D1446454

HF COMMUNICATION

Science & Technology

HF COMMUNICATION

Science & Technology

John M. Goodman

VNR VAN NOSTRAND REINHOLD
New York

Library of Congress Catalog Card Number 91-8065
ISBN 0-442-00145-2

Manufactured in the United States of America

Published by Van Nostrand Reinhold
115 Fifth Avenue
New York, New York 10003

Chapman and Hall
2-6 Boundary Row
London, SE1 8HN, England

Thomas Nelson Australia
102 Dodds Street
South Melbourne 3205
Victoria, Australia

Nelson Canada
1120 Birchmount Road
Scarborough, Ontario M1K 5G4, Canada

16 15 14 13 12 11 10 9 8 7 6 5 4 3 2 1

Library of Congress Cataloging-in-Publication Data

Goodman, John M.
HF communications : science and technology / John M. Goodman
p. cm.
Includes bibliographical references and index
ISBN 0-442-00145-2
1. Radio, Short wave. I. Title.
TK6553.G65 1991
621.384' 151--dc20 91-8065
CIP

To

My Wife JANE,

and

Children

MARY JANE, JOHN, and JENNY

PREFACE

The radiowave component of the electromagnetic spectrum has had a major influence on modern society, most especially in areas of wireless communication, radio and television broadcasting, remote sensing, and radar. Even though radiowaves may be generated naturally and have existed since the dawn of time, the ability to intentionally generate and exploit them has been a 20th century phenomenon. This generation and exploitation capability grew significantly following World War I, it accelerated during World War II with the development of radar applications and, from 1960 to the present, satellite technology has challenged the telecommunications specialist with a host of potential applications for radiowaves. Work in the HF portion of the radio spectrum has figured prominently in the development of various radio applications from the very beginning.

With the advent of artificial earth satellites, the communication discipline entered a new era with vastly improved coverage and better reliability for both radio and television signals. Even so, HF radio, which requires no costly space segment, has remained a major component of the hierarchy of military communication systems essential for maintenance of requisite command and control functions.

In the non-military and commercial world, constraints imposed by security, electronic countermeasures (ECM), non-optimum terrain, communication suite balance, asset mobility, and related factors are largely unimportant. As a consequence, satellite communication has an insurmountable edge over HF over most of the globe. Nevertheless, in remote areas where satellite terminals, undersea cables, and radio relay are either nonexistent or sparse, HF will remain the communication system of necessity. Furthermore because of relatively low capital investment, HF will continue to be an important medium of communication in the developing countries for some time to come. Since geosynchronous satellite coverage is not reliable in the near-polar regions, HF service is important there as well.

HF has suffered from an inferiority complex for a number of years, having been afflicted with the presence of new and advanced satellite communication schemes. Since, for simplicity in argument, investment in communication technology is a zero-sum-game, then investment in satellite-based technology has naturally led to a diminution in HF capability. In fact, for a number of years, not only did HF become degraded relative to satellite communication, but it also suffered degradation with respect to its previous capability. This diminished capability which was induced by neglect, and has led to lack of operator training, inadequate replacement schedules, and a reduction in necessary R&D. This has been unfortunate component in the set of negative perceptions about HF. Communication by HF radio does have its difficulties at times, but many of them can be overcome or dealt with sensibly.

To a certain degree things are changing and we explore this development within the text. It is the author's intent that a greater appreciation for the realities of HF communication may be derived and that many of the myths may be appropriately suppressed if not dismissed.

Chapter 1 is an introduction and contains all of the usual background material one would expect in a text of this type. A certain amount of time is devoted to historical aspects, not simply out of personal taste, but rather because it gives us some insight about the future. Hopefully it will be instructive, but the impatient reader might wish to only scan this portion of the introduction.

In Chapter 2 we examine the complicated inter-relationships between the sun, the interplanetary medium, the near-earth plasma encompassed by the magnetosphere, and the ionosphere which is of primary concern to HF specialists. The general topology of the magnetosphere and the underlying solar source is also described. We look at the basic solar structure with particular emphasis on the origins of various disturbances and sunspots, the area and number of which are utilized in a number of HF propagation codes.

In Chapter 3 we look at ionospheric layer formation and examine the radiations which are important in this process. Using the continuity equation to provide an intuitive physical understanding, the general ionospheric structural behavior is discussed, along with implications for HF propagation. The formation of the E, F1, and F2 regions are characterized in terms of their daily, seasonal, and solar epochal dependencies . In addition, certain phenomena such as spread F, sporadic E, and traveling ionospheric disturbances (TIDs) are described. Various ionospheric models are identified. Their utility in the prediction process is outlined, but we reserve for Chapter 5 a more thorough discussion of HF propagation prediction methods.

Chapter 4 provides basis for much of the book, given the fact that an understanding of HF propagation effects is central to performance prediction, frequency management strategy, and the specification of certain design factors which are embodied in operational systems. Radiowave phenomena are covered in general along with a brief review of non-ionospheric propagation features including those associated with spacewaves and surface waves. Ionospheric propagation is clearly a major emphasis, and we outline the main features of the Appleton-Hartree formalism as it relates to the HF regime. Effects specific to the magnetoionic medium are covered including those impairments associated with ionospheric irregularities and disturbances. Path loss methods are examined for point-to-point skywave links. Special limiting cases of near-vertical-incidence-skywave (NVIS) and multiple hop propagation are discussed along with ducted and chordal mode possibilities. The actual path description is an important ingredient for specification of the complete HF channel. The ionospheric sub-channel exhibits considerable variability but its properties are commensurate with the picture provided by climatological models which are described in chapter 3. However, specific phenomenological details and propagation effects cannot be adequately represented by median models, and we shall highlight specific propagation effects by showing examples. We gain further insight by reviewing specified ray tracing methods. Major elements of the Chapter include fading and multipath phenomena and methods for mitigation of these effects. We conclude the Chapter with an review of channel models and simulators.

Chapter 5 is devoted to methods for prediction of HF communication system performance, and a brief history of graphical and computer methods is included. To illustrate existing methodology, a particular CCIR model designed for determination of MUF and field strength is outlined. The IONCAP program is representative of state-of-the-art computer methods, and a short review of IONCAP and similar programs is given. Within the context of HF performance prediction, HF noise and interference have great importance. Accordingly, a treatment of noise models and their application in prediction codes is found in the Chapter. Recognizing a trend in recent years, we also include a section on small microcomputer models which are used in lieu of large mainframe programs. The current and future needs for prediction services are outlined.

Chapter 6 contains a comprehensive treatment of various Real-Time-Channel-Evaluation (RTCE) schemes including ionospheric sounding. Both Vertical-Incidence-Sounding (VIS) and Oblique-Incidence-Sounding (OIS) techniques are discussed in the context of RTCE. A detailed discussion of chirp sounding is given in view of a continuing interest in these devices for frequency management. Modern dual-purpose sounders, having the potential to satisfy both scientific and certain operational requirements, have also been developed and are outlined. The possibility of sounder networking holds a considerable amount of charm in view of the possibility for global real-time assessment. This is especially true for OIS-based systems. Of major interest in this Chapter are various update schemes which are hybrid combinations of propagation prediction models and RTCE devices. A thorough discussion is provided and a number of examples are given.

Chapter 7 deals with system adaptivity and its evolution against a background of evolving technologies. We begin by reviewing some definitions of adaptive HF. We take note of the fact that a clear definition is rather elusive being relative to the vantage points of the specific developers. However it is clear that adaptive HF is distinct from some systems which are intrinsically robust. We define adaptive HF and compare that with other definitions put forward by government agencies and supporting contractors. Because the term adaptive HF is used in so many ways, we shall endeavor to develop the material in terms of modern HF as an umbrella under which adaptive HF will be contained. Within this chapter we will examine new technologies which might be ingredients of modern (and perhaps adaptive) HF. Many of these technologies involve component development while others involve disturbance mitigation schemes using signal processing, or involve the implementation of new procedures for resource management. The hierarchy of adaptive HF approaches are described and some specific realizations are given. A special section dealing with nuclear phenomenology and its impact on HF communication network design is provided in the Chapter. Newly emerging technologies, and the extent to which they impact the development of improved HF radios and the utilization of the HF spectrum, will be described. To make order out of potential chaos, the need for industry and government standards is obvious. Steps in this direction are outlined.

The final component of the text is an epilogue which might have been alternatively entitled *Putting it all together.* The major conclusions contained within the entire monograph are provided in this chapter along with a discussion of what the future might bring.

It is impossible to acknowledge everyone who has contributed in the development of this text material. However there are a number of present and former colleagues who have contributed significantly through the simple process of passing along their personal insights during informal discussions. I would like to express my appreciation to all those individuals who permitted me to utilize published materials including photographs and certain line drawings to better elucidate the text. Hopefully I have suitably referenced all such contributions and I apologize for any oversights. Finally, I would like to express my appreciation to Mrs. Jean Ware, for artwork assistance and for preparation of the camera-ready manuscript.

John M. Goodman

CONTENTS

HF COMMUNICATION

Science & Technology

1

INTRODUCTION

HISTORICAL REVIEW, CURRENT ROLES, FUTURE TRENDS

"In 1901 Marconi first succeeded in sending signals across the Atlantic Ocean. This was an achievement of outstanding and technical importance. There was much in the way of theory to deter anyone contemplating such an experiment. But Marconi, a true experimenter, was not to be deterred by (existing) theories."

Excerpt of a tribute to Marconi by E.V. Appleton[1]

1.1 CHAPTER SUMMARY

We begin our journey into the domain of modern HF communications with an abbreviated history of the *wireless* experiments which were conducted at the turn of the century. We will also make note of some of the key activities which have brought us to where we are today. The treatment is not meant to be thorough, but it is instructive, allowing the reader to get a feel for the excitement and controversy which was generated by the pioneering efforts of Marconi and others. The author believes that historical studies are too often relegated to dusty archives, and that experience is lost in the quest to reexamine old physics with new instruments in the hope that new physics will emerge. In most applications, this is simply not the case. As in any field, progress in the HF-related disciplines is built upon a foundation of advances over a period of many years. In most instances, while the advances may be significant, they are incremental in nature; in a few cases the events are monumental and shake the foundations of the subject. Along with the revolution in modern physics which occurred during the first part of the 20th century, exciting events were taking place in the fields of radiowave propagation and upper atmospheric physics. In Section 1.2 we note some of these events.

---

1. Radio broadcast by E.V. Appleton on 20 July 1937, the day in which Marconi died. The quotation was reported in a paper by W.J.G. Beynon [1975] commemorating contributions by Marconi to radio science and ionospheric studies [Checcacci, 1975].

An examination of the chronology of major events shows that studies of geomagnetism, the ionosphere, and HF propagation have been closely intertwined. Early in the 20th century the synergistic relationship between (radio) engineers and (ionospheric) physicists was a natural one, analogous in many respects to the joint endeavors of experimentalists and theorists who were responsible for development of the new physics. However, specialization led to an erosion in these relationships. Ionosphericists no longer depend exclusively upon high frequency radiowaves to diagnose the ionospheric plasma characteristics. Also, with the advance of radio system engineering involving new technologies for generating, transmitting, receiving, and processing of radio signals, the ionosphere is becoming a less important factor in the development of modern HF communication systems.

Following a short discourse on the history of HF communication, we examine the sensitivity of HF to variabilities in the ionospheric medium, and we will outline both the inherent strengths and weaknesses of ionospheric radiowave propagation. The weaknesses were believed to have been overcome by the development of satellite communication systems, but space systems are known to have vulnerabilities of their own, especially in an environment which is *stressed* [2].

To place HF in perspective as far as the ionosphere is concerned, we have included a short discussion of the various ionospheric effects which are encountered throughout the radio spectrum. For communication beyond the horizon using HF, the ionosphere is a necessary ingredient in the total system design. It is a satellite which *doesn't fall down* even though it may appear to be unreliable at times. Coping with ionospheric variability at HF is a traditional problem. Many of the historical concerns about HF propagation , especially that element which depends upon the ionosphere, are being overcome although they cannot be totally removed from consideration. Many of the countermeasures to propagation effects are mentioned, and we conclude with a summary of current HF system needs and some future requirements.

1.2 HISTORICAL PERSPECTIVES

As we go through this brief summary of the history of HF, it will become evident that HF radio development and early ionospheric studies have been closely related. The first few chapters will emphasize certain aspects of the ionospheric medium itself, along with the general properties of radiowave

---

2. The (communication) environment may be *stressed* in a number of ways, the most important being those which degrade system performance including: enhancement in the background radio noise, deliberate jamming, and propagation disturbances. Although propagation disturbances may derive from natural and man-made causes, military telecommunication specialists generally reserve the word *stressed* to describe nuclear-disturbed media.

propagation which are important for an understanding of HF communication theory and practice. Later chapters will presume a basic appreciation for details of the ionospheric *personality* as well as the solar-terrestrial environment within which HF radiowaves must propagate.

Historically both HF radio and the ionosphere were exciting curiosities, and early investigations were directed toward development of a basic understanding of both. As knowledge grew, it became evident that the ionospheric medium was a perturbing factor and would introduce a basic limitation in the ability to exploit evolving radio technology. As a result, the perception of the role of HF in providing communication service has changed over the years. The most serious difficulties resulted from the emergence of satellite communication technology. Even in the military world, HF was regarded as merely a back-up to SATCOM. A number of events changed the perception of the value of HF as a viable communication tool. We shall find that this has had little to do with any improvement in knowledge of the ionosphere or even in the development of any truly exciting new techniques. It has had more to do with the reluctant acceptance by communication planners that satellites have their own intrinsic limitations in a stressed environment. A redefinition of the role which HF can best play in the requisite mix of communication systems has also been a key factor in the rebirth of HF. This redefinition amounts to a restriction in the requirements which are placed upon it. To paraphrase a nursery rhyme, "when HF is good, it is very very good, but when it is bad it is horrid." This *personality* is exploited in modern HF systems in a variety of ways. In later chapters we shall examine both the ionospheric *personality* and the exploitation techniques. To fully appreciate the changing perceptions of HF, let us briefly review its early history. For those interested in further information, the author has compiled a set of references and a bibliography both of which appear at the end of the chapter.

The role of geomagnetism in understanding the nature of ionospheric personality is important, if not central, in many HF applications. This is especially true for systems which exploit the skywave mode of propagation at high latitudes. The geomagnetic field, as represented by the position of a compass needle, was observed to undergo transient fluctuations as early as the 1700s. The Swedish scientist Celsius, famous for development of the centigrade temperature scale, discovered that *magnetic storms* were not isolated like ordinary tropospheric weather; rather they had a global characteristic. Celsius also discovered the correlation of magnetic activity with optical auroras, and in 1741 he determined that auroral forms were aligned with the magnetic field orientation. The first theory of magnetism was developed by Poisson in 1824 and the first systematic measurements were made by Gauss. It was Gauss, in 1839, who postulated the existence of ionized regions in the upper atmosphere. (Now, of course, a gauss is defined as a basic unit of magnetic field strength.)

Although the relationship is not precise, sunspots are generally regarded as an important factor in the field of HF predictions, especially for climatological predictions of behavior. There are well established circumstantial correlations between the sunspot number and ionospheric disturbances which are useful. The first telescopic observations of sunspots were made in 1611 by a number of observers, the most famous of whom was Galileo. Sunspots have been monitored continuously since that time although a pronounced minimum was observed between about 1645 and 1715 (the Maunder minimum). Hardly any aurora were observed during that period. Thus there appeared to be a relationship between the absence of sunspots and the lack of observable auroras. Samuel H. Schwabe, a German observer, discovered a ten year variation in the number of sunspots and reported his findings in 1843. More systematic observations over many decades have shown the average cycle to be eleven years rather than ten. Small transient magnetic variations on the earth were shown to exhibit the same periodicity.

According to Chapman [1968], in his book *Solar Plasma, Geomagnetism, and Aurora*, the first picture of the so-called auroral oval centered about the geomagnetic pole, was drawn by Elias Loomis of Yale University in 1860; in 1878 Balfour Stewart suggested that ionization in the upper atmosphere would account for small fluctuations in the geomagnetic field. By 1892 Stewart had identified the existence of an electrified layer in the upper atmosphere. At about the same time A. Schuster associated this electrified layer as the origin of electric currents which cause compass variations. These variations were observed to be most intense near the auroral zone and the geomagnetic equator. Schuster also developed a dynamo theory to explain the diurnal component of these same currents, and he associated those currents with tidal motions of the neutral atmosphere.

The most remarkable feature of the upper atmosphere is the visible aurora, a luminous display which appears in the nocturnal sky in the polar regions. It is generically called *Aurora Polaris*, but is termed *Aurora Borealis* in the northern hemisphere and *Aurora Australis* in the southern hemisphere. It is now known that auroras may exist at any time of day, but they cannot be seen in the presence of competing sunlight. Auroral forms were studied extensively by C. Stormer who by 1911 had developed one of the earliest explanations of auroral zone formation. One of Stormer's most significant contributions was his theory of the motion of charged particles in the geomagnetic field.

In the years following the work of Stewart and Schuster, the development of magnetic activity indices continued. Although indices were first being published by 1885 (in the yearbook of the Greenwich Observatory), the first step toward defining a geomagnetic index at an international level was made in 1905 [Mayaud, 1980]. Magnetic activity indices were employed by Bartels in 1932 in connection with his discovery that the mysterious M-regions on the sun were associated with 27-day recurrence cycles of magnetic activity. These 27-day cycles were also shown to be related to the period of solar rotation.

Many years later, magnetic indices similar to those used by Bartels were used to explain the terrestrial effects caused by high speed solar wind streams associated with the appearance of coronal holes.

Half a century before Stewart made his suggestion about the existence of the ionosphere, the English physicist Michael Faraday developed a theory of the electromagnetic field; and 17 years before Stewart's announcement, Maxwell had predicted the existence of radiowaves (or more properly, electromagnetic waves). Maxwell's work dealt with the speed at which magnetic disturbances travel. His equations are the cornerstone of EM theory. His predictions about radiowaves could not be verified at that time. Confirmation of the theory was later made by the German physicist Heinrich Hertz.

In 1887 Hertz developed the first radio transmitter and he also built the first loop receiver. With this simple equipment he was able to determine the basic transmission properties of radiowaves. Hertz generated radio frequencies between 31 and 1250 MHz (note: Megahertz and not Megacycles per second) in his early experiments and this limited the range of his coverage because of the earth's curvature. Hertz tuned his system by changing the transmitter antenna configuration but this was not a very selective technique. (The Hertzian type antenna is still employed in modified form in TV reception. Household radios use a modified type of antenna invented by Guglielmo Marconi.) In 1892, Prof. E. Branly of France developed the first radio detector and one year later the Russian experimenter A.S. Popoff recorded radiowaves emanating from lightning. Popoff modified the Hertzian antenna so that the upper half was a tall tower and the lower half was replaced by a connection to the earth. This was thought to be superior to the Hertzian antenna since the resulting radio beam more closely conformed to the earth's surface. This antenna was to become known as the Marconi antenna because Marconi was the first to exploit it commercially.

The Italian inventor Guglielmo Marconi, while working in England in 1896, succeeded in sending the first short distance (2 miles) radio telegraphic signals. He used a Branly-Popoff type receiver and a Hertzian type transmitter. Since Marconi did his work in England, England became the world center for *wireless* as it was called at the time. Also, in that year, Marconi brought his equipment to the U.S. in order to report the international yacht races off the New Jersey coast. The use of his equipment was witnessed by officers of the U.S. Navy and it led to a test using a shore station and two ships. Gebhard [1979] reports that the USS Massachusetts was able to receive transmissions from the USS New York over a distance of 46 miles. The first official use of wireless by the U.S. Navy occurred in 1899 when a message was sent from the USS New York to a navy facility in New Jersey requesting refueling services in the Navy yard. In 1900, the U.S. Navy began negotiations with the Marconi Company for the purpose of obtaining radio equipment which could be installed on its ships and shore facilities. These negotiations failed because the equipments could not be obtained unconditionally. By 1900, the respective navies of the United States, Great Britain, France, Germany, and Sweden adopted radio for use by their fleets. The early Marconi-inspired tests, although important, did not challenge the existing rectilinear propagation theory.

In 1901 Marconi transmitted the first long distance (trans-Atlantic) signals from a site at Poldu, England to Newfoundland. It is thought that he used a radiofrequency of 313 kHz, a frequency which was appropriate for ionospheric bounce which was unknown at the time. This experiment of Marconi created a puzzle since the earlier work of Hertz had conclusively demonstrated that radiowaves travel in straight lines unless they are deflected by an obstruction. Marconi's work triggered additional theoretical work in EM wave propagation, and he received the Nobel prize for physics in 1909 as a result of his monumental achievement.

In 1902, Kennelly and Heaviside independently postulated the existence of conducting strata in the upper atmosphere which could guide radiowaves for great distances. For many years, what we now refer to as the ionosphere was called the Kennelly-Heaviside layer in their honor. (In some articles, the region is termed the Heaviside layer.) Marconi had no particular liking for the Kennelly-Heaviside layer, and as late as 1924 was quoted as saying that perhaps it didn't exist. Nevertheless the existence of the layer is the basis upon which long-haul HF communication is based.

Following the Marconi experiment, interest arose for the use of wireless by the navies of many countries. At about that time a number of commercial companies were established, and by 1904 interference problems could no longer be ignored. The first attempt at international radio regulation, in 1906, failed; but in 1912 regulations assigning stations specific frequencies and call signs went into effect.

Many important achievements were made during the period between 1910 and 1920. We mention the names of Pierce (who examined the interference effects of ground and skywaves), Eccles (who was the first to describe the effect of plasma on EM waves), DeForest (who made many contributions to radio technology and suggested that the reflecting layer in the upper atmosphere was about 62 miles high), Fuller (who collaborated with DeForest), and Watson (who verified the theoretical predictions of Rayleigh and Poincare which indicated the inadequacy of diffraction as the cause of long distance radio propagation). The following events, during the same period, were noteworthy: the Americans, British and Germans established trans-Atlantic radio service; the Americans succeeded in establishing trans-Pacific radio service; the vacuum tube was developed; Marconi built the first directional radio station; and massive projects using long wave radio telegraph were initiated.

Following World War I more improvements were made in radio apparatus, and both theoretical and experimental studies continued. Eckersley conducted important direction-finding work, in proving proved that radiowaves transmitted over a long distance were indeed downcoming waves. Other researchers emerged in the twenties: Larmore, who developed the famous formula relating transmitter frequency to plasma frequency; Swann, who first speculated about receiving backscatter echoes from the ionosphere; Pickard, who made the first quantitative measurements of natural fading phenomena; Appleton, who made substantial contributions in magneto-ionic theory and conducted A-O-A measurements; Breit and Tuve, who developed their well-known theorem and who conducted landmark radio pulse experiments; and a number of others who performed radio direction finding experiments (including Barnett, Smith-Rose, and Barfield). In 1924, Appleton and Barnett unequivocally proved the existence of the ionosphere, and new techniques for probing this region were developed.

The twenties saw contributions from scientists at the newly established Naval Research Laboratory in the United States. In 1926, Taylor and Hulburt described properties of the radio-reflecting layer such as day-night variations, seasonal variations, latitude effects, and electromagnetic properties. They gave estimates of the conducting layer electron density. They also described the new phenomenon of *skip distance*, observed *round-the-world* propagation by multiple hops, and encountered the phenomenon of ionospheric *splash-back*. This latter effect eventually led to the development of Over-the-Horizon (OTH) radar. Some of the ionospheric phenomena which were uncovered during the early HF radio experiments are depicted in Figure 1-1.

The Carnegie Institute and NRL collaborated in a series of landmark experiments in the late twenties which led to the invention of radar[3]. The NRL-Carnegie team developed an echo interference technique to distinguish between phase path length changes of 20 meters (1/2 wavelength) by feeding a small amount of signal from the transmitter crystal to the receiver to serve as a phase reference. In-phase signals caused the oscillograph spot to be displaced upward, and out-of-phase signals did the opposite. Unfortunately, NRL is located near Bolling AFB and aircraft were constantly taking off and landing. On these occasions, the oscillograph spot moved violently. Clearly more than ionospheric signals were being directed to the receiver. The concept of radar was born.

As a direct consequence of the problems encountered by Tuve of Carnegie Institute and his NRL colleagues, Lt. W.S. Parsons wrote a letter to his superiors in the U.S. Navy Bureau of Ships in 1931 suggesting that the technique might be usefully employed in the detection and location of enemy aircraft

----

3. Attribution for the invention of radar is still a matter of some controversy, with legitimate claims being made by groups in three countries: the United States, England, and Germany. Radar development in the U.S. was stimulated by the ionospheric reflection experiments, whereas British radar development was a direct result of meteorological programs led by R.A. Watson-Watt. Early German radar research has been chronicled by Reuter [1971]. The origins of radar have been discussed by Hill [1990].

approaching U.S. ships.

One of the most prominent scientists involved in early investigations of the ionosphere was Sidney Chapman, who in 1931 published a paper dealing with the Kennelly-Heaviside layer, and who, like Hulburt before him, provided a fundamental basis for our current understanding of the ionosphere. To this day the Chapman hypothesis for ionized layer formation is a useful model, especially for the lower layers of the ionosphere.

From the theoretical vantage point, Larmore, Lorentz, Appleton, and Hartree provided a clear understanding of radiowave propagation in magneto-ionic media, and the so-called Appleton-Hartree formula for the radio refractive index is of fundamental importance in the analysis and prediction of media effects. Appleton published a complete theory of radio propagation in magnetoionic media, such as the Kennelly-Heaviside layer, in 1932. He also coined the term *ionosphere* to describe the ionized strata upon which HF skywave propagation depends.

There is clearly more to the early history of HF than has been provided here. This sketch will give the reader an outline of the past before we cover current technology and speculate about the future.

1.3 CURRENT UTILIZATION

With the advent of satellites, HF was ignored because of ionospheric variability, frequency management problems, and inadequate bandwidth. The recent history of HF has been uncertain, at least from the perspective of its capability and survivability in a nuclear environment. This situation is changing.

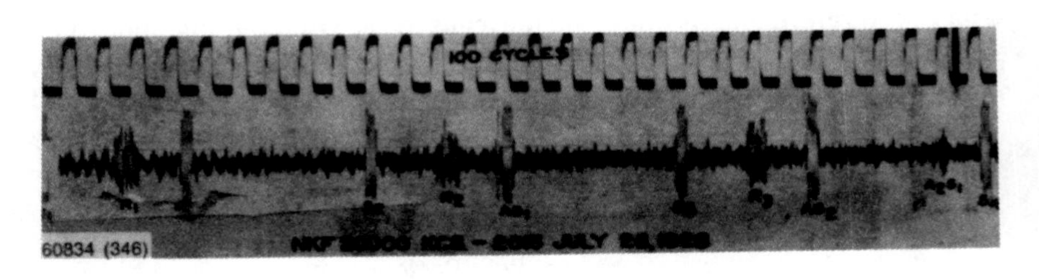

Fig.1-1. This recording of *round-the-world* radio transmissions was made in 1928. Such phenomena were first reported in 1926 [Taylor and Hulburt,1926]. The NRL transmitter pulses are given by S1,S2, and S3; *Splashback* echoes are indicated by R1, R2, and R3; AS1 is the once-round-the-world signal associated with S1; AS2 is the once-round-the-world signal associated with S2; and A2S1 is the twice-round-the-world signal associated with S1. At the top of the photograph is a record of 100 Hertz timing signals. The observation of *splashback* lead ultimately to Over-the-Horizon or OTH radar technology. Indeed studies of this type were central in the development of radar itself. [From Gebhard, 1979]

Figure 1-2 is a cartoon which depicts the *roller-coaster* perception of HF following World War II. Clearly this picture is influenced by military concerns, but it is nonetheless an indicator of the interest (and financial investment) in high frequency technology over the years.

HF systems are a major component in the command, control, and communication (C^3I) disciplines within the military. In many instances, these systems are relatively old and are to be replaced by a new class of radios exploiting solid state and other modern technologies.

1.3.1 Communication

The earliest form of long distance terrestrial communication involved either longwave or HF bands, and to the present, civil and military use of such techniques is considerable. Overuse of the HF spectrum (3-30 MHz) has led to some interesting frequency management challenges. HF is used for tactical and strategic military purposes, and for international broadcasting by organizations such as the British Broadcasting Corporation (BBC), the Voice of America (VOA), Radio Liberty, Deutsche Welle, Radio Nederland, and Radio Moscow. HF, or shortwave, exploitation is relatively common in remote areas and is a leading method for communication in developing countries. The advantages of HF communication arise from its relative simplicity, its ability to provide near global connectivity at low power without relay, its ease of proliferation, and its moderate cost. The apparent disadvantages are directly related to ionospheric variability, and in many applications the performance of HF communication using the skywave method is a mirror of this variability. Nonselective fading caused by solar disturbances, and selective fading arising from multipath and multimodal interference leads to a reduction in reliability for point-to-point links. Specific countermeasures involving propagation diversity or real-time system adaptivity may bound the impact of these difficulties. Propagation disturbance effects are further exacerbated by time varying congestion and environmental noise effects under conditions of fixed frequency assignments.

Recently, the HF broadcast community has invested in applied propagation modeling to better estimate coverage in *target areas* of interest as well as in the sound quality of its programs. System performance tools such as IONCAP [Teters et al., 1983] have been modified to provide a better picture of broadcast coverage. This discipline is similar to OTH radar, and many of the modeling and propagation assessment measures bear a striking resemblance to area coverage concepts used in OTH radar analysis.

New technology is improving the capability of HF communication systems, especially in digital transmission. *Rake*[4] processing schemes lead to elimination of multipath distortion, and equalization methods compensate for intersymbol interference effects. The development of novel adaptive procedures and the use of robust system architectures is providing a more favorable view of HF, even for stressed environments. Though high altitude nuclear effects (HANE) may limit skywave coverage for several hours, the use of networking techniques coupled with mixed media approaches is a powerful countermeasure. In addition, the ionosphere is actually a robust component of the system; it repairs itself, something that other components of the total system cannot easily do.

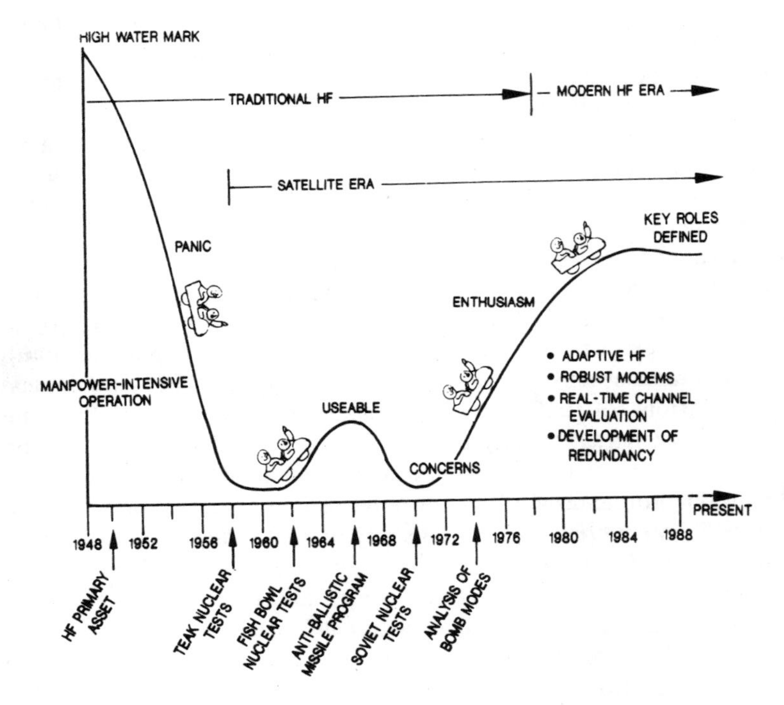

Fig.1-2. The HF *Roller Coaster* ride since the end of World War II.

---

4. *Rake* is a processing technique designed to compensate for multipath fading effects by the process of "raking" together all multipath components which are encountered over a signal path. It belongs to a class of matched filter techniques which may be used to dispose of selective fading. Suitable modification of the concept, adaptive equalization, permits elimination of intersymbol interference. (See Chapter 7 for more discussion.)

HF radio has always constituted a major fraction of all intercontinental communications although the emergence of SATCOM has served to diminish the role of HF in recent years. This de-emphasis has been most pronounced in the nonmilitary segment of the industrialized world, although HF will retain its importance for communication in remote areas and in locations where accessibility to commercial cable, radio relay systems, and satellite service is limited. Indeed HF is expected to be a major method for providing communication service within developing countries for some time to come.

1.3.2 HF Surveillance and Direction-Finding Systems

Over-the-Horizon radar (OTHR) concepts arose in the late 1920s in connection with HF communication experiments which clearly exhibited the *splashback* of echoes from Beyond-Line-Of-Sight (BLOS) sea clutter into the radio receiver. These experiments were first conducted by scientists at NRL. The technology has been pursued vigorously in the United States only since the late 1950s, and OTHR programs currently exist in a number of countries (see Section 6.2.2.2 on backscatter sounding). OTHR system development, along with other radar surveillance programs in the U.S., received a boost with the advent of the Strategic Defense Initiative (SDI) program.

OTH radars may employ ionospheric forward scatter or backscatter geometries. Operating in the HF band these systems are vulnerable to ionospheric variations which may seriously degrade OTH radar usefulness both for fleet defense and for measuring ocean surface conditions in real-time at ranges from 600 to 2200 nautical miles. This reduction in performance is largely restricted to high latitude surveillance missions which are influenced by auroral effects in the ionosphere.

The U.S. Air Force has developed an OTH Backscatter radar (OTH-B) and the U.S. Navy is developing its own system version which has the virtue of being relocatable (ROTHR). Other OTH radar projects are underway in the United Kingdom, Australia, and China.

Over-the-Horizon backscatter radars operate on the principle of ionospheric bounce for coverage, and backscatter for target detection and ranging. Signal processing allows the extraction of targets such as aircraft from unwanted clutter. Some separation between transmit and receive antennas is employed in OTH-B in order to permit utilization of a duty cycle approaching 100%, and an FM/CW waveform is used to provide some processing gain, to reduce high peak power components, and to suppress interference. Ionospheric sounding is employed to monitor the ionosphere and this data is used to assist in system frequency management and control. It should be recognized that target range is ultimately determined by a measurement of time delay between the transmission of the radar signal and the radar echo. Since the ionospheric reflection height and layer *tilts*[5] influence the result, a method is required to solve for the essential ionospheric properties in order to resolve this range *registration* problem.

Because of the extensive use of the HF band, there has been a considerable investment in various HF emitter location disciplines. This capability is important for both civil and military purposes. A treatise on the subject was published by Gething [1978] and there is a wealth of information available on various aspects of this technology. Abstracts of the literature published from 1899 to 1965 have been compiled by Travers and Hixon [1966].

There are several methods for deriving geolocation information at HF and various so-called *fix algorithms* attempt to exploit attributes of the schemes. Multiple sites may be used to take advantage of signal time-of-arrival differences (called TDOA) or bearing angle-of-arrival differences (called AOA). There are also frequency difference methods and hybrid approaches. Perhaps the most intriguing class of methods from the point of view of the ionosphere, is called Single-Station-Location (SSL). This procedure implies that a signal source is located from information derived from a single site. Clearly this implies some intelligence about the ionospheric height as well as how the HF signal travels from the signal transmitter (or emitter) to the receiver. It presents a challenge not only for the system designer but also for the ionospheric specialist. Recent tests have examined this concept for both short and long range emitters [Goodman and Uffelman, 1983; McNamara, 1988], and relevant ray tracing schemes have been discussed by Baker and Lambert [1989].

1.3.3 Spectrum Congestion

Even though the capital investment in HF radios may have diminished relative to other categories of communication equipment, the need for communication, in general, has increased dramatically. As a consequence, the use of HF radio has not decreased in absolute terms. For example, demand for HF spectrum space has been especially strong in Europe region even though satellite service is obviously available. Over-utilization of HF in the region has caused legitimate concern among users of present systems and among architects of new systems yet to be deployed. The congestion problem, as a function of day and nighttime conditions, is illustrated in Figure 1-3.

---

5. The height of maximum electron concentration of the ionosphere forms a distorted spherical surface. At any point on this surface we may contruct a tangent plane. The term ionospheric *tilt* refers to the angle between this plane and the local horizontal.

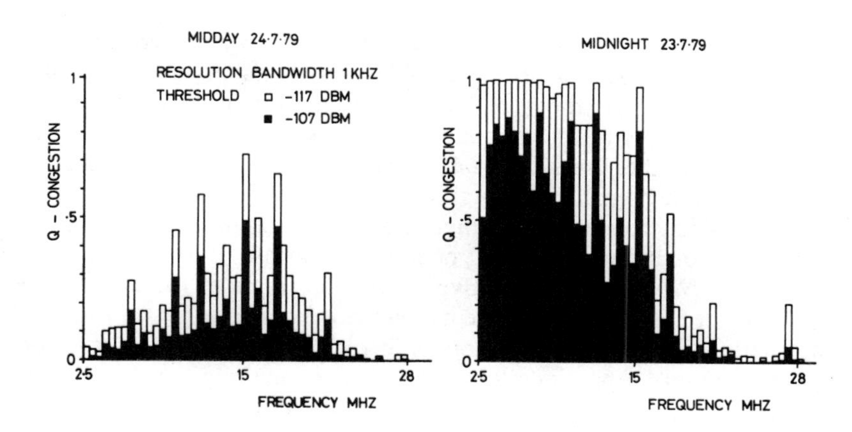

Fig.1-3. Spectral congestion across the HF band. The index Q is the probability of finding, at random, two 100 Hz bandwidth regions within each 50 kHz slice of spectrum such that the average interference level exceeds a defined threshold. For this illustration Q is much larger during the nighttime than daytime. Nocturnal congestion is most pronounced at the low part of the HF band, since the ionosphere does not provide support in the upper part. The relatively high *spikes* are the congestion from the broadcast bands. (from Dutta and Gott [1982];IEE, London, by permission)

With the anticipated growth of HF for tactical military use, and with the possibility that a new class of robust spread-spectrum systems may be developed and used, there is a concern that the full benefit of HF may not be realized (especially for disadvantaged users). This concern has provided the impetus for a re-examination of how the HF spectrum is utilized. Obvious technological remedies such as frequency sharing are being examined. The *pooling* or sharing of a set of frequencies by users within an organized group or administration should reduce the integrated requirements of all members of the group. This in turn should reduce the pressure for assignments within allocated bands. Furthermore, since each member in the group potentially has a greater range of frequencies available (on a time-shared basis), then each member has a higher degree of propagation support and better reliabilities should result. However, to successfully implement such a system it is necessary to have central management control, and the dissemination of frequency management instructions should occur in near real-time. This would not have been possible a few years ago but new technological advances in computers and real-time-channel-evaluation may make the approach achievable. A number of system architectures employing the frequency pooling concept have been conceived.

1.3.4 International Coordination Efforts

The frequency allocation process at HF and the studies which accompany it are of special significance. An advantage of HF systems is the inherent ability of signals in the band to propagate great distances without the necessity for intermediate relay and re-transmission. This property may pose an interference problem for other users if not taken into account properly. The purpose of HF spectrum management is to allocate resources (i.e., frequencies) to satisfy various general telecommunication service requirements[6]. Such a process must be orchestrated globally, or at least regionally.

Periodically World Administrative Radio Conferences (WARCs) are convened by the International Telecommunications Union (ITU) to review the international frequency allocation table and to examine other issues of importance to users. Approximately 150 countries maintain membership in the ITU and the group is charged with the responsibility for specification of radio regulations, operational criteria, frequency utilization, and for consideration of related technical matters. The central purpose of the WARC is to distribute spectral space in an equitable fashion so that all specified telecommunication services (such as HF communication) may be accommodated. It also provides rules and guidelines for the assignment of frequencies within allocated space by various administrations so that interference may be minimized. To perform this function effectively, due consideration is given to relevant propagation phenomena as well as to system requirements, priorities, and specifications.

The ITU also convenes topical conferences for examining specific issues such as maritime-mobile and aeronautical-mobile usage. The International Radio Consultative Committee (CCIR) is a technical arm of the ITU. This organization is composed of a number of Study Groups organized at both national and international levels. The efforts of Study Group 6, which deals with Propagation in Ionized Media, is of special significance. Study Group 6 is presently composed of six working groups as follows: (WG-1) Ionospheric Properties and Propagation, (WG-2) Operational Considerations, (WG-3) System Design Factors, (WG-4) Natural and Man-made Noise, (WG-5) Field Strength Below 1.6 MHz, and (WG-6) Field Strength Above 1.6 MHz. The deliberations of each of these working groups is of interest to HF system architects and users (with the possible exception of WG-5). The *Recommendations and Reports of the CCIR* are based upon Plenary Assemblies held every four years; Oslo (in 1966), New Delhi (in 1970), Geneva (in 1974),

---

6. Frequency allocations are made on the basis of the following service types: Fixed, Mobile, Aeronautical Mobile, Maritime Mobile, Land Mobile, Broadcasting, Amateur, and Standard Frequency. (See Section 5.3.4)

Kyoto (in 1978), Geneva (in 1982), Dubrovnik (in 1986), and Dusseldorf (1990) The so-called *Green Books*, which are based upon the deliberations of these assemblies, are published by the CCIR and contain many reports of major interest to users of the HF spectrum. Specific references to these reports will be given where appropriate[7].

1.4 HF: CONTEXT OF ELECTROMAGNETIC & RADIO SPECTRA

For electromagnetic waves, the relationship between the electric (and magnetic) vector oscillation frequency and the wavelength of the radiation is:

$$f\,L = 300 \tag{1-1}$$

where f is frequency in Megahertz (MHz), L is the wavelength in meters, and the constant 300 is free space velocity of light in megameters/sec. The velocity of light is traditionally given in MKS units, but the expression given here is more convenient for immediate conversion between frequency and wavelength within the HF band.

The electromagnetic spectrum is composed of a number of major components which overlap. At the highest frequency portion of the spectrum, the behavior of the electromagnetic radiation is particulate in nature, and at the lower frequency end the properties are more nearly (if not exclusively) wave-like in nature. As illustrated in Figure 1-4, the electromagnetic spectrum may be organized into the following segments: Gamma radiation, X-rays, Ultraviolet light, Visible light, Infra-red light, and Radiowaves. The radio spectrum breakdown is shown in Figure 1-5. We see that the HF band (decametric waves) is defined as Band 7 and that it extends from 3 to 30 MHz in frequency and has a wavelength span between 10 and 100 meters. These wavelength dimensions have particular significance in the design of transmitting and receiving antennas. However, the propagation effects at HF are best understood by considering the frequency domain (if we exclude certain ducting and scattering problems).

----

7. The proceedings of the XVIIth Plenary Assembly held in Dusseldorf, Germany in 1990, was printed and made available to the public near the time of publication of this book. The reader should consult with the Dusseldorf documents to ascertain any changes which have been made to various recommendations and reports since the Dubrovnik Plenary held in 1986. It should be noted that other study groups (SGs) may be of interest to HF workers. The reports of SG5: Radiowave Propagation in Non-ionized Media are important since ground constant and surface wave propagation studies are examined. Aside from SG5 and SG6, the others of interest will include: SG1 (Spectrum Management Techniques), SG3 (HF Fixed Service, currently defunct), SG9 (Fixed Service, general), and SG10 (Broadcasting Services - Sound). Barclay [1990] has discussed the "The Working of the CCIR". Documents may be ordered by writing the ITU, c/o the General Secretariat, Sales Section, Place des Nations, Geneva, Switzerland.

To the military, the HF band covers the range between 2 and 32 MHz. In a later chapter we shall also discuss the concept of frequency extension, a process in which the enhanced *skywave* properties which might be encountered in a nuclear-disturbed environment are exploited. This HF extension into the lower VHF band is defined by the disturbance model used, and the concept is of importance in certain strategic communication scenarios.

1.5 HF: GENERAL PROPERTIES AND PROPAGATION MODES

The term *propagation Mode* is used in so many contexts that it may be confusing at times. Other terms are also used such as approach, technique, or method. We will suggest use of the latter terms instead of *mode* in certain instances, while reserving the term *mode* for special circumstances of interest in the HF skywave context. The following hierarchy is proposed by the author as a matter of convenience:

1. *Propagation Approach*: a term used to differentiate between various path types: the earth-space path, the line-of-sight path, the surface wave path, and the class of media-dependent paths. We would use the term *technique* to distinguish between the media-dependent schemes. (See #2 below.)

2. *Propagation Technique*: a term used to differentiate between various media-dependent schemes such as meteor scatter, troposcatter, ionoscatter, skywave, etc.

3. *Propagation Method*: a general term which may be used as a replacement for *approach* or *technique*.

4. *Propagation Mode*: a term used to differentiate between various propagation categories of the skywave technique. For example, the multiplicity of skywave paths would be characterized by mode descriptors such as 1F2 for the single-hop F2 mode, 2Es for the 2-hop sporadic E mode, etc. Certain modes may be further differentiated by consideration of low and high angle (Pederson) rays, and by specification of either the ordinary or extra-ordinary waves.

HF communication is enabled by a receiver which performs an operation on specified radiowave properties or characteristics such as field intensity, phase, or polarization. Successful operation on these properties depends upon factors which can be controlled, including the transmitted waveform, antenna characteristics, and transmitter power; but operation upon these parameters is constrained by certain factors which cannot be controlled, such as specified propagation effects identified with the channel. Some of these effects may be

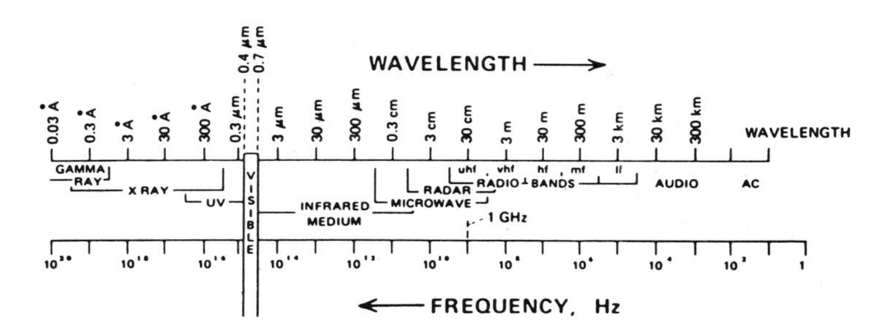

Fig.1-4. The electromagnetic spectrum. (Adapted from Figure A-1 of the *Air Force Handbook*, [Jursa, 1985], AFGL, NTIS, Springfield VA.)

BAND	FREQUENCY	WAVE LENGTH (Free Space)	BAND
ULF	Below 6 Hz	≥ 50 Mm	
ELF	6 - 3000 Hz	50,000 - 100 km	
VLF	3 - 30 kHz	100 - 10 km	4
LF	30 - 300 kHz	10 - 1 km	5
MF	300 - 3000 kHz	1 - 0.1 km	6
HF	3 - 30 MHz	100 - 10 meters	7
VHF	30 - 300 MHz	10 - 1 meters	8
UHF	300 - 3000 MHz	100 - 10 cm	9
SHF	3 - 30 GHz	10 - 1 cm	10
EHF	30 - 300 GHz	1 - 0.1 cm	11

KEY

Mm = megameter = 1,000,000 meters
km = kilometer = 1000 meters
cm = centimeter = 0.01 meter
Hz = Hertz = 1 cycle/sec
kHz = kilohertz = 1000 Hz
MHz = megahertz = 1,000,000 Hz
GHz = gigahertz = 1000 MHz

Fig.1-5. The radio spectrum

estimated and found to be insignificant for certain applications, but other effects are unpredictable and may seriously degrade communication.

Later chapters will deal with the major propagation issues. In this introduction media refractivity and its variations are of major concern. This property of the medium influences the wave phase and group velocities, signal or waveform integrity, the field strength, the polarization, and the direction of signal arrival. There has been a great deal of research directed toward characterization of the propagation medium and its impact on radiowaves.

To analyze the potential performance of a radiocommunication system, it is necessary to consider first the relevant propagation approaches. Preliminary questions include the following: Is the path between transmitter and receiver unobstructed so that it may be regarded as essentially rectilinear? If the path is obstructed by terrain, and the path distances are relatively small, can communication be achieved by the process of diffraction? May connectivity be achieved by surface or ground wave? For long-haul connectivity, do we need to consider the potential of tropospheric scatter or meteor scatter? Do we need to be concerned with ducted modes within the troposphere or ionosphere? We know that the ionosphere provides *beyond-line-of-sight* (or BLOS) connectivity to great distances. Is this achieved, as it is at HF, by reflection from the E and F regions of the ionosphere, or is it the result of propagation within the earth-ionosphere waveguide, as it is at VLF and below? Many of these issues are immediately resolved by specification of the frequency and path geometry. The frequency and path geometry are usually dictated by the communication requirement, although there is typically some flexibility in terms of the frequency selection, following consideration of platform constraints, waveform requirements, etc.

Of major concern to the communication architect is the *precariousness* of the communication channel in terms of propagation effects, both to benign and nuclear disturbances. Several questions arise when considering this class of problems. For example: How many outages would be expected during specified reckoning times, and is this behavior consistent with the quality of communication service required? Can these outages be tolerated or can they be mitigated in ad-hoc fashion by corrective action in the field? What is the potential for incorporating a greater degree of robustness in the system through clever engineering or by applications of modern technology? Propagation support is clearly a major concern when addressing a specific communication requirement. Usually there is no clearly unassailable answer. For the military, there is an additional concern about electronic countermeasures (ECM), including jamming which complicates the problem. In the final analysis, the prudent military planner should always choose a balanced mix of assets to satisfy requirements. We shall see that this will increase the probability for avoidance of propagation effects during stressed conditions, and that it will increase the necessary resource investment by the enemy to launch an ECM attack. In any case, it is important to know the strengths and weaknesses of all bands within the radio spectrum to analyze the problem adequately.

Returning to the problem of determining the type of propagation support which specified radiofrequencies will provide, we may simplify the situation by considering four basic propagation approaches: earth-space, Line-of-Sight (LOS), surface (or ground) waves, and finally an approach that includes all techniques which are media-dependent. These media-dependent techniques include HF-VHF meteor scatter, ionoscatter, VHF-UHF troposcatter, VHF-UHF (atmospheric) ducted modes, HF (ionospheric) ducted and chordal modes, long wave guided modes, MF-HF skywave, etc.

When considering HF, two of the approaches will be most helpful, the surface wave approach and the ionospherically-dependent approach. Using the latter, we are principally interested in the skywave technique, although meteor scatter is an important consideration in the context of *extended HF* for strategic communication requirements. Ducted and chordal mode propagation are not sufficiently prevalent for reliable communication, but they may be important for an understanding of occasional ultra-long range communication via the ionosphere. Ionoscatter is a vague term, but at HF we shall use it to refer to scatter from ionospheric irregularities including field-aligned ionization in the auroral zone and plasma plume formations near the magnetic equator. It is well known that the existence of spread-F and related ionospheric inhomogeneities are associated with so-called above-the-MUF propagation effects.

Ionoscatter should not be confused with the scatter from individual electrons in the ionosphere called Thomson or incoherent scatter. The Thomson scatter cross section is sufficiently small to be unimportant as a means for communication. Thomson scatter has nevertheless been successfully exploited as an ionospheric diagnostic, and it has made a major contribution to our understanding of upper atmospheric structure and dynamics.

Figure 1-6 gives a crude synopsis of the major propagation approaches and techniques, and the primary uses of the radio spectrum between VLF and SHF.

Examining the HF band, we shall find, for long ranges involving BLOS propagation, that ionospheric bounce or some alternate form of media interaction is required for range extension. That is, ground wave and direct (rectilinear) waves are precluded, and this is true during both day and night. At the shorter ranges (0-200 km), we see that HF may provide ground wave (or surface wave) support in addition to skywave support. The ground wave support is greater over the oceans than over the ground, and it diminishes rapidly with increasing radio frequency, and is effectively limited to vertical polarization. This polarization selectivity has implications for antenna selection. *Extended-line-of-sight* (or ELOS) coverage capability, arising from contributions by groundwave and *line-of-sight* (or LOS) components, is important in tactical military maritime-mobile communications. *skywave* support for short ranges (0-500 km) is limited to the lower portion of the HF band and the scheme is termed *Near-Vertical-Incidence-Skywave* (or NVIS).

1.6 PROPAGATION FACTORS AT HF

The ionosphere is the medium of primary significance for HF skywave propagation and its behavior defines the temporal and geographical properties of HF system performance for a significant majority of the cases of interest. Figure 1-7 serves to remind us of the basic ionospheric structure exhibiting the various regions and layers of interest. The physical basis for this structure will not be discussed here. It will be covered in detail in Chapters 3 and 4. We shall simply assert that it results from a very complex interaction of the multicomponent solar electromagnetic and corpuscular flux with the earth's neutral atmosphere. The fact that the earth possesses a magnetic field also complicates the picture.

Separation of the environment (especially the ionosphere) into two states, benign and disturbed, is qualitatively simple but quantitatively difficult. There are a number of disturbing influences which occur quite naturally and these disturbances give rise to a continuum of effects upon C^3I systems.

We know that solar flares, geomagnetic storms, and nuclear detonations will produce perturbations. Disturbances may also be produced by releases of chemical reagents, and nonlinear RF interactions with the ionosphere. The magnitudes of the introduced effects vary widely and may lie beneath the upper limit of variability associated with *unknown* sources. It is also true that certain regions of the ionosphere are always disturbed. The auroral zone is a noteworthy example. The equatorial region could also be placed in this category depending upon the observable involved.

BAND	FREQUENCY	PROPAGATION (Principal Modes)	USES
VLF	3-30 kHz	Waveguide Groundwave	Navigation Standard frequency Standard time
LF	30-300 kHz	Waveguide Groundwave	Broadcasting (BCST) Navigation
MF	300-3000 Hz	Groundwave Skywave (E)	International distress AM BCST
HF	3-30 MHz	Groundwave Skywave (E,Es,F1.F2)	International BCST Amateur Service Citizens OTH radar
VHF	30-300 MHz	Line-of-sight (LOS) Es Scatter	Television (TV) FM BCST
UHF	300-3000 MHz	LOS	SATCOM BCST Radar Navigation TV
SHF	3-30 GHz	Troposcatter LOS	SATCOM BCST Radar Navigation TV

Fig.1-6. The radio spectrum and its utilization. The ULF and ELF bands (below VLF) and the EHF band (above SHF) are not shown. *Communication* is understood to be an important *use* in all of the bands shown even though a specific form of service may not be specified in the listing. It will be noted that at HF, the principal modes of propagation involve reflection or refraction from ionospheric layers. However groundwave *approaches* are also useful over relatively short unobstructed distances and for oversea paths where losses are reduced. At VHF, scatter from meteor trails is an important *technique*.

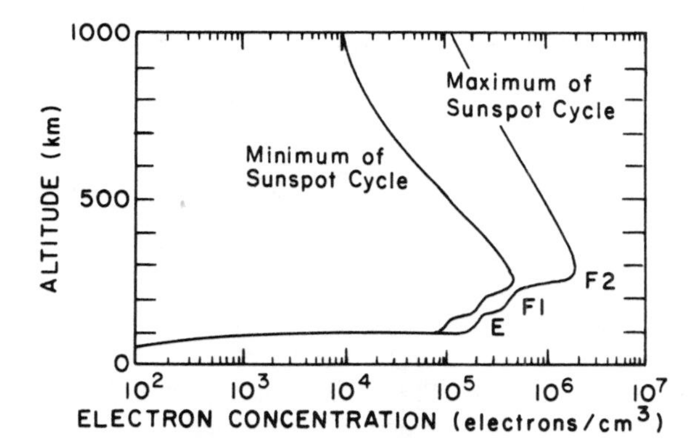

Fig.1-7. The daytime ionosphere for solar maximum and minimum conditions. Higher portions of the HF band are supported near the distribution peak.

A convenient definition of the benign ionosphere state is that ionosphere which is undisturbed by solar flares, large geomagnetic storms, and <u>known</u> manmade (including nuclear) events. In short, the smooth ionosphere which is based upon quiet geomagnetic and non-erratic solar conditions will be termed the benign ionosphere provided nuclear effects may be ignored. The benign ionosphere, so defined, is exemplified in well-known mean morphological models of electron concentration. This definition admits to a wide variation in solar activity, provided it is not impulsive in nature. Nevertheless, the removal of obvious disturbing influences still leaves a substantial residue of benign ionospheric variability unaccounted for. A large portion of this variability is the result of the superposition of a large class of *traveling ionospheric disturbances* (TIDs). Other sources of variability include *sporadic-E*, *spread-F*, and scintillation-producing irregularities. Discussion of these and other sources of variability is found in chapter 3. Success in modeling these sources has been achieved in recent years, at least from the statistical point of view.

The HF band is by far the most sensitive to ionospheric effects. Indeed, HF radiowaves experience some form of almost every conceivable propagation mode including groundwave, line-of-sight and earth-space modes, reflected modes, refracted modes, ducted modes, chordal (earth-detached) modes, and scatter modes. The major ionospheric layers possess characteristic plasma frequencies which lie within the HF band, and as a result, ionospheric interaction is most pronounced. Accordingly, most of the traditional ionospheric diagnostic systems exploit this feature and use HF waveforms to probe the ionosphere and determine its structure. Figure 1-8 depicts the various possibilities. Other (non-HF) methods have been employed as well. For example, in-situ methods such as satellite- and rocket-borne plasma probes have been effective in scientific studies; and Thomson scatter radars provide considerable detail, especially about the topside ionosphere. Nevertheless, HF methods, and especially HF ionosonde instruments, form the backbone for ionospheric characterizations which have been most useful in HF propagation forecasting and assessment. Furthermore HF wideband and narrowband channel probes operated over oblique paths may be used to estimate the properties of an actual HF circuit. The popularity of oblique-incidence-sounders (OIS) and similar devices attests to the advantage of sampling the environment with a frequency (and a geometry) which is similar to, if not the same as, that which the HF system will use itself.

Propagation factors at HF have been well chronicled. Comprehensive treatments have been published by the CCIR in the form of the so-called *Green Books* [CCIR, 1986a]. The CCIR has also documented various field strength prediction methods for use in the HF band [CCIR,1986b;1986c]. In view of the renewed interest in HF, research in the area has increased, and recently a book emphasizing HF system applications has been published [Maslin,1987].

One of the major problems which arise in connection with HF system performance is the variability in coverage and reliability for a fixed transmitter site and a specified frequency. This variability mimics the ionospheric variability itself, and recently schemes have been developed to monitor the ionosphere and adjust certain system parameters in near-real-time to compensate for the system effects. The value of these *Real-Time Channel Evaluation* (RTCE) schemes, which may be central in specified adaptive-HF methodologies, will be addressed later.

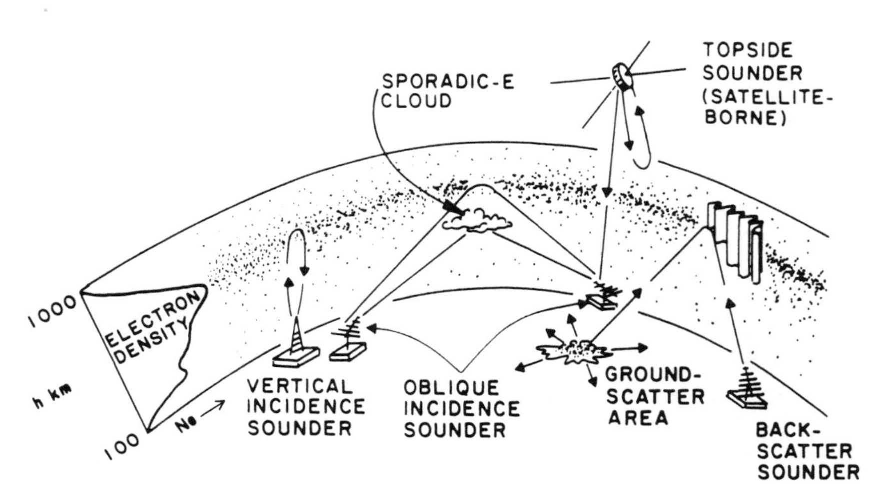

Fig.1-8. Sounding and ionospheric diagnostic techniques which exploit the HF band. (By permission, Goodman and Aarons [1990], copyright 1990 IEEE.)

Included among the quasi-global disturbance phenomena which may influence long-haul HF systems are *Sudden Frequency Deviation* (SFD) and *Short Wave Fades* (SWF), both of which occur within minutes of the appearance of an x-ray flare on the solar surface. These events are significant only on the sunlit portion of the ionosphere and the effects are diminished as the ionospheric distance from the subsolar point increases. Less immediate but near-term phenomena associated with energetic solar protons are also encountered. *Polar cap absorption* events (or PCAs) are perhaps the most catastrophic events in connection with HF radio propagation in the high latitude zone. Attenuation over skywave circuits in excess of 100 dB is sometimes encountered. These absorption conditions may last from hours to days. Fortunately, they are rare events which are not typically encountered at solar minimum and are observed approximately once a month at solar maximum.

Probably the most interesting phenomenon to be encountered at HF is the ionospheric storm which gives rise to a hierarchy of effects at midlatitudes. Although the time history of ionospheric storms is also of importance in earth-space satellite applications, it may be devastating in the HF band since it may limit ionospheric support in the normally-propagating higher frequencies causing nonabsorptive *blackout* of HF trunks in the affected area.

An ionospheric storm is an ionospheric manifestation of the geomagnetic storm whose basic phenomenology has been fully described by Akasofu [1977]. At HF we are principally interested in the diminutions of the F2 region electron density which is highly correlated with the temporal structure of the main phase of the geomagnetic storm. Maximum Observable Frequency (MOF) reductions of up to 50% may be observed for the day of the disturbance with full recovery occurring after several days. There is a tendency for more ionospheric storms to be observed during solar maximum conditions.

Figure 1-9 is a tabulation of ionospheric effects, the magnitudes, durations, occurrence rate, and related factors.

There is clearly a need to study the ionosphere and its coupling to the magnetosphere above and the troposphere below in order to more fully develop the insight required to specify the radiowave propagation effects introduced by these various media. This is of special significance in the development of HF forecasting schemes, HF being the most precarious of all the RF bands. This venture must be closely allied to studies of effects associated with both the sun and the interplanetary medium, so that adequate predictive and forecasting technology may be realized. Current empirical models of the ionosphere are inadequate except perhaps for system design guidance and more effort should be spent in this area. Models are especially poor over oceanic areas and in zones where experimental observations are sparse. Therefore those HF users who are required to operate long-haul oceanic circuits may not be particularly well served.

IONOSPHERIC DISTURBANCES

DISTURBANCE	PROPAGATION EFFECTS	TIME AND DURATION	APPROX OCCURRENCE FREQUENCY SOLAR MAX	SOLAR MIN	Probable Cause
a) Sudden Ionospheric Disturbance (SID)	In sunlit hemisphere, strong D layer absorption (shortwave fades), anomalous VLF-reflection, F-region effects.	All effects start approx simultaneously. Duration ~1/2 hour.	2/Week	2/Year	Enhanced solar X-ray and EUV flux from solar flare.
b) Polar Cap Absorption (PCA)	Intense radiowave absorption in magnetic polar regions. Anomalous VLF-reflection.	Starts a few hours after flare. Duration one to several days.	1/Month	0	Solar protrons 1-100 MeV.
c) Magnetic Storm	F-region effects; increase of foF2 during first day, then depressed foF2, with corresponding changes in MUF.	May last for days with strong daily variations.	26/Year	22/Year	Interaction of solar low energy plasma (solar wind) with earth's magnetic field, causing ener getic electron pre cipitation, auroral effects, heating, and TID generation.
d) Auroral Absorption (AA)	Enhanced absorption along auroral oval in areas hundred to thousand kilometers in extent. Sporatic K may give enhanced MUF.	Complicated phenomena lasting from hours to days.	Essentially omni-present		Precipitation of elec trons with energies a few tens of keV within an oval region equatorward of the polar cap.
e) Travelling Ionospheric Disturbances (TID)	Changes of foF2 with corresponding changes of MUF sometimes periodic.	Typically the periods are from tens of minutes to hours.	Essentially omni-present with larger scales enhanced during magnetic storms		Atmospheric waves.

Fig.1-9. Ionospheric disturbances which influence radio systems. (By permission, Goodman and Aarons [1990], copyright 1990 IEEE.)

The data void associated with oceanic areas may be filled (through estimation) by physical models in some instances, but such measures generally lead to predictions which have marginal utility in most operational scenarios. In addition, the more sophisticated scientific models, whether they be empirical or physical in nature, may overburden the computational capacity of an operational system. As a result, future trends may be directed toward the development of more simplified operational models which may be updated with real-time observables. It is emphasized, however, that users must recognize that the scientific models have to precede the operational models to fully identify the relationships involved.

Most of the macroscopic features of the ionosphere are reasonably well understood although certain details remain as perplexing problems. However, for short-term prediction or forecasting capability, we encounter far more serious deficiencies and the day-to-day and other short-term temporal variations in electron density contribute to these difficulties. There are also some uncertainties associated with the geomorphology of ionospheric inhomogeneities of all scales, although considerable progress has been made during the last decade through comprehensive experimental and theoretical studies. In short, the irregular properties of the ionosphere are clearly inadequate for purposes of command and control from the point of view of phenomenology and driving or triggering functions. This deficiency affects propagation assessment for transionospheric paths as well as ionospherically-reflected modes in profound ways. The definition of the irregular ionosphere and our ability to predict its impact upon radiowave systems in near-real time could be the most important contribution in the decade of the 1990s.

Within the HF portion of the radio spectrum, the prediction and analysis of solar flares and specifically the various ionospheric disturbances produced by solar flares present difficulties. There also appear to be ionospheric irregularities which are only indirectly related to the sun. Such irregularities - including tilts, layer height variations, TIDs, spread F, and sporadic E - are difficult if not impossible to predict from first principles. Direct observation (remote sensing) and extrapolation techniques may be the only solution to such problems. These topics will be covered later.

1.7 THE TRIALS AND TRIBULATIONS OF HF

1.7.1 The Tribulations

HF communication supporters have proclaimed that the ionosphere is a satellite which doesn't fall down! With all of the problems which the ionosphere may introduce (in terms of variability and so on), the statement does have an element of truth[8]. Indeed, there is reason to expect that the ionosphere will always be available to permit HF skywave propagation. Restoration of the normal ionosphere is even expected to occur following a massive use of nuclear weapons. The time might extend to a few hours or days, but the prognosis is good. The same cannot be said of satellites. Direct physical destruction coupled with nuclear *electromagnetic pulse* (EMP) attack could render SATCOM subordinate to HF in both trans- and post-attack periods.

The primary advantages of HF are apparent upon review of the attributes given in Table 1-1. Although telecommunications architects generally relegate HF to a *backup* status, in real-world tactical situations which require quick setup and utilization, it is HF which generally provides the most reliable capability.

----

8. It may be said that the ionosphere, although variable, is virtually indestructable. Therefore an investment in communication systems which exploit the ionosphere has been viewed to be a prudent one by military planners. The development of major systems such as ELF-Seafarer, VLF-TACAMO, and a variety of *strategic* HF systems has been the result.

TABLE 1-1. THE ADVANTAGES OF HF UTILIZATION

ECONOMY: Equipment, physical plant, and the cost per circuit mile are relatively low.
AVAILABILITY: Many communication requirements may be satisfied with existing assets. Incremental improvements possible with appliques in some instances, and the technology required to upgrade HF is now *on-the-shelf,* and advanced technology R&D shows promise.
RELAY-INDEPENDENCE: Communication is possible without relays over short, intermediate, and global distances, provided multiple hops are allowed.
COMMUNICATION ADEQUACY: Ability exists to provide 2-way voice and data communication with adequate bandwidth for most applications.
FLEXIBILITY: HF systems are easily proliferated and netted.
FUNCTIONALITY: Ability exists to facilitate non-time-critical functions such as recall, redirection, reconstitution, attack assessment, and civil defense.

1.7.2 The Trials

The problems which arise with HF propagation are well known and have been mentioned earlier. Primary problems include absorption, selective fading, and dispersion. If these were the only problems, the situation would be acceptable since there are approaches for coping with these degradation effects. In some instances it may even be possible to compensate for the impairments which are bound to arise. Table 1-2 is a list of unfavorable features attributed to HF communication systems.

The negative aspects of vintage HF are largely the result of factors which can be controlled. The decade of the 1980s saw a revival of HF because of an awareness of this fact. Indeed, within the military, programs for the improvement of HF resource management and for replacement of old equipment are well underway. Methods for improving HF have been known for many years. Although many of these proposed improvements have been made attractive by modern technology and microcomputers, the implementation of most of them has been simply a matter of resolve.

1.8 SOLUTIONS TO HF PROBLEMS

One of the most fascinating developments in the field of HF in recent years has been the investment in advanced HF concepts which take advantage of ionospheric diversity paths to improve system performance. This topic will be discussed in detail in Chapter 7, along with techniques termed *adaptive HF.*

Incorporated within the umbrella of *adaptive HF*[9] schemes are those components listed in Table 1-3.

TABLE 1-2. LISTING OF HF UTILIZATION DISADVANTAGES

1. HF has a restricted range of coverage if single-hop coverage is required for reduced distortion. The coverage *footprint* around each transmitter has a radius of about 4000 km for an isotropic antenna. Global connectivity using single-hop links requires netted operation.
2. Because of its dependence on ionospheric support, HF *skywave* generally exhibits a lower intrinsic reliability than satellite telecommunication methods. (Naturally, this is untrue if the satellite system is the object of direct physical, electromagnetic pulse (EMP), or electronic warfare (EW) attack. Satellite communication (SATCOM) must contend with other forms of environmental impairment: rainfall attenuation, ionospheric scintillation, spacecraft charging, etc.)
3. Although technology is available to improve HF reliability markedly, much of the existing inventory is archaic and utilizes rather unimaginative architectures. These older radios are characterized by small *mean-time-between-failure* (MTBF) and a large *mean-time-for-repair* (MTFR).
4. Most pre-1990 equipment was based on technologies of the 1950-1970 period. This necessitated procedures for operation which were manpower-intensive. Since much of the current inventory is of this *older* vintage, there is still the need for skilled operators for optimum performance. This increases training cost.
5. Pre-1990 HF equipment is non-adaptive, and utilization is exacerbated by the use of frequency management procedures which are rigid and based upon inadequate long-term prediction methods.
6. Although some systems have the potential to be quasi-adaptive, this capability is limited, and it features non-organic sounding equipments which are costly and require manual interpretation.
7. The HF spectrum is highly congested, complicating the frequency management problem.

----

9. *Adaptive HF*, discussed in chapter 7, is a valiant attempt to optimize information rate or the system reliability in a time-varying environment. To accommodate this feature, it is necessary for the system to possess agility across a wide range of system parameters. Sufficiency involves *real-time channel evaluation* (RTCE) and a method for conveyance of channel intelligence to the system controller (typically a microprocessor).

TABLE 1-3. ADAPTIVE HF ATTRIBUTES AND CHARACTERISTICS

1. Real-time sounding to determine optimum transmission frequencies. The sounding instrumentation may be either organic to the system or external (adaptive frequency diversity).
2. Real-time measurement of channel occupancy and interference to allow for adaptive selection of transmission frequencies from the observed set of *open* channels (adaptive frequency diversity).
3. Real-time measurement of system performance with allowance for adjustment of system parameters such as: power level, error-correction code rate, and effective transmission rate. The availability of modern equipment such as microprocessor-controlled modems is implied (adaptive time diversity).
4. Advanced signal processing countermeasures to compensate for selective fading effects or to resolve multipath thereby achieving *implicit* diversity gain (adaptive equalization or *Rake* schemes are candidates).
5. Adaptive antenna arrays to *null-out* jammers (adaptive space and/or mode diversity).
6. Real-time adaptive relay capability. Network reorganization would be accomplished automatically through application of link-level RTCE information (adaptive path and/or site diversity)

Other features of modern HF systems, which may or may not be classified as adaptive in nature, may involve the concept of frequency extension and the exploitation of spread spectrum technology. Adaptive features are typically regarded as measures which are applicable in the context of a *MUF-seeking* HF system architecture.

MUF-seeking[10] strategies are wholly consistent with the traditional view that optimal performance is achieved at the highest frequency which will support connectivity between two terminals. Indeed, the ionospheric absorption losses are less at the highest frequencies, the multipath probability is generally reduced, noise levels are diminished, and congestion difficulties are more manageable. *MUF-seeking* concepts require the selection of a rather small frequency window for each designated path, and this selection is decidedly time-dependent over a variety of scales which depends on solar activity and other imprecisely predicted parameters. That is why Real-Time-Channel-Evaluation (RTCE) concepts are essential in most adaptive HF designs. Some

---

10. The term *MUF-seeking* is attributed by the author to R. Bauman of NRL in connection with the development of strategic HF system concepts for the U.S. Navy in the late 1970s and early 1980s. *MUF-seeking* systems belong to the class of adaptive systems which may be personified as *smart* since the derived channel information (or *intelligence*) is exploited.

modern HF designs exploit *robust*[11] concepts which may be *non-MUF-seeking* in nature. Such systems are typically characterized by low data rates and are specialized. More will be said about this class of systems in chapter 7.

Modern HF radio systems will require a high degree of automation in order to take advantage of adaptive HF technologies. In order to reduce conflicts between multiple users of a network, a combination of discipline and central control will be necessary. Frequency pooling will be necessary ultimately to increase the frequency management flexibility and maintain circuit reliabilities. Currently most sounding functions are performed external to the system using separate transmitters, receivers, and antennas. Future systems will likely provide all requisite RTCE capability in a manner which is organic to the system.

1.9 REFERENCES

Akasofu, S., 1977, *Physics of Magnetospheric Substorms*, D. Reidel Publishing Co., Boston.

Baker, D.C. and S. Lambert, 1989, "Range Estimation for SSL HFDF by Means of a Multiquasiparabolic Ionospheric Model," *IEE Proceedings*, Vol.136, Pt.H, No.2, pp.120-125.

Barclay, L.W., 1990, "The Working of the CCIR", *Electronics and Communication Engineering J.*, December issue, pp.244-249.

Beynon, W.J.G., 1975, "Marconi, Radio Waves and the Ionosphere," *Radio Science*, 10(7):657-664.

CCIR, 1986a, *Recommendations and Reports of the CCIR, 1986; Volume VI: Propagation in Ionized Media*, Proceedings of Meeting held in Dubrovnik, Yugoslavia, ITU, Geneva.

CCIR, 1986b, "CCIR Interim Method for Estimating Skywave Field Strength and Transmission Loss at Frequencies between the Approximate Limits of 2 and 30 MHz," Report 252-2, (See also Supplement to 252-2), in *Recommendations and Reports of the CCIR, 1986; Volume VI: Propagation in Ionized Media*, ITU, Geneva.

---

11. *Robust* systems incorporate a variety of measures which taken together allow the system to satisfy mission requirements without consideration of detailed channel information. A properly engineered system would allocate sufficient margin(s) to overcome any anticipated system impairments. *Robust* systems exploit diversity without characterizing it with RTCE. Some examples of robust measures include: low-rate coding, non-coherent detection, higher power, relay-capable networking, netted groundwave operation, and reduced signalling rate. *Robust* systems are not personified as being *smart*; however it may be smart to use such an architecture when the environment is changing more rapidly than may be accounted for in an adaptive approach. The nuclear-disturbed environment is an example.

CCIR, 1986c, "Simple HF Propagation Prediction Method for MUF and Field Strength," Report 894-1, in *Recommendations and Reports of the CCIR, 1986; Volume VI: Propagation in Ionized Media*, ITU, Geneva.

Chapman, S., 1968, *Solar Plasma, Geomagnetism and Aurora*, Gordon and Breach, New York.

Dutta, S. and G. Gott, 1982, "HF Spectral Occupancy," in *HF Communication Systems and Techniques*, (Second Conference), IEE, Savoy Place, London.

Gebhard, L.A., 1979, *Evolution of Naval Radio-Electronics and Contributions of the Naval Research Laboratory*, Naval Research Laboratory Report 8300, USGPO, Washington DC.

Gething, P.J.D., 1978, *Radio Direction Finding*, Peter Peregrinus Ltd., IEE, Stevenage, Herts, England.

Goodman, J.M. and D. R. Uffelman, 1983, "On the Utilization of Ionospheric Diagnostics in the Single-Site Location of HF Emitters," presented at the NATO-AGARD 32nd Symposium on *Propagation Factors Affecting Remote Sensing by Radio Waves*, NATO-AGARD-CP-345, Tech. Edit. and Reproduction Ltd., UK.

Goodman, J.M. and Jules Aarons, 1990, "Ionospheric Effects on Modern Electronic Systems", *Proc. IEEE*, 78(3):512-526.

Hill, R.D., 1990, "Origins of Radar", in *EOS: Trans. American. Geophys. Union* 71(27):781.

Jursa, A.S. (editor), 1985, *Handbook of Geophysics and the Space Environment*, AFGL, Bedford MA, NTIS, Springfield, VA.

Maslin, N., 1987, *HF Communications: A Systems Approach*, Plenum Press, New York and London.

Mayaud, P.N., 1980, *Derivation,Meaning, and Use of Geomagnetic Indices*, Geophysical Monograph 22, American Geophys. Union, Washington DC.

McNamara, L.F., 1988, "Ionospheric Limitations to the Accuracies of SSL Estimates of HF Transmitter Locations," in *Ionospheric Structure and Variability on a Global Scale and Interaction with the Atmosphere and Magnetosphere*, NATO-AGARD Conference, Munich FRG.

Reuter, F., 1971, "Funkmess," Westdeutscher Verlag, Opladen.

Taylor, A.H. and E.O. Hulburt, 1926, "The Propagation of Radio Waves over the Earth," *Physical Rev.*, February issue.

Teters, L.R., J.L. Lloyd, G.W. Haydon, and D.L. Lucas, 1983, "Estimating the Performance of Telecommunication Systems Using the Ionospheric Transmission Channel: Ionospheric Communications Analysis and Prediction Program User's Manual," (IONCAP User's Manual), NTIA Report 83-127, NTIS, Springfield, VA.

Travers, D.N. and S.M. Hixon, 1966, "Abstracts on the Available Literature on Radio Direction Finding 1899-1965", Southwest Research Institute, San Antonio, TX (U.S. Navy Bureau of Ships Report).

1.10 BIBLIOGRAPHY

Atherton, W.A., 1987-89, "Pioneers," a series of articles found in successive issues of *Electronics and Wireless World*, Reed Business Publishing Ltd., Sutton, Surrey, UK.

Checcacci, P.F.(editor), 1975, Special Issue of Radio Science in commemoration of Marconi, *Radio Sci.*, 10(7):655-761.

Chiles, James R., 1987, "The Road to Radar," in *Invention & Technology*, (Spring issue).

Dunlap, O.E., 1944, *Radio's 100 Men of Science*, Harper & Bros., New York.

Gilmore, C. Stewart, 1982, "William Altar, Edward Appleton, and the Magneto-Ionic Theory," *Proc. American Philosophical Soc.*, 126(5):395.

Gordon, William E., 1985, "A Hundred Years of Radio Propagation," *IEEE Trans. A/P*, Vol.AP-33, No.2, p.126.

Ivall, Tom, 1985, "Radar in Retrospect," in *Electronics & Wireless World*, pp. 74-75 (August).

Kraus, John D., 1985, "Antennas Since Hertz and Marconi," *IEEE Trans. A/P*, Vol.AP-33, No.2, p.131.

Kniestedt, J., 1989, "A Century Ago Heinrich Hertz Discovered Electromagnetic Waves," *Telecommunications J.*, 56(6):376-380.

Mitra, S.K., 1947, *The Upper Atmosphere*, The Royal Asiatic Society of Bengal, Calcutta.

Morgan, P.F.A., 1986, "Highlights in the History of Telecommunications," in *Telecommunications J.*, 53(3):138-148.

Nagle, John J., 1984, "The Development of Amateur SSB: A Brief History," *Ham Radio Magazine*, pp. 12-23 (September).

Ratcliffe, J.A.(editor), 1974, special issue of JATP to commemorate the 50th anniversary of the early work of Appleton, Barnett, Breit and Tuve, *J. Atmospheric Terrest. Phys.*, 36:2069-2313.

Skolnik, M.I., 1986, "Radar Research and Development at NRL," publication number 0073-2630, NRL, Washington, DC.

Smith, K.L., 1988, "A Radiant Century," *Electronics and Wireless World*, pp. 1061-1062.

Waynick, A.H., 1974, "Fifty Years of the Ionosphere. The Early Years-Experimental," *J.Atmospheric.Terrest.Phys.*, 36:2105-2111.

2

THE SOLAR-TERRESTRIAL ENVIRONMENT

HF SYSTEM IMPACT

2.1 CHAPTER SUMMARY

In this chapter we will turn our attention to the solar-terrestrial environment and discuss those interactions which occur in the ionosphere, a medium which is essential for HF skywave propagation. We will examine the solar structure and the origin of solar activity, magnetospheric topology, and solar-terrestrial relationships in general. We will cover solar activity cycles and indices, and we will discuss geomagnetic activity and the various indices which represent this important parameter. We shall conclude with a discussion of recent progress in the development of solar-terrestrial prediction systems and will identify those services which are available for forecasting purposes.

2.2 INTRODUCTION

The reader will find many excellent texts which deal almost exclusively with solar and magnetospheric physics, and the author will not attempt to duplicate them either in terms of scope or rigor. For those interested in detailed discussions of some of the topics to be covered reference should be made to the bibliography found at the end of the chapter.

The importance of solar activity as it relates to HF communication is well known. It is recognized that the sun exhibits sudden outbursts of energy that are called solar flares, and that these events may play havoc with the performance of certain radiowave systems including commercial television. Even in the HF band which is most sensitive to solar effects, these impulsive events are typically short-lived; the adverse effects disappearing in less than an hour. Less well-known influences of the sun include those changes which are related to significant enhancements in the extent and magnitude of the visible aurora. These events affect high latitude communication, but the disturbances are actually global in nature with ionospheric storms introducing significant alterations in HF coverage and rendering performance predictions uncertain. We shall find that solar influences on the ionosphere, and HF communication systems, may generally be characterized as immediate or delayed, with the long-term occurrence of these categories following an 11-year cycle. A well-known index of solar activity which exhibits the cyclic pattern is the sunspot number. This index roughly characterizes the number of spots on the visible solar disk, and is proportional to that component of solar activity which most severely influences radiowave systems, and especially HF systems. The 11-

year solar cycle is not subject to precise characterization in terms of onset, duration, or magnitude; and its direct influence on the ionosphere is not always clear. Nevertheless, we use indices of solar activity in all current long-term prediction programs. This will be discussed in Chapter 5.

The ionosphere owes its existence to the sun, but it would clearly exist even in the absence of the 11-year solar cycle of sunspots. Indeed, the ionosphere possesses some rather interesting features even during periods of few sunspots. Without the 11-year modulation of activity, the ionosphere would possess a reasonably deterministic variability which is associated with the local solar zenith angle (including diurnal and seasonal effects which are largely controlled by geometry rather than by fundamental physics). Moreover even the benign ionosphere would be characterized by relatively unpredictable variations which arise because of the constitution and dynamics of the underlying neutral gas. In fact there are a host of temporal fluctuations which originate from sources other than the sun including neutral atmospheric *weather* patterns and turbulence. This residual class of fluctuations poses interesting challenges for ionospheric forecasting specialists who have long concentrated on the more obvious and direct association between the sun and the ionosphere. Invoking impulsive and long-term variations in solar activity, additional contributions to ionospheric variance include non-vanishing changes in median properties which are highly correlated with the median activity, as well as a collection of disturbances which are directly related to solar flares and geomagnetic storms. HF channel properties exhibit variations which are highly correlated with the ionospheric behavior.

2.3 THE SUN AND ITS INFLUENCE

Following the tradition of many monographs dealing with HF radiowave propagation and texts on magnetospheric and ionospheric physics, we shall include a review of the nature of solar activity. However, since we are ultimately interested in HF communication and not solar physics, our treatment will exclude the more esoteric topics and consider mainly those issues which will provide necessary insight for the HF communicator.

2.3.1 Solar Structure and Irradiance Properties

A qualitative picture of the modes of outward energy flow from the sun is given in Figure 2-1. The seat of solar activity lies within the central core region. The process by which the energy is generated within the core is similar to the mechanism exploited in the detonation of fusion weapons, but in the case of the sun these reactions are hidden from us because they are contained by the enormous gravitation pressure of the overlying solar layers.

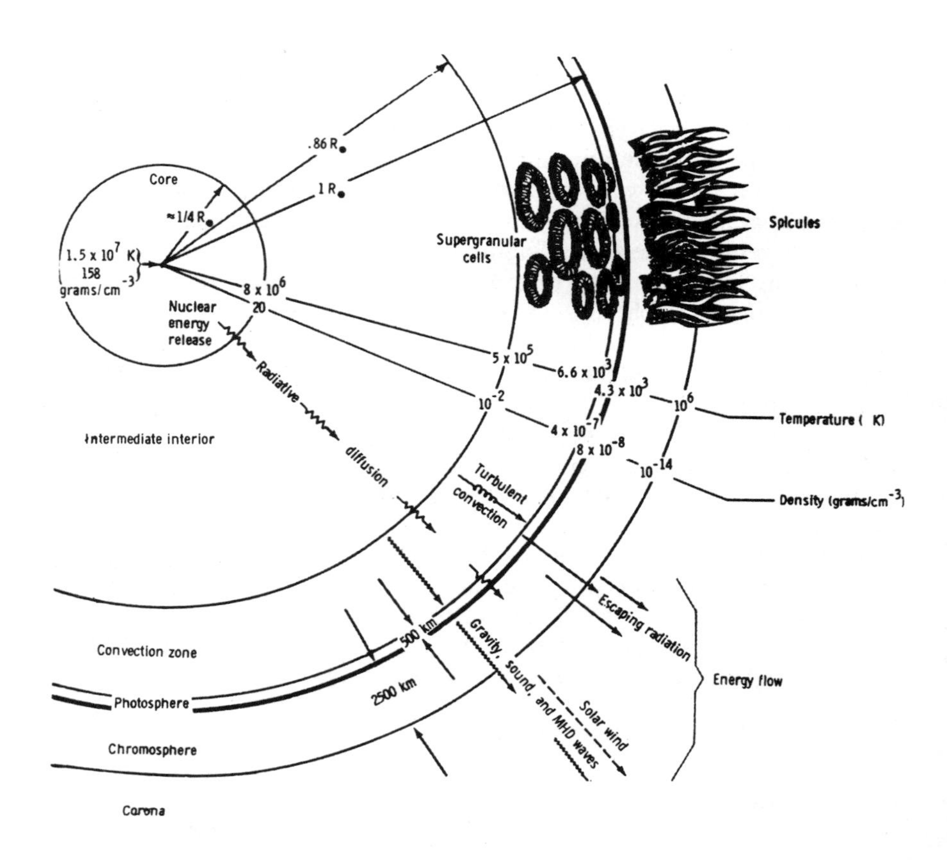

Fig.2-1. A model of energy flow within the sun. This qualitative description depicts the energy transitions from the fusion source to a hierarchy of exiting electromagnetic and particulate flux. (From USAF *Handbook of Geophysics and the Space Environment*, 4th Edition, Jursa [1985]; After Gibson [1973] with some modifications.)

Figure 2-2 depicts the complex structure of the sun and shows a number of the features which have been studied by solar scientists as well as by engineers involved in the prediction of ionospheric impact on terrestrial systems.

Solar physicists now have a good understanding of many of the basic characteristics of the sun, including its average temperature, mass, size, constituents, etc. The sun is about 93 million miles from the earth, it has a mass of about 330 thousand earths, it is a gaseous body, and it rotates (from left to right as viewed from the earth) with a period of about 27 days.

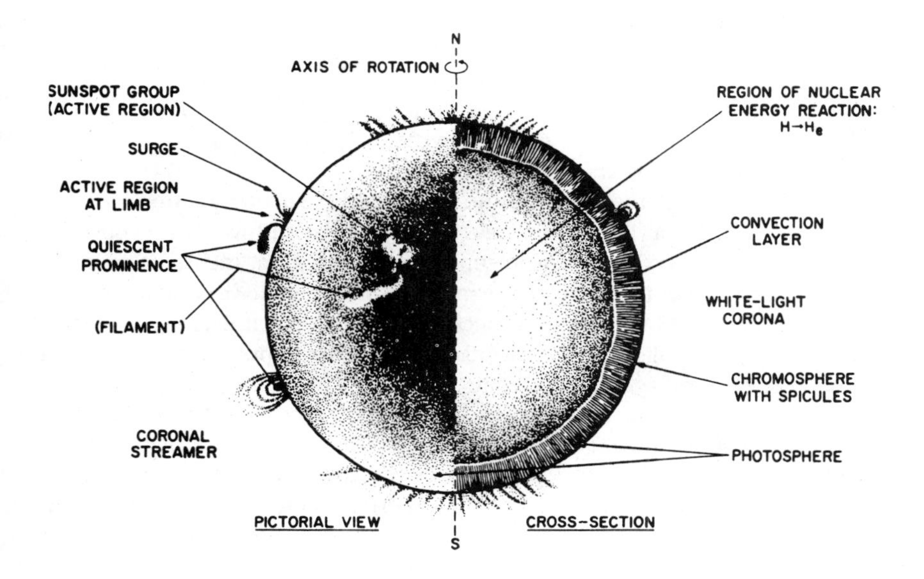

Fig.2-2. Principal features of the sun. Coronal holes, an important source of enhanced solar wind are not depicted. (From USAF *Handbook of Geophysics and the Space Environment*, 2nd Edition, Valley [1965])

The sun is composed of 90% hydrogen, 9% helium which is formed by the fusion of hydrogen, and 1% heavier trace elements. The energy-producing fusion reactions occur in the central core region, generating a temperature of approximately 10 million degrees Kelvin. The energy from the fusion reactions takes 1 million to 50 million years to reach the solar surface.

As observed in Figure 2-1, the temperature of the solar regions diminishes from millions of degrees in the core to its minimum value of several thousand degrees in the lower chromosphere. It rises again to roughly a million degrees in the tenuous solar corona. The equivalent *blackbody*[12] temperature of the

----

12. The blackbody temperature derives from Planck's radiation law. The brightness (or radiance per unit bandwidth) at radio wavelengths is given by the Rayleigh-Jeans approximation:

$$B = 2kTf^2 / c^2$$

where k is Boltzmann's constant, c is the velocity of light (m/s), f is the frequency (Hz), T is the absolute temperature (deg K), and B is expressed in watts/m^2 per Hz per sterad. The power received from the sun in a specified frequency band is called the power flux density, $F = B\Omega$, where Ω is the solid angle subtended by the sun. The power flux density (or irradiance per unit bandwidth) is normally expressed in terms of solar flux units (sfu). One(1) sfu is 10^{-22} watts/m^2 per Hz. The total solar irradiance is about 1370 watts/m^2 .

sun is about 6000 degrees Kelvin for electromagnetic (and radio) emissions having wavelengths less than 1 cm. For longer wavelengths, the equivalent temperature is much higher and is variable. For wavelengths in the range of 10 meters the temperature may range from 10^6 deg K for the benign sun to 10^{10} deg K during highly active periods. In the lower part of the HF band (i.e., 100 meters) the situation is far more pronounced.

At one time it was thought that the integrated electromagnetic flux from the sun as reckoned at a fixed distance from the solar surface was a constant. However, even correcting for earth-sun distance variation, the *solar constant*, which is a measure of the total solar irradiance at a distance of one astronomical unit (1 a.u.), has been found to fluctuate slightly. Its value is approximately 1370 watts/m^2 . Figure 2-3 illustrates the relative importance of selected bands in the electromagnetic spectrum.

The largest contribution to solar irradiance variability (in percentage rather than absolute terms) may be found on either side of the visible, near-infrared, and near-ultraviolet portions of the spectrum. These regions of greatest variability have the most profound effect on the constitution of the upper atmosphere and the ionosphere. Unfortunately (for the purpose of observational science, but fortunately for purposes of biological safety) a large proportion of the high energy component corresponding to extreme-ultraviolet (EUV), ultraviolet (UV) and x-rays is strongly absorbed in the atmosphere and cannot be observed at the earth's surface.

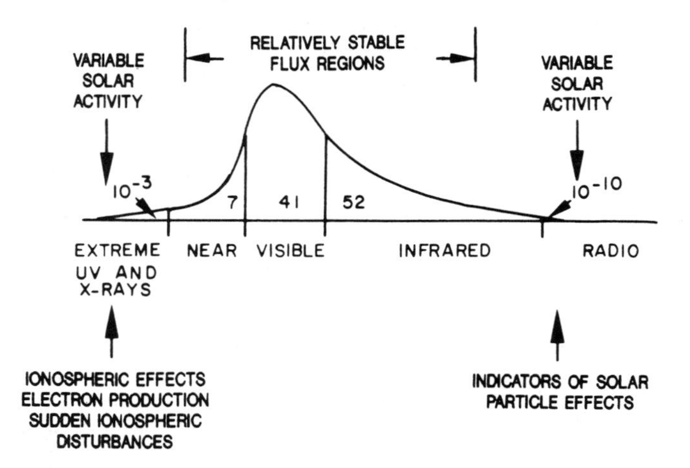

Fig.2-3. Solar spectral distribution from the high energy x-ray band through the radio band (not to scale). The numbers represent percentages of the *solar constant* H where $H \approx 1370$ watts/m^2 .

EUV and x-ray components are quite influential in photoionization processes, and the UV component is associated with ozone layer production. The low energy component of spectral irradiance corresponding to the radiowave region provides information about energetic particle events and about the general level of solar activity. In fact radio data at 10.7 cm (2800 MHz) may be used as a measure of solar activity since such data is more reliable (and perhaps more meaningful) than the well-known sunspot number.

Solar physics is a complex subject and many of the fundamental solar phenomena are still incompletely understood. Therefore, at the present time, the important components of electromagnetic and corpuscular flux exiting the sun would appear to exhibit a degree of chaotic behavior. This chaotic behavior is exclusive of more familiar tendencies toward regularity in the temporal pattern of radiation; especially those phenomena associated with the solar rotation period (27 days), and the 11-year solar cycle of sunspot population. As would be expected, the regular patterns are more amenable to prediction. Nevertheless, the chaotic behavior generally controls the short-term environment making near-term forecasts of solar behavior difficult. Moreover, a clear relationship between a particular solar event and its properties (the cause) and the terrestrial disturbance (the effect) is generally lacking. This has a profound effect upon our ability to predict ionospheric behavior and, of course, HF communication performance.

Because of the circumstantial relationship existing between sunspot number and ionosphere state, considerable effort has been directed toward the development of sunspot number prediction methods. Various prediction methods have been reviewed by Withbroe [1989]. In recent years neural networks have been used to provide estimates of the maximum number of sunspots and when this maximum will occur [Koons and Gorney, 1990]. Predictions of the number of sunspots and other measures of solar activity are covered in sections 2.3.3 and 2.5.

2.3.2 On the Origins and Nature of Solar Activity

Solar physics provides a basis for our comprehension of solar phenomena, and we use this knowledge to explain previous solar events or to forecast future activity. The relationship between sun and earth is of profound importance for a variety of reasons seemingly unrelated to HF communication issues raised in this book. Still, an unequivocal linkage between certain solar phenomena and the ionospheric state has been established, largely the result of ionospheric measurements using HF sounders and probes. Friedman's *Sun and Earth* [1985] is recommended for additional reading.

2.3.2.1 *Introductory Remark*

Sunspots are related to solar activity, and have been monitored for several centuries, revealing a certain periodicity or cyclic behavior. The technical literature suggests that large numbers of sunspots may lead to difficulties in systems which exploit radiowaves, so we will review this correlation.

In the parlance of military planners involved in the command and control of forces, we speak of C^3I (for Command, Control, Communication, and Intelligence) as a *force multiplier*. In military journals, communication is referred to as the "glue" which holds it all together. In recent years, the tendency toward automation has led to the incorporation of a fourth C, standing for Computers, leading to the term C^4I. The notion of C^3I difficulty is naturally a relative term dependent upon the sensitivity of the system to various ionospheric effects which may be introduced by solar disturbances or related events. The ionosphere, whether significant to a particular system or not, is profoundly effected by sunspots, at least circumstantially. Most ionospheric models reflect this fact. Certainly the quiet mean morphological models are constructed with sunspot activity as a major component. The average electron densities increase with increasing solar activity, and other characteristics of the total ionospheric personality including features such as sporadic E, scintillation, and spread F also track sunspot number in a general way. However, if we examine this correlation closely, the relationships become difficult to determine. Specifically, we find that direct relationships between sunspot activity and short-term ionospheric behavior are disguised by unpredictable variations. Some reasons for this will be discussed in Chapter 3.

2.3.2.2 *Solar Magnetic Field, Differential Rotation, and Sunspots*

To study the origin of sunspots, it is necessary, first, to examine the magnetic field structure on the sun, because, in the absence of the solar magnetic field, current theory does not explain the generation of sunspots or their cyclic behavior.

The sun's field is oriented N-S in its quiescent configuration (sunspot minimum), and its intensity is little more than that of the earth's magnetic field, being approximately 1 gauss. However, the sun differs from the earth, where the source of the field is within the metallic core, because the solar field is confined near the surface. The field is *frozen-in* to the surface plasma which can move, transporting the field lines with it. In short, the magnetic field is generally too weak to extricate itself from the control of the highly ionized solar plasma. Since the sun and its surface plasma rotates about its N-S axis, co-rotation of the surface magnetic field also occurs. However, since the sun is a fluid, this rotation is not uniform as a function of solar (or heliographic) latitude. Indeed, the solar surface rotates differentially, with the equatorial region moving more rapidly than higher heliographic latitudes.

This causes the solar magnetic field to become wrapped around the sun over a period of time. It also increases the equatorward magnetic field. Eventually the neighboring stretched field lines become intertwined because of turbulent motion originating in the underlying convection zone. Figure 2-4 shows how this happens.

The twisted field lines are hidden below the visible surface and the most intense regions are associated with local magnetic fields of about 4000 gauss. Such fields exert enormous magnetic pressure on the surrounding plasma. As the magnetic pressure begins to exceed the plasma pressure, the fields penetrate the surface and appear as bipolar loops. This phenomenon arises first at a solar latitude of about 40 degrees where the field line stretching and convergence is most intense. At the points where the field lines protrude from the surface, the magnetic field intensity is so large that energy is prevented from reaching the surface. These points of opposite polarity are several thousand degrees cooler than their surroundings and appear as dark spots on the photosphere.

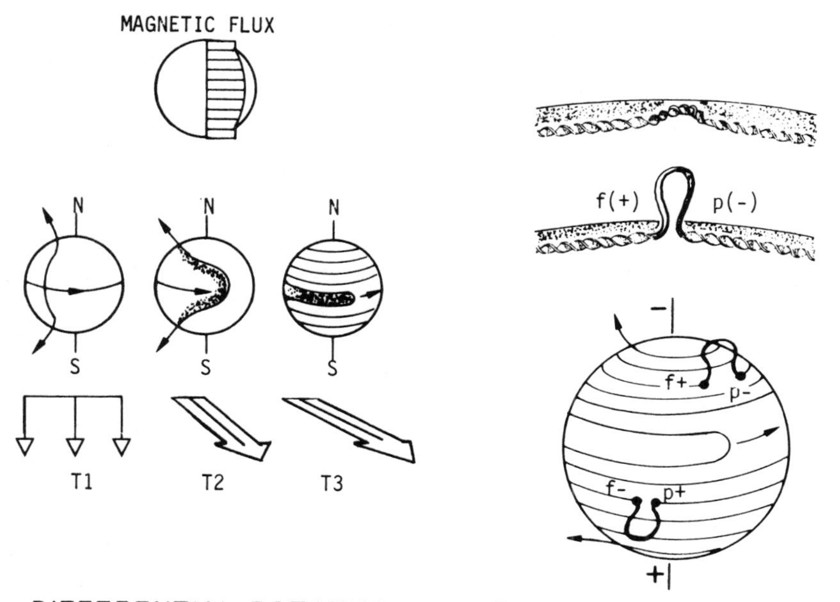

Fig.2-4. Effect of differential rotation on the sun. (a) Development of east-west component of the surface field as the field lines become stretched out between times τ_1, τ_2, and τ_3. This brings lines of magnetic flux closer together. (b) Formation of kinks in the plasma field configuration leading to development of bipolar sunspots. An eventual reversal of the field at the poles results from the effective poleward migration of *following* spots which have an algebraic sign opposite to that of the pole in its hemisphere. (Adapted from Gibson [1973].)

Sunspot pairs usually occur in large groups and are contained within rather long-lived (bright calcium plage) regions. The *preceding* sunspots of the sunspot pairs have the same polarity as the pole in their hemisphere, whereas the *following* sunspots have the opposite polarity. Because of differential rotation, the *following* sunspots lag the overall group motion and form distended unipolar regions which gravitate toward the pole. As a result the latitude of maximum stress moves equatorward and the polar fields become eroded. At sunspot maximum, the polar fields have become completely neutralized. Beyond this point the pole reversal process begins ,and the amount of the sunspots, now being formed near the low latitude region of limited differential rotation, begins to wane. By the time the polarity of the magnetic field has completely reversed, no sunspots are evident. At about the time of the solar minimum, the field lines which were once intertwined return to mostly longitudinal configuration. It takes about 11 years for this process to be completed, and it takes 22 years for the original magnetic configuration to recur. This process is shown in Figure 2-5. From the figure, we see that the spots first start to appear below 40 degrees latitude , both north and south. The maximum solar activity (shown here by the sunspot area index) occurs several years after sunspots first appear near 40 degrees heliographic latitude and a several years before the last sunspots appear near the equator. The latitude-time presentation of sunspot number is sometimes referred to as a *butterfly* diagram.

2.3.2.3 *Active Regions, Coronal Holes, and the Solar Wind*

Active areas on the sun are the regions where there arise many phenomena whose form depends upon the region of the spectrum being monitored. Sunspots are best observed in the visible wavelengths, whereas disturbances in the coronal area overlying the disk are best examined in the soft x-ray band with satellite or rocket-borne instruments which are not effected by atmospheric absorption. X-ray emissions are not observable at ground level, and white light observations of the tenuous corona are made extremely difficult by the overwhelming brightness associated with the disk area.

The earliest measurements of the solar corona were made during solar eclipses or by using a special instrument called a coronagraph. By superposing successive limb scans, it has been possible to reconstruct an image of coronal disturbances in the visible part of the spectrum. While this method does not generate a frozen picture in time (i.e., a snapshot), it allows coronal observations to be mapped over the entire disk area. The first direct x-ray and UV maps of coronal disturbances were made using rocket probes, but it remained for instruments aboard the Skylab satellite to produce the first comprehensive observations of these disturbances and call attention to so-called coronal holes.

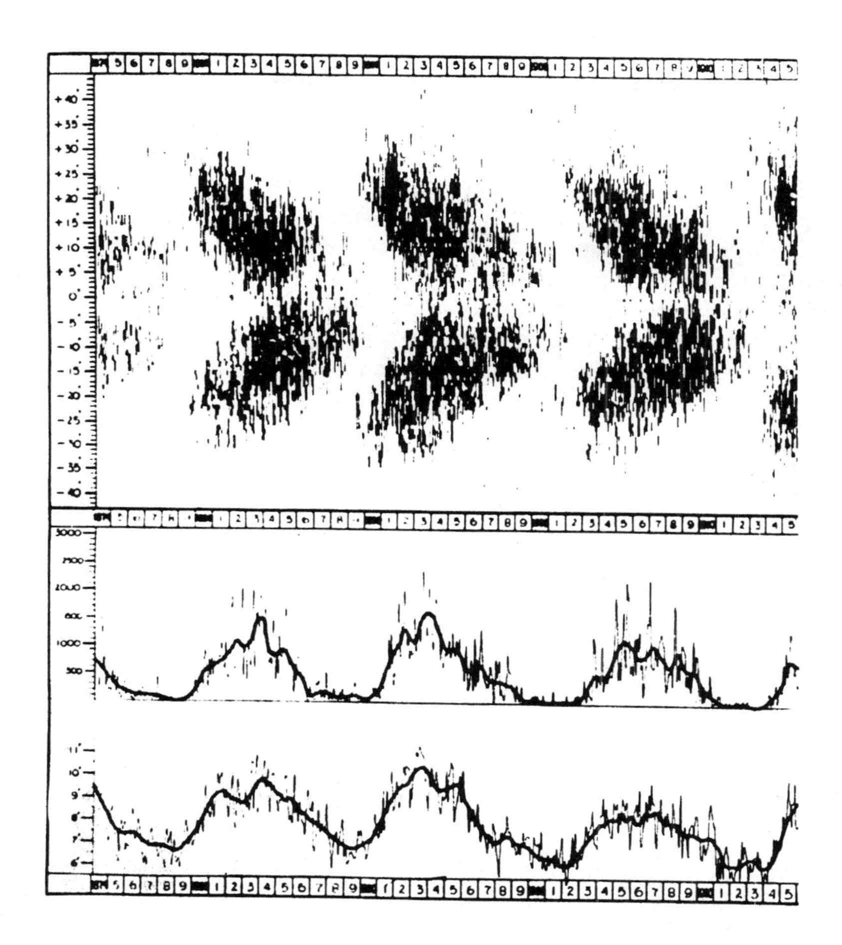

Fig.2-5. This is the so-called Maunder *butterfly* diagram which shows the migration of sunspots from high latitudes to low as the solar cycle progresses. Also given are the area of sunspots and a measure of magnetic activity. [From Chapman, 1968].

What is a coronal hole? Employing a coronagraph, the visible manifestation of a coronal hole is a lack of coronal brightness in certain regions surrounding the disk; this diminution usually arises and persists near the solar poles. These regions, termed coronal holes, are not devoid of plasma but the density is much less than that found in the surrounding gas. These holes are coupled to underlying unipolar active regions where the field lines are nearly radial. Such a configuration allows plasma to escape the sun and propel itself into space. The observation of coronal holes near the north and south poles of the sun is not surprising, since field lines are naturally vertical in those regions. The existence of coronal holes at low latitudes is a direct result of the generation of bipolar sunspots, progressively in the equatorward direction, and the growth of large unipolar regions. Figure 2-6 shows a coronal hole which was observed from the Skylab x-ray telescope. This hole extended from the north polar region into the southern hemisphere, and was persistent in this general form for over six 27-day rotations of the sun. Through a hole of this type, solar plasma has an escape route similar to that which it has from the polar regions. Plasma which escapes from the sun carries a signature with it, the embedded magnetic field which may be either sunward or anti-sunward. This coronal magnetic field is transported by the expanding corona into interplanetary space along distended spiral arms which are called Archimedes spirals. These spirals resemble a rotating gardenhose. They appear as spirals because the magnetic field is dominant within the corona (causing an initial corotation of the exiting plasma) but with increasing distance from the sun, eventual dominance of the magnetic field energy by the kinetic energy of the plasma will cause the plasma to fall behind with respect to the rotating disk. We refer to the transported solar magnetic field as the interplanetary magnetic field (IMF) and the accompanying solar plasma is called the solar wind. Usually the predominant magnetic field polarity within the large unipolar regions in the southern and northern solar hemispheres have opposite signs. Opposing fields from the large unipolar regions tend to reconnect at a great distance from the sun producing a neutral sheet in the neighborhood of the ecliptic plane.

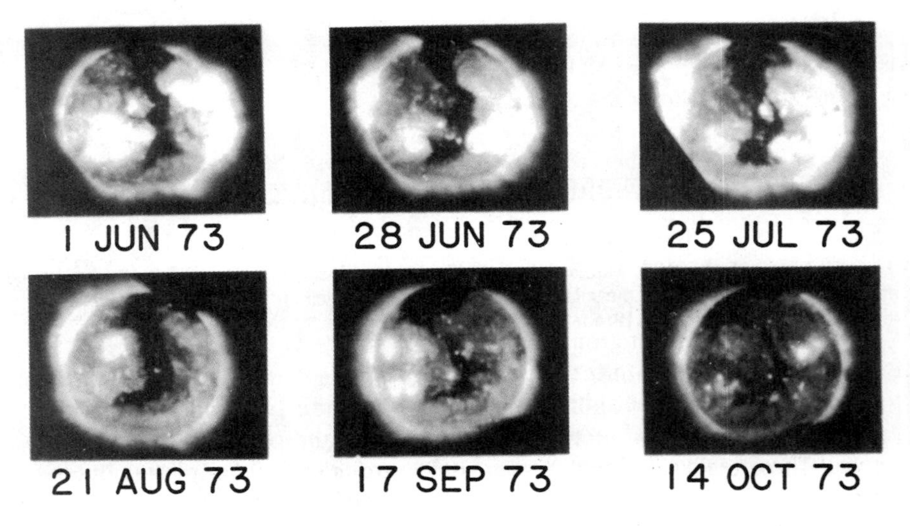

Fig.2-6. An example of a coronal hole observed by Skylab. This event occurred during the decreasing portion of solar cycle 20. Each photograph is separated from its neighbor by 28 days, the period of a solar rotation. (Selected photographs are taken from Figure 1-29 of the USAF *Handbook of Geophysics and the Space Environment*, 4th Edition, Jursa [1985])

Looking down on the pole, a four-sector structure of the IMF is observed, with the magnetic field polarity in adjacent sectors being reversed. This interesting feature is the result of a latitudinal undulation in the neutral current sheet which, under quiescent conditions, would reside in the neighborhood of the ecliptic. The solar wind speed is greatest away from sector boundary crossings. Wind speeds may vary from 700 km/sec during disturbed times and within the center of a sector, to 300 km/sec in the neighborhood of a sector boundary crossing. Greater wind speeds cause more significant ionospheric effects.

Measurement of the distributions of the IMF and the solar wind speed would be useful for the derivation of indices for estimating ionospheric effects. Unfortunately these measurements are difficult to make without the aid of satellite-borne magnetometers and the placing of special particle detectors at distances from the earth which are not appreciably influenced by the earth's magnetosphere. Some research has suggested that scintillation of selected interstellar radio sources (using measurements made by ground-based radio telescopes) may be associated with the plasma density inhomogeneities within the solar wind.

2.3.2.4 *Sunspots and Solar Activity Indices*

Sunspots are visual manifestations of other effects which are important to HF propagation. Sunspots themselves are not important; they are merely circumstantially related to ionospheric behavior. Nevertheless, sunspots have been monitored for centuries and have proved to be a useful if imprecise index.

The most common index of solar activity is based upon a count of the number of sunspots on the solar disk. The fundamental index is the relative sunspot (or Wolf) number. which is reckoned daily. It is given by the following relationship developed by Rudolf Wolf who was the first director of the Swiss Federal Observatory in Zurich:

$$R = k(10g + s) \tag{2-1}$$

where k is a correction factor dependent upon the observatory, g represents the number of sunspot groups, and s is the number of individual spots. To illustrate a peculiarity in the construction of the Wolf number, we need only to compute R for some simple cases. If, for example, only one sunspot were to be observed, then R would be 11 and not 1. By definition each sunspot must belong to a group even though the population may be small. If another spot is observed within the same group, then R becomes 12. On the other hand if two isolated sunspots are observed (belonging to different groups), then R equals 22. The net effect of this construction is to enhance R for low sunspot counts and to limit its growth as the count increases. There is no unassailable reason why this rather artificial recipe for R should properly represent solar activity.

For many years the Wolf number was compiled from measurements compiled at Zurich. Until 1981 when it was discontinued, it formed the basis for many solar and ionospheric studies. After 1981 the Zurich number R_Z was replaced by the International sunspot number R_I. About 25 stations are involved in the construction of R_I.

Records of daily and averaged sunspot numbers are archived by the World Data Center A for Solar-Terrestrial Physics[13] through the National Geophysical Data Center located in Boulder, Colorado. Another index called the *American Relative Sunspot Number* R_a which is compiled by amateur and professional observers is also available. Details about these indices are given in a government publication explaining the available data reports [NOAA, 1987].

Another index of solar activity used by many because of its ease of determination and its power as a representative index of solar activity is the Ottawa 10.7 cm (2800 MHz) solar flux ϕ. This index is expressed as a monthly mean value in units of 10^{-22} Watts/m² per Hz. Stewart and Leftin [1972] have compared the Ottawa flux index with the sunspot number and have derived the following relationship:

$$\phi_{12} = 63.7 + 0.728\, R_{12} + 8.9 \times 10^{-4}\, R_{12}^{2} \tag{2.2}$$

where ϕ_{12} and R_{12} are the 12-month running mean values of ϕ and R respectively. In this expression, R corresponds to the Zurich sunspot number R_Z but if the relationship were a general one, it would be acceptable to replace it with the newer international number R_I. Several HF propagation codes contain a relationship similar to Equation 2.2 permitting the use of either the 10.7 cm flux index or the sunspot number as a parameter.

The National Geophysical Data Center (NGDC) publishes *Prompt*[14] and *Comprehensive* data reports dealing with solar effects and related geophysical phenomena. Solar activity bulletins are also published by the NOAA/NGDC and may be obtained through its offices in Boulder, Colorado. In addition,

---

13. Solar and geophysical data are archived at World Data Centers for Solar-Terrestrial Physics. These centers are located in Boulder, Colorado, USA (WDC-A), Izmiran, USSR (WDC-B), Tokyo, Japan (WDC-C1), and Slough, UK (WDC-C2). The WDC-A for Solar-Terrestrial Physics is one organizational component of the National Geophysical Data Center (NGDC) of the National Environmental Satellite, Data, and Information Service (NESDIS), in the National Oceanographic and Atmospheric Administration (NOAA) of the U.S. Department of Commerce.

14. Two separate reports of solar data are issued by the US Department of Commerce in Boulder, Colorado. These reports: *Part I (Prompt Reports)* and *Part II (Comprehensive Reports)* are a subset of the data archived at NGDC which is collocated with WDC-A for Solar-Terrestrial Physics. The *Prompt* reports contain the most recent data (1-2 months of publication date), and the *Comprehensive* reports contain data which is 6 months old. (See Section 2.5)

circulars of the CCIR and monthly issues of the *ITU Telecommunication Journal* contain tables of solar activity indices. For the user with more immediate requirements, two other services are available: one a public bulletin board, and the other an interactive system called SELDADS which contains a more comprehensive data base for user access. Sunspot numbers are also broadcast on station WWV, and more detailed data sets may be accessed directly over an SESC satellite broadcast channel for those with appropriate terminals (see Section 2.5.3).

2.3.2.5 *Long Term Variations in Solar Activity*

Figure 2-7 gives the range of variability of the sunspot number for a period of 170 years. The plot exhibits daily and long term variability. Prior to 1880, there is an obvious quantization effect shown by the distinct levels which the sunspot number assumes.

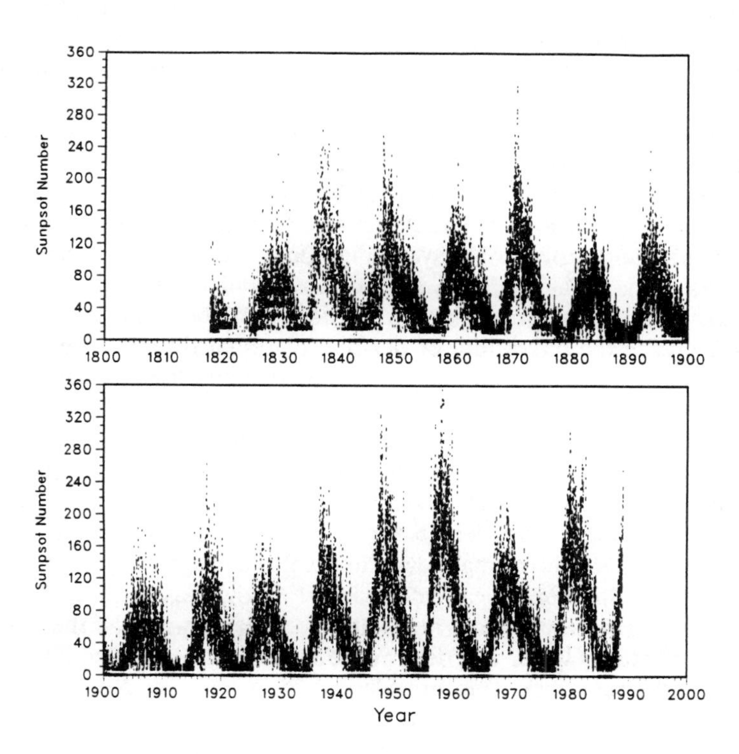

Fig.2-7. Variation of the daily sunspot number between January 1818 and January 1989. Solar cycles 6 through 22 (incomplete) are shown. (From *Transactions of the American Geophysical Union, EOS*, Vol.70, No.32, Aug.8,1989; reproduced by permission AGU, Washington DC)

Recently there has been a perception that the long-term trend involves a grouping of cycles in a ramping type of behavior. For example, solar cycles 16-19 appear to be one such group, and cycles 20 through 22 (the current cycle) appear to be another.

Long term trends in solar activity are well correlated with median properties of the ionosphere. Because of this, median ionospheric models based upon a long term (running average) in solar activity have been developed. Applications of these models include ionospheric and communication performance predictions. The application of solar activity in ionospheric models is discussed in Section 2.3.4.

2.3.2.6 *Short Term Solar Activity Variations*

Observations of the sun have been made for many years, with results being archived in various World Data Centers. In the United States, the National Geophysical Data Center issues reports of solar observations. Figure 2-8 shows a set of solar activity characterizations which are used by scientists and forecasters. For June 16, 1989 the solar disk is shown for the following observational formats: an image in Hydrogen-Alpha light, a magnetogram showing regional clusters of positive and negative polarity, the distribution of sunspots, and a depiction of the solar corona in three spectral regions. These kinds of pictures change dynamically, and display short-term variability.

The solar electromagnetic and particulate flux reaching the earth exhibits considerable short term variability, and the (long term) time-averaged behavior tracks the general tendency, but not the detailed morphology, of solar active regions and sunspots. This narrow bandwidth behavior is well known. As we increase the bandwidth of our observational filter, we begin to see more irregular behavior. Indeed, in the time domain, the relevant intervals over which the output of the filter will be noticeably different ranges from minutes to years. If we take the Wolf sunspot number as a gauge of solar activity, we find that an 11-year filter (which smoothes out the 11-year cycle) yields an average value of R between 50 and 100, a 1-year filter yields an average value ranging between 5 and 150, a 1-month filter yields a range between 2 and 175, and a single-day filter yields a range between 0 and 350. The corresponding nonoptical categories of activity associated with radio flux, ultraviolet emission, and x-rays will be correlated with R if the temporal filters are relatively large. R is determined by spatially averaging information over the entire solar disk. This procedure yields a value for R which does not vary significantly over a day. Flare activity and its resulting radiation, on the other hand, are not spatially-averaged and may vary extensively from hour-to-hour. The short-term activity, which may be related to short-term HF effects, is correlated with a class of ionospheric disturbances called SID or Sudden Ionospheric Disturbances. SID effects will be discussed in Section 2.5.2 and in Chapter 3.

Most of the ionospheric and HF propagation prediction programs in use today rely upon sunspot number, and it is presumed that a running 12-month average of R_Z or R_I is to be used. Such programs were designed for the estimation or study of long-term behavior. The underlying ionospheric data base was constructed using a median representation for the various ionospheric parameters. This suggests that ionospheric and HF propagation predictions based upon projections of R_{12} would not be possible since the smoothed values themselves would not be available. For purposes of HF propagation forecasts (i.e., short-term predictions), an estimated R, reckoned over a smaller averaging interval, may be developed . A running 27-day average might be considered but, in practice, an average over the last 5 days is sufficient to eliminate much of the more erratic (and misleading) sunspot behavior while retaining some of the near-term flavor. The use of daily values of sunspot number from any single observatory is certainly to be discouraged.

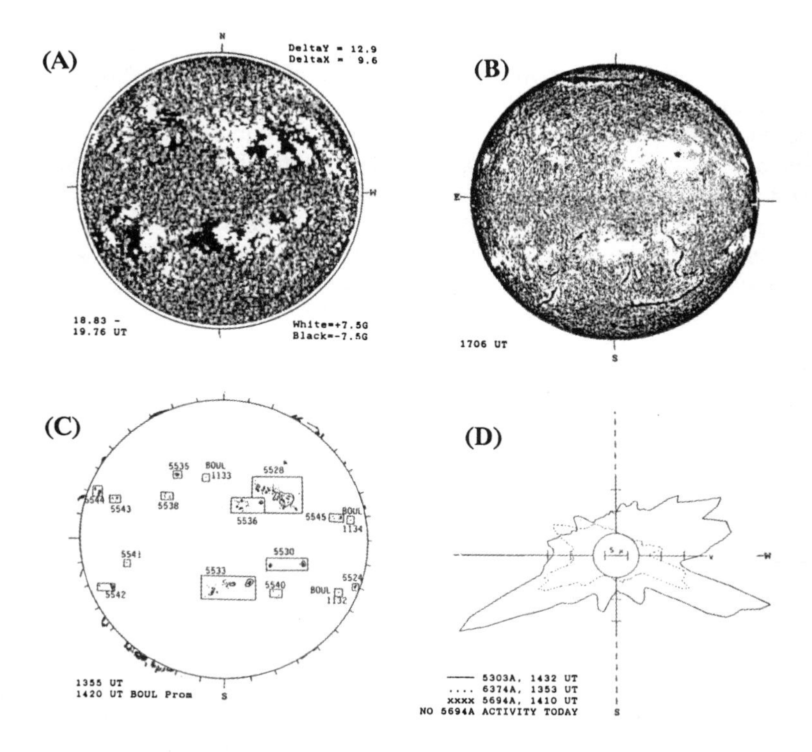

Fig.2-8. (A) A magnetogram from Mount Wilson Observatory showing regions of opposite polarity. (B) The solar disk as observed in H-alpha at Boulder. (C) The Boulder observation of the distribution of sunspots. (D) An observation of the solar corona in three emission lines (FeXIV, FeX, and CaXV). (From *Solar-Geophysical Data*, No.540, Part 1, National Geophysical Data Center, NOAA, Dept. of Commerce, Boulder Colorado.)

In addition to solar flares and related impulsive phenomena which have durations of the order of minutes to hours, there is a longer term variability in solar activity which can be found by observing the sun through a hypothetical filter with a 1-day inverse bandwidth. This variability is thought to be most pronounced during the epoch of sunspot minimum and, on the earth, it introduces a geomagnetic activity cycle of 27 days. This 27 day period is correlated with the solar rotation period, but is significant only if a distinct (longitudinally-isolated) solar active region with a lifetime in excess of 27 days exists. If the lifetime is much smaller than the solar rotation period, then recurrence is impossible. Also, if multiple active regions are distributed over the solar disk, then recurrence phenomena will be averaged out, even if the active regions are long-lived. Recurrence, when observed, can be used to predict future effects on the ionosphere and on HF propagation conditions. We have already seen from Figure 2-5 that the long-term trends in solar and magnetic activity are correlated. Figure 2-9 illustrates the effect of the 27 day solar rotation period on geomagnetic activity.

There is a greater likelihood that active regions will be isolated at solar minimum than at solar maximum. For this reason, we expect recurrence phenomena to be more pronounced at the minimum of the solar cycle. Nevertheless, if an especially active longitude is persistent, it may still introduce a a resolvable 27 day modulation in solar activity even when the average levels are high. This situation was quite evident during 1990 if the solar activity is characterized by the observed 10.7 cm solar flux. From Figure 2-10, we note a steady background level of 150 solar flux units for the first six months. Quite obvious is an oscillatory component of activity, with a 27 day period, with a rectified amplitude of about 40% of the background level. We shall return to this in Section 2.3.3.

The most well-known solar event responsible for HF disturbance is the solar flare. These events trigger many of the short duration ionospheric events called Sudden Ionospheric Disturbances (SID), and are closely related to other solar features which are responsible for certain geomagnetic substorm phenomena, enhanced auroral activity, and ionospheric storms. Sunspot occurrence is closely associated with the observation of solar flares. In general the number of flares observed per solar rotation N_F is proportional to the sunspot number.

$$N_F = a(R - 10) \tag{2.3}$$

where R is the smoothed sunspot number over the 27-day rotation period, and a is a "constant" which ranges between 1.5 and 2. Thus for $R = 110$ (near solar maximum), the value for N_F is about 200. This implies that 7 flares/day will be observed on a global basis.

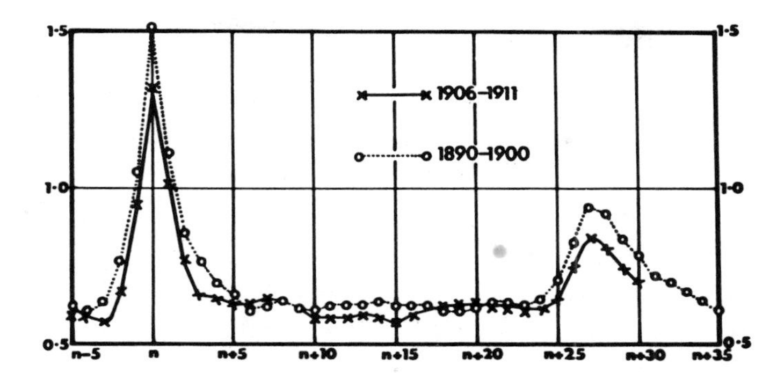

Fig.2-9. Variation in the geomagnetic activity index compared with an observed 27-day periodicity in the solar activity based upon by Chree and Stagg [1927]. Geomagnetic character figures C (which range from 0-2) are determined for a set of disturbed days, are averaged for various lags and plotted. The 0 lags yielded high means (by the selection process) but a secondary peak is observed at 27 days. Character indices are not generally used by propagation specialists or ionosphericists, but are employed by solar physicists for the study of various recurrence phenomena (See Fig.2-13). The various Character indices C (used here),Ci, and C9 are described by Mayaud [1980]. It is interesting that the C9 index was developed by Bartels [1951] to provide an emphasis on 27-day recurrence phenomena and possible absence of relationship with sunspot numbers. Solar M Regions, now associated with coronal holes, were discovered by use of C indices. (Figure from Chapman [1968], by permission, Gordon & Breach)

Flares have been classified in terms of the solar surface area which is enclosed as observed in the hydrogen α line of the solar spectrum. Subflares cover about 2 square degrees or less but the largest class of flares may cover about 25 square degrees of the solar surface. Another optical designation provides a qualitative indication of the brightness of the flare: F=faint; N=normal; B=bright. The most important flare classification for association with ionospheric effects is the flare strength as measured in the x-ray band. Table 2-1 shows the x-ray classifications in the 1-8 Angstrom band.

The number of x-ray flares observed during solar cycle 21 is given in Figure 2-11a. The total number was 172. More optical flares were observed than x-ray flares, differing by about an order of magnitude. More x-ray flares were observed following sunspot maximum than at the peak itself. Figure 2-11b shows the number of magnetic storm days/month which also exhibit no preferential concentrations at sunspot maximum. Figure 2-12 shows that flare activity and storm occurrence were not strongly correlated with sunspot number, at least during solar cycle 19. Definitive patterns linking sunspot activity with ionospheric storms or solar flares (and thus Sudden Ionospheric

Disturbances) has been elusive since observable relationships seem to vary from cycle-to-cycle. We shall return to this matter in Section 2.4.

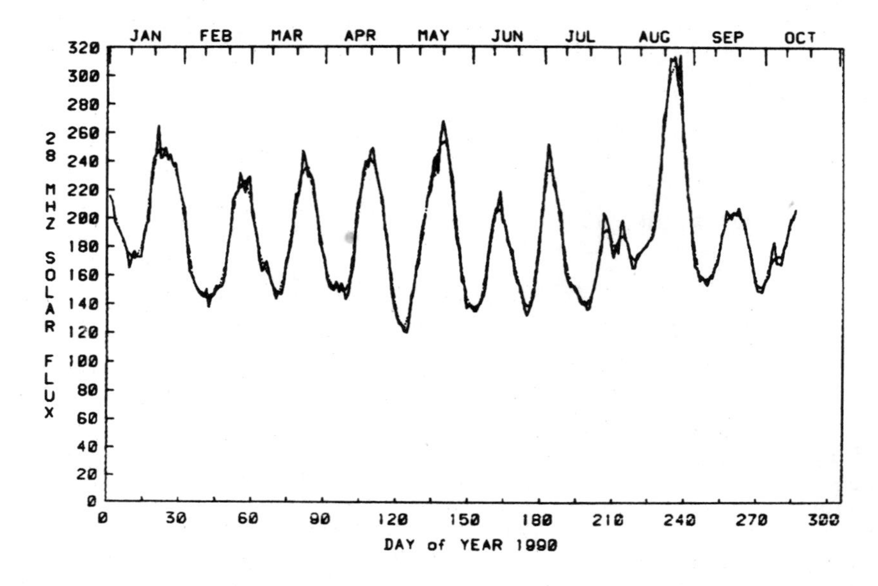

Fig.2-10. Variation of the 2800 MHz (10.7 cm) solar flux during 1990 showing evidence of a 27-day recurrence in solar flux. This is somewhat unusual in view of the fact that the background level of solar activity was so high. Active longitudes are normally masked since they are so numerous and lack a consistent phase relationship with each other. Accordingly, one would not expect the sunspot count to present as coherent a picture as this since solar disk (i.e., hemispherical) averaging is more complete. The daily average and 5-day running average (partially obscured but appearing as a superimposed dotted line) are shown. (Raw data was obtained from the NOAA-USAF Space Environment Services Center, Boulder.)

TABLE 2-1. CLASSIFICATION OF X-RAY FLARES

CLASS OF X-RAY FLARE	*E*, FLUX LEVEL @ 1 AU.[15]
C	$10^{-6} < E < 10^{-5}$
M	$10^{-5} < E < 10^{-4}$
X	$E > 10^{-4}$

---

15. The x-ray flux is referred to that value which would be observed at 1AU (the earth's distance from the sun). The units are watts/m² .

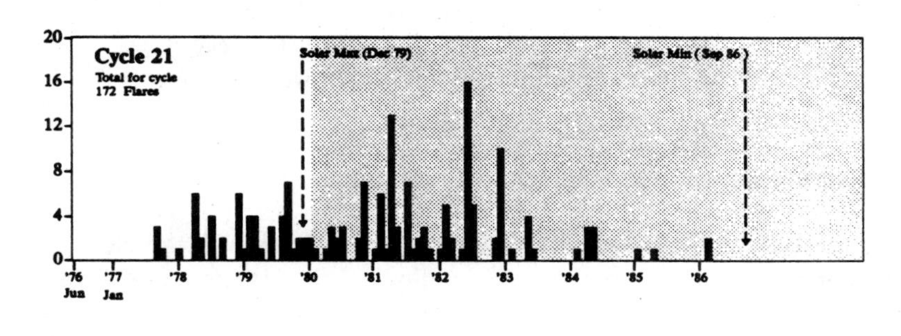

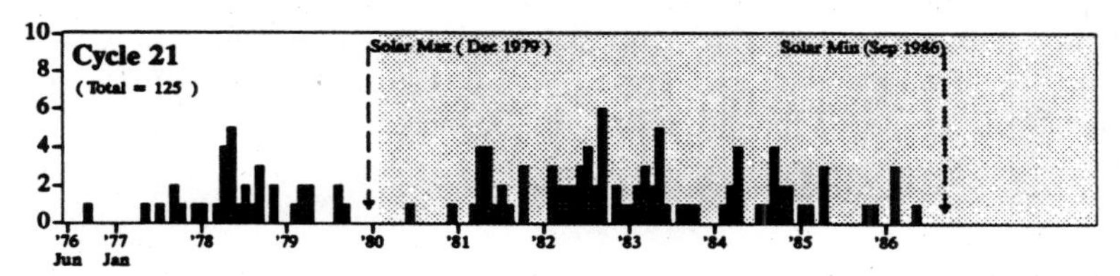

Fig.2-11. (a) X-ray flares/month for solar cycle 21. (b) Magnetic storm days/month for solar cycle 21 where a storm day is defined for the instance for which Ap $\geq$50. (From Hildner [1990])

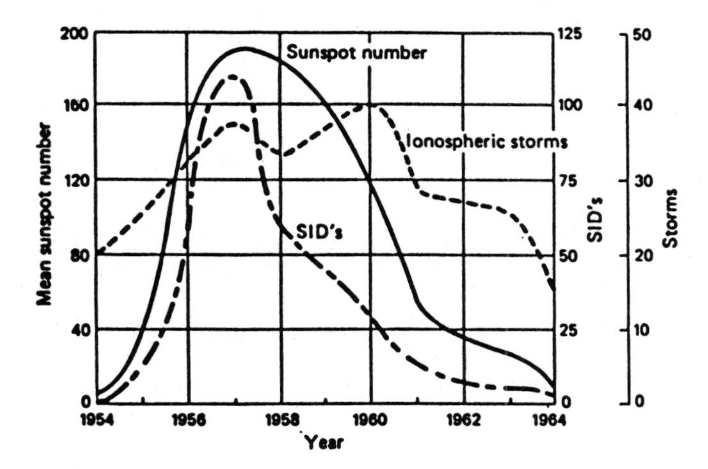

Fig.2-12. Comparison of Sunspot number, the number of ionospheric storms, and number of Sudden-Ionospheric-Disturbances for solar cycle 19 (1954-1964). (From Jacobs and Cohen [1979].)

X category x-ray flares may cause significant absorption of HF signals on the sunlit side of the earth. This will be discussed in Section 2.5.2. Figure 2-13 shows a significant x-ray flare with a flux of more than 10^{-3}watts/m^2 as observed by instruments on the GOES-7 satellite. This event was associated with a host of highly energetic particles which were ejected from the sun during the flare.

2.3.3 Prediction and Measurement of Solar Activity

The prediction of solar activity should not be confused with the measurement of solar activity for the purpose of geophysical disturbance predictions. The measurement of solar features and events has led to algorithms for estimating the probability that magnetospheric, ionospheric, and HF system disturbances might occur as the consequence of solar activity, either immediately or after some specified time delay. These may be termed forecasts in view of the short-term nature of the situation, and because near-real-time solar data is involved in the process.

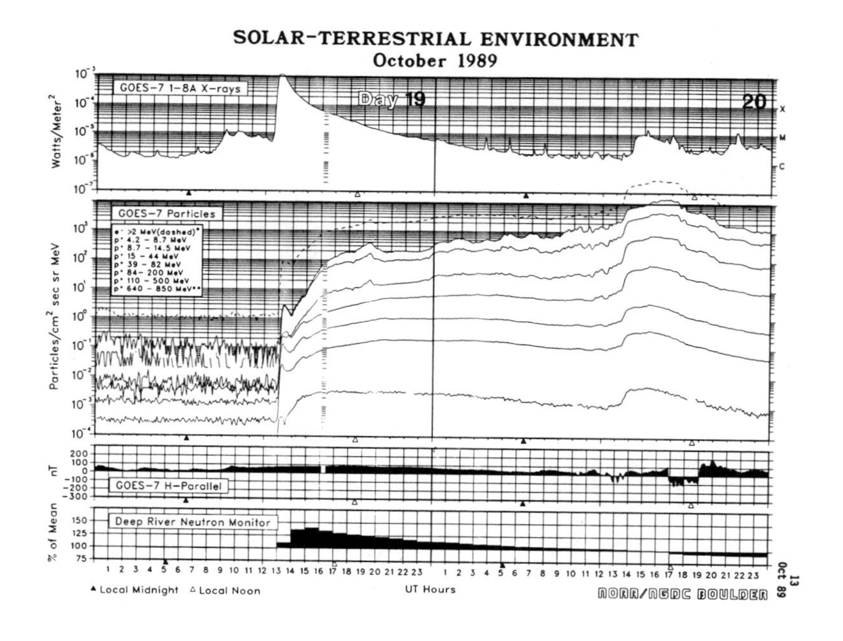

Fig.2-13. The 1-8 Angstrom flux as observed by instruments on GOES-7 on October 19, 1989. Also given is the particle flux enhancement, the horizontal component of the magnetic field at the satellite altitude, and neutrons monitored at ground level. [From Solar-Geophysical Data, Prompt Reports, Nov. 1989, No.543, Pt.1, National Geophysical Data Center, NOAA, Boulder, CO.]

A true long-term prediction of a solar-driven geophysical disturbance involves a projection of the solar activity for some future time, followed by the application of this information in a algorithm which relates solar activity to the geophysical or propagation effect of interest. A knowledge of the long-term behavior of solar activity, specifically cycle amplitude versus phase, enables one to predict the amplitude of the solar cycle at some future time. Such predictions are always in the form of smoothed estimates of solar activity, whether represented by sunspot number or solar flux at 10.7 cm. The use of these solar predictions can only lead to smoothed representations for the hierarchy of geophysical effects. In the context of HF communication the geophysical effects of interest involve the ionosphere. Typically a correspondence is made between running *averages* of sunspot number and *median* properties of the ionosphere. This has important implications in connection with the use of models for HF communication performance evaluation. A long-term prediction algorithm may model the relationship between an ionospheric property and solar activity with great precision, but a prediction of the property is no better than the prediction of solar activity. Moreover there is no reason to expect that an *instantaneous* measurement of a specified ionospheric parameter will agree with this prediction, since it is a median value.

We have already mentioned that the 27-day recurrence of active regions on the sun might provide a basis for updates of the predictions of geophysical disturbances which are otherwise based upon long-term trends. Persistence of features on the sun coupled with solar rotation creates the possibility for of predictable recurrence in the amplitude of parameters such as the solar wind speed which is associated with coronal holes, and solar flux, including corpuscular and electromagnetic components. Some of these recurring parameters are more important than others. Coronal hole recurrence (associated with an enhanced solar wind) is significant parameter. The latter phenomenon might allow predictions of up to six months to be made. This has been suggested by Sheeley and his colleagues at NRL [Sheeley et al., 1976, 1978; Bohlin, 1977]. Figure 2-14 shows the excellent correlation between the appearance of coronal holes, solar wind velocity, and magnetic disturbance.

2.3.4 The Application of Solar Activity Indices in Models

All HF propagation prediction programs allow for indirect modification of the underlying ionospheric parameters by changing the solar activity index. The convenience afforded by a single solar index *driver* seems clear enough, but even this simplified approach may present a problem. With the plethora of solar indices available, some confusion has arisen regarding the proper choice of index and the way it is applied in specified programs. These indices are distinguished principally by the epochs over which they are reckoned or averaged, or the observation network which contributes to construction of the index.

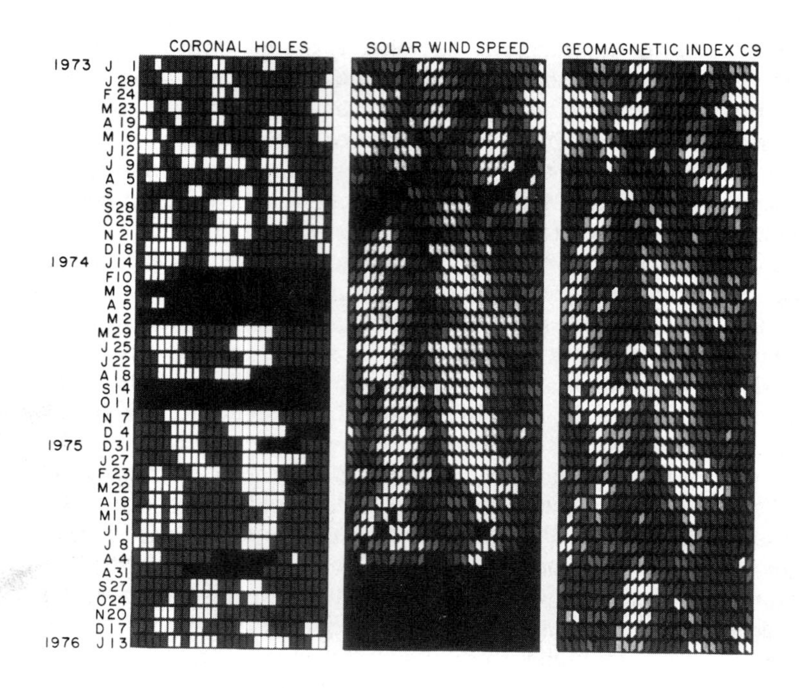

Fig.2-14. A comparison of coronal holes with the solar wind speed and geomagnetic activity index C9 for the 1973-1975 period. The comparison was made by Sheeley et al. [1976]. The data are arranged in 27-day sequences (in the rows) in a manner suggested by Bartels to correspond to solar rotations. A strong correlation of the features is evident. (Figure provided by courtesy N. Sheeley, NRL.)

A majority of the programs specify the sunspot number while others specify the 10.7 cm radio flux in *solar flux units*[16]. Since these indices are closely related to solar observables, they are termed *direct* indices. The matter is complicated by the introduction of global or regional *ionospheric indices* which act as proxies for the sunspot number.

In recent years, a system-specific *pseudoflux* parameter has been developed as a method to update models. The *pseudoflux* accounts for any departure

----

16. The 2800 MHz radio flux from the sun (10.7 cm) is measured in solar flux units (*sfu*'s). An *sfu* has the units 10^{-22} watts/m² per Hz (see Section 2.3.2.4).

between actual observation and model predictions, with the predictions being based upon direct indices or their ionospheric proxies. *Pseudoflux* is an *ersatz* index masquerading as *direct* index, and it is model-specific. In the literature, the term *pseudoflux* is not used universally. For example it may be referred to as *pseudo-sunspot number, effective solar flux* or *effective sunspot number.* It has been shown that *ersatz* indices based upon near-real-time ionospheric data outperform their direct index counterparts. The CCIR ionospheric coefficients which are exploited in the most popular models are based upon sunspot number, a direct index.

A few comments on CCIR recommendation 371-5 [CCIR, 1986a, 1988] which concerns the use of solar indices for long term predictions are now in order. The reader is referred to Section 2.3 (and especially sub-section 2.3.2.4) and Section 2.5 dealing with solar activity predictions and services. Table 2-2 is a list of indices and their characteristics. We shall conclude with a few remarks about short-term applications.

The CCIR recommends use of the index R_{12} for all ionospheric predictions $\geq$ 12 months in the future. The recommendation further stipulates that F-layer predictions for periods up to six months in the future (and possibly up to 12) be computed using the indices R_{12} or ϕ_{12}. The F-layer parameters suitable for prediction under these recommendations are foF2 and M(3000)F2, which are discussed more fully in Chapter 3. On the other hand. the E and F1 layer parameters (specifically foE and foF1) are controlled more directly by EUV flux from the sun, but EUV data are not readily available. Fortunately, the 10.7 cm (2800 MHz) solar radio flux, from which the index ϕ_{12} is constructed, has been shown to be fairly well correlated with EUV radiation, and the radio flux value is monitored routinely by the Algonquin Radio Observatory in Ottawa, Canada. Accordingly the CCIR recommends the use of the index ϕ_{12} based upon the Ottawa measurements, made at 1700 UT every day, for prediction of foE and foF1 for periods up to six (and possibly 12) months in advance.

Since both R_{12} and ϕ_{12} have equal standing for intermediate term F-layer predictions (i.e., up to 12 months), it is not surprising that a relationship between the two has been sought. The result is Equation 2.2 which is repeated below:

$$\phi_{12} = 63.7 + 0.728\, R_{12} + 8.9 \times 10^{-4}\, R_{12}^{2} \tag{2.2}$$

where ϕ_{12} and R_{12} are the 12-month running mean values of ϕ and R respectively. It should be recognized that if a prediction of foF2 is required for the particular month, a twelve month average of R or ϕ_{12} (centered about that month) is still recommended. This is done to smooth-out solar activity variabilities which are not represented in the long-term data base which characterizes the ionosphere. The 12-month smoothing associated with solar flux and/or sunspot number relies on the availability of actual data, so ϕ_{12} and R_{12} will only exist for previous months. In fact, if the current month is designated by the sequence index m, then the last month for which R_{12} may be computed without a change in recipe is m-6, or six months prior to the current month. The nature of how R_{12} is constructed will be discussed shortly.

TABLE 2-2. LIST OF SOLAR ACTIVITY INDICES (& Proxies)

INDEX	DESCRIPTION
R	*Generic Sunspot Number.* Corresponds to either a daily value or a monthly mean. For a specified observatory, *R* is just the Wolf number.
R_a	*American Relative Sunspot Number.* Developed from observations made by mixture of amateur and professional astronomers in north America.
R_b	*Boulder Sunspot Number.* Developed by SESC in Boulder CO based upon the Solar Optical Observing Network (SOON).
R_I	*Monthly Mean International Relative Sunspot Number.* (Current construction, from 1982 to the present.) Values are developed by taking the monthly average of the weighted daily values....formed from sunspot counts which are reckoned daily at ≈ 25-30 observatories worldwide. Values may be Provisional or Final. [Sunspot Index Data Center (Brussels); A. Koeckelenbergh]
R_Z	*Monthly Mean Zurich Sunspot Number.* (used prior to 1982) The recipe is similar to that for R_I
R_{12}	Twelve-Month Running Mean of Sunspot Number R_I (see Eqn.2-4.)
ϕ	*2800 MHz (10.7 cm) Solar Flux.* Measured at Ottawa. It may correspond to either a daily value or monthly mean. The daily values are deduced at 1700 UT (local noon). Values may be Observed or Adjusted.
ϕ_{12}	*Twelve-Month Running Mean of Solar Flux,* ϕ
IG	*Liu Solar Activity Index.* An ionospherically-derived solar activity index. [Liu et al., 1983]
IG_{12}	*Twelve-month Running Mean of IG.* [Smith, 1986]
I_{F2}	*Minnis Solar Activity Index.* An ionospheric index based upon noontime foF2 values. The index is based on noontime foF2 values [Minnis, 1964]. Predictions of this index are prepared by the United Kingdom for six months into the future based upon a method developed by Smith [1968].
I_{F212}	*Twelve-month Running Mean of* I_{F2}
T	*Australian Ionospheric index.* Similar to I_{F2} except all hours of the day are used in its construction [IPSD, 1968].
T_{12}	*Twelve-month Running Mean of T.*

OTHER

Pseudoflux parameter [NRL workers]

Effective Sunspot Number [AFGWC] in connection with ICED model

Separate relationships have also been determined between R_{12} and the indices IG_{12}, I_{F212}, and T_{12}, and these are provided in Recommendation 371-5.

It is worth noting how R_{12}, the twelve-month running-mean sunspot number, is constructed. We have already indicated that if the latest month for which monthly mean data is available is labeled m, it is possible to develop a 12-month running mean value for R for any preceding month n provided $n \leq m - 6$. The expression for R_{12} is as follows:

$$R_{12} = (1/12)\left[\sum_{n-5}^{n+5} R_k + \tfrac{1}{2}\,(R_{n+6} + R_{n-6})\right] \tag{2.4}$$

where R_k is the monthly mean sunspot number for the kth month, and R_{12} corresponds to the running mean index for the month for which $k = n$. It is seen by this recipe that 13 months are actually involved in the smoothing process, with the first and thirteenth being weighted by a factor 0.5. We note that Equation 2.4 implies operation on monthly means. The peculiar form of Equation 2.4 is dictated since only monthly means are used, with individual daily values not being (directly) involved in the running-mean calculation. For example if we wish to compute R_{12}, the process may be represented by taking the arithmetic average of two 12-term sequences as shown below:

Yr N ————> increasing time ————> Yr N+1

$$\Sigma R^- = R_{Jan} + R_{Feb} + R_{Mar} + R_{Apr} + R_{May} + R_{Jun} + R_{Jul} + R_{Aug} + R_{Sep} + R_{Oct} + R_{Nov} + R_{Dec}$$

$$\Sigma R^+ = + R_{Feb} + R_{Mar} + R_{Apr} + R_{May} + R_{Jun} + R_{Jul} + R_{Aug} + R_{Sep} + R_{Oct} + R_{Nov} + R_{Dec} + R_{Jan}$$

Months	R_{Jan}	$+R_{Feb}$	$+R_{Mar}$	$+R_{Apr}$	$+R_{May}$	$+R_{Jun}$	$+R_{Jul}$	$+R_{Aug}$	$+R_{Sep}$	$+R_{Oct}$	$+R_{Nov}$	$+R_{Dec}$	$+R_{Jan}$
(times)	1	2	2	2	2	2	2	2	2	2	2	2	1

In this illustration ΣR^+ and ΣR^- are offset sums of 12 contiguous monthly mean sunspot numbers. Offset to the future by one month implies a (+) sign; offset to the past by one month implies a (-) sign. Equivalently, we find the following relationship to hold: $R_{12}(\mathrm{Jul}) = (1/24)[\,\Sigma R^+ + \Sigma R^-\,]$. The procedure defined by Equation 2.4 and illustrated above is certainly appropriate for *hindcasting*. In *hindcasting* we evaluate ionospheric properties, channel parameters, and system performance under previously observed conditions using direct indices such as R or ϕ as a basis. For current and future conditions, we must use another approach.

As noted in the preceding paragraph, while R_{12} is recommended for use in predictions, Equation 2.4 above is not an appropriate way to construct R_{12} in such situations. In fact, Equation 2.4 suggests a primary role for R_{12} in *hindcasting* rather than prediction. *Hindcasting* involves an archival function.

In an operational environment, we are in the realms of real-time index assessment and index prediction. To estimate current and future system behavior, we will be forced to make system performance predictions based upon indices which are themselves predictions. and no method for making such predictions is totally unassailable. There is an internationally accepted procedure, however. This is the so-called McNish-Lincoln objective method [1949] which has been improved upon by Stewart and Ostrow [1970]. This procedure is mentioned in *Explanation of Data Reports*, a supplement to the *Solar-Geophysical Data Reports* [NOAA, 1987]. An excerpt from CCIR Recommendation 371-5 describing the method is provided as follows:

> "First a mean cycle is computed from all past values of R_{12} starting from the sunspot minimum of each cycle through eleven years thereafter. For prediction of a value in the current cycle, the first approximation is the value of the mean cycle at the stated time of the minimum. This estimate is improved by adding a correction proportional to the departure of the last observed value for the current cycle from the mean cycle. .... The statistical uncertainty is small for the first months after the last observed value but becomes large for predictions 12 months or more in advance. As soon as a minimum is identified, new correction factors can be computed by including the observed values for the preceding cycle, for application to the new cycle."

As indicated, the McNish-Lincoln procedure is essentially an update of the mean of previous cycles (No.8-20, or 13 cycles) obtained through application of a correction term based upon an evolution of the current cycle. This correction term is related to the amount of departure of current values from those which correspond to the 13-cycle mean. Using this method, it is possible to forecast monthly mean sunspot numbers, as well as the 12-monthly running means for the present and future. However, it means that the current 12-monthly value of R (i.e., the estimated R_{12} for the present month) will be based upon equal amounts of past data and *future data*, with the latter being developed using the McNish-Lincoln procedure. The same approach could be applied to indices such as ϕ and others listed in Table 2-2.

For HF communication applications, some workers have suggested smoothing epochs much shorter than twelve months, or even one month. Some improvement in MOF prediction performance (using MUF prediction programs such as MINIMUF) has actually been achieved using 5-day and 7-day averages of the daily sunspot number or the 10.7 cm flux. These near-term procedures and results are covered more fully in Chapter 6. Use of such small smoothing epochs should be avoided whenever possible, but the necessity to obtain near-term prediction of system performance, and the ready availability of daily values for R and ϕ is certainly a temptation.

The use of the ionospheric indices in the manner just described for short-term prediction is much more defensible. This is because estimations of the future ionospheric state are more closely correlated with the current ionospheric state than they are to the current state of the sun! Still near-term forecasts of solar activity will influence the answers which are derived on the basis of ionospheric indices alone. The ultimate process should no doubt involve an hierarchy of solar-terrestrial data sources, and both solar and ionospheric indices should be involved.

In conclusion, it is noted that smoothed values of R and ϕ are recommended for use in propagation prediction models. The indices R_{12} and ϕ_{12} are the most appropriate since the ionospheric data contained in propagation models has been developed and organized based upon 12-month running mean solar activity indices. Generally speaking R_{12} should be used for F2 layer predictions, and ϕ_{12} should be used for E and F1 layer predictions. Regardless of how these indices are constructed (either directly or by the McNish-Lincoln method), there are difficulties if the predictions are needed for more than a year in advance. In this situation use of R_{12} is as good as anything.

2.4 THE MAGNETOSPHERE AND GEOMAGNETIC STORMS

We next turn our attention to the magnetosphere and, in particular, to the coupling of the solar wind, the magnetosphere, and the ionosphere. Much is unknown about these coupling processes, but several studies scheduled for 1990-2000 will improve our understanding. We will identify the factors which appear at present most important, and which are thought to have a significant bearing on HF propagation in the ionosphere. Additional readings appear in the bibliography.

2.4.1 The Geomagnetic Field

To obtain an understanding of the magnetosphere, we must first describe the geomagnetic field. The earth's magnetic field is an important feature since it generally prevents a direct encounter between the ionosphere and energetic particles of solar origin, and especially solar wind streams. A geographically localized but significant exception to this pattern arises in the polar ionosphere where profound solar-terrestrial effects are observed. Since the geomagnetic field is an efficient deflector of the solar wind, why are parameters of the solar wind significant in the morphology of the magnetosphere and the ionosphere beneath it?

Figure 2-15 depicts the field surrounding a bar magnet. This field resembles the magnetic field of the earth in many respects. The longitudinal field lines are aligned with the axis of the magnet at the ends (poles), and the transverse field lines define an equatorial plane bisecting the magnet. If the field around this bar magnetic were to represent the first-order field of the earth, then we see that the polar field line orientation is nearly vertical while the equatorial field lines are horizontal. This is a good model but there are some differences. First, the geomagnetic field is not purely dipolar, and secondly the axis of the best fit dipole does not correspond precisely to the rotational axis of the earth.

The geomagnetic field is generated by several sources and current systems located within the earth, the ionosphere, and the magnetosphere. The internal sources include a field produced by currents flowing near the earth's core at a depth of about 3000 km. This component dominates all other sources below about five earth radii. The geomagnetic field may be adequately represented by a magnetic dipole tilted with respect to the earth's rotational axis. Some local anomalies result from direct magnetization of crustal material, but these are generally averaged out at ionospheric heights. The effects of ionospheric/magnetospheric current system sources depend upon the heights being analyzed, but these components are usually small below a few earth radii.

The simplest approximation to the field is an earth-centered dipole directed southward and inclined at about 11.5 degrees to the earth's rotational axis. Thus the North pole is 78.5°N, 291°E, and the South pole is 78.5°S, 111°E. This model can be improved by displacing the dipole a distance equal to 0.0685 R_e toward 15.6° N and 150.9° E, where R_e is the earth radius. This modification places the North pole at 81°N, 84.7° W and the South pole at 75° S, 120.4° E. However, there is considerable wander in the precise coordinate placement if the model is slightly changed because of longitude sensitivity at high latitudes.

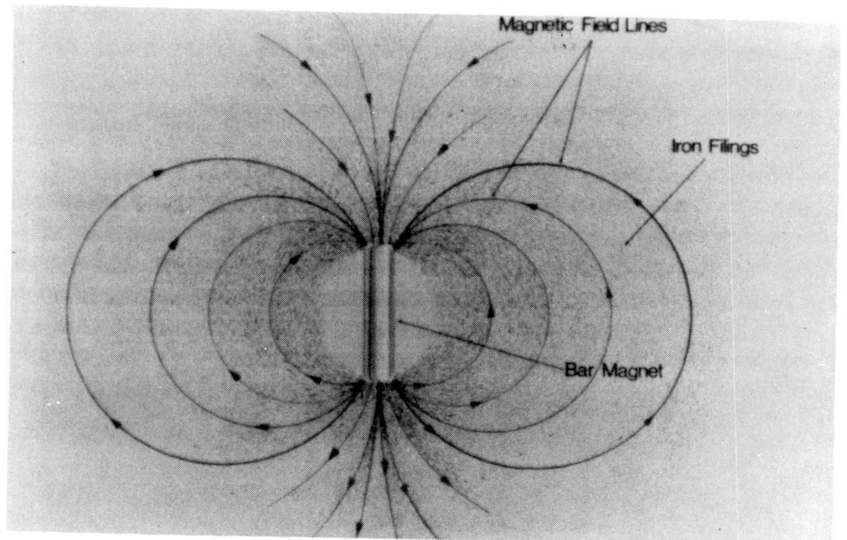

Fig.2-15. The magnetic field surrounding a bar magnet. In many respects the geomagnetic field mimics this pattern. Differences are introduced by local anomalies in the earth's crust and current systems in the atmosphere.

There are also secular variations of the field associated with gradual reduction in the dipole field strength, a migration of regional anomalies, a northward movement of the dipole, and other variations. Some approximation methods have been based upon the fact that the geomagnetic field decreases in intensity with the inverse cube of geocentric distance, and these methods extrapolate surface values to ionospheric heights. Such approximations tend to emphasize local effects, but the availability of surface magnetic field properties makes the use of such approaches very tempting. Maps of surface values of the total magnetic field, the azimuthal variation of the compass (declination), and the inclination of the magnetic field from the horizontal (dip) may be found in a number of sources.

Figure 2-16 shows the conventions associated with measurements of the geomagnetic field. Units vary depending upon application. The primary transformations are given in Table 2-3.

There are a number of representations of the geomagnetic field. A description of the methods has been given by Knecht and Shuman [1985]. One of the methods which is most physically attractive for demonstrating ionospheric-magnetospheric interactions is one for which the field is modeled in a so-called *B-L* coordinate frame (see Fig.2-17). In this system, the field may be exhibited in curves of constant magnetic field intensity *B* and curves of constant *L*.

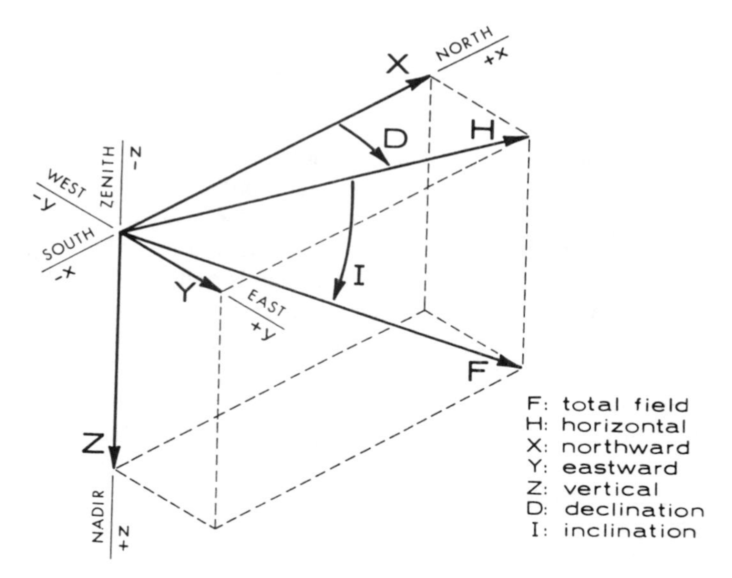

Fig.2-16. Conventions used in geomagnetic field measurements

TABLE 2-3. MAGNETIC FIELD UNITS

MAGNETIC INDUCTION (B)

1 Gamma = 10^{-5} Gauss = 10^{-9} Tesla = 1 nT

MAGNETIC INTENSITY (H)

1 Gamma = 10^{-5} Oersted [CGS]

1 Gamma = $10^{-2}/(4\pi)$ Ampere Turns/meter [MKS]

AMPLITUDES OF THE VARIOUS FIELDS

1.	Earth surface	1/2 Gauss	5×10^4 nT
2.	Benign Solar Field	1 Gauss	10^5 nT
3.	Disturbed Solar Field	10^4 Gauss	10^9 nT
4.	Solar Wind		~ 6 nT
5.	Secular Field Decay @ the equator		~ 16 nT/yr
6.	Sq Field Variations from Equatorial currents		0-50 nT
7.	Lunar-Solar Tidal Variations		~ 3 nT
8.	Geomagnetic Storms (for $K_p = 9$) @ midlatitude station		~ 10^3 nT

In the *B-L* system, a particular magnetic shell is characterized by a unique *L* value corresponding to the normalized geocentric distance of the field vector over the equator. Thus, $L = 2$ corresponds to a field line which reaches its maximum height over the geomagnetic equator at 2 R_e, where R_e is the earth radius and normalization factor. This system is quite useful in the study of particles trapped in the magnetosphere such as those found in the Van Allen radiation belts. The terrestrial footprint of a specified field line will occur at two points. These are called conjugate points.

The ionospheric plasma and the resultant HF skywave phenomena are best characterized in terms of geomagnetic coordinates rather than geographic coordinates. Accordingly emphasis should be placed on determining the geomagnetic coordinates of HF terminals as well as their location in a standard geographical framework. As we have shown earlier, the simplest representation of the geomagnetic field is one in which the magnetic field is assumed to be a tilted dipole. This geomagnetic reckoning has greatly assisted in the understanding of ionospheric phenomena, especially those effects which are largely under geomagnetic rather than solar control.

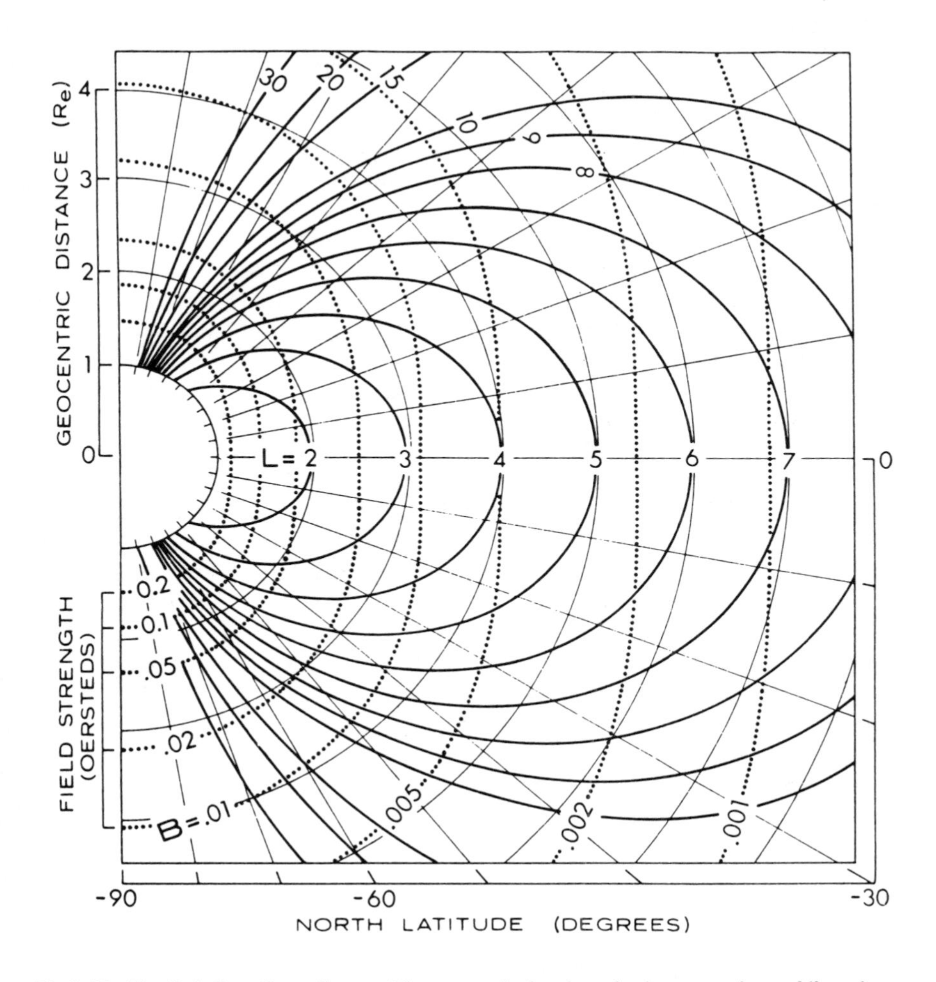

Fig.2-17. The *B-L* Coordinate System. The curves depicted are in the magnetic meridian plane. The *L* parameter is related to the height of the field line over the magnetic equator. (See text). *B* is the magnetic field intensity in Oersteds. (From Chapt.4 in the USAF *Handbook on Geophysics and the Space Environment*, 4th Edition, Jursa [1985])

The dipole method leads to a coordinate system of Geomagnetic Latitudes and Longitudes, and it is appropriately termed the Geomagnetic Coordinate System. Figure 2-18a is a map of Geomagnetic Latitudes for a dipole model based on the 1965 IGRF (International Geomagnetic Reference Field) [CCIR, 1980]. Other methods include those which replace Geographic (or even Geomagnetic) Latitude with Dip Latitude or with Invariant Latitude [Jensen et al., 1960]. The Corrected Geomagnetic Coordinate System (CGCS) is a refinement of the Geomagnetic Coordinate system and like the *B-L* system mentioned earlier is useful for the study of conjugate point phenomena. [Knecht and Shuman, 1985].

Current methodology uses a representation of the field in terms of a multipole expansion of the magnetic scalar potential function in which the coefficients are based upon a least squares approach to provide a best fit to the field data. This method is now well established and the models, and coefficients, for computing the field are widely available. The internationally accepted model of the geomagnetic field is the IGRF (mentioned above) with the 1985 version being the most accurate. Previous models have been developed at 5-year intervals beginning in 1965. Coefficients for these models are available from the World Data Center A for Rockets and Satellites at Beltsville, Maryland.

From Figure 2-18a we see that the geomagnetic latitude lines are shifted southward (with respect to the geographic latitudes) in the American sector. Thus, phenomena observed at a fixed geographic latitude but under geomagnetic control will be different in the American sector from those in the Eurasian sector. This effect is especially important when we examine phenomena at the higher geographic latitudes. It is instructive to view the world in geomagnetic coordinates rather than in geographic coordinates to get a better feel for the phenomenological distortion which geomagnetic control may introduce. Figure 2-18b is such a display. It shows the world (as well as positions of many of the active geomagnetic observatories) in a Mercator representation of geomagnetic coordinates.

Another useful coordinate is the Magnetic Latitude, as opposed to Geomagnetic Latitude. This might be more properly called the Dip Latitude since it is based upon a transformation of observed dip angles by the following formula:

$$\text{Tan}\ \Theta = 0.5\ \text{Tan}\ I \qquad (2.5)$$

where Θ is the magnetic latitude and I is the dip angle.

Figure 2-19 is a plot of Magnetic (Dip) Latitude. Notice that the magnetic latitude lines passing through the United Kingdom and northern Europe cut through the middle of the continental United States.

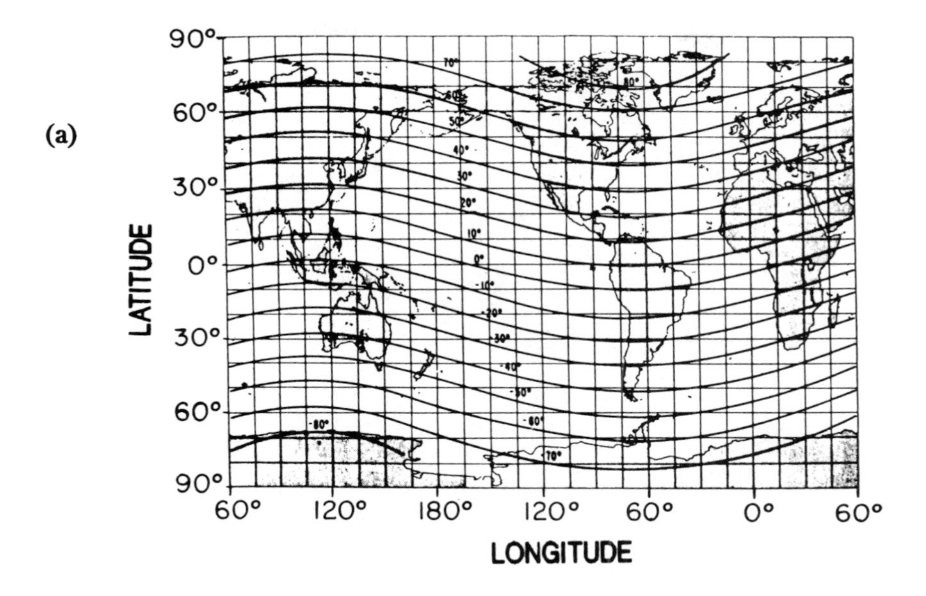

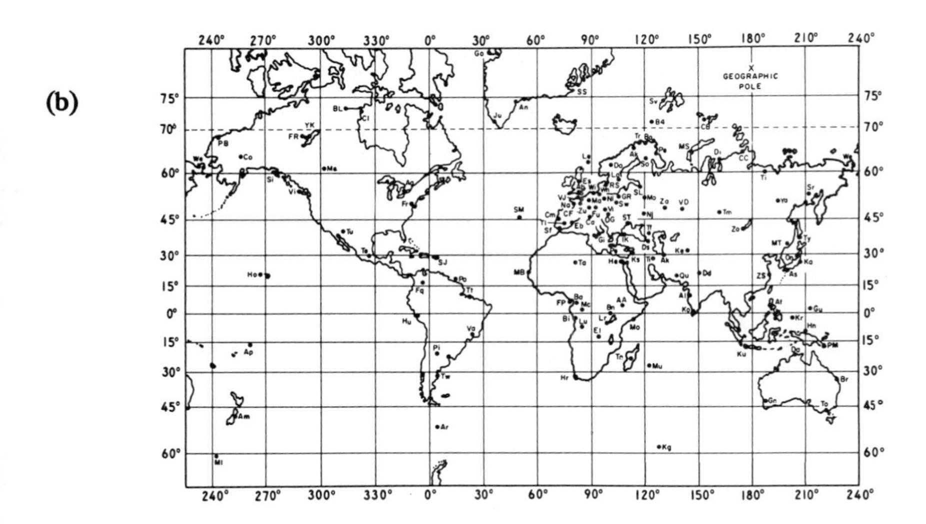

Fig.2-18. (a) Geomagnetic Latitudes [CCIR, 1980]. (b) Mercator representation of the world in Geomagnetic Latitude format. (From Knecht and Shuman [1985])

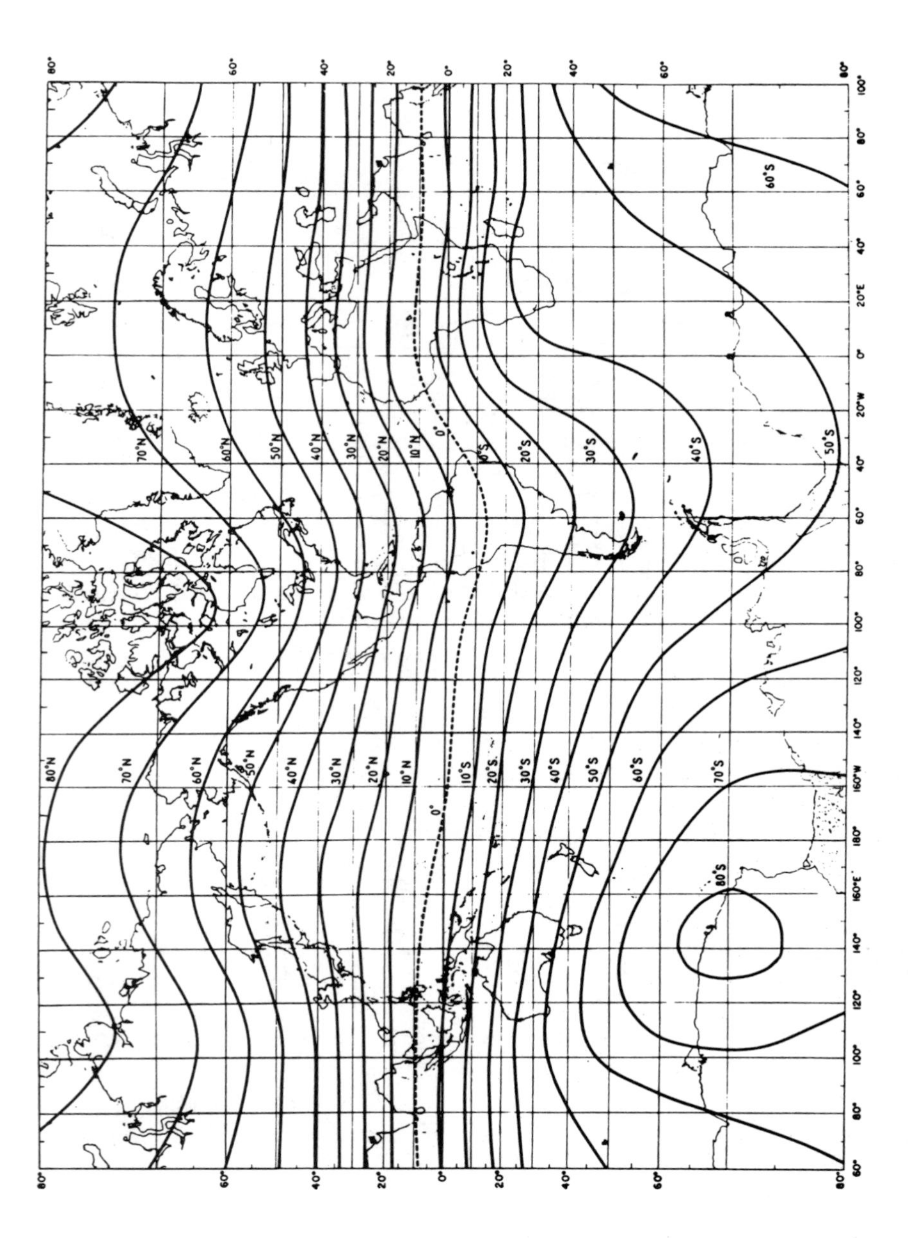

Fig.2-19. Magnetic Latitudes. The dashed line is the Dip equator. This is where the magnetic field is horizontal to the earth's surface. (From USAF *Handbook of Geophysics and the Space Environment*, 2nd Edition, Valley [1965]; p.11-15)

2.4.2 Magnetospheric Topology

The geomagnetic field and the solar wind interact like a blunt object in a supersonic flow field. The earth's field is compressed on the sunward side and distended on the anti-sunward side, giving rise to a characteristic shape resembling a comet (See Figure 2-20). Within the magnetosphere, solar wind particles are generally excluded, being deflected by the severely distorted geomagnetic field. A collisionless bow shock is formed upstream of the magnetospheric boundary (or magnetopause), and the region between the shock boundary and the magnetosphere is termed the magnetosheath. The magnetosheath is the region of closest approach for the deflected solar wind particles. Within the magnetosphere the motion of plasma is governed by the geomagnetic field. This is a primary property. Since ion-neutral collisions are not insignificant within the ionosphere and may restrict geomagnetic control of plasma motion, we do not regard the ionosphere as part of the magnetosphere, and since the geomagnetic field vanishes above the magnetopause, the magnetosheath is not part of the magnetosphere either.

Solar wind particles are typically denied entry to the ionosphere because of the geomagnetic field interaction just mentioned. However, there are some exceptions. Particles may gain entrance through the polar cusp regions. During energetic particle events, this process is enhanced and polar cap absorption (PCA) is the result. Also, because of a process called magnetic merging (of the IMF with the geomagnetic field) magnetosheath plasma may be temporarily captured by the plasma sheet.

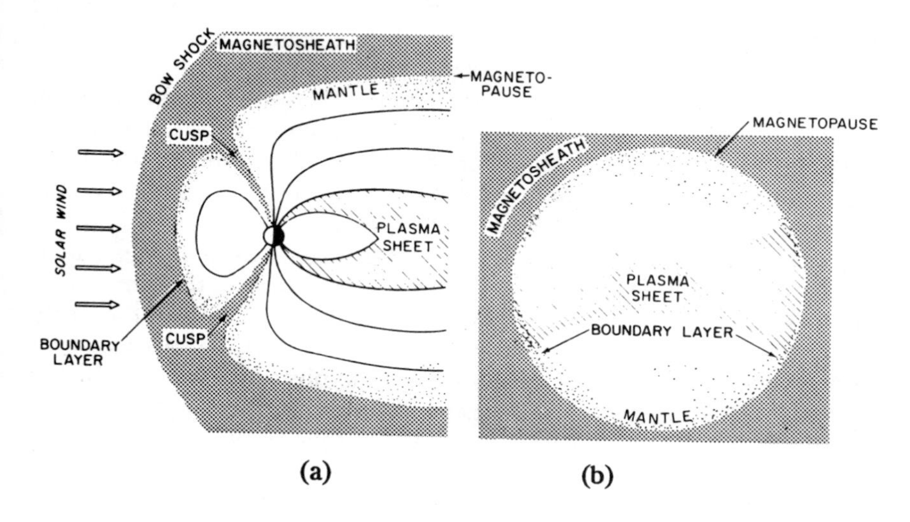

Fig.2-20. A depiction of the magnetospheric regions as related to the solar wind, the bow shock, the magnetosheath, and the ionosphere. From Hill and Wolf [1977].

Another region of interest is the plasmasphere which serves as a reservoir for ionospheric replenishment during the night and acts as a *sink* for electrons during the daytime. A very important property of the plasmasphere is its closed field lines. The plasmasphere contains the Van Allen radiation belts. The poleward boundary of the plasmasphere (called the plasmapause) maps into an ionospheric region called the high latitude trough. Electron concentrations are relaxed in this region, leading to a reduction in the range of critical frequencies suitable for exploitation in HF skywave propagation. Poleward of the plasmapause, the geomagnetic field lines are no longer closed, but are stretched out well into the magnetotail. This region of open field lines is called the plasma sheet and it has important implications for HF communication at high latitudes. Disturbances within the plasma sheet produce enhanced auroral activity.

2.4.3 Magnetic Storms and the Ionosphere

It is now widely believed that auroral activity is embodied in the auroral substorm concept described by Akasofu [1964], and that the auroral substorm is only one manifestation of a general process called a magnetospheric substorm [Akasofu, 1968, 1977]. The most dramatic consequence of the magnetospheric substorm is the aurora.

While the solar wind blows smoothly through the magnetosheath, the topology of the magnetosphere is not disturbed. The plasma sheet is calm and auroral displays are subdued as long as this quiet condition exists. When sunspot activity is at high levels the probability for disturbed aurora and geomagnetic storms is increased. Evidently the enhanced solar wind which exists when active regions (and sunspots) are most prevalent disturbs the magnetospheric boundary as well as the plasma sheet within the magnetotail. Another source of the solar wind is the coronal hole, and this solar feature can occur even during solar minimum conditions. This explains why ionospheric disturbances are sometimes observed to occur in the absence of any apparent solar activity (as measured in terms of sunspots).

A crude picture of the magnetic substorm process is given in Figure 2-21. An important factor in the generation of substorm energy is the direction of the Interplanetary Magnetic Field (IMF) along the dipole axis of the earth. When it is directed southward, as shown in Figure 2-21, the plasma sheet becomes pinched driving ionization toward the polar regions. Recall that Sheeley et al.[1976] associated high speed solar wind and geomagnetic activity (See Figure 2-14). Others have shown that this relation is not true if a substorm index such as AE (discussed below) is used [Arnoldy, 1971]. The current view is that the IMF direction is the more fundamental discriminant.

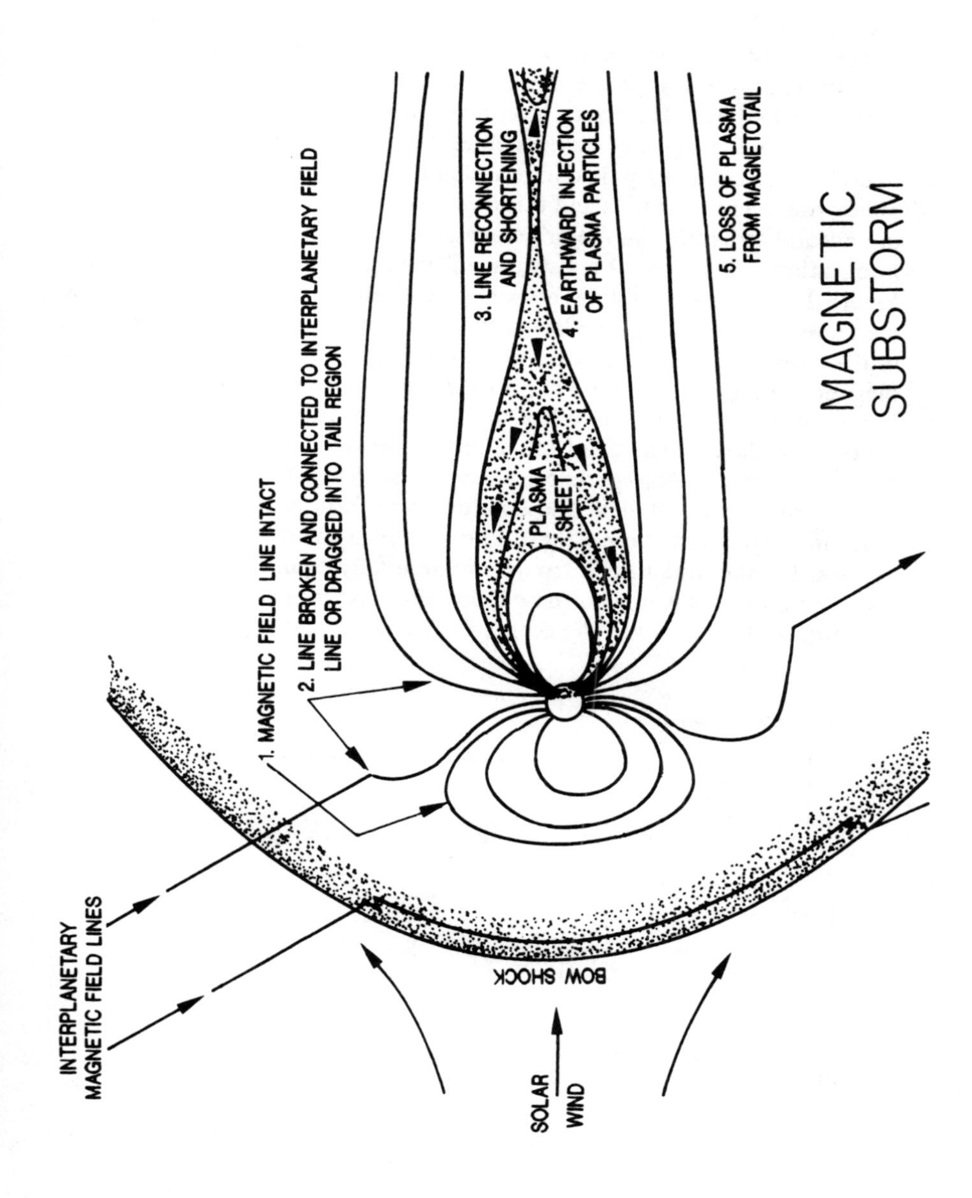

Fig.2-21. Anatomy of a magnetospheric substorm

The solar wind exhibits fluctuations when high speed wind streams mix with the ambient solar wind. These fluctuations may be transferred to the N-S component of the IMF. A southward fluctuation may introduce a series of substorms until the excess energy associated with differential merging is exhausted. Differential merging is a process in which the dayside merging rate is in excess of the nightside merging rate. It is a phenomenon which is thought to build up stresses along field lines within the plasma sheet in the tail region. Substorms occur in groups and this process serves to produce a pulsation of the auroral structures. A major geomagnetic storm may be regarded as the integration of a series of substorms. The sequence of events is shown in Figure 2-22. As the field lines associated with the IMF turn southward, they merge with geomagnetic field lines inside the bow shock region (in the vicinity of the cusp) and flux is transferred from the dayside to the nightside of the magnetosphere. This buildup in potential energy lasts for less than an hour. Subsequently the energy is released as open field lines in the magnetotail become pinched and some of them become reconnected. As the field lines snap into a dipolar configuration, energetic particles are injected toward the earth, forming auroral arcs. These auroral arcs form beautiful nocturnal displays in regions in the northern and southern hemispheres located just poleward of the midlatitude trough but generally equatorward of the polar cap. These regions were shown by Feldstein to be represented best by ovals anchored at their center by the geomagnetic pole.

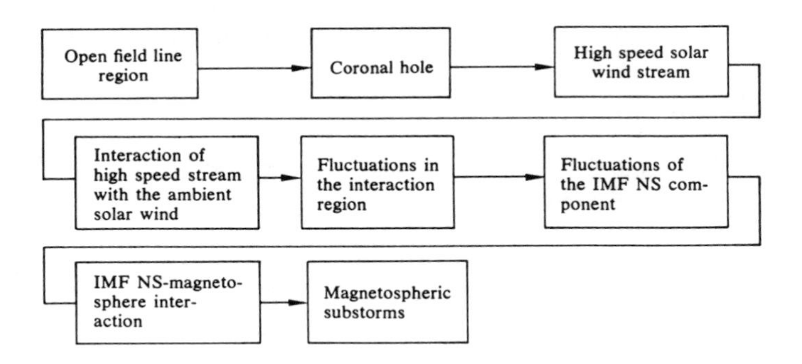

Fig.2-22. The association of open field lines, coronal holes and geomagnetic disturbances. (From Akasofu [1977]; by permission D.Reidel Publishing Co., Boston)

Since the auroral oval is a geomagnetic feature, an interesting diurnal pattern emerges because of the departure of geomagnetic from geographic coordinates. Figure 2-23 shows how the steady state auroral oval appears to move as a function of local time. Notice how Iceland becomes alternately an auroral and nonauroral station even though the magnetic activity is taken to be fixed.

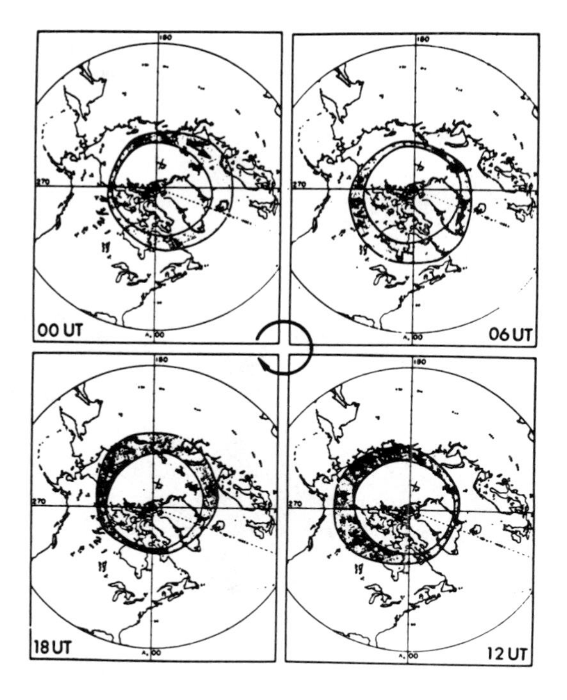

Fig.2-23. The diurnal motion of the auroral oval in geographic coordinates.

Our understanding of the dynamics of the aurora and its location has been greatly enhanced by the use of optical imagers from the Defense Meteorological Satellite Program (DMSP). City lights are easily seen in Figure 2-24a along with the auroral image providing registration of the auroral zone. This picture was obtained in 1974. More recently, striking images of the enhanced auroral zone have been obtained over the northern and southern hemispheres (see Figures 2-24b and 2-24c). It has been found that the auroral oval thickness increases as the magnetic activity increases and the diameter of the oval expands equatorward. This process is depicted in Figure 2-25. A quantitative measure of the oval parameters may be obtained through an association with magnetic activity indices. More will be said about this in the next section.

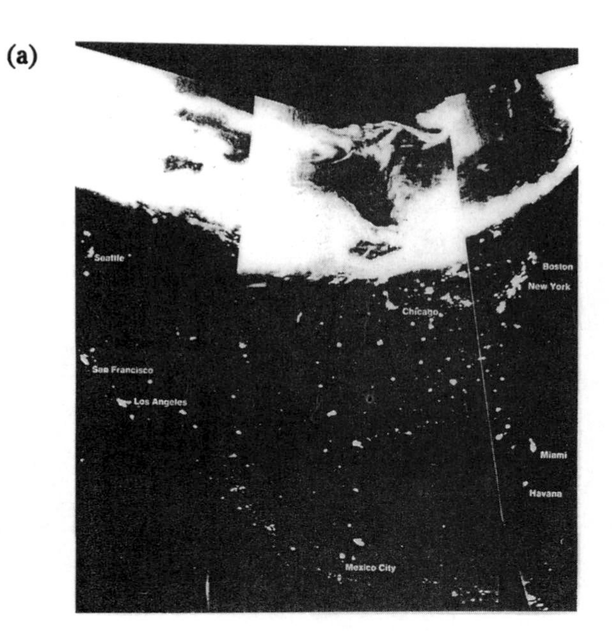

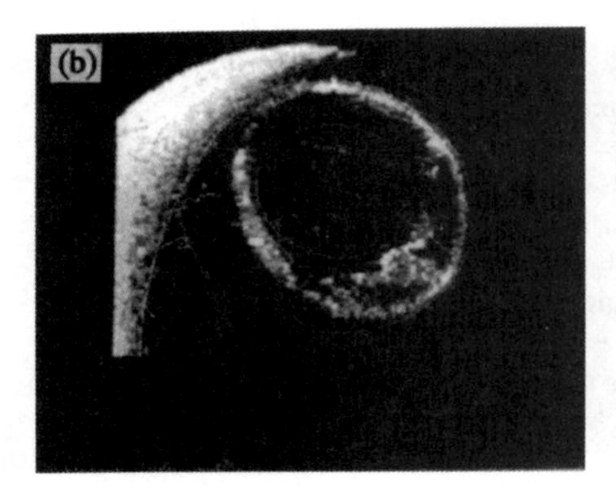

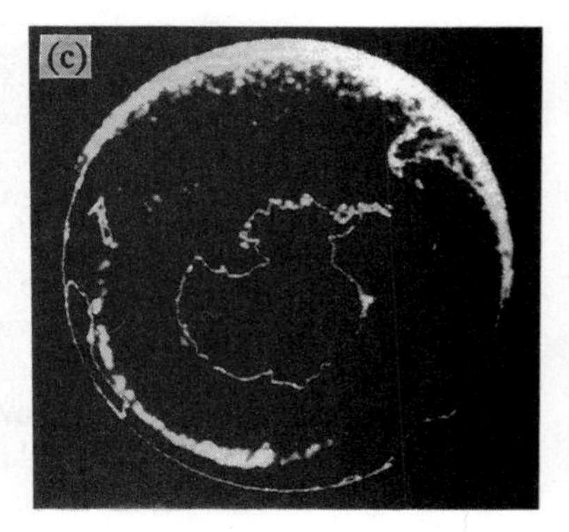

Fig.2-24. (a) Auroral image obtained from the Defense Meteorological Satellite (DMSP). The composite view was recorded on April 14, 1974. DMSP satellites are maintained in a moderately low earth orbit, and are continuing to take images of this type to satisfy DoD and civilian requirements. (b) Auroral image of North America obtained with University of Iowa instrumentation aboard the Dynamics Explorer satellite (DE1) of on November 8, 1981, a period of enhanced solar and magnetic activity. (c) Image by DE1 over obtained by DE1 of the southern hemisphere on March 13, 1989. This is the largest auroral zone ever recorded by the satellite. The DE1 orbits at 20,000 km altitude. (Sources: 1974 image [Stern, 1978]; 1981 DE1 image courtesy Whalen et al. [1985] from the *Air Force Handbook* [Jursa, 1985]); 1989 DE1 image courtesy Allen et al. [1989] from *EOS*, by permission AGU, Washington, D.C.)

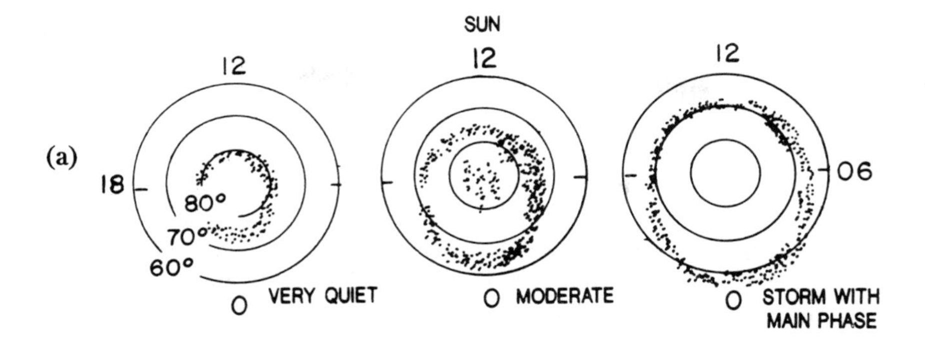

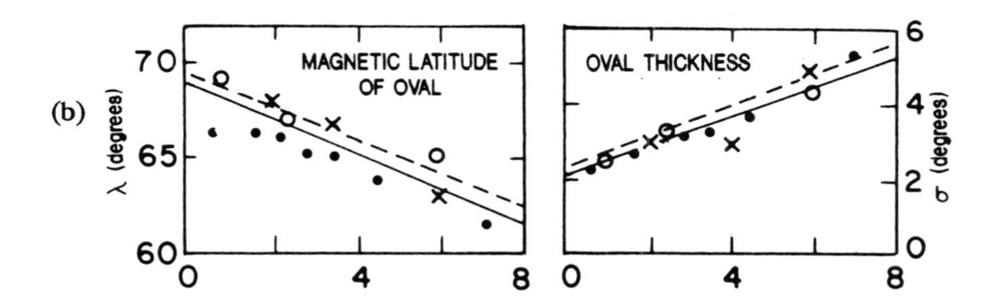

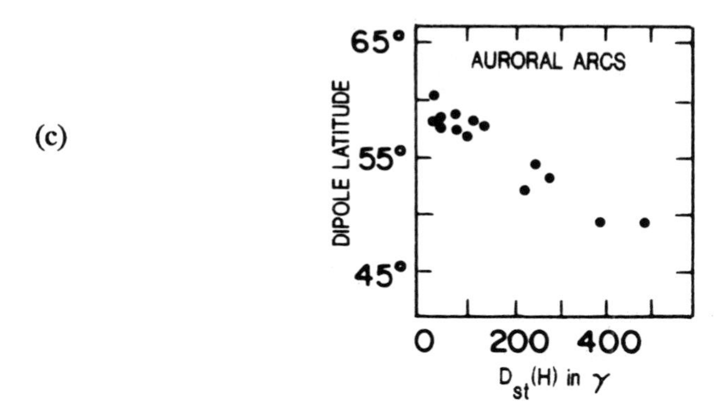

Fig.2-25. (a) Illustration of the descent of the auroral oval with increasing magnetic activity. (b) Median latitude and oval thickness as a function of *Kp* index. (c) Dipole latitude as a function of the magnetic index *Dst*.

A geomagnetic storm develops as the direct result of the energy transferred from the solar wind to the magnetosphere in a series of substorm events. In macroscopic terms we may regard the geomagnetic storm as composed of two parts: (1) an initial (short-lived) positive phase associated with an increase in the horizontal component of the magnetic field, which is followed shortly by enhanced auroral displays and (2) a main negative phase (or bay) in the horizontal field intensity which may last for several days. The initial phase is associated with short-lived enhancements in electron concentration at ionospheric heights, while the main phase is associated with large scale diminutions in electron concentration. These effects are often called ionospheric storms to emphasize the ionospheric disturbances involved. By monitoring the total electron content of the ionosphere we can see how an ionospheric storm is almost a one-to-one mapping of the geomagnetic storm-time profile (Figure 2-26). Since the parameter foF2 is of fundamental importance in sustaining ionospheric skywave support in long-haul HF communications, an ionospheric storm may have a substantial negative influence on performance within a designated communication zone.

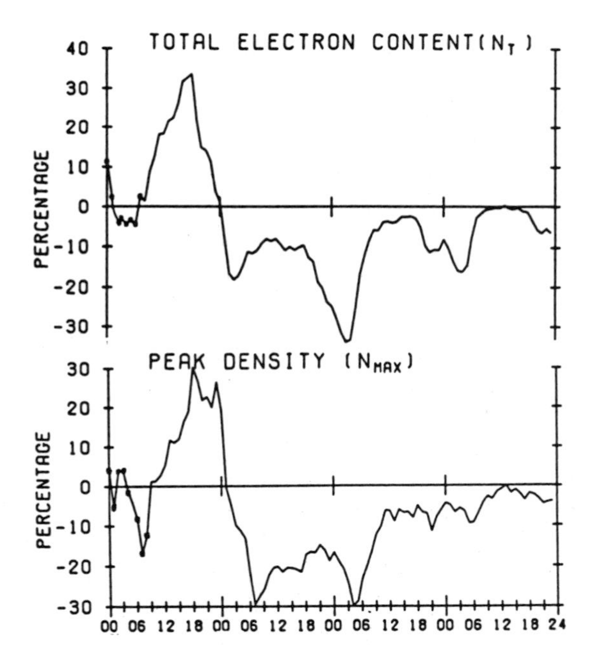

Fig.2-26. Storm-Time variation of the total electron content and the parameter foF2

2.4.4 Geomagnetic Activity Indices

Indices are useful for empirical modeling as well as for forecasting because they provide a convenient parameter set which may be used for driving the model. We have seen that sunspot number R or the flux index ϕ are convenient if not totally representative of solar activity. The magnetic activity also lends itself to the development of a wide range of index representations. Moreover, the magnetic activity indices are organized and smoothed in a variety of ways which have the potential for confusing the user who is not an ionospheric specialist. Mayaud [1980] has discussed the array of indices in his book *Derivation, Meaning, and Use of Geomagnetic Indices*. He traces the history of magnetic index development from the earliest forms to those of the present. Table 2-4 is a list of the current indices sanctioned by the International Association of Geomagnetism and Aeronomy (IAGA).

TABLE 2-4. INDICES OF MAGNETIC ACTIVITY

K-index	Three-hourly quasi-logarithmic index is a measure of the irregular variations of the horizontal field component of regular magnetograms at a specified station. Within the U.S. K_{fr} values from the station at Fredericksburg Virginia is used as a reference for the American midlatitudes. Values of K run from 0-9 with 9 corresponding to the most disturbed condition. K_p is derived from 12 stations between geomagnetic latitudes of 48 and 63 degrees. Most of these stations are in Europe.
Dst-index	Hourly index associated with low latitude magnetic activity. It is designed to be a measure of the ring current in the magnetosphere; viz, above the geomagnetic equator at about 5.6 earth radii. *Dst* stands for "Disturbance amplitude storm time" and is measured in nT. Four midlatitude stations are used in its construction [Sugiura,1964].
AE-index	Auroral electrojet activity index. It is derived from indices from a number of auroral stations at various longitudes. It is an hourly index.
Q-index	High latitude index with 15 minute time resolution. Related to oval position.
A-index	Equivalent Amplitude *A*-index. It is a daily index and linear version of *K*. A_p is a planetary value similar to K_p. The index ranges between 0 and 400. Three-hourly K_p may be converted to three-hourly a_p indices which are averaged to yield a single daily *A* index.
aa-Index	Three-hourly indices computed from *K* indices of two nearly antipodal magnetic observatories (invariant latitude = 50 degrees). The index provides a measure of global activity.

The most widely used index is K_p. It is used for ionospheric predictions. However, if we want a simple daily average for the magnetic activity, the fact that K_p is quasi-logarithmic makes it a mathematically poor choice. Even so, a number of studies have used the sum of the eight 3-hourly values of K_p to represent the smoothed daily behavior. The *A*-index is a better choice for use in averaging. Table 2-5 gives a transformation between *A* and *K*

TABLE 2-5. TRANSFORMATION FROM *K*-INDEX to *A*-INDEX

K-INDEX	*A*-INDEX
0	0
1	3
2	7
3	15
4	27
5	48
6	80
7	140
9	240
10	400

Magnetic field data may be obtained from publications and bulletins issued by the International Service of Geomagnetic Indices (ISGI) or the International Association of Geomagnetism and Aeronomy (IAGA) by writing the publications office of the International Union of Geophysics and Geodesy (IUGG) located in Paris, France. The World Data Centers (WDC) also maintain archives of geomagnetic data. Subcenters of WDC-A (USA) are located in Boulder, Colorado (NGDC) and Greenbelt, Maryland (NSSDC). Bulletins issued by NOAA/NGDC are also mailed to interested users, and NOAA publishes *Prompt Reports* and *Comprehensive Reports* of *Solar-Terrestrial data* (including magnetic activity indices). Generally speaking the K-index (or its A equivalent) is available from the same sources which issue sunspot number reports and advisories.

2.4.5 Relationship Between Geomagnetic and Solar Activity

As was indicated earlier, the geomagnetic field is composed of an internal main field, which very roughly resembles the field resulting from an embedded bar magnet, and an external field generated by currents which flow within the ionosphere and magnetosphere. The main field varies quite slowly, while the external field exhibits variation over time scales of the order of hours, as 10% of the main field component. Although global indices such as K_p and *Dst* increase during active periods, the activity levels observed at individual stations are observed to be preferentially larger for high latitudes. than those for other latitude regimes. This is a manifestation of the relative significance of high latitude current systems during disturbed periods. It has long been known that the high latitude atmosphere is heated during periods of enhanced magnetic activity. To the extent that solar activity is correlated with magnetic activity, this high latitude heating effect also increases with sunspot number. Moreover, sunspot number R_{12} is itself correlated with the Ottawa flux index ϕ_{12} (see Equation 2-2) which is directly related to solar EUV, a predominant equatorial heat source. A major consequence of this dual-component heating is an increase in the atmospheric scale height ($H = kT/mg$) resulting in an expansion of the thermospheric neutral gas profile. The orbit of SKYLAB, a scientific satellite which monitored solar activity during the seventies, decayed well before of its projected end because of solar-related increases in satellite drag. This phenomenon was caused by enhanced (and poorly predicted) levels of solar activity which was composed of the two terms discussed above. Satellite drag studies by Jacchia [1977] showed this to be a general result, and solar activity has been found to be a major station-keeping consideration for low-orbiting platforms. Figure 2-27 shows the relative importance of solar flux index ϕ_{12} and magnetic activity index A_p in the decay of Explorer 9.

Magnetic activity is related to the global atmospheric heating process indirectly. Geomagnetic energy is released at high latitudes preferentially through processes such as joule heating. This excess heat causes the auroral atmosphere to heave thereby providing an enhanced and relatively localized source of potential energy. This source decays through the production of atmospheric waves which transport energy away from the auroral zone. These waves may be observed at middle latitudes as large-scale TIDs if appropriate ionospheric measurements are made. But the propagating waves suffer dissipation as the result of kinematic viscosity in the upper atmosphere, and this viscous damping heats the thermosphere at distances far removed from the auroral source. The significance of magnetic activity as a major engine for driving thermospheric circulation may be seen from Figure 2-28. At high magnetic activity, it appears that the dominant source for defining the mean meridional wind pattern above 100 km derives from the high latitude region. At low magnetic activity, the dominant source is near the subsolar point (i.e., the equatorial region).

As indicated above, there is a general correlation between long-term averages in magnetic activity and solar activity, but the correlation is not precise and is sometimes rather low. Figure 2-29, for example, shows that the number of days for which A_p exceeds a value of 50 maximizes during the descending phase of the sunspot cycle with a secondary maximum evident slightly before the peak in solar activity.

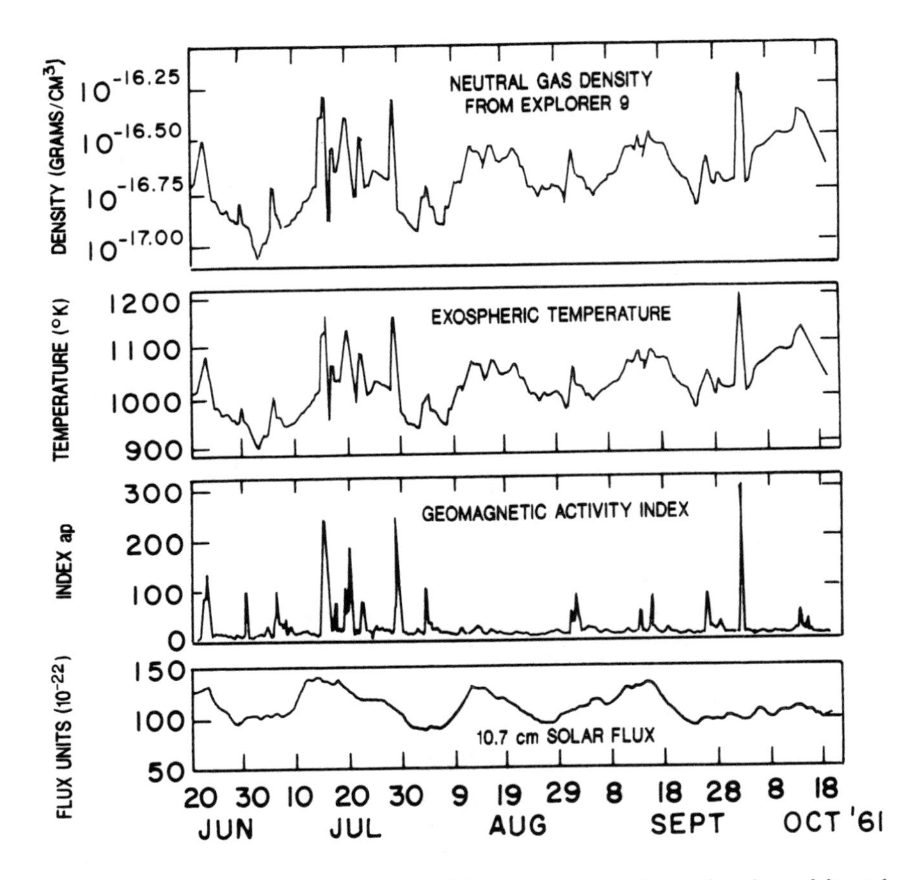

Fig.2-27. These curves exhibit the response of the atmosphere to changes in solar activity and magnetic activity. Data correspond to the period between June and October 1961. The *smoothed* values of gas density (curve a) and exospheric temperature (curve b) appear to follow the trend associated with the 10.7 cm. solar flux (curve c). Shorter period variations in temperature and density appear to be correlated with geomagnetic index (curve d). The density and temperature curves were derived from satellite drag data. [Jacchia, 1977].

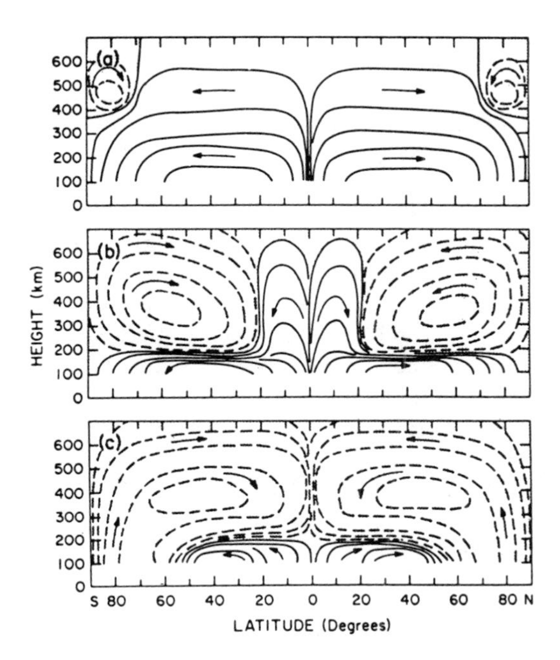

Fig.2-28. Diagram showing the mean meridional circulation in the thermosphere ($h \geq 100$ km) for equinox conditions and for three levels of geomagnetic activity. The top set of curves correspond to low activity, the middle plot corresponds to moderate activity, and the bottom set of curves correspond to a disturbed condition (i.e., a substorm) [Roble,1977].

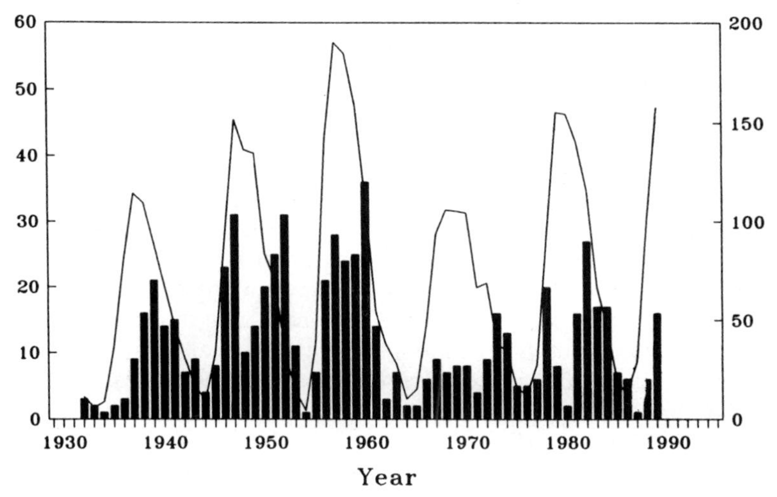

Fig.2-29. Comparison of the number of geomagnetically disturbed days with sunspot number [from Hildner, 1990].

2.4.6 Establishment and Use of Real-Time Geomagnetic Data

Magnetometer stations have been established at many locations, and have been quite useful in long-term aeronomic studies as well as in short-term experiment campaigns. A list of stations providing data to the various World Data Centers are given in the compilation by Shea et al. [1984]. Recently attention has been given to a more current assessment of geomagnetic conditions, and the proposed INTERMAGNET program is a program which would exploit satellite connectivity to provide near real-time geomagnetic data to Geomagnetic Information Nodes (GINs) for analysis and timely dissemination of data to users [Green, 1990]. The establishment of such a network, if it could be realized, would be consistent with the creation of other real-time ionospheric data networks and solar monitoring systems.

2.5 SOLAR ACTIVITY PREDICTION METHODS AND SERVICES

2.5.1 Long-Term Predictions

As has been shown throughout this chapter, the prediction of solar activity is like predicting tropospheric weather more than a few days in advance. It is very difficult. In the case of our atmospheric environment, theoretical models predict the climate, but day-to-day forecasts and even five-day outlooks are principally based upon updates of empirical models or rules. These updates are now largely based upon satellite pictures which allow for forecasting based upon persistence and the dead reckoning of a particular event. Examples include the motion of a weather front or the path of a hurricane. In the solar-terrestrial environment, the problem of forecasting ionospheric effects has recently been attacked by direct assessment or nowcasting methods involving networks of ionospheric sounders which generate pseudoflux indices to drive empirical models. Satellite imagery also allows for a near real-time examination of the auroral zone. However, in the world of long-term ionospheric predictions, the sun still retains our interest as the ultimate source of energy from which the condition of the ionosphere is derived.

Ionospheric properties are at least circumstantially related to the sunspot number. This is the basis for a number of ionospheric models and HF propagation prediction methods. Although the eleven year solar cycle has been observed for over 200 years, there is still no reliable method for predicting the peak amplitude of an upcoming cycle, or precisely when it will occur. Harmonic analysis and statistical methods have both been employed and only the latter has been found to have modest utility. Once a cycle has started, some success has been achieved in the prediction process. Irreducible errors in the prediction R_{12} of 10 are typical based upon studies undertaken by the CCIR in the 1950s. A standard technique for the prediction of solar indices is the

Lincoln-McNish method. A discussion of the choices of indices available for long-term prediction is found in CCIR Recommendation 371-5 [CCIR, 1986a].

Other indices have also been used to represent solar activity. A general discussion of these indices is given in Section 2.3.4 above. I_{F2} [Minnis,1964] is based on noontime values of the ionospheric parameter foF2 at a specified set of sites; and the Australians have developed a *T* index [IPSD, 1968] based upon the average value of the monthly median values of foF2 for each hour. Thirty stations are involved in the construction of the *T* index. Liu et al. [1983] have also developed a solar activity index based on ionospheric data. Smoothed versions of this index, IG_{12} have been shown to be more accurate indicators than R_{12} [Smith, 1986]. Comparisons of various indices have shown that ionospherically-derived indices are generally better at predicting future ionospheric events than solar-derived indices.

Measured and predicted values of R_{12}, IG_{12}, I_{F2}, and ϕ_{12} are published in monthly issues of the Telecommunications Journal of the ITU; and circulars are available through the CCIR.

2.5.2 Short-Term Predictions

Short-term predictions of solar activity, which may be used to estimate the future state of the ionosphere, must be distinguished from solar measurements which may be used to forecast the future state of the ionosphere. In the former case we are generally dealing with statistics whereas in the latter case we may attempt to solve the problem somewhat deterministically. Both approaches are used with varying degrees of success. It must be said that the short-term prediction of solar activity itself is basically a subjective exercise, and success is only modest at best.

Using various satellite and ground-based measurement schemes, it is possible to obtain a considerable amount of information about radio, optical, and x-ray signatures of solar flares, the nature of coronal holes, energetic particle flux, the interplanetary magnetic field configuration, and other parameters. These observations allow us to estimate certain special events in the ionosphere. These include: Polar Cap Absorption (PCA), and geomagnetic storm effects with some degree of confidence and lead time, and Sudden Ionospheric Disturbances (SID) with considerable confidence but with virtually no lead time. The most well-known SID is the short-wave-fade (SWF), affecting the daytime ionosphere. The SOLRAD satellite was used by the U.S. Navy in the 1970s to relate solar X-ray and EUV flux to SWFs. Successful demonstration of the capability to detect the onset of SWF and to forecast other ionospheric effects led to the development of the PROPHET system by workers at the Naval Ocean Systems Center in San Diego, California. The prediction of SID will not be discussed any further in this section (see Chapter 3).

To gain an appreciation for the panorama of solar-terrestrial effects see Figure 2-30.

Radio flux measurements provide a key to the prediction of PCA. It is now possible to predict the likelihood that a particular flare will produce a PCA and the maximum absorption to be expected during the event [Smart and Shea, 1979].

By far the most interesting prediction problem is that of forecasting various effects associated with geomagnetic storms. This is because a sufficient lead time should exist to adapt to the predicted effect, if necessary. Given the existence of a geomagnetic storm, the variation of many of the ionospheric parameters as a function of storm-time is moderately well-known. Consequently, the occurrence probability of a geomagnetic storm as derived from solar observations is an important factor to determine.

There are two principal categories of magnetic storms relative to their sources on the sun. The first is associated with a solar flare and the second is related to coronal holes and High Speed Solar Wind Streams (HSSWS). Sheeley et al. [1976, 1978] has examined the latter.

The probability that a solar flare will give rise to a magnetic storm is dependent upon the flare energy and the position of the flare on the solar disk. The time lag between the flare and the magnetic storm commencement is of the order of 2 to 3 days but is highly variable.

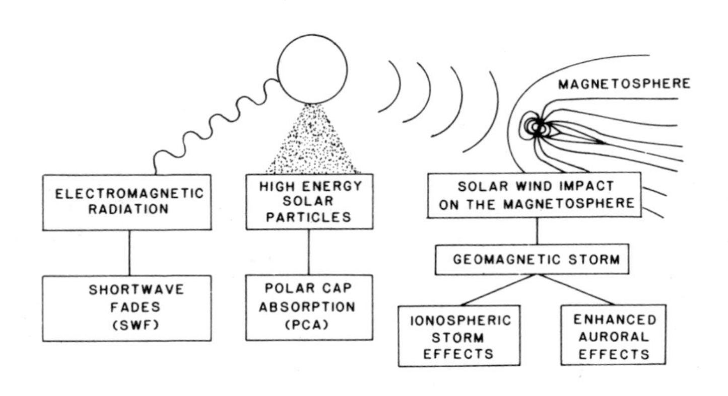

Fig.2-30. Hierarchy of Solar-Terrestrial Effects

As has been indicated, the solar rotation period of 27 days provides a natural recurrence period for isolated active regions. Recurrent storms are now associated, for the most part, with coronal holes and associated HSSWS which interact strongly with the magnetosphere. The speed of the recurring HSSWS is much higher than the average solar wind. It takes 2 to 3 days for a geomagnetic storm to commence following the passage of a coronal hole over the central meridian of the sun. The most surprising aspect of this phenomenon is the persistence of recurrence. Bohlin [1977] showed that these features have been observed to persist for up to 10 rotations. This offers exciting forecasting prospects for HF communicators. No current method used for HF propagation performance calculations accounts for this long-term effect. The uncompensated prediction algorithms will therefore generate overestimates of the MUF for several days every month at a recurrence period of 27 days. This is precisely the wrong direction we wish to go if we are relying on predictions to request frequency authorizations.

2.5.3 Prediction Services and Products

Information on short-term predictions may be found in CCIR report 727-2 [CCIR, 1986b]. Other relevant information may be found in the *Prediction Group Reports* published in connection with the 1979 Solar-Terrestrial-Physics Predictions Workshop [Donnelly, 1980].

2.5.3.1 *International URSIgram*[17] *and World Day Service (IUWDS)*

Information on data exchange for short-term forecasting, and the transmission of disturbance warnings is contained in Annex 1 to CCIR Recommendation 313-5 [CCIR, 1986c]. A permanent service, the IUWDS was established by URSI in association with the IAU[18] and the IUGG[19]. Solar-geophysical and radio propagation data are collected and rapidly exchanged through use of Regional Warning Centers (RWCs). Ultimately all data is transmitted to the

---

17. URSIgrams are broadcasts of scientific information of direct interest to radio communicators and ionospheric specialists involved in forecasting effects on radio systems. The information contained in URSIgrams is now organized under the aegis of the International URSIgram and World Day Service (IUWDS) so that summary data may be exchanged rapidly. The data exchange takes place at Regional Warning Centers (RWCs). There are nine RWCs which provide data to the IUWDS World Warning Agency located in Boulder, Colorado. At Boulder, a decision concerning the issuance of a geophysical alert is made.

18. IAU, International Astronomical Union

19. IUGG, International Union of Geophysics and Geodesy

IUWDS World Warning Agency in Boulder Colorado for action as necessary. Such action might be the issuance of a GEOPHYSICAL ALERT to members of IUWDS by teletype or similar means. The IUWDS sponsors the periodic Solar-Terrestrial Workshops [Donnelly, 1980; Simon et al., 1986]. The most recent workshop was held in Australia in 1989.

2.5.3.2 *The Space Environment Services Center*

In the USA, space environment forecasts are available on a 24 hour basis through the US Air Force (AFGWC) and the Department of Commerce (NOAA-SEL). As noted above, short-term reports and advisories are also issued by Regional Warning Centers (RWCs), but the definitive advisories originate from the World Warning Agency located in Boulder, Colorado. NOAA also offers two interactive services: SELDADS which allows access to many data sets, and a public bulletin board system (PBBS) which provides a subset of available data. SELDADS is also available via satellite link using the WESTAR satellite. Station WWV broadcasts solar activity data and propagation bulletins at 18 minutes past each hour. These bulletins contain the daily solar flux level, the three-hourly Boulder *K*-index, the daily *A*-index, and a commentary on recent HF propagation conditions, and a statement about what may be expected in the near future.

The USAF Global Weather Central (AFGWC) of the Air Weather Service (AWS) has for many years joined with NOAA and its Space Environment Laboratory (SEL) in connection with activities of the joint Space Environment Services Center (SESC). The SESC is a 24-hour real-time space environment and forecasting center which is located at Boulder Colorado. Independently, the U.S. Air Force operates two space forecast facilities: one located at Offutt AFB in Nebraska and the second located at Falcon AFB in Colorado.

A discussion of SESC products has been prepared by Joselyn [1988], and Table 2-6 is a listing of the data types available from SESC. Recently, SESC sponsored a conference and workshop [SESC, 1990] at which current and future products were described.

The Solar Observing Network is composed of three types of sites and the World Warning Agency (WWA) located at SESC in Boulder. The oldest type is the Regional Warning Center (RWC) which operates during regular working hours and prepares messages about major optical flares and solar radio bursts. Additionally, the RWCs develop daily solar and geomagnetic forecasts which are summarized by the WWA. A second type of site is run by the Air Force. These sites are geographically configured so that at least one views the sun at any given time; this enables 24-hour coverage of the sun in the optical and radio wavelengths. These Air Force sites form the primary resource for the Solar Observing Network , and are organized within an umbrella system called the Solar Electro-Optical Network (SEON) which is itself composed of two sub-nets: The Solar Optical Network (SOON) and the Radio Solar Telescope Network (RSTN). A third type of sites is the so-called contributing type which provides data of a supplemental character, or other data, on an intermittent basis. The Canadian site in Ottawa which provides the 2800 MHz (10.7 cm) solar flux data. SEON data is virtually real-time.

For investigators interested in direct access to solar images, SESC has developed the Space Environment Lab Solar Imaging System (SELSIS). This is a database of solar images in digital format available from the upgraded SELDADS (viz., SELVAX) over computer network such as SPAN or Internet. Five SELSIS observing stations provide solar images and magnetograms on a continuous basis. Data similar to that illustrated in Figure 2-8 is provided.

The Magnetometer Observing Network is composed of two types of sites: the first belonging to the Remote Geophysical Observing Network (RGON), and the second is a set of auxiliary sites. A complete set of data from the RGON may be relayed to Boulder in near real-time via GOES. Data from auxiliary stations which are not within the footprint of GOES, provide summary data to Boulder every 90 minutes. Another source of geomagnetic data is International URSIgram and World Day Service (IUWDS). The IUWDS sites provide 24 hour summaries of geomagnetic activity in the form of indices such as the daily A-index and 3-hourly K-indices. Section 2.4.6 describes a real-time magnetometer network called INTERMAGNET.

The Ionospheric Network provides data to Boulder in near real-time (from 12 total electron content sites), on an hourly basis (from 8 sites reporting foF2, fmin, and foEs), and in a daily time frame (from 12 Soviet and 3 non-Soviet sites reporting foF2, fmin, M3000, and foEs; and 17 sites providing only foF2). This network is currently being upgraded by replacement of the older variety of analog sounders (which must be hand scaled) with a new class of digital sounders which allow autoscaling. In chapter 6, we shall discuss ionospheric sounding in more detail.

2.5.3.3 *Other Prediction Services and Systems*

Recently Damboldt [1991] has described a new system for distribution of near-real-time solar-terrestrial data using BTX, a German videotext service. The service is telephonically available to worldwide users with computers, modems, and special decoding equipment (required in some instances). An interesting feature of the BTX is the inclusion of some real-time data obtained by the Deutsche Bundespost Telekom at the Regional Warning Center (RWC) located at Darmstadt in Germany. It consists of HF propagation data from 24 distant HF transmitters, additional critical frequency data, and data from a magnetometer.

Another short-term service is available as a product line of GEC-Marconi [1991]. Forecasts are issued by TELEX at 1600 GMT, by recorded message (telephone), and over PRESTEL, a British videotext service. The service, reportedly the only forecasting service in the UK, is being augmented with a bulletin board feature, which will allow the user to access a data base of archived data.

TABLE 2-6. ACTIVITIES AND DATA SETS HELD BY SESC

RESOURCES

* Solar Observing Network
* Magnetometer Observing Network
* Ionospheric Network
* GOES Satellites (currently GOES-6/7)
* NOAA-10 Satellite

PRODUCTS AVAILABLE

* SGARF: Solar and Geophysical Activity Report and Forecast [Daily at 2200 UT]
* SGAS: Solar and Geophysical Activity Summary [Daily at 0030 UT]
* SRS: Solar Region Summary [Daily at 0030 UT]
* Preliminary Report and Forecast of Solar-Geophysical Activity [Weekly on Wed]
* WWV Message @ 2.5, 5, 10, 15, and 20 MHz [Hourly: 18 minutes past each hour]
* Real-Time Alerts

SYSTEMS FOR DISSEMINATION OF DATA

* SELDADS: Space Environment Laboratory Data Acquisition and System. [24 hours/day]
* PBBS: Space Environment Laboratory Public Bulletin Board System [24 hours/day] Computer/Modem: (303-497-5000)
* Satellite Broadcast: SELDADS (subset) available in continental US, Hawaii, Alaska, and Canada.
* Telephone: Queries [24 hours/day] (303-497-3171)
  WWV Message: (303-497-3235)
* Teletype: SGARF, SGAS, SRS, Alerts, 27-day predictions

2.5.4 Data Archives and Publications

The National Geophysical Data Center (NGDC) for Solar-Terrestrial Physics (or equivalently WDC-A/STP) in Boulder distributes a pair of *yellow* reports called *Solar-Geophysical Data: Part I (Prompt Reports)* and *Solar-Geophysical Data: Part II (Comprehensive Reports).* These reports contain information on solar and magnetic activity and related geophysical data sets. They are not real-time but are useful for analysis and hindcasting. The NGDC, which is operated under NOAA of the Department of Commerce, also issues a monthly *Geomagnetic Indices Bulletin* and a *Solar Indices Bulletin* which are somewhat more timely than the *yellow* reports. These reports and bulletins may be obtained for a modest fee from NGDC, Solar-Terrestrial Physics Division, 325 Broadway, Boulder CO 80303. Another product being prepared by NGDC at NOAA is a special compilation of geomagnetic and solar-terrestrial physics data on compact disk (CD-ROM) format suitable for use with an IBM PC/AT (R) compatible machine with EGA color graphics.

The National Space Science Data Center (NSSDC) for Rockets and Satellites (or equivalently WDC-A/R&S) is located at NASA's Goddard Space Flight Center in Greenbelt, MD. It maintains an archive of data obtained largely from satellites. Data are available in a variety of forms. Details are found in WDC-A/R&S report 88-01 [Horowitz and King, 1988].

The Air Force Global Weather Central (AFGWC) maintains a dynamic (running 45-day) database called the Astrogeophysical Database (AGDB). After 45 days, these data sets form part of an archived climatic database system. Both data bases are the responsibility of the AFGWC Space Environment Support System (SESS). Developed by AWS at Scott AFB in the sixties, the SESS function has grown significantly over the years with expanded operations in the near real-time environment and with an archival function to satisfy climatological requirements. This database is described in a government report [USAFETAC, 1986]. Questions concerning this database should be addressed to: QL-A, USAFETAC, Federal Bldg., Asheville NC 28801-2723.

2.5.5 International Programs and Activities

International programs have added much to our knowledge of the solar-terrestrial environment over the years. The International Geophysical Year (IGY) and International Magnetospheric Study (IMS) are good examples of how international cooperation in research can assist in the basic understanding of global phenomena. In the present context, the Solar-Terrestrial Energy Program (STEP) for 1990-1995 has the goal of advancing the quantitative understanding of the coupling mechanisms which transfer energy and mass from one region of the solar-terrestrial system to another. STEP involves ground-based, rocket, satellite, and balloon experiments, as well as theoretical studies and simulations. A host of projects are orchestrated under STEP, sanctioned by the Scientific Committee on Solar-Terrestrial Physics (SCO-STEP) of the International Council of Scientific Unions (ICSU). A summary of STEP and its initial research projects is available [STEP, 1990]. Of some interest is Geospace Environment Modeling (GEM), a program of solar-terrestrial research sponsored by the National Science Foundation [NSF, 1988]. SCOSTEP publishes the *STP Newsletter*, distributed by WDC-A for Solar-Terrestrial Physics, which provides up-to-date information on STEP, coordinated programs, and future meetings. A separate *U.S. STEP Newsletter* is published by NASA's Space Science Data Center (WDC-A/R&S).

2.6 REFERENCES

AGU, 1989, cover illustration of EOS, *Transactions of the American Geophysical Union*, Vol.70, No.32, Aug. 8th issue.

Akasofu, S.I., 1964, "The Development of the Auroral Substorm, *Planetary Space Science*, 12:273.

Akasofu, S.I., 1968, *Polar and Magnetospheric Substorms*, D. Reidel Publ. Co., Dordrecht, Holland.

Akasofu, S.I., 1977, *Physics of Magnetospheric Substorms*, D. Reidel Publishing Co., Boston.

Allen, J., H. Sauer, L. Frank, and P. Reiff, 1989, "Effects of the March 1989 Solar Activity", *EOS, Transactions of the AGU*, November 14, p.1479.

Arnoldy, R.L., 1971, "Signature of the Interplanetary Medium for Substorms", *J.Geophys.Res.*, 76:5189.

Bartels, J., 1951, "Tagliche Erdmagnetische Charakterzahlen 1884-1950 und Planetarische Driestundliche Erdmagnetische Kennziffern K_p 1932/33 und 1940-1950", *Abh. Akad.Wiss.Gottingen Math.Phys.Kl.*, 7:28.

Bohlin, J.D., 1977, "Extreme Ultraviolet Observations of Coronal Holes. 1. Locations, Sizes, and Evolution of Coronal Holes, June 73-Jan 84", *Solar Physics*, 51:377-398.

CCIR, 1980, "CCIR Atlas of Ionospheric Characteristics", Supplement No.3 to Report 340, (based upon CCIR Mtg.in Kyoto, 1978), ITU, Geneva; p.29. (Map extracted fr. W.H. Campbell [1972], ERL 244-SEL 23, USGPO, Washington, DC.)

CCIR, 1986a, "Choices of Indices for Long Term Ionospheric Predictions", Rec.371-5, in *Recommendations and Reports of the CCIR, 1986: Propagation in Ionized Media*, Vol. VI, pp. 40-46, ITU, Geneva.

CCIR, 1986b, "Short-Term Prediction of Solar-Induced Variations of Operational Parameters for Ionospheric Propagation",Report 727-2, *Recommendations and Reports of the CCIR, 1986: Propagation in Ionized Media*, Vol. VI, pp. 86-93, ITU, Geneva.

CCIR, 1986c, "Exchange of Information for Short-Term Forecasts and Transmission of Ionospheric Disturbance Warnings", Rec. 313-5, in *Recommendations and Reports of the CCIR, 1986: Propagation in Ionized Media*, Vol. VI, pp. 81-85, ITU, Geneva.

CCIR, 1988, draft modification of CCIR Recommendation 371-5, in Compilation of "Green Book" modifications submitted by Study Group 6.

Chapman, S., 1968, *Solar Plasma Geomagnetism and Aurora*, Gordon and Breach, New York, London and Paris; p.28-32 [references have been made specifically to Figs.1.12 and 1.11 in Chapman's text.]

Chree, C. and J.M. Stagg, 1927, "Recurrence Phenomena in Terrestrial Magnetism", in *Phil.Trans.Roy.Soc.* (A), 227:21-62.

Damboldt, T., 1991, "Near Real-Time Information about the Solar-Terrestrial Environment and HF Propagation with BTX", *5th International Conference on HF Radio Systems and Techniques*, conference proceedings, pp. 103-107, IEE, Savoy Place, London, UK.

Donnelly, R.F.(editor), 1980 "Prediction Group Reports", in *Solar-Terrestrial Predictions Proceedings*, Dept. of Commerce, US Government Printing Office, Washington, DC.

Friedman, H., 1985, *Sun and Earth*, Scientific American Books, New York.

GEC-Marconi, 1991, brochure passed out at the 5th IEE conference of *HF Radio Systems and Techniques* held in Edinburgh, Scotland, 22-25 July 1991, GEC-Marconi Research Centre No.M214 dtd.1989.

Gibson, E.G.,1973, "The Quiet Sun", NASA SP-303, National Aeronautics and Space Administration, USGPO, Washington, DC.

Green, A.W, 1990, "Intermagnet, A Prospectus", proposal for global real-time digital geomagnetic observatory network, distributed at *First SESC Users Conference*, Boulder, CO., 15-17 May.

Hildner, E., 1990, "Whither Solar Cycle 22", paper presented at *First SESC Users Conference*, Boulder CO., 15-17 May.

Hill, T.W. and R.A. Wolf, 1977, "Solar Wind Interactions", in *The Upper Atmosphere and Magnetosphere*, Studies in Geophysics, National Research Council, NAS, Washington, D.C.

IPSD, 1968, "The Development of the Ionospheric Index T", IPS-R11, Ionospheric Prediction Service, Sydney, Australia.

Jacchia, L., 1977, "Thermospheric Temperature, Density and Composition: New Models", Smithsonian Astrophysical Observatory Report 375.

Jacobs, G. and T.J. Cohen (Editors), 1979, *The Shortwave Propagation Handbook*, Cowan Publishing Corp., Port Washington, NY.

Jensen, D.C., R.W. Murray, and J.A. Welch Jr., 1960, "Tables of Adiabatic Invariants for the Geomagnetic Field 1955.0", Air Force Special Weapons Center, Kirtland Air Force Base, New Mexico.

Joselyn, J.A., 1988, "General Real-Time Support of WITS Campaigns by the Space Environment Services Center", in *World Ionosphere/Thermosphere Study: WITS Handbook*, Vol.1, edited by C.H. Liu and Belva Edwards, pp.214-224, SCOSTEP, University of Illinois, Urbana, IL.

Jursa, A.S.(Scientific Editor), 1985, *Handbook of Geophysics and the Space Environment*, Air Force Geophysics Laboratory, Air Force Systems Command, U.S. Air Force, NTIS, Springfield, VA.

Knecht, D.J. and B. M. Shuman, 1985, "The Geomagnetic Field", in *Handbook of Physics and the Space Environment*", edited by A.S. Jursa, AFGL, available through NTIS, Springfield, VA.

Koons, H.C., and D.J. Gorney, 1990, "A Sunspot Maximum Prediction Method Using a Neural Network", *EOS: Trans.AGU*, 71(18):677.

Liu, R.Y., P.A. Smith, and J.W. King, 1983, "A New Solar Index Which Leads to Improved foF2 Predictions Using the CCIR Atlas", *Telecommunication J.*, 50(VIII):408-414.

Mayaud, P.N., 1980, *Derivation, Meaning, and Use of Geomagnetic Indices*, American Geophysical Union, Washington, DC.

McNish, A.G. and J.V. Lincoln, 1949, *Trans. AGU*, 30:673-685.

Minnis, C.M., 1964, "Ionospheric Indices" in *Advances in Radio Research*, Vol. II, edited by J.A. Saxon, Academic Press, London and New York.

NOAA, 1987, *Solar-Geophysical Data, explanation of data reports*, No.515, July (Supplement), National Geophysical Data Center (NGDC), NOAA, Dept. of Commerce, Boulder, CO.

NOAA, 1989a, *Solar-Geophysical Data*, No.540, Part 1, NGDC/NOAA, Dept. of Commerce, Boulder, CO.

NOAA, 1989b, *Solar-Geophysical Data, Prompt Reports*, No.543, Part 1, NGDC/NOAA, Dept. of Commerce, Boulder, CO.

NSF, 1988, *Geospace Environment Modeling (GEM)*, prepared for National Science Foundation by the GEM steering committee chaired by J. Roederer, report based upon workshop held at University of Washington, Seattle Washington (August 6-8, 1987).

Roble, R.G., 1977, "The Thermosphere", in *The Upper Atmosphere and Magnetosphere*, Studies in Geophysics, National Research Council, NAS, Washington, DC.

SESC, 1990, *Space Environment Services Center Users Conference*, Boulder, CO., 15-17 May.

Shea, M.A., S.A. Militello, H.E. Coffey, and J.H. Allen, 1984, *Directory of Solar-Terrestrial Physics Monitoring Stations*, Edition 2, MONSEE Special Publication No.2, published jointly by AFGL (Hanscom AFB, Bedford, MA.) and WDC-A for Solar-Terrestrial Physics (NGDC, NESDIS, NOAA, Boulder, CO.) under aegis of SCOSTEP, AFGL-TR-84-0237, Special Report No.239.

Sheeley, N.R. Jr., J.W. Harvey, and W.C. Feldman, 1976, "Coronal Holes, Solar Wind Streams, and Recurrent Geomagnetic Disturbances. 1973-1976", *Solar Physics*, 49:271.

Sheeley, N.R. Jr., and J.W. Harvey, 1978, "Coronal Holes, Solar Wind Streams, and Geomagnetic Activity During the New Sunspot Cycle", *Solar Physics*, 59:159-178.

Simon, P.A., G. Heckman, and M.A. Shea (Editors), 1986, *Solar-Terrestrial Predictions*, Proceedings of a Workshop at Meudon, France, June 18-22, 1984, published cooperatively by NOAA (Boulder, CO.) and AFGL (Hanscom AFB, Bedford, MA.)

Smart, D.F., and M.A. Shea, 1979, "PPS76 - A Computerized Event Mode; Solar Proton Forecasting Technique", in *Solar-Terrestrial Predictions*, Vol.1, Prediction Group Reports, ERL, NOAA, Commerce Dept., Boulder, CO.

Smith, P.A., 1968, "An Ionospheric Prediction System based on the Index I_{F2}", *J. Atmospheric Terrest. Phys.*, 30:177-185.

Smith, P.A., 1986, "Some Techniques Used to Predict Solar Activity Through the 11-year Cycle", in *Solar-Terrestrial Predictions*, Proceedings of a Workshop at Meudon France 18-22 June 1984, edited by P.A. Simon, G. Heckman, and M.A. Shea, published by NOAA (Boulder, CO.) and AFGL (Hanscom Field, MA).

STEP, 1990, *Solar-Terrestrial Energy Program*, prepared by the STEP steering committee chaired by G. Rostoker, under the aegis of SCOSTEP (J. Roederer), Geophysical Institute, University of Alaska, Fairbanks, AK.

Stern, D.P., 1978, *Solar-Terrestrial Programs: A Five Year Plan*, NASA, Goddard, Greenbelt, MD.

Stewart, F.G. and M. Leftin, 1972, "Relationship Between the Ottawa 10.7 cm solar flux and Zurich Sunspot Number", *Telecommunication J.*, 39:159-169.

Stewart, F.S. and S.M. Ostrow, 1970, "Improved Version of the McNish-Lincoln Method for Prediction of Solar Activity", *Telecomm. J.*, 37:228-232.

Sugiura, M., 1964, "Hourly Values of Equatorial Dst for the IGY", *Annals Int. Geophys. Yr.*, 35:49.

USAFETAC, 1986, "SESS: USAFETAC Climatic Database: Users Handbook No.3", Report No. USAFETAC/UH-86/003USAF, Environmental Technical Applications Center, Asheville, NC.

Valley, S.L. (editor), 1965, *Handbook of Geophysics and Space Environments*, Air Force Cambridge Research Laboratories, Office of Aerospace Research, USAF, Hanscom AFB, Bedford, MA.

Whalen, J.A., R.R. O'Neil, and R.H. Picard, 1985, "The Aurora", in *Handbook of Geophysics and the Space Environment*, Air Force Geophysics Laboratory, Air Force Systems Command, USAF, NTIS, Springfield, VA.

Withbroe, G.L., 1989, "Solar Activity Cycle: History and Predictions", *J.Spacecraft and Rockets*, 26:394.

2.7 BIBLIOGRAPHY

Akasofu, S. and S. Chapman, 1972, *Solar-Terrestrial Physics*, Oxford University Press, Ely House, London.

Al'pert, Ya.L., 1974, *Waves and Satellites in the Near-Earth Plasma*, Consultants Bureau (a Division of Plenum Press), New York.

Basu, S., J. Buchau, F.J. Rich, E.J. Weber, E.C. Field, J.L. Hechscher, P.A. Kossey, E.A. Lewis, B.S. Dandekar, L.F. McNamara, E.W. Cliver, G.H. Millman, J. Aarons, J. Klobuchar, and M.F. Mendillo, 1985, "Ionospheric Radio Propagation" in *Handbook of Geophysics and the Space Environment*, edited by A.S. Jursa, AFGL, available through NTIS, Springfield, VA.

Burke, W.J., D.A. Hardy and R.P. Vancour, 1985, "Magnetospheric and High Latitude Ionospheric Electrodynamics", in *Handbook of Geophysics and the Space Environment*, edited by A.S. Jursa, AFGL, available through NTIS, Springfield, VA.

Carovillano, R.L., J.F. McClay, and H.R. Radoski (Editors), 1968, *Physics of the Magnetosphere*, D. Reidel Publishing Co., Dordrecht, Holland.

Chamberlain, J.W., 1961, *Physics of the Aurora and Airglow*, Academic Press, New York and London.

Chang, C.C. and S.S. Huang (Editors), 1965, *Proceedings of the Plasma Space Science Symposium*, Gordon & Breach Science Publishers, New York.

Chapman, S., 1968, *Solar Plasma Geomagnetism and Aurora*, Gordon and Breach, New York, London, and Paris.

Donnelly, R.F. (Editor), 1980, *Solar-Terrestrial Predictions Proceedings*, in four Volumes: Vol.1, "Prediction Group Reports"; Vol.2, "Working Group Reports and Reviews; Vol.3, "Solar Activity Predictions"; Vol.4, "Prediction of Terrestrial Effects of Solar Activity", Dept. of Commerce, U.S. Government Printing Office, Washington, DC.

Frazier, K., 1980, *Our Turbulent Sun*, Prentice-Hall Inc., Englewood Cliffs, NJ.

Hundhausen, A.J., 1972, *Coronal Expansion and the Solar Wind*, Springer-Verlag, Berlin, Heidelberg, and New York.

Johnson, F.S., 1961, *Satellite Environment Handbook*, Stanford University Press, Stanford, CA.

Kane, S.R. (Editor), 1975, *Solar Gamma-, X-, and EUV Radiation*, International Astronomical Union, Proceedings Symposium No. 68, D. Reidel Publishing Co., Dordrecht, Holland and Boston.

Mayaud, P.N., 1980, *Derivation, Meaning, and Use of Geomagnetic Indices*, American Geophysical Union, Washington, DC.

McCormac, B.M.(Editor), 1972, *Earth's Magnetospheric Processes*, D. Reidel Publishing Co., Dordrecht, Holland and Boston.

Mitra, A.P., 1974, *Ionospheric Effects of Solar Flares*, D. Reidel Publishing Co., Dordrecht, Holland and Boston.

Newkirk, G. (Editor), 1974, *Coronal Disturbances*, International Astronomical Union, Proceedings of Symposium No. 57, D. Reidel Publishing Co., Dordrecht, Holland and Boston.

Nishida, A., 1978, *Geomagnetic Diagnosis of the Magnetosphere*, Springer-Verlag, New York, Heidelberg, and Berlin.

Olson, W.P. (Editor), 1979, *Quantitative Modeling of Magnetospheric Processes*, Geophysical Monograph 21, American Geophys. Union, Washington, DC.

Ratcliffe, J.A., 1972, *An Introduction to the Ionosphere and the Magnetosphere*, Cambridge University Press, London and New York.

Rich, F.J. and Su. Basu, 1985, "Ionospheric Physics", in *Handbook of Geophysics and the Space Environment*, edited by A.S. Jursa, AFGL, available through NTIS, Springfield, VA.

Roederer, J.G., 1970, *Dynamics of Geomagnetically Trapped Radiation*, Springer-Verlag, New York, Heidelberg, and Berlin.

Roederer, J.G. (Editor), 1988, *Solar-Terrestrial Energy Program: Major Scientific Problems*, Proc. SCOSTEP Symposium during XXVII COSPAR Plenary held at Espoo, Finland, SCOSTEP Secretariat, Univ. of Illinois.

Rosen, A. (Editor), 1976, *Spacecraft Charging by Magnetospheric Plasmas*, American Institute of Aeronautics and Astronautics, MIT Press, Cambridge, MA. and London.

Schulz, M. and L.J. Lanzerotti, 1974, *Particle Diffusion in the Radiation Belts*, Springer-Verlag, New York, Heidelberg, and Berlin.

Svestka, Z., 1976, *Solar Flares*, D. Reidel Publishing Co., Dordrecht, Holland, and Boston.

Svestka, Z. and P. Simon (Editors), 1975, *Catalog of Solar Particle Events 1955-1969*, D. Reidel Pub. Co., Dordrecht, Holland, and Boston.

3

THE IONOSPHERE AND ITS CHARACTERIZATION

"There may possibly be a sufficiently conducting layer in the upper air. If so, the waves will, so to speak, catch on to it more or less. Then the guidance will be by the sea on one side and the upper layer on the other."

Oliver Heaviside[20]

3.1 INTRODUCTION

The main features of the ionosphere are now well-known and understood even though the details require more research and debate. Much of our understanding of the fundamental structural integrity of the ionosphere has been derived from an analysis of vertical incidence sounders which operate at HF. Indeed the nomenclature for the layers of ionization had been developed in the twenties and thirties from interpretations of ionograms based upon a rather incomplete theory of radiowave propagation through ionized media. Since HF signals are so profoundly influenced by the ionosphere, it is not surprising that HF sounders have been used to deduce properties. On the other hand, we should not be surprised to find that immediate inferences drawn from HF ionograms are not always correct about detailed structure. Improved sensors and analysis techniques have provided the necessary information and have extended our understanding into the realm of ionospheric inhomogeneities and short-term temporal fluctuations.

Median structure is important in a number of applications involving the HF spectrum. One significant example is HF broadcasting, not only by the military but also by groups such as the Voice of America, the British Broadcasting Corporation, Radio Liberty, Radio Free Europe, Radio Moscow, Deutsche Welle, and others. Because of the nature of civilian broadcasting, it is necessary to plan for coverage many months in advance. Because of sky-wave variability introduced by the ionosphere, it is not possible to predict events with absolute certainty, but under certain conditions it is possible to

----

20. Quotation by Heaviside appears in an article entitled "The Theory of Electric Telegraphy" in the 10th edition of the Encyclopedia Britannica published in 1902. A similar suggestion was made by Authur Kennelly in the same year. To honor the two individuals, the upper atmospheric layer was for many years labelled the Kennelly-Heaviside layer. For additional information regarding the contributions of Heaviside, the book *Oliver Heaviside: Sage in Solitude* by Paul J. Nahin may be consulted (1990, IEEE Press, New York).

compute the coverage and reliability contours for a particular situation using median properties of the ionosphere. Typically these median representations are parameterized in terms of a set of future conditions, many of which are known and one or two of which must be predicted. The known conditions which are not directly related to the ionosphere include time(s) of broadcast, transmission frequency(ies), the transmitter power-gain product, the coverage area required, and related factors. These are not of interest to us here. The unknown conditions involve the ionospheric personality which is typically a function of sunspot number (or its equivalent) as well as magnetic activity index. If we can successfully predict the solar and magnetic indices for the conditions stated, then we can specify a median ionosphere. Finally, if we can do this we are in a good position to solve the median coverage prediction problem. Near-term prediction of coverage may be an improvement over a median representation, or it may not be. This depends upon the relationship between near real-time indicators of solar activity and the extent to which they are correlated with the ionospheric morphology. Ultimately ionospheric indicators must serve as replacements for the solar indices. More will be said about this matter later.

There are many excellent sources of information about the ionosphere, from both a theoretical and experimental perspective. Books by Rishbeth and Garriott [1969], Davies [1965, 1969, 1990], Ratcliffe [1972], and Giraud and Petit [1978] should be consulted. Volumes such as the revised U.S. Air Force *Handbook of Geophysics and the Space Environment* [Jursa, 1985] is quite useful for background information as well as a more detailed exposition of many aspects not specific to HF propagation and systems. As an alternative to the Air Force handbook which is rather bulky, a condensed volume, *Introduction to the Space Environment*, has been written by Tascione [1988]. Kelley [1989] has published a book emphasizing ionospheric plasma physics and electrodynamics, and a practical handbook on the ionosphere has been developed by McNamara [1991]. For those interested in historical aspects of ionospheric physics and research, special issues of *Radio Science* [Checcacci, 1975] and the *Journal of Atmospheric and Terrestrial Physics* [Ratcliffe, 1974] should be examined. The *Radio Science* issue commemorated the centennial of the birth of Guglielmo Marconi, a scientist who made monumental contributions to radio and its commercial development. The *Journal of Atmospheric and Terrestrial Physics* issue was entitled "Fifty Years of the Ionosphere" and chronicles ionospheric research from the first steep-incidence observations of radiowaves reflected from the ionosphere in December 1924.

3.2 FORMATION OF THE IONOSPHERE

The ionosphere exhibits a multilayered structure although only one of these layers (i.e.,the F-layer) is typically dominant in terms of its contribution to the total population of electrons or electron content. Figure 3-1 depicts the various ionospheric layer distributions for solar maximum conditions , and a drawing exhibiting the F1/F2 and D/E region bifurcations as one proceeds from the nightside to the dayside hemisphere.The F-layer domination does not always translate into a one-to-one correspondence with radiowave propagation effects. Lower layers, because of geometrical factors or collisional processes, may have a major impact on HF system operation. In some instances the impact may be controlling. The normal D-region, which is characterized by radiowave absorption rather than refraction, is the lowest sensible layer having an impact on HF signals and acts as a power robbing attenuator. This feature is especially enhanced during the daytime when the sun's photoionization flux is greatest. Additionally, patches of sporadic E ionization may blanket the upper ionospheric layers and serve as the primary reflecting layer. This has both favorable and negative implications for HF coverage and will be discussed fully in the propagation section contained in Chapter 4.

We have just mentioned that several distinct regions or layers of ionization may exist in the upper atmosphere, but we have not mentioned how this ionization is formed or how it is maintained. Let us discuss this briefly. The sun exerts a number of influences over the upper atmosphere, but the interactions of most importance for our discussion are photodissociation and photoionization. These processes were mentioned in Chapter 2. Figure 3-2 depicts the neutral atmosphere, its various regions and the depth of penetration of the various components of solar flux.

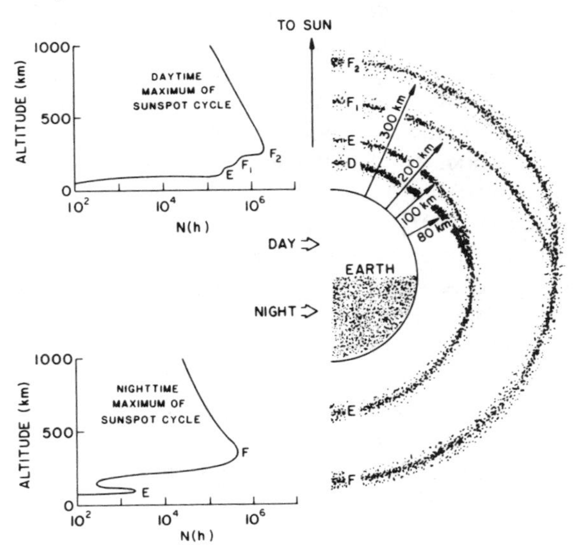

Fig.3-1. Electron density distributions for the ionosphere during solar maximum conditions, day and night. The right-hand-side shows the diurnal layering for a mid-latitude slice through the earth (not to scale).

In the lower atmosphere, species such as N_2, O_2, He and H_2 dominate the constituents even though other species such as helium, water vapor, carbon dioxide, nitric oxide, and trace element gases are influential in specific contexts. In the upper atmosphere, however, molecular forms are dismembered by incoming solar flux into separate atomic components. This process is so efficient that we refer to the distribution of neutral species in a vast segment of the upper atmosphere (i.e., above 200 km) as a monatomic gas. In the lower atmosphere (i.e., below roughly 200 km), the gas is largely polyatomic. This has implications for the lifetime of ion-electron pairs created through photoionization.

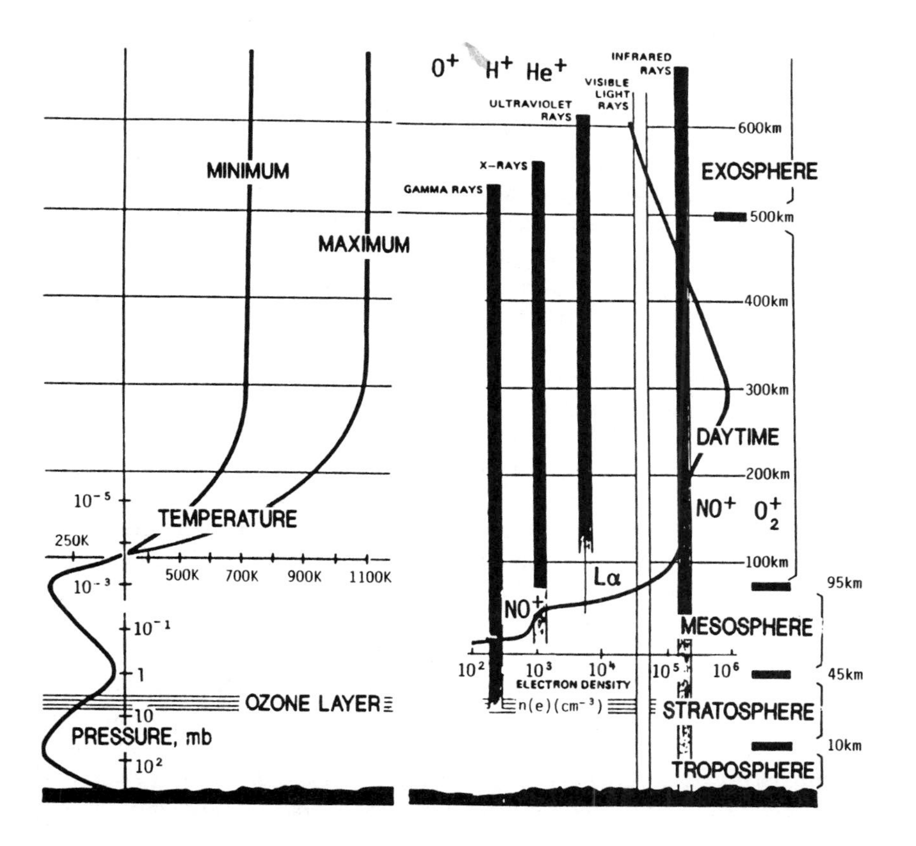

Fig.3-2. Various atmospheric and ionospheric layers, average neutral density profile, the designation of primary ionic species, and the depth of penetration of solar radiation by category. (From National Research Council study [NRC,1981])

We shall also see in the altitude regime of primary interest to us (i.e., above 90 km), that collisions become a rarity and mixing of the various species becomes unimportant in comparison with diffusive forces. As a consequence a phenomenon called diffusive separation occurs, with constituents of the neutral gas seeking their own unique height distributions dictated by their atomic masses, the gas temperature, and the acceleration of gravity. Figure 3-3a shows a height profile of ionic species in the upper atmosphere, and Figure 3-3b contains typical distributions of midlatitude electron density for daytime and nighttime under solar maximum and minimum conditions.

It may be seen that monatomic oxygen in its ionized form is the constituent of major interest for HF skywave concerns. This is because O^+, the majority ion between roughly 180 and 1000 kilometers, is dominant between about 200 and 500 km, a height domain which figures prominently in the determination of HF coverage. As a consequence of this fact we shall give considerable attention to the interaction of solar ionizing flux with atomic oxygen. From Figure 3-3a, we also note that NO^+ is the major ionic constituent below roughly 180 km even though (neutral) nitric oxide itself is not an important ingredient in the atmosphere. This is the result of a chemical reaction between N_2^+ (or N_2) and O (or O^+) which produces N and NO^+. The presence of NO^+ in the ionosphere is important since it provides a major vehicle for electronic loss through the process of recombination below 200 km. The other vehicle for loss in this region is O_2^+.

In outlining the ionization process we shall restrict our attention to atomic oxygen for simplicity. The oxygen atom consists of 8 protons (which are positively charged) and 8 neutrons within its nucleus, and 8 (bound) electrons in its extranuclear region. We may picture the extranuclear electrons much like satellites circling a massive planet in various orbits or shells. The electrons which are closest to the oxygen nucleus are more tightly bound while those which are in the outer shells are more susceptible to outside influence. The sun's radiation is such an influence. Extreme Ultraviolet (EUV) radiation from the sun interacts strongly with various atmospheric constituents, and it is a primary force in the development of the ionosphere. Photoionization is best understood by recognizing the duality of electromagnetic radiation. In lower frequency regimes (within the radio spectrum), EM radiation is dominated by wavelike properties such as interference and diffraction, while at higher energies EM radiation begins to exhibit corpuscular properties such as the exertion of pressure. When corpuscular processes are important, we use the term photon to identify individual radiation elements. Although the notion is far from precise, it is convenient to associate these corpuscles or photons with the stripping of electrons from the outer shells of oxygen atoms in a process analogous to a collision. To visualize the interaction between the EUV radiation and the oxygen atom, it is convenient to regard energetic solar flux in this pseudo-corpuscular role. Electrons are tied or bound to a region surrounding the oxygen nucleus by electrostatic forces which may be overcome by the collision of a photon in the EUV range. The photon transfers its energy to the electron as excess kinetic energy. If conditions are right, the excess kinetic energy will exceed the binding energy for the electron and it will be removed from the influence of the nucleus. This process is called photoionization, and it results in a positively charged atom (i.e., an oxygen atom minus one negative charge) and a free electron. It is important to note that the full ensemble is still neutral as before. Most of the significant ionization is produced by radiation corresponding to wavelengths less than 1026 Angstrom units. This includes EUV and x-rays, with the latter radiation responsible for D-layer enhancements in electron concentration.

(a)

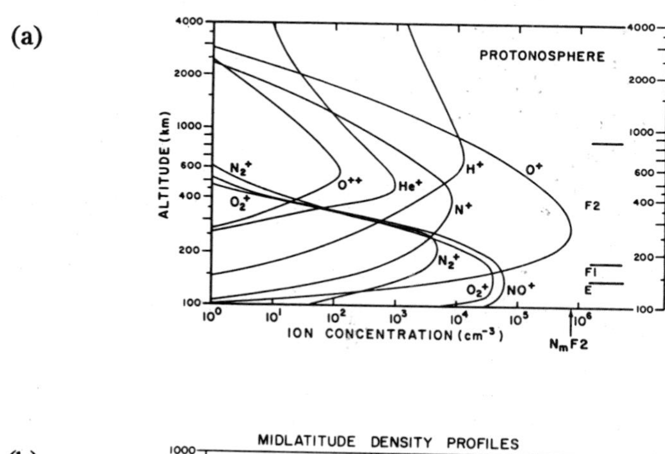

(b)

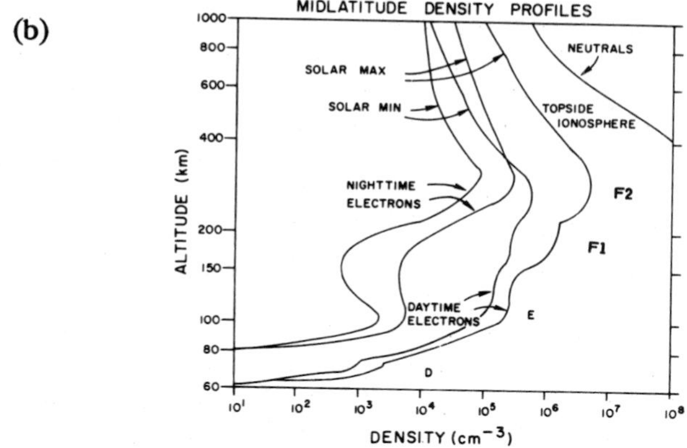

Fig.3-3. (a) Profiles of ion concentrations, as a function of height, for daytime conditions [NRC,1977]. (b) Electron density distributions for day-night and solar maximum/minimum conditions [Jursa,1985]. Note: Plots a and b were obtained under differing conditions

What has changed? We still have no net charge, but individual free charges within the medium are now subject to the influence of electromagnetic forces and fields such as those associated with radiowaves. The significance of the photoionization event lies in the generation of *free* electrons, and the significance is profound. There were free molecules before, and both free positive ions and electrons afterwards. Are we concerned with the free positive ions? The answer is yes, but the amount of interest is slight. Since a nucleon (either proton or neutron) is 1,840 times more massive than an electron, and since there are 16 nucleons in an oxygen atom, the ratio of a singly-charged oxygen atom to a free electron 8 x (1840) or (2.944) x 10^4. Clearly the light (free) electrons will have a greater mobility than the heavy ions and will respond freely and strongly to RF fields. By comparison, the ions will remain relatively motionless for radio frequencies of interest (i.e., 3-30 MHz), and except for collisional effects, we shall generally ignore their presence in the ionosphere.

3.3 PROPERTIES OF THE IONOSPHERIC REGIONS

Table 3-1 provides information about the various ionospheric layers, the altitude ranges of each, the principal ionic constituents, and the means of formation. A comment is appropriate here on the nature of ionospheric layering with some emphasis on the historical distinctions made between the words *layer* and *region* as they pertain to the ionosphere. Often the terms are used interchangeably. If one term were to be preferred, the most appropriate would be the term *region* since it properly does not leave the impression that sharp discontinuities in electron density prevail at well-defined upper and lower boundaries. The most obvious exception is sporadic E ionization which, owing to the sharp density gradients defining its boundaries, is commonly referred to as the sporadic E layer. Since the sporadic E layer is generally restricted in geographical extent, it is sometimes called a sporadic E patch. There are additional situations for which the restrictive term *layer* is appropriate. For example, the normal E region may occasionally be characterized by an electron density profile displaying a degree of boundary sharpness. The context of discussion is important.

The ionospheric electron density distribution is logically evaluated first in terms of its height profile, followed by its geographical variability. Even though there is substantial evidence for an ionospheric height distribution, or profile, which is variable with multiple ionization peaks, the most fundamental understanding about HF skywave properties comes from a simple picture of an ionosphere dominated by a single region, or layer, having a distinct maximum in electron density.

The ionosphere is often described in terms of its component regions or layers with the earliest designations being historically traced to E.V. Appleton. These were the so-called E and F regions. These designations were based upon data obtained from a myriad of radiowave measurement schemes involving vertical and oblique geometries, pulsed and cw systems, variable and constant frequencies, and a variety of observables such as polarization and signal time delay. These early crude measurements often exhibited evidence for a distinct intermediate layer in addition to regions E and F. This led to the notion of an F1 and an F2 layer comprising the F region. This layering evidence was most apparent in HF radio signals owing to their strong interaction with the ionosphere in locations where the profile exhibits a discontinuity in its height derivative. Indeed, we still refer to the E, F1, and F2 layers of the ionosphere, even though there is occasionally only one distinct maximum exhibited throughout the whole of the ionosphere. Certainly the term region would be preferred and justifiable. There is nonetheless strong evidence for a distinct E layer since a valley is often observed between the E and F regions based upon incoherent backscatter records and rocket flights. There is no such evidence for the F1 *layer* however; as a result, it is probably more appropriate to use the term F1 *ledge*. For the D-region, there is no consistent tendency for a maximum to be produced within the 70-90 km height domain. Consequently, the term D *region* should be preferred over the term D *layer*. In short, exclusive of sporadic E, the term *region* is the least misleading (and more general) suffix to use. The term *layer* should be reserved for the more obvious situations where an emphasis on a real tendency for layering exists. Fortunately, the context of discussion reduces the likelihood of confusion.

Simple layering occurs as the result of two factors. First, the atmospheric neutral density decreases exponentially with altitude, while the solar ionizing flux density increases with height above sea level. This leads to the formation of single region for which the ionization rate is maximized, and ultimately results in a layer having the so-called Chapman shape. This shape is based upon a simple theory advanced by Sidney Chapman [1931] (see Figure 3-4). We observe nonetheless a degree of structure in the ionosphere which suggests of more than one layer. One cause for multilayer formation is the existence of a multicomponent atmosphere, each component of which possesses a separate height distribution at ionospheric altitudes. But there are other factors. Solar radiation is not monochromatic as suggested in simple Chapman theory, and it has an exoatmospheric energy density which is not distributed evenly in the wavelength domain. Furthermore its penetration depth and ionization capability depend upon wavelength and atmospheric constitution. All of this results is a photoionization rate which is a structured function of altitude. A very readable account of the ionospheric layering process is given by Ratcliffe [1972].

TABLE 3-1. IONOSPHERIC PROPERTIES: MID-LATITUDES

Region	Height Range (km)	Nmax Range (No/m^3)	f_p(MHz) [Note 1]	Major Ingredient	Basis of Formation
D [Note 2]	70-90	10^8-10^9	[Note 3]	NO^+, O_2^+	L-alpha x-rays
E	90-130 hmax=110	$\approx 10^{11}$ [Note 4]	night:≈ .5 day:≈ 3.0	O_2^+, NO^+	L-beta x-rays Chapman form
E_s	90-130	[Note 5]	[Note 5]	Metallic Ions	Wind Shear
F1 [Note 6]	130-210	$\approx 2 \cdot 10^{11}$	day:≈ 4.5	O^+, NO^+	He II line UV Chapman form
F2	200-1000 [Note 7] hmax≈300	$\approx 10^{12}$ [Note 8]	day: ≈ 10 [Note 8]	O^+, N^+	[Note 9] Diffusion

COMMENTARY

Note 1: The plasma frequency $f_p = 0.9\, N^{\frac{1}{2}} \times 10^{-5}$. Plasma frequencies of interest fall in the HF band.

Note 2: The D region is bounded from above by the E region (having similar properties) and the C region below. The C region is produced by galactic cosmic rays. The D region generally departs from Chapman-like rules. Still it exhibits strong solar control.

Note 3. Plasma frequency is unimportant for HF interactions.

Note 4: Strong solar zenith angle control. Normal E is not important following sunset.

Note 5: Electron densities vary widely and may be $\approx 10^{12}$ or so. E_s occurs in a variety of shapes, but generally takes the form of thin patch-like concentrations of electrons. Thought to be associated with wind shear.

Note 6: Merges with the F2 layer at night. The peak of F region production occurs near the peak of ionization of the F1 region. Since the F1 region shows no clear layering, its peak is often termed the F1 ledge.

Note 7: The peak may range between roughly 250 and 500 km.

Note 8: Peak values strongly influenced by the movement term in the continuity equation. Non-Chapmanlike.

Note 9: Limited loss rates due to dominance of monatomic ions. Primary ionization source arises from vertical transport with production being a secondary source.

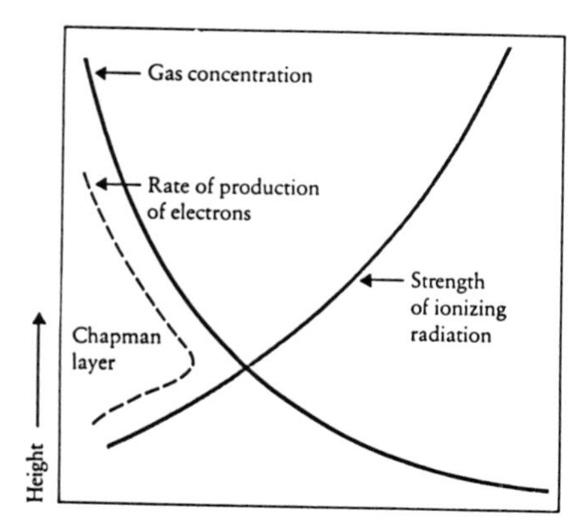

Fig.3-4. A representation of how the Chapman profile of ionization production is developed as the sun's radiation enters a neutral gas with an exponentially increasing density (with penetration depth).

3.3.1 Properties of the Chapman Layer

One of the basic tenets of Chapman theory is that solar radiation will descend to an altitude for which the total number of molecules P (populating a column of unit cross sectional area directed toward the sun) is equal to the inverse of the absorption (or interaction) cross section σ. In other words $P = 1/\sigma$. The peak in ionization will be produced in the neighborhood of that altitude, and the concept is valid for oblique solar illumination as well as for the case in which the sun is overhead. Nevertheless it is convenient to look at the production rate in terms of its deviation from the peak value which would result if the sun were overhead. For this it is useful to define a reduced height z corresponding to the normalized departure of the height to be reckoned from the (possibly hypothetical) height which would result if χ, the solar zenith angle, were equal to 0. Thus,

$$z = (h - h_0)/H \tag{3.1}$$

where h_0 is the peak height for vertically-incident radiation from the sun, and H is the neutral scale height given by the following expression:

$$H = kT/mg \tag{3.2}$$

where k is Boltzmann's constant, T is the absolute gas temperature, m is the molecular mass, and g is the acceleration of gravity. The scale height H is a convenient parameter since it may be used as a measure of layer thickness for an equivalent fixed density slab. More importantly, it has a physical meaning. If the atmosphere is in diffusive equilibrium governed by the force of gravity and the gas pressure gradient, we have:

$$N = N_0 \exp(-h/H) \tag{3.3}$$

$$H = -\{(1/N)(dN/dh)\}^{-1} \tag{3.4}$$

where N_0 is the molecular density at some reference height.

Figure 3-5 depicts the production rate curves associated with an ideal Chapman-like production profile and a range of solar zenith angles. The production q is greatest for $\chi = 0$ (overhead case corresponding to $q = q_0$), and we see that q_{max} decreases in magnitude and occurs at increasing (reduced) heights as χ becomes larger (moves toward the horizon). The rate of production may be written as:

$$q = q_0 \exp\{1 - z - \sec\chi \exp(-z)\} \tag{3.5}$$

where q_0 corresponds to the maximum production rate for $\chi = 0$ and is greater than or equal to q_{max}. At altitudes well above the peak in q, the rate of electron production drops off in a roughly exponential fashion imitating the exponential decrease in gas pressure with height. All of this is intuitively pleasing. In order to relate Chapman production curves to actual electron density distributions, we must examine loss processes and certain dynamical factors.

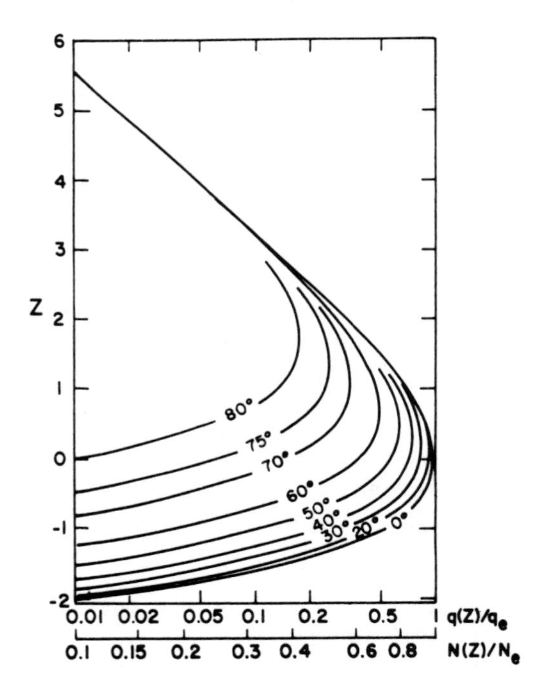

Fig.3-5. Curve illustrating the rate of electron production as a function of reduced height $(h-h_0)$ and for selected values of solar zenith angle χ. (From Davies [1965], p.16)

3.3.2 The Continuity Equation

The equation which expresses the time rate of change of electron concentration N_e is called the continuity equation:

$$dN_e/dt = q - L(N_e) - \text{div}(N_e V) \tag{3.6}$$

where N_e is the electron density, $L(N_e)$ is the loss rate which is dependent upon the electron density, div stands for the vector divergence operator, and V is the electron drift velocity.

The divergence of the vector in equation 3.6 is the so-called movement term. Simply put, the continuity equation says that the time derivative of electron density within a unit volume is equal to the number of electrons which are generated within the volume (through photoionization processes), minus those which are lost (through chemical recombination or attachment processes), and finally adjusted for those electrons which exit or enter the volume (as expressed by the movement term). To first order, the only derivatives of importance in the divergence term are in the vertical direction, since horizontal N_e gradients are generally smaller than vertical ones and horizontal velocities are relatively small in comparison with vertical drift velocities. Hence, the vector velocity $V = (V_x, V_y, V_h) \approx (0,0,V_h)$. We thus replace div $(N_e V)$ with $d/dh(N_e V_h)$ where V_h is the scalar velocity in the vertical direction. We rewrite equation 3.6 as follows:

$$dN_e/dt = q - L(N_e) - d/dh(N_e V_h) \tag{3.7}$$

Now let us look at some special cases. If $V_h = 0$ (no movement), then the time variation in electron concentration is controlled by a competition between production q and loss L. At nighttime, we may take $q = 0$, and this results in

$$dN_e/dt = - L(N_e) \tag{3.8}$$

In principle, there are two mechanisms to explain electron loss: one defined by attachment of electrons to neutral atoms (in the upper ionosphere), and the second defined by recombination of electrons with positive ions (in the lower ionosphere). The attachment process is proportional to N_e only, while recombination depends upon the product of N_e with N_i, where N_i is the number of ions. The process of attachment involves radiative processes and results in an extremely low cross section (probability of occurrence). We may ignore it in practical situations and take recombination as the only major source for electron loss. Since $N_e = N_i$, the recombination process proceeds as N_e^2. Recombination is very rapid in the D and E regions.

In the vicinity of local noon, $dN_e/dt = 0$ and we may analyze the quasi-equilibrium conditions suggested by Equation 3.7 when the LHS = 0. The two main types of equilibrium processes are given in Table 3-2.

TABLE 3-2. TYPES OF EQUILIBRIUM PROCESSES

Continuity Equation: $q = L(N_e) + d/dh(N_e V_h)$

I. Photochemical Equilibrium: [Production balanced by loss]

$$L(N_e) >> d/dh(N_e V_h)$$

2. Drift Equilibrium: [Production balanced by drift]

$$L(N_e) << d/dh(N_e V_h)$$

The equilibrium processes identified in Table 3-2 are the dominant possibilities during daytime when photoionization is significant. During nocturnal hours, equilibrium is seldom achieved at F region heights, although it is approached in the period before sunrise.

The reader should not be misled by the simplicity of the continuity equation. The generic terms: production, loss, and movement represent a host of complex photochemical and electrodynamic processes which exhibit global variations and are influenced by nonstationary boundary conditions within the atmosphere and the overlying magnetosphere. Notwithstanding these complications, the equation provides a remarkably clear view of the basic processes which account for ionospheric behavior. In fact, the relative contributions of terms in the continuity equation will account for the majority of the anomalous ionospheric properties; that is, those ionospheric variations which depart from a Chapman-like characteristic. (Some of these anomalies are discussed in Section 3.3.7.) This is especially true for the F2 layer for which the movement term attains paramount status. In the E and F1 regions where the movement term is small compared with production and loss (through recombination), photochemical equilibrium exists in the neighborhood of midday. All of this has had a significant bearing on the development of ionospheric models, on the prediction process, and on the manner in which HF frequency management systems have evolved. Indeed, as it relates to the F region of the ionosphere, it may be said that the existence of a nonvanishing divergence term in the continuity equation has been the primary impetus for the development of statistical modeling approaches. Nevertheless, noble efforts to account for all terms in the continuity equation through physical modeling continue to this day.

The underlying assumptions used by Chapman in his theory of layer production are in substantial disagreement with observation. The Chapman layer was based upon an isothermal atmosphere, and it is well known that the atmosphere has a scale height kT/mg which varies with height. This is principally because of the altitudinal variation of temperature (see Figure 3-2), but been taken to correct this flaw in the original theory and further extensions have been made over the years [Rishbeth and Garriott, 1969a]. Still the Chapman profile remains as a benchmark for use in ionospheric modeling and analysis.

Even though the Chapman profile is relatively simple, it does not readily admit to analytical treatment for the solution of radio propagation integrals. For this reason before the availability of modern computational methods, many alternative profiles had been suggested to handle various portions of the ionosphere. They included the following: linear, exponential, parabolic, height-squared, cosine, and secant-squared profiles. These simplified alternatives are described by Davies [1969]. Current approaches avoid such simplifications and rely upon numerical integration schemes which use electron density profiles which are even more complex (and realistic) than the Chapman description.

3.3.3 The D-Region

The D-region[21] is responsible for most of the nondeviative absorption encountered by HF signals which exploit the skywave mode. In most instances, D region absorption is a primary factor in the determination of the lowest frequency which is useful for communication over a fixed skywave circuit. Details of D-region electron concentration are sketchy in comparison with information available about the E and F regions, principally because of the difficulty in making diagnostic measurements. In addition, analysis is hampered because many photochemical processes with poorly defined reaction rates take place in the D region. Over 100 reactions have been compiled by Ogawa and Shimazaki [1975]. Fortunately, details of the D region electron density distribution are not as important at HF as they are in the longwave regime. We are principally interested in the integrated features which will give us a picture of the absorption which HF signals experience during a complete passage through the region.

---

21. The D region has insufficient ionization for introducing a strong refractive interaction with radiowaves in the HF band. This may be seen from inspection of the most rudimentary form of the ionospheric refractive index n (see Equation 3.9). It is convenient to express electron density in terms of the plasma frequency f_p where $f_p = 9 N_e^{0.5}$. A typical daytime D region electron density is 2×10^8 electrons/m^3. This indicates a D region plasma frequency of roughly a tenth of a MHz. At the low end of the HF band (i.e., 3 MHz), we have $n \approx 1$. Thus refractive bending is negligible. On the other hand absorption (which in the upper D region is proportional to the product of the electron density and the collision frequency) is not at all negligible. Assuming a fixed collision frequency, absorption at a fixed radio frequency will be proportional to the electron population of the region traversed. Thus one would expect the amount of absorption to roughly replicate the time variation of the D region electron density. Ionospheric D region absorption is measured with instruments called Relative Ionospheric Opacity Meters (Riometers). The matter of absorption will be discussed more in Chapter 4.

Table 3-1 shows that the D region lies between 70 and 90 km. In fact, the upper and lower levels are not precisely defined. Some investigators place the lower boundary of the D region at 50 km to account for the contribution of galactic cosmic rays in the neighborhood of 50-70 km. This altitude regime, termed the C region, is not produced by solar radiation. It exhibits different characteristics from the region between 70 and 90 km. Specifically, a minimum in electron concentration is observed during solar maximum conditions for the lower portion (region C) while the reverse is true in the upper portion (region D). This can be explained if we assume that the galactic cosmic ray source is partially diverted from the earth by an increase in the interplanetary magnetic field (IMF) which occurs during solar maximum conditions. (It is fortuitous that the C region is denoted by the letter *C* which is just before *D* in the alphabet and that cosmic rays are the primary ionization source.) Even in the normal D region above 70 kilometers, particle interactions may be important after geomagnetic storms because particle precipitation events may lead to an enhancement of electrons.

The main influence of the D region on HF systems is absorption. It is therefore regarded as a region without much virtue. Higher layers introduce refractive effects which are responsible for the ionospheric bounce phenomenon. If we wish to take advantage of skywave absorption to enhance the relative importance of groundwave signaling through reduction in cochannel interference, then the D region does have its positive benefits. Also HF communication schemes may take advantage of D region absorption to achieve a limited degree of privacy. We will say more about the absorption introduced by the D region in Chapter 4.

3.3.4 The E-Region

The normal E region exhibits a Chapman-like characteristic with the peak in electron density occurring at local noon. As a matter of convenience we shall use notation which has evolved from ionospheric sounding studies to specify the electron concentration.

3.3.4.1 *On Plasma Frequencies, Critical Frequencies and Electron Densities*

If a plane wave is vertically incident on the ionosphere from below, it will interact with the electron gas and will be reflected from a level at which the refractive index $n = 0$. In general we have:

$$n = [1 - (f_p/f)^2]^{\frac{1}{2}} \tag{3.9}$$

where f_p is the plasma (resonance) frequency and f is the radio frequency. The plasma frequency f_p is defined for any point in the ionospheric plasma and is proportional to the square root of the electron density. More precisely we have:

$$f_p = 9\, N_e^{\frac{1}{2}} \tag{3.10}$$

where f_p is expressed in Hertz and N_e is expressed in electrons/m^3. For a typical E region electron density of 10^{11}, we have f_p = 2.85 MHz.

From Equation 3.9, we see that the reflection of the plane wave occurs at the point for which the radio frequency matches the local plasma frequency. If the layer has a defined peak in electron concentration N_{max}, then the corresponding plasma frequency profile will also have a maximum. The maximum plasma frequency in the layer (having a distinct peak) is called the critical or penetration frequency. This is because if the radiofrequency transmitted from below exceeds this critical value, called f_c, then the radiowave will exit the layer and not return. The frequency f_c is the highest one, vertically incident on the ionosphere, which has the possibility to be reflected and monitored with a ground receiver which is colocated with the transmitter.

The expression for electron density in terms of its equivalent plasma frequency is especially convenient in the context of HF propagation studies and ionospheric modeling which depends so fundamentally upon a large archive of ionospheric critical frequency data derived from sounders. In fact f_c is determined directly from a vertical incidence ionogram if the ordinary wave trace $h'(f)$ can be identified. Here h' is the virtual height or time delay of reflection and f is the sounder frequency. Specifically, for the E layer, $f_c E = foE$ and it is the frequency at which $d(h')/df$ asymptotically tends to infinity along the ordinary wave trace. (In this expression *foE* is the ordinary ray critical frequency. Both ordinary and extraordinary rays propagate in the ionosphere, but the ordinary ray properties are more conveniently related to electron density. This matter will be discussed more thoroughly in Chapter 4.)

In an idealized Chapman layer for which photochemical equilibrium has been established, the following equation represents the electron density distribution as a function of reduced height z:

$$N_e(z) = (q_0/\alpha)^{\frac{1}{2}} \exp[\,(1 - z - e^{-z} \sec \chi)] \tag{3.11}$$

where α is the recombination coefficient, χ is the solar zenith angle, and q_0 is the maximum production rate in the layer[22]. It has been found that values for the recombination coefficient α are of the order of 1.6×10^{-13} m^3/sec and values for the peak production rate q_0 are of the order of 4.7×10^9m^{-3}/sec. The rate of electron removal is $\alpha N_e^{\,2}$ and the layer governed by Equation 3.11 is referred to as an alpha-Chapman layer.

The maximum rate of electron production q_0 occurs only for the overhead sun. However, it may be shown that actual maxima for other zenith angles are

---

22. In Equation 3.11, the function *exp* [ ] implies $e^{[\ ]}$ where e = 2.73. The function *sec* χ corresponds to the secant of the solar zenith angle.

simply related by this expression[23]:

$$q_{max} = q_0 \cos \chi \qquad (3.12)$$

It is noteworthy that N_{max} may not be obtained from Equation 3.11 by setting the reduced height $z = 0$. To do so would generate an underestimate for N_{max} (see Figure 3-5). N_{max} is obtained by setting $z = \ln \sec \chi$ in Equation 3.11 [Chapman, 1931]. Converting N_{max} to the critical frequency nomenclature, we obtain [in MKS units]:

$$f_c(\text{Hz}) = 9\,(q_0/\alpha)^{\frac{1}{4}}\,(\cos \chi)^{\frac{1}{4}} \qquad (3.13)$$

and for the E region,

$$\text{foE(Hz)} = 9\,(q_0/\alpha)^{\frac{1}{4}}\,(\cos \chi)^{\frac{1}{4}} = k\,(\cos \chi)^{\frac{1}{4}} \qquad (3.14)$$

where k is a constant of proportionality which is dependent upon the sunspot number. As a representative calculation, taking $\alpha = 1.6 \times 10^{-13}$ m^3/sec and $q_0 = 4.7 \times 10^9$ m^{-3}/sec corresponding to a moderate value of the sunspot number ($R \sim 60$) [Allen,1965], we find that $k \sim 3.8 \times 10^6$. For $\chi = 0$ we have $foE = 3.8$ MHz. In general we have $foE \sim k(\cos \chi)^n$ where n tends to 0.25. Some workers have found $n = 0.3$, or more, for the diurnal dependence but the long term seasonal dependence was close to 0.25. A value for n in excess of 0.25 might be expected if the E layer scale height gradient is considered [Rishbeth and Garriott, 1969].

The solar activity dependence of the ratio of peak production to the effective loss (recombination) coefficient has been studied by a number of workers, and the results enable values of foE to be deduced. The generally accepted values may be found in a number of texts [Davies, 1965; Jursa, 1985]

$$foE(\text{MHz}) = 0.9\,[\,(180 + 1.44\,R)\cos \chi\,]^{\frac{1}{4}} \qquad (3.15a)$$

$$= 3.3\,[\,(1 + 0.008\,R)\cos \chi\,]^{\frac{1}{4}} \qquad (3.15b)$$

where foE is given in MHz and R is the running 12-month sunspot number.

Figure 3-6 contains an E region critical frequency map for solstice conditions in 1958. Solar maximum conditions existed in June 1958 with the monthly mean value of R being 172 and the yearly mean value being 185. Using $R = 185$ in Equation 3.15 above, we have $foE = 4.1$ MHz. We see that this is only slightly less than the maximum value observed on the map where $\chi \sim 0$. Figure 3-6 also shows the seasonal variation of foE for all of 1958 for a midlatitude site near Washington, D.C. Note the strong solar control with foE virtually vanishing during nocturnal hours.

The E region height varies diurnally but in a manner exhibiting rough symmetry about local noon. An average value of 110 km is typical, and this value is used in most empirical models.

----

23. The function $\cos \chi$ corresponds to the cosine of the solar zenith angle.

3.3.5 The F1 Region

The F1 region is not unlike the E region in the sense that it obeys many of the features of Chapman theory. We look for a relation of the form:

$$foF1(\mathrm{Hz}) = 9\,(q_0/\alpha)^{\frac{1}{4}}\,(\cos \chi)^{\frac{1}{4}} \tag{3.16}$$

where the value of q_0/α is specific to the F1 region in this case. Ratcliffe and Weekes [1960] find $q_0/\alpha = 500(1+0.016R) \times 10^{20}$ meter^{-6}. Inserting these values into Equation 3.16 and setting $R = 0$ and $\chi = 0$, we have $foF1 = 4.25$ MHz. The currently accepted recipe for $foF1$ comes from Hargreaves [1979]:

$$foF1(\mathrm{MHz}) = 4.25\,[\,(1 + 0.0015\,R)\cos \chi]^{\frac{1}{4}} \tag{3.17}$$

We have tacitly assumed that the power n to which $\cos \chi$ is raised is 0.25 to conform with an equilibrium Chapman layer. However, values considerably less than 0.25 have been observed near solar minimum and also in the summertime. Allen [1948] and Davies [1969] take $n = 0.2$ as a better fit to data. Figure 3-7 shows the solar zenith angle control of $foF1$ under sunspot maximum and minimum conditions.

The height of the F1 ledge, $hF1$, is taken to be range between 180 and 210 km. From Chapman theory we anticipate that $hF1$ will be lower in summer than in winter and will be higher at midlatitudes than at low latitudes. Unfortunately the reverse is true. Explanations for this behavior may be found in a detailed study of scale height gradients, a nonvanishing movement term (as expressed in the continuity equation), or gradients in upper atmospheric chemistry.

3.3.6 The F2 Region

The F2 region is the most prominent layer in the ionosphere. and it generally is the most important one in the application of skywave propagation to HF communication. This significance arises as a result of its height (the highest of all the component layers) and of course its dominant electron density. It is also characterized by large ensembles of irregularity scales $\{\delta L\}$ and temporal variations $\{\delta T\}$. The F2 region is a vast zone which eludes prediction on the microscale ($\delta L < 1$ km) and mesoscale ($1 \leq \delta L \leq 1000$ km) levels, and even provides challenges to forecasters for global and macroscale ($L >$ 1000 km) variations. This is largely because of the elusive movement term in the continuity equation. There are also a host of so-called anomalous variations (the subject of Section 3.3.7) which make the F2 region a most exciting region to study.

The F2 layer is much more prominent than the F1 ledge although it does exhibit considerable variability which is not easily accounted for in terms of external solar and magnetospheric influences. It possesses a climatology with features similar to normal weather. Both categories of weather, the F region ionosphere as well as the troposphere, present a challenge for the long term forecaster. In the intermediate and short-term domains, which is the province of forecasting rather than pure prediction, the state of the art in applied meteorology far exceeds the state of the art in ionospheric weather prediction, which may be referred to as *ionospherology*. This is largely because in-situ and remote sensing systems for meteorological use are superior to those for *ionospherological* use. We shall return to this topic in Chapter 6.

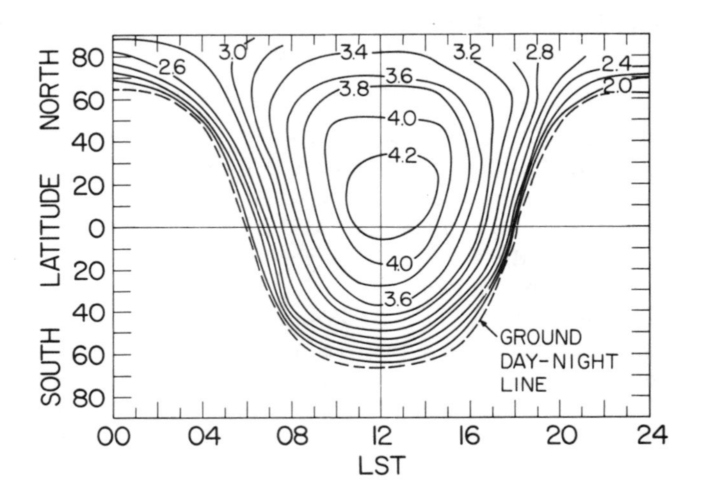

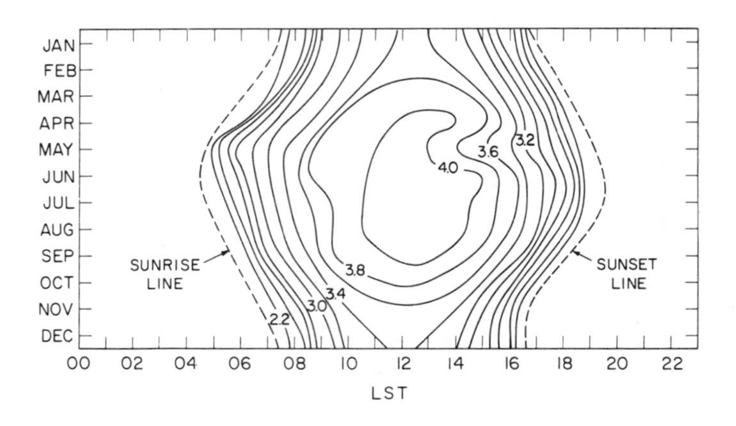

Fig.3-6. (a) Map showing the latitudinal and diurnal behavior of the critical frequency *foE* for summertime solar maximum conditions. (b) Critical frequency foE contours for Fort Belvoir during 1958 showing seasonal variations. (From Davies, [1965].)

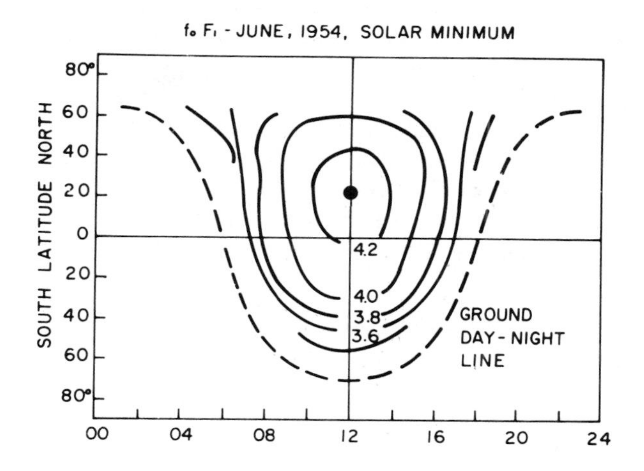

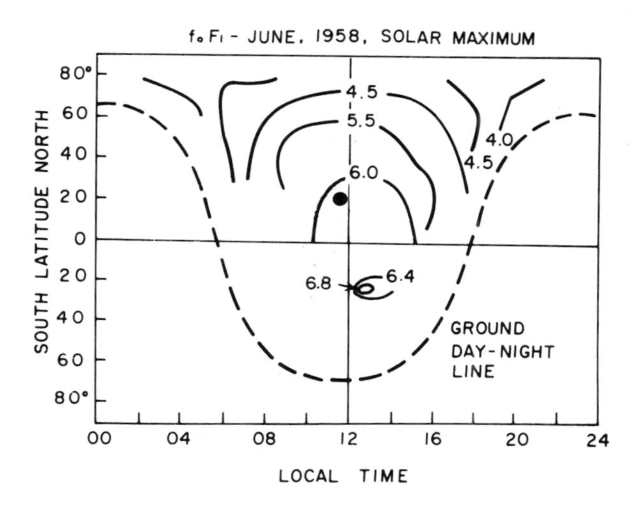

Fig.3-7. Maps showing the geographical variation of the F1 region critical frequency *foF1* (MHz) for the June 1954 solar maximum (top) and the June 1958 solar minimum (bottom). (From Jursa, [1985].)

The ionosphere has been examined with a variety of sensing systems, but the most valuable tool for providing a long-term global picture of the key parameters, such as peak densities and layer heights, has been the vertical incidence sounder. From measurements taken over several solar cycles and from a large number of ground observatories, a picture of the basic features of the ionosphere has been determined. The vast network of sounder stations and a host of dedicated scientists, who have assembled and analyzed the vertical incidence sounder records, have led to the development of a comprehensive database which has been archived by the National Geophysical Data Center (NGDC) in Boulder, Colorado[24]. This has allowed scientists to characterize the F2 region behavior in a meaningful way. A sample ionogram is shown in Figure 3-8. More will be said about the past, current, and future exploitation of sounders in Chapter 6.

As in the E and F1 regions, we may conveniently specify the behavior of the F2 region in terms of a critical frequency rather than the underlying peak electron density. We have:

$$N_{max}F2 = 1.24 \times 10^{10}\,(foF2)^2 \quad (3.18a)$$

$$= 1.24 \times 10^{10}\,(fxF2\text{-}0.5f_B)^2 \quad (3.18b)$$

where $foF2$ is the ordinary ray critical frequency, $fxF2$ is the extraordinary critical frequency, and f_B is the electron gyrofrequency.

As a consequence of Equations 3.18a and 3.18b, we may interpret the electron density in terms of either $foF2$ or $fxF2$ provided we know the gyrofrequency f_B. This is fairly well known since it depends only upon the magnetic field strength which is adequately modeled to the level of accuracy required. The electronic gyrofrequency f_B is proportional to the product of the magnetic field and the charge to mass ratio (e/m). Since e/m is fixed for an electron, the value of f_B is easily obtained. We have:

$$f_B\,(\text{MHz}) = 2.8\,B(\text{Gauss}) \quad (3.19)$$

where B is the order of 1/2 Gauss leading to a value of 1.4 MHz as the value for f_B.

Maps of the total magnetic field intensity (or induction) are found in various handbooks [Jursa, 1985]. Figure 3-9 gives the total magnetic field in gamma (1 gamma = 10^{-5} Gauss). Note the reduction in field over South America where the field is 0.24 Gauss.

----

24. The National Geophysical Data Center (NGDC) has produced a series of CD-ROMs containing Solar-Terrestrial Physics data. CD-ROMs containing the ionospheric database, labeled NGDC-05/2 and NGDC-05/3, contain 101 stations, 1520 station years, and more than 680 Mbytes of data. It is anticipated that the second CD-ROM will be updated with additional data from countries holding data not yet archived at World Data Center A in Boulder, specifically ionospheric data from the Soviet Union.

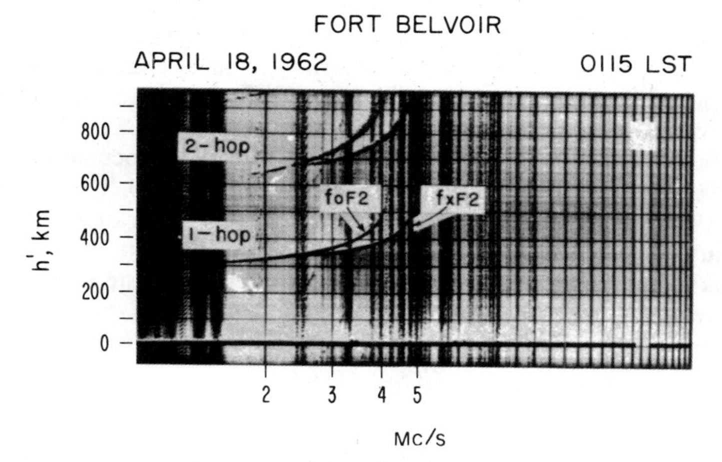

Fig.3-8. Ionogram which is typical of the nighttime at middle latitudes. Two pairs of traces on the h' (f) record are shown. The lower pair corresponds to the first hop. A separation of about 1 MHz is seen between the fx and the fo curves as dh'/df approaches infinity. This separation $(fxF2\text{-}foF2) = 0.5\, f_B$ where f_B is the gyrofrequency in the F region neighborhood experienced by the radiowave.

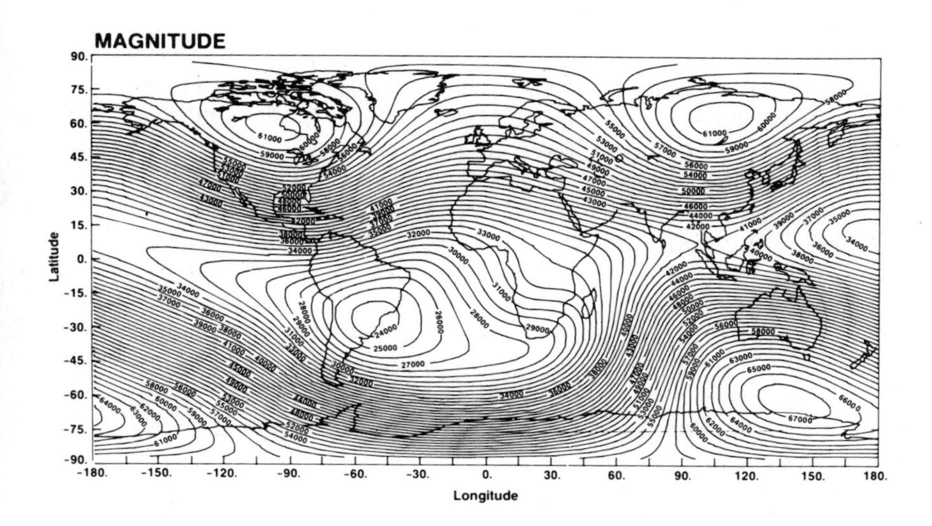

Fig.3-9. World map of the electron gyrofrequency in gamma units. 1 gamma = 10^{-5} Gauss. (From Jursa [1985].)

As part of the *Recommendations and Reports of the CCIR* [CCIR, 1986a], the CCIR publishes its *CCIR Atlas of Ionospheric Characteristics* which includes global maps of F2 layer properties for sunspot numbers of 0 and 100, for every month, and for every even hour of Universal Time [CCIR, 1986b]. This *atlas* first appeared in connection with the Xth Plenary documentation [CCIR, 1966] but has been updated or supplemented in 1970, 1974, 1978, 1982, and 1986. The latest version will be based upon deliberations of the XVII Plenary held in Dusseldorf, Germany in 1990 (see footnote on page 15 of this book). The CCIR atlas uses a term, *EJF*, referring to the Estimated Junction Frequency on an ionogram. It corresponds closely to the *MUF* for specified skip distances. The following paragraph outlines the differences.

> For an oblique wave refracted from a single layer, two distinct paths or trajectories may be observed, one corresponding to the high ray and the other corresponding to the low ray. This phenomenon arises provided the wave frequency is greater than the overhead (midpath) critical frequency. The two rays may be clearly observed on an oblique-incidence ionogram. The point at which the high and low rays merge is called the Junction Frequency (*JF*) and is the classical equivalent of the Maximum Observable Frequency (*MOF*). It is possible for either the high or low rays to extend beyond the junction frequency to create a type of *nose extension*. The highest frequency observed on an oblique ionogram may be called an Operational *MOF*. The junction frequency (or the classical MOF) is important since it may be directly related to the electron density profile and N_{max} in particular through classical propagation theory. The Estimated Junction Frequency (or *EJF*) is a modeled parameter corresponding to a median value for *JF*. The median value of the F2 region *EJF* for a skip distance *d* is termed *EJF*(d)*F2*. This corresponds to the Maximum Usable Frequency, *MUF*(d)*F2*, which is a more commonly used term. If $d = 0$, then we have *EJF(zero)F2* as the parameter to be mapped. This is simply the *MUF* for vertical incidence. Consequently we have *EJF(zero)F2* = *MUF(zero)F2* = *fxF2* = *foF2* + 0.5 f_B. (See Equations 3.18a and 3.18b. More discussion of these concepts is found in Chapters 4 and 5.)

Such maps are derived from coefficients based upon data obtained from a number of ionosonde stations for the years 1954-1958 as well as for the year 1964. This set of coefficients is sometimes identified by an ITS prefix but is known more generally as the CCIR coefficients. The numerical procedure used to produce the maps of ionospheric parameters from the CCIR coefficients was published by Jones and Gallet [1960, 1962a, 1962b, 1965]. The methodology is thought to be basically sound. Largely because of the paucity of data over oceanic areas, a number of errors in values of *foF2* have been uncovered. These errors are particularly striking in the Pacific basin and in the southern hemisphere. A method for improving the basic set of coefficients by adding theoretically-derived data points has been discussed by Rush et al. [1983] and new maps were developed in a followup paper [Rush et al., 1984]. These new coefficients have been sanctioned by URSI and thus are referred to as URSI coefficients. The CCIR coefficients are currently the set which is used in specified HF prediction methods [CCIR, 1986c; 1986d].

Figure 3-10a is a global representation of the F region parameter *EJF (zero)F2* for December at 2000 hours UT and for nearly solar maximum conditions (i.e., $R = 100$). Also shown, in Figure 3-10b, is a map depicting the contours of constant solar zenith angle for the same month.

Several interesting features emerge from this map. From Figure 2-16 and Figure 3-9 we note the departure of the geomagnetic dip equator from the geographic equator, especially in the vicinity of 60 degrees West Longitude. The *EJF(zero)F2* contours appear to track this behavior. More generally, in the low latitude region, the *EJF(zero)F2* contours exhibit a global symmetry about the geomagnetic equator with *islands* of excess ionization apparent on either side. This is a manifestation of the Appleton or equatorial anomaly which will be discussed below. By inspection of the solar zenith angle chart, we see a tendency for the F region to grow rapidly at ionospheric sunrise with a more gradual decay in the dusk sector. The effect is not nearly so dramatic as that exhibited by the E and F1 regions.

3.3.7 Anomalous Features

The F2 layer of the ionosphere is probably the most important region for HF radiowave systems since it provides us with the capability to achieve the greatest skywave propagation range (by a single hop) at generally the highest allowable frequency (in relation to underlying layers). Unfortunately the F2 layer exhibits the greatest degree of typically unpredictable variability because of the so-called movement term in the continuity equation. As indicated previously, this term represents the influences of ionospheric winds, diffusion, and dynamical forces. The Chapman description for ionospheric behavior depends critically upon unimportance of the movement term. Consequently many of the attractive, and intuitive, features of the Chapman model are not observed in the F2 region. The differences between observation and those features which would be predicted on the basis of a hypothetical Chapman description have been termed anomalies.

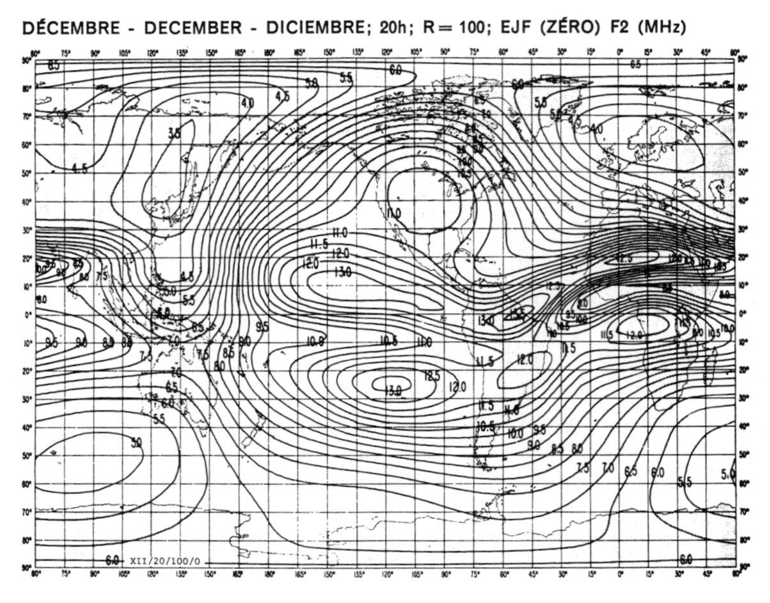

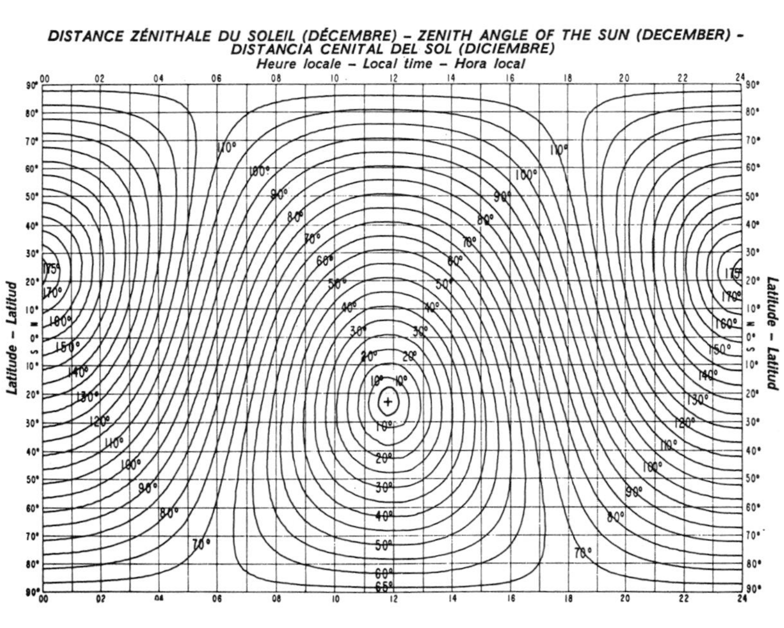

Fig.3-10. (a) Map of *EJF(zero)F2* for December at 2000 hr UT and R = 100. (b) Solar Zenith Angle for December parameterized in local time. This plot may be used as an overlay for the map above (From CCIR [1966])

The following list represents the major forms of anomalous behavior in the F2 layer: diurnal, Appleton, December, winter, and the F region trough. A few comments are provided for each major form.

The Diurnal Anomaly. The diurnal anomaly refers to the situation in which the maximum value of ionization in the F2 layer will occur at a time other than at local noon as predicted by Chapman theory. On a statistical basis, the actual maximum occurs typically in the temporal neighborhood of 1300 to 1500 LMT. Furthermore there is a semidiurnal component which produces secondary maxima at approximately 1000-1100 LMT and 2200-2300 LMT. The presence of two maxima (one near 1000 and the other near 1400) may give the appearance of a minimum at local noon. This feature, when observed, is called the midday biteout. It is noteworthy that dual daytime maxima may be reversed in amplitude with the morning peak being larger than the afternoon peak. This situation is depicted in Figure 3-11a (midlatitude case) and Figure 3-11b (magnetic equator). A striking factor in both regions is the strong symmetry of the electron density contours up to altitudes of 180 km or so. This indicates Chapman-like behavior for altitude regimes containing the E and F regions. It also exhibits the diurnal anomaly for the F2 region; the anomalous behavior is clearly an increasing function of altitude.

Appleton Anomaly. This feature is symmetric about the geomagnetic equator and goes by a number of names including : the geographic anomaly, the geomagnetic anomaly and the Appleton anomaly, as well as the equatorial anomaly. (The author prefers the term Appleton anomaly to give credit to E.V. Appleton [1954] who performed its first thorough examination.) The Appleton anomaly is associated with the significant departure in the latitudinal distribution of the maximum electron concentration within 20 to 30 degrees on either side of the geomagnetic equator. The phenomenon is described as an *equatorial fountain* initiated by $E \times B$ (Hall) plasma drift which is upwards during the day since the electric field associated with the equatorial electrojet is eastward at that time. The displaced plasma gradually removed from electrodynamic forces as the electrojet decays is now subject to downward diffusion when the atmosphere begins to cool. This diffusion is constrained along paths parallel to B which maps to either side of the geomagnetic equator. The poleward extent of the anomaly crests is increased if initial Hall drift amplitude is large. This anomalous behavior accounts for the valley in the *EJF(zero)F2* parameter (with peaks on either side) seen at the geomagnetic equator in Figure 3-10a. Figure 3-12 illustrates the Appleton anomaly based upon data derived from topside sounder studies. Note carefully the plasma frequency contour labeled hmF2 for this is the lowest height which may be monitored by topside ionosonde and, of course, is the only height of any interest to the HF propagation specialist. Magnetic inclination I is of interest as well as the hmax curve (as a function of geographic latitude) shown on the lower section of the figure. Recall from Equation 2.4 that the magnetic latitude ϴ may be derived from I (i.e., Tan ϴ = 0.5 Tan I).

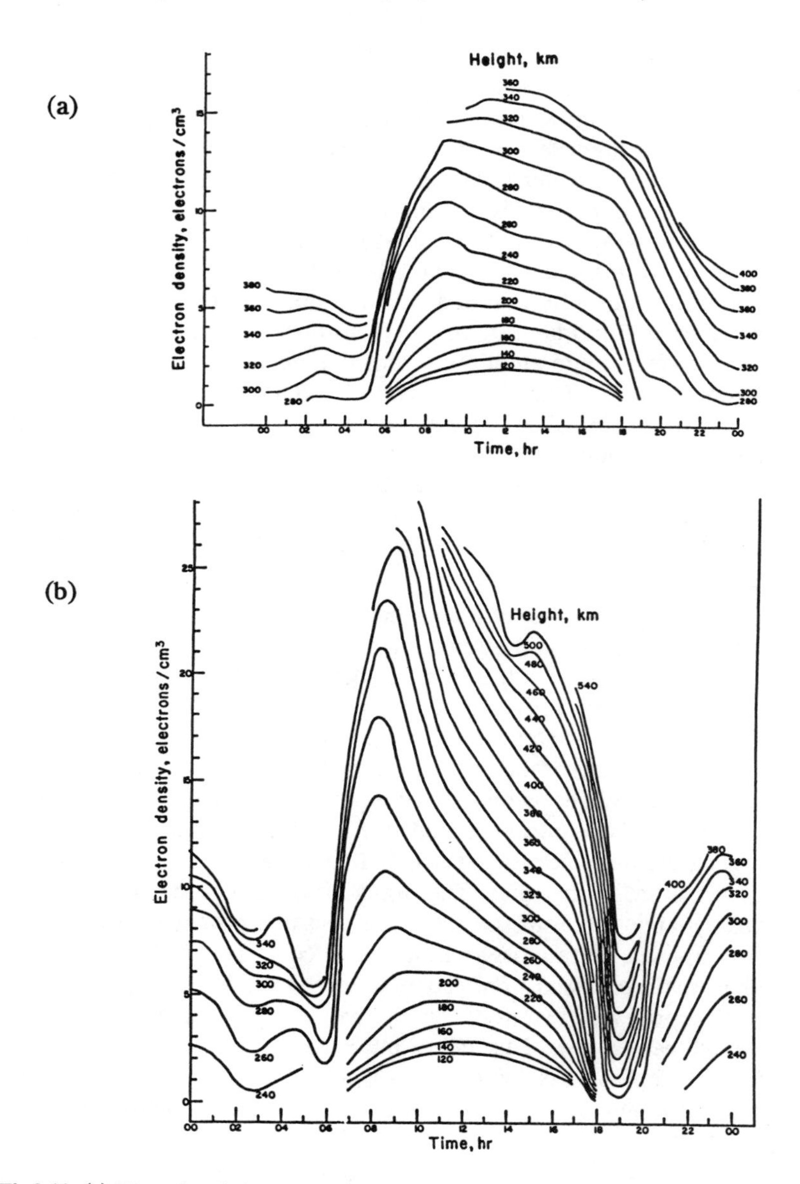

Fig.3-11. (a) Diurnal variation of the electron density at Washington DC. (b) Diurnal variation of the electron density at Huancayo Peru. Data from Washington and Huancayo were obtained in April 1958, and the electron densities are reckoned at fixed altitudes. The number densities are plotted in units of 10^5 electrons/cm^3. (From Newell [1966].)

The December Anomaly. This phenomenon refers to the fact that the electron density at the F2 peak over the entire earth is 20% higher in December than in June, even though the solar flux change due to earth eccentricity is only 5% (with the maximum in January).

Winter (Seasonal) Anomaly. This is the effect in which the noontime peak electron densities are higher in the wintertime than in the summertime despite the fact that solar zenith angle is smaller in the summer than it is in the winter. This effect is modulated by the 11-year solar cycle and virtually disappears at solar minimum. Figure 3-13 exhibits this effect quite nicely.

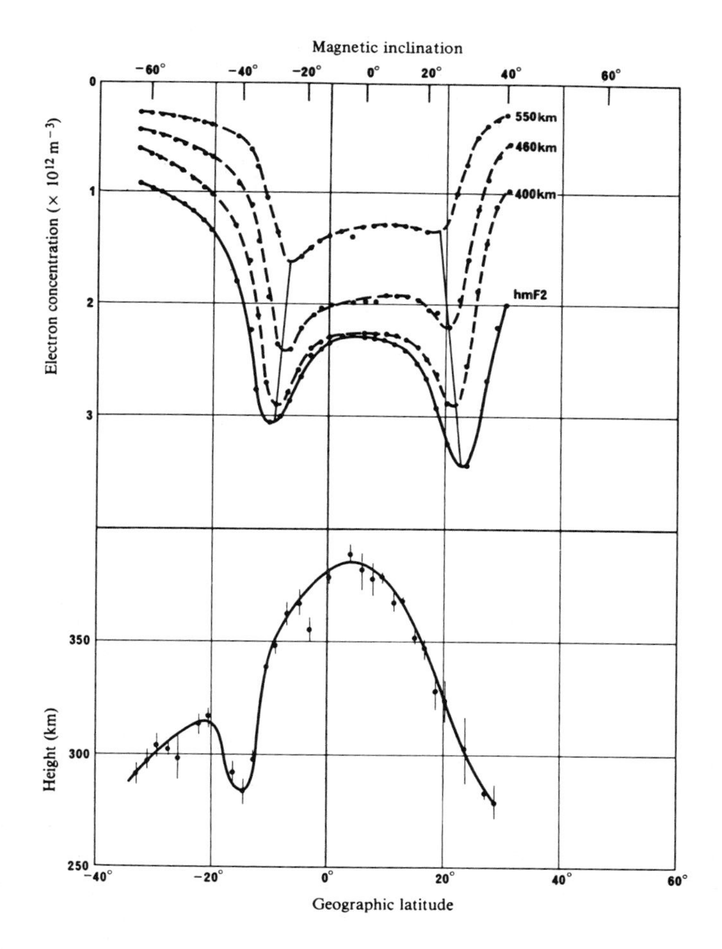

Fig.3-12: Appleton Anomaly: [top] electron concentration; [bottom] height of the F2 maximum hmF2. (From CCIR [1986e].)

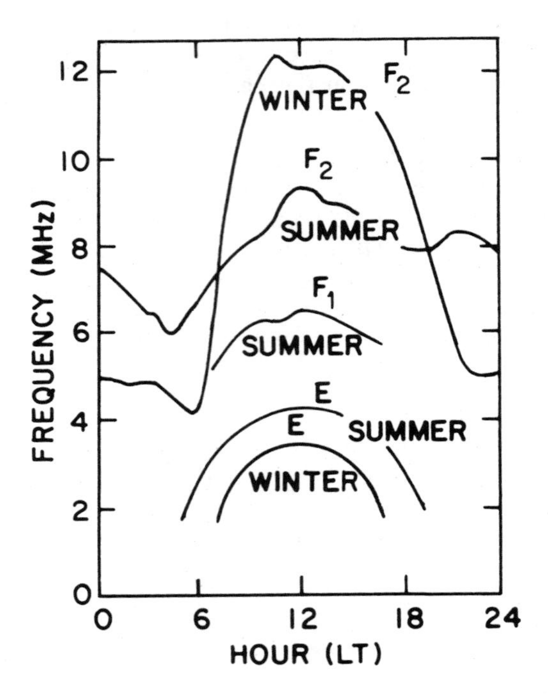

Fig.3-13. Mean diurnal variation in *foE*, *foF1*, and *foF2* with solar maximum conditions assumed (from Jursa [1985].)

The F-Region (High Latitude) Trough. This is representative of a number of anomalous features which are associated with various circumpolar phenomena including particle precipitation, the aurora, etc. The trough is a depression in ionization which extends from 2 to 10 degrees equatorward of the auroral oval. This region is though to be associated with a *mapping* of the plasmapause onto the ionosphere along field lines on the night side of the earth.

At least one group of investigators is convinced that the F region winter anomaly is the result of a seasonal variation in the proportion of atomic oxygen to molecular neutrals [Alcayde et al., 1974]. When this ratio is large, as it is in the wintertime, production increases at the expense of loss through recombination. It may still be said, however, that there is not yet an unassailable argument to fully explain the various anomalous features including the winter anomaly. It is likely than a number of the possible causes listed in Table 3-3 may act together in varying degrees to produce the observed effect.

TABLE 3-3. EXPLANATIONS FOR ANOMALOUS FEATURES

a. Changes in loss rate of electrons associated with changes in the relative concentration of atomic oxygen and molecular nitrogen.
b. Changes in the reaction rates resulting from changes in upper atmospheric temperature.
c. Production of ionization resulting from energetic particles entering the magnetosphere.
d. Movement of ionospheric plasma caused by neutral atmospheric winds (or by electric fields originating in the lower ionosphere) which might transport the peak to locations where the loss coefficient has different values.

3.3.8 Long-Term Solar Activity Influence

There is a clear tendency for the ionospheric critical frequencies to increase with sunspot number. Figure 3-14 shows the long-term variation of *R*, *foF2*, foE, and the riometer absorption level (at 4 MHz) for noontime conditions. A slow 11-year modulation in the ionospheric parameters is evident. After smoothing, the results correlate well with sunspot number. Superimposed on this solar epochal variation is an annual variation, with D-region absorption and *foE* exhibiting summertime maxima while *foF2* exhibits a wintertime maximum.

Davies and Conkright [1990] have discussed recent trends in ionospheric data management stressing digital archival and retrieval methods. Figure 3-15 is derived from a long-term (1938-1990) database of $N_{max}F2$. It shows the noon and midnight variation of the F2 region electron density observed near Washington D.C. in comparison with sunspot number over a period of more than 50 years.

The slow but definite dependence upon mean sunspot number is illustrated in Figure 3-16. This plot is rather unusual since it represents the running 12-month averages of the ionospheric parameters as well as the sunspot number. This naturally disguises (i.e., averages) the seasonal effects which caused the oscillations observed in Figure 3-14. This was, of course, the intent. A residual effect may still be seen upon careful examination, however. These data have also been analyzed by Davies [1965] who suggests that there is a tendency for the critical frequency *foF2* to flatten for *R* in excess of 150. This has implications for ionospheric modeling during epochs of high solar activity.

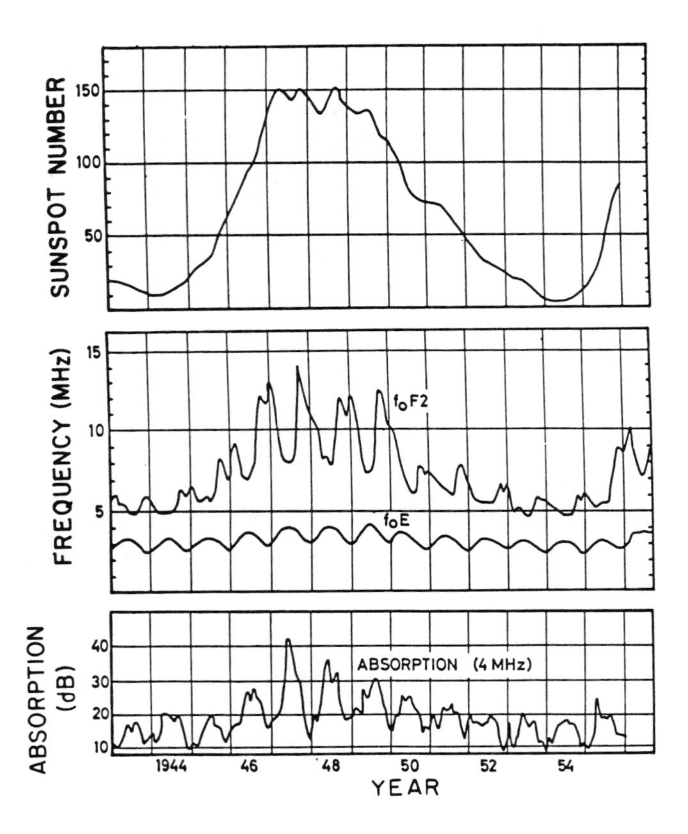

Fig.3-14: Long term variations in the noontime values of parameters *foE*, *foF2*, and the amount of absorption at 4 MHz. Data were obtained at Slough in the United Kingdom. The *R* index is shown also. (From Leid [1967].)

3.3.9 Sporadic E

3.3.9.1 *General Characteristics*

Even though the normal E region is Chapman-like in nature, anomalous forms of ionization are often observed with a variety of shapes and sizes. These anomalous ionization forms are termed sporadic E because they are observed to occur quasi-randomly and generally defy deterministic prediction methods. Although sporadic E is comprised of an excess of ionization (against the normal E region background) it does not appear to be strongly tied to solar photoionization. It does exhibit tendencies which have been examined statistically, and at least three different types of sporadic E ionization have been discovered with distinct geographical regimes. These are low latitude (or equatorial), midlatitude (or temperate), and high latitude ionization.

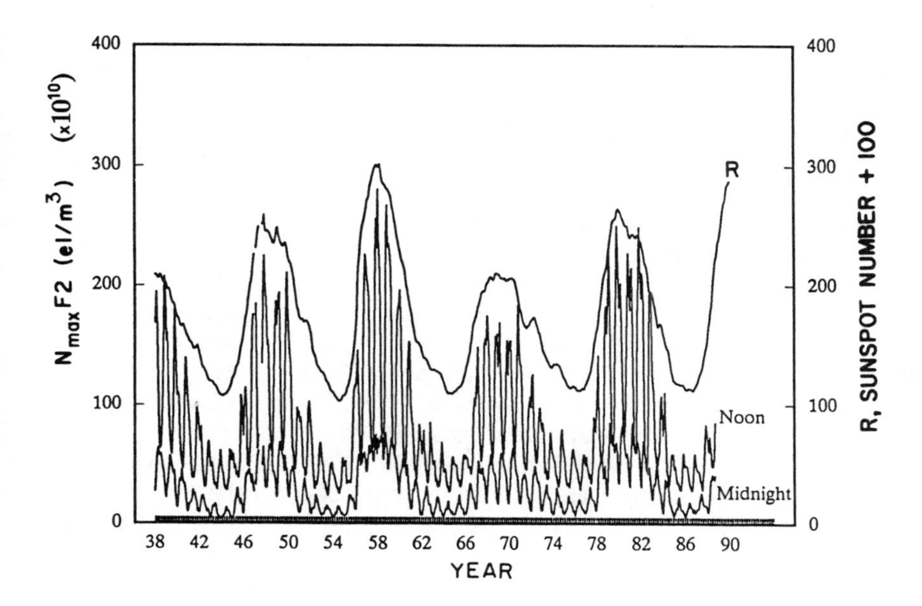

Fig.3-15. Maximum electron density of the F2 region, $N_{max}F2$, near Washington D.C. for the period between 1938 and 1990. (From Davies and Conkright [1990].)

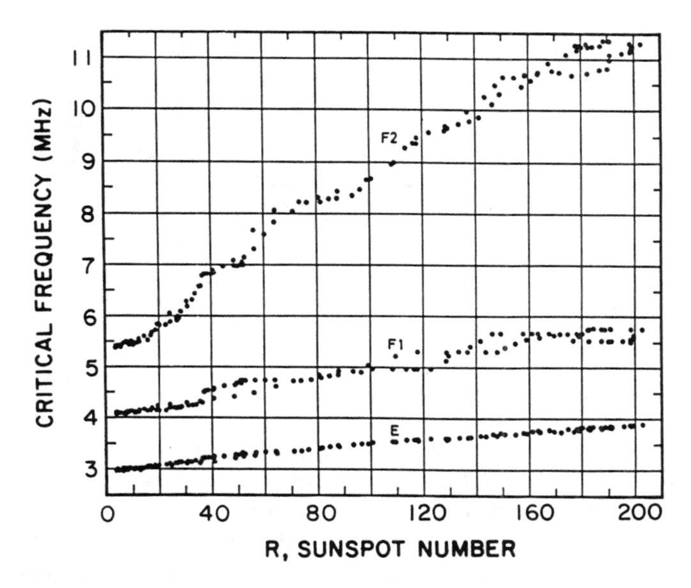

Fig.3-16. Variation of the parameters *foE*, *foF1*, and *foF2* derived from the Ft.Belvoir ionosonde vs *R*. The ionospheric parameters were 12-month running averages as was the sunspot number. Only noontime values were used. (From Leid [1967].)

Figure 3-17 contains a set of Thomson scatter profiles of the ionosphere obtained at Arecibo, Puerto Rico. These profiles show the F region and the normal E region (separated by a valley of ionization) along with a sharp sporadic E echo. One of the features of sporadic E is its limited vertical thickness and horizontal extent .

Sporadic E has been observed during rocket flights with layer thicknesses of the order of 2 kilometers being observed (See Figure 3-18). Furthermore some studies have suggested that the ensemble of midlatitude sporadic E patches may be distributed with systematic separations which are multiples of 6 km [Seddon, 1962]. Meteor trail observations, which are used to examine neutral wind properties in the upper atmosphere, have suggested that wind shears exist in the vicinity of the E region with vertical separations of 6 km, and that these shears have a horizontal extent of 100 - 200 km [Millman,1959; Greenhow and Neufeld, 1959]. Whitehead [1961, 1967, 1970] has suggested that wind shears in the upper atmosphere are responsible for the formation of sporadic E at midlatitudes. Axford [1961] and Hines [1964] have also made contributions to our understanding of the formation of midlatitude Es.

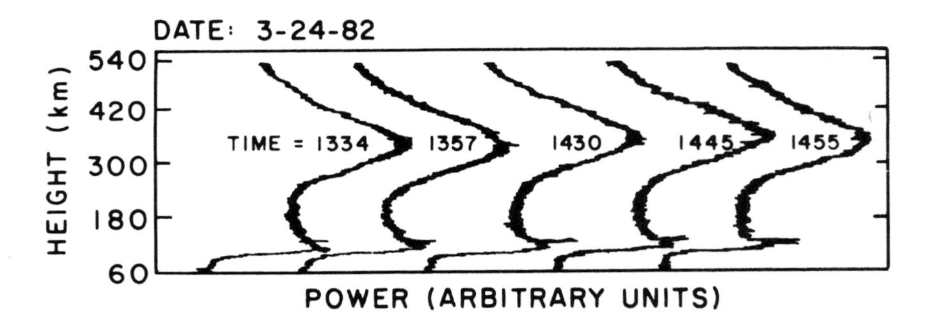

Fig.3-17. Thomson scatter power profiles of the ionosphere. These data were obtained at the Arecibo facility located in Puerto Rico. Clearly evident in the E-F valley and persistent sporadic E.

3.3.9.2 *Formation of Midlatitude Sporadic E*

It should be recalled from examination of photochemistry in the ionosphere that molecular ions such as those which exist in the E region introduce relatively rapid electron loss by recombination. At the same time it is recognized that an enormous number of meteors suffer their demise in the neighborhood of the E region. This meteoric debris is largely comprised of metallic ions which are monatomic. Their presence has been confirmed by mass spectroscopy measurements using rockets [Narcisi and Bailey, 1965]. These atomic species include iron, sodium, magnesium, etc.

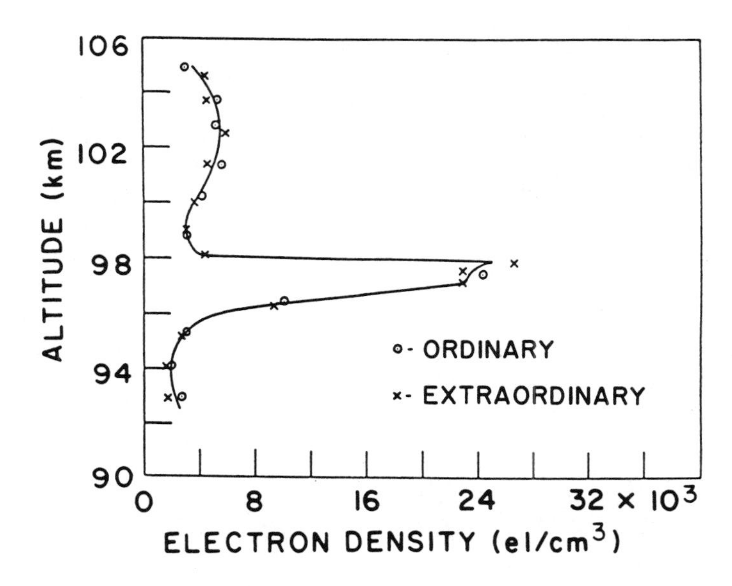

Fig.3-18. Sporadic E observed in connection with a rocket flight over Fort Churchill, Canada on February 4, 1958 at 0018 LST. (From Seddon [1962].) By permission of Macmillan Book Company, New York.

The influx of this foreign mass of metallic ions when distributed over the whole of the E region is still insufficient to overwhelm the omnipresent molecular species such as NO^+ which are in a state of photochemical equilibrium were it not for a mechanism which preferentially concentrates the meteoric ions. Wind shear is this mechanism.

Again, looking at the continuity equation, we are reminded that there is a movement term which must be considered, especially in the F region which is dominated by atomic ions such as O^+. That movement is important in the context of monatomic gas is a direct consequence of the fact that monatomic ions have relatively long lifetimes in a sea of free electrons (in comparison with molecular ions). In short, we effectively have two versions of the continuity equation operating in the E region: one for the generally dominant molecular gas, and a second for the meteoric species. In the first instance, movement is vanishingly small; in the second, the divergence of particle flux may be important.

The existence of well-defined neutral gas wind shears in the E region neighborhood is well documented. Sporadic E is also well defined, and may sometimes resemble thin sheets of ionization while at other times it may be patchy. Like neutral wind shears, Es ionization is observed in the altitude region between 100 and 125 km. As indicated above, Whitehead's wind shear theory would appear to explain the strong correlation between the neutral gas shears and the appearance of Es. The origin of the neutral shear itself is of some interest. Gossard and Hooke [1975] contend that the basic theory involving buoyancy waves and atmospheric tides provides the answer, and it is closely related to the theory which underlies the development of other ionospheric irregularities such as TIDs. The ultimate process is one in which there is a corkscrew propagation of atmospheric gravity waves and atmospheric tides resulting in a rotation of wind velocity as a function of altitude. Another interesting feature of gravity waves is the downward phase progression along with an upward flow of energy parallel to these surfaces of constant wave phase. The velocity rotation effect can cause the wind to change direction over an altitude of only a kilometer or so, sufficient to trap meteoric ions at an intermediate point having zero velocity. This buildup in a narrow region is sufficient to generate an intense sporadic E patch.

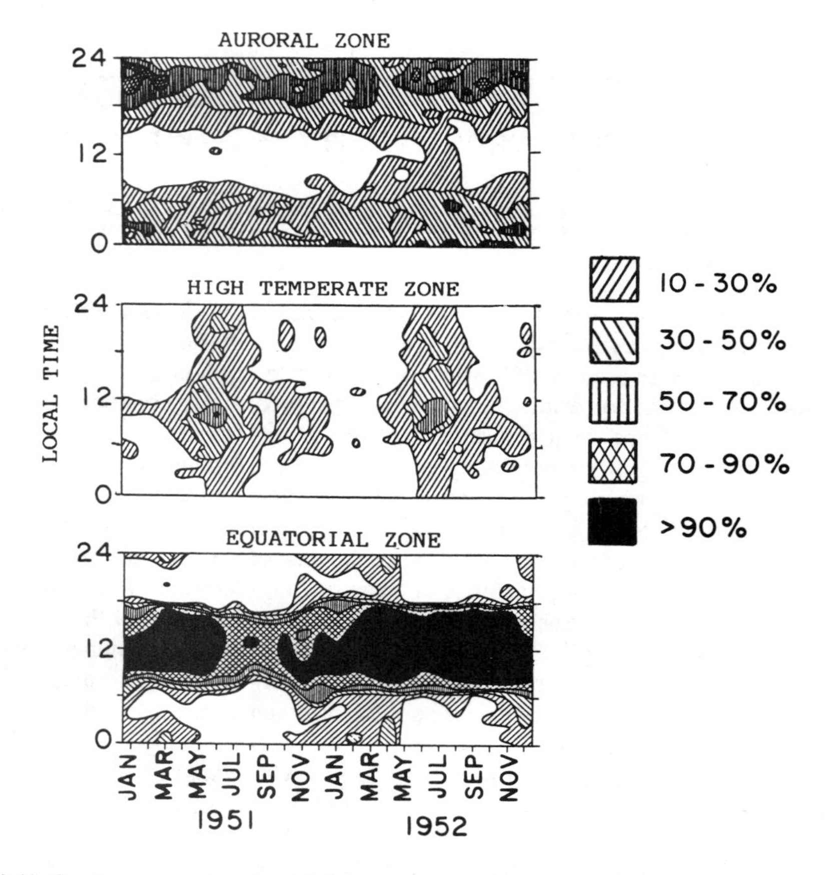

Fig.3-19. Graph representing the global, seasonal, and diurnal variation of sporadic E. This picture is illustrative of three distinct zones for Es in the high latitude (auroral), temperate, and equatorial regions. (From Davies [1965].)

3.3.9.3 *Global Morphology*

The reader is referred to a text edited by Smith and Matsushita [1962] to get a flavor of some of the early work on sporadic E. Whitehead [1970] has made a significant contribution to knowledge of the worldwide behavior, and Matsushita and Smith [1975] have reviewed the general subject. World maps of sporadic E have been published by Smith [1957,1976,1978] to allow communicators to assess the probability of oblique-incidence Es modes at VHF. Similar graphical presentations have been published by the CCIR [1986b] for solar maximum and minimum conditions. A chart which has appeared in many texts (e.g., Davies [1965]) summarizes the global probability that Es will exceed 5 MHz (Figure 3-19).

The general properties of sporadic E is well documented in various reports and recommendations of the CCIR [1986e,1986f,1986g].

We have indicated that midlatitude sporadic E is probably the result of wind shear. The high latitude sources are evidently of two types depending upon whether the observation is made in the neighborhood of the auroral oval or within it (i.e., in the polar cap region). It has been found that auroral Es is basically a nocturnal phenomenon and is associated with the optical aurora in a complex fashion. Because of its proximity to the seat of auroral substorm activity, it is not surprising to find some correlation between auroral Es and some appropriate magnetic index. Indeed, it has been found that auroral Es is positively correlated with magnetic activity. On the other hand, polar cap Es may be relatively weak , and is negatively correlated with substorm activity. Turning equatorward, it has been found that equatorial Es is most pronounced during daylight hours and evidence points to the equatorial electrojet as the responsible agent for sporadic E occurrence at low latitudes. More will be said about sporadic E and its effect on HF in later chapters. Figure 3-20 is an oblique-incidence ionogram exhibiting sporadic E.

3.3.9.4 *Current Views Concerning Es Structure*

Paul [1986] has examined sporadic E using an advanced digital sounder, and concludes that nonblanketing Es may be associated with a number of thin patches of ionization having substantial tilts. Rapid temporal variations in the Es as observed from a vertical incidence sounder may be associated with tilt variations rather than temporal variation in the Es ionization density. Paul concludes that nonblanketing Es is the manifestation of a geometric rather than an optical property. He also laments the fact that existing data banks are inadequate for detailed study of temporal and spatial properties of Es. Some support for this picture has been obtained by From and Whitehead [1986] based upon HF radar experiments. These authors indicate that *spread Es* consists of small clouds which reflects signals for only a few seconds. These cloudlike features may be related to the same kinds of structures observed by

Paul. Other classes of midlatitude Es have also been observed by From and Whitehead based upon their HF radar approach. *Totally reflecting Es*, or *blanketing Es*, is associated with a relatively smooth horizontal sheet of ionization having a scale size of ~ 1000 km, but composed of small scale structure or ripples. Representative values for the wave period, horizontal scale, vertical amplitude and average speed of these superimposed ripples are 5 - 10 minutes, 30 - 60 km, 1 km, and 100 m/sec respectively. Finally *partially reflecting Es* is thought to be related to an ensemble of ionization clouds.

In order to understand the climatology of Es, it will be essential that instruments having time resolutions of the order of seconds be employed as sensors. It would also be useful to employ antenna arrays which would allow the direction(s) of signal arrival to be obtained. The implications of Es presence in HFDF applications is enormous, and the implications in certain communication scenarios is significant as well. For more details, refer to Chapter 6.

3.4 THE HIGH LATITUDE IONOSPHERE

From a morphological point of view, the high latitude region is the most interesting part of the ionosphere. It has been said that the auroral zone and associated circumpolar features are our windows to the distant magnetosphere. It is also a constant reminder of the enormous amount of corpuscular energy which emanates from the sun. For HF users it is a special challenge because of the variety of (typically) deleterious effects which are associated with the region. It also provides scientists and propagation specialists with an excellent laboratory for detailed examination of pathological conditions. In particular, the high latitude environment may mimic some conditions which arise during nuclear disturbances. In this section we will outline the major ionospheric features of the high latitude region. For more details about storm-time phenomena, the reader may refer to Section 3.6.

The close relationship between the high latitude ionosphere and the magnetosphere has already been covered in Chapter 2; it will only be summarized in this section. For a description of the high latitude propagation channel and communication aspects, the reader is referred to Chapter 4. To prepare for that account, we provide a preview of the phenomena responsible for the channel personality. To do this, we must discuss the morphology of certain HF propagation parameters which may be used to define the high latitude ionosphere, be it benign or disturbed.

There are many good references dealing with the high latitude region. CCIR Report 886-1 [CCIR, 1986h] discusses the properties of the region which are important in the context of telecommunications. Report 886-1 also provides an excellent bibliography. Other useful information is contained in Chapters 8,9 and 12 of the previously-cited Air Force Handbook [Jursa, 1985], and in NATO-AGARD Conference Proceedings #295 [Schmerling,

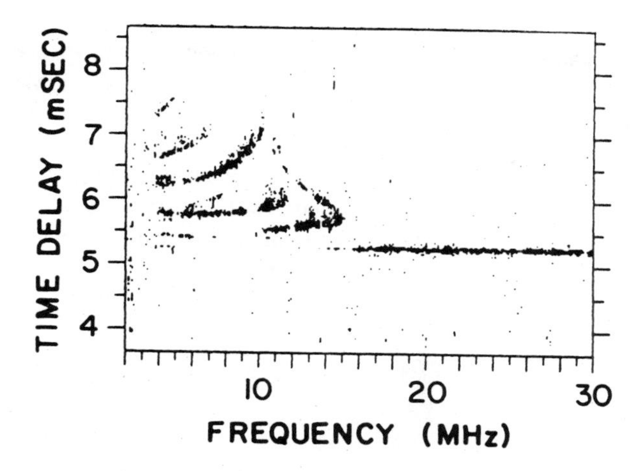

Fig.3-20. Oblique-incidence ionogram for an intermediate range midlatitude path exhibiting the effect of sporadic E ionization, Es. The maximum frequency is seen to be at least 30 MHz (the limit of the ionogram sweep). Thus, using the secant law (see Chapter 4), we find that *foEs*(MHz) ≥30 cos ϵ where ϵ is the ray elevation angle.

1981] and #382 [Soicher, 1985]. Other accounts are contained in a special edition of *Radio Science* [Hunsucker and Greenwald, 1983] and in papers prepared by Hunsucker and his associates [Hunsucker, 1983; Hunsucker and Bates, 1969].

3.4.1 The High Latitude Region: Where Is It?

When we refer to the high latitude region, we refer to that portion of the ionosphere which is characterized by the hierarchy of phenomena which are largely orchestrated by magnetospheric and interplanetary events (of a corpuscular nature) rather than solar (electromagnetic) flux variations. As was mentioned in Section 2.4.1 of Chapter 2, there are a number of coordinate systems which may be used when studying this region. For high latitude phenomena, geomagnetic coordinate systems are to be preferred. (Note that geomagnetic time approximates regular local time at great distances from the pole. It is reckoned by extending a great circle from the subsolar point through the magnetic pole.) Corrected Geomagnetic Latitude (CGL), along with the orthogonal coordinate: Corrected Geomagnetic-local Time (CGLT), is an improved version of the geomagnetic coordinate representation which was derived from a dipole model approximation. The Corrected Geomagnetic Coordinate system (CGCS), described in Chapter 2, may be obtained from geomagnetic coordinates by means of tables developed by Gustafsson [1970]. Figure 3-21 compares the CGL (i.e., corrected geomagnetic latitude) with normal geographic coordinates for the northern hemisphere.

In Section 2.4.3 of Chapter 2 we saw that a fixed site under the auroral oval at night will be equatorward during the day. Also we saw from Figure 2-25 the extent to which the oval thickens and descends as K_p increases. Most of the circumpolar phenomena are coupled to the auroral oval, at least circumstantially. Thus all features tend to move equatorward with the oval boundary as activity increases. Hunsucker [1983] has examined the salient features and they are depicted in Figure 3-22. In Figure 3-23, from Bishop et al.[1989], many of the same features are depicted (in a mercator format employing geographic coordinates) compared with worldwide features.

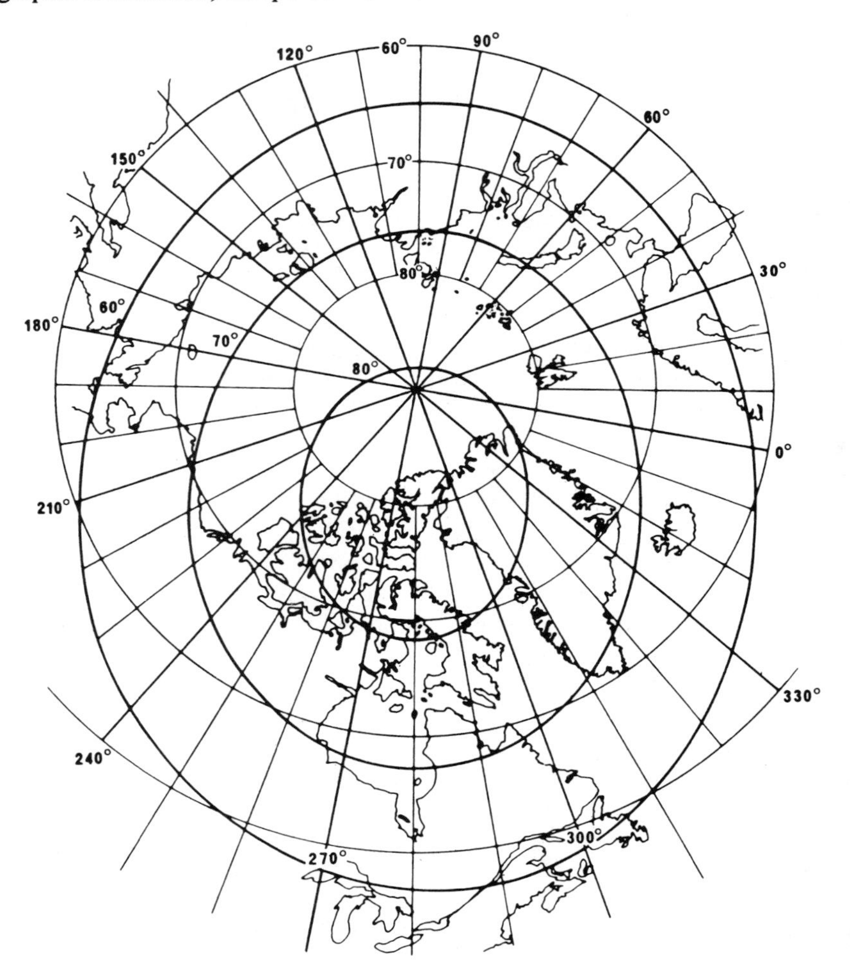

Fig.3-21. Corrected Geomagnetic Coordinates and Geographic Coordinates. (From CCIR Report 886-1, CCIR [1986h].)

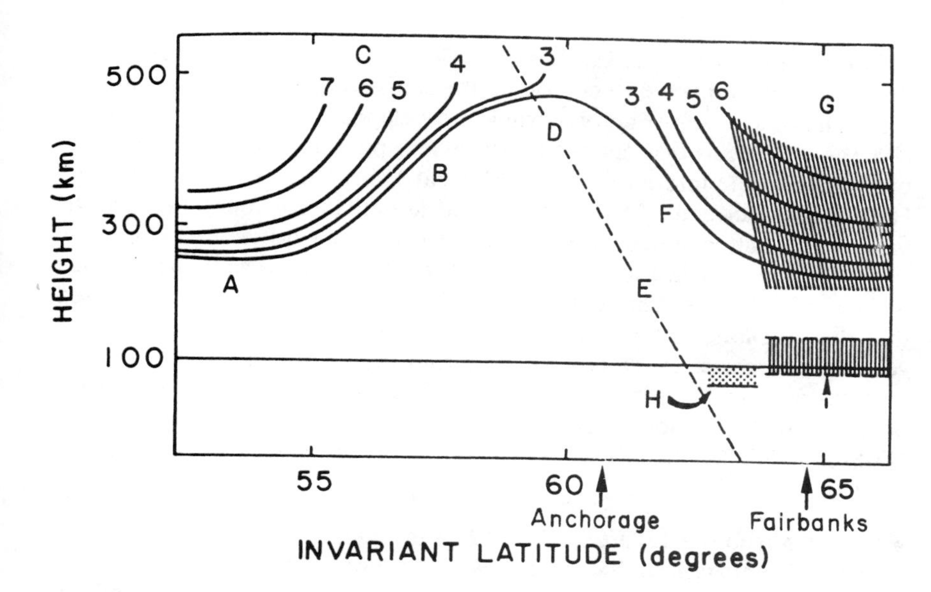

Fig.3-22. Idealized picture of plasma frequencies in a N-S plane. E = equatorward of trough, B = equatorward edge of trough, C = plasma frequency, D = trough minimum, E = plasmapause field line, F = poleward edge of trough, G = F region blobs, H = enhanced D region absorption, I = E region irregularities. (From CCIR Report 886-1, After Hunsucker [1983].)

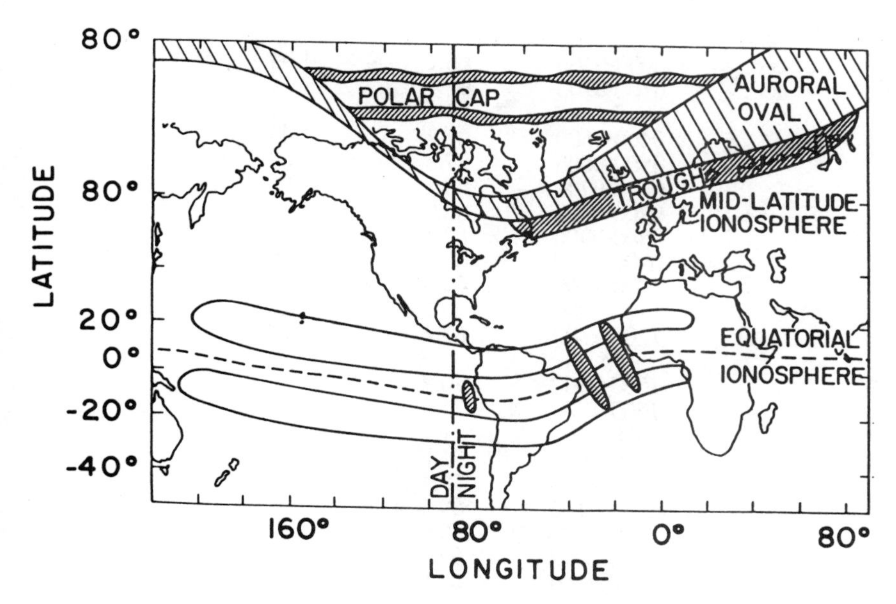

Fig.3-23. Various ionospheric regions. (From Bishop et al.[1989].)

The magnetic activity index K_p is generally available and is typically used as the parameter of choice to determine the statistical position of the auroral zone. The concept of the auroral oval was developed by Feldstein and Starkov [1967] on the basis of a set of all-sky camera photographs which were obtained during the International Geophysical Year. Other models exist (e.g., Hardy and MacKeen [1980]) but the Feldstein picture is found in most HF models which attempt to include auroral effects in some way.

One of the most fascinating properties of the various circumpolar features is their latitudinal motion as a function of magnetic activity. We have noted that the ionospheric plasma is best organized in terms of some form of geomagnetic coordinates. However we find that the high latitude plasma patterns are not fixed in that frame of reference either. The equatorward boundary of the region of precipitating electrons has been deduced from DMSP satellite instruments and it takes the form [Gussenhoven et al.,1983]:

$$L = L_0 + a K_p \tag{3.20}$$

where CGM coordinates are used, and L_0 is the equatorward boundary when $K_p = 0$. It is emphasized that L_0 and a are functions of time. Table 3-3 shows the relationship observed between Magnetic Local Time (MLT), a, and L_0. Given K_p it is possible to deduce L from Equation 3.20.

TABLE 3-3. EQUATORWARD BOUNDARY OF THE AURORAL OVAL

MLT	L_0	a
0000-0100	66.1	-1.99
0100-0200	65.1	-1.55
0400-0500	67.7	-1.48
0500-0600	67.8	-1.87
0600-0700	68.2	-1.90
0700-0800	68.9	-1.91
0800-0900	69.3	-1.87
0900-1000	69.5	-1.69
1000-1100	69.5	-1.41
1100-1200	70.1	-1.25
1200-1300	69.4	-0.84
1500-1600	70.9	-0.81
1600-1700	71.6	-1.28
1700-1800	71.1	-1.31
1800-1900	71.2	-1.74
1900-2000	70.4	-1.83
2000-2100	69.4	-1.89
2100-2200	68.6	-1.86
2200-2300	67.9	-1.78
2300-2400	67.8	-2.07

Movement of the visible aurora has been studied by Chubb and Hicks [1970] who found that per unit increase in K_p the equatorward boundary moved equatorward 1.7 degrees in the daytime and about 1.3 degrees in the nighttime sector. Using the index Q, Starkov [1969] finds:

$$L = 72^\circ - 0.9^\circ Q - 5.1^\circ \cos[(360^\circ/24)t_{CGLT} - 12^\circ] \qquad (3.21)$$

where t_{CGLT} is the corrected geomagnetic local time, and $Q > 1$.

Consistent with the picture of median oval location and thickness presented in Figure 2-25, Kamade and Winningham [1977] have found an equatorward movement of the auroral oval by between 1 and 3 degrees during auroral substorms. They have also discovered that the thickness of the oval is closely tied to the amplitude of the southward component of the interplanetary magnetic field (IMF). Specifically, an increase of 2 nanoTesla (nT) in the southward IMF causes the latitudinal extent of the oval to increase by 1 degree. From Table 2-3, it is noted that the scalar IMF as manifested within the solar wind is about 6 nT.

The U.S.Air Force prepares daily summaries of the index Q in order to provide a basis for various analyses of the high latitude ionosphere. Since Q, viewed as a parameter, defines the shape and location of the auroral zone, it is a convenient index for transmission to communication facilities and forecasting facilities. Its utility is dependent upon timeliness and accuracy. As originally designed by Feldstein, the Q index defines only a statistical relationship between the oval position and magnetic activity which could be parameterized by the planetary index K_p. Nevertheless, the Feldstein oval concept has been shown to have some utility under real-time circumstances. Satellite imagery is used to deduce an effective index, $Q_E(F)$, based upon Feldstein oval concept. An alternative effective index, $Q_E(D)$, based upon the Hardy model. For those instances in which satellite data is unavailable, Q_E is estimated from K_p. For the Feldstein oval, the following relationship obtains:

$$K_p = Q - 2, \quad 3 < Q < 8 \qquad (3.22a)$$

$$K_p = Q/3\,, \quad Q < 3 \qquad (3.22b)$$

and for values of K_p equal to or exceeding 6, Q is set to 8.

3.4.2 Major Features of the High Latitude Zone

Table 3-4 summarizes the major propagation issues for HF signals traversing the high latitude zone. It borrows from CCIR Report 886-1 [CCIR,1986h] and other generally accepted sources of information. It starts with information about location of specified circumpolar features and is followed by a summary of the magnetic index dependence, region-specific data, and concludes with signal effects.

TABLE 3-4. HIGH LATITUDE HF PROPAGATION ISSUES

OVAL LOCATION

Variable. An annular region which is eccentric with respect to the geomagnetic pole being distended ~ 4 degrees in the midnight sector compared to noontime. The equatorward boundary ranges between 67 and 70 degrees in the morning sector and 68-72 in the evening sector (CGM). The annular region expands and moves equatorward as magnetic activity increases.

TROUGH LOCATION

Tied to the oval position. Generally located at ~ 60-65 degrees magnetic latitude. The trough expands and moves poleward when the oval does the same. It contracts and moves equatorward as the oval moves equatorward.

POLAR CAP LOCATION

Poleward of the oval.

INDEX RELATIONS

All circumpolar features move poleward as the magnetic activity diminishes. There are several indices used: Kp, AE, and Q. The equatorward boundary of the oval is given by $Lo + a\,Kp$ where Lo is the unperturbed position, Kp is the planetary magnetic activity and a is a constant which ranges between -1.35 and -2.15 in the Northern hemisphere. A correspondence between Q and Kp exists with $Q = 3$ representing an undisturbed condition.

D-REGION EFFECTS

HF absorption occurs over the entire polar region (within the polar cap principally) due to proton precipitation, and in the auroral zone due to the precipitation of electrons.

AURORAL ABSORPTION

Composed of an intense bursty variety centered in the midnight sector, and a more continuous variety in the morning sector. Substorms seem to trigger a growth of the morning auroral absorption events; the midnight-centered variety is correlated with visual aurora. The absorption zone, like the oval, moves equatorward with Kp. It approaches a geomagnetic latitude of about 60 degrees for Kp ~ 7. It has a statistical width of about 5 - 8 degrees of latitude.

PCA EVENTS

Polar Cap Absorption (PCA) events generate the most intense periods of HF absorption. The altitude region with PCA is between 45 and 75 km. PCA is a result of the precipitation of MEV protons which originate from the sun and is correlated with intense x-ray flares. Following an SID event, PCA may occur 1/4 hour to an hour or so later depending on the energy of the protons. Some predictions are possible because of a non-vanishing propagation time delay from the associated flare region on the sun and the ionospheric *target*. PCA protons are restricted to the field lines which map into the polar cap region which, like the auroral zone, expands and contacts.

TABLE 3-4. HIGH LATITUDE HF PROPAGATION ISSUES {Continued}

PCA EVENTS (Continued)

HF absorption levels may be hundreds of decibels and may last for several days. According to Bailey [1964], the number of PCA hours is given by $N(\text{hrs}) \sim -70 + 4.4\ R_z$ where R_z is the Zurich sunspot number.

NORMAL E

The normal E layer is produced by solar illumination and when the sun is above the horizon. The normal E layer ionization is termed *solar E* and is augmented by so-called *auroral E.*

AURORAL E

Auroral E (or Ea) is caused by precipitation. The ordinary ray critical frequency associated with the auroral E, *foEa*, is generally more spread (or diffuse) than the solar-driven critical frequency *foE*. Auroral E is roughly 4-5 degrees wide. It occurs at a greater height during active periods than during quiet times.

AURORAL Es

Associated with auroral activity. Altitude is between 90 and 130 km. Appears to be correlated with discrete auroral arcs within the oval. Generates rapid fluctuations on HF signals. Values for *foEs* can exceed 15 MHz at times. Although reduced in amplitude in wintertime, it is an important mode at HF for communication in the region. Auroral foEs is positively correlated with magnetic activity.

POLAR Es

Negatively correlated with magnetic activity. It is decidedly weaker than the auroral counterpart. It is generally characterized by bands of ionization which extend across the cap in the midnight-noon direction.

TROUGH

Alternative terms: high latitude trough, midlatitude trough. The region is *underpopulated* because the sources of ionization are diminished. It is equatorward of the precipitation zone. It does not receive electrons through the process of antisunward drift, and it is not replenished by the plasmaspheric reservoir during nocturnal hours. The trough is a persistent feature at night but is usually filled-in by solar photoionization in daytime. The trough has a significant effect on HF trunks which would utilize *control points* located in the trough. The primary effect is loss of (refractive) support.

AURORAL F

Low energy electrons are deposited in the F region in the latitudes between the poleward part of the oval and a portion of the polar cap region. Penetration frequencies are generally no less than 3 MHz in this region but it is highly variable. The precipitation is quite intense in the noontime sector generating irregularities of all scales. These dayside cusp inhomogeneities drift equatorward in nocturnal hours and the dominant scales are small. F Region irregularities are an important issue for satellite scintillation, but they are also related to HF fading and signal fluctuation events.

TABLE 3-4. HIGH LATITUDE HF PROPAGATION ISSUES {Continued}

GRADIENTS

Horizontal gradients have been known to generate bearing fluctuations of 10-20 degrees or more. This effect is most severe if the HF ray trajectories are roughly tangent to the oval region. MOF *nose extension* may result from Non-Great-Circle (NGC) propagation. Often NGC and normal modes exist at the same time generating multipath. Broad-beam antennas may also accommodate sidescatter modes. Such NGC modes may dominate in local wintertime.

FADING

The signal levels may exhibit rapid variations as a result of the pathological nature of the region. Fading results from a variety of selective and nonselective events. It may result from an interference phenomenon or it may be dissipative in nature. An absorption event generally exhibits a modest short-term variability (being the order of minutes) but its long-term pattern depends upon the nature of the source region: polar cap, auroral zone, and high midlatitudes. Midlatitude absorption may arise from x-ray flares, but auroral and polar forms are due to particle bombardment or precipitation. Absorption events are usually non-selective or relatively wideband phenomena.

Fluctuations in signal level may be the result of interference of several modes or even micromultipath. Fading may be relatively slow and quasiperiodic, as in the case of Faraday fading, or it can be irregular and rapid. Fading effects are discussed in Chapter 4. It is also a subtopic within CCIR Report 266-6 [CCIR, 1986i]. Rapid fading is sometimes called scintillation.

DOPPLER

Scattering centers in the ionosphere as well as the normal layers are typically in more rapid motion in the high latitude region than at midlatitudes. Doppler shifts are, of course, dependent upon the frequency used as well as the drift velocities associated with the ionosphere. Drift velocities vary but may be of the order of 1 km/sec at times. Typical spectra are of the order of 8 Hz with the greatest values being associated with NGC modes.

TIME SPREAD

The auroral zone admits to a wide variety of conditions from benign (like midlatitude) to disturbed. Thus time delay spread, not unlike Doppler spread, is difficult to classify precisely. Values between 100 microseconds and 2 milliseconds have been observed for selected modes.

TABLE 3-4. HIGH LATITUDE HF PROPAGATION ISSUES
{Continued}

POLARIZATION

In theory, a vertical antenna should perform better than a horizontal one at high latitudes since for quasitransverse (QT) propagation at low angles, the X-mode is horizontal and is strongly absorbed. In practice, a horizontal antenna performs adequately. High latitude NVIS systems exploit horizontal antenna structures since vertical whips have nulls in the zenith direction.

DISTORTION

Interference between the two sidebands of an AM transmission makes AM vulnerable to distortion. SSB or CW transmissions are not as vulnerable.

3.4.3 Concluding Remark

Auroral physics is an exceedingly rich and complex subject. Not all phenomena in the high latitude region are understood at this time, and insufficient data are available to fully characterize those factors for which a general understanding exists. This situation limits our ability to model the high latitude zones. Certainly, it is an impediment to prediction of HF effects. Simple climatological models for the region do exist and more complex codes are under development. It is indeed unfortunate that for the most part, phenomenological models of the ionosphere which are in common use for HF propagation work do not contain any significant information about the region above 60 degrees geomagnetic. The equatorial region also requires further exploration, but large uncertainties associated with system performance at high latitudes make the development of more efficient auroral models a matter of some priority. CCIR Report 1012 covers some aspects of HF propagation modeling for high latitudes [CCIR, 1986j]. In this section, we have attempted to lay out some of the features of most concern to HF workers. Those interested in more detail may consult Whalen et al. [1985] and other references listed at the end of the chapter.

3.5 IONOSPHERIC RESPONSE TO SOLAR FLARES

We have already mentioned solar flares in Section 2.3.2.6. Now we shall take note of a special class of effects called Sudden Ionospheric Disturbances (SID). These constitute those events which arise as a result of the atmospheric interaction with electromagnetic flux from solar flares. A book by Mitra [1974] is an excellent treatise on the ionospheric effects of solar flares.

We recognize that the sun is the ultimate source for a large variety of ionospheric and magnetospheric effects. Many of these are of some importance in HF work, while others are only of marginal interest. Figure 2-30 exhibited the hierarchy of solar-induced ionospheric effects. We have ignored many interesting phenomena, but we have accented the major effects of interest for HF systems.

There are many types of SID which are observed, and many are only of importance to longwave systems or have limited diagnostic application. At HF the most important form is the Short-Wave-Fade (SWF), although a relatively small and sudden enhancement in *foF2* may also be observed. The SWF which affects the entire sunlit ionosphere is by far the most important form of SID. An x-ray flare generates a significant increase in D layer ionization content with a temporal pattern which mimics that of the flare itself. This results in an increase in the product of the electron density and the collision frequency. It is the growth of this product which accounts for the absorption of HF signals. See Chapter 4 for a discussion of absorption theory and its application to HF channel modeling.

3.6 THE IONOSPHERIC STORM

In Section 2.4.3 we commented on geomagnetic substorms, their origins in the magnetosphere, their association with the auroral zone, and we provided some insight into the ionospheric response. In this section we look at the subject again with more emphasis upon the temporal and spatial effects which ionospheric storms introduce on certain HF system parameters including the Maximum Observable Frequency (*MOF*) over a given skywave link.

The ionospheric storm is the ionosphere's response to a geomagnetic storm, and it may be the most important solar-induced phenomenon from the point of view of HF propagation impact. While the associated effects may often be quantitatively unpredictable over large geographical areas, even the most accurate prediction of stormtime effects will not provide an unassailable basis for improved HF communication, although it will help. Ionospheric storms are known to introduce substantial and enduring diminutions in F region electron concentrations which will have the effect of reducing the availability of the upper portion of the HF band. This is true even during the daytime. Whether a specified event admits to a successful short-term forecast or not, loss of ionospheric support over a wide range of assigned frequencies cannot be easily overcome. And this is precisely the problem in many HF point-to-point or netted configurations. Unless a sufficient set of frequencies is available to the system manager, communication by skywave may be prevented.

At midlatitudes the ionospheric storm signature is one in which the F region ionization momentarily increases in the dusk sector following storm commencement (SC), after which it decreases dramatically. The initial short-lived enhancement is observed in *foF2* records and it is correlated with the initial positive phase of the geomagnetic storm. The main phase of the geomagnetic storm is correlated with a concomitant *foF2* diminution, which is the deleterious effect noted in the previous paragraph. This reduction in *foF2* may last for a day or longer. It is thought that the initial enhancement in *foF2* is a result of electrodynamic forces while the long-term reduction in *foF2* is associated with ionospheric heating through dissipation of storm-induced gravity waves. This heating effect will cause the thermosphere to expand, and ionospheric loss rates will increase.

Figure 3-24 shows the storm-time variation in the *MUF(zero)F2* and *MUF(3000)F2* parameterized in terms of the season of occurrence for a midlatitude site [Dominici, 1975]. An enhancement in the *MUF* is observed to occur within the first 12 hours of SSC, in concert with the initial positive phase of the associated geomagnetic storm, but it lasts only a few hours. The major effect is associated with the main phase condition, during which time the *MUF* diminution may exceed 10% of the undisturbed *MUF* for a duration of about 24 hours, depending upon the season. This effect starts roughly 12 hours after SSC and may persist for several days.

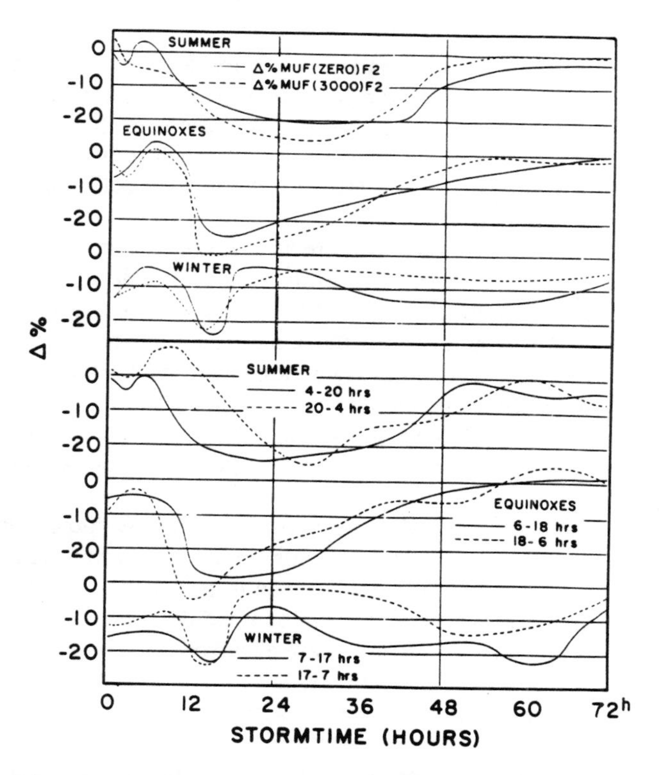

Fig.3-24. Storm-time variation of *MUF(zero)F2* and *MUF(3000)F2* for summer, equinoctial, equinoctial, and winter ionospheric storms. (From Dominici [1975].)

3.7 IONOSPHERIC CURRENT SYSTEMS

There are four principal current systems in the ionosphere which give rise to relatively rapid fluctuations in the geomagnetic field. They are: the ring, magnetopause, atmospheric dynamo, and polar current systems. The first two owe their existence to the occurrence of the same phenomena which are associated with magnetic storms, and they occur at magnetospheric distances. The polar and atmospheric dynamo currents, on the other hand, occur at lower ionospheric heights in the neighborhood of 110 km.

The atmospheric dynamo theory was first suggested by Balfour Stewart in his work on magnetic fluctuations. There are excellent accounts of the theory in Hines [1963], Maeda and Kato [1966], Rishbeth and Garriott [1969], and Ratcliffe [1972]. Borrowing principally from the latter two references, the main features of the theory are given below:

3.7.1 Main Features of Atmospheric Dynamo Theory

The Dynamo region is located in a rather restricted altitude neighborhood centered at roughly 110 km where the collision frequencies are in excess of the electron and ion gyrofrequencies. The plasma present in this region contributes to a relatively large DC conductivity. The region is restricted in height because of the lack of electrons below 110 km, and because collision rates are not sufficiently high above 110 km.

Atmospheric tides are driven by both the sun and the moon. In the upper atmosphere, the solar tidal force, the result of heating within the middle atmosphere, is more substantial than the lunar force, which is the result of gravitational forces. This is quite unlike the situation for the ocean tides. The resulting force is therefore predominantly periodic with a 24 hour period and a strong semi-diurnal (12-hour) component. The lunar period corresponds to 24.8 hours (i.e., a lunar day). The atmospheric tide manifests itself in gas motion which is largely horizontal in nature with a periodicity of 12 hours (not 24 hours).

The motion of the tide-driven *dynamo region* across the geomagnetic field induces EMFs within the region, and currents are driven by this force. This current flow associated with this effect gives rise to so-called Sq (solar-quiet) magnetic variations. In connection with the lunar tidal forces, there are also Lq (lunar-quiet) magnetic variations which are observed.

Conductivity is not uniform; consequently the induced electric fields $\boldsymbol{V} \times \boldsymbol{B}$ will be variable and a distribution of space charge is developed. This results in a current flow modification. The total field which drives this current is given by $\boldsymbol{E} = \boldsymbol{V} \times \boldsymbol{B} - \text{grad}\,\phi$ where ϕ is the potential set up by the space charge.

The electrostatic field developed within the dynamo at 110 km is transferred to F region heights along field lines which serve as equipotentials. (The parallel conductivity is much greater than the transverse conductivity above the dynamo region.) In the F region, the electrostatic field forces the plasma to drift (note: there is no current). This drift is $V = B^{-2}(E \times B)$. Notice that the neutral wind cannot move the plasma across magnetic fields since the gyrofrequency far exceeds the ion-neutral collision rate in the F region. Thus there is no induced electric field to augment the electrostatic polarization field. If I is the magnetic dip angle and the eastward electric field is E_{EAST}, we find that the vertical component of the drift will be $W = (E_{EAST}/B)\cos I$. The equatorial electrojet is a current system centered over the geomagnetic equator. It flows in the eastward direction during the day and for sites in the neighborhood of the equator it produces substantial variations in the magnetic field intensity. It is associated with an anomalously high value of the conductivity over the magnetic dip equator where field lines are horizontal. Instabilities develop in this region and are correlated with observations of an equatorial variety of sporadic E.

3.7.2 The Ring Current System

The ring current is generally associated with the main phase of a geomagnetic storm but it is only enhanced during that event. Trapped particles mirror back and forth between northern and southern hemispheres undergoing gyromotion along the field lines as they move. Mirroring times are the same for both ions and electrons within the same L shell. It may be 0.5 seconds or more depending on the L shell selected. At the same time, a charge separation occurs with electrons drifting eastward and protons drifting westward resulting in a *ring current* to the west. This phenomenon is greatly enhanced during storms because of an increase in the population of energetic particles in the Van Allen belts at the time.

3.7.3 The Magnetopause Current System

Magnetopause currents are associated with a compression of the geomagnetic field. These currents are most evident during events when the solar wind pressure is greatly enhanced. During sudden storm commencements (SSCs), the magnetosphere is abruptly compressed to an altitude at which the solar wind pressure balances the magnetic pressure.

3.7.4 High Latitude Current Systems

The polar current system is not driven by the atmospheric dynamo. Rather it is ultimately due to magnetospheric electric fields. The solar wind interacts with the magnetospheric plasma causing a circulating plasma flow pattern in two zones resembling vortices, one located in the dawn and the other in the dusk sector, when viewed in the ecliptic plane. The flow in each vortex is oriented such that plasma motion is sunward near the earth and antisunward at greater magnetospheric heights (see Figure 3-25). This plasma circulation pattern will be mapped out of the ecliptic and poleward ultimately being *reflected* down into the ionosphere to an altitude for which the gyrofrequency becomes comparable to the collision rate. The reflection process effectively reverses the orientation of the plasma flow tendency as it maps into the polar ionosphere. In a plan view in which we look down at the north polar region, the dawn circulation is counterclockwise and the dusk circulation is clockwise. The altitude regime for which the gyrofrequency is comparable to the collision frequency is different for electrons and ions (between roughly 80 km for electrons and 140 km for ions). As a consequence, electrons will undergo motion independent of ionic motion between 80 and 140 km.. This plasma separation, in which electrons are relatively free to move while the heavy ions are collision dominated, corresponds to a (positive) current flow in a direction opposite from the (negative) electron flow. This current is sunward and is termed Sq^{p}.

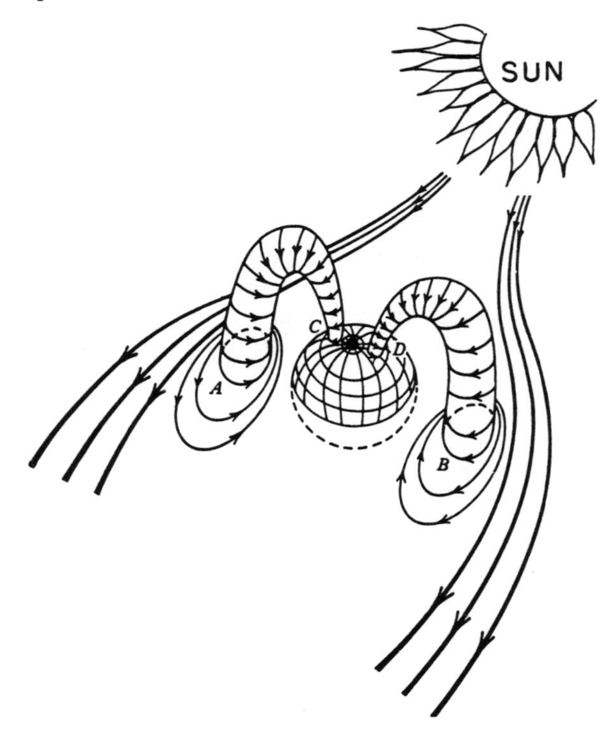

Fig.3-25. Schematic diagram of ionospheric and magnetospheric plasma flow. These patterns may be associated with currents in those instances for which collisions become important. For example, in regions *C* and *D*, the antisunward plasma flow over the polar cap generates a current flow toward the sun. This is because the ions are not as free to move in the low ionosphere. (Figure adapted from Ratcliffe [1972].)

There are other current systems which are closely tied to the polar regions and may be especially enhanced during active periods. During substorms, for example, there are two systems called DP1 and DP2. The DP2 corresponds to an enhanced version of the Sq^p system. The DP1 current system is associated with current flow toward the west (in both northern and southern hemispheres) and lies in the auroral oval. The oval feature is distended equatorward on the midnight side and the DP1 system is more enhanced at that time as well, being termed the polar electrojet. While DP1 is virtually omnipresent, substorms drive additional energetic particles into the auroral zone thereby increasing the conductivity even more. This causes the appearance of brilliant auroral arcs and radio aurora. The term *auroral zone* is a generic description of a latitude region where an observer is apt to see optical aurora overhead. This obviously will vary depending upon the level of general magnetic activity but may be put at a geomagnetic latitude of about 68 degrees. Of course, observers at lower latitudes may observe aurora at low elevation angles.

Brekke [1980] provides an excellent synopsis of the major features of relevant high latitude current systems. It has been thought for some time that magnetic activity fluctuations in the auroral zone were associated with strong currents. The earliest suggestions were made by Birkeland [1908], and the upgoing and downgoing legs of the auroral current system (i.e., the field -aligned components) are called Birkeland currents to recognize his contributions to auroral physics. The horizontal currents associated with the Birkeland system were studied by Harang [1946] and he determined that it was comprised of two branches, one moving eastward in the PM sector, and the other moving westward in the AM sector. These branches are also termed the auroral (or polar) electrojets and are associated with so-called Hall currents. The Harang discontinuity is the nocturnal boundary between the branches.

The following picture has emerged in connection with the Birkeland current system. In the evening sector, the downgoing leg is on the equatorward side of the oval (called Region 2) and the upgoing leg is on the poleward side (called Region 1). The situation is reversed in the morning sector. Thus the field-aligned Birkeland currents are actually nearly-vertical current sheets. These sheets are closed by a meridional current within the E layer (called the Pederson current) which connects Regions 1 and 2. The amount of Joule heating in the auroral ionosphere Q is proportional to the height-integrated Pederson conductivity times E^2, but Cole [1971] has found that Q is simply proportional to the square of the magnetic field fluctuations. At the same time the magnetic field fluctuations are proportional to the Hall (electrojet) currents. Consequently, we find that the Joule heating is equal to the square of the height-integrated Hall conductivity times the square of the E field. The electric fields driving the auroral currents are largely meridional, forming eastward and westward electrojets perpendicular to E.

3.8 IONOSPHERIC MODELS

As in many areas of geophysical study, ionospheric modeling may assume a number of forms ranging from the purely theoretical to the totally empirical. Approaches may also include a combination of these forms, although empirical models dominate the field. Recent developments include allowance for adaptivity within the models to accommodate exploitation in the near-real-time environment for special applications.

Although the distribution of electrons in the ionosphere is important in many specific applications such as HF skywave propagation, it should be recognized that ionospheric modeling is actually a field unto itself. For a specified application such as the prediction of HF communication reliability, the ionospheric model is actually a submodel of the full program (or code) which may include within its modular construction the following components: a front-end user interface, an executive program, a propagation submodel, an ionospheric submodel, a noise and interference submodel, various antenna and equipment-specific submodels, special purpose applications, output data and graphics management. Not all HF prediction codes contain the full range of these component features, but it is clear that the ionospheric submodel must be included and prominent.

The well-known HF performance prediction code IONCAP is not an ionospheric model, even though it is often mentioned in a number of general articles on ionospheric modeling. By the same token, a well-known ionospheric model such as the International Reference Ionosphere (IRI) is not an HF performance prediction model. In this section we refer to the narrow but important field of ionospheric modeling. We shall leave special applications of ionospheric models, for which the ionospheric representation takes the form of a submodel, to a later chapter on HF prediction methods.

The author began development of an ionospheric model thesaurus and user's guide several years ago [Goodman, 1982] but the project was left uncompleted because of funding limitations. Recently several reports have been issued which fill a distinct void in this area, Bilitza [1989,1990] and Secan [1989]. The report by Secan is a very detailed exposition of currently available empirical models and those which he terms semiempirical models, which reflect a theoretical flavor in their makeup. Bilitza's report is not limited to models but includes a description of the worldwide ionospheric data base as well. Other worthwhile surveys of ionospheric modeling have been written by Kohnlein [1978], Davies [1981], Rush [1986], and Dudeney and Kressman [1986]. Reviews principally dedicated to theoretical approaches have been prepared by Schunk [1983] and Rawer [1984] while global scale modeling using both empirical and theoretical methods are described by Schunk and Szuszczewicz [1988].

Ionospheric models have been required for a number of applications aside from HF communications. The most notable example is the study of area coverage and transmission frequency requirements for over-the-horizon backscatter radar (OTH-B). Ionospheric impact has been studied by authors including Trizna and Headrick [1981] and Millman [1978]. More recently Schleher [1990] and Szuszczewicz [1988] have looked at ionospheric models for test and evaluation of the Air Force OTH-B system. To satisfy the study requirements, ray tracing must be exploited in conjunction with selected models. Therefore the models should include faithful reproductions of ionospheric electron density profiles including effects of gradients and tilts. Not all models satisfy this requirement.

3.8.1 Theoretical Models of the Ionosphere

Ionospheric models based upon data, and called empirical models, are generally preferred over theoretical models in many practical applications since they are require less computational power and speed. In addition, they are perceived to be more accurate than their theoretical counterparts over regions for which a significant amount of data has been gathered. Given the additional computational overhead as well as the negative perceptions, why would one still consider theoretical approaches in ionospheric modeling? Some of the attributes of theoretical models are given in Table 3.5 below:

TABLE 3-5. ATTRIBUTES OF THEORETICAL MODELS

1. They provide a basis for understanding the physical processes involved.

2. They are useful for providing insight in circumstances which are beyond the scope of earlier physical evidence.

3. They allow for the extrapolation of experimental results into areas of space and time using factors which are not adequately derivable from the available data base.

These are indeed powerful arguments for the use of theoretical approaches, especially for those circumstances in which ionospheric data sets are sparse. Very little ionospheric data has been obtained over ocean areas because of a natural limitation on the number of sounder stations. Mapping procedures generally perform best if data sets are uniform, and this dictates the *invention* of data at selected grid points. Traditionally this has been accomplished by spatial extrapolation of observed data using generally accepted empirical rules. One would suppose, therefore, that the vast ocean areas might provide a good test for the partial use of *theoretical extrapolation* as opposed to *experimental extrapolation*. Rush et al. [1983,1984,1989] under the aegis of URSI have used this approach to improve the set of ionospheric coefficients which are used to map the global variation of the ionospheric parameter *foF2*. These new coefficients, termed URSI coefficients, were generated as a possible replacement for the current set of coefficients which are used to derive the maps found in CCIR Report 340 [CCIR, 1986b] and which form the basis for almost all statistical models in use today. Surprisingly, a comparison of the two sets of coefficients against a unique set of data shows only a slight improvement overall if we use the semitheoretical URSI coefficients instead of the CCIR coefficients. The theoretical approach appears to be best over the Pacific Ocean zone, as would be expected.

Figure 3-26a is a contour map for the parameter *foF2* based upon CCIR recommendations (circa 1986); Figure 3-26b shows *foF2* based upon the newer URSI coefficients. Maps of parameters such as *foF2* represent the best view of how the selected parameter varies worldwide at a specified sunspot number, universal time, and date. A departure from reality would be expected for a variety of reasons including those on the following list:

1. Modeled parameter is based upon median data.
2. Mapping algorithm relies upon spatial correlation.
3. Data base is globally non-uniform.
4. Theory is incomplete at high latitudes.
5. Sunspot number (as a driver) is only a circumstantial index.
6. Parameters needed to drive the model may be unavailable.
7. Parametric relationships are typically approximations.

For long-term predictions, theoretical models and semiempirical approaches are ultimately predictions of predictions. In other words, we must predict the driving functions or parameters (such as K_p and R_I) which are themselves used to predict the ionospheric parameters. For HF communication predictions, we must go one step further, of course, and translate these predictions into communication performance estimates for a specified application.

To be fully successful, theoretical models must take the mutual coupling among the neutral atmosphere, the ionosphere and the magnetosphere into account. This makes the problem rather complex but significant progress in the development of such models has been reported [Schunk et al., 1986], [Roble et al., 1988], [Schunk, 1988]. The main successes of theoretical models have been in the auroral regions, but more work is needed to match prediction with reality.

The Penn State MK-1 model developed by Nisbet [1971] was probably the first largely physical model of the ionosphere to be programmed on a digital computer. Over the years other models have included such factors as neutral winds [Strobel and McElroy, 1970], solar flare effects [Oran and Young, 1977], as well as electric and magnetic field effects [Stubbe, 1970].

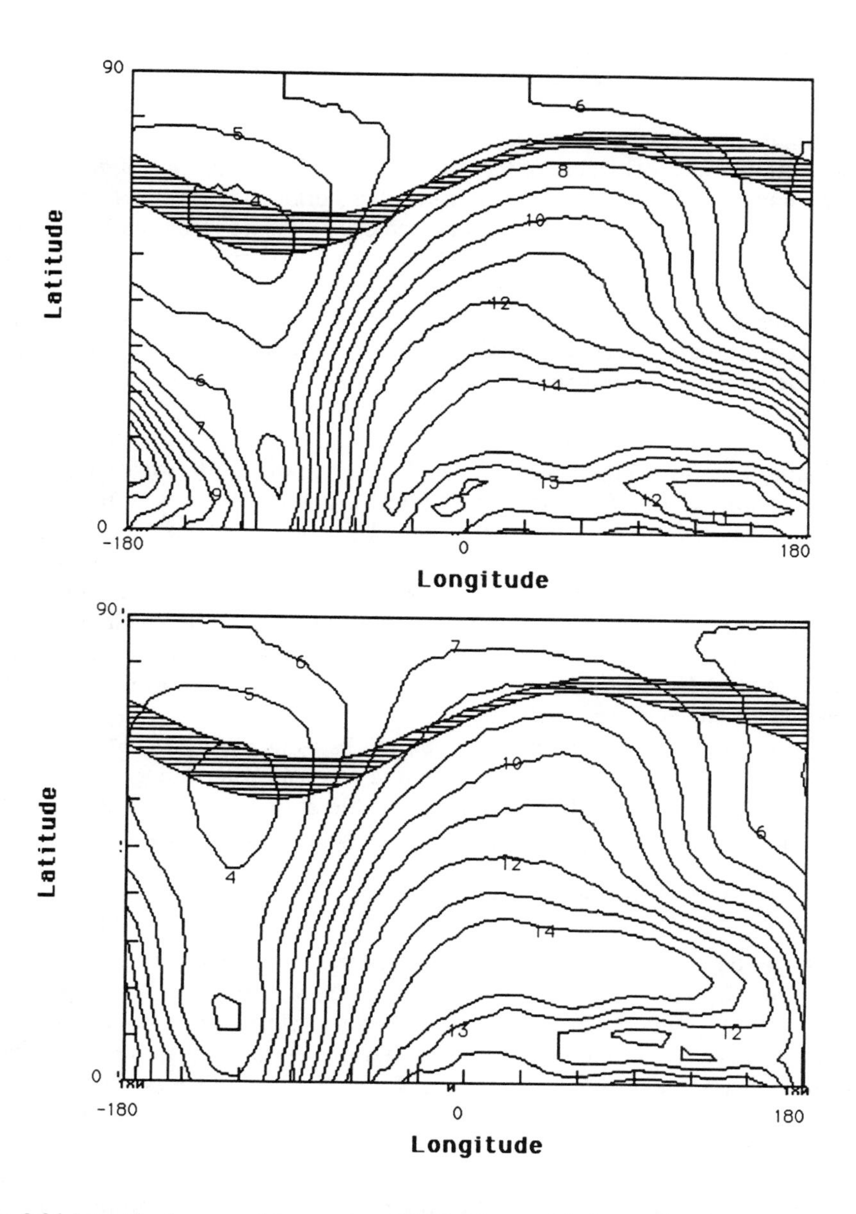

Fig. 3-26. Ionospheric maps of the parameter *foF2* with a modeled auroral oval superimposed (Feldstein Q = 3, shaded region). The northern hemisphere is shown for two sets of ionospheric coefficients: (a) [top] CCIR-1971 coefficients, (b) [bottom] URSI-1988 coefficients.

Purely theoretical models generally require a substantial amount of computer time to generate a global representation of the ionosphere. Models which are mainly empirical but have been constructed using theoretical notions (i.e., profile shapes, temporal variabilities, etc.) are termed semiempirical. Such models have some of the speed advantages of empirical models and the physical appeal of theoretical versions. Furthermore, it may be shown that reductions in computation time can be achieved in some instances if a novel suggestion made by Batten et al.[1987] is followed. These workers have suggested that theoretical data bases be constructed and parameterized using procedures similar to those previously used in the exploitation of empirical data. Once these theoretical data bases are formed through calculation, one achieves a reduction in computation time for future applications. Computational overhead is required to generate the theoretical data base, but this should be compared with the time, and cost, of compiling the equivalent number of empirical data points through observation. Anderson et al.[1985] have computed a grid of theoretical profiles in connection with the Semi-empirical Low-latitude Ionospheric Model (SLIM). The best-fit Chapman layer approximations to the theoretical profiles were deduced and specified by a set of coefficients. The resultant *data* set is then used in much the same way as observational data. In view of the manner in which models such as SLIM and the Penn State models have been developed, they properly belong to the subclass of semi-empirical models. Nonetheless, they do possess some of the positive attributes of theoretical models.

The components of theoretical models are listed in Table 3-6 and the main difficulties encountered in the application of theoretical modeling are provided in Table 3-7. The difficulties given in Table 3-7 are primarily associated with purely theoretical models rather than the semi-empirical ones.

3.8.2 Empirical Models of the Ionosphere

A vast amount of data has been accumulated over the years using ionospheric sounders. On space platforms other instruments such as in situ plasma probes and topside sounders have provided additional information about electron concentration above the peak of F2 ionization. Occasional sources of data include incoherent scatter radars and, of course, rocket probes. These data have been used to generate empirical models of the distribution of electrons in the ionosphere.

Most empirical models are climatological in nature and represent the global distribution in a statistical manner. Although some exceptions exist, most ionospheric parameters are represented in models as median values. Very little direct information on variability is contained in the models. The models are typically statistical descriptions of the ionosphere which is characterized by the set of standard profile parameters listed in Table 3-8 and depicted in Figure 3-27. The functional relationships which are accounted for in the models include: geographic coordinates, geomagnetic coordinates, time, solar zenith angle, seasonal effects, solar activity, and magnetic activity.

TABLE 3-6. COMPONENTS OF THEORETICAL MODELS

1. EQUATIONS
   - Continuity equation
   - Momentum equation
   - Energy equation
2. INPUT DATA
   - Solar extreme ultraviolet flux
   - Particle precipitation
3. BOUNDARY CONDITIONS
   - Ionosphere-magnetosphere
   - Atmosphere-ionosphere
4. INTERACTION CROSS SECTIONS
   - Collisions
   - Chemical reactions
5. SUBMODELS
   - Neutral atmospheric constituents
   - Neutral winds
   - Magnetic fields
   - Electric fields
   - Solar flares

TABLE 3-7. DIFFICULTIES IN APPLICATION OF THEORETICAL MODELS

1. The accuracy is strongly dependent upon current physical understanding of the various geophysical processes. For simplicity some effects may be excluded.
2. Accuracy requires a significant degree of precision in the specification of a large number of necessary parameters used in the model.
3. Average boundary conditions in the models are based upon observation and may be subject to an amount of uncertainty.
4. EUV Flux necessary in the model is not routinely available.
5. Even though better physical insight is achieved, theoretical models do not perform as well as physical models in data-rich regions from which the empirical models were partially derived.

TABLE 3-8. PARAMETERS OF EMPIRICAL MODELS

MODELED PARAMETER	DESCRIPTION
foE	Critical Frequency of the Normal E Region
foEs	Critical Frequency of the Sporadic E Layer
foF1	Critical Frequency of the F1 "Ledge"
foF2	Critical Frequency of the F2 Region
hmE	Height of the E Region
hmF1	Height of the F1 "Ledge"
hmF2	Height of the F2 Region
ymE	Semi-thickness of the E Region
ymF1	Semi-thickness of the F1 Region
ymF2	Semi-thickness of the F2 Region

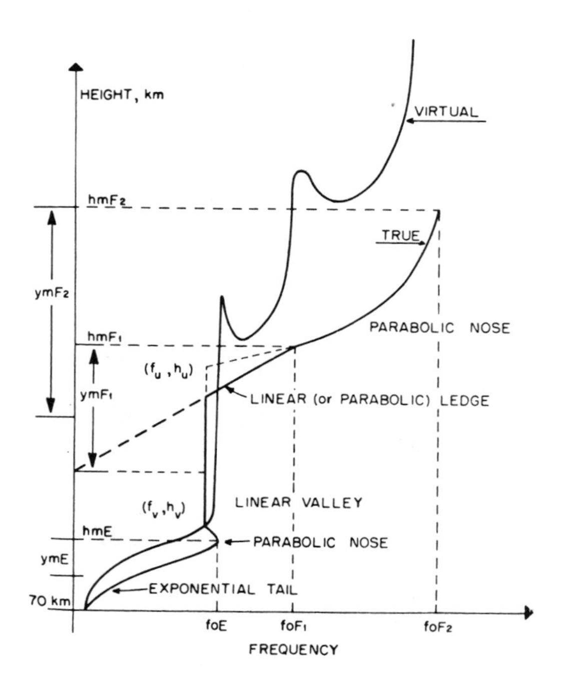

Fig.3-27. Schematic representation of ionospheric profile parameters of importance.

The single most important (ionospheric) parameter in connection with an examination of HF skywave communication performance is *foF2*, corresponding to the ordinary ray critical frequency of the F2 region. We have noted previously that the peak F2 region electron density is proportional to the square of *foF2*. It is generally observed in using vertical incidence ionospheric sounders. Figure 3-28 is a daytime nighttime ionogram showing *foF2* and other parameters of importance. We observe a number of so-called ionogram *traces*. Generally the trace which produces the highest frequency return is the mode corresponding to interaction of the extra-ordinary ray (X-mode) with the F2 layer. In the near neighborhood (but below the X-mode asymptote) we find the trace which corresponds to the ordinary ray (O-mode) *reflection* from the same layer. The X-mode asymptote is termed *fxF2* and the O-mode asymptote is called *foF2*. We shall return to ionograms in Chapters 4 and 6.

Table 3-9 is a listing of empirical and semiempirical models with associated references. The Chapman profile is listed for comparison even though it is quasi-theoretical in nature.

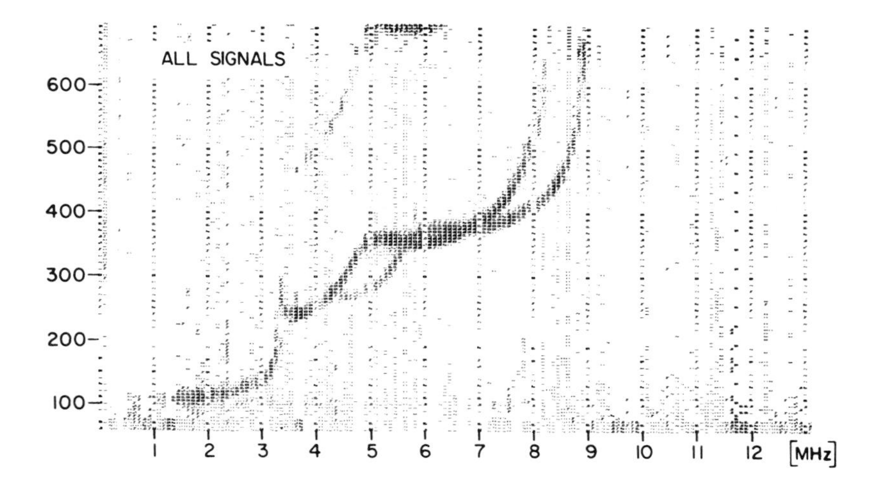

Fig.3-28. Vertical incidence sounder recordings (ionograms) showing *foF2*, *fxF2* and other parameters.

TABLE 3-9. LISTING OF EMPIRICAL IONOSPHERIC MODELS

MODEL	REFERENCES	TYPE
CCIR foF2	[Jones & Obitts,1970]	D
URSI foF2	[Rush et al., 1984]	D
CCIR foE	[Leftin,1976]	D
CCIR foF1	[Rosich and Jones,1973]	D
CCIR M3000	[Jones et al.,1969]	D
CCIR Atlas	[CCIR,1986b;Rpt 340-5]	D
Ching-Chiu	[Ching and Chiu, 1973] [Chiu,1975]	E
NRL:Ching-Chiu	[Keskinen & Fedder,1988]	E
RBTEC	[Flattery & Ramsay,1976]	E
BENT	[Bent et al.,1976]	E
Elkins-Rush Polar	[Elkins & Rush,1973]	E
Miller-Gibbs	[Miller & Gibbs,1978]	E
NRL Global: RADARC	[Thomason et al.,1979]	E
Chatanika: AMBCOM	[Vondrak et al.,1978]	E
ICED	[Tascione et al.,1986,1988]	E
	[Daniell et al., 1990]	E
IRI	[Rawer,1981][Bilitza,1986]	E
Dudeney Profile	[Dudeney,1978]	P
Bradley-Dudeney	[Bradley & Dudeney,1973]	P
Chapman	[Chapman,1931]	P
DH Profile	[Damen & Hartranft,1970]	P
Penn State:Mk1-3	[Nisbet,1971][Lee,1985]	SE
AWS 4-D Analysis	[Secan & Tascione,1984]	SE
SLIM	[Anderson et al.,1985]	SE
HLISM	[Daniell et al.,1988]	SE
FAIM	[Anderson et al.,1989]	SE
DETMOD	[Secan,1985]	SE

E	Empirical
SE	Semi-Empirical
P	Profile
D	Ionospheric Coefficients and Maps

Empirical models of the ionosphere form the basis for all HF prediction methods in current use. One of these, the Ching-Chiu model, being a phenomenological description, is one of the simplest models to use since it does not employ a spherical harmonics expansion method. Unfortunately, because of the simplistic construction of the Ching-Chiu model, it only represents the F region parameters crudely. Still, the Ching-Chiu model has distinct advantages in the many applications for which relatively accurate electron density distributions are not necessary. The Ching-Chiu model has even been used in a MUF prediction program [Daehler et al., 1988].

Brown et al. [1990] have examined six ionospheric models as predictors of the total electron content, which is the integral of the electron density distribution of the ionosphere. The models were: BENT, FAIM, ICED, IRI, Penn State Mk-3, and a model called Damen-Hartranft-Ramsay (or DHR). The profile in the DHR model is similar to the DH profile listed in Table 3-9, but revised to account for the F1 layer. In the study, the DHR model used the URSI-88 coefficients as a basis for near global mapping and the underlying methodology is similar to that found in RBTEC, which is also listed in Table 3-9. The main purpose of the study was to examine TEC prediction performance against a large data base. However, a comparison of the prediction performance of *foF2* was also made and is significant for this book. Table 3-10 has been constructed from Table 6 given in Brown et al. [1990].

TABLE 3-10. YEARLY MEAN RMS DEVIATIONS OF *foF2*

(Wallops Island: 38.71 deg.N, 289.1 deg.E; Brown et al. [1990])

MODEL	SOLAR MAXIMUM (MHz)	SOLAR MINIMUM (MHz)
BENT	0.55	0.48
DHR	0.55	0.59
ICED	0.80	0.46
IRI	0.85	0.44
FAIM	1.88	0.88
Penn State Mk-3	2.32	0.99

Models in Table 3-10 have been ordered on the basis of decreasing performance at solar maximum, generally thought to be the most stressed condition. We see that for this particular site, the top four models perform equally well under solar minimum conditions, but the IRI model is a marginal winner. For solar maximum conditions, the BENT and the DHR models are clear winners. The fact that the revised URSI coefficients are employed in DHR should not be an advantage over the other models which employ the older CCIR coefficients, since Wallops Island is in a data rich region where distinctions between the results produced by application of the two sets of coefficients would be expected to be slight. Indeed the BENT model appears to be the best selection for Wallops Island under solar minimum conditions. Naturally for specialized applications, such as the correction of orbit determination errors introduced by ionospheric TEC effects, the BENT model might not be the best choice. In fact Bilitza et al. [1988] suggest that the IRI model might be best for this purpose.

For HF applications, most investigators are concerned with the ionospheric submodels within a small group of computer codes or structures such as IONCAP, AMBCOM, RADARC, and PROPHET, which represent four rather distinct methodologies in HF modeling for the benign environment.

Although there are programming convenience issues involved, each methodology should admit to the introduction of common ionospheric submodels. To the author's knowledge, no comprehensive comparisons of this type have been made. The ionospheric submodels presently contained within IONCAP, HFMUFES, and RADARC are similar in many respects. The computer codes for these three models have been developed with HF applications in mind, and the underlying ionospheric databases are virtually identical, both in content and structure. PROPHET, is a U.S. Navy product containing a number of applications other than HF, and is based upon a unique data base. It will be noted that a given program such as IONCAP may produce different results under the same parametric specifications (i.e., sunspot number, time-of-day, etc.). This may arise because of minor modifications in the ionospheric submodel, or alterations in the application involved. Thus, the user must be wary of comparisons made between models unless the version numbers are specified. Configuration control and management is a major concern to DoD and commercial users of HF propagation and modeling codes. The ionospheric profile recipe for IONCAP is shown in Figure 3-30. We shall return to IONCAP and to some of the other pertinent models when we discuss performance prediction in Chapter 5.

3.8.3 Improvements in Modeling

In the context of HF communications, there have been a number of modeling improvements since the original CCIR *Atlas of Ionospheric Characteristics* (CCIR Rpt 340) was published [CCIR, 1966]. This atlas contains information about the parameters *foF2* and *M(3000)F2* which have been mentioned previously. The parameter *foF2*, the ordinary ray critical frequency, is proportional to the square root of the peak F2 layer electron density; *M(3000)F2* is the normalized *MUF* for a skip distance of 3000 km, with 1/*foF2* being the normalization factor. Although the raw data used to form the atlas is voluminous, the data is still sparse over ocean areas. As a consequence, mapping procedures have undergone a number of iterations over the years.

The original procedure followed by Jones and Gallet [1965] is outlined by Suchy and Rawer [1987]. The first step involved a seventh-order Fourier analysis of the data from each designated station (thus generating 15 Fourier coefficients). Taking each coefficient in turn, a Legendre analysis was applied, with the result that 15 distinct worldwide representations were produced. This ultimately enabled worldwide maps of *foF2* and *M(3000)F2* to be constructed. This procedure works well in data-rich regions where it is possible to construct a uniform spacing of grid points. Due to the lack of data over the oceans, it was necessary to extrapolate data to grid points for which data were unavailable. This procedure was not completely successful. One of the first improvements was one in which the geographic latitude was replaced by a modified dip latitude (called MODIP). This procedure, suggested by Rawer [1963], better accounted for the geomagnetic control especially in regions where data had to be invented to make the Legendre procedure work efficiently. A more recent improvement in the development of the coefficients used in the ionospheric atlas formulation is due to Rush et al.[1986] as discussed in Section 3.8.1.

Upgrades to various models are continually being made. For example, the Penn State model has been restructured to run on a personal computer [Nisbet and Divany, 1987] and the same strategy is being used by other developers. This is placing the previously main frame models in the hands of the actual users. This procedure has been accelerated for models specializing in HF work, a matter we shall return to in Chapter 5.

The most significant improvements in empirical ionospheric modeling (to be distinguished from HF performance prediction modeling) have been promoted by sponsorship from the U.S. Department of Defense[25]. This is not surprising in view of the large number of applications of ionospheric specification in radiowave systems used by the military. The AWS-4D concept and the ICED model are examples of the interest expressed by the services.

The original ICED model was intended principally to be a northern hemispheric ionospheric specification model to serve the requirements of the US Air Force. Thus it was only a regional model, intended for use at midlatitudes but extending into the auroral zone. It was designed to allow for recovery of some of the dynamic features embodied in auroral climatology which are smeared out in most mapping procedures. The model as described by Tascione et al.[1988] is driven by an effective sunspot number SSN_{eff} and an index derived from auroral oval imagery, Q_{eff}. The effective sunspot number, is not based on solar data at all, but is derived from ionospheric data extracted from the US Air Force real-time ionosonde network. This effective sunspot number is similar to the pseudoflux concept used in HF predictions of the MUF and is described in Chapter 6. It is also reminiscent of the ionospheric T-index discussed in CCIR Recommendation 371-5 [CCIR, 1986k]. Important modifications to ICED include incorporation of versions of SLIM, FAIM and the Ching-Chiu model. To accommodate problems encountered outside the American sector, a new model called Global ICED is envisioned. This version would involve a greater emphasis upon physical principles, rather than purely climatological ones, while still preserving the capability to be driven by real-time data. Global ICED is to be organized into five latitudinal regions the

---

25. The Air Weather Service (AWS) of the Department of the Air Force has promoted the development of various space models. Those of interest include the following: the solar wind transport model (SWIM), the magnetospheric specification and forecast model (MSFM), the thermosphere-ionosphere general circulation model (TIGCM), the parameterized real-time ionospheric specification model (PRISM), and an ionospheric forecast model.

boundaries of which are to be dynamically restructured in accordance with observations. The regions are: northern and southern high latitudes, northern and southern mid-latitudes, and low latitudes. The global ionospheric specification will be based upon the following near-real-time data sets: ionosonde measurements of ionospheric parameters; the TEC; in-situ plasma density, temperature, and composition; and satellite-based measurements of ultraviolet airglow and auroral emissions. For a synopsis of the methodology in Global ICED, the reader should consult a paper by Daniell et al.[1990].

3.9 IONOSPHERIC PREDICTIONS

Ionospheric predictions influence several disciplines including those associated with HF radiowave systems. As we have seen, long-term predictions are generally based upon predictions of driving parameters such as sunspot number, the 10.7 cm solar flux, magnetic activity indices, etc. Unfortunately these parameters are not easy to predict. Moreover, the functions relating these parameters with the ionosphere are imprecise. Therefore long-term predictions needed for system design are subject to a considerable amount of uncertainty. To first order the uncertainty in the median value of *foF2* for a particular time and location is proportional to the uncertainty in the sunspot number. Referring to Figure 3-15 which shows *foF2* versus sunspot number for Ft Belvoir Virginia, we estimate that the slope of *foF2* for local noon is about 6 MHz/200. Thus an uncertainty of ± 40 would correspond to an error in the prediction mean noontime value of *foF2* of ± 1.2 MHz. Long-term prediction methods are discussed in CCIR Rec. 371-5 [CCIR, 1986k], and the preferred method is evidently the Lincoln-McNish procedure [Stewart and Ostrow, 1970]. Some of the other methods are discussed by Withbroe [1989]. A prediction method exploiting neural network theory [Koons and Gorney, 1990], uses a training sequence of solar cycles 7-21, and predicts a cycle #22 maximum of 194 ± 26 to occur in March of 1990. There are also other indices which may be used for the purpose of long-term ionospheric prediction, but the 12-month running average sunspot number performs as well as any other solar-derived index. There is some evidence to suggest that ionospherically-based indices may perform somewhat better, but not by a significant margin. However, in the short-term forecasting arena, ionospherically-derived indices have a clear edge.

In addition to the uncertainty in the mean parameters, we must account for the fact that ionospheric parameters have real distributions, and with the possible exception of *foE*, the spread of these distributions is such that errors about the mean may be a dominant source. Ionospheric predictions in the short and intermediate terms provide the most exciting challenge for the ionospheric researchers. The internationally-sanctioned Middle Atmospheric Program (MAP) and the World Atmospheric Gravity Wave Study (WAGS) are providing valuable information about the interaction between the ionosphere and the lower atmosphere including its undulations. A survey of the effects of TIDs on radiowave system and a description of WAGS may be found in a review paper by Hunsucker [1990].

The field of ionospheric predictions is undergoing continuous evolution with the introduction of new scientific methods and instruments which are providing fresh insight. Additional discussions of this topic appear in Chapters 5 and 6.

3.10 REFERENCES

Alcayde, D., P. Bauer, and J. Fontanari, 1974, *J. Geophys. Res.*, 74:629.

Allen, C.W., 1948, "Critical Frequencies, Sunspots, and the Sun's Ultraviolet Radiation", *Terrest. Magn. Atmospheric Elec.*, 53:433-438.

Allen, C.W., 1965, "The Interpretation of the UV Solar Spectrum", *Space Sci. Rev.*,4:91-122.

Anderson, D.N., M. Mendillo, and B. Herniter, 1985, "A Semi-Empirical Low-Latitude Ionospheric Model", AFGL TR-85-0254, Hanscom AFB, MA.

Anderson, D.N., M. Mendillo, and B. Herniter, 1987, "A Semi-Empirical Low Latitude Ionospheric Model", *Radio Science*, 22:292.

Anderson, D.N., J.M. Forbes, and M. Codrescu, 1989, "A Fully Analytic, Low and Middle Latitude Ionosphere Model", *J. Geophys. Res.*, 94:1520.

Appleton, E.V., 1954, "The Anomalous Equatorial Belt in the F2 Layer", *J. Atmospheric Terrest. Phys.*, 5(5,6):348-351.

Axford, W.I., 1961, "Note on a Mechanism for the Vertical Transport of Ionization of the Ionosphere", *Can.J.Phys.*, 39:1393-1396.

Battan, S., D. Rees, and J. Fuller-Rowell, 1987, "A Numerical and Data Base for Vax and Personal Computers for the Storage, Reconstruction, and Display of Global Thermospheric and Ionospheric Models", *Planet. Space Sci.*, 35:1167-1179.

Bent, R.B., S.K.Llewllyn, G. Nesterczuk, and P.E. Schmid, 1976, "The Development of a Highly-Successful Worldwide Empirical Ionospheric Model and its Use in Certain Aspects of Space Communications and Worldwide Total Electron Content Investigations", in *Effect of the Ionosphere on Space Systems and Communications, IES'75*, edited by J.M. Goodman, USGPO, available through NTIS, Springfield, VA, pp.13-28.

Bilitza, D., 1986, "International Reference Ionosphere: Recent Developments", *Radio Science*, 21:343-346.

Bilitza, D., K. Rawer, and S. Pallaschke, 1988, "Study of Ionospheric Models for Satellite Orbit Determination", in *The Effect of the Ionosphere on Communication, Navigation and Surveillance Systems, IES'87*, J.M. Goodman (Editor-in-Chief), USGPO, No.1988-195-225, available through NTIS, Springfield, VA.

Bilitza, D., 1989, "The Worldwide Ionospheric Data Base", NSSSDC/WDC-A-R&S 89-03, National Space Science Center/ World Data Center A for Rockets and Satellites, Goddard Space Flight Center, Greenbelt, MD.
Bilitza, D., 1990, "Solar Terrestrial Models and Application Software", NSSDC/WDC-A-R&S 90-19, NASA, Greenbelt, MD.
Birkeland, K., 1908, "The Norwegian Aurora Polaris Expedition: 1902-03", Christiania, Norway.
Bishop, G.J., J.A. Klobuchar, A. E. Ronn, and M.G. Bedard, 1989, "A Modern Trans-Ionospheric Propagation Sensing System", in *Operational Decision Aids for Exploiting or Mitigating Electromagnetic Propagation Effects*, NATO-AGARD-CP-453, Specialised Print. Services Ltd., UK.
Bradley, P.A. and J.R. Dudeney, 1973, "A Simple Model of the Vertical Distribution of Electron Concentration in the Ionosphere", *J. Atmospheric Terrest. Phys.*, 35:2131-2146.
Brekke A., 1980, "Currents in the Auroral zone Ionosphere", in *The Physical Basis of the Ionosphere in the Solar-Terrestrial System*, AGARD-CP-295, Tech. Edit. & Reprod. Ltd., London, pp.13.1-13.9.
Brown, L.D., R.E. Daniell, M. W. Fox, J.A. Klobuchar, and P. Doherty, 1990, "Evaluation of Six Ionospheric Models as Predictors of TEC", in *Effect of the Ionosphere on Radiowave Signals and System Performance, IES'90*, J.M. Goodman (Editor-in-Chief), USGPO, available through NTIS, Springfield, VA.
Burke, W.J., D.A. Hardy and R.P. Vancour, 1985, "Magnetospheric and High Latitude Ionospheric Electrodynamics", Chapter 8, in *Air Force Handbook of Geophysics and the Space Environment*, AFGL, NTIS, Springfield, VA.
CCIR, 1966, "CCIR Atlas of Ionospheric Characteristics", Report 340, General Assembly held in Oslo, ITU, Geneva.
CCIR, 1986a, *Recommendations and Reports of the CCIR, 1986: Vol.VI (Propagation in Ionized Media)*, XVIth Plenary Assembly in Dubrovnik, published under aegis of ITU, Geneva.
CCIR, 1986b, "CCIR Atlas of Ionospheric Characteristics", Report 340-5 in *Recommendations and Reports of the CCIR, 1986, Vol. VI, (Propagation in Ionized Media)*, XVIth Plenary Assembly held in Dubrovnik, ITU, Geneva.
CCIR, 1986c, "Simple HF Propagation Prediction Method for MUF and Field Strength", Report 894-1, in *Recommendations and Reports of the CCIR, 1986, Vol. VI, (Propagation in Ionized Media)*, XVIth Plenary Assembly held in Dubrovnik, ITU, Geneva.
CCIR, 1986d, "CCIR Interim Method for Estimating Skywave Field Strength and Transmission Loss at Frequencies Between the Approximate Limits of 2 and 30 MHz", Report 252-2, in *Recommendations and Reports of the CCIR, 1986, Vol. VI, (Propagation in Ionized Media)*, XVIth Plenary Assembly held in Dubrovnik, ITU, Geneva.

CCIR, 1986e, "Ionospheric Properties", Report 725-2, in *Recommendations and Reports of the CCIR, Volume VI (Propagation in Ionized Media)*, XVIth Plenary Assembly, Dubrovnik, ITU, Geneva.

CCIR, 1986f,"Method of Calculating Sporadic E Field Strength", Rec. 534-2, in *Recommendations and Reports of the CCIR: Vol.6 (Propagation in Ionized Media)*, XVIth Plenary Assembly, Dubrovnik, ITU, Geneva.

CCIR, 1986g, "VHF Propagation by Regular Layers, Sporadic E, and Other Anomalous Ionization", Report 259-6, in *Recommendations and Reports of the CCIR: Vol.6 (Propagation in Ionized Media)*, XVIth Plenary Assembly, Dubrovnik, ITU, Geneva.

CCIR, 1986h, "Special Properties of the High Latitude Ionosphere Affecting Radiocommunications", Report 886-1, in *Recommendations and Reports of the CCIR: Vol.6 (Propagation in Ionized Media)*, XVIth Plenary Assembly held in Dubrovnik, ITU, Geneva.

CCIR, 1986i, "Ionospheric Propagation and Noise Characteristics Pertinent to Terrestrial Radiocommunication Systems Design and Service Planning", Report 266-6, in *Recommendations and Reports of the CCIR: Vol.6 (Propagation in Ionized Media)*, XVIth Plenary Assembly held in Dubrovnik, ITU, Geneva.

CCIR, 1986j, "Operational Modelling of HF Radio Propagation Conditions at High Latitudes", Report 1012, in *Recommendations and Reports of the CCIR: Vol.6: Propagation in Ionized Media)*, XVIth Plenary Assembly held in Dubrovnik, ITU, Geneva.

CCIR, 1986k, "Choice of Indices for Long Term Ionospheric Predictions", Rec. 371-5, in *Reports and Recommendations of the CCIR: Vol VI (Propagation in Ionized Media)*, XVIth Plenary Assembly held in Dubrovnik, ITU, Geneva.

Chapman, S., 1931, "The Absorption and Dissociative or Ionizing Effect of Monochromatic Radiation in an Atmosphere on a Rotating Earth", *Proc. Physical Soc*, 43:26.

Checcacci, P.F. (Guest Editor), 1975, special publication commemorating the centennial of the birth of Marconi, *Radio Science*, Vol.10, No.7.

Ching, B.K. and Y.T. Chiu, 1973, "A Phenomenological Model of Global Ionospheric Electron Density in the E-,F1-, and F2-Regions",*J. Atmospheric Terrest. Phys*, 35:1615-1630.

Chiu, Y.T., 1975, "An Improved Phenomenological Model of Electron Density", *J. Atmospheric Terrest. Phys*, 37:1563-1570.

Chubb, T.A., and G.T. Hicks, 1970, "Observations of the Aurora in the Far Ultraviolet from OGO 4", *J.Geophys.Res.*, 75:1290-1311.

Cole, K.D., 1971, "Electrodynamic Heating and Movement of the Thermosphere", *Planet. Space Science*, 19:59-75.

Daehler, M., M.H. Reilly, F.J. Rhoads, and J.M. Goodman, 1988, "Comparison of Measured MOFs with Propagation Model Forecasts", in *Effect of the Ionosphere on Communication, Navigation and Surveillance Systems, IES'87*, J.M. Goodman (Editor-in-Chief), USGPO 1988-195-225, available through NTIS, Springfield, VA.

Damon, T.D., and F.R. Hartranft, 1970, "Ionospheric Electron Density Profile", Tech.Memo.70-3, Tech.Supp.Center, Aerospace Environ. Supp. Center, Scott AFB, IL.

Daniell, R.E. Jr., L.D. Brown, D.N. Anderson, J.J. Sojka, and R.W. Schunk, 1988, "A Real-Time High-Latitude Ionospheric Specification Model", paper presented at the *1988 Cambridge Workshop on Theoretical Geoplasma Physics* (in press).

Daniell, R.E.Jr., D.W. Decker, D.N. Anderson, J.R. Jasperse, J.J. Sojka, and R.W. Schunk, 1990, "A Global Ionospheric Conductivity and Electron Density (ICED) model", in *Effect of the Ionosphere on Radiowave Signals and System Performance, IES'90*, J.M. Goodman (editor-in-chief), USGPO, available through NTIS, Springfield, VA.

Davies, K., 1965, *Ionospheric Radio Propagation*, NBS Monograph 80, USGPO, Washington, DC.

Davies, K., 1969, *Ionospheric Radio Waves*, Blaisdel Publishing Company, Waltham, Massachusetts.

Davies, K., 1981, "Review of Recent Progress in Ionospheric Predictions", *Radio Science*, 16:1407-1430.

Davies, K. 1990, *Ionospheric Radio*, IEE Electromagnetic Wave Series 31, Peter Peregrinus Ltd., IEE, London.

Davies, K., and R. Conkright, 1990, "Some Recent Trends in Ionospheric Data Management at World Data Center-A", in *Effect of the Ionosphere on Radiowave Signals and System Performance, IES'90*, J.M. Goodman (editor-in-chief), USGPO, available through NTIS, Springfield, VA.

Dominici, P., 1975, "Magnetic Storms and Ionospheric Forecasting over Italy", *Radio Science*, 10(7):699-703.

Dudeney, J.R., 1978, "An Improved Model of the Variation of Electron Concentration with Height in the Ionosphere", *J. Atmos. Terrest. Phys.*, 40:195-203.

Dudeney, J.R., and R.I. Kressman, 1986, "Empirical Models of the Electron Concentration of the Ionosphere and Their Value for Radio Communication Purposes", *Radio Science*, 21:319-330.

Elkins, T.J. and C. Rush, 1973, "A Statistical Predictive Model of the Polar Ionosphere", in *Air Force Surveys in Geophysics*, No.267, AD766240, Air Force Geophysics Laboratory, Hanscom Field, MA, pp.1-100.

Feldstein, Y.I. and G.V. Starkov, 1967, "Dynamics of Auroral Belts and Polar Geomagnetic Disturbances", *Planet. Space Sci.*, No.15, p.209.

Flattery, T.W. and A.C. Ramsay, 1976, "Derivation of Total Electron Content for Real-Time Applications", in *Effect of the Ionosphere on Space Systems and Communications, IES'75*, J.M.Goodman (Editor), USGPO, available through NTIS, Springfield, VA., pp.336-344.

From, W.R., and J.D. Whitehead, 1986, "Es Structure using an HF Radar", *Radio Science*, 21(3):309-312.

Giraud, A. and M. Petit, 1978, *Ionospheric Techniques and Phenomena*, D. Reidel Publishing Co., Dordrecht (Holland), Boston, and London.

Goodman, J.M., 1982,"A Survey of Ionospheric Models: A Preliminary Report on the Development of an Ionospheric Model Thesaurus and User's Guide", NRL Rpt. 4830, Naval Research Lab, Washington, D.C.

Gossard, E.E. and W.H. Hooke, 1975, *Waves in the Atmosphere*, Elsevier Scientific Pub. Co., Amsterdam, Oxford, and New York.

Greenhow, J.S., and E.L. Neufeld, 1959,*J. Geophys.Res.*, 64(12):2129.

Gussenhoven, M.S., D.A. Hardy, and N. Heinemann, 1983, "Systematics of the Equatorward Diffuse Auroral Boundary", *J.Geophys.Res.*, 88:5692.

Gustafsson, G., 1970, "A Revised Corrected Geomagnetic Coordinate System", *Arkiv fö r Geofysik*, 5:595-617.

Harang, L.M., 1946, "The Mean Field Disturbance of Polar Geomagnetic Storms", *Terrest. Magnet. Atmosph. Elec.*, 51:353.

Hardy, D.A. and R. MacKeen, 1980, "An Algorithm for Determining the Boundary of Auroral Precipitation using Data from the SSJ/3 Sensor", AFGL-TR-80-0028, ADA084482, DTIC, Cameron Station, Alexandria, VA.

Hargreaves, J.K., 1979, *The Upper Atmosphere and Solar-Terrestrial Relations*, Van Nostrand Reinhold, New York.

Heaviside, O., 1902, "The Theory of Electric Telegraphy", *Encyclopedia Britannica*, 10th edition,

Hines, C.O., 1963, "The Upper Atmosphere in Motion", *QuarterJ.Meteorolog. Soc.*, 89:1-42.

Hines, C.O., 1964, "Minimum Vertical Scales in the Wind Structure above 100 Kilimeters", *J.Geophys.Res.*, 69:2847-2848.

Hunsucker, R.D. and H.F. Bates, 1969, "A Survey of Polar and Auroral Region Effects on HF Propagation", *Radio Sci.*, 4:347-365.

Hunsucker, R.D., 1983, "Anomalous Propagation Behavior of Radio Signals at High Latitudes", in *Propagation Aspects of Frequency Sharing, Interference and System Diversity*, H. Soicher (editor), AGARD-CP-332, available from NTIS, Springfield, VA.

Hunsucker, R.D. and R.A. Greenwald (Editors), 1983, *Radio Science*, Special Issue on "Radio Probing of the High Latitude Ionosphere and Atmosphere: New Techniques and New Results", Vol.18, No.6.

Hunsucker, R., 1990, "Atmospheric Gravity Waves and Traveling Ionospheric Disturbances: Thirty Years of Research", in *Effect of the Ionosphere on Radiowave Signals and System Performance, IES'90*, J.M.Goodman (Editor-in-Chief), USGPO, available through NTIS, Springfield, VA.

Jones, W.B., R.P.Graham, and M. Leftin, 1969, "Advances in Ionospheric Mapping by Numerical Methods", ESSA Tech. Rpt. No. ERL-107-ITS-75, USGPO, Washington, D.C.

Jones, W.B. and D.L. Obitts, 1970, "Global Representation of Annual and Solar Cycle Variation of foF2 Monthly Median 1954-1958", OT/ITS Res. Rpt. 3, COM 75-11/43/AS, available through NTIS, Springfield, VA.

Jones, W.B. and R.M. Gallet, 1960, "Ionospheric Mapping by Numerical Methods", *Telecomm. J.*, 27(12):280-282, ITU, Geneva.

Jones, W.B. and R.M. Gallet, 1962a, "Representation of Diurnal and Geographic Variations of Ionospheric Data by Numerical Methods", *Telecomm. J.*, 29(5):129-147, ITU, Geneva.

Jones, W.B. and R.M. Gallet, 1962b, "Methods for Applying Numerical Maps of Ionospheric Coefficients", *J. Research of NBS*, 66D(6):649-662.

Jones, W.B. and R.M. Gallet, 1965, "Representation of Diurnal and Geographical Variations of Ionospheric Data by Numerical Methods", *Telecomm. J.*, 32(1):18-28, ITU, Geneva.

Jursa, A.S. (Scientific Editor), 1985, *Handbook of Geophysics and the Space Environment*, Air Force Geophysics Laboratory, Air Force Systems Command, U.S. Air Force, National Technical Information Service (NTIS), Springfield, VA.

Kamide, Y. and J.D. Winningham, 1977, "A Statistical Study of the Instantaneous Nightside Auroral Oval: The Equatorward Boundary of Electron Precipitation as Observed by thw ISIS 1 and 2 Satellites", *J.Geophys.Res.*, 82:5573-5583.

Kelley, Michael C., 1989, *The Earth's Ionosphere, Plasma Physics and Electrodynamics*, Academic Press Inc., San Diego CA.

Keskinen, M.J. and J. A Fedder, 1988, "Approximate Analytical Formulae for Electron Density and Collision Frequency in the Natural and Nuclear-Disturbed Ionosphere and Inner Magnetosphere", NRL Memorandum Rpt. 6134, Naval Research Lab., Washington, D.C.

Kohnlein, W., 1978, "Electron Density Models of the Ionosphere", *Reviews of Geophysics*, 16:341-354.

Koons, H.C., and D.J. Gorney, 1990, "A Sunspot Maximum Prediction Using a Neural Network", *EOS*, Trans. AGU, Vol. 71, No.18.

Lee, S.C., 1985, "The Penn State Mark III Ionospheric Model: An IBM XT Computer Code", Rpt. No. PSU CSSL SCI 482, Penn State University, PA.

Leftin, M., 1976, "Numerical Representation of Monthly Median Critical Frequencies of the Regular E Region (foE)", Office of Telecomm. Rpt. 76-88, NTIS No.PB-255-484, Springfield, VA.

Lied, F. (editor), 1967, *High Frequency Radio Communications with Emphasis on Polar Problems*, AGARDograph 104, NATO, published by Technivision, Maidenhead, England.

Maeda, K. and S. Kato, 1966, "Electrodynamics of the Ionosphere", *Space Sci. Rev.*, 5:57-79.

Matsushita, S. and E.K. Smith (Editors), 1975, *Radio Science*, Special Issue, Vol.10, No. 3.

McNamara, L., 1991 (in press), *User's Guide to the Ionosphere*, Orbit Book Company Inc., Malabar, FL.

Miller, D.C. and J. Gibbs, 1978, "Ionospheric Modeling and Propagation Analysis", Rome Air Development Center, Report No.RADC-TR-78-163, ADA062998, Hanscom Field, MA.

Millman, P.M., 1959, *J. Geophys. Res.*, 64(12):2122.

Millman G.H., 1978, "Ionospheric Propagation Effects on HF Backscatter Radar Measurements", in *Effect of the Ionosphere on Space and Terrestrial Systems, IES'78*, J.M. Goodman (Editor), USGPO, Washington, D.C., pp.211-218.

Mitra, A.P., 1974, *Ionospheric Effects of Solar Flares*, D.Reidel Pub.Co., Dordrecht (Holland) and Boston.

Narcisi, R.S., and A.D. Bailey, 1965, "Mass Spectrometric Measurements of Positive Ions at Altitudes from 64 to 112 km", in *Space Research V*, edited by D.G. King-Hele, P. Muller, and G. Righini, North Holland Publishing Company, Amsterdam, pp. 753-754.

Nahin, Paul J., 1990, *Oliver Heaviside: Sage in Solitude*, IEEE Press, New York.

Newell, H. (editor), 1966, "Ionospheres and Radio Physics", NASA Report SP-95, Scientific and Technical Information Division, NASA, USGPO, Washington, D.C.

Nisbet, J.S., 1971, "On the Construction and Use of a Simple Ionospheric Model", *Radio Science*, 6:437.

Nisbet J.S. and R. Divany, "Penn State Mk III Model", abstract in *Solar-Terrestrial Models and Applications Software*, NSSDC/WDC-A-R&S 90-19, by D. Bilitza, NASA, Greenbelt, MD.

NRC, 1977, *The Upper Atmosphere and Magnetosphere*, National Research Council, National Academy of Sciences, Washington, DC.

NRC, 1981, *Solar-Terrestrial Research for the 1980's*, National Research Council, National Academy Press, Washington, DC.

Ogawa. T. and T. Shimazaki, 1975, *J. Geophys Res.*, 80:3945.

Oran, E.S. and T.R. Young, 1977, "Numerical Modeling of Ionospheric Chemistry and Transport Properties", *J. Phys. Chem.*, 81:2463.

Paul, A.K., 1986, "Limitations and Possible Improvements of Ionospheric Models for Radio Propagation: Effects of Sporadic E Layers", *Radio Science*, 21,(3):304-308.

Ratcliffe, J.A., 1972, *An Introduction to the Ionosphere and the Magnetosphere*, Cambridge University Press, London and New York.

Ratcliffe, J.A. (Editor), 1974, "Fifty Years of the Ionosphere", *J.Atmospheric Terrest.Phys.*, Vol.36, Special Issue.

Ratcliffe, J.A. and K. Weekes, 1960, "The Ionosphere", Chapter 9 in *Physics of the Earth's Upper Atmosphere*, edited by J.A. Ratcliffe, Academic Press, New York.

Rawer, K., 1981, "International Reference Ionosphere - IRI 79", World Data Center A Report UAG-82, J.V. Lincoln and R. Conkright (editors), NOAA, Boulder, CO.

Rawer, K., 1984, "Analytical Description of Profiles Through Planetary Atmospheres", *Acta Astronaut.*, 11, pp.607-608.

Rawer K., 1963, paper appearing in *Meteorological and Astronomical Influences on Radio Wave Propagation*, B. Landmark (editor), Academic Press, New York, p.221.

Rishbeth, H. and O.K. Garriott, 1969, *Introduction to Ionospheric Physics*, Academic Press, New York and London.

Roble, R.G., T.L. Killeen, N.W. Spencer, R.A. Heelis, P.H. Reiff, and J.D. Winningham, 1988, "Thermospheric Dynamics During November 21-22, 1981: Dynamics Explorer Measurements and Thermospheric General Circulation Model Predictions", *J. Geophys. Res.*, 93:209-225.

Rosich, R.K. and W.B. Jones, 1973, "The Numerical Representation of the Critical Frequency of the F1 Region of the Ionosphere", Office of Telecomm. Rpt. 73-22, NTIS No. COM-75-10813, available through NTIS, Springfield, VA.

Rush, C.M., M. Pokempner, D.N. Anderson, F.G. Stewart, and J. Perry, 1983, "Improving Ionospheric Maps Using Theoretically Derived Values of *foF2*", *Radio Science*, 18(1):95-107.

Rush, C.M, M. Pokempner, D.N. Anderson, J. Perry, F.G. Stewart, and R. Reasoner, 1984, "Maps of foF2 Derived from Observations and Theoretical Data", *Radio Science*, 19:1083-1097.

Rush, C.M., 1986, "Ionospheric Radio Propagation Models and Predictions-A Mini-Review", *IEEE Transactions on Antennas and Propagation*, Vol.AP-34, pp.1163-1169.

Rush, C., M. Fox, D. Bilitza, K. Davies, L. McNamara, F. Stewart, and M. Pokempner, 1989, "Ionospheric Mapping: An Update of foF2 Coefficients", *Telecommunication Journal*, March, ITU, Geneva.

Schleher, J.S., 1990, "The Impact of the Solar Cycle on Over-the-Horizon Radar Systems", in *Effect of the Ionosphere on Radiowave Signals and System Performance*, IES'90, J.M. Goodman (Editor-in-Chief), USGPO, Washington, DC. (available through NTIS, Springfield VA.)

Schmerling, E. (Editor), 1981, *The Physical Basis of the Ionosphere in the Solar-Terrestrial System*, NATO-AGARD-CP-295, Tech.Edit. & Reprod. Ltd., London.

Schunk, R.W., J.J. Sojka, and M.D. Bowline, 1986, "Theoretical Study of the Electron Temperature in the High Latitude Ionosphere for Solar Maximum and Winter Conditions", *J. Geophys. Res.*, 91:12041-12054.

Schunk, R.W., 1988, "A Mathematical Model of the Middle and High Latitude Ionosphere", *Pageoph*, 127:255-303.

Schunk, R.W. and E. P. Szuszczewicz, 1988, "First Principle and Empirical Modeling of the Global Scale Ionosphere", *Annales Geophysicae*, 6:19-30.

Secan, J. and T.F.Tascione, 1984, "The 4-D Ionospheric Objective Analysis Model" in *Effect of the Ionosphere on C3I Systems, IES'84*, J. M. Goodman (editor-in-chief), USGPO, available through NTIS, Springfield ,VA, pp.336-345.

Secan, J.A., 1985, "Development of Techniques for the Use of DMSP SSIE Data in the AWS 4D Ionosphere Model", Final Report, AFGL-TR-85-0107(I), Air Force Geophysics Lab, Hanscom Field, MA.

Secan, J.A., 1989, "A Survey of Computer-Based Empirical Models of Ionospheric Electron Density", Northwest Research Associates Report NWRA-CR-89-R038, Bellevue, WA.

Seddon, J.C., 1962, "Sporadic E as Observed with Rockets", in *Ionospheric Sporadic E*, pp.78-88, Macmillan Company (Pergamon Press), New York.

Smith, E.K. 1957, "Worldwide Occurrence of Sporadic E", NBS Circular 582, Dept. of Commerce.

Smith, E.K., 1976, "World Maps of Sporadic E (foEs > 7 MHz) for Use in Prediction of VHF Oblique-Incidence Propagation", OT Pub. 76-10, NTIS, Springfield VA 22161.

Smith, E.K., 1978, "Temperate Zone Sporadic E Maps (foEs > 7 MHz)", *Radio Sci.*, 13(3):551-575.

Smith, E.K. and S. Matsushita (Editors), 1962, *Ionospheric Sporadic E*, Pergamon Press, MacMillan Company, New York.

Soicher, H. (Editor), 1985, *Propagation Effects on Military Systems in the High Latitude Region*, NATO-AGARD-CP-382, Specialised Print. Services Ltd., UK.

Stewart, F.G., and S.M. Ostrow, 1970, "Improved Version of the McNish-Lincoln for Prediction of Solar Activity", *Telecomm. J.*, 39:159-169.

Starkov, G.V., 1969, "Analytical Representation of the Equatorial Boundary of the Oval Auroral Zone", *Geomag. & Aeron.*, 9:614.

Stroble, D.F., and M.B. McElroy, 1970, "The F2 Layer at Middle Latitudes", *Planet. Space. Sci.*, 18:1181.

Stubbe, P., 1970, "Simultaneous Solution of the Time-Dependent Coupled Continuity Equations, and the Equations of Motion for a System Consisting of a Neutral Gas, an Electron Gas, and a Four-Component Ion Gas", *J.Atmospheric Terrest. Phys.*, 32:865.

Suchy, K. and K. Rawer, 1987, "Improvements in Empirical Modeling of the Worldwide Ionosphere", AFGL-TR-87-0109, AFGL (now Phillips Laboratory), Hanscom AFB, MA.

Szuszczewicz, E., 1988, "OTH-B-PROP: A Computer Model for the Test and Evaluation of OTH-B Propagation Characteristics in the Global Scale Ionosphere", unpublished briefing material, developed for the Air Force Test and Evaluation Center (9 August 1988), SAIC, McLean, VA.

Tascione, T., 1988, *Introduction to the Space Environment*, Orbit Book Company, Malabar, FL.

Tascione, T.F., H.W. Kroehl, B.A. Hausman, and R.C. Cregier, 1986, "A Technical Description of the Ionospheric Conductivity and Electron Density Profile Model (ICED, Version 1986-II)", Hqrtrs Air Weather Service, Scott AFB, IL, (Unpublished)

Tascione, T.F., H.W. Kroehl, R. Creiger, J.W. Freeman, R.A. Wolf, R.W. Spiro, R.V. Hilmer, J.W. Shade, and B.A. Hausman, 1988, "New Ionospheric and Magnetospheric Specification Models", *Radio Science*, 23:211-222. (also in proceedings of IES'87).

Thomason, J., G.Skaggs, and J.Lloyd, 1979, "A Global Ionospheric Model", Rpt. No. 8321, Naval Research Laboratory, Washington, D.C.

Trizna, D. and J. Headrick, 1981, "Ionospheric Effects on HF Over-the-Horizon Radar, in *Effect of the Ionosphere on Radiowave Systems, IES'81*, J.M. Goodman (Editor-in-chief), USGPO, Washington, D.C., pp.262-272.

Vondrak, R.R., G. Smith, V.E. Hatfield, R.T. Tsunoda, V.R. Frank, and P.D. Perreault, 1978, "Chatanika Model of the High Latitude Ionosphere for Application to HF Propagation Prediction", Rpt. No.RADC-TR-78-7, Rome Air Development Center, Hanscom Field, MA.

Whalen, J.A., R.R. O'Neil, and R.H. Picard, 1985, "The Aurora", Chapter 12 in *Handbook of Geophysics and the Space Environment*, AFGL, NTIS, Springfield, VA.

Whitehead, J.D., 1961, "The Formation of Sporadic E in Temperate Zones", *J. Atmospheric. Terrest. Phys.*, 20:49.

Whitehead, J.D., 1967, "Survey of Sporadic E Processes", in *Space Research VII*, edited by R.L. Smith-Rose, North-Holland Publishing Company, Amsterdam, pp.89-99.

Whitehead, J.D., 1970, "Production and Prediction of Sporadic E", *Reviews Geophys.Space Phys.*, 8:65-144.

Withbroe, G.L., 1989, "Solar Activity Cycle: History and Predictions", *J. Spacecraft and Rockets*, 26:394.

3.11 BIBLIOGRAPHY

Banks, P.M. and G. Kockarts, 1973, *Aeronomy*, in two parts (A&B), Academic Press, New York and London.

Checcacci, P.F. (Editor), 1975, *Radio Science*, Vol.10, No.7, Special Issue.

Donnelly, R.F. (Editor), 1980, *Solar-Terrestrial Predictions Proceedings*, in four Volumes: Vol.1- "Prediction Group Reports", Vol.2- "Working Group Reports and Reviews", Vol.3- "Solar Activity Predictions", Vol.4- "Prediction of Terrestrial Effects of Solar Activity", Dept. of Commerce, US Government Printing Office, Washington, D.C.

Fleagle, R.G. and J.A. Businger, 1963, *An Introduction to Atmospheric Physics*, Academic Press, New York and London.

Goodman, J.M. (editor), 1975 *Effect of the Ionosphere on Space Systems and Communications, IES'75*, USGPO, Stock Number 008-051-00064-0, NTIS, Springfield, VA.

Goodman, J.M. (editor), 1978, *Effect of the Ionosphere on Space and Terrestrial Systems, IES'78*, USGPO, Stock Number 008-051-00069-1, NTIS, Springfield, VA.

Goodman, J.M. (editor-in-chief), 1982, *Effect of the Ionosphere on Radiowave Systems, IES'81*, USGPO, 1982-0-367-001:QL-3, NTIS, Springfield, VA.

Goodman, J.M. (editor-in-chief), 1985, *Effect of the Ionosphere on C3I Systems, IES'84*, USGPO, 1985-0-480-249:QL-3, NTIS, Springfield, VA.

Goodman, J.M. (editor-in-chief), 1988, *Effect of the Ionosphere on Communication, Navigation and Surveillance Systems, IES'87*, USGPO: 1988-195-225, NTIS, Springfield, VA.

Goodman, J.M. (editor-in-chief), 1990, *Effect of the Ionosphere on Radiowave Signals and System Performance, IES'90*, USGPO, NTIS, Springfield, VA.

Hines, C.O., I. Paghis, T.R. Hartz, and J.A. Fejer (Editors), 1965, *Physics of the Earth's Upper Atmosphere*, Prentice-Hall, Englewood Cliffs, NJ.

Mitra, A.P., 1974, *Ionospheric Effects of Solar Flares*, D. Reidel Publishing Co., Dordrecht (Holland) and Boston.

Mitra, S.K., 1947, *The Upper Atmosphere*, The Royal Asiatic Society of Bengal, Calcutta, India.

Nicolet, M., 1964, *Contributions to the Study of the Structure of the Ionosphere*, (Thesis, University of Brussels, 1945), NASA Technical Translation F-253, NASA, Washington, DC.

Ratcliffe, J.A., 1960, *Physics of the Upper Atmosphere*, Academic Press, New York and London.

Smith, E.K. and S. Matsushita (editors), 1962, *Ionospheric Sporadic E*, Pergamon Press, the Macmillan Company, New York.

Svestka Z., 1976, *Solar Flares*, D. Reidel Publishing Co., Dordrecht (Holland) and Boston.

Thrane, E. (Editor), 1964, *Electron Density Distribution in the Ionosphere and Exosphere*, NATO-AGARD, North-Holland Publishing Co., Amsterdam and John Wiley & Sons, New York.

Verniani, F. 1974, *Structure and Dynamics of the Upper Atmosphere*, Elsevier Scientific Publishing Co., Amsterdam, Oxford and New York.

Volland, H., 1984, *Atmospheric Electrodynamics*, Springer-Verlag, Berlin, New York, Heidelberg, and Tokyo.

Whitten, R.C. and I.G. Poppoff, 1965, *Physics of the Lower Ionosphere*, Prentice Hall Inc., Englewood Cliffs, NJ.

Yeh, K.C. and C.H. Liu, 1972, *Theory of Ionospheric Waves*, Academic Press, New York and London.

4

HF PROPAGATION AND CHANNEL CHARACTERIZATION

$$\text{Curl } \mathbf{E} = -(1/c)\, d\mathbf{B}/dt$$

$$\text{Curl } \mathbf{B} = (\mu/c)(4\pi \mathbf{I} + d\mathbf{D}/dt)$$

$$\text{Div } \mathbf{D} = 4\pi\rho$$

$$\text{Div } \mathbf{B} = 0$$

Maxwell's Equations, James Clerk Maxwell[26]

4.1 INTRODUCTION

We begin with a discussion of the theoretical basis for our understanding of HF radiowave propagation in the ionosphere. We will be brief in this treatment and turn rapidly to the more practical issues which have confronted radio amateurs and system engineers since the advent of HF radio. We will discuss the nature of the HF channel itself. The ionospheric subchannel varies widely in a manner which is roughly commensurate with the picture provided by mean climatological models as described in Chapter 3. However, specific phenomenological details and propagation effects cannot be adequately represented by median models, and we shall highlight specific propagation effects by examples. We will gain further insight by showing results obtained through ray trace methods for some special cases of interest. These techniques, now facilitated by the availability of high speed computer assets, are compared with approximate graphical procedures used in earlier years. Finally, it must be recognized that two of the major factors for determining system performance are the interference environment and the ambient noise level, both of which are components of the total HF channel description. Available noise and interference models are mentioned along with an indication of their deficiencies. Specific outputs from these models are given in Chapter 5, which deals with the prediction of system performance. Channel occupancy measurement as a scheme for assisting in real-time frequency management is covered in Chapter 6.

----

26. Vector version of Maxwell's equations. While at Edinburgh Academy in 1865, Maxwell predicted electromagnetic wave propagation at the speed of light. The existence of such waves within the radio regime was established by Heinrich Hertz in the latter part of the 19th century. The era of ionospheric radio propagation was inadvertently ushered in by Marconi in 1901. Refer to Chapter 3 for more discussion of the early history of ionospheric studies.

4.2 HF PROPAGATION: FUNDAMENTAL PROPERTIES

We have indicated that HF is the most precarious of bands for propagation within and through the ionosphere. In this chapter, we outline its complexity, and emphasize all possible ionospheric modes belonging to the skywave class (or method). Surface wave propagation and Line-of-Sight (LOS) methods, as opposed to skywave methods, do not involve the ionosphere directly, and consequently do not suffer from the same vagaries. Conventional treatment usually suffices for LOS paths; whereas surface wave paths, which can be mathematically complex for situations of mixed media and rugged terrain, do not suffer from magnetoionic effects. We shall see that HF coverage arising from (hypothetical) half-space isotropic radiation is typically composed of two distinct zones: the first being a circle of roughly fixed radius with transmitter at the center (i.e., the ground wave component), and the second being a highly variable and distended annulus (i.e., the skywave component). The outer boundary of the annulus may extend around the world (in the upper portion of the HF band) while the inner boundary may be the transmitter itself (in the lower part of the HF band). In the classical description involving a single-frequency of transmission, we have a donut-type coverage for the skywave zone with the inner boundary defined by a skip distance which may be greater than the coverage provided by the ground wave. This leads to a silent zone where no coverage is provided (see Figure 4-1). By changing frequencies, we may reduce or even eliminate the silent zone.

We shall not elaborate on the fundamental radiowave properties. Excellent references are available which cover the material is great detail. However, there are several texts which treat ionospheric radiowave propagation with a special emphasis, and these deserve mention here. In particular we draw attention to the work of Ken Davies who has previously published two books on ionospheric radio propagation [Davies, 1965; 1969], and who has recently written an updated book [Davies, 1990]. An earlier text by John Kelso [1964] is also a good source, and a treatise by Budden [1961] has many advocates.

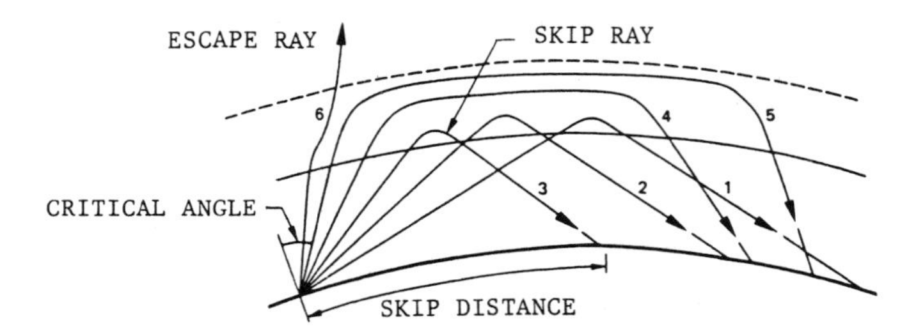

Fig. 4-1. Simplified geometry of HF coverage, illustrating the silent zone at a fixed frequency. This region is located beyond the point where the ground wave signal is no longer discernible but within the so-called skywave skip distance from the transmitter.

For those interested in the esoteric features of the subject of radiowave propagation, *Lectures on Magnetoionic Theory* by Budden [1964] should be examined. The influence of the ionosphere on propagation is also nicely treated by Boithias [1987], and a compact discussion of the subject is found in the popular American Radio Relay League book on antennas [ARRL, 1988].

Below is an abbreviated discussion of radiowave properties and their definitions. The treatment is intentionally superficial at this point, since we intend to limit distractions generated by a premature discussion of side issues. Amplification will be provided in special sections. The reader is referred to an IEEE standards document containing definitions of terms associated with radio wave propagation: Standard 211-1990; [IEEE,1991]. Definitions used in this Chapter are consistent with, if not identical to, this Standard.

4.2.1 Field Strength

The strength of a wave field is diminished as the wave progresses from the source (transmit antenna) into free space because the original energy must be distributed over an increasingly larger surface area. The strength is measured in terms of a voltage between two specified points on the wave surface and it is sometimes called the field intensity. Preference is given to the term field strength since field intensity has an implication of power in some contexts [IEEE, 1991]. The units are volts/meter. Since the field strength is generally small at distances of practical interest for HF communication, the field strength is usually measured in either millivolts/meter (mV/m) or microvolts/meter (μV/m). The specified points are along the electric field vector which, for a locally plane wave, lies on the surface of constant wave phase and is transverse to the direction of ray (phase) propagation. In free space, the field strength drops off as $1/R$. Thus if the field strength is 200 mV/m at 1 km from the transmitter, it will be reduced to 100 mV/m at 2 km, 50 mV/m at 4 km, 25 mV/m at 8 km, etc.

4.2.2 Power Density

The power density, or power flux density, of a radiowave is a measure of the energy the wave, and is an important characteristic for assessing the performance of communication links. In accordance with IEEE-Std.211-1990, the power density is defined as the time-averaged Poynting vector amplitude. Power density is proportional to the square of the field strength and the two parameters are related as follows:

$$P = E^2/Z \tag{4.1}$$

where E is field strength in volts/meter (V/m), P is power density in watts per square meter (W/m^2), and Z is the free space impedance (377 Ohms). The basic concept is shown in Figure 4-2. If the power density at 1 km from the transmitter is 1 mW/m^2, then it is diminished to 0.25 mW/m^2 at a distance of 2 km. At a distance of 1000 km (a factor of 10^3 increase in distance), the power density will drop to a value of (1 mW/m^2 x 10^{-6}. Because of the low power levels involved, it is conventional to use decibel units to specify the power density. If *P*1 and *P*2 represent two power densities, then the power difference expressed in decibels is simply 10 log (*P*1 / *P*2). Noting that power decreases with the square of distance, the power density loss is simply *L*(dB)= 20 log *d*(km). Thus, proceeding from 1 to 1000 km we should observe a 60 dB reduction in power density because of the distance factor alone. Expressed in terms of dB with respect to a watt (not milliwatt) per square meter, the power density at 1000 km is - 90 dBW/m^2.

The term basic transmission loss (in dB) is sometimes used to express the free space loss between two isotropic antennas. It can be used to determine the reduction in power density and may be expressed as a function of frequency f and distance d:

$$L_b\,(f,d) \;=\; 32.45 \;+\; 20 \log f(\text{MHz}) \;+\; 20 \log d(\text{km}) \qquad (4.2)$$

The penalty paid for traversing the first kilometer is 32.45 dB regardless of the frequency. There is an additional free space loss associated with the frequency term in Equation 4.2. At 3 MHz this added loss is 9.54 dB; at 30 MHz it is 29.54 dB.

In addition to the basic transmission loss, there are other loss terms for propagation in media such as the ionosphere, and these are discussed below.

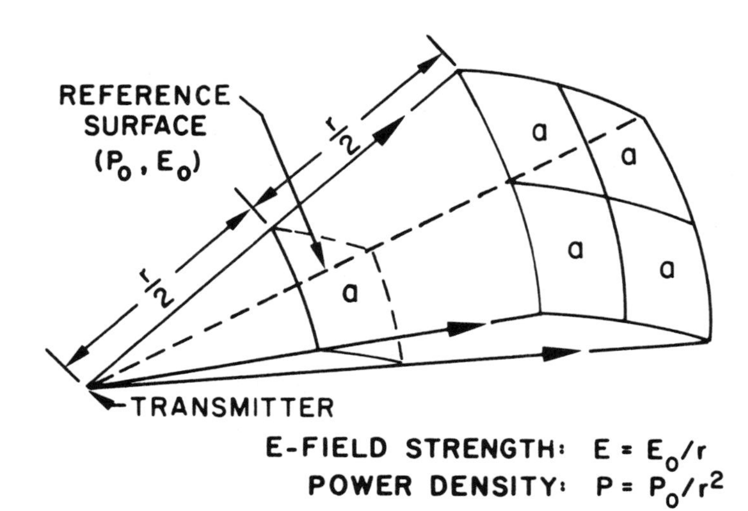

Fig.4-2. Spatial spreading of energy as a function of distance from a transmitter. It is seen that energy is diminished in proportion to the area over which it is delivered. Since the surface area of an expanding sphere is proportional to the square of the sphere radius r, an isotrope located at the center of the sphere will have its power reduced by a factor $1/r^2$ at the surface.

4.2.3 Polarization

This is a property which is generally computed from the orientation of the electric field vector E. In general, it is defined by the locus of the tip of the electric field vector in a plane which is perpendicular to the wave normal. If E is parallel to the surface of the earth, the wave is said to be horizontally polarized. If E is parallel to the earth's radius vector, then the wave is said to be vertically polarized. Horizontally and vertically polarized radiowaves are said to be linearly polarized, and for HF paths sufficiently removed from the earth the generic term is appropriate when used to analyze propagation effects. However, for radiowave transmission close to the earth's surface (viz. surface or ground waves), the distinction between vertical and horizontal may be significant because of an important boundary condition associated with electric fields in the vicinity of conductors. This condition is that an electric field vector in contact with a perfect conductor must be perpendicular to the conductor surface; this implies that horizontally polarized radiowaves cannot propagate along a conductor. Since the earth, although generally a good to fair conductor, is not a perfect conductor, vertically polarized waves are somewhat tilted in the direction of wave propagation. This foot dragging effect is more exaggerated if the ground is poorly conducting and also indicates of a greater amount of penetration into the earth and more dissipative losses. Similarly, we find that horizontally polarized radiowaves will propagate over earth, but not a great distance. Horizontally polarized signals are extinguished rapidly over the ocean because of the excellent conductivity of salt water.

Linearly polarized radiowaves may be decomposed into two waves of equal amplitude having electric field vectors which rotate in opposite directions. These component waves may propagate as independent waves in the ionosphere since in a magnetoplasma they encounter slightly different refractive indices. As a result, path splitting effects are introduced and may be observed as the ordinary (O) and extra-ordinary (X) traces on ionograms. For situations in which the spatial (and time-delay) separations of the O- and X-mode components are unresolved, the two components combine to form a resultant quasilinear mode. The electric field vector orientation of this resultant mode may be quite different from its initial direction. This effect is called Faraday rotation and over time the rotation angle changes introduce a slow regular fading effect on signals which propagate within the ionosphere, provided the receiver antenna is itself linearly polarized. This is because the Faraday rotation which occurs over an ionospheric path is directly proportional to the electron content over the path involved, and the effective electron content is time-varying. A rotating electric field vector is elliptically polarized in general. We shall return to the Faraday effect when we take up fading phenomena.

4.2.4 Wavelength

This is the dimension over which E field properties are recurrent. ($Lf = 300$ where L is wavelength (meters) and f is the radio frequency (MHz).) This property is useful for antenna sizing and for assessing the interaction with media containing irregularities of various scales. The term is interchangeable with frequency according to the formula given above, but the wavelength interpretation is preferred by radio amateurs.

4.2.5 Frequency

The rate of oscillation of E field properties. (Units are in cycles/second or Hertz (Hz).) The frequency sensitivity of media effects, including attenuation, absorption, scattering, and refraction, is substantial and typically plays a significant role in system concept definition. (See the term wavelength above.)

4.2.6 Phase Velocity

This term is associated with the speed (and direction) of the surface of constant wave phase. (The speed is given by $v_{\phi} = c/n$ where c is the free space speed of light and n is the refractive index). In the ionosphere, the (phase) refractive index is less than unity; consequently the phase surface may advance faster than the speed of light c. We shall see shortly that ionospheric critical frequencies exist in the HF band for which $n = 0$. This suggests an infinite phase velocity but in practice corresponds to a simple π phase reversal indicating the reflection condition.

4.2.7 Group Velocity

Speed with which a signal travels in media (given by the expression $v_g \approx n\, c$). Since $n \leq 1$ in the ionosphere, we see that the signal velocity will be less than the speed of light. This retardation effect is directly proportional to the electron content along the ray trajectory. Retardation is greatest near the critical frequency, a fact which is illustrated by ionosonde recordings of virtual *reflection* height versus transmission frequency.

4.3 HF RADIOWAVE PHENOMENA

Upon observation of the specified wave properties, it is found that changes often arise as functions of space and time. These variations may be large or small, fast or slow, and the importance is often dependent upon system factors. In the present context, a primary reason for wave property variation is the dependence of refractive index upon media properties. We emphasize the ionospheric plasma properties at HF because of its obvious relationship with ionospheric skywave propagation. There are, however, other factors which should be mentioned. These include effects which arise as a result of refractive index fluctuations in the nonionized tropospheric and middle atmospheric regions. The reader is referred to a treatment by Ippolito [1986] of tropospheric effects, principally emphasizing satellite communications. Phenomena of interest to HF specialists are identified below.

4.3.1 Attenuation

The attenuation of a radiowave corresponds to additional energy loss it suffers from factors other than the range spreading loss (R^{-2}). This attenuation could be the result of ionospheric absorption or the dissipative effects associated with imperfect ground, foliage, and atmospheric factors. Ionospheric absorption occurs predominantly in the D-region where refractive bending is negligible at HF, and this form of attenuation is termed nondeviative. Section 4.4 deals with the attenuation arising from ground and vegetation, while Section 4.5 covers ionospheric absorption.

4.3.2 Reflection

The process of reflection is the result of a substantial change in media properties occurring over a relatively short distance. This distance is sufficiently small so that wave penetration is effectively prevented, thereby allowing dissipative losses to be ignored, and the ray trajectory obeys the rule: $i = r$ where i and r are the incidence and refraction angles respectively. The term ionospheric reflection is often used in the literature in reference to skywave modes of propagation involving *reflection* from the E, F1, and F2 layers. This is most often used in connection with the so-called *mirror* model, which is a convenient (but inexact) representation of wave interaction process. The actual process is more properly seen to be refraction (see Section 4.3.3).

4.3.3 Refraction

Refraction corresponds to a nonvanishing change in media properties leading to a perturbation in the ray direction. Media penetration occurs and under suitable conditions Snell's Law is taken to be valid: $n_1 \sin \theta_1 = n_2 \sin \theta_2$ where n_1 and n_2 are the indices of refraction for the exited (region 1) and entered (region 2) media, respectively, and θ_1 and θ_2 are defined with respect to a line orthogonal to the (plane) interface between the two homogeneous, isotropic, and semi-infinite regions. Details of wave propagation in homogeneous media is covered in Kelso [1964]. In place of the term *refracted*, the more descriptive word *bent* is often used. A technique known as ray tracing is used in practical assessments of HF coverage provided by the skywave channel. We shall cover ray tracing in Section 4.10.

4.3.4 Diffraction

Diffraction is a wave property which is associated with the reradiation of a wave when it encounters a surface or an obstacle. This theory is employed to determine field properties within or in the neighborhood of shadow zones which are implied by simple ray theory. At HF, the interest in diffraction theory generally arises in connection with applications involving groundwaves, line-of-sight paths, and spacewaves. Specific applications include the impact of irregular terrain on antenna patterns, and of mixed media on groundwave coverage. In a nuclear environment, the diffraction of signals around a fireball is of relevance for determination of connectivity degradation.

4.3.5 Fading

Fading is a process in which the signal level varies in the time domain. Rapid fading is typically termed scintillation, but fading may range from seconds to hours. The longest duration may arise from relatively nonselective processes such as SWF or PCA events. The shorter duration events are generally caused by multipath interference effects. Other possibilities include the Faraday effect and above-the-MOF operation. Multipath and related fading phenomena are covered in Section 4.13. Faraday fading is discussed in Section 4.5.1.

4.3.6 Scattering

Scattering is a process that arises when the refractive properties correspond to a distribution of scales, a portion of which are less than or commensurate with the wavelength of the radiowave. For a plane wave traversing an inhomogeneous medium, the separation between strong and weak scatter is the condition that the rms phase fluctuations be $\geq \pi$ (strong) or $< \pi$ (weak) over a Fresnel zone. This process is also associated with a rapid fading effect called scintillation.

4.3.7 Dispersion

Dispersion is a process in which the frequency, phase, and temporal structure of a signal are modified by the presence of the magnetoionic medium (ionosphere). Since phase and group velocities are frequency-dependent in the ionosphere, the media-generated excess doppler and time delay will vary across the bandwidth. This distortion causes narrow pulses (or wideband signals) to suffer more distortion than cw (or narrowband signals). Distortion of wideband signals is greatest near the Maximum Operating Frequency (MOF) of a link, where the group-path time-delay slope (versus frequency) is steepest. MOF dispersion effects will also exaggerate the time delay jitter on spread spectrum systems which use frequency hopping. This distinction between wideband and narrowband effects implies specification of MOF-seeking architectures for conventional narrowband systems, but non-MOF-seeking architectures for wideband and spread-spectrum signaling. (See Section 4.14 for more information on wideband modeling, and Section 7.6 for discussion of methods for coping with ionospheric effects such as dispersion.

4.3.8 Doppler Shift and Spread

A change in radio frequency is introduced by the presence of free electrons in the ionosphere, if the electrons are in motion. Of importance is the mean motion of the electron population and the distribution of velocities. Reflection from well defined ionospheric layers will introduce a Doppler shift; scattering from an ensemble of irregularities will introduce a spread. In practice, the latter effect is more significant in the design of systems.

4.3.9 Group Path Delay (see Section 4.2.7)

Group-path delay corresponds to an increase in the transit time for a signal traversing the ionospheric medium. The amount of delay is directly proportional to the total electron content along the ray trajectory. Reflection from a distended region will give rise to time-delay-spread of the pulse. This spread effect is a type of multipath.

A classic treatment of the various ionospheric effects upon earth-space propagation was published by Lawrence et al. [1964] and this treatment has been updated by Flock [1987]. Recently Goodman and Aarons [1990] have re-examined the effects from the point of view of modern electronic systems which might have to cope with exaggerated solar activity in the 1989-91 period. Chapter 10 in the Air Force *Handbook of Geophysics and the Space Environment* [Jursa, 1985] is a good source of information as well. One of the most important parameters in the estimation of ionospheric effects for earth-space paths at VHF is the electron content of the ionosphere. Climatological models compute the Total Electron Content (TEC)[27] along the vertical, thus

---

27. The electron content, or EC, is the integral of an electron density distribution, typically along the vertical direction. Several terms are used. The total electron content of the ionosphere, or TEC, is given by $\int N dh$. The slant total electron content, or STEC, is the total content along a straight line from an earth terminal through the ionosphere, with no restriction on the path orientation. The STEC is given by $\int N ds$ where ds is the differential path element. The TEC is the fundamental parameter of interest to ionosphericists. The STEC is the parameter of concern to propagation specialists. The modifying annotation *T* for *total* refers to situations for which there is total ionospheric penetration. If the path does not penetrate, the *T* is dropped. The path EC, or PEC, is a more general expression and is equivalent to STEC for rectilinear earth-space paths at VHF and above. Since HF paths are curved, the PEC integral requires ray tracing to evaluate properly.

requiring the user to determine the appropriate slant value (STEC) given the actual ray zenith angle. Naturally, a reduction in the STEC is to be stipulated if the ray trajectory terminates within the sensible ionosphere. The frequency dependence of various effects is given in Table 4-1. We use the generic term electron content (EC). At HF special care must be taken (see Footnotes 27 and 28.)

One of the nice features of earth-space propagation at VHF and above is that rectilinear propagation of ray sets may be assumed. In short, the system (including its antenna electrical boresight and beam pattern) will determine the geometry of the relevant ray trajectories, and the medium is assumed to be unimportant in this regard. This convenience allows us to easily associate the STEC with the various effects. At HF, the path involved is not rectilinear but is curved and is determined to some extent by the medium. Thus it is not generally a trivial matter to find an appropriate value of the electron content along the path (PEC). Moreover the possibility of path splitting makes an analysis of the effects even more complicated.

TABLE 4-1

EFFECTS RELATED TO ELECTRON CONTENT [28]

EFFECT	UNITS	FORMULAE
Faraday Rotation	Radians	$2.97\,(10^{-2})\,f^{-2}\,H_L$ EC
Group Path Delay	Seconds	$1.34\,(10^{-7})\,f^{-2}$ EC
Phase Advance	Radians	$8.44\,(10^{-7})\,f^{-1}$ EC
Doppler Shift	Hertz	$1.34\,(10^{-7})\,f^{-1}$ d/dt EC
Time Dispersion	Seconds/Hertz	$-2.68\,(10^{-7})\,f^{-3}$ EC
Phase Dispersion	Radians/Hertz	$-8.44\,(10^{-7})\,f^{-2}$ EC

Path splitting effects are particularly troublesome when assessing the impact of Faraday rotation at HF. In general, the amount of Faraday rotation diminishes with increasing frequency (see Table 4-1), provided the PEC is fixed. However we shall find that the PEC increases with f (for HF propagation below the MOF) at a greater pace than the intrinsic decrease in Faraday rotation per unit of EC (essentially f^{-2}). Thus we will look for more Faraday

---

28. It should be noted that HF ray trajectories of interest in terrestrial skywave communication do not penetrate the entire ionosphere. On the other hand, the path integral has both *upleg* and *downleg* components, and paths may exhibit considerable obliquity. It is also worth noting that *high* rays traverse the upper F region over a considerable fraction of their trajectories. The units of EC are electrons/m^2 , and H_L is that component of the geomagnetic field which lies along the geomagnetic field (ampere turns/meter).

fades near the upper part of the HF band than in the lower part. Whether this is good or bad depends upon the system implementation. Ray tracing methods are used to provide the best estimates of these effects at HF.

4.4 NON-IONOSPHERIC PROPAGATION REGIMES

Figure 4-3 depicts propagation approaches employed in HF operations. We are concerned principally with spacewave or groundwave approaches when the distance between terminals is small, although an exception arises for heavily forested and/or rugged terrain. As the figure suggests, the HF band interacts very strongly with the ionospheric plasma, admitting to a full array of physical mechanisms which will support connectivity between two designated points. This array, including the family of single- and multiple-hop modes along with scatter, ducted, and chordal modes, is augmented by other nonionospheric approaches such as groundwave, spacewave or Line-of-Sight (LOS).

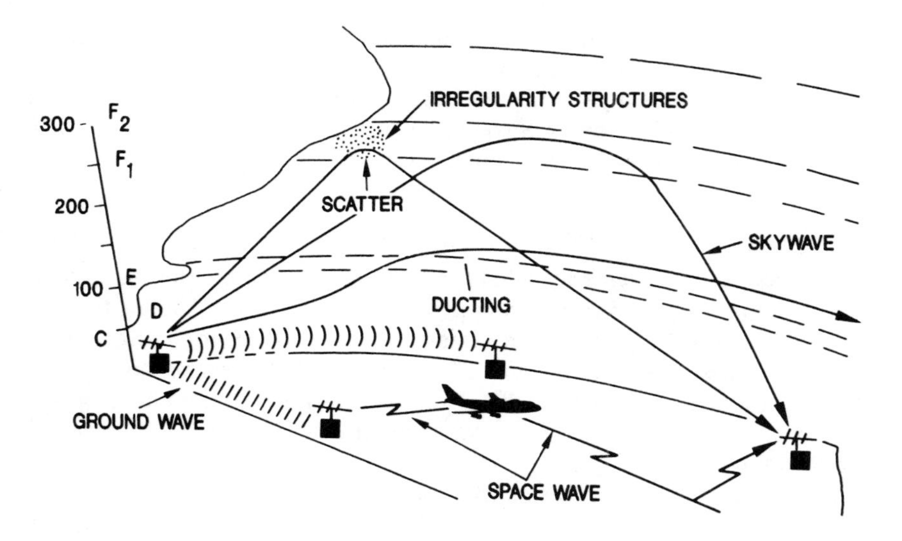

Fig.4-3. Cartoon depicting the various propagation approaches which are exploited at HF. The most prominent approach is associated with the set of skywave modes. These fall into four basic groups: those associated with regular refraction from the ionospheric layers E, F1 and F2; those associated with "reflection" from sporadic E ionization; those associated with ducted and chordal modes of propagation, and those associated with scatter modes which might figure prominently in above-the-MOF connectivity. Other non-skywave possibilities include groundwave, spacewave, and line-of-sight. A combination of these approaches is also possible in some scenarios.

Next we will examine some of the major features of the nonionospheric propagation regimes[29].

4.4.1 Spacewave

The spacewave constitutes the superposition of all signals types which may reach a receiver (or target area) under the condition that the transmitter and receiver are within Line-of-Sight (LOS). Besides the direct or LOS signal this generally implies the inclusion of any earth-reflected signal of significance, but under specific conditions we must consider secondary ionospheric modes as well. Earth-reflected and LOS signals are the types which are usually considered spacewave communications. These types will support a relatively high signal bandwidth, as compared to ionospheric modes, and in many practical instances have comparable path lengths. If the spacewave is an admixture of LOS and ionospheric skywave, the situation should be avoided not only because of the obvious reduction in composite signal bandwidth, but also because of the enormous multipath separation (in time) which may limit data rates.

Figure 4-4a illustrates the limiting condition for LOS propagation between two terminals with antennas located at heights h_a and h_b, and Figure 4-4b shows the conventional geometry for the composite spacewave signal. To see if two terminals satisfy the LOS condition (referring to Figure 4-4a), we may use the expression given in Equation 4.3 twice, once for d_a and once for d_b, and then add the two.

$$d(\text{km}) \quad = \quad 4.124\ \text{h(meters)}^{\frac{1}{2}} \tag{4.3}$$

In equation 4.3, we use the so-called 4/3 earth approximation which allows us to assume rectilinear propagation provided the radius of the earth is increased by a third to account for atmospheric refraction. If an airborne receiver is located at an altitude h of 10 km, then its LOS distance to a ground terminal (at sea level) is 412.4 km. If two aircraft are communicating with each other at this altitude, the maximum separation allowing LOS communication is about 825 km.

One of the potential problems with spacewave communication is the interference between the two modes. In an airborne case for which the platforms are at 10 km altitude but are separated by 500 km in ground range, we see that the difference between the direct and reflected paths is 36 meters. If

---

29. Propagation in nonionized media is treated by the CCIR [1986a] in a separate volume of its *Recommendations and Reports.* Reference to particular CCIR reports or recommendations will be provided at appropriate places in the book. Additional texts of general interest contained within the body of CCIR [1986a] include: *Calculation of free-space attenuation, The concept of transmission loss for radio links,* and *Definitions of terms relating to propagation in nonionized media.*

horizontal polarization is employed, then the reflected signal will undergo π-phase reversal (at least approximately), whereas vertically polarized signals will not. At 2 MHz the radio wavelength is 150 meters, and 36 meters corresponds to a phase difference of roughly $\pi/2$ in the absence of any phase reversals introduced by the reflection process. On the other hand, at 30 MHz (a wavelength of 10 meters), the phase difference amounts to several multiples of π. As the separation between the two aircraft approaches the LOS limit, the path length differences grow smaller and we find a minimal phase difference between the interfering signals at 2 MHz. For these geometries and in the low frequency case, phase interference effects will serve to approximately cancel the two signals if horizontal polarization is employed. Naturally there are other situations for which the signals will add in-phase.

A degree of protection may be achieved against fading skywave signals if airborne platforms are equipped to exploit both direct-skywave and earth-reflected-skywave signals. Diversity gain may be achieved by isolating the two independently-fading signals and combining them after detection. This situation would arise when the airborne platform is Beyond-Line-of-Sight (BLOS) and the spacewave scheme cannot be invoked. The process is most effective if the skywave path is near the horizon and the aircraft altitude is sufficiently high. This increases the possibility that the ionospheric portions of the competing paths are sufficiently different to introduce uncorrelated signal fluctuations. A procedure similar to this has been used to eliminate the deleterious effect of satellite scintillation on earth-space links involving aircraft flying over the ocean.

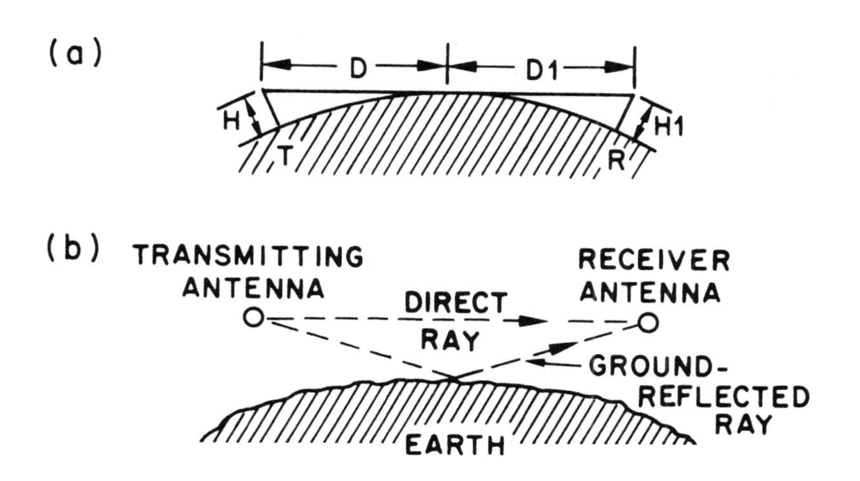

Fig.4-4. (a) Distance of terminals at heights h_a and h_b from the radio horizon. The largest LOS distance from *a* to *b* is the sum of the two horizon distances. (b) Geometry associated with a general spacewave signal and equal to the combination of direct or LOS signal and an earth-reflected signal.

4.4.2 Groundwave

The groundwave propagation method has special relevance in short-haul HF communications. A groundwave exists in the absence of the ionosphere as the result of a source in the vicinity of the earth's surface [IEEE STD, 1990]. It is comprised of a *Norton* surface wave and a space wave. Although the definitions for ground and surface waves have more than subtle differences, they are often used interchangeably, especially for BLOS communication ranges.

Surface waves are exploited for intra-task-force (ITF) communications since terminal separations in a task force are generally less than a few hundred kilometers. Over arbitrary paths, the use of vertical polarization is the only practical approach, but even this polarization suffers high attenuation over poor ground. Oversea paths have diminished losses because of increased surface conductivity. Also, the wave front obtains a forward tilt as it progresses, and its magnitude is established by the surface conductivity (Figure 4-5).

The most obvious property of groundwave propagation is its tendency to be guided by the earth's surface. This tendency is influenced by tropospheric refractivity which effectively reduces earth curvature (increases the earth's radius R_e) by 1/3. The use of the 4/3 earth radius approximation (where we let R_e = 8500 km instead of 6370 km) begins to lose its validity in the lowest part of the HF band, and the standard earth radius is most appropriate in VLF band [CCIR, 1986b]. It has been shown that the effective-earth-radius-method (ERRM) depends upon the radio path altitude, and is only appropriate below about 1 km.

The attenuation experienced by the propagating wave depends upon the ground constants, siting factors, and a number of system specifications such as the polarization, the antenna height, and the radiofrequency employed. Other factors which influence the attenuation include terrain roughness for overland paths and sea state for oversea paths. The effect of obstacles must also be considered. Diffraction effects over both smooth earth and irregular terrain are discussed in CCIR Report 715 [CCIR, 1986c].

Fig.4-5. The surface wave is elliptically polarized because of the complex dielectric constant of the earth. The wave front tends to ***drag its feet*** as the conductivity decreases, an effect which is directly tied to wave dissipation; i.e., the transfer of energy from the wave to the ground.

In an effort to explain the landmark achievement of Marconi, Zenneck [1907] postulated the existence of a surface-type wave which would be constrained to follow the earth's curvature over a considerable distance. The study of ground wave propagation began with Sommerfeld [1909] whose work supported Zenneck's theory, but it languished in a sea of mathematical complexity for several years. Moreover, Sommerfeld's original paper was mistaken about the existence of an *enduring* surface wave [Weyl, 1919] [Burrows, 1936] [Niessen, 1937]. Suitable extensions and modifications by Norton [1936, 1937, 1941] made the theory of ground wave more applicable to the communication problem, and Norton's corrections were supported by others [Burrows, 1937]. An historical background (up to about 1970) may be found in accounts by Wait [1964] and Barrick [1970]. Some of the mathematical and physical aspects are discussed by Wagner [1977] with particular stress on the effective decomposition of the ground wave into surface and space wave components. For issues of ground wave propagation over rugged terrain, a report by Fulks [1981] should be consulted. This report also discusses the effect of atmospheric refraction on ground wave propagation and the influence of obstacles. A commentary on the recent history of ground wave theory and modeling has been published by Roy et al.[1987].

A number of models for ground and surface wave propagation have been developed and the CCIR has published a series of *Groundwave Propagation Curves for Frequencies between 10 kHz and 30 MHz* as Annex I of CCIR Recommendation 368-5 [CCIR 1986d]. These curves have been generated using the program GRWAVE [Rotheram, 1981], which is described in CCIR Report 714-1 [CCIR, 1986b]. Figures 4-6 and 4-7 exhibit the field strength versus distance curves for a short vertical dipole with an isotropic radiated power of 1 kW under two conditions. The first condition corresponds to an oversea path (best case: $\sigma = 5$ siemens/meter, $\epsilon = 70$) and the second corresponds to relatively poor ground ($\sigma = 10^{-4}$ siemens/meter, $\epsilon = 3$). The units for conductivity are sometimes given as mhos/meter, but this is now archaic[30].

One of the more useful concepts for point-to-point communication circuits is that of basic transmission loss L_b. It corresponds to the loss between isotropic antennas expressed in decibels. The transmission loss between half-wave dipoles which exhibit a degree of directivity is 4.3 dB smaller than that between isotropes if the intervening medium is free space. Figure 4-8 gives L_b parametrically in f(MHz) as a function of distance for a smooth oversea path. The attenuation effect is most pronounced at the high part of the HF band, especially for paths over land.

---

30. Resistance may be expressed in ohms and conductance was originally expressed in mhos, a contrivance to emphasize the the inverse relationship between the two. Conductance is now expressed in siemens. The terms used to designate conductivity are siemens/meter (S/m).

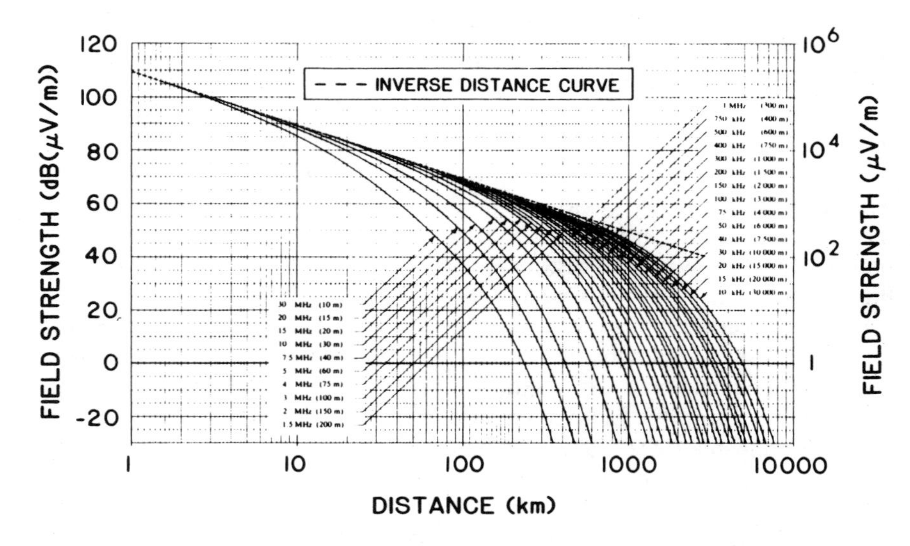

Fig.4-6. Groundwave propagation curves: for sea and average salinity conditions, T = 20 ° C, σ = 5 S/m, ϵ = 70. (From Rec.368-5 [CCIR, 1986d].)

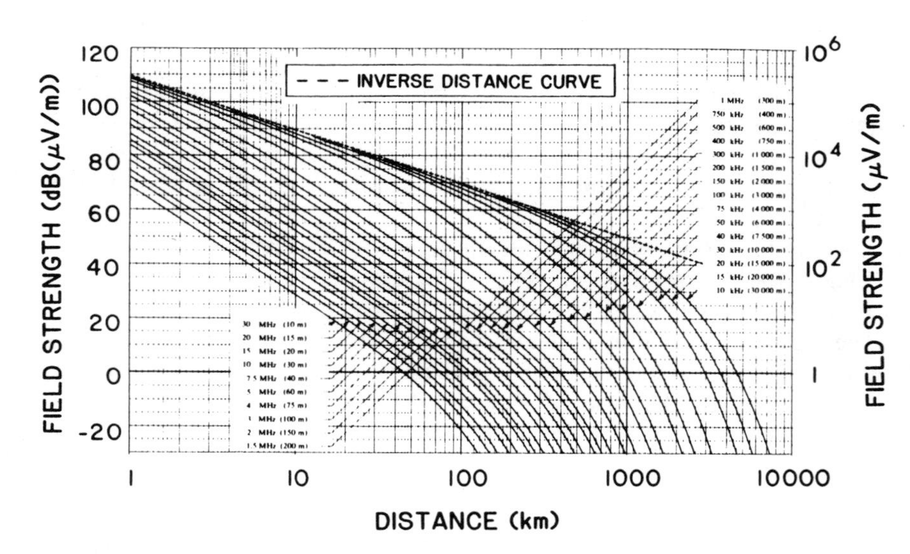

Fig.4-7. Groundwave propagation curves: for very dry conditions, σ = 0.0001 S/m, ϵ = 3. (From Rec.368-5 [CCIR, 1986d].)

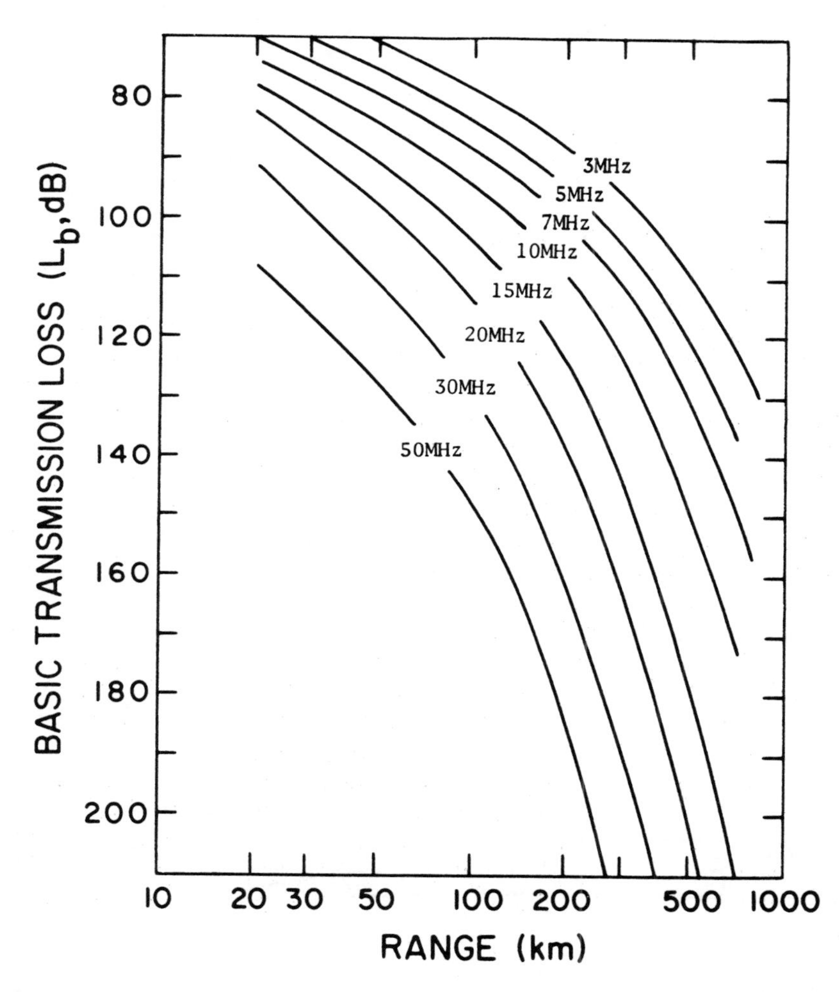

Fig.4-8. Basic transmission loss for wave propagation over a smooth sea, and using the 4/3 earth radius approximation. Conditions: σ = 4 S/m and ϵ = 80, R_e = 4/3. (From Roy et al.[1987].)

Figure 4-9 shows the ground wave loss at 30 MHz for land, marsh and sea [Roy et al., 1987]. Rice [1951] investigated propagation over slightly rough surfaces and this theory was extended by Barrick [1970], who examined the effect of sea state on the basic transmission loss, and found that an additional loss is introduced when the wind speed increases. Figure 4-10 shows a rather flat effect (i.e., fixed added loss) out to roughly 100 km, but the effect changes rapidly beyond that point. A rather curious effect may be observed in the lowest part of the HF band where sea state conditions may actually reduce the amount of roughness loss (see Figure 4-11).

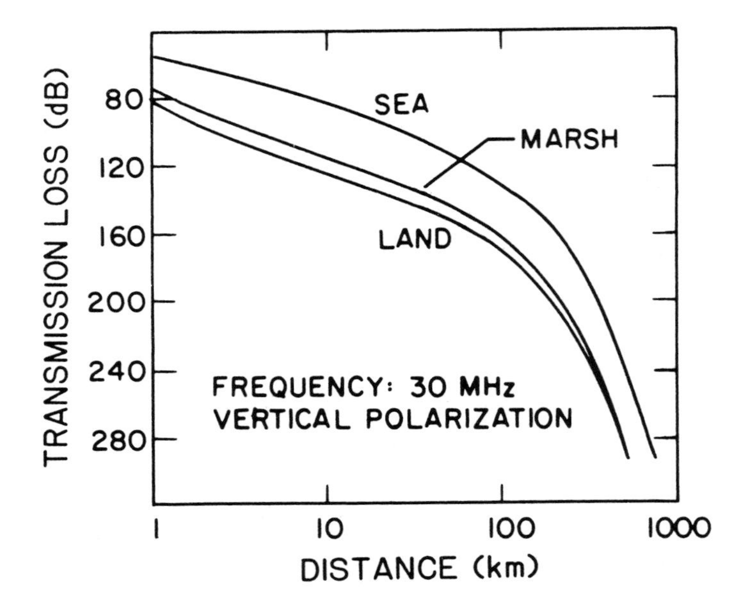

Fig.4-9. Surface wave loss at 30 MHz over sea, marsh, and land. (From Roy et al.[1987].)

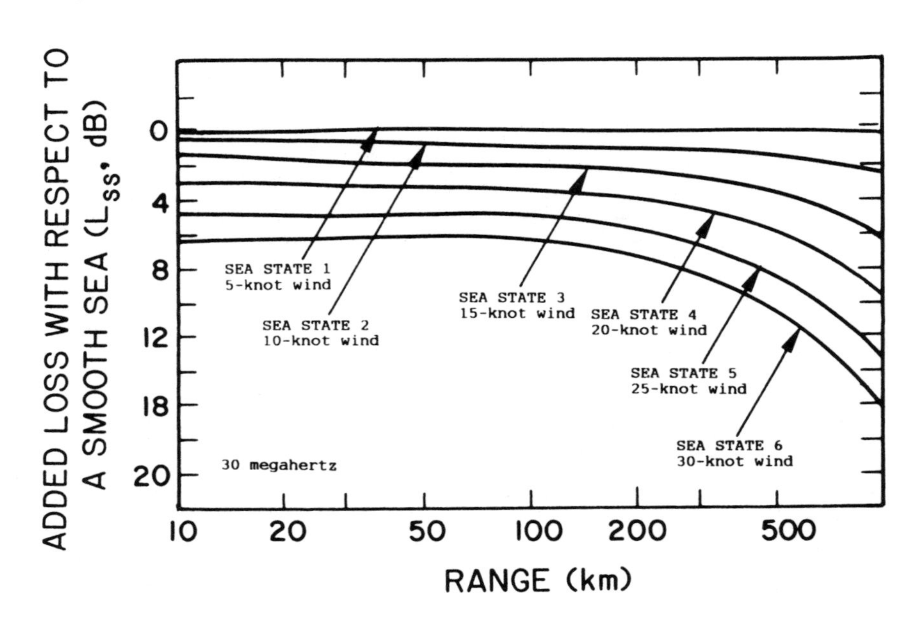

Fig.4-10. Added loss as a function of sea state at 30 MHz. Antennas are near the surface and a Phillip's isotropic ocean-wave spectrum is used. (From Roy et al.[1987], after Barrick [1970].)

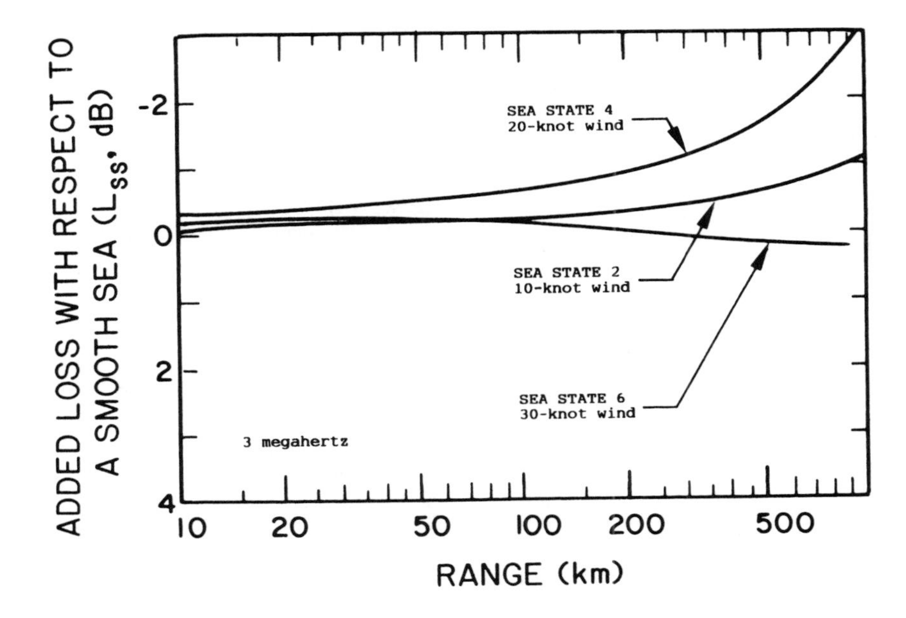

Fig.4-11. Additional transmission loss due to sea state effects at a frequency of 3 MHz. (From Wagner [1977], after Barrick [1971b].)

Mixed path conditions may also exist, especially near littoral areas. General procedures are outlined in CCIR Recommendation 368-5 [CCIR 1986d]. The reader is also referred to the Army *GW Book* which is a series of ground wave propagation charts [ESEIA, 1984]. Table 4-2 is a list of various ground wave propagation models.

Methods for examination of ground wave propagation over inhomogeneous terrain include the Millington method (from Table 4-2) and other methods such as one developed by Suda [1956]. The Millington procedure and a graphical approximation [Stokke, 1975] are outlined in Annex II to CCIR Rec. 368-5. The field strength in the neighborhood of obstacles and surface ridges has also been studied rather extensively. The phenomenon of obstacle gain is a real one and may be examined by consideration of knife-edge diffraction [Fulks,1981]. It may be shown that the diffracted field behind a ridge drops off as $r^{-1/2}$ rather than r^{-2} (which is the generally accepted loss rate for ground waves over flat earth). Even in front of the ridge there is some gain experienced since the field diminishes as r^{-1} in that region.

TABLE 4-2. LISTING OF SURFACE WAVE MODELS

NAME (Type) OF MODEL	REFERENCES
(Argo) Rough Sea-Surface	[Roy et al., 1987]
Booker-Lugananni	[Booker and Lugananni, 1978]
	[Barrick, 1971a, 1971b]
Diffraction Propagation	[Blomquist and Laddell, 1975]
	[Blomquist, 1975]
ECAC	[ECAC, 1975]
Graphical Mixed Path	[CCIR, 1986d] [Stokke, 1975]
GROUNDWAVE (GDWAVE)	[Roy et al., 1987]
GRWAVE	[Rotheram, 1981]
GWAPA	[DeMinco, 1986]
GWSNR	[Berry, 1978]
GWSNR2	[Roy et al., 1987]
Knife-edge Obstacle	[Furutsu and Wilkerson, 1971]
	[Furutsu, 1962]
Levine	[Levine, 1978]
	[Barrick, 1971a, 1971b]
(Levine) Rough Terrain	[Roy et al., 1987]
(Levine) Vegetation	[Roy et al., 1987]
Lustgarten-Madison (EPM-73)	[Lustgarten and Madison, 1977]
Millington Mixed Path	[Millington, 1949]
MRC Ground Wave	[Fulks, 1981]
Multiple Knife Edges	[Deygout, 1966]
Multiple Rounded Obstacles	[Assis, 1971]
NBS Diffraction Method	[NBS, 1967]
NBS Ground Wave Program	[Berry & Chrisman, 1966]
Rounded Obstacle	[Dougherty,1969]
TIREM	[ECAC, 1988] [Morcerf, 1991]

A program called WAGNER, developed by R. Ott at ITS, may be used over rough terrain although it is has not been fully validated at HF. An improved version of the WAGNER model has been developed by R. Bevensee of Lawrence Livermore National Laboratory (LLNL), and a computer program called GWAPA[31] has been developed [DeMinco, 1986]. Another popular program developed under the aegis of ECAC is a ground wave code called TIREM. This computer code has been incorporated within the Network Assessment Model (NAM) developed by the U.S. Army Signal Corps to support the Army IHFR effort. The reader is apprised that neither GWAPA

---

31. Ground Wave Automated Performance Analysis Program. This is a user-friendly program that can predict propagation loss, electric field strength, received power, SNR, and antenna factors over lossy earth. Smooth and irregular earth propagation loss can be used over homogeneous or mixed paths. Antenna performance can also be deduced. (adapted from abstract of NTIA Report 86-209 [DeMinco, 1986].)

nor TIREM[32] are universally available, being controlled by certain military agencies and their designated contractors.

MRC Ground Wave, developed by Mission Research Corporation, is based upon Bremmer's method (also employed by CCIR [1986d]), but it does not include atmospheric refraction. The MRC program is similar to NUCOM, a nuclear effects code. NUCOM has been embodied in yet another HF code, HFNET, developed for the Defense Nuclear Agency (DNA) and used in the analysis of HF networks in a nuclear-stressed environment (see Section 7.8.2).

The Naval Ocean Systems Center has developed a FORTRAN 77 computer program called GDWAVE to execute the GROUNDWAVE model. GROUNDWAVE is a composite of several ground wave submodels, each of which is appropriate for a particular situation. These sub-models include the Booker/Lugananni model[33], the ECAC model[34], the Lustgarten/Madison model[35], the Argo model[36], and several models by Levine[37] (see Table 4-2).

---

32. Terrain Integrated Rough Earth Model. The TIREM model selects an appropriate mode of propagation based upon terrain profile, and computes path loss between two terminals. The propagation regimes include line of sight (LOS), diffraction, and troposcatter. The TIREM program applies between 20-30 MHz. (from "Catalog of Computer Program Abstracts, Vol I; Analysis Capabilities", ECAC Handbook 81-047, Electromagnetic Compatibility Analysis Center, Annapolis, MD.)

33. The Booker and Lugananni model [1978] is based on earlier work of Barrick [1970, 1971a, 1971b], being an empirical fit to Barrick's curves. It is designed to handle the propagation loss over a smooth sea from 2 to 30 MHz.

34. The ECAC model, a term used by Roy et al [1987], is designed to compute transmission loss over a smooth spherical earth of any type (i.e., land or sea). It accounts for losses of vertically or horizontally polarized radio waves between 1 and 30 MHz.

35. This model is commonly referred to as EPM-73. It is similar in many respects to the ECAC model, although it incorporates a troposcatter region in addition to reflection and diffraction regions.

36. An empirical fit to the rough sea model of Norton [1957].

37. Levine [1978] models rough terrain through use of a specified roughness scale and an rms wave height, and rough sea surfaces are modeled using a so-called Phillips spectrum. A simple model for surface vegetation effects is incorporated.

4.4.3 Ground Constants

Ground constants play a substantive role in the groundwave coverage. They are also of interest in the performance analysis of HF antennas which are located on the earth or near its surface. Models such as GROUNDWAVE use frequency-invariant values for both the dielectric constant ϵ and conductivity σ. Hagn et al.[1982] have shown this to be in error; Figures 4-12 and 4-13 show the frequency dependence of ϵ and σ for specified types of terrain. An analysis by Roy et al.[1987] indicates that GROUNDWAVE (with fixed ground constants) may severely underestimate the attenuation in the lower part of the HF band over poor earth (i.e., 20 dB errors may arise at 3 MHz).

Electrical characteristics of the surface of the earth are discussed in CCIR Report 229-5 [1986e]. Unfortunately the CCIR still recommends inaccurate curves for the ground constants at HF in Rec. 527-1 [CCIR, 1986f]. However, the reader is referred to this report and its antecedents for an introductory commentary on the various factors which determine the effective ground constants, and the methodologies for measuring the ground constants. (Use the curves found in Figures 4-12 and 4-13 as a replacement.)

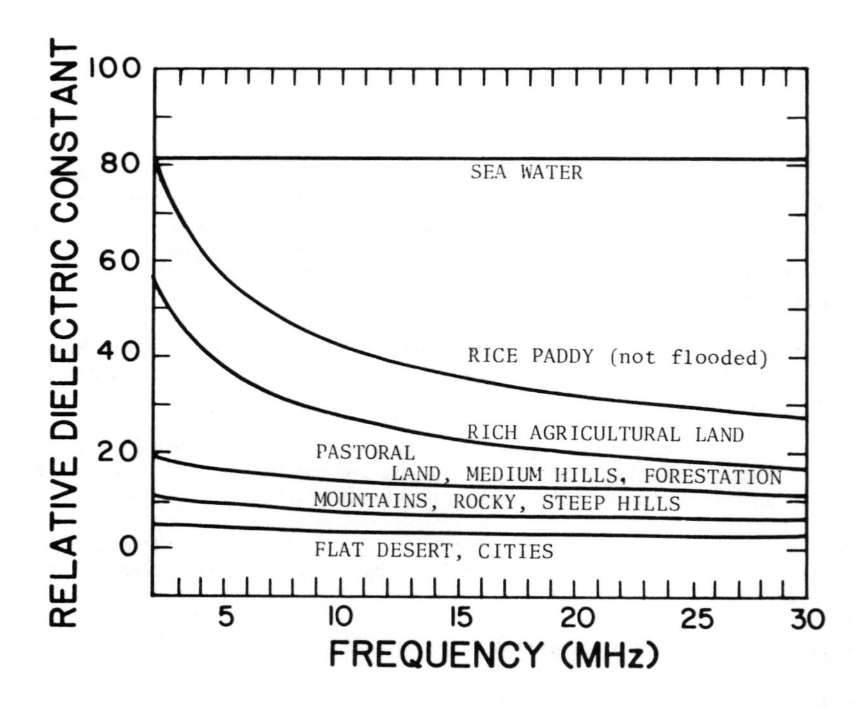

Fig.4-12. Relative dielectric constant for specified terrain types. (From Hagn [1988a], by permission.)

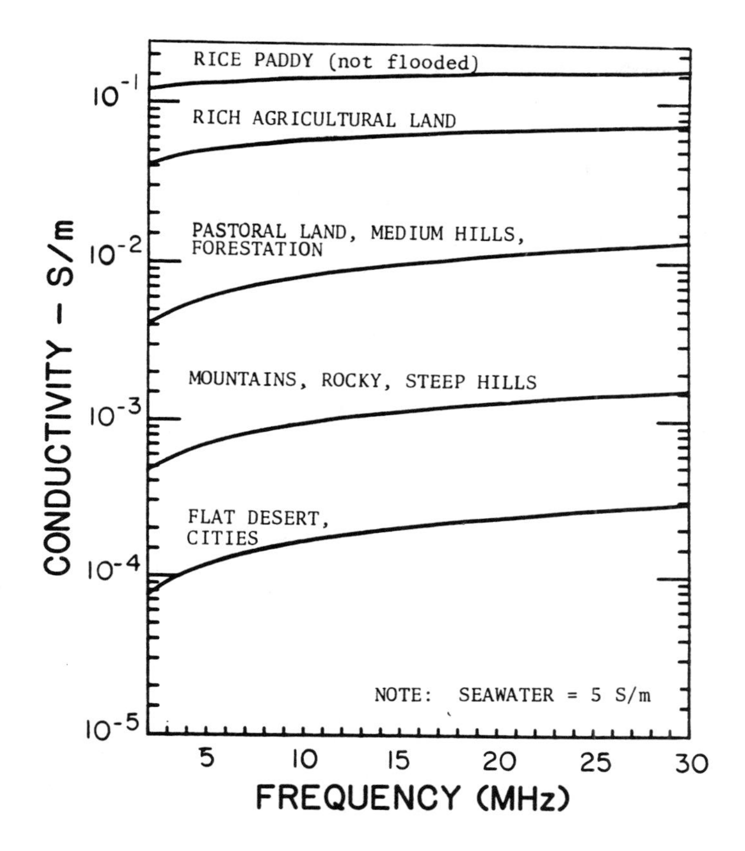

Fig.4-13. Ground conductivity for various terrain types (From Hagn [1988a], by permission.)

The factors which specifically relate to soil properties include: soil type, moisture content, and temperature. As we have already noted, the radio frequency is an important factor as well, and this is principally because of the presence (or absence) of moisture in the soil. Thus the most obvious ground factor is the intrinsic ability of the soil to absorb moisture and retain it over time. Accordingly soil type and drainage conditions must be taken into account, and also precipitation occurrence in an area. Soil such as loam may exhibit a two-orders-of-magnitude variation in conductivity between wet and dry conditions.

Measurements of ground constants are important to provide the best possible data for specified antenna designs. The SRI Open-Wire-Line (OWL) measurement scheme [Hagn and Gaddie, 1984; Hagn, 1990] has proven to be a successful method for application in the HF band. This particular method has been used by Hagn of SRI to develop generic curves of the relative dielectric constant, ϵ_r, and the conductivity, σ, for a variety of soil types.

These generic curves, given in Figures 4-12 and 4-13, have been transformed into analytical form for ease in computation [Sailors,1984]. The equations for both σ and ϵ_r take the form $A = Bf^M$ where f is the frequency in MHz. Corrected tables of B and M are found in a paper by Hagn [1985]. For seawater we have no frequency dependence (i.e., $M = 0$); consequently, the conductivity = 5 siemens/meter (S/m), and the relative dielectric constant = 81. In this instance the ground constants (expressed by A) are equal to the appropriate values of B.

The dissipation factor DF (often called the loss tangent) associated with surface wave propagation effects is proportional to the ratio of the conductivity σ to the permittivity (or dielectric constant) ϵ. Specifically, we have:

$$\mathrm{DF} = (2\pi f)^{-1}\sigma/\epsilon \tag{4.4}$$

where DF » 1 corresponds to a lossy conductor, DF « 1 corresponds to a lossy dielectric, and DF ≈ 1 corresponds to a semiconductor. It is of some interest to note that the transition from a lossy conductor to a lossy dielectric occurs at HF (or somewhat below) for most ground types [Hagn, 1988a].

The skin depth SD (or penetration depth) is an important factor in the spread of HF radiowaves and in the effectiveness of groundwave propagation. If the penetration is significant, as in desert regions, the properties of lower strata will influence the effective ground constants to be used in the calculations. On the other hand, if the skin depth is small compared to the water table then the effect of the subsurface water on radiowave dissipation will be unimportant. Values for skin depth are provided in CCIR Recommendation 527-1 [1986f], and have been recompiled by Hagn [1988a]. Figure 4-14 gives the skin depth for a variety of soil types and conditions. Note that for a frequency of 3 MHz, the SD values may range from 100 meters for a flat desert to about 10 cm for sea water. Also we see that the skin depth decreases with increasing frequency. For sea water at $f = 30$ MHz, the skin depth SD ≈ 4 cm. For rich agricultural land, we encounter intermediate values for skin depth; SD ranges from 1.5 meters at 3 MHz to 0.4 meters at 30 MHz.

Another property of interest is the effective wavelength of the HF radiowave in the ground. This property is needed to assess the effectiveness of buried antenna segments. For flat desert, the wavelength is 60 meters at 2 MHz (compared to 150 meters for free space) and 6 meters at 30 MHz (compared to 10 meters for free space). For seawater the in-media wavelength is 1 meter at 2 MHz and about 25 cm at 30 MHz.

4.4.4 Terrain and Vegetation Effects

A discussion of the impact of terrain features and vegetation on radio propagation is given in CCIR Report 236-6 [1986g]. High attenuation rates are expected in the upper part of the HF band in forested regions, especially if trees are covered with wet snow. Raindrops on leaves will also introduce dissipation. In a forest environment wave propagation between two terminals may be accomplished in several ways. This has been modeled as propagation within a lossy dielectric slab [Wait et al.,1974]. The modeled path may be direct (i.e., through the forest along the shortest path), or appear as a so-called lateral wave (i.e., along the top of the trees), or some hybrid combination. Lateral waves are described fully by Tamir [1977], and they rely upon continuous leakage of wave energy from the treetop boundary to the terminals (and vice-versa). For short distances, the direct path is of most significance especially for the lower part of the HF band. Lateral waves at 30 MHz, however may be quite important in comparison with direct waves for intermediate to longer distances. The SRI OWL kit has been used to measure the effective conductivity of living vegetation and soil in jungle environments [Hagn, 1973] [Doeppner et al.,1974].

4.4.5 On the Use of Ground Constant Data

We have already seen that ground constant data is a major consideration in the determination of surface wave coverage. This may be inferred from the curves in Figures 4-6 and 4-7 corresponding to radiation from a short vertical dipole and an isotropic radiated power of 1 kilowatt. The curves are parameterized in frequency and show the field strength $\mu V/m$ on the right-hand-scale or alternatively dB $\mu V/m$ on the left-hand-scale. Taking 10 MHz as an example, it may be seen that for the oversea case (where $\epsilon = 80$ and $\sigma = 1/4$ S/m) the field strength over a 100 km path is +60 dB; for the overland case (for which $\epsilon = 4$ and $\sigma = 10^{-3}$ S/m is assumed) the field strength is -10 dB. This difference of 70 dB is fully attributable to surface wave dissipation differences. The effects are reduced for frequencies below 10 MHz but are increased for frequencies above 10 MHz, provided ground constants are fixed. The dissipation effect in this example corresponds to very poor ground. Figure 4-9, illustrating the surface wave losses at 30 MHz, shows that good ground will reduce the losses. We see that the difference in transmission loss between oversea and overland cases for a 100 km path may be no more than 40 dB.

We have noted that ground constants are important in deriving the performance of HF antennas regardless of whether or not surface wave propagation is exploited. For example, the antenna models within IONCAP, derived from earlier subroutines due to Lucas and Haydon [1966] (i.e., ITSA-1) and Barghausen et al., 1969] (i.e., ITS-78), require the computation of power gain, radiation resistance, and antenna efficiency. The first two terms are critically dependent upon ground influence. The IONCAP procedure may be derived from the cited references and may be found in an unpublished draft report by Lloyd et al.[1978]. This report is not generally available and must be requested from the authors directly or from the Institute for Telecommunication Sciences, Department of Commerce, Boulder Colorado. IONCAP and other models such as HFMUFES and PROPHET will be discussed in Chapter 5.

All of these models contain antenna pattern information and either directly or indirectly make use of ground constant data. In a PC version of IONCAP developed by F. Rhoads [1990], one may enter either the appropriate ground constants or the ground type. This is illustrated in Table 4-3.

IONCAP and similar propagation programs allow the user to specify standard antenna types. For more arbitrary antenna types, a method-of-moments technique has been developed by Lawrence Livermore National Laboratory (LLNL) under the sponsorship of the U.S. Navy. Over the years all the services have been involved in the development of the basic methods employed. The Numerical Electromagnetic Code (NEC) [Burke and Poggio, 1977] is a computer code designed for analysis of the electromagnetic response of antennas and other metal structures. NEC allows the insertion of ground parameters (i.e., σ and ϵ) for the neighborhood of the antenna, the specification of ground parameters for a second region not in the vicinity of the antenna, or the specification of a ground screen. A mini-computer version of NEC, called MININEC, has been developed [Logan and Rockway, 1986].

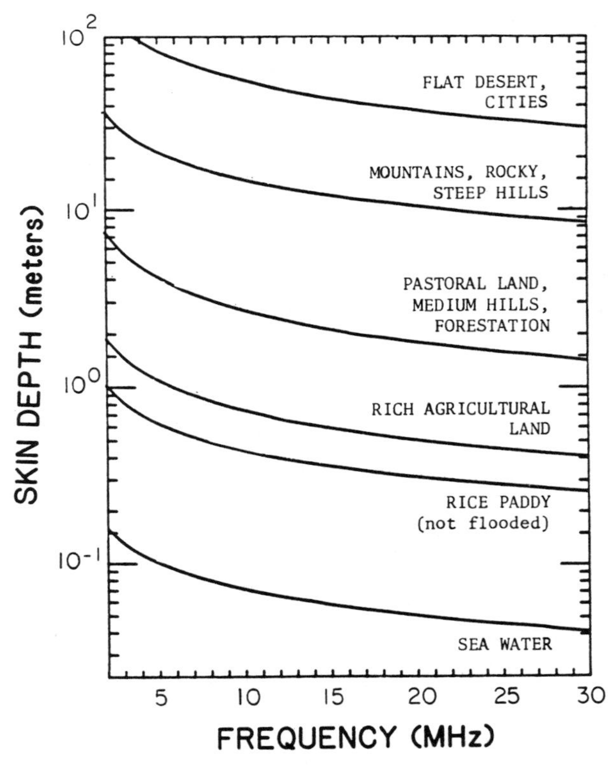

Fig.4-14. Skin depth (meters) versus frequency for selected terrain categories. (from Hagn [1988a], by permission.)

TABLE 4-3. ENTRY OF GROUND CONSTANT DATA IN IONCAST

Case 1: User is prompted to specify transmitter (or receiver) ground conductivity by the Ground type (G), by inserting new Values (V), or <R>eturn for acceptance of defaults. In this example new values: $\sigma = 0.001$, $\epsilon = 4$ are entered.

{Note: The page number (p. 141) corresponds to information to be found in the IONCAP user's manual [Teters et al., 1983]}

```
------------------------------------------------------------
XMTR GROUND CONDUCTIVITY = 0.0100 (MHOS/METER)
     RELATIVE DIELECTRIC =  10.    (REL FREE SPACE)
============================================================
 Do you wish to enter Ground Conductivity
   and Relative Dielectric Constant by:

    V       <V>alue
    G       <G>round type  (Page 141)
    Ret     <R>eturn and accept values above

  Enter letter in left column for your choice?                      v

GROUND CONDUCTIVITY AT XMTR= 0.010   (MHOS/METER  page 141)?  0.001

RELATIVE DIELECTRIC AT XMTR=10.000   (RATIO  page 141)?  4
```

Case 2: Here the user specifies that <G>round type is the method of data entry. After pressing <G>, the user is prompted for the type of ground. The desired type is entered.

```
============================================================
 Do you wish to enter Ground Conductivity
   and Relative Dielectric Constant by:

    V       <V>alue
    G       <G>round type  (Page 141)
    Ret     <R>eturn and accept values above

  Enter letter in left column for your choice?                      g

GROUND TYPE AT XMTR=:
============================================================

              Type Ground Around Node Location
  ---------------------------------------------------------------
   S          <S>eawater
   G          <G>ood ground     (Midwest, Eastern USA, Europe, etc.)
   F          <F>resh water
   P          <P>oor ground or Sea Ice  (Mountains, Desert, Permafrost, etc
   I          <I>ce Cap      (Greenland, Antarctica, Glaciers, etc.)

  Enter letter on left for your choice?                             g

------------------------------------------------------------
```

NEC is the most advanced computer code available for the analysis of thin wire antennas, and is designed for detailed analysis including the possibility of nonradiating networks, transmission lines, perfect and imperfect conductors, lumped element loading, and perfectly or imperfectly conducting ground planes. Antenna excitation options include either an incident plane wave or an applied voltage source. A number of output options exist including: gain, directivity, and power budget [Logan and Rockway, 1986]. NEC requires a main frame computer, and should be dedicated for the larger antenna problems.

MININEC is a simplified version of NEC designed principally for exploitation by a microcomputer. The BASIC language is used, and the program is optimized for small problems; e.g., less than 10 wires in the antenna system. In addition, it is required that the antenna be placed in free space or over ground which is perfectly conducting. Because of the relative simplicity of the program, but principally because the host computer may be a microcomputer, MININEC has attracted a large number of users. Other improvements in MININEC are also being made to ease its use. One such program called MN has been reported [White, 1990]. Development of the basic NEC methodology is being pursued by workers at the Navy Post Graduate School (NPGS), the Naval Ocean Systems Center (NOSC), and Lawrence Livermore National Laboratory (LLNL) with LLNL having the leading developmental role. The current version of the mainframe code is NEC-3 [Burke and Poggio, 1981; Breakall et al., 1985; and Burke, 1983]. The NEC code has been shown to perform best when observed data rather than predicted ground constant values are used. An organization which serves as a forum for international development in this general area as well as for similar topics is the Applied Computational Electromagnetics Society (ACES).

4.4.6 Siting Considerations and Antenna Selection

Antenna siting factors and choice of antenna are major considerations for missions which exploit HF radio principles and engineering concepts. The choice of antenna may depend upon a number of factors such as directivity and gain desired, the platform (or station) properties and constraints, the coverage requirements, low-probability-of-intercept (LPI) and anti-jam (AJ) requirements, and anticipated propagation parameters. The siting factors include not only the ground constants which were discussed above, but also the terrain slope in the region of antenna system placement, and the nature of any obstacles or terrain irregularities in the near field of the antenna. Another siting factor is the nature of atmospheric, galactic, and man-made or vehicular noise. The last factor may be most significant for reception of signals characterized by low signal strength. Aspects of noise will be summarized in Chapter 5 where the CCIR 322 model will be discussed in connection with the prediction of HF system performance.

An understanding of certain HF antenna siting factors generally requires the use of the Fresnel zone concept. The Fresnel zone plays a significant role in the formation of the first major lobe in the radiation pattern, and thus the ultimate downrange coverage. For the application of terrestrial HF communication, the Fresnel zone is a region in front of the antenna which represents a phase departure of at most 180 degrees (for all contributing reflected rays) of an advancing plane wave with respect to a reference ray. It is presumed that the plane wave advances in the direction of a reference ray. It is desirable to have a site which is locally smooth over this Fresnel zone, but this may be impossible, especially for low radiation angles. An approximate relation stipulating maximum allowable vertical departures from the mean terrain profile may be deduced from ray optics. Utlaut [1962] indicates that the size of tolerable terrain irregularities is equal to $\frac{1}{4} h_A$ where h_A is the antenna height.

The Fresnel zone is approximated by an ellipse located in front of an antenna with the semimajor axis along the direction of propagation, The near point, far point, and width of the ellipse all increase with decreasing radiation angle and frequency. For a radiation angle of 10 degrees and a frequency of 15 MHz, the Fresnel zone is characterized by a transverse dimension (width) of about 150 meters, a near edge of about 5 meters, and a far edge of 400 meters.

There are distinct advantages to locating an antenna such that the boresight (or principal direction of propagation) is along a downward slope. Such a configuration effectively increases the antenna height and lowers the minimum radiation (or elevation) angle, a desirable goal for long distance communication. It also has the effect of reducing the Fresnel zone dimension, which softens the ground homogeneity requirement somewhat. Naturally, if one can locate an antenna on high ground such as a high cliff, it would be possible to achieve the very lowest possible elevation angles for signal launch, and thereby obtain the greatest skywave coverage.

The radiation patterns for standard HF antennas may be found in various handbooks such as the *American Radio Relay League Antenna Book* [ARRL, 1977] and the *Field Antenna Handbook* [ECAC. 1984]. Pamphlets such as the *HF Communications Data Book* developed by Rockwell-Collins [1978] are also helpful to the HF novice. For those interested in tactical antennas, a compendium of articles has been published by the Armed Forces Communications Command [Christinsin, 1986]. For those interested in air-to-ground communication antennas, Chapter 7 in a book by Maslin [1987] is suggested.

Antenna pattern measurements are required for nonstandard antennas or for standard antennas which are located over irregular terrain, or are deployed in regions of unspecified and/or variable soil categories. Workers at SRI have developed a full-scale airborne pattern measurement system called XELEDOP [Hagn and Harnish, 1986]. This system allows measurement of the absolute gain and directivity of an antenna.

As a matter of convenience, vertical monopoles are used for many tactical applications. This type of antenna is precisely the wrong choice for high radiation angles. The best choice for a short range skywave communication antenna is a horizontal dipole aligned broadside to the direction of propagation. On the other hand, the monopole is an excellent choice for long-distance (low radiation angle) communication if rather low gain is acceptable. The monopole also provides a uniform azimuthal pattern. A ground screen composed of several radials tends to produce the lowest radiation angle for the monopole (whip antenna).

4.4.7 Earth Reflection

Earth reflection is important for that class of antennas which have low directivity, such as those which are associated with mobile services and tactical military communications. It is also important when examining the impact on coverage by multiple hops. Details are found in CCIR Report 1008 [1986h].

The (specular) reflection coefficient for vertical polarization is less than it is for horizontal polarization. At the so-called Brewster angle, the amplitude (or modulus) of the complex reflection coefficient is a minimum. Most applications of interest correspond to low grazing angles. In such instances, the reflection coefficient approaches a value of -1. This implies that the direct and reflected fields have roughly equal magnitudes and have nearly a 180 degree phase difference, since the path difference is negligible.

At very low grazing angles, diffraction becomes the dominant process for defining wave properties and geometric optics is discarded. The angle in milliradians below which diffraction is dominant is $[2100/f(MHz)]^{1/3}$. This limiting angle is 0.51, 0.34, and 0.24 degrees at 3 MHz, 10 MHz, and 30 MHz respectively.

4.5 SKYWAVES AND THE APPLETON-HARTREE EQUATION

In this section we shall outline the main points associated with propagation in the ionosphere. The basic properties of such propagation result from an application of the magnetoionic theory. We shall not cover the basic theory in any detail nor derive the relevant equations. Many excellent texts cover these subjects, and most have been cited in Section 4.2. We shall, however, identify the important relationships and provide examples which are pertinent at HF.

One of the primary characteristics of long-haul HF communication is its dependence upon the ionospheric subchannel. The skywave method is employed for link connectivity Beyond-Line-of-Sight (BLOS) and this involves reflection from the ionosphere which is composed of one or more layers. Ionospheric properties are addressed in CCIR Report 725-2 [CCIR, 1986g] and in Chapter 3 of this book. An interesting series of articles have been published by McNamara and Harrison [1985-1986] in the *Australian Electron-*

ics Monthly on skywave propagation from a communicator's perspective. Chapter 10 (Section 10.4 was written by McNamara) of the *Air Force Handbook* [Jursa, 1985], is recommended reading.

It is well known that the ionosphere is composed of several distinct layers which are capable of reflecting HF skywave signals thereby providing terrestrial communication by ionospheric bounce. The term skywave is a rather curious term, but it concisely describes the method of propagation by which signals originating from one earth terminal arrive at a second terminal by reflection (or more properly refraction) from the ionosphere. Originally the term skywave was used to "designate the sound of distant cannons that propagated through the upper reaches of the atmosphere and returned to earth by virtue of the refraction of sound waves downward by the atmosphere" [Ames et al., 1977]. Figure 3-3b depicts the E, F1, and F2 layers for a midday midlatitude ionosphere and for two epochs of solar activity. Even though the electron densities for the layers underlying the F2 peak of ionization are relatively small (as shown in the illustration), their impact may at times be dominant. Such is the nature of HF propagation.

The general expression for the index of refraction for a radio wave propagating through a magnetoionic medium is given by the Appleton-Hartree dispersion equation:

$$n^2 = (n_r - i\, n_i)^2 \tag{4.5a}$$

where n_r and n_i are the real and imaginary components of the refractive index respectively.

$$n^2 = 1 - \frac{X}{(1-iZ) - \dfrac{Y_T^2}{2(1-X-iZ)} \pm \left[\dfrac{Y_T^4}{4(1-X-iZ)^2} + Y_L^2\right]^{\frac{1}{2}}} \tag{4.5b}$$

In Equation 4.5b, the terms X, Y_L, Y_T and Z are conveniently defined as follows:

$$X = f_p^2/f^2 = N_e\,(q^2/\epsilon_0 m)(2\pi f)^{-2} \tag{4.5c}$$

$$Y = f_g/f = (\mu_0\, H)(|q|/m)(2\pi f)^{-1} \tag{4.5d}$$

$$Y_L = -(\mu_0\, H \cos\theta)(e/m)(2\pi f)^{-1} \tag{4.5e}$$

$$Y_T = -(\mu_0\, H \sin\theta)(e/m)(2\pi f)^{-1} \tag{4.5f}$$

$$Z = v/2\pi f \tag{4.5g}$$

where N_e = number of electrons (m^{-3})

q = electronic charge (-1.6×10^{-19} coulombs)

m = rest mass of an electron (9.1091×10^{-31} kg)

ϵ_0 = free space permittivity ($10^{-9}/36\pi$ farads/meter)

μ_0 = free space permeability (henries/meter)

ν = electron/neutral collision rate (sec^{-1})

H = magnetic field intensity (ampere turns/m) = B/μ_0

and where f, f_g and f_p are the radio-, gyro- and plasma-frequencies respectively (Hz). Notice that Y and the gyrofrequency f_p are always positive but Y_L and Y_T depend upon the sign of the particle charge. For electrons, this sign is negative, thus canceling the negative signs given in the expressions for Y_L and Y_T.

Several approximations to Equation 4.5b may be made in order to simplify the analysis. A thorough discussion of the many interesting features suggested by the Appleton-Hartree equation is beyond the scope of this book. The interested reader will find necessary details elsewhere (e.g., Chapters 4 and 5 in Davies [1969] and Chapters 2 and 4 in Kelso [1964]). We shall, however, make the most obvious simplifications to demonstrate various well-known HF propagation effects.

We shall make the following assumptions in turn: (a) no collisions, (b) no magnetic field, and (c) no collisions and no magnetic field. When collisions are ignored, then $\nu = 0$ and Z vanishes. With no magnetic field $f_g = 0$ and all terms involving Y vanish. Naturally we cannot employ assumption (a) if we wish to estimate the impact of absorption; and assumption (b) is invalid if we are estimating Faraday fading phenomena. In fact we should exercise caution in the use of simplifications to the general form since a number of factors which are presumed to be side issues may become important under actual conditions of operation. This having been said we may write down the following approximations to the Appleton-Hartree equation:

Case 1: No Collisions (No Absorption)

$$n^2 = 1 - \frac{X}{1 - \frac{Y_T^2}{2(1-X)} \pm \left[\frac{Y_T^4}{4(1-X)^2} + Y_L^2\right]^{\frac{1}{2}}} \qquad (4.6)$$

Case 2: No Magnetic Field (no anisotropy)

$$n^2 = (n_r - i\, n_i)^2 = 1 - \frac{X}{(1-iZ)} \qquad (4.7)$$

Case 3: No collisions & No Magnetic Field (no anisotropy, no absorption)

$$n^2 = 1 - X \qquad (4.8)$$

4.5.1 The Faraday Effect and Related Phenomena

Using the case 1 result (i.e., nonzero magnetic field but no collisions), we observe from Equation 4.6 that the refractive index is bivalued. This means that two characteristic waves will propagate. Two limiting situations of interest in connection with propagation in a magnetoionic medium are: (1) when the propagation has a significant component along the direction of the external magnetic field (QL or Quasi-Longitudinal) and (2) when propagation is nearly transverse to the magnetic field (QT or Quasi-Transverse). Without going through the details, it may be shown that QL propagation constitutes a major fraction of all practical HF system geometries. There are some special cases, and we shall discuss them at an appropriate time.

Under QL conditions, we have a major simplification in the arithmetic for case 1 conditions; viz., $Y_T = 0$. Thus Equation 4.6 may be written:

$$n^2 = 1 - \frac{X}{(1 \pm Y_L)} \qquad (4.9)$$

Following the convention in Davies [1969], we associate the (+) sign with the ordinary wave vector (O-mode), and the (-) sign with the electric vector of the extraordinary wave vector (X-mode). Thus the refractive indices for the two modes are:

$$n_o^2 = 1 - \frac{X}{(1 + Y_L)} \quad \{\text{O-Mode}\} \qquad (4.10a)$$

$$n_x^2 = 1 - \frac{X}{(1 - Y_L)} \quad \{\text{X-Mode}\} \qquad (4.10b)$$

It may be shown that a linearly polarized radiowave can be represented by two oppositely-rotating circular waves. Conveniently, for ionospheric propagation in the direction of the magnetic field, the two characteristic modes, having indices of refraction described in Equations 4.10a and 4.10b above, are also circularly polarized and rotate in the opposite sense. Even in the QL approximation the result is reasonably accurate. Therefore if we wish to investigate the polarization aspects of a linearly-polarized radiowave in the ionosphere, we need only to compute the effects upon the two characteristic (O and X) modes and combine them appropriately.

Making appropriate substitutions in Equations 4.10a and 4.10b, and noting that for $\delta \ll 1$, then $(1 + \delta)^{-1} \approx 1 - \delta$ and we have:

$$n_O \approx 1 - (f_p/f)^2 [1 - (f_g \cos \Theta / f)] \tag{4.11a}$$

$$n_X \approx 1 - (f_p/f)^2 [1 + (f_g \cos \Theta / f)] \tag{4.11b}$$

Clearly if the magnetic field were to vanish, then $f_g = 0$ and the two indices would be identical. In this case $n = n_O = n_X$ would describe the refractive index in an isotropic plasma. Of course if $f_p = 0$, then $n = 1$ independent of the magnetic field. In the general situation both indices of refraction are less than unity suggesting phase velocities, V_X and V_O in excess of c. However, since the index of refraction for the X-mode is slightly smaller than the O-mode, the X-mode phase advance is somewhat more rapid. Since $V_X > V_O$, then at a given distance *s*, the electric vector of the extraordinary wave would have appeared to have rotated less than that of the ordinary wave. Upon combination of the two modes, assuming each has the same amplitude, one finds that the net rotation of the resultant linearly polarized radiowave is in the clockwise sense. It should be noted that if the direction of propagation is antiparallel to $\boldsymbol{H}$ then $\cos \Theta = \cos \pi = -1$ and the resultant rotation would be counter-clockwise. Naturally there are intermediate values of Θ as well. When Θ is close to $\pi/2$, and for certain other geometrical conditions, the QL approximation breaks down.

The natural sense of rotation of electrons in a magnetic field is clockwise when viewed in the direction of the field. This is in the same sense as the E vector for an extraordinary wave directed along the magnetic field but reckoned at a fixed point in space. In order for the latter to be true, it may be shown that the X-mode must trace out a left-handed helix as one moves physically along the direction of propagation and in the direction of the magnetic field. Now the X-mode always rotates in the same direction as the natural motion of electrons (which is fixed for a given external magnetic field), so even if the direction of radio propagation is reversed, polarization rotation is not reversed. Since the resultant polarization angle does not unwind when the paths are reversed, we have a non-reciprocal situation. Figure 4-15 illustrates the situation.

The amount of Faraday rotation Ω introduced over a given distance is given by the average of the X- and O-mode rotation angles or:

$$\Omega = \tfrac{1}{2} (\Omega_+ - \Omega_-) \tag{4.12}$$

where the (+) sign refers to the O-mode. The individual electric vector angles are proportional to $(\pi f/c) n$. Subtracting n_X from n_O and multiplying by $(\pi f/c)$ we obtain the following result:

$$\Omega = (\pi/c)\,(f_p/f)^2\, f_g\, (\cos\theta)\, s \qquad (4.13)$$

where s is the distance traversed. This equation must now be translated into more convenient terms and we must generalize the result for the situation for which the various parameters vary over s.

After traversing a distance ds in a magnetoionic medium, the Faraday rotation, under conditions of QL propagation and for $f \gg f_p > f_g \gg \nu = 0$, is given by:

$$d\Omega = 2.97 \times 10^{-2}\, f^2\, N_e\, H \cos\theta\; ds \qquad (4.14)$$

where MKS units are used.

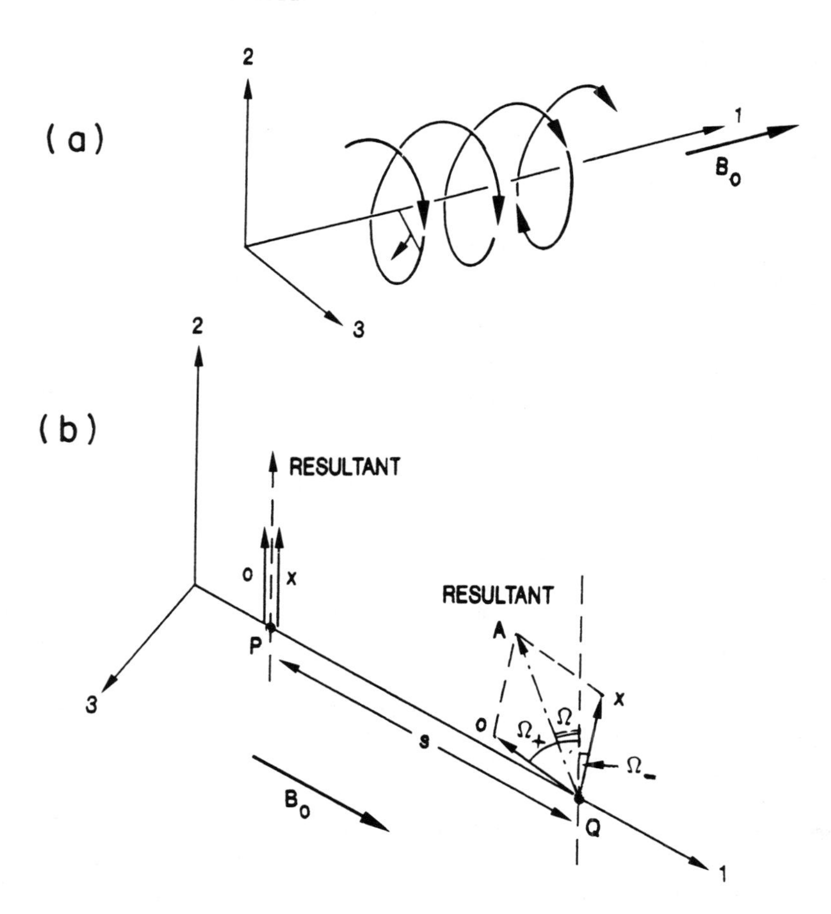

Fig.4-15. Geometrical Considerations in Faraday Rotation. (a) Radiowave advances in the 1 direction which is aligned with the magnetic field **B** = μ**H**. (b) The resultant electric vector rotates as the wave progresses over the distance s.

The amount of Faraday rotation of a linearly polarized radiowave over an actual path is given by the following equation:

$$\Omega = 2.97 \times 10^{-2} f^{2} \int_{\text{path}} H \cos\Theta \, N_e \, ds \qquad (4.15)$$

where Ω is in radians, H is in ampere-turns/meter, Θ is the angle between the ray path and the magnetic field H, N_e is the electron density (el/m^3), f is the radio frequency (Hz) and s represents displacement along the path.

For the case of rectilinear propagation, computation based upon Equation 4.15 is almost trivial since $H \cos\Theta$ is slowly-varying over the path and may be removed from the integral and replaced by its mean value. Using this procedure we may separate magnetic field effects from electron density effects, at least to first order. At HF, ray trajectories are curved in regions where most of the Faraday rotation would be encountered, and this complicates the picture considerably. Indeed, one finds that over a fixed link the higher frequencies will suffer more Faraday rotation than the lower ones despite the f^2 term in Equation 4.15. This is because the higher frequencies are refracted from greater ionospheric heights where the electron population is maximized. This result surprises those principally interested in earth-space aspects of ionospheric propagation. Figure 4-16 is germane to the discussion which follows, and is a preview of a propagation feature which is unique to HF under the condition that the transmitted frequency f is greater than the overhead critical frequency f_c but less than the maximum observable frequency for the path (or MOF). From the figure we see that two distinct ray paths exist between points A and B. The lower ray travels a shorter distance, suffers less dispersion, and is appropriately called the low ray. The upper ray, called the high ray or the Pederson ray by its discoverer, spends a considerable amount of time in the more dense portion of the refracting layer. Faraday rotation is greater for the Pederson ray than for the low ray, everything else being equal.

One way of looking at the Faraday effect is to define a Faraday bandwidth B_F. This is equivalent to the polarization bandwidth B_p for the case in which the polarization angle equals π radians (or 180 degrees), an amount which constitutes a full fade interval. Epstein [1967] defines a coherent polarization bandwidth $B_p = \frac{1}{2} B_F$ on the basis of engineering considerations. In order not to introduce confusion, we shall use Epstein's definition for the polarization bandwidth. A computer simulation shows how B_p should vary as a function of frequency at fixed azimuth, and azimuth at fixed frequency (Figure 4-17). Part a of Figure 4-17 exhibits the rather substantial range of B_p values for the low ray as compared to the high ray. One also observes the strong aspect sensitivity with less Faraday rotation (i.e., larger B_p) along an E-W baseline where, in most circumstances, $\cos\Theta$ should be minimized.

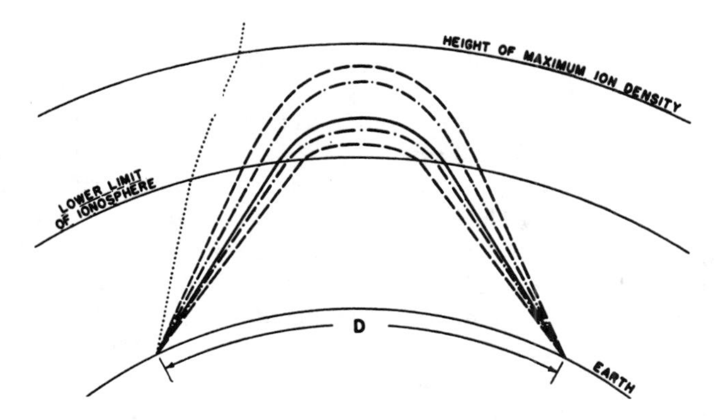

Fig.4-16. The existence of high and low rays in the ionosphere. The highest frequency which may propagate between the points A and B corresponds to a single ray which is the superposition of the high and low rays in a limiting case. This limiting frequency is sometimes referred to as the MUF, or maximum usable frequency. The MUF is usually defined as a monthly statistic and reckoned for each hour of the day. A single observation is referred to as the MOF, or the maximum observable frequency. In this illustration, the two dashed lines (----) correspond to the high and low rays at a frequency considerably less that the MUF. The two dot-dashed curves (·—·—·) correspond to a frequency closer to the MUF. The solid line (—) corresponds to the MUF, while the dotted line (· · · · ·) corresponds to a frequency above the MUF which penetrates the ionosphere. (Original curve from Davies [1965].)

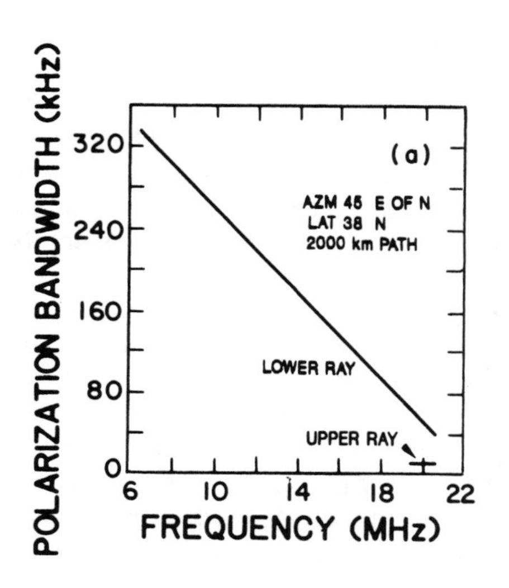

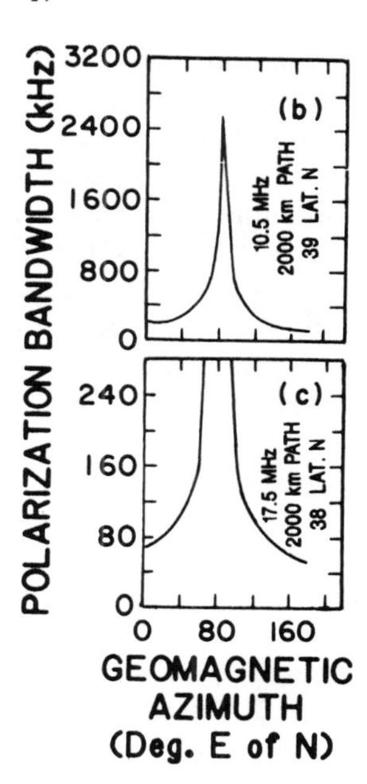

Fig.4-17. Polarization bandwidth in the ionosphere: (a) as a function of frequency, (b) as a function of azimuth with f = 10.5 MHz, (c) as a function of azimuth with f = 17.5 MHz. (From Epstein [1967], after Ames et al. [1977].)

Epstein and Villard [1968] of Stanford Research Institute (now SRI International) studied the problem of Faraday fading with an emphasis on the fade rates at fixed frequency. They concluded from observations over midlatitude paths that the polarization null rate increased as the ratio of the carrier to the MOF increased, and as the total electron content of the ionosphere increased (i.e., time-of-day effect). They observed that the average daytime polarization null rate was 0.5 nulls/minute, although it was often observed to be higher. Nighttime rates were generally around 0.1 nulls/minute. Clearly, a fixed frequency system will not encounter a significant amount of signal fluctuation (arising from the Faraday effect) during nocturnal hours. Of course, very slow and deep fades present a different kind of challenge. As a rule of thumb, one may take B_F to be 400 kHz for E-W paths and 100 kHz for N-S paths, based upon the work of the SRI group. Therefore, if the frequency range 15-20 MHz is supported for a daytime N-S path , we would observe roughly 25 fades (nulls) over this interval. For an E-W path about 6 fades would be observed.

Demonstration of the rate-of-change of Faraday rotation can be made with a sounder. The only problem is that sounder sweep rates may be sufficiently long that temporal effects become mixed with frequency effects. Sounders and related instruments are covered in Chapter 6, but suffice it to say that an ionospheric sounder is nothing more than a radar which is specially configured to determine ionospheric properties over a range of contiguous frequencies. For an oblique sounder, the transmitter and receiver are generally separated over a considerable distance, and the conventional output is a display of signal time delay (on the y-axis) versus frequency of transmission (on the x-axis). Figure 4-18 demonstrates the rate-of-change of the Faraday rotation angle with respect to transmission frequency, $d\Omega/df$. The path is generally E-W, and we observe roughly 6 nulls, certainly consistent with the suggestion made in the previous paragraph. Also shown (Part b of Figure 4-18) is an obscuration of Faraday fading (in the integrated signal) when more than one mode is present.

At a more practical level, the Faraday effect, unlike phase path and absorption effects, may lead to a type of nonreciprocal behavior for a path. As we have seen, Faraday rotation depends upon $\cos \theta$ (including the algebraic sign). As reckoned from the transmitter (be it A or B), polarization rotation from A to B will be in the opposite sense to that from B to A. Although the two-way rotation is in a consistent direction and the magnitude of rotation is the same, nonreciprocal effects are introduced because of the interaction of the oppositely-directed waves with the antennas at each end. A discussion of this effect is found in Davies [1969; Chapter 13] and in Budden and Jull [1964].

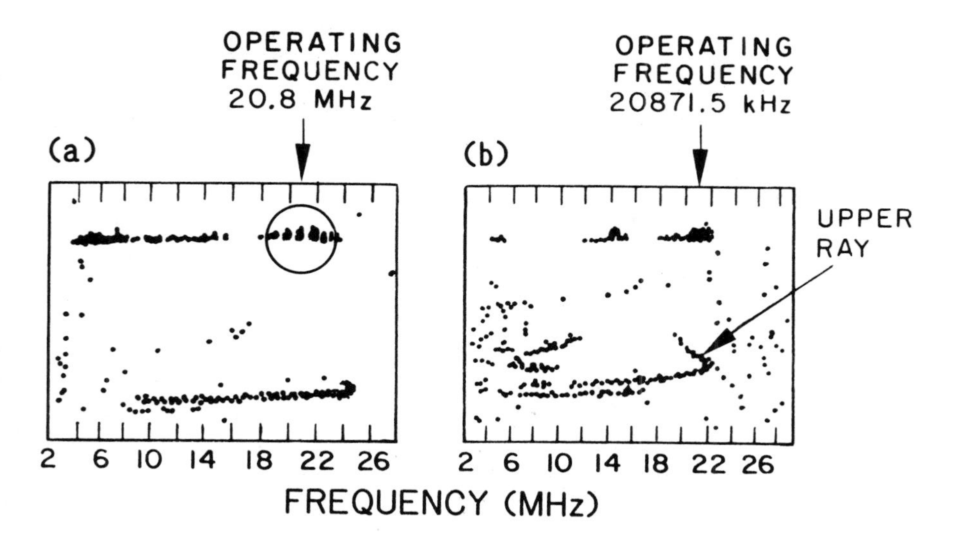

Fig.4-18. (a) Chirpsounder record which illustrates some features of Faraday rotation. Transmitter at Langley AFB. Receiver at Scott AFB. Time : 1030 EST on 2 February 1981. (b) Chirpsounder record of Langley AFB to Scott AFB on 3 February 1981 at 1410 EST. The AGC indicator at top exhibits high signal level but no appreciable null structure near the MOF. (Courtesy BR Communications.)

Before leaving Case 1 (nonvanishing magnetic field effects and no collisions) we shall examine a few salient features of O- and X-mode reflection from the ionosphere at vertical incidence. We shall make use of this information in section 4.6.1 for NVIS communication, and in Chapter 6 when sounder methods for RTCE are discussed. We may rewrite Equation 4.6 as follows:

$$n^2 = 1 - \frac{X(1-X)}{2(1-X) - Y_T^2 \pm [\, Y_T^4 + 4(1-X)^2\, Y_L^2\,]^{\frac{1}{2}}} \qquad (4.16)$$

The reflection condition corresponds to the situation for which $n^2 = 0$. For this to be true we have the following possibilities [Davies, 1965; pp. 71-73]:

$$X = 1 \qquad (+\text{ sign}) \qquad (4.17a)$$

$$X = 1 - (Y_T^2 + Y_L^2)^{\frac{1}{2}} = 1 - Y \qquad (-\text{ sign}) \qquad (4.17b)$$

$$X = 1 + (Y_T^2 + Y_L^2)^{\frac{1}{2}} = 1 + Y \qquad (-\text{ sign}) \qquad (4.17c)$$

The solution $X = 1$ corresponds to reflection as though the magnetic field were not involved, and is associated with the O-mode. Two solutions (Equations 4.17b and 4-17c) are obtained upon use of the (-) sign which appears in equation 4.16. Reflection at $X = 1 - Y$ corresponds to the situation for which the transmitting frequency is greater than the electron gyrofrequency, and Y is less than unity. This corresponds to one branch of two X-modes which may propagate, and is the only extraordinary mode at HF frequencies of interest. (The other X-mode corresponds to the $X = 1 + Y$ solution and it arises when $f < f_g$. Since f_g is below the defined HF band, we shall not consider this X-mode branch, sometimes called the Z-mode, any further.) Assuming a simple layer having a maximum plasma frequency f_{pmax}, we may derive the following relationships:

$$f_{pmax}^2 = f_o^2 \tag{4.18a}$$

$$f_{pmax}^2 = f_x^2 - f_x f_g \tag{4.18b}$$

$$f_x - f_o \approx \tfrac{1}{2} f_g \tag{4.18c}$$

where f_o and f_x are the ordinary and extraordinary wave critical frequencies respectively, and Equation 4.18c is true if the O- and X-mode criticals are sufficiently large. (Note: The critical frequency for a specified layer and mode is the frequency that will enable a vertically-incident radiowave to penetrate the layer if the frequency is exceeded.) The O- and X-mode criticals differ by roughly 0.5 f_g regardless of the value of f_{pmax} provided it is not in the lowest portion of the HF band.

4.5.2 Absorption in the Limit of No Magnetic Field

We will now develop a practical expression for computing the amount of ionospheric absorption under the presumption that the earth's magnetic field may be ignored. This assumption allows sufficiently accurate results in most instances. We shall use the result expressed in Equation 4.7 (i.e., corresponding to case 2).

A radio wave which is attenuated along the y-axis takes the form:

$$E = E_o e^{-\alpha y} = E_o \exp[-(2\pi f/c)\, n_i\,]\, y \tag{4.19}$$

where n_i is the imaginary part of the refractive index and $\alpha = (2\pi f/c)\, n_i$ is the absorption coefficient for the medium. The real and imaginary parts of Equation 4.7 are easily determined as:

$$n^2 = (n_r^2 + n_i^2) + i(2\, n_i n_r) = [1 - X/(1+Z^2)] + i\,[XZ/(1 + Z^2)] \tag{4.20}$$

Considering the imaginary part and solving for n_i we have:

$$n_i = (1/2\, n_r)\, [XZ/(1 + Z^2)] \tag{4.21}$$

where n_r is the real part of the refractive index (associated with bending) and the other terms are defined in Equation 4.5. This indicates that the absorption may be altered to some extent by ray bending. The absorption index α is $(2\pi f/c)$ times the RHS of Equation 4.21, and may be simplified as:

$$\alpha = (q^2/2\epsilon_o m c\, n_r)\,(N_e \nu)\,[(2\pi f)^2 + \nu^2]^{-1} \tag{4.22}$$

For HF propagation, most absorption occurs in the D-region where the amount of bending is insignificant. Thus $n_r \approx 1$ and equation 4.22 may be adjusted to read:

$$\alpha \approx (q^2/2\epsilon_0 mc)\,(N_e \nu)\,[(2\pi f)^2 + \nu^2]^{-1} \qquad (4.23)$$

The absorption coefficient as written above gives the amount of absorption in nepers/meter if MKS units are employed. A more practical version of 4.23 may be derived by using decibel units (based upon the common log of a power ratio) instead of nepers (based upon natural log of a voltage ratio) recognizing that 1 neper = 20 $\log_{10}$e = 8.68 dB. Making the appropriate substitutions for the constants, and computing the absorption over a km rather than a meter, we have:

$$\alpha \approx 4.6 \times 10^{-2}\,(N_e \nu)\,[(2\pi f)^2 + \nu^2]^{-1} \text{ dB/km} \qquad (4.24)$$

A further simplification in the expression for α may be made if $2\pi f$ is large in comparison with the collision rate ν.

$$\alpha \approx 1.16 \times 10^{-3}\,(N_e \nu)/f^2 \text{ dB/km} \qquad (4.25)$$

The imaginary part of the refractive index is associated with wave dissipation, which corresponds to a conversion of wave energy into heat. There are two types of absorption in connection with ionospheric propagation: deviative and nondeviative. Deviative absorption, arises near the apogee of a ray trajectory or at any position along a path for which significant bending occurs. For most practical situations, absorption which arises under this restricted definition is not significant. Indeed, most of the ionospheric absorption occurs within the D-region where propagation is essentially rectilinear, but where collision rates are high and the electron densities are sufficient. We refer to this significant form of absorption as the nondeviative type. The mathematical consequence of this is a simplification in the procedure for estimating the amount of absorption.

Taking the D-region electron density to be 10^9 per cubic meter and the collision rate to be 10^7 per second, then $N_e \nu$ is 10^{16} Hz and we find from equation 4.25 that the absorption at 10 MHz is slightly more than a 0.1 dB/km. Even for effective D-region paths of about 20 km, the absorption loss will be modest for this specification of $N_e \nu$. Unfortunately the product may sometimes reach about 10^{17}, and the radiofrequency may be 3 MHz for short-range (tactical) near-vertical-incidence-skywave (NVIS) links. This implies a loss of $(1.16 \times 10^{-3})\cdot(3 \times 10^6)^{-2}\cdot(10^{17}) \approx 12.9$ dB/km. Integrating this specific rate over an effective path of say 10 km (an NVIS path), we encounter roughly 130 dB of absorption. The absorption (A) in dB increases in proportion to the path length; whereas basic transmission loss (L_{bf}) in dB increases with the logarithm of distance. Thus, assuming path equivalence for upleg and downleg segments, the two-way value for L_{bf} is only 3 dB more than the one-way value; but the two-way value of absorption is twice the one-way dB value, which may be hundreds of dB. The two-way path will thus experience 260 dB of absorption. Figure 4-19 indicates a general range of values for N_e, ν and the product $N_e \nu$.

Methods for predicting absorption at HF involve assumptions of electron density distribution and electron collision rates in the D-region. Since these distributions (especially of N_e) depend upon solar activity, solar zenith angle and other factors, procedures established for estimating absorption loss take a generally different form from that given in equation 4.25. The procedure established by the CCIR (i.e., Rpt.252-2; [CCIR, 1970]) and incorporated in most computer methods is similar to the relation given below (see Figure 4-20) [Lucas and Hayden, 1966; Hayden, 1979]:

$$A(\text{dB}) = 677.2 \sec\phi \sum_{j=1}^{n} I_j \left[(f + f_H)^{1.98} + 10.2 \right]^{-1} \quad (4.26a)$$

where

$$I = \text{Greater of } 0.1 \text{ and } [1 + 0.0037\, R_{12}] [\cos 0.881\, \chi]^{1.3} \quad (4.26b)$$

$$n = \text{the number of hops}$$

and where ϕ is the ray zenith angle, f_H is the electron gyrofrequency (Hz), R_{12} is the 12-month running-average sunspot number, χ is the solar zenith angle, and f is the radio frequency (Hz). Various nomograms have been constructed as a matter of convenience for those without computers (see Davies [1965]).

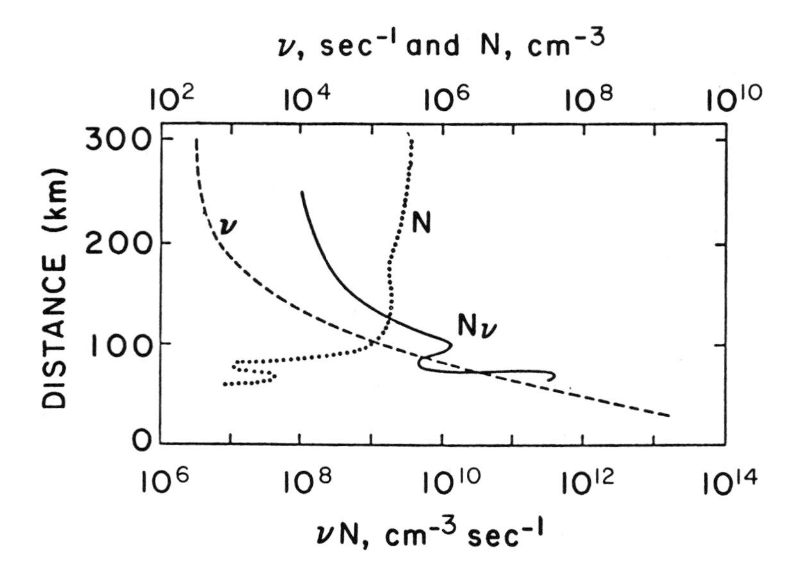

Fig.4-19. Variation in electron density and electron collision rate with altitude. Also shown is the product of N_e and ν. (From Davies [1965].)

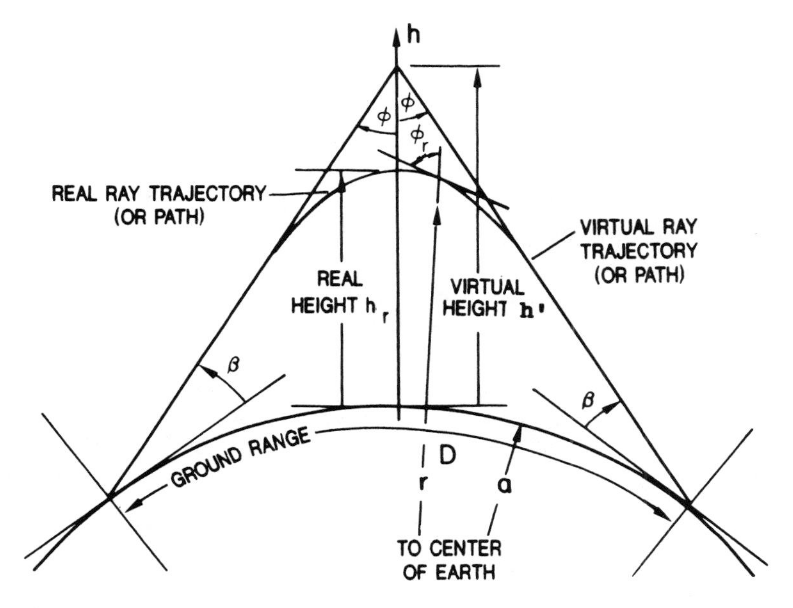

Fig.4-20. Geometry of ray propagation over a curved earth. β is the ray elevation angle, ϕ is the ray zenith angle. The virtual height h' is shown as the distance to the intersection of the virtual rays.

The term I (Equation 4.26b) is sometimes called the absorption index and it represents the frequency- and (radio) path-independent absorption strength. For a sunspot number of 100 and an overhead sun ($\chi = 0$), we have $I = 1.37$. The ionospheric sunset value for χ (call it χ_{min}) such that I must default to 0.1 is 100.55. For $R_{12} = 0$, χ_{min} is 100.11. This implies that in the vicinity of the terminator[38] the sunspot number alters the effective solar illumination time. The effective solar illumination time near the terminator is augmented by about a minute for solar maximum compared to solar minimum.

A somewhat different, and more detailed, expression for ionospheric absorption is given by Davies [1990]. That expression is based upon formulas of Barghausen et al [1969] and Shultz and Gallet [1970]. The expression takes the so-called winter anomaly absorption factor into account, and requires an explicit account of geographic latitude, the solar zenith angle at local noon, and the collision frequency.

---

38. The terminator is the line separating day from night on the earth's surface. At ionospheric heights, the solar illumination is more persistent, a geometrical fact which enables solar zenith angles well in excess of 90 degrees to be effective. The terminator is well known for so-called "grey line" propagation characteristics. Radio amateurs have noted that long distance "grey line" communication may often be achieved when the radio path is aligned with the day-night terminator.

The winter anomaly of absorption corresponds to a wintertime increase in absorption which arises at middle latitudes and which appears to be associated with a warming of the underlying stratosphere. The phenomenon is not readily predictable but nevertheless the monthly median absorption for fixed solar zenith angles has been observed to increase by about a factor of 2 (in dB) during the course of these events. Thus if the monthly median absorption at 10 MHz is 2 dB, we could expect the absorption to increase to 4 dB during winter anomaly events. In addition to the winter anomaly, another interesting departures from a Chapmanlike picture in the amount of absorption is associated with a geomagnetic anomaly around the magnetic equator. In this instance, a zone of enhanced absorption is observed to occur in the vicinity of 25 degrees if data are organized in terms of modified dip angle. An actual reduction in absorption is observed at the dip equator [George, 1971]. An examination of these anomalous absorption features may be found in work by Appleton and Piggott [1954]. This behavior is reminiscent of what is observed in connection with the F-region peak electron density at low latitudes (i.e., the Appleton anomaly); however, in this case we are dealing with D-layer total electron content effects. Davies [1990] indicates that the effect is probably related to large scale circulation in the mesosphere.

Absorption takes a variety of forms depending upon the source of the excess ionization responsible for the disturbance. The preceding discussion refers principally to normal absorption which is introduced by nondisturbed solar illumination. Focusing on the absorption index I in Equation 4.26b, we see that there is strong solar control with symmetry about local noon. Of course, in any real application, the term local must be defined appropriately. In general local time for an oblique path suffering absorption must be computed at one or more control points[39] which correspond to ray path intersections with the D-region. It should be recognized that the strict solar control of absorption is met only approximately. In fact, owing to the effect of recombination (which increases with N^2), there is a delay in the absorption peak.

---

39. The term *control point* is generally used to designate the apogee of a single-hop ray trajectory involving the F2 layer (and sometimes the E and F1 layers). In practice, when observing in a plan view, one takes the midpoint of a great circle path connecting two terminals to be the control point, although it need not be the case if gradient effects are to be considered. The control point concept is a convenient one, and is based upon the notion that most of the refraction (i.e., bending) occurs near the apogee of the ray trajectory. Thus, ionospheric properties associated with this region are of utmost importance. The control point concept may also be used to identify locations where most of nondeviative absorption occurs in the D layer. Given the ray launch angle, it a simple matter to estimate the ionospheric D layer piercing point. In general, provided the ionospheric E and F layers will support ionospheric bounce, there will be two D layer control points for every single hop path, one for upward entry and the other for downward exit.

The amount of delay is given by the following expression:

$$\delta T = (2 \alpha N)^{-1} \tag{4.27}$$

where α is the recombination coefficient. This relationship has been used by several investigators to deduce an effective recombination coefficient for individual ionospheric layers [Whitten and Poppoff, 1965]. From Figure 4-19, if we take $N = 10^5/\text{cm}^3$ and $\alpha = 10^{-6}$ cm^3/sec [Mitra, 1963], then the amount of time delay is approximately 17 minutes following local noon.

Martyn's absorption theorem states that for a flat earth and no magnetic field, the amount of absorption A_{ob} along an oblique path may be related to the vertical A_v amount by the formula:

$$A_{ob} = A_v \sec \phi \tag{4.28}$$

where ϕ is the ray zenith angle and the absorption is in dB. This relationship, which has its counterpart in a spherical geometry, is useful when it is necessary to extrapolate absorption measurements made at one location to an operational region for which measurements are not available.

Figure 4-21 depicts several interesting types of absorption. They include the normal type (discussed above), and short-wave-fades (SWFs) which are associated with x-ray flares. Both of these types affect only the sunlit side of the earth. Two other types are also shown, and both are specific to high latitude morphology. These phenomena, arising from particle precipitation, will be discussed later.

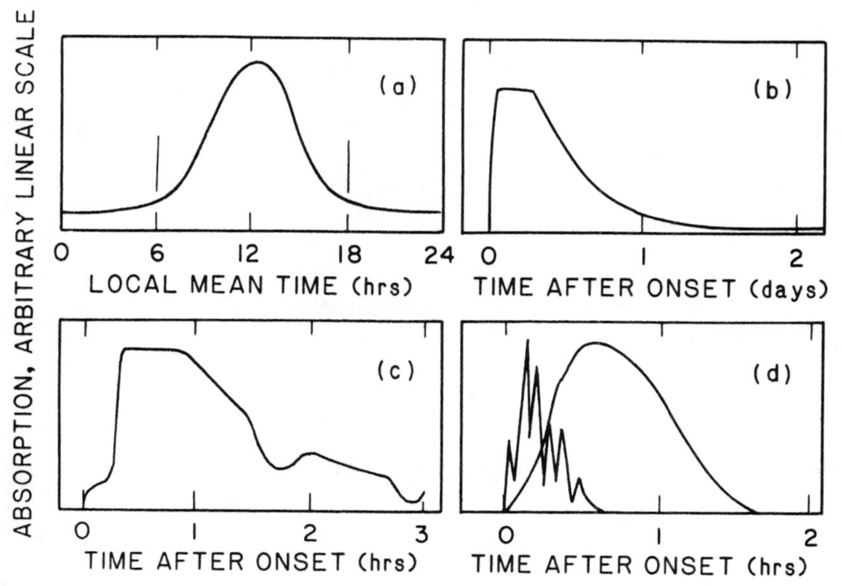

Fig.4-21. Categories of absorption. (a) normal, (b) short-wave-fade, (c) polar-cap-absorption, and (d) Auroral.

4.5.3 Propagation when the Magnetic Field and Collisions are Ignored

A magnetoionic medium such as the ionosphere interacts strongly with HF radiowaves. Generally, the ionosphere exerts the following influences: dispersion, absorption, birefringence, and anisotropy. Absorption is caused by collisions, and the presence of the geomagnetic field introduces anisotropy and birefringence. Under certain circumstances path splitting between the characteristic modes may result. With no collisions and no magnetic field we are left with dispersive effects which are introduced by the plasma. Even though this case (i.e., No. 3; Equation 4.8) would appear to be unrealistic at first glance, there is a class of problems which may be analyzed under this approximation. Obvious examples include the determination of gross refractive errors and group path delay effects associated with signals traversing the ionosphere.

Equation 4-8 is repeated here in order to discuss it in the current context.

$$n^2 = 1 - X = 1 - (f_p/f)^2 \tag{4.8}$$

A graph of n^2 versus X is called a dispersion curve, and Figure 4-22 is the simplest non-trivial example. As is obvious from the equation (and the figure) the square of the refractive index is always equal to or less than 1, a value it assumes at $X = 0$. If a radiowave is entering the ionosphere from below, we generally encounter a monotonically increasing function of f_p. As a result of this behavior n^2 will decrease proportionately from an initial value of unity to a critical value at which reflection occurs. When this point is reached, the ray retraces its steps. At normal incidence, reflection occurs at $n^2 = 0$ and happens when $X = 1$. At other than vertical incidence, reflection will arise when $n^2 = \sin^2 \phi$ where ϕ is the ray incidence angle.

The dispersion phenomenon arises because the refractive index of the ionosphere is a function of the radiofrequency employed. This dispersive behavior will lead to signal distortion. In general the phase velocity of a radiowave is different from the group velocity. We have:

$$V_\phi V_g = c^2 \tag{4.29}$$

where c is the free space speed of light, V_g is the group velocity of the radiowave, and V_ϕ is the phase velocity. In the ionosphere, the following two relations correspond to the phase and group velocities respectively.

$$V_\phi = c/n \tag{4.30a}$$

$$V_g = c/n_g = nc \tag{4.30b}$$

where n is understood to be the phase refractive index.

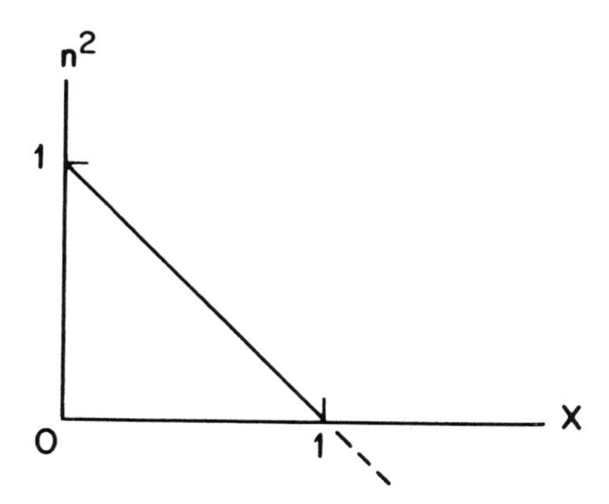

Fig.4-22. Dispersion curve for the situation of no magnetic field and no collisions.

The group refractive index n_g is simply the inverse of n. As is evident, if the wave front advances more rapidly, then the velocity of a wave group (i.e., a signal) will be reduced in order to satisfy equation 4.29. Since a pulse may be represented by a Fourier spectrum the bandwidth of which is inversely related to the pulse duration, and since the ionosphere is dispersive, it is clear that a very narrow pulse may exhibit significant distortion while a long pulse may exhibit little or no distortion. A useful picture of ionospheric pulse distortion of this type has been discussed by Millman [1967].

4.6 NEAR-VERTICAL-INCIDENCE-SKYWAVE (NVIS)

Vertical-incidence skywave propagation is associated with the following conditions: (a) radiowave propagation along a zenith-nadir line, and (b) $f < f_{pmax}$. As we have noted previously, ionospheric reflection from a specified layer and mode of propagation occurs provided the transmitted frequency is less than the critical frequency f_c for the layer and mode in question. Generally the largest of the critical frequencies is associated with the F2 layer although exceptions may arise. For the F2 layer the ordinary wave critical frequency is designated *foF2*; for the extraordinary wave it is *fxF2*. Communication coverage by this scheme is crudely uniform provided the terminals are contained within a roughly circular region surrounding transmitter and if we use a frequency generally smaller than *foF2*. This particular geometry has received a considerable amount of attention in the literature since it is applicable to conventional ionospheric sounders which have been operating for roughly half a century. From the point of view of remote sensing for ionospheric characterization this geometry has much to offer, since it is possible to conveniently colocate transmitters and receivers, although the latter may require some protection. The reader may wonder why the umbrella-type coverage associated with this geometry has any merit, considering the availability of line-of-sight systems (at HF, VHF, and UHF), ground wave systems (at MF and HF), and satellite systems which have a proven reliability. In this section we shall discuss the properties of a HF communication scheme called Near-Vertical-Incidence-Skywave (NVIS), and we shall identify some of its advantages and disadvantages. Figure 4-23 illustrates the NVIS geometry.

Lecture notes developed by Hagn [1988b] for an AFCEA short course on military uses of the HF spectrum [Goodman, 1983-1988] were used in the preparation of this section. Useful information has also been extracted from certain technical notes [Sedgwick, 1985] and reports. Although military exploitation of the NVIS scheme dominates the field, there are some interesting civilian applications.

4.6.1 Requirements for Use of NVIS

Line-of-sight (LOS) communication between terminals at heights h_1 and h_2 may be possible if they are separated by a distance d as determined by an appropriate form of Equation 4.3:

$$d(\text{km}) = 4.124 \cdot (\sqrt{h_1} + \sqrt{h_2}) \qquad (4.31)$$

where h_1 and h_2 are in meters and the 4/3 earth approximation is used. Equation 4.31 is valid for a spherical earth characterized by a smooth terrain. If the terminals are located at heights of 25 meters, then the LOS range between the two terminals will be about 40 km (or 25 statute miles). Unfortunately terrain is not always smooth and obstacles (both natural and man-made) may block the LOS path. Mountainous regions are natural barriers to LOS communication and such features are generally not conducive to ground wave propagation. Communication by ground wave or LOS is also inhibited by high signal attenuation if one or both terminals are located within a forested area or a region of heavy wet vegetation, such as a jungle. Lateral attenuation through vegetation has been shown by Hagn and Barker [1970] to be an increasing function of frequency:

$$\alpha_L \text{ (dB/m)} \approx 0.009 f(\text{MHz}) + 0.1 \qquad (4.32)$$

At 10 MHz, the amount of attenuation experienced by a radiowave upon emergence from 1 km of vegetation is of the order of 1000 x 0.19 = 190 dB. At 3 MHz the attenuation is reduced to 127 dB, but at 30 MHz it is 370 dB. No wonder that NVIS is an attractive alternative, since the free space loss is a fairly uniform value for ranges of interest, being (from Equation 4.2, taking 600 km to be the ionospheric virtual path) 97 dB at 3 MHz, 108 db at 10 MHz, and 117 dB at 30 MHz. Of course, frequencies in excess of 10 MHz have limited practical application for NVIS, and 30 MHz is clearly impossible except for oblique skywave paths.

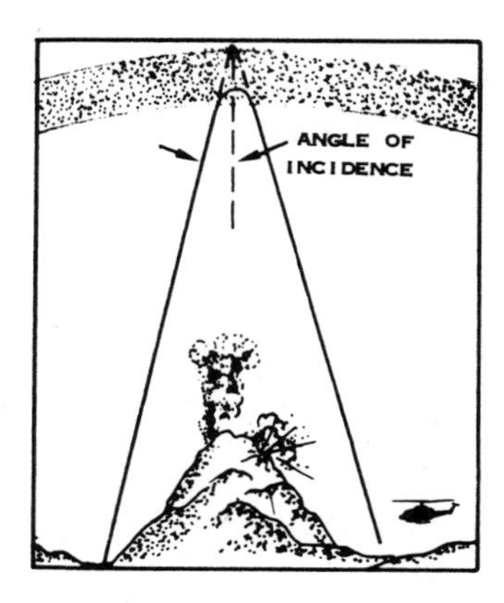

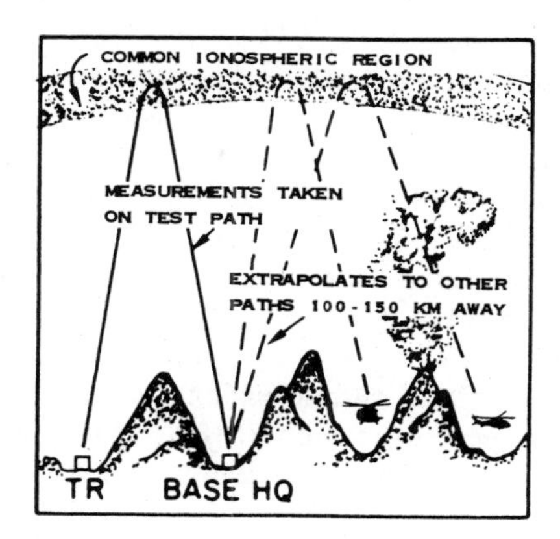

Fig.4-23. Near-vertical-incidence-skywave communication (NVIS) geometry.

At 10 MHz, propagation through a 500 meter stretch of vegetation is approximately 95 dB. The impact of forested environments upon LOS and ground wave communication links depends critically upon antenna height and configuration (i.e., polarization). For air-to-ground links significant fluctuations in signal level are observed in the direction of forested regions as aircraft move (azimuthally) with respect to the ground terminal. These fluctuations may lead to significant impairments in communication performance.

Specific military applications require NVIS alternatives. In particular, it may not be possible to assume control of the high ground to achieve an unobstructed LOS path. NVIS allows the tactical communicator to launch HF signals up-and-over an obstacle such as a mountain to achieve valley-to-valley connectivity. It has also been shown that groundwave jammers and emitter location systems may be partially thwarted through use of skywave communication. There are special missions which also dictate the use of an alternative to LOS and groundwave communication. One is the so-called nap-of-the-earth (NOE) operation which requires low flying aircraft such as helicopters to fly close to the surface of the earth using the local terrain features as cover from enemy radar and other location measures.

A rich history exists for NVIS propagation but very little of it has been treated in the open literature. One notable exception is the early work of the 1920s and 1930s which dealt with magnetoionic theory and vertical-incidence sounding to study the ionosphere. Work in this area still continues, but the communication aspects have not been widely reported. A survey of the literature emphasizing equatorial NVIS conditions was published by Hagn and Posey [1966]. Except for the early activity, much attention has been directed toward military aspects of the subject, a fact which probably accounts for the limited coverage. During World War II, jungle communications were improved through use of NVIS schemes [Aikens et al., 1944]. During the period of the Korean and Malaysian wars, and as an indirect consequence of those conflicts, several studies were undertaken to overcome the problem of mountainous terrain and jungle attenuation. A noteworthy study of NVIS was published [Egli and Lacy, 1952], and a number of antenna aspects were examined [Shirley, 1952; Moore 1958] during this period. In addition, Piggott [1959] computed median NVIS field strengths for jungle regions. Activity accelerated in the 1960s, during the Viet Nam era with additional antenna studies being conducted [Busch, 1963; Ho and Ranvier, 1963; Bergman and Barnes, 1967]. Hagn and his coworkers performed a comprehensive investigation in Thailand throughout the 1960s and this work has been summarized in an open-source report [Hagn and Barker, 1970]. This same group also expanded on dipole orientation theory for equatorial environments and developed a sounder technique [Hagn, 1964; Hagn et al., 1966]. The 1970s saw a number of studies directed toward the exploitation of NVIS for NOE communications [Brune and Reilly, 1975; Pinson et al., 1977; Tupper and Hagn, 1978; Holmes, 1979; Burgess, 1979; Maslin, 1979; Brune and Ricciardi, 1979]. Work has continued into the 1980s with some emphasis on spread spectrum NVIS.

4.6.2 NVIS Propagation Factors and Constraints

We tend to think of the HF radio spectrum as a long-haul band, with the upper portion being used to provide coverage (or connectivity) to extended distances of up to 4000 km by a single ionospheric bounce (or hop) and to far greater distances by multiple hops. In achieving these distances we find that a penalty is paid in short-range coverage. In fact, since the maximum frequency (f_{max}) supporting communication between two points separated by a distance d_0 is always in excess of the overhead critical frequency, we conclude that no coverage is provided at f_{max} for $d < d_0$. This is the so-called skip distance. As we reduce the separation distance d_0 between the two terminals, the value for f_{max} diminishes. At $d_0 = 0$ the value for $f_{max} = f_{pmax}$ where f_{pmax} is the maximum plasma frequency of a specified layer above the transmitter (and receiver) and is called the critical frequency f_c. In most applications $f_c = foF2$ since the electron density of the F2 maximum is generally in excess of both the E and F1 layer values. Thus if we optimize propagation to fulfill a long-haul mission, we generate skip zones and limit short range coverage to that which may be achieved by LOS and ground wave techniques. On the other hand, if we transmit at a frequency equal to or less than the overhead critical frequency, we generate an umbrella-like local coverage, without a skip zone, but also without long range capability.

Figure 4-23 shows the geometry associated with NVIS propagation. Note if transmission frequencies exceed the overhead critical frequency, taken to be *foF2*, we generate a skip zone for terrestrial coverage. At the same time we provide transionospheric coverage above the transmitter through an ionospheric iris. The iris may be defined by a zenith angle ϕ_I which diminishes to 0 degrees for $f < foF2$ but expands to 90 degrees (i.e., the whole sky) if $f \gg foF2$. We say that breakthrough has occurred when f exceeds $foF2$. The iris half-angle is given by the approximate expression:

$$\phi_I = \cos^{-1}(foF2/f) \tag{4.33a}$$

or equivalently

$$\phi_I = \sec^{-1}(f/foF2) \tag{4.33b}$$

where a locally plane stratified ionosphere is assumed and we ignore anisotropic propagation. Geomorphology of the breakthrough phenomenon has been examined by the author [Goodman, 1979].

It is clear that a different type of frequency management is required for the NVIS scheme. For conventional long-haul HF requirements, MOF-seeking architectures are typically employed as a means to limit multimode effects and to reduce the f^2 absorption. This dictates the selection of a frequency which to a first approximation is given by the following relation:

$$f \dashrightarrow f_c \sec\phi > f_c \quad \text{(MOF-Seeking, Long-Haul HF)} \tag{4.34}$$

In relation 4.34, ϕ is the ray zenith angle, and where the arrow signifies that the optimum frequency is approached from the low frequency end. For NVIS we are even more encouraged to use a MOF-seeking frequency management philosophy since the requisite transmission frequencies are more severely effected by absorption. Thus we require that the following rule be satisfied for successful NVIS operation:

$$f \dashrightarrow f_c \quad \{\text{MOF-Seeking, NVIS HF}\} \tag{4.35}$$

The probability of ionospheric support for NVIS communication depends critically upon the overhead value of $foF2$. As we have seen this varies considerably during the day and drops to relatively low values at night reaching its minimum before sunrise. The daytime variability is a problem but is not insurmountable. The nocturnal reduction is a serious problem. Figure 4-24 is an example from Rufenach and Hagn [1966] of the percentage of time a specified operating frequency will exceed the MOF (i.e., the overhead $foF2$) as a function of local time. We see that there is a sizable probability that NVIS will not be available during nocturnal hours, at least at HF.

The daytime variability of $foF2$ must be accounted for as well. Vertical-incidence sounders may be exploited in connection with NVIS to provide a real-time estimate of $foF2$, which is the MOF for this geometry. It could be worse. We note that the MOF for an oblique path depends upon both $foF2$ and the height of the refracting layer. Thus estimates of MOF variability for oblique paths must account for two functions which may or may not be independent. If sounders are unavailable, predictions of $foF2$ alone may suffice for the NVIS mode. (See Chapter 5 for a discussion of prediction methods.)

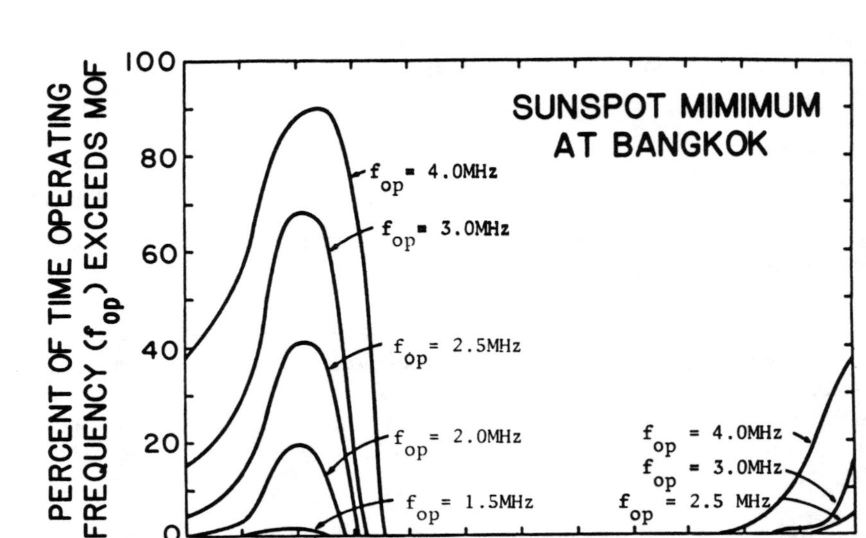

Fig.4-24. Percentage of time that the operating frequency exceeds the overhead value of foF2 on a path of less than 50 km. (From Rufenach and Hagn [1966].)

It may be critical that the transmission frequency be as close to *foF2* as possible in order to reduce the absorption which has its maximum value during the daytime hours. In short, one requires very good predictions. As noted in Figure 4-25, the predicted range of available frequencies (between the LUF and the MUF) is quite limited, especially in the temporal period near dawn. Naturally the same situation occurs with the observed range (between the LOF and the MOF).

One of the fundamental problems with NVIS, aside from those already mentioned, is the spectrum congestion which exists in the band of interest. This leads to unintended interference and a further reduction in channel availability. To make matters worse atmospheric and man-made power-line noise in the lower part of the HF band is enhanced relative to the upper part. (See Chapter 5 for a discussion of noise level prediction.)

Since NVIS requires a vertically-launched group of rays, it is clear that appropriate antennas must be selected to communicate efficiently. Tactical communicators have a tendency to use the ubiquitous whip antenna which is precisely the wrong antenna for NVIS. It has been shown that the effectiveness vertical whips deployed on jeeps may be increased by bending them backward and away from the jeep, thereby minimizing the effect of the overhead null.

4.6.3 Advantages of NVIS

We have noted some requirements for alternatives to LOS and ground wave schemes in section 4.6.1, and NVIS has been suggested as a replacement. NVIS is not without its own propagation peculiarities and these were outlined in Section 4.6.2. Despite the path loss associated with absorption at the low end of the available spectrum and the danger of MOF exceedance at the other, NVIS does have a role to play. Some of the main advantages are given in Table 4-4.

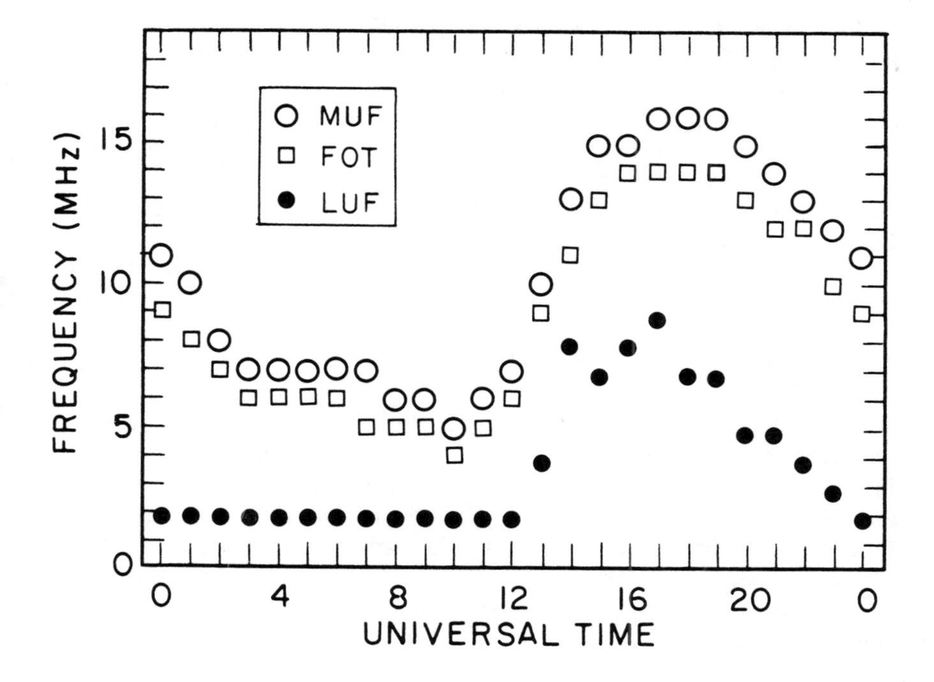

Fig.4-25. Plot of MUF, LUF, and FOT for an NVIS path over Washington DC, based upon the IONCAP predictions [Teters et al., 1983]. (R_z= 200, Washington site at 38.82N, 77.02W.)

TABLE 4-4. PRINCIPAL ADVANTAGES OF NVIS PROPAGATION

1. Omni-directional communication
2. Generally constant signal level over coverage area
3. No skip zone
4. Local terrain effects are minimized
5. Effective for valley-to-valley and NOE communications
6. Electronic warfare

4.7 OBLIQUE-INCIDENCE-SKYWAVE

In this section we shall be dealing with a geometry of fundamental importance to HF communications and related disciplines. Typically the conveyance of information is necessary over significant ground ranges and this may require access to frequencies in excess of the local overhead critical frequency. This means that the upwardly-directed rays which would normally provide NVIS coverage (discussed in the previous section) will be lost in outer space. On the other hand, the use of frequencies f such that $f > f_c$ will serve to enhance BLOS coverage, and system performance will actually improve with range out to 3000 km or so. This result is somewhat surprising to people not familiar with HF.

4.7.1 Introduction and Definition of Terms

Before embarking on the general subject of oblique-incidence-skywave, it is important to distinguish between several terms which are sometimes confused, even by otherwise competent experts. The two terms of prime concern are the Maximum Usable Frequency (MUF) and the Maximum Observable Frequency (MOF). These terms are not the same, but are used in some texts interchangeably. Fortunately, the context often clarifies the intent of the author allowing the reader to make a mental adjustment. The difficulty generally arises in discussions about the process of updating MUFs for use in near real-time applications. In the real-time limit, an update procedure may actually involve replacement of a MUF value by an observed MOF. On the other hand, if an update becomes aged the MUF is a better estimate of the maximum support which will be available over a given link. Updating procedures are discussed more fully in Chapter 6. Below we define several terms which are used with oblique propagation. For the most part these definitions are consistent with those contained within CCIR Recommendation 373 [CCIR, 1978; 1986j]. Generally, MUF refers to a median representation of an ensemble of MOF values, although CCIR definitions tend to obscure the distinction between the two. This will be clarified as we go along. (Note: the source of the definition will be appended. If no source is provided, the definition is suggested by the author of this book.)

4.7.2 Transmission Frequency Definitions

A number of definitions for extrema in the set of HF propagation frequencies has been developed over the years, and this has led to some confusion in usage. Table 4-5 contains a list of terms which we shall define in this section.

TABLE 4-5. TRANSMISSION FREQUENCY TERMS

TERM	ABBREVIATION	SOURCE
Basic MUF		[CCIR,1986j]
Classical MUF		[CCIR,1978]
Standard MUF		[CCIR,1978]
Est.Junction Frequency	EJF	[CCIR,1978]
Operational MUF		[CCIR,1986j]
Estimated MUF	EMUF	
Optimum Working Frequency	OWF	[CCIR,1986j]
Frequency Optimum de Travail	FOT	
Estimated FOT	EFOT	
FOT Band		
Maximum Observable Frequency	MOF	[CCIR,1978]
Estimated MOF	EMOF	
Lowest Usable Frequency	LUF	[Davies,1965]
Lowest Observable Frequency	LOF	

1. *Basic MUF*: The Basic Maximum Usable Frequency is the highest frequency by which a radio wave can propagate between given terminals, on a specified occasion, by ionospheric refraction alone [CCIR, 1986j]. The word *usable* in MUF generally implies a median or suitable long-term representation. The Basic MUF is one intended output of well-known prediction programs. (The basic MUF incorporates two other terms which are no longer recommended by the CCIR: the Classical MUF and the Standard MUF. These terms are given below to provide continuity with older texts and documents. The following definitions 1a and 1b require a more explicit understanding of ionospheric propagation than does the accepted definition.)

1a. *The Classical MUF, or Junction Frequency (JF)*: The Classical Maximum Usable Frequency is the highest frequency that can be propagated in a particular mode between specified terminals by ionospheric refraction alone. In an oblique ionogram it is the frequency at which the high and low angle rays merge into a single ray [CCIR, 1978].

1b. *The Standard MUF, or Estimated Junction Frequency (EJF)*: The Standard Maximum Usable Frequency is an approximation to the Classical MUF. It is obtained by application of the conventional transmission curve to vertical incidence ionograms, together with use of a distance factor, or optionally, by some other suitable theoretical method [CCIR, 1978].

2. *Operational MUF*: The Operational Maximum Usable Frequency is the maximum frequency that would permit acceptable operation of a radio service between given terminals at a given time under specified working conditions. This includes antenna specification, emission class, information rate, and required signal-to-noise ratio [CCIR, 1986j]. Again, this definition corresponds to a median representation, and not to individual values. Furthermore the Operational MUF cannot be deduced unequivocally from sounder data.

3. *Estimated MUF (or EMUF)*: The Estimated Maximum Usable Frequency is an approximation to the Basic MUF (not the Operational MUF). It is typically obtained by suitable conversion of vertical incidence ionograms to accommodate the general case involving oblique geometry. This definition is similar to that of the (now defunct) standard MUF or EJF. For a high angle (medium range) path involving the F2 layer, a conversion is of the form: EMUF $\approx$ $<foF2>$ sec ϕ where the corner brackets imply median value, ϕ is the local ray zenith angle, and *foF2* is the ordinary ray critical frequency which is scaled from an ionogram. (See Equation 4.41.) For an NVIS path involving the F2 layer, EMUF $\approx$ $<foF2>$.

4. *Optimum Working Frequency (OWF) or Frequency Optimum de Travail (FOT)*: (The term FOT is used most often). The FOT is the lower decile of the daily values of the Operational MUF at a given time over a given period, usually a month. That is, it is the frequency that is exceeded by the operational MUF during 90 % of the specified period [CCIR, 1986j]. Emphasis is on operational in this definition.

5. *Estimated FOT (or EFOT)*: An estimate to the FOT obtained by taking 85% of the Basic MUF. This definition is a convenient one used in the computation of FOT in a number of simpler programs which provide basic rather than operational MUFs.

In connection with analysis of most favorable bands for transmission over specified links using oblique ionograms as a basis, the term FOT band has been used specifically for links exploiting F2 layer propagation. (To understand this term fully the reader must refer to Definitions 6 and 7.) The author suggests the following term to define a band of frequencies which might logically be accessed in a frequency management strategy which is MOF-seeking.

6. *FOT Band*: The FOT Band is the region on an oblique ionogram which satisfies two conditions: no discernible multipath and high signal strength. This definition may be made more precise by defining discernible multipath and a threshold minimum detectable signal. Often one defines discernible multipath as those secondary returns (or echoes) which are no more than 10

dB below the principal return in the frequency domain of interest. In many practical applications, the principal 1-hop return on an ionogram is the F2 layer low ray. At the high frequency end, the FOT Band is typically limited by the contribution of the single-hop F2 layer high ray which introduces a multipath component. At the low frequency end, it is limited by multiple hop F2 layer modes, F1 modes, E modes, and sporadic E. The second condition of high signal strength tends to place the center of the FOT band in the range 80%-90% of the MOF although conditions will vary. The fact that EFOT and the center of the FOT Band are ≈ 85% of the Basic MUF and the MOF respectively is only a coincidence. The center of the FOT Band is dependent upon a mode and signal level analysis of a single oblique-incidence-ionogram; EFOT for the F2 layer, on the other hand, depends almost exclusively on the statistics of F region variability over a full month.

7. *Maximum Observable (or observed) Frequency (MOF)*: The Maximum Observable Frequency is the highest frequency at which signals are observable on an oblique-incidence ionogram [CCIR, 1978]. In connection with this definition the CCIR indicates that the MOF is defined on a per-mode basis. When used without qualification the term implies simply the highest of all propagation frequencies on a given path. It is emphasized that the MOF refers to observables from an oblique sounder system, but not necessarily a telecommunication system using the same path.

8. *Estimated Maximum Observable (or Observed) Frequency (EMOF)*: The Estimated Maximum Observable Frequency is an approximation to the Maximum Observable Frequency for a specified path, and is derived by a suitable transformation of the overhead critical frequency, obtained from a vertical incidence ionogram, to account for the oblique geometry of the specified path. For high angle (medium range) paths involving the F2 layer, this transformation takes the form EMOF $\approx$ *foF2* sec ϕ where ϕ is the ray zenith angle. This definition is the observable equivalent of the EMUF above. Here no ensemble median is used, however.

9. *Lowest Usable Frequency (LUF)*: The Lowest Usable Frequency is the minimum frequency below which the (circuit) reliability is unacceptable [Davies, 1965]. This naturally suggests that the LUF is strongly dependent upon system parameter specification or the signal-to-noise ratio (SNR). Other factors which are important in determination of the LUF include: ionospheric absorption, E-layer occultation or screening, system power-gains product, as well as the required SNR for a specified reliability at a given grade of service. Discussions of reliability determination and the LUF will be given later in this chapter.

10. *Lowest Observable Frequency (LOF)*: The Lowest Observable Frequency is the minimum frequency detectable on an oblique-incidence ionogram.

The Ionospheric Prediction Service of Australia in the context of its ASAPS prediction program uses other terms such as ALF, for absorption limiting frequency, and BUF, for best usable frequency (see Chapter 5).

4.7.3 Discussion of Transmission Frequencies

All estimated values of the maximum transmission frequencies (i.e., EMUF, EMOF, and the EJF) which are derived from vertical-incidence data must rely upon a form of Martyn's Equivalence Theorem called the secant law for extrapolation to the oblique path. Martyn's theorem and other fundamental relations are examined in Section 4.8.

It should be noted in all definitions involving oblique-incidence-sounding measurements such as MOF, LOF and FOT-Band that the antenna employed in the sounding plays a key role. It is typically assumed that an antenna providing constant gain over the range of operational signal arrival angles is rather optimum. Otherwise, the sounder may suppress modes which are of importance for the actual telecommunication system. Using this strategy, it is still important to know the pattern of the actual antenna so that constant-gain sounder assessments will not lead to an overly optimistic (and false) picture of the actual propagation and circuit performance. Naturally, it would be preferable to use the same antenna for both sounder and telecommunication systems, but this is not always possible.

It is also noted that the Classical MUF (now defunct but embodied generally within the context of the Basic MUF) is independent of power, whereas the Operational MUF does depend on power through the designation of a required SNR. The Classical MUF and Basic MUF may be deduced through a consideration of ray optics alone without detailed system considerations. They may be determined from ensembles of oblique-incidence soundings, provided of course that the sounder system is well calibrated and has sufficient sensitivity (not a big challenge). Furthermore these (nonoperational) MUFs exclude non-classical layer refraction phenomena such as scatter or off-great-circle reflections which may introduce nose-extensions on individual ionograms. Such increases are often observed and may permit connectivity at higher than normally predicted frequencies.

From inspection of the definitions, we note that there are two versions of the highest and lowest transmission frequencies. There are those which contain a *U* and those which contain a *O*. For a moment we shall concentrate on the MUF and the MOF as canonical examples. The MUF represents the tendency (i.e., median) over a monthly period for a specified hour. It may be computed from sounder data (by deducing the median value of the MOF for example) or it may be predicted. Prediction is typically based upon a driving parameter such as a smoothed sunspot number (to be specified) in conjunction with an archive of median data. In either case variability information is disguised. The MOF is an individual value. An ensemble of MOF values contains variability information as well as statistical summary information such as average and median values. For purposes of HF frequency management the median is significant since this is the manner in which the long term data base has been assembled. The MUF and EMUF may be deduced in accordance with the following formulas:

$$\mathrm{MUF} = \langle \mathrm{MOF} \rangle \tag{4.36}$$

$$\mathrm{EMUF} = \langle foF2 \rangle \sec \phi \tag{4.37}$$

where the corner brackets imply that an ensemble median is to be deduced. In principle it is possible to predict an individual value of either MOF or foF2 for a specific time (and condition) in the future. Routinely these predictions are simply the predicted monthly medians for the conditions specified. Consider the following hypothetical exercise in prediction as a preview of some of the issues we shall address more fully in Chapter 5.

It is March 5, 1991 and we wish to predict the MUF for 0600 LMT for June 22, 1991 over a specified link for which values of the MOF are being obtained routinely. (The routine observations may be from sounders of the Chirp type for example.) Since the time period between today's date and the future date is 109 days there is no unassailable possibility for providing a forecast based upon a 27-day persistence of solar activity (even though prediction gain has been postulated by some workers based upon recurrence principles). Thus we are forced to use long-term prediction methods which lose track of the day of the month. The long-term prediction method of choice relates the monthly median for a specified hour with a 12-month running average of the sunspot number. Consequently our best guess for the MUF to be expected on June 22, 1991 is no different than it would be if today's date were March 8; or if the target date were June 14, 1991 for that matter. Only the link geometry, the LMT, and the solar activity index matter for the prediction method. The generic 0600 LMT estimate of the median will depend upon the model being used as well as the predicted solar activity (typically some effective sunspot number). There are two components of error in our predicted MUF for June 22. The first component is associated with a departure of the 0600 LMT observation (from a sounder, say) from the observed June median. This is an unavoidable ionospheric variability term. A second component is associated with the departure of the predicted median from the observed median. This is a measure of the performance of the model as well as our ability to predict sunspot numbers. It is virtually impossible to eliminate intrinsic (short-term) ionospheric variability effects since they exhibit such limited temporal correlation properties. One method for reducing prediction errors involves the insertion of near-real-time data which may assist in removing some of the bias errors in the predicted median. The process usually involves updating the model at a certain time (by adjusting model coefficients to force the model to match experience). The updated model is then exploited for a near-term forecast. We use the term forecast here since we have now inserted real information rather than resorting only to the discipline of statistics.

4.8 VERTICAL AND OBLIQUE PROPAGATION RELATIONSHIPS

There are a number of situations for which information derived over a vertical path must be applied to an oblique path. For transionospheric propagation in the high frequency approximation (where $f \gg f_p$) it involves the transformation of locally-derived ionospheric profiles $N(h)$ to oblique profiles $N(s)$ where h and s are measured along the vertical and oblique paths respectively. On a differential basis, this transformation is trivial for a flat earth since we may write:

$$ds = dh \sec \phi \tag{4.38}$$

where ϕ is the ray zenith angle which for a flat earth is constant. However, for a curved earth the local ray zenith angle becomes smaller as we move along the ray trajectory (See Figure 4-20). In short, the result of a calculation involving Equation 4.38 depends upon distance along the path. This would appear to be only a minor annoyance since Equation 4.38 is simply modified by noting the earth-centered surface angle which is subtended by the oblique path. If this angle is δ, and the elevation angle is β, then:

$$ds = dh \csc (\beta + \delta) \tag{4.39}$$

where $\pi/2-\beta-\delta$ is simply the local ray zenith angle ϕ'. Nevertheless, for the general long-haul HF problem (where $\delta \neq 0$), in which the ionosphere is not flat, the ray trajectory is not rectilinear, and earth curvature cannot be ignored, complications arise. Before considering this matter any further, we will now write down a set of equivalence theorems which will be of considerable importance in associating parameters which define vertical-incidence and oblique-incidence ray trajectories. We shall first take the simplest of cases and take the ionosphere and the earth's surface to be parallel planes. We shall refer to Figure 4-26.

4.8.1 The Secant Law and Other Useful Relationships

HF signal parameters are strongly dependent upon the properties of the medium which is traversed and the radiofrequency which is used. Estimates of ionospheric properties and/or certain transmission frequencies are required for prediction methods used in spectrum planning, or channel assessment methods which are the keys to successful real-time frequency management. If (oblique) path sounding is unavailable, data derived from vertical incidence sounders (VIS) might be the only available source of propagation data. This VIS data may be part of an archive, or it may be in real-time form. Some means for conversion of VIS information is needed so that it might be applicable to an equivalent oblique communication path. As suggested in the introductory commentary (Section 4.8), there exist some theorems which provide a convenient, albeit approximate, way to perform this transformation.

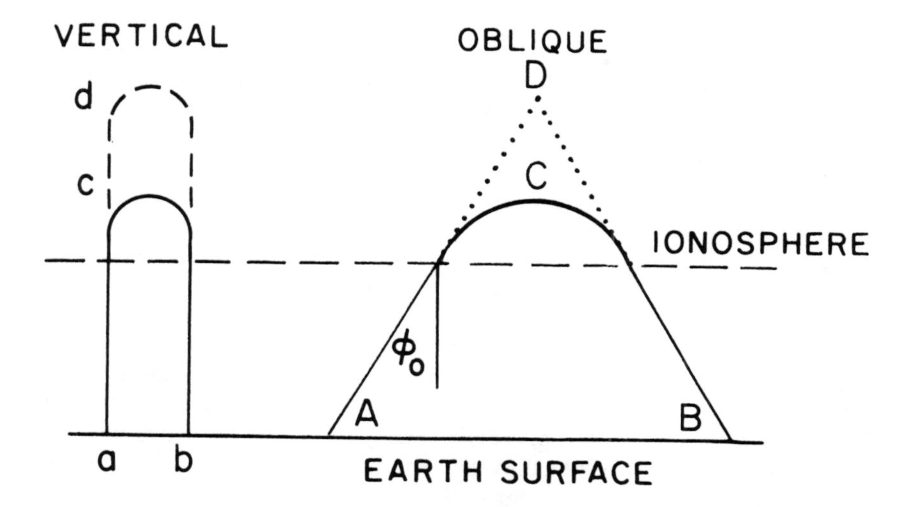

Fig.4-26. Geometry for ray propagation. Earth and ionosphere are plane parallel surfaces. Terminals (endpoints) of the oblique path are A and B, its real apogee is C, and its virtual apogee (triangle vertex) is denoted by D. For the vertical path, the terminals a and b are colocated. The true point of reflection is at c, and the virtual point of reflection is at d. The angle of incidence (i.e., ray zenith angle) is termed ϕ, and is equal to $\frac{1}{2}$ the vertex angle at D.

4.8.1.1 *The Secant Law*

The secant law is known as the first part of Martyn's equivalence theorem. Two rays, one incident obliquely while making an angle ϕ with the vertical, and the other vertically incident, will have the same real height of ray reflection if the radio frequencies obey the following relation:

$$f_o = f_v \sec \phi \tag{4.40}$$

4.8.1.2 *Martyn's Equivalent Path Theorem*

The virtual heights of reflection for the oblique path and the equivalent vertical path are the same, provided the secant law is valid. From Figure 4-26, this implies that the oblique frequency f_o will penetrate just as deeply in the ionosphere as the vertical frequency f_v. Naturally, if we use f_v for the same oblique launch angle such that $f_v = f_o$, then the depths of penetration will not be equal.

4.8.1.3 *Breit and Tuve Theorem*

The time required for a wave to propagate an actual oblique trajectory is the same as the time for a ray to traverse a (virtual) triangular path at the free space speed of light. In Figure 4-26, this implies that $\delta T(A \rightarrow C \rightarrow B) = \delta T(A \rightarrow D \rightarrow B)$ where δT is the propagation time.

4.8.1.4 *Martyn's Absorption Theorem*

The absorption A_o along an oblique path is related to the absorption A_v along the vertical path by the relation (see Section 4.5.2):

$$A_o = A_v \sec \phi \tag{4.41}$$

where absorption is typically reckoned in dB units. The application of Equations 4.40 and 4.41 in the conversion of locally measured quantities to an oblique path (for which the properties are graphically removed) is rather direct. For example if f_v is equal to the ordinary ray critical frequency for the F2 layer, we note that $f_v = f_c$(O-mode, F2) = foF2 and,

$$f_o = \text{foF2} \sec \phi \dashrightarrow \text{(EMUF or EMOF)} \tag{4.42}$$

A brief comment should be made at this point about Equation 4.42 since we have introduced the terms, EMUF and EMOF for the first time. The right-hand-side of Equation 4.42 is EMUF if foF2 represents an ensemble median. Typically, such a median refers to a specified month and hour, and is reckoned from N data points, where N is the number of days in the month. More precisely, we have:

$$\text{EMUF} = \langle \text{foF2} \rangle \sec \phi \tag{4.43}$$

where as before the corner brackets represents the ensemble median over N days. If a single value of foF2 is used in the deduction, then the right-hand-side of Equation 4.42 is EMOF. A number of studies have been undertaken to compare predicted MUFs (such as those obtained from programs like IONCAP) with observations (accommodated through the use of sounders). Clearly the appropriate correspondence is between the predicted MUF and the $\langle$MOF$\rangle$, where a suitable number of MOF values are used in the computation of the ensemble median. (It is noteworthy that medians are involved here and not means.) This is a good test provided, of course, that the data involved in the $\langle$MOF$\rangle$ computation are independent of the archived data (i.e., coefficients) used in the model prediction of the MUF.

Figure 4-27 illustrates the relationship between refractive index and the electron density profile for both oblique and vertical paths in the ionosphere, under the no-field assumption (i.e., $n^2 = 1 - X$). For the vertical case we see that when the transmitted frequency f is identical to the critical frequency f_c (i.e., $f \equiv f_c$ and $X = 1$), then $n \rightarrow 0$ precisely where N_e is maximized. Reflection occurs at this point and for all frequencies such that $f \leq f_c$. As f becomes very large, then $f \rightarrow 1$ everywhere. For the oblique geometry, we see for the electron distribution depicted that a radiowave with frequency f will be refracted downward for a launch elevation given by:

$$\beta = \cos^{-1}[1 - (f_c/f)^2]^{\frac{1}{2}} \tag{4.44}$$

For f_c = foF2, and noting that $\pi/2-\beta = \phi_I$, we see that Equation 4.44 is equivalent to Equations 4-33a and 4.33b. Thus in the limit of plane geometry, the launch elevation β may be used to define the border of an ionospheric iris for the specified values of foF2 and f. If we fix the ratio foF2/f, we may define this limiting elevation angle β, termed β_I. For values of $\beta > \beta_I$, energy at a frequency f will be lost into space. It is noteworthy that a specified value of β_I also may be related to a specific propagation distance given a virtual height of reflection. It then carries the additional significance that the corresponding value of f is the Basic MUF for that propagation distance. (Note we use MUF and not MOF under the presumption that foF2 is a modeled median value.) Although earth curvature and a spherically stratified ionosphere will introduce some modifications in these notions, the principles remain the same. Figure 4-28, although specific to a real (not even spherically stratified ionosphere) and a curved earth, shows the ionospheric iris in connection with the breakthrough phenomenon. Of course, ray tracing is employed instead of a mirror model. We shall return to Figure 4-28 and the issue of ray tracing later.

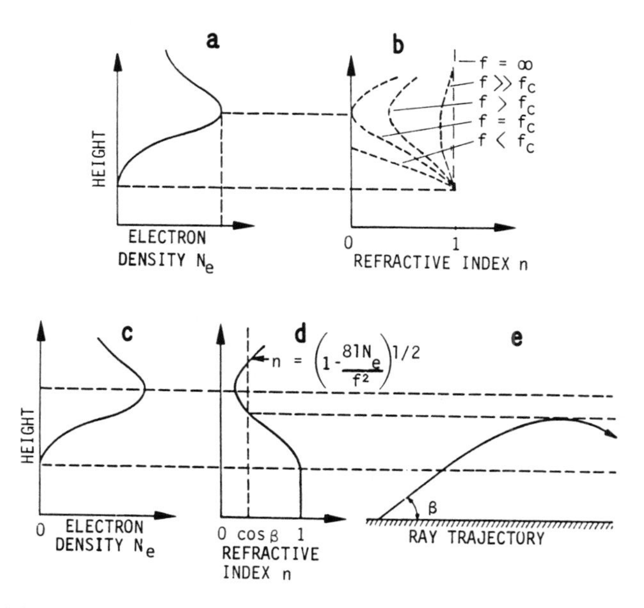

Fig.4-27. Relationships between ray path, refractive index, and electron density

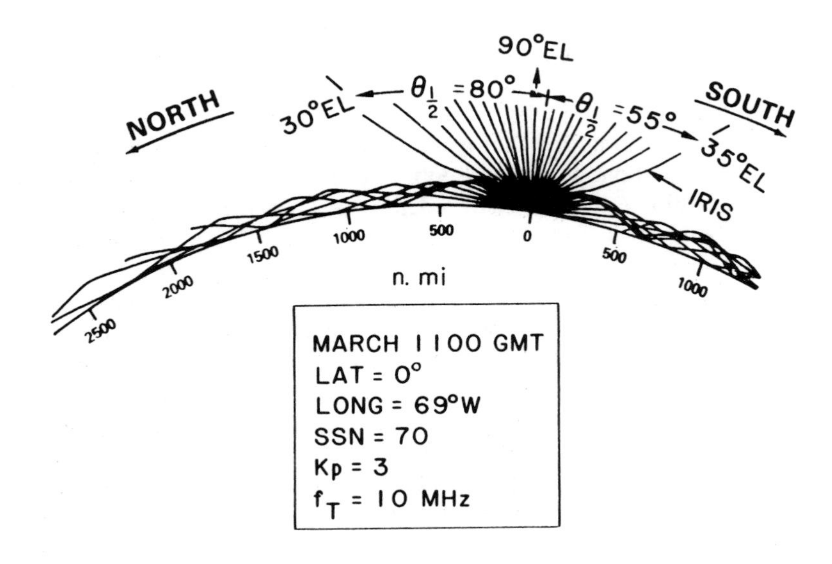

Fig.4-28. Raytracing through the ionosphere showing the iris effect.

The secant law is not strictly valid for a curved earth. A correction factor is generally applied to account for this, and we rewrite Equation 4.40 as follows:

$$f_o = f_v (k \sec \phi) \tag{4.45}$$

where ($k \sec \phi$) is termed the corrected secant ϕ [Lucas and Haydon, 1966] and k is a numerical correction factor. The correction factor k depends not only on distance of propagation but also the ionospheric layer profile. This has been discussed by Smith [1939]. A simple representation depending only upon distance is a practical approximation (see Figure 4-29). We see that k may range between unity for distances of 250 km or less to almost 1.2 at the limiting range for 1 hop F2 layer propagation. This simplifies the construction of transmission curves similar to the one given in Figure 4-30. Transmission curves for a fixed distance of propagation provide the user with a graphical way to transform a vertical-incidence ionogram to an oblique-incidence ionogram. Figure 4-31 shows how this is done. We shall return to the issue of transmission curves in Chapter 6.

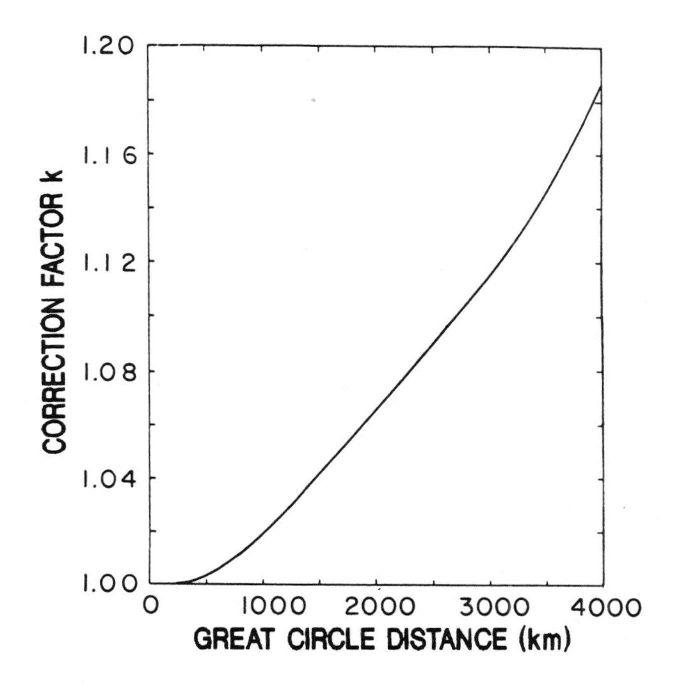

Fig.4-29. Ionospheric curvature correction factor. This factor is used in the equation $f_o = f_v\, k$ sec ϕ (secant law). (From Lucas and Haydon [1966], after N. Smith [1939].)

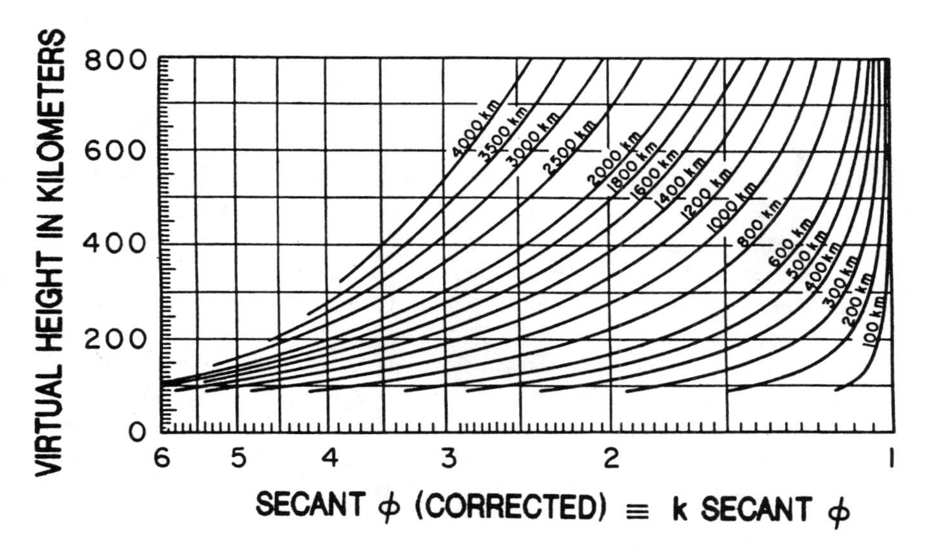

Fig.4-30. Logarithmic transmission curves: virtual height of reflection versus secant ϕ (corrected) and parameterized in terms of the propagation range [Davies, 1965].

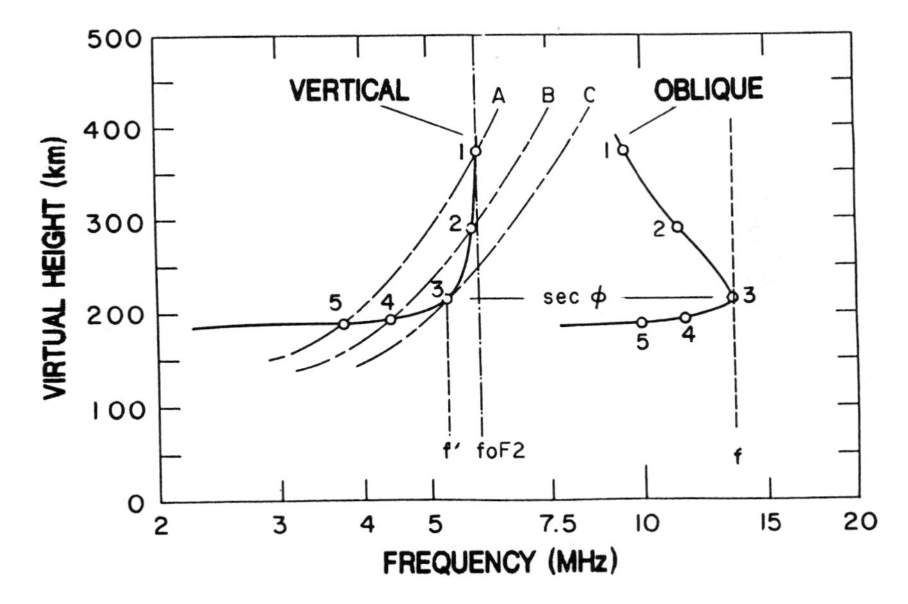

Fig.4-31. Illustration of the transformation from VIS to OIS geometry. Association of the numbers on the two curves indicates the $k \sec \phi$ dependence upon the height. The curve C is a transmission curve just tangent to the vertical-incidence ionogram at point 3. This corresponds to the MUF. Another curve marked B intersects at two locations: 2 and 4. Points 2 and 4 may be associated with High and Low rays respectively. (From Lucas and Haydon, 1966.)

4.9 PROPERTIES OF OBLIQUE PROPAGATION

Of some interest are depictions of oblique ray trajectories under specified conditions involving launch or elevation angle β, transmission frequency f, and terminal separation distance or ground range. We have the possibilities:

1. Fixed terminal positions, variable f and β (Figure 4-16)
2. Fixed f, variable ground range and β (Figure 4-32)
3. Fixed β, variable f and ground range (Figure 4-33)

Curves of this type are found in many texts including that of Davies [1965].

The situation depicted in Figure 4-16 is one of the more practical situations encountered. It corresponds to a fixed link without restrictions on frequency or elevation angle, and illustrates the possible mechanisms for connectivity using one-hop single-layer modes.

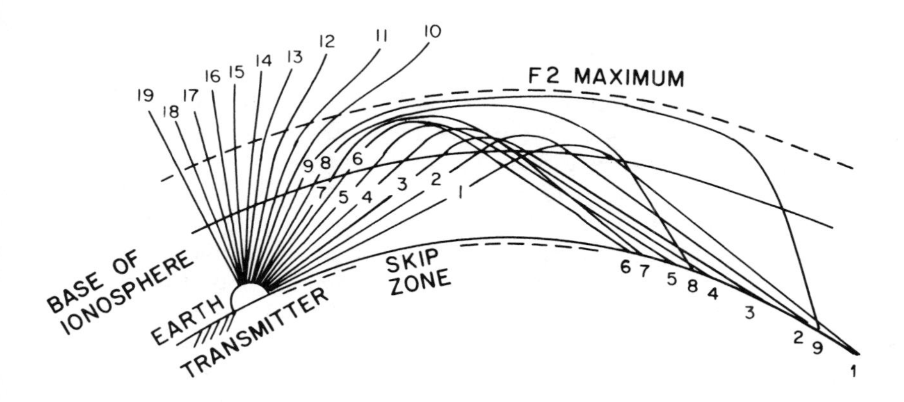

Fig.4-32. Ray trajectories for fixed frequency transmission. (adapted from Hortenbach [1986].)

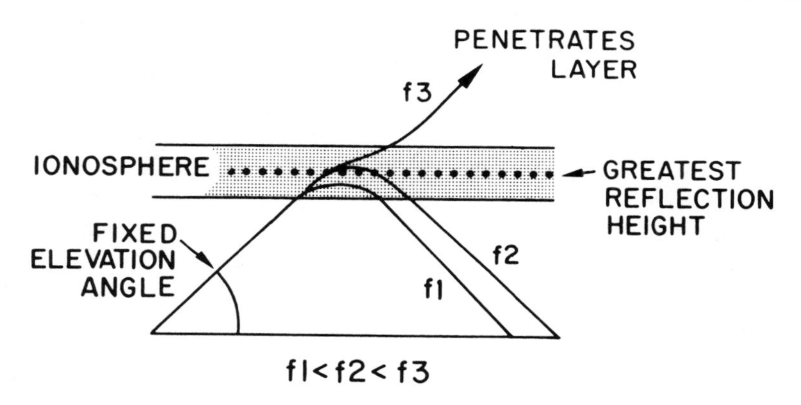

Fig.4-33. Ray trajectories for fixed launch angle

Even though the situation has been highly simplified by ignoring the magnetic field, a number of possibilities still exist for connectivity between points A and B. At the (basic) MUF [40], we presume the existence of only one path. At frequencies below the MUF, the ray path bifurcates into a low ray

---

40. Refer to Section 4.7 for more discussion about maximum transmission frequencies. The MUF, as used in this discussion, is taken to be the so-called classical MUF as opposed to the operational MUF, the latter selection being favored by communication specialists. In either case the MUF is the point on an oblique ionogram where the high and low angle rays merge. The fact that we use MUF instead of MOF indicates that a median representation serves as the basis of reckoning. Still the comment is applicable to *classical* or *operational* MOFs observed over a specified link at a specified time.

and a high ray. The high ray, sometimes called the Pederson ray[41], is launched at an elevation higher than is the MUF ray and penetrates more deeply into the ionosphere before being refracted downward again. This high ray encounters more electrons in its traverse of the ionosphere, is more strongly dispersed, and suffers more Faraday rotation than a low ray at the same frequency. The low ray exhibits the least dispersion, possesses the largest field strength, and will support communications relatively more effectively. There are occasions, however, at low elevation angles when the low ray is obscured by the earth leaving only the high ray to provide connectivity. Even though we have used the term MUF is this discussion, we may replace MUF by MOF in order to analyze real-time situations (for which ionograms are available).

Figure 4-32 is a rather complicated picture requiring individual rays to be indexed. In this instance, the transmission frequency is fixed but elevation is allowed to vary. This provides a crude view of the one-dimensional terrestrial footprint of fixed frequency transmissions. At the highest elevation angles, the signal escapes and no energy returns to earth (paths 19-10). At a certain elevation angle one encounters the ionospheric iris limit which was discussed above (between paths 9 and 10). At this limit, the impact of the ionospheric layer on the ray trajectory begins to become significant. As we reduce the elevation slightly below the iris limit, rays will be returned to earth but at distended distances (paths 9, 8, and 7). As we continue to move below the iris limit, we reach a minimum ground range from the transmitter corresponding to the skip range, or skip distance (path 6). Unlike NVIS propagation, this minimum distance d_{min} will not be achieved at the highest elevation since, in the present instance, f exceeds the overhead critical frequency. It will also not be achieved at the lowest elevation angles. The value of d_{min} is actually achieved at an intermediate elevation angle which is in the near neighborhood of the critical angle β_I (which is the complement of the layer iris half-angle ϕ_I) but slightly below it . The impact of the ionosphere is maximized for this situation in the sense that the largest refraction is observed. As the elevation is reduced even further, the refracted rays become less influenced by the ionosphere since the ray paths encounter less dense portions of the electron density profile. As a result, the ray trajectory intersections with the earth

---

41. For realistic layer types, such as ones with a quasiparabolic shape, it may be shown that there are two ray launch angles possible for a specified terminal separation and a fixed frequency. Using a form of Snell's Law, Kelso [1964] develops an analytic expression for horizontal ray displacement , initial ray zenith angle ϕ_o, the radio frequency f, the critical frequency f_c , the layer "bottom" height h_o , and a layer maximum $h = h_o + z_m$, where z_m is the layer semi-thickness. Provided reflection is permitted and ionospheric support for a specified range occurs, there are always two rays connecting any two points within the specified range provided f exceeds f_c . To deduce the highest elevation angle still permitting reflection we take $f/f_c \cos\phi_o = 1$. As an aid in visualization, the reader is referred to Figure 5-10 in Kelso [1964].

move outward again to increasingly greater ranges (paths 5, 4, 3, 2, and 1). Accordingly d_{min} represents the skip distance for specified frequency and it is also the distance where high and low rays concentrate resulting in so-called MUF-focusing. For real-time applications, we would more properly use the term MOF-focusing. It may be shown that d_{min} at a specified frequency and ionospheric layer is given as:

$$d_{min} = 2\,h[(f/f_c)^2 - 1]^{\frac{1}{2}} \qquad (4.46)$$

where f is the transmitted frequency, f_c is the critical frequency, and h is taken to be the virtual height of reflection (i.e., $h = h'$). This relation was originally derived by Appleton and Beynon [1948] for the case of a thin layer and a flat earth. Naturally, if f is close to f_c, then d_{min} is close to zero.

The maximum distance of propagation by ionospheric reflection (or ground range), d, is limited as a result of earth curvature. As a good approximation we have the following:

$$d_{max} = [8R_e\,h]^{\frac{1}{2}} \qquad (4.47)$$

where R_e is the earth radius. Over that distance, if we accept the validity of the secant law, then we may associate a maximum frequency with d_{max}. This is given by:

$$f_{max} = f_c\,(\sec\phi)_{max} = f_c\,[R_e/2h]^{\frac{1}{2}} \qquad (4.48)$$

where f_c is taken to be the ordinary ray critical frequency for the layer in question. The value of ϕ in Equation 4.48 is the earth-curvature corrected value of the ray zenith angle. To a good approximation the quantity $[2h/R_e]^{\frac{1}{2}}$ is the complement to ϕ_{max}, and it equals the angle which the ray makes with the ionospheric layer at the point of reflection, or $\pi/2 - \phi_{max}$. This ionospheric angle, β' is equal to the ray elevation (or radiation) angle β for a flat earth. For a curved earth, of course, it is increased by the earth angle between the ray launch point and the ionospheric intersection point (see Figure 4-20). Table 4-6 gives some estimates of d_{max}, $(\sec\phi)_{max}$, and $\pi/2 - \phi_{max}$ for specified reflection heights. The value $(\sec\phi)_{max}$ is the largest possible MUF-factor for the conditions specified.

The values of $\beta' = \pi/2 - \phi_{max}$ in Table 4-6 indicate the difficulty encountered in trying to launch an HF signal into an ionospheric duct. Clearly, it may be necessary to use elevated antennas (to allow negative radiation angles), exploit possible atmospheric superrefractivity conditions (which will increase the effective value for R_e), or invoke some ionospheric scattering/refraction process so that favorable conditions for ducted mode injection will occur. We will return to the topic of ducted and chordal mode propagation in Chapter 4 (viz., Section 4.12).

TABLE 4-6
LIMITING PARAMETERS FOR IONOSPHERIC REFLECTION
(Thin Layer, Curved Earth Model)

Heights (km) → / Earth Radii (km) ↓		100	150	200	250	300	350	400
d_{max}	R_e = 6370	2257	2765	3192	3569	3909	4223	4516
	8500	2608	3194	3688	4123	4516	4879	5214
$(\sec \phi)_{max}$	R_e = 6370	5.6	4.6	4.0	3.6	3.3	3.0	2.8
	8500	6.5	5.3	4.6	4.1	3.8	3.5	3.3
$\pi/2-\phi_{max}$	R_e = 6370	10.2	12.4	14.4	16.1	17.6	19.0	20.3
	8500	8.8	10.8	12.4	14.0	15.2	16.4	17.6

Notes: 1. R_e = 8500 km corresponds to a 4/3 earth radius approximation.
2. $(\sec \phi)_{max}$ is the largest possible MUF factor.

From Table 4-6, we see that the effective earth radius selected is quite important in these limiting conditions which correspond to effective radiation angles of zero degrees. The artifact of 4/3 R_e has derived from a construction whereby refracted rays appear as straight lines over an earth with a modified curvature. (For median conditions of refractive index, this may be accomplished by supposing that the normal earth radius is increased to 1.33 R_e = 8500 km [Hall, 1979].) Above a few degrees of elevation, the impact of tropospheric refraction starts to become unimportant, and this is especially true if the transmitter is elevated (such as in a high flying aircraft or aerostat), a factor which reduces the refractivity experienced by the radiowave. Although tropospheric refractivity is intrinsically frequency independent, long wavelengths (in the lowest part of the HF band) cannot fully experience the scale height gradient which leads to tropospheric refraction, and which formed the basis for the 4/3 R_e concept in the first place. Furthermore, actual patterns of deployed antennas indicate that launch angles below 3 degrees or so cannot be realized in most practical situations. Boithias [1987] takes R_e = 6800 km rather than the 4/3 earth approximation. Use of this would naturally lead to different values in Table 4-6.

Pictures are often better than tables in providing insight. Since we are not usually dealing with limiting cases as suggested in Table 4-6, we have provided Figure 4-34 which shows two typical MUF factor curves which one would expect for the E and F layers. The implication is clear. For a common propagation distance (i.e., ground range), the lower layers require much reduced values of critical frequency (or electron density) to produce the same MUF as that which would be associated with the upper layers. This partially explains why regions E and F1 may exhibit a communication importance which is disproportionately higher than would be expected on the basis of the electron density profile. An illustration of the combined effects of communication range, ionospheric height, and other factors is given in Figure 4-35. From this figure one may estimate the propagation coverage for arbitrary radiation angles (or elevation angles) and specified ionospheric layer heights. One can also deduce the range of effective reflection heights (and radiation angles) which would terminate at a specified range. It should be recognized however that the results obtained are only approximate. Detailed coverage patterns can only be derived with ray tracing through realistic ionospheric representations.

Figure 4-33 exhibits several ray trajectories for a fixed elevation angle. As may be seen, the frequency dependence of the refractive index leads to more penetration for the higher frequencies. For those rays when are reflected back toward earth, their apogees move outward in range and upward in altitude as the transmission frequency is increased. Ionospheric penetration will eventually occur at a frequency consistent with the secant law. Even so, the path is modified from its launched trajectory. In the limit where $f \rightarrow \infty$, the ray trajectory is naturally rectilinear. As f is reduced, refraction becomes evident and eventually the ray will be returned to earth providing communication connectivity. How the individual set of rays will behave is critically dependent upon the initial ray launch angle. For elevation angles well below the complement to the iris half-angle (i.e., for $\beta < \beta_I$), which may be defined for the given ionospheric conditions and frequency employed, we are dealing with a set of low rays. For elevation angles in the neighborhood of $\beta_I - \beta'$ (where β' is a moderately small number) and above, we are dealing with a set of high rays. At an intermediate elevation angle, we are dealing with a single ray, and the frequency involved for that condition is the MUF (in the monthly-median context) or the MOF (in the real-time context).

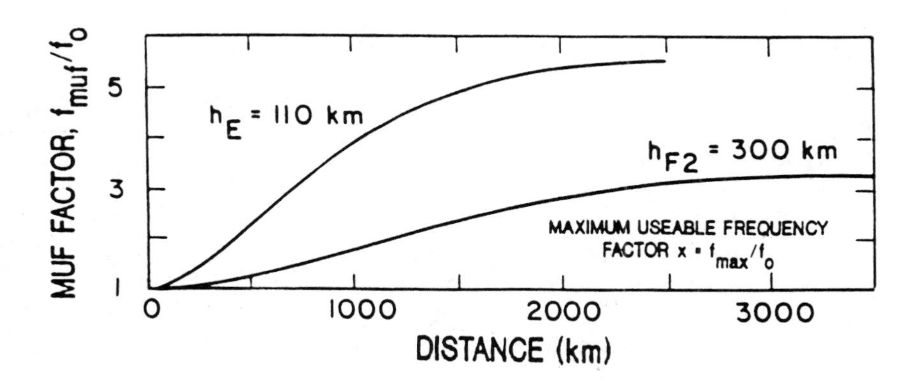

Fig.4-34. MUF curves for ionospheric reflection from the E and F2 layers with an effective E-layer height h_E = 110 km, and an F-layer effective height h_{F2} = 300 km. (from Appleton and Beynon, 1948].)

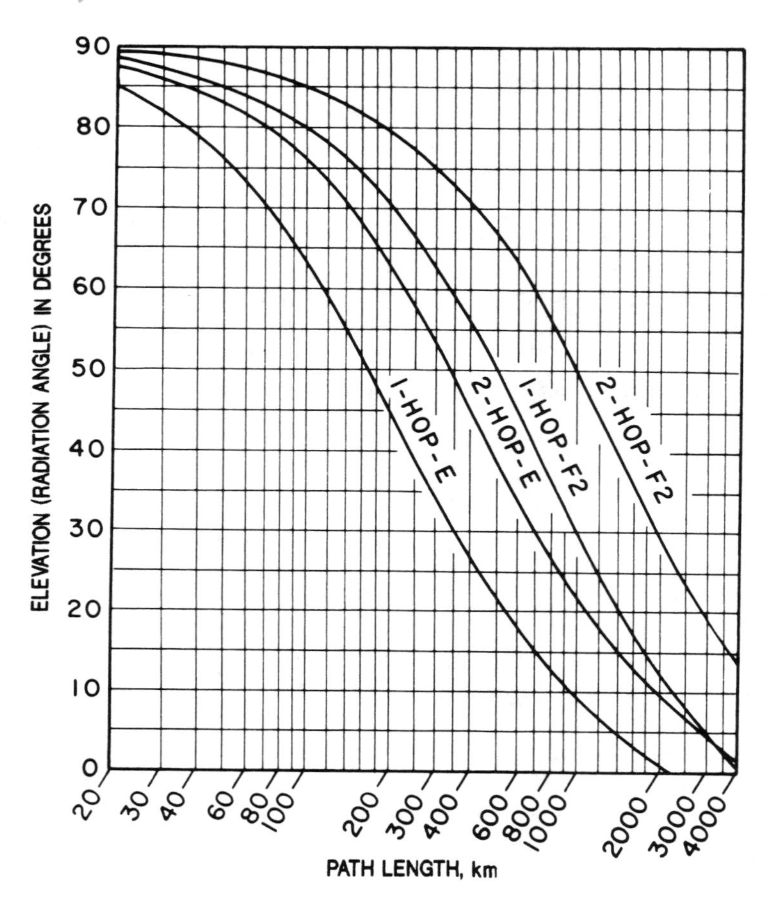

Fig.4-35. Propagation geometry. Ray elevation (radiation) angle versus path length (ground range) for 1- and 2-hop E and F2 layer reflected modes. The mirror model is used and the E and F2 layer virtual heights are assumed to be 105 and 320 km respectively. (from Davies [1965].)

As illustrated in Figures 4-16 and 4-32, the notion of high and low rays are important concepts for oblique HF propagation and they are often misunderstood. In the limit, when the two rays coincide, we have focusing of the two types of rays and an enhancement of the signal level. We looked at this effect from two vantage points: fixed range and fixed frequency. The fixed range case is the one most often encountered in practice, and is the most easily visualized since we may see the effect quite directly from oblique-incidence-ionograms (see Figure 4-36). It is the fixed-frequency, variable-range case requires more discussion. It also has significant practical application in air-ground communications and in other situations in which terminal separations are variable.

One of the best schemes developed for picturing the relationship between elevation and ground range is embodied in the transmission diagram which is a plot of ray elevation angle β versus ground range d [Hayden, 1979]. Figure 4-37 shows how choice of frequency impacts the function $\beta(d)$ for a two-layer model including an E-region and an F2-layer. Three frequencies are examined to produce three distinctly different $\beta(d)$ functions. In Figure 4-37a, the transmission frequency is less than both the E-region and F-region critical frequencies (i.e., $f <$ foE $<$ foF2). In this example, the propagation mode is simply NVIS under E-region control. In Figure 4-37b, the transmission frequency is an intermediate one (i.e., foE $< f <$ foF2). Under this intermediate frequency condition, the propagation mode is essentially F2-layer-controlled NVIS for the short ranges, but for the longer ranges it is conventional oblique propagation dominated by the E-layer. Since $f >$ foE, the E region interaction produces an E-region MUF and, of course, a skip distance d_{min}. Finally in Figure 4-37c, the transmission frequency is in excess of critical frequencies for the E and F2 layers (i.e., $f >$ foF2 $>$ foE). In this instance there will be two skip distances and two MUFs if we examine the full range of elevation angles.

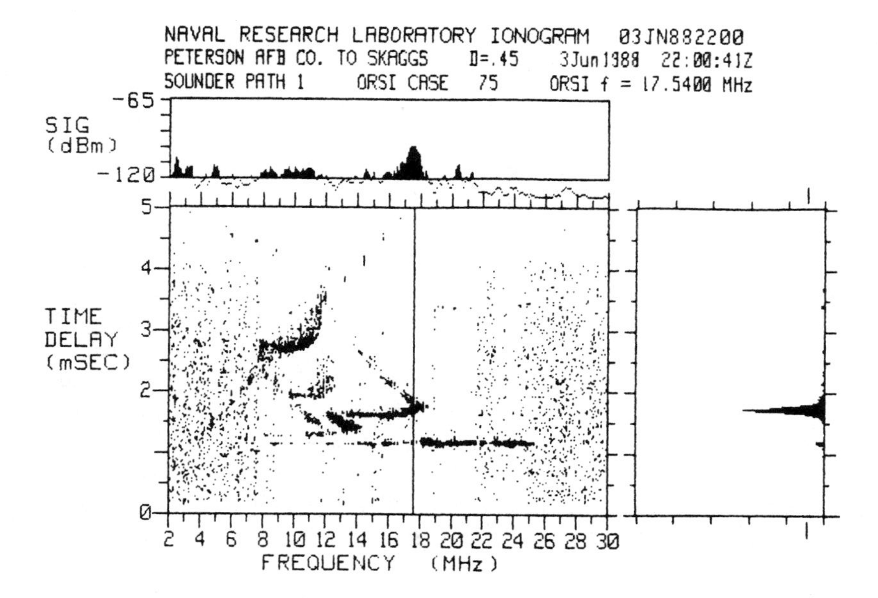

Fig.4-36. Ionogram showing high and low rays merging at the MOF. The example corresponds to a path between Peterson AFB in Colorado and a site near San Francisco, California. The vertical line on the ionogram is placed near the MOF (but slightly below). The upper curve shows that the relative signal level is maximized at 17.54 MHz. The plot on the right shows that the predominant signal source is a single mode at about 1.75 milliseconds time delay. This data was obtained by NRL investigators. GPS was used to provide absolute timing information. (Figure provided courtesy M. Daehler, Naval Research Laboratory, Washington, DC.)

For larger and larger values of transmission frequency (such that $f >$ foF2 » foE but not illustrated in Figure 4-37), the MUF will be defined by the F2 layer, and there will be only a single skip distance (at approximately 4000 km). Hayden [1979] used raytracing methods to generate more realistic results for spherical geometry. We have modified Hayden's original work only slightly to emphasize the points of interest here.

We see from Figure 4-37a that if the frequency is quite low (i.e., $f <$ foE) then the ground range d increases monotonically with decreasing elevation angle. For higher values of f (i.e., $f >$ foF2 and depicted in Figure 4-37c), monotonicity is totally destroyed, and two critical elevation angles are noted, β_I(F2) and β_I(E). One corresponds to the F2 layer iris and the other corresponds to the E-region iris. We see that for elevation angles in excess of β_I(F2), there will be no terrestrial communication possible. Furthermore for $\beta < \beta_I$(E), the E-region will control the maximum range for communication coverage. This is called E-layer screening, and naturally requires that f exceed foE, but it is manifested when this excess is not too large. We thus anticipate E-layer screening for some elevation angle if f satisfies the relation: foE $< f <$ foF2. However it is also seen to be true when the following holds: $f >$ foF2 $>$ foE. Note that the maximum range is roughly 2000 km for the conditions assumed, this being the approximate geometrical limit for a horizontal (β = 0) ray which intersects the E region at 110 km.

We have just discussed the notion of the focusing of high and low rays to generate a MUF-focusing effect. This is a very real situation and is depicted in terms a spatial snapshot derived from a rapidly moving receiver in Figure 4-38. In the spatial neighborhood of the high intensity signal we observe rapid oscillations introduced as a result of the interference of high and low rays. This picture is highly variable in both space and time as a result of ionospheric inhomogeneities and temporal variability of the electron density profile. Another way of looking at this variability (on a time scale for which small scale effects are averaged out) is to examine the diurnal pattern of skip distance contours using frequency as a parameter. Figure 4-39, for example, shows how one-hop F layer modes provide a decidedly different diurnal coverage pattern for designated frequencies. This is the result of both ionospheric height and critical frequency fluctuations during the day. The curves represent the MOF and correspond to two days of data [Croft, 1982]. We see that the F-layer skip distance at 12 MHz ranges from roughly 800 km during the midday period to more than 2800 km during the night. This is a skip range differential of more than 1000 kilometers, with a change rate near the dawn-dusk terminators of the order of 1000 km/hour or more. It is obvious that a MOF-seeking philosophy of frequency management may be subject to some problems in successful implementation unless predictions (based upon MUF) are sufficiently accurate or unless a control function can be implemented to track the skip zone in conjunction with the respective ranges of telecommunication customers from the transmitter.

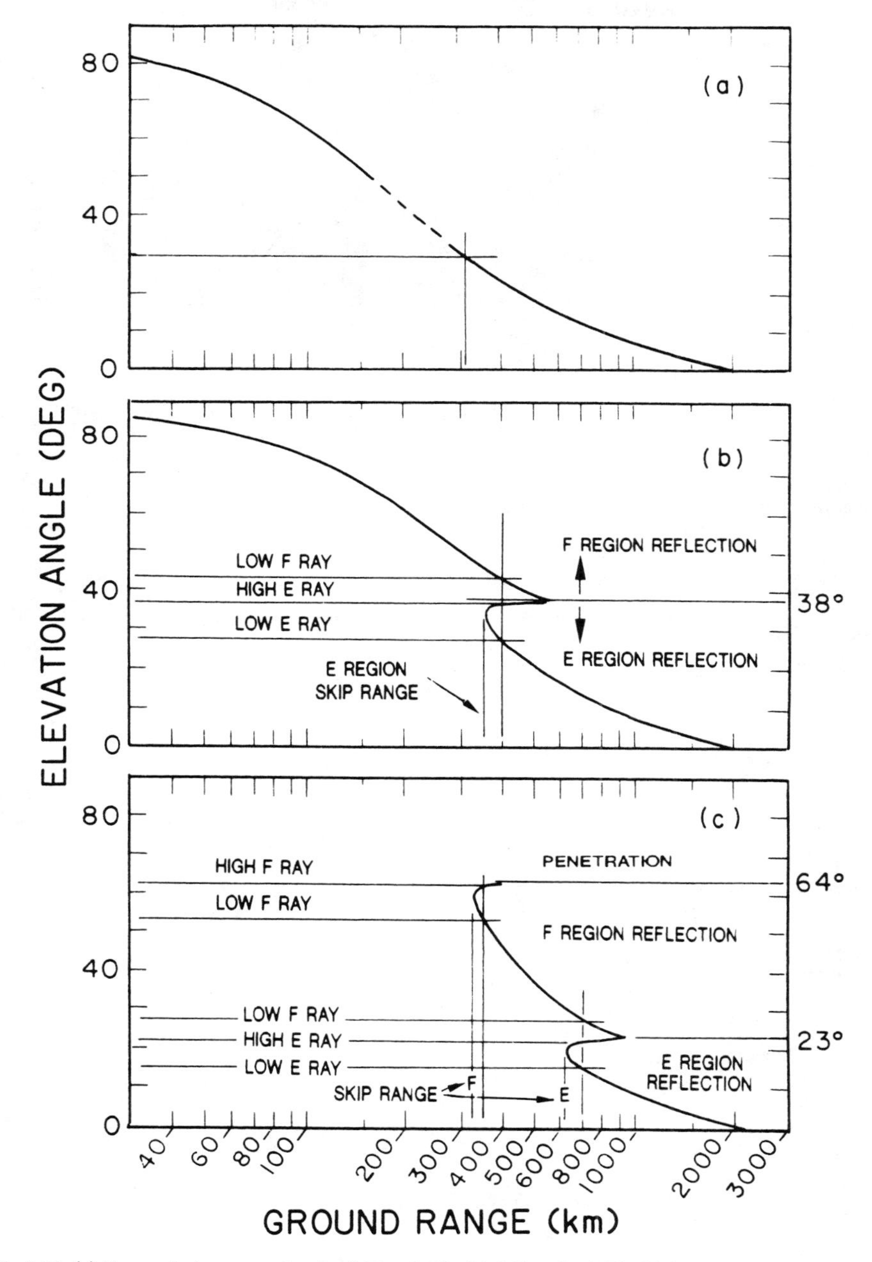

Fig.4-37. (a) Transmission curve for f < foE < foF2. (b) foE < f < foF2. (c) foE < foF2 < f. Figure adapted from Hayden [1979].

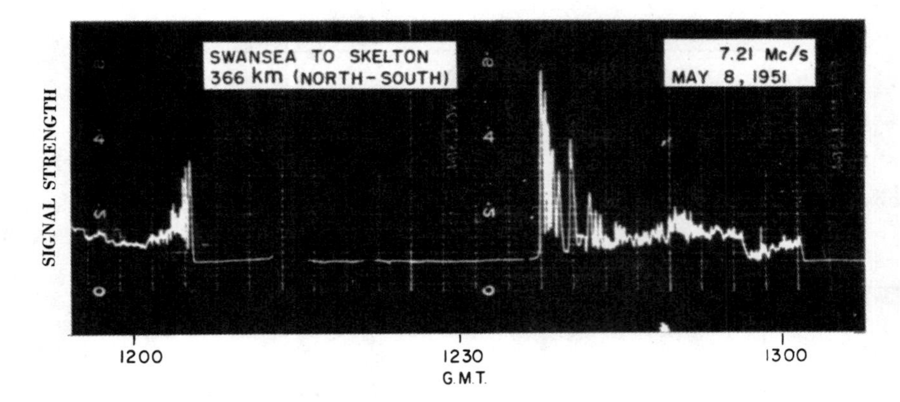

Fig.4-38. MOF focusing. In this illustration the path length and frequency are fixed. Since the actual MOF is time-varying, the skip zone boundary will move, leading to a distinctive fading pattern as the receiver moves in and out of the boundary. The focusing gain may be as much as 6-9 dB. The effect reduced by dispersive effects especially for wideband signaling, and it may be ignored when f is more than a few kHz away from the MOF. Although not depicted in this picture, MOF-focusing is a factor only within a few kilometers of the skip boundary [Davies, 1965]. (Figure courtesy of Ken Davies.)

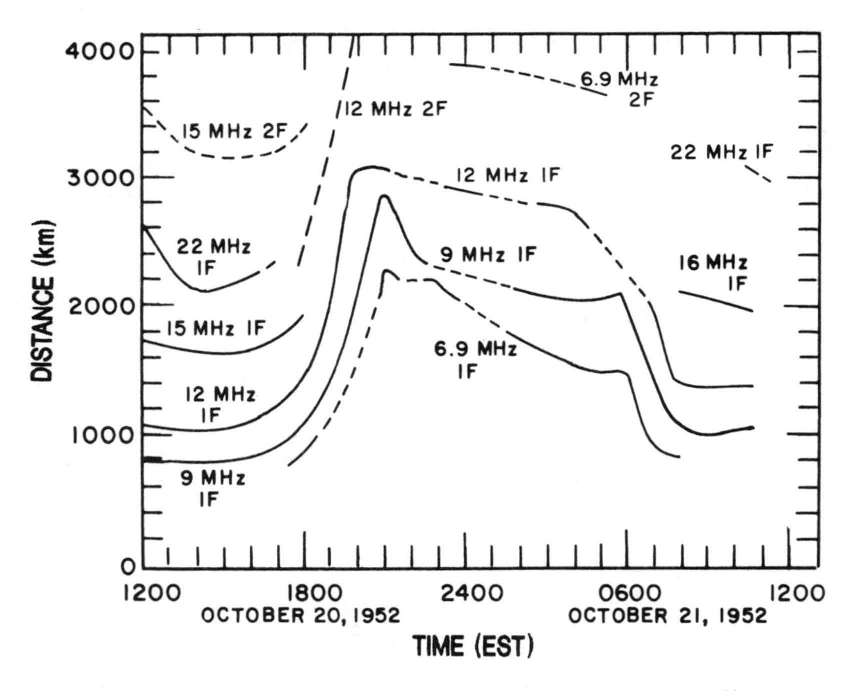

Fig.4-39. Diurnal variation of the skip distance (from Ames et al. [1977].)

4.10 HF RAYTRACING TECHNIQUES

4.10.1 Rationale

Raytracing is an alternative to simpler schemes which involve the concept of virtual geometry. We have emphasized the virtual methods, not only because of historical precedence, but also because of the satisfactory performance of graphical methods and preprocessed analysis tools (i.e., nomograms) based upon such methods. Although a number of factors are involved, system performance methods using more complex raytrace algorithms have not achieved widespread use. Moreover, raytracing methods offer substantial improvement over virtual approaches only if the ionospheric models (or data representations) are appropriate for the physical situation involved. Models such as IONCAP use the virtual methods (sometimes referred to as virtual raytracing) rather than true raytracing to test for the ray paths which should be considered in performance calculations. Another HF model, AMBCOM, uses a raytracing approach. IONCAP and its predecessors such as HFMUFES use the simpler virtual methods to minimize computational load. These latter codes are also well matched to the median data base and simplistic profile shape which is used in the calculations. There is considerable justification for the IONCAP approach for system planning activities such as frequency (resource) management. Raytracing methods are, in principle, well matched to actual environments. This would suggest that such methods have a major role to play in real-time applications for which current ionospheric data are available. This would surely be the case if the computational constraints imposed by the raytracing method were not too severe. Possibly limiting constraints might be: data assimilation requirement (i.e.,input data rate), physical properties of computer required for the application (such as size, power requirement, etc.), and delivery speed of the software application (i.e., output data rate).

It is not clear to the author that raytrace methods are complex (by inference) any more than nonraytracing methods are simple. Indeed, virtual methods have a rich history, and evolutionary growth has led to rather artificial constructions one purpose of which is to provide the analyst with a set of graphical tools. The development of transmission curves for converting vertical information to the oblique geometry is an example. Raytracing approaches are intrinsically more appealing since they are more intuitive. It should be recognized, however, that ray optics has its own shortcomings. For example, in the region of a caustic (i.e., where rays cross each other) ray theory cannot provide an estimate of signal strength.

The choice between raytracing and virtual methods is not just a matter of personal taste but is driven by application. Raytracing is needed to fully examine details of HF propagation. For example, if the distribution of electrons does not fall within the guidelines suitable for use of virtual methods, raytracing will be required. Such conditions arise whenever the ionosphere exhibits a non-stratified behavior or if the actual $N(h)$ profiles depart from standard forms. Propagation effects might include focusing caused by large scale ionospheric structures including the effects of refraction arising from tilts and gradients over the full beam of the transmitting antenna. Raytracing is also essential for a precise assessment of magnetoionic path splitting effects, Faraday rotation, ionospheric Doppler, and angle-of-arrival information. A special issue dealing with ray tracing was published in *Radio Science* [1968], and AGARD CP-13 [Jones, 1969] contains a number of insightful articles on oblique raytracing.

4.10.2 Analytic Raytracing Methods

For quite simple models of the ionosphere, it is possible to determine the range to an HF emitter analytically. This analytic raytracing procedure was originally developed at Stanford Research Institute by Croft [Croft and Gregory, 1963; Croft and Hoogasian, 1968], but operational improvements have recently been made [Milsom, 1985]. It is termed the Quasi-Parabolic (QP) method and its successful operation requires an admixture of simple ionospheric layer representations (parabolas, linear segments, etc.) which do not necessarily arise in nature. An improved ionospheric model has been developed by Dudeney [1978] but this increased level of sophistication requires numerical ray tracing. To avoid numerical raytracing complexities while retaining some degree of ionospheric realism, Multi-Quasi-Parabolic (MQP) models have also been developed [Woyk,1978; Baker and Lambert, 1989]. The success of the general QP approach had been limited to a rather unrealistic ionospheric model [Bradley-Dudeney,1973] but MQP appears to have a broader application and retains the elegance and speed of an analytic process. Unfortunately it works only if the magnetic field effects can be ignored.

Because of their speed, analytic methods have found considerable application in the solution of problems for which the impact of tilts and gradients must be found. Large scale tilt effects have been studied by Nielson [1968]. Another example is the refractive effect of a TID which can be relatively localized and time-varying. Ray trajectory effects of TIDs have been studied by George [1972a, 1972b] from the aspect of HF-SSL systems, and by Croft [1972] from the aspect of OTH radar.

4.10.3 Numerical Raytracing Methods

Conventional wisdom says that numerical procedures, while more friendly in terms of the magnetoionic complexities which may be addressed, are too computationally time-consuming for many purposes. Examples include tactical direction finding (HFDF) and OTH radar surveillance systems. It is taken for granted that real-time systems will not admit to anything but a fast analytic algorithm. Although speed is very important, it is not necessarily true that computational schemes (even those which involve consideration of the magnetic field) need to be slow. The issue of ray homing is critical in many applications and several techniques for solving this problem have been reported (e.g., Rao et al.,1976].

An essentially exact solution for the determination of the path followed between a transmitter and a receiver involves the numerical solution of ray trace equations of a form suggested by Lighthill [1965] and Budden [1985]. This form is a set of eight coupled, first order, nonlinear differential equations. The standard computer program for the solution of the raytrace equations in spherical coordinates [Haselgrove, 1954] is the Jones-Stephenson program [Jones, 1968] [Jones and Stephenson, 1975]. The time dependence of the ionosphere is usually neglected, an excellent approximation over the time scale of the propagation, so that the problem reduces to the solution of six equations for the position and wave vector coordinates.

If magnetic effects are neglected, the solution of the raytrace equations is considerably simplified. The six first order equations are equivalent to three second order equations for the position coordinates (similar to Lagrange's equations in classical mechanics), and it has recently been shown [Reilly and Strobel,1988] that an exact solution of these equations for a ray path increment is possible for the case of no magnetic field if the index of refraction is expanded in a Taylor series about the beginning of the increment. The net result is that a numerical solution of six first order equations for a ray path increment, each requiring four function evaluations in the typical fourth order Runge-Kutta numerical solution, is replaced by an exact solution of two second order equations (one of the coordinates is used as the independent variable in the integration), each requiring one function evaluation. There seems to be an improvement in accuracy over the Jones-Stephenson program results for no magnetic field, but, more significantly, the computations are completed an order of magnitude faster. Ionospheric tilts and inhomogeneities fit naturally into a ray trace program. Reilly and Strobel [1988] use their program to demonstrate the importance of ionospheric tilt effects in single-site location algorithms by means of calculations for a typical sunrise transition region.

As indicated above, one argument against taking on the extra complexity of a raytrace program for selected applications is that ionospheric information is incomplete, and the use of a climatological ionospheric model typically makes errors in the range of 10-25% for bottomside ionospheric specification at any single time. Hence, the increased accuracy of the propagation model is presumed to be unnecessary, in view of the inaccuracy of the ionospheric model. Nevertheless, there is strong evidence that: (1) sounder updates of an ionospheric model can significantly enhance the accuracy of ionospheric specification; (2) it is incorrect to generally neglect ionospheric tilts and inhomogeneities; and (3) magnetic effects are significant. These elements are properly included in the framework of a raytrace program. As regards magnetic effects, it has been estimated that the error from the neglect of the magnetic field is comparable with a 15% error in ionospheric specification at 10 MHz [Reilly, 1990]. At higher frequencies the error from neglecting the magnetic field diminishes (with a dependence inversely proportional to the frequency) relative to the ionospheric specification error. Consequently a raytrace program should be one important ingredient in a system which requires rather exacting ray trajectory specification. New three-dimensional algorithms operate rather efficiently and even include magnetic field effects. The added complexity is easily handled by modern computers, even PCs. A raytrace program also has value in the simulation and interpretation of complex signal structures encountered in HFDF and OTHR systems, and in the development of techniques for handling them. As a matter of fact, the interpretation of ionograms is aided significantly by the use of raytrace methods.

4.10.4 Magnetic Field Influence: The Spitze and Other Things

Inclusion of the magnetic field in numerical raytracing allows for examination of anisotropy and birefringence. A glimpse of this may be found in a discussion by Davies [1990]. At vertical incidence, one finds that the O and X modes deviate significantly from one another. In the magnetic meridian, and at a frequency less that the critical frequency, the poleward excursion of the O mode is much greater than the equatorward deviation of the X mode at the point of reflection. On the other hand, at a given height (but prior to reflection), the X mode has the greater deviation. Such deviations may be considerable, up to about 50 kilometers. Accordingly the notion of vertical incidence sounding might appear to be strange. Perhaps the NVIS term would be more appropriate to describe all short-range skywave applications whether the intent is for overhead coverage or not. VIS naturally involves such an intent. Nevertheless, in order to avoid a semantic struggle, we will still refer to a vertically firing array or other suitable antenna as capable of supporting vertical incidence skywave propagation if that is the intent.

Oblique propagation in a magnetic field is not without its interesting features. A spitze is a discontinuity in the ray trajectory which arises during propagation in the magnetic meridian plane. Davies [1990] discusses this phenomenon as well as other features for meridional and zonal propagation. Of importance in RTCE systems is the ray separation for X and O modes. For propagation in the magnetic meridian plane, the O and X rays remain in that plane. On the other hand, the two magnetoionic modes are deflected in opposite directions for zonal propagation. However, to first order, the deviations for the upgoing (prereflection) and downgoing (postreflection) legs for each mode are equal and opposite. Ultimately neither the O nor the X modes will experience any substantial deviation out of the zonal plane because of the compensation effect of the upgoing and downgoing deviations. Nevertheless, at the point of reflection, the maximum separation between the modes will occur. This separation may be as much as a kilometer.

4.11 PROPAGATION LOSS CONSIDERATIONS

Equation 4.2 defined the Basic Transmission Loss (in dB) the free space loss between two isotropic antennas. It is used to determine the reduction in power density as a function of both frequency and propagation distance: For convenience we repeat it,

$$L_{fs}(f,d) = 32.45 + 20 \log f(\text{MHz}) + 20 \log d(\text{km}) \qquad (4.2)$$

where the last term on the right hand side is called the distance loss L_d. Figure 4-40 is a nomogram which may be used to deduce $L_{fs}(f,d)$.

We have shown that for ionospheric skywave modes, we have additional losses due to absorption and focusing (actually divergence or convergence) as the rays propagate. Also, for multiple hops, losses introduced as a result of ground reflection will arise. The magnitude of this loss is nominal for sea reflection (being typically much less than 1 dB) but may be significant for ground reflection (ranging between 1 and 10 dB). Figure 4-41 [Davies, 1965] shows the losses associated with two extremes of media reflection: poor ground and sea under the assumption of random polarization.

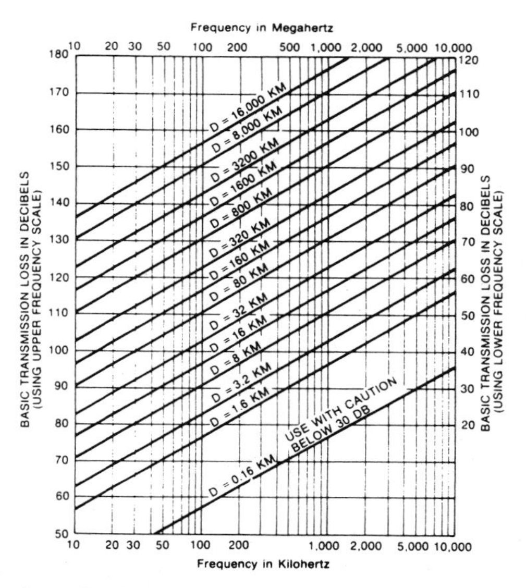

Fig.4-40. Nomogram for estimating the spreading loss between two terminals for specified distances as a function of f.

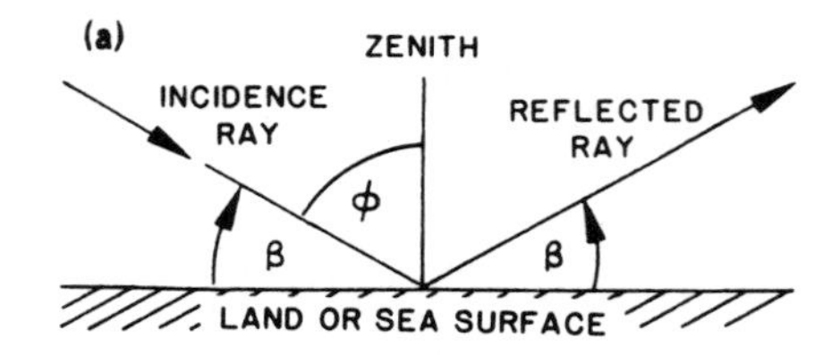

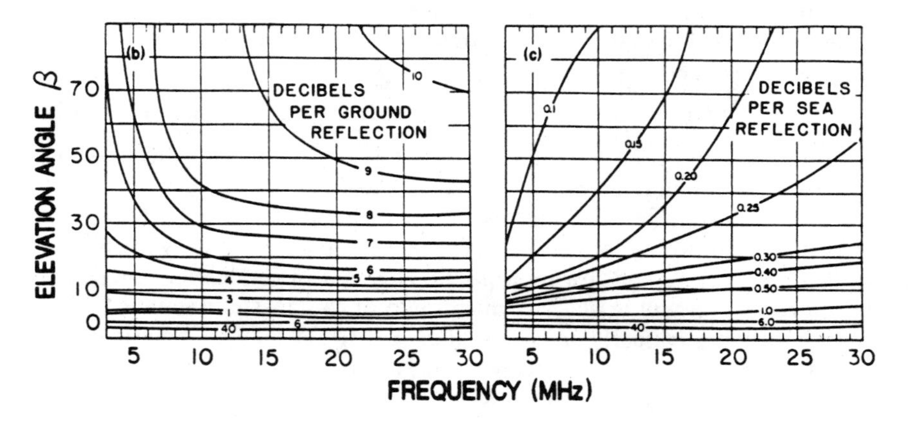

Fig.4-41. [Left] Reflection loss over poor ground (ϵ = 4, σ = 0.001 S/m). [Right] Reflection loss over sea (ϵ = 80, σ = 5 S/m). Adapted from Davies [1965].

We have discussed the ionospheric absorption process in Section 4.5.2. The CCIR method is convenient to use since we need only specify the sunspot number, solar zenith angle, ray elevation, and the transmitter frequency. Several nomogram procedures have been developed for use at HF [Davies, 1965; Hayden,1979]. Equation 4-26 is reproduced below:

$$L_i \text{ (dB)} = 677.2 \sec\phi \sum_{j=1}^{n} I_j \left[(f + f_H)^{1.98} + 10.2 \right]^{-1} \qquad (4.26a)$$

where

$$I = \text{Greater of } 0.1 \text{ and } [1 + 0.0037\, R_{12}]\,[\cos 0.881\chi]^{1.3} \qquad (4.26b)$$

$$n = \text{the number of hops}$$

and where ϕ is the ray zenith angle, f_H is the electron gyrofrequency (Hz), R_{12} is the 12-month running-average sunspot number, χ is the solar zenith angle, and f is the radio frequency (Hz). We will be dealing with losses other than those due to absorption. Because of this and to be more consistent with accepted terminology emphasizing ionospheric absorption loss, we have replaced the term A by L_i. Otherwise the current version of Equation 4-26a is the same as the previous one.

The divergence and convergence processes are best accommodated through the mechanism of ray tracing as we have discussed in Section 4.10. In practice, however, the detailed effects of focusing and defocusing are consolidated into an additional (excess) system loss term L_x along with other factors which
are also difficult to assess. The array of excess losses may include: loss due to sporadic E, L_{Es}; polarization mismatch loss, L_p; above-the-MUF loss, L_M and other factors (see Equations 4.50 and 4.51). To partially account for the fact that the transmitted signals experience r^2 spreading losses only approximately over the actual path of propagation, it suffices to take d in Equation 4-2 to be some effective distance. Taking h' to be the virtual height of reflection, D to be the ground range of propagation, and R_e to be the earth radius, we may compute the effective distance of propagation from geometrical considerations. Since we use virtual height in the calculation, then d is a virtual distance. In this way we take the spreading loss to be proportional to d^2.

$$d = [2 R_e^2 + 2 R_e h' + h'^2 - (2 R_e^2 + R_e h') \cos (D/2 R_e)]^{\frac{1}{2}} \qquad (4.49)$$

We use the result of Equation 4.49 as a replacement for d in Equation 4-2 (or the nomogram in Figure 4-40) to obtain the effective distance loss. For large ground ranges the ionospheric height takes on less importance as a contributor to the effective distance of propagation d. Indeed if D is in excess of 3000 km, we may take $d \approx D$. Of course, at the short ranges such as those associated with NVIS modes, the ionospheric height is the principal component of d. Figure 4-42 exhibits this clearly.

Although other consolidations exist, components in the excess system loss typically include all other than the free space spreading term, the ionospheric absorption loss, and the ground reflection losses. In general we have:

$$L(\text{db}) = 32.45 + 20 \log f(\text{MHz}) + 20 \log d(\text{km}) + L_i + L_g + L_x \qquad (4.50)$$

where L_x is the excess system loss, f is in MHz, d is in km, and the first three terms on the right hand side corresponds to the free space loss for which the path is a modified distance to account for propagation in the ionospheric medium [i.e., $L_{fs}(f,d)$ with d = virtual distance of propagation]. With respect to the excess system loss, there is no common view as to the totality of components which are important to consider. Including terms found in various sources, L_x may be written as:

$$L_x = L_m + L_p + L_{focus} + L_{Es} + L_c + L_{fade} \qquad (4.51)$$

where
- L_m is above-the-MUF loss
- L_p is the polarization mismatch loss
- L_{focus} is the focusing loss
- L_{Es} is excess loss due to sporadic E
- L_c is correction to absorption loss for E-region modes for which absorption is incomplete.

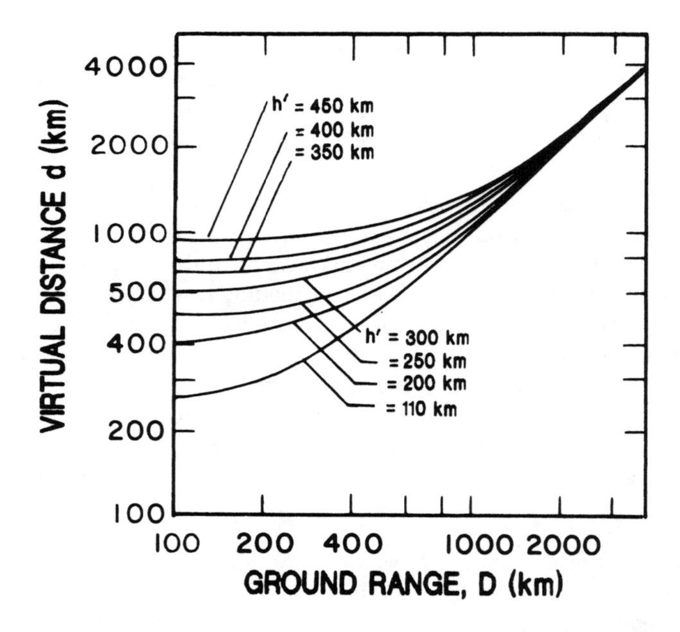

Fig.4-42. Effective distance of propagation *d* as a function of virtual height of reflection *h'* and the ground range (great circle distance) *D*.

Section 5.13.3 describes a CCIR procedure for estimating field strength. The various loss factors such as the polarization mismatch loss [Davies, 1990] are considered in connection with long path and short path models for propagation. In view of its significance in the estimation of field strength in the neighborhood of the MUF, we shall now examine the term L_m.

4.11.1 Above-the-MUF Loss[42]

Above-the-MUF loss accounts for the fact that a limited amount of ionospheric support exists even when the predicted support (from a hypothesized laminar ionosphere) vanishes. In CCIR Report 252-2 [1986k], it has the form:

$$L_m(\text{dB}) \quad = \quad 130\,\{(f/f_{bMUF}) - 1\,\}^2 \qquad (4.52)$$

----

42. Above-the-MUF loss, as used in this section, is really an attempt to characterize the situation for an instantaneous value of basic MUF, or MOF. It is likely that noninstantaneous values of the basic MUF, such as the monthly medians which are provided in statistical models, will lead to different mathematical relationships than observed by Wheeler [1966] and others. There is some reason to believe that the current CCIR model, given in Equation 4.52, should be modified so that the loss is related to the frequency difference between the *f* and the MOF rather than the departure of their ratio from unity [Hagn, 1991].

where f_{bMUF} is the basic MUF. and f is a transmission frequency in excess of the basic MUF. If f is 1.1 times the basic MUF, for example, then the loss associated with selection of this frequency is a nominal 2.6 dB. On the other hand, if $f = 1.4\ f_{bMUF}$, then $L_{fm} \approx 21$ dB; $f = 1.6\ f_{bMUF}$, then $L_{fm} \approx 47$ dB. It is thought that the necessity for including the effects of above-the-MUF propagation (and a loss term) is brought about by ionospheric scatter.

Since the shortwave broadcast community uses high gain antennas and moderately high power transmissions, the details of above-the-MUF loss takes on added significance, although not for the usual reasons. The most important consideration involves the unintended interference which is implied, rather than the increased capability to convey meaningful traffic or intelligible programming. Scatter modes associated with above-the-MUF propagation are unlikely to be acceptable for broadcasts since they would be fading channels. Nevertheless, such modes, if not properly accounted for, can still interfere with other broadcasts or services. From a communication perspective, diversity techniques may provide sufficient gain to overcome a portion of the difficulty and allow support of low data rate (i.e., manual Morse) communications in some instances.

There are currently thought to be two distinct mechanisms which will support above-the-MUF propagation: scatter from ionospheric inhomogeneities and two-hop ground backscatter. Early theory pertaining to ionospheric scatter is due to Phillips and Abel [1958] and Wheeler [1966]. Recent studies have suggested that the ground backscatter component may be significant [Gibson and Bradley, 1991]. There is also a dependence on path length. An unassailable theory is not yet available, although there is experimental evidence that ionospheric scatter dominates for small ratios f/f_{bMUF}, and that the ground backscatter mode dominates for the largest ratios. The CCIR is actively investigating the matter in connection with the 1990-1994 study period.

4.12 Long Distance Propagation by Unconventional Modes

A discussion of long distance ionospheric propagation without intermediate ground reflection is contained in CCIR Report 250-6 [CCIR, 1986L]. The reader is also referred to books by Gurevich and Tsedilina [1985] and Whale [1969] which cover long distance radio propagation. The former book deals specifically with HF. The matter is of definite interest to HF broadcasters since long distance propagation of this type would reduce the total number of stations required to provide global coverage. At issue is the reliability of the process involved, and its efficiency when the process is active. There are a number of ways for HF signals to propagate to considerable distances, and Figure 4-43 depicts the principal mechanisms.

It has been known for some time that round-the-world (RTW) HF signals could be detected (see Figure 1-1). In retrospect, and given our current could be detected (see Figure 1-1). In retrospect, and given our current understanding of ionospheric hop propagation, RTW signal detection should not be too surprising since there exist an infinite number of great circle paths which may be drawn through a single point. Furthermore the concept of focusing at the antipode has been verified [Bold, 1957, 1972]. Aside from this rather unique geometry, long-range and nonantipodal paths have also been studied. Although such long-range paths are not necessarily RTW in the strict sense of the word (RTW being a distance equal to the earth's circumference) the term seems to be used predominantly for any path which is at least trans-antipodal. NRL workers found that signals originating from European radio stations and monitored at Washington, D.C. (a nominal northeast path as viewed from the Washington terminal) were sometimes received from the southwest [Taylor and Young, 1928]. This seemingly spurious transantipodal signal was most evident during the morning and over the September-December period. This corresponds to a situation for which most of the nocturnal (and least absorbing) path was over the southern hemisphere. Wrong-Way paths between New York and Berlin were observed by Mögel [1934] and found to be more pronounced at solar maximum and whenever the direct (shorter) path was in sunlight.

Based upon earlier work by Rawer [1948], a convenient expression for the antipodal focusing gain has been developed by Hortenbach and Rogler [1979] and it is shown graphically in Figure 4-44. The antipode is located at a distance of 20,000 km, and it is seen that the influence of focusing may be felt within a rather large spatial neighborhood. For example at 16,000 km the gain is 12 dB. These authors also show that a further reduction in propagation loss arises as a result of the elimination of ground reflection loss terms (which may otherwise be several dB/hop). This assumption is permissible if rays are allowed to propagate for great distances without intermediate ground reflection using a chordal mode (see Figure 4-45). Thus the total propagation loss may be seen to be much reduced from that which would be anticipated on the basis of multihop propagation alone. Chordal modes might normally be expected to occur during nocturnal hours because of the elimination of any possible channeling between the F-layer and the E-layer (virtually non-existent). This would also explain the observation of higher signal levels over longer nocturnal paths as compared to shorter daytime paths which are influenced by D-layer absorption. If ducted modes could be introduced over daytime paths, it would be possible to reduce absorption loss. On the other hand, there are nonvanishing amounts of duct injection and ejection loss to consider.

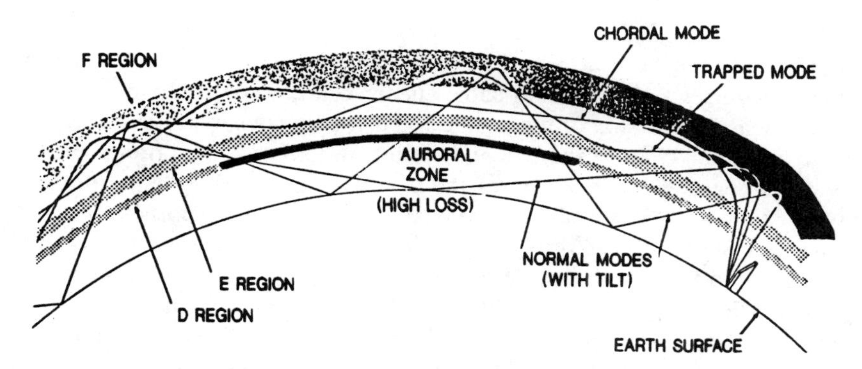

Fig.4-43. Principal mechanisms for long distance propagation. Two of the paths which are shown (i.e., chordal and trapped) are accomplished without intermediate ground reflection. Two others correspond to normal paths experiencing multihop transmission losses which may be estimated through conventional analyses.

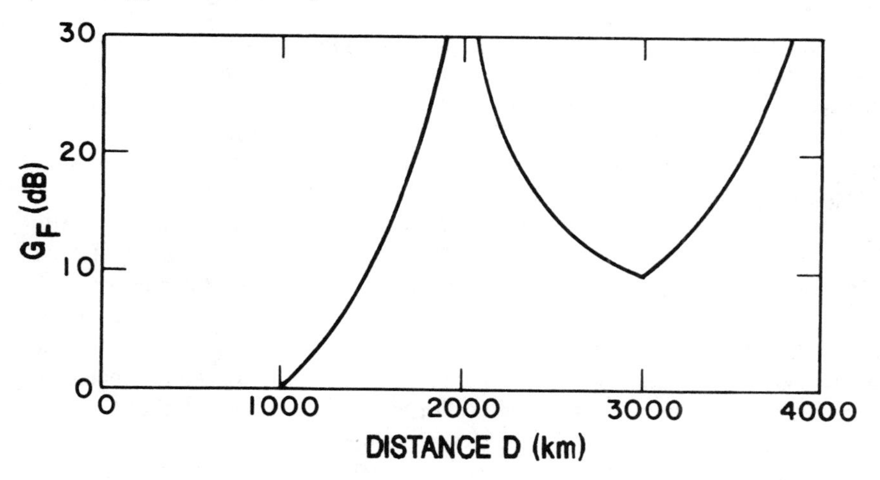

Fig.4-44. Antipodal focusing gain (after Hortenbach and Rogler, [1979]).

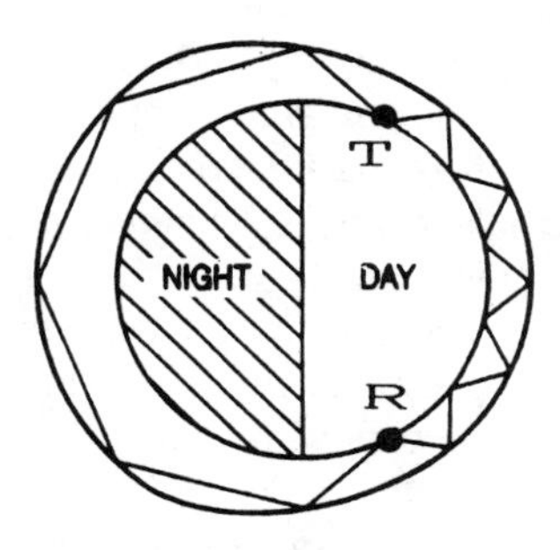

Fig.4-45. Long distance propagation modes using a virtual reflection model

For RTW signals, certain peculiarities have been observed. First, it has been found that there are no distinguishing features which can be used to separate oversea from overland paths unequivocally. Second, there is little or no frequency dispersion of the signal. Finally, many observations have been made and the weight of the evidence points to antenna reception angles of about 20 degrees, far in excess of the several degrees which is required to optimize a standard multihop pattern. A glancing wave hypothesis was advanced by Schmidt [1936] to account for the high elevation angles of the nondispersed RTW signals. There are, however, conflicting accounts about the elevation angle required. Reports of abnormally high signal levels launched over low-elevation RTW paths have appeared from occasionally in papers given at engineering symposia. Bain [1963, 1965] argues that "super modes" may be launched from elevated sites thereby enabling a low elevation signal to be transmitted efficiently. Although such an hypothesis is attractive, several attempts to verify the notion have been largely unsuccessful [Engel, 1967; Whale, 1969].

Predictability is at the core of any practical application, and long distance propagation generally requires special circumstances which limit its general utility. Nevertheless, these special circumstances do occasionally arise and lead to interesting results. Recently Lane and Richardson [1990] have analyzed radio broadcasts by a Deutsche Welle station from Kigali, Riwanda to Washington, D.C., a transequatorial path of some 12,000 km. During the sunset transition periods, signal levels were observed to be substantially higher than predicted by the IONCAP program. (IONCAP may treat long distance propagation in accordance with two options: the first corresponds to a decomposition of the entire path into a sequence of ray hops; the second is a so-called long-path model which assigns a significant role to ionospheric scatter. Without some ad hoc modification, treatment is not given to ducted or chordal mode options. IONCAP will be discussed in greater detail in Chapter 5.) The Kigali to Washington results are consistent with observations of Hortenbach and Rogler [1979] who detected unexpectedly high signal intensities over a transantipodal path when twilight occurred at either end of the path. These high signal intensities are thought to be associated with natural ionospheric ducts accessed by the coupling and decoupling action of layer tilts which are exaggerated near the geomagnetic equator in the temporal neighborhood of sunset. Figure 4-46 shows how the Appleton anomaly would be expected to influence HF signals traversing the equator.

Observations of long distance propagation have been made for several years and a variety of mechanisms have been suggested. Most familiar is the notion of so-called grey line propagation for which signals are efficiently transmitted in the North-South direction along the day-night terminator. More precisely, optimal RTW signal reception is achieved when the angle of intersection of the radiowave path and the plane of the terminator is a minimum [Yefimuk et al., 1982]. Pederson (high) ray propagation has been suggested by Fulton et al. [1960], and Muldrew and Maliphant [1962] have found evidence for Pederson rays over daytime summertime paths. Other workers have proposed that the long distance echoes are the result of tilt-supported modes, and similar to the Kigali to Washington result mentioned above, many observations have been made for the transequatorial geometry [Stein, 1958; Southworth, 1960; Gerson, 1968]. A case for chordal hops in conjunction with tilts has been suggested by Albrecht [1957, 1959], and a similar mechanism was favored by Fenwick and Villard [1963] to explain RTW echoes. Guided propagation, as opposed to a simplistic ray hop theory not involving tilts, has been advocated by some workers and approaches have been divided between those favoring a ray tracing procedure and those who have carried out a full wave analysis [Chang, 1971a, 1971b].

Ray tracing approaches provide the analyst with a convenient picture for deriving insight and it is also possible to examine the effect of tilts and gradients in the ionosphere. Such studies have been undertaken by Grossi and Langworthy [1966, 1968], and work has been described in a series of Soviet papers by Chvojková [1965, 1974a, 1974b, 1976] and Tsedilina [1975]. A computational study of long-range HF ducting was reported by Toman and Miller [1977], and Toman [1979] has reviewed the topic rather thoroughly. Advocacy for ray tracing procedures is presented by Carrara et al.[1970] for HF waves emanating from satellite platforms.

Some effort is usually required to distinguish between various alternatives to explain RTW signal properties. Accordingly some workers have examined only those HF frequencies in excess of the basic MUF as defined by a standard multihop path. This is because guided modes have been shown to propagate in this above-the-MUF regime. By convention they are termed whispering gallery modes, which is a term borrowed from the field of acoustics.

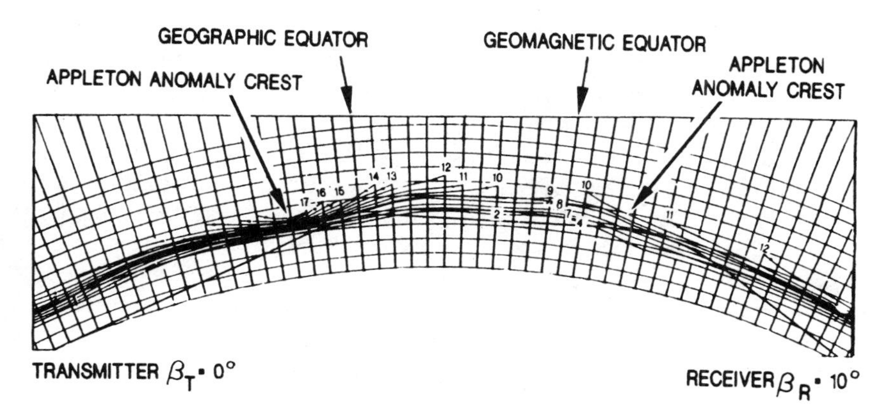

Fig.4-46. Effect of the Appleton anomaly on transequatorial propagation (from Davies [1965]).

The main features of the whispering gallery are given in Table 4-7 [CCIR, 1986L]. Both E and F region whispering galleries exist, but only the F region variety is effective for RTW propagation since collisions will significantly degrade the E-region whispering gallery.

An adiabatic invariant method has been pioneered by Gurevitch and Tsedilina [1985] to study guided waves in the ionosphere. Tichovolsky [1984] has applied the method to the problem of ground-to-ground, ground-to-satellite, and satellite-to-satellite communication. Using a method developed by Fishchuk and Tsedilina [1980], Tichovolsky has produced simulated ionograms. Figure 4-47 is an example of an RTW ionogram under the presumption of no absorption loss.

TABLE 4-7

PROPERTIES OF LONG-RANGE WHISPERING GALLERY MODES

1. Possible during any phase of the solar cycle.
2. Most favorable launch (elevation) angle is roughly 7-8 degrees.
3. Requisite launch angles are not reachable from ground sites without alteration by refraction or scatter.
4. The whispering gallery MUF, f_{wg} obeys the rule: $f_{wg} \approx 2$ MUF where MUF refers to an equivalent multihop path.
5. Whispering gallery and multihop modes may coexist.
6. Path loss, including absorption, is independent of terminal separation and is the order of 125-138 dB in the upper part of the HF band.

As indicated in Table 4-7, there is no terrestrial access to a whispering gallery unless the effective launch angle can be minimized by the process of scattering or refraction by features which exist in the auroral zone and in the neighborhood of the Appleton anomaly crests. Of course it is also possible to locate a transmitter at a great height to allow for nearly grazing incidence with the ionosphere. The existence of these natural features as candidates for inducing ducted and chordal mode support for communication has been suggested by Gerson [1979]. The full range of natural injection and ejection mechanisms includes: ionospheric tilts, sporadic E, auroral zone, high latitude trough, Appleton anomaly crests, Appleton anomaly trough, and possibly even meteor trails. Artificial methods such as the creation of plasma clouds by seeding the ionosphere with readily ionizable constituents or the creation of plasma holes by the addition of chemical reagents have been mentioned. Perhaps the most prominent suggestion refers to various nonlinear phenomena embodied in HF heating technology. HF heating as a general topic has received widespread attention, notably in connection with long-range HF propagation in articles by Gurevich and Tsedilina [1975], Elkins [1979], and Sales [1979].

Before leaving this general area of long distance propagation, it should be mentioned that HF signals may also be trapped in magnetospheric ducts. This phenomenon is associated with long-delayed echoes. Finally, it is recognized that a large number of terms are found in the open literature which describe various facets of the phenomena attributed to ultra long distance HF propagation. They include: trapping, ducting, channeling, guided modes, chordal modes, whispering gallery, earth-detached modes, and so on.

4.13 MULTIPATH AND FADING PHENOMENA

4.13.1 Introduction

In this section, multipath phenomena will be briefly reviewed, the impact on HF communications will be assessed, and various mitigation or avoidance schemes will be listed and described. In the process we shall examine HF channel models.

Multipath arises when the received signal obeys one of the following conditions: (1) the signal is nondispersively distended by multiple reflections from the ionosphere with or without ground reflections (more than 1 hop or layer is involved); or (2) the signal is distorted by the superposition of multiple and near-equal amplitude sources within a single layer. Fading is involved in both cases. Although amplitude fading is an important deleterious effect attributed to ionospheric multipath, the most serious impact of multipath in digital systems is intersymbol interference (ISI). It is also worth noting that ionospheric irregularities within a single layer are generally in pseudorandom motion, and we must be prepared to cope with Doppler spreads when micro-multipath returns are evident.

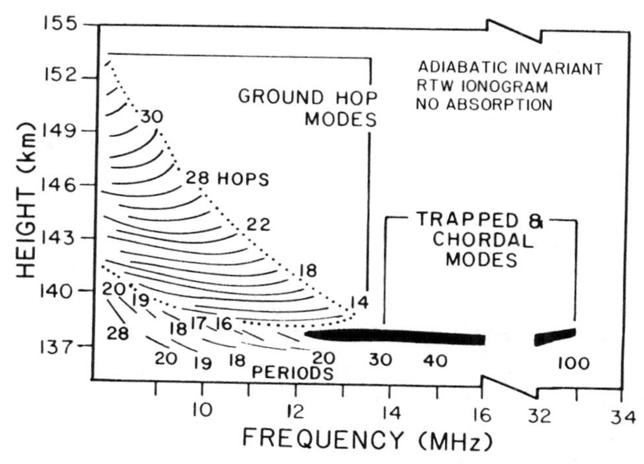

Fig.4-47. Simulated absorptionless ionogram for RTW conditions. The source is elevated, and radiates due east from a transmitter located at 42N and 120E (from Tichovolsky [1984]).

4.13.2 Description of the Phenomenon

Multipath may be regarded as a general term which describes the transfer of energy from the transmitter to the receiver by different paths nearly simultaneously. The degree of simultaneity is a function of system resolution. Multipath may be introduced as a result of scatter from multiple patches of ionization within the same layer, by reflection from one or more distinct layers (multimode), or by multiple ionospheric bounces from a specified layer or layers (multihop). The definitions are given below:

Multipath: The transfer of energy from a transmitter to a receiver by different radio paths nearly simultaneously. This is a general term.

Multimode: A multipath condition in which RF energy is received from two (or more) ionospheric layers simultaneously, such as sporadic E, normal E, F1, and F2. It is also possible to have a set of these individual modes in combination. These are referred to as mixed modes. Multimode specification also includes the possibility of a multihop mode.

Multihop: A multipath condition in which the radio wave undergoes multiple bounces from a specified ionospheric layer. The multihop condition is included under the general definition of multimode propagation.

Scatter: A condition in which the transmitted signal is reradiated (scattered) either from small-scale inhomogeneities or individual electrons. The former condition may lead to signal spread such as spread F, while the latter condition is associated with an incoherent low-level re-radiated power process (Thomson scatter). Thomson scatter is not significant for HF communication.

The multipath resulting from multiple hops has been examined statistically by Croft [1982] among others, and is found to depend upon the transmitter to receiver distance. From Figure 4-48 we see that the time delay distributions are widest for the shortest links and that they also exhibit the largest median values. This result would be expected on the basis of simple geometrical considerations.

Bailey [1959], using oblique-incidence-sounder data sets, has developed an empirical relationship between multipath spread, communication path length, and the MOF-normalized transmitter frequency. The normalized transmitter frequency is called the Multipath Reduction Factor or MRF. Several conclusions are evident upon inspection of the MRF curves shown in Figure 4-49a. Of greatest significance is the fact that the delay spread is smallest near the MOF or when the MRF is close to unity. Also, we see that if we specify a fixed delay spread (of about 1 ms), then the MRF is minimized at a range of about 2000 kilometers. A second curve by Salaman [1962] gives the maximum value of the multipath spread as a function of the path length (i.e., Fig. 4-49b). We see that this upper limit exhibits a valley in the 2×10^3 to 10^4 kilometer region and rises somewhat at greater distances. It is safe to say that multipath is most pronounced for NVIS (and medium range tactical circuits up to 500 km) but is reduced for most long-haul circuits of interest. The reader is reminded that these curves are empirical and are not universally applicable. They do however indicate the general trends. Other workers have found maximum multipath spreads in excess of those suggested by Bailey and Salaman. Values of 10 ms have been suggested by Otten [1962] and 12 ms has been observed by Schmidt [1960]; however these values are at frequencies well below the MOF.

Figure 4-49 or equivalent information may be useful in estimating the maximum transmission rate which is likely to be supported over a specified link. For example, a 200 km circuit may experience a maximum time delay spread, δT_{max}, of 8 ms. This is about the largest amount of multipath one should reasonably expect to experience as a result of multihop/multilayer propagation, even for NVIS circuits. The order of magnitude of the largest allowable transmission rate is $\approx 1/\delta T_{max}$. Hence we should expect to transmit symbols no faster than about 125 sec^{-1}. In practice there is a difference in the relative amplitudes of the contributing multipath returns which compose the maximum spread. As a result, a more liberal value 200 bits/sec is often quoted to be the upper limit.

At a distance of 3000 km, Figure 4-49b suggests that $1/\delta T_{max} \approx 333$ bps; however from Figure 4-49a, if we use a transmission frequency near the MUF (say with a MRF $\approx$ 0.9) then $1/\delta T_{max} \approx 2000$ bps. Even higher values are possible under certain circumstances. Nevertheless, prudence would suggest that 200 bps should be used for planning purposes.

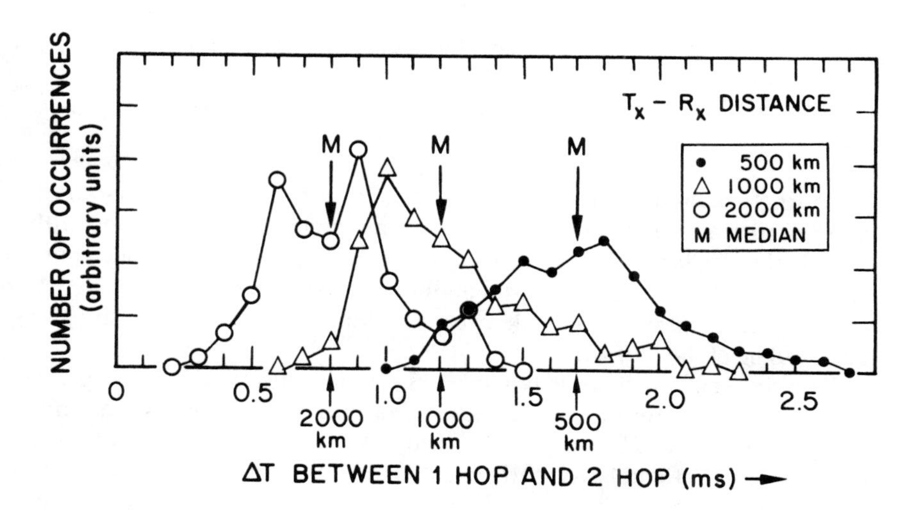

Fig.4-48. Multipath spread distributions corresponding to one and two hops and parameterized in terms of link distance. (Courtesy T. Croft, SRI)

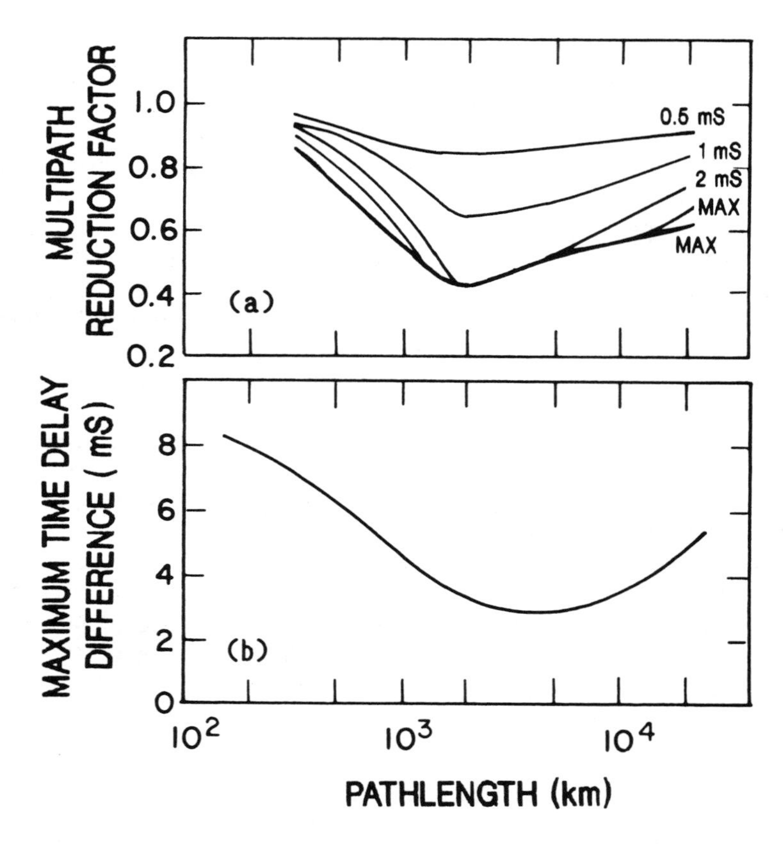

Fig.4-49. (a) Multipath Reduction Factor versus path length. (b) Maximum time delay difference versus path length.

If a single discrete mode were to be isolated for communication purposes, the transmission rate would be limited by micromultipath associated with ionospheric roughness. This spread may range between 10 and 100 μsec. For $\delta T_{max} = 100\ \mu$sec, we have a maximum transmission rate $\approx$ 10 kilobits/sec. It is worth noting that the small scale structure within a given layer may still give rise to fading of the diffraction type.

We have seen that short-haul and NVIS circuits may experience the greatest number of discrete multipath echoes, including the groundwave mode which is not observed for long-haul circuits. When any two or more of these modes are present and have nearly equal amplitudes, conditions are exist for marked interference fading which tends to be selective in its character. This means that subbands within a normal 3 kHz channel may fade independently. In fact frequencies differing by as little as 0.1 kHz may undergo independent fading processes. We will say more about fading statistics later.

4.13.3 Wideband Examination of Multipath

Wagner et al. [1988] have developed a wideband probe having an effective instantaneous bandwidth of 1 MHz, and measurements have been made over midlatitude and auroral circuits. This system provides an effective pulse width of 1 microsecond, and thus allows the observer an extraordinary view of ionospheric microstructure. Wagner and his colleagues examine the time-varying HF channel in two ways: (1) by constructing a time history of the coherent pulse response, and (2) by computing the channel scattering function. The coherent pulse response is simply the in-phase component of the received signal amplitude at a specified frequency, while the scattering function (for a specified ionospheric mode) is a plot of Doppler frequency versus time delay spread.

Figure 4-50 shows the in-phase component of the channel pulse response. It is characterized by a multipath spread does not obscure the individual ionospheric echoes. In this illustration, any echo spread is the result of refraction rather than a distended ionospheric multipath structure. The rather well-defined pulse print indicates the specular (nondiffuse) nature of the echoes.

The disturbed auroral zone may generate a pulse response which has decidedly more personality. Figures 4-51a and 4-51b give the mode pulse response and the corresponding scattering function for a 1 hop F1 transauroral mode of propagation. We note that this single mode is spread over roughly 100 microseconds and the scattered signal exhibits a relatively large Doppler spread. Since no coherent pulse print structure is evident, this auroral channel is diffuse in nature.

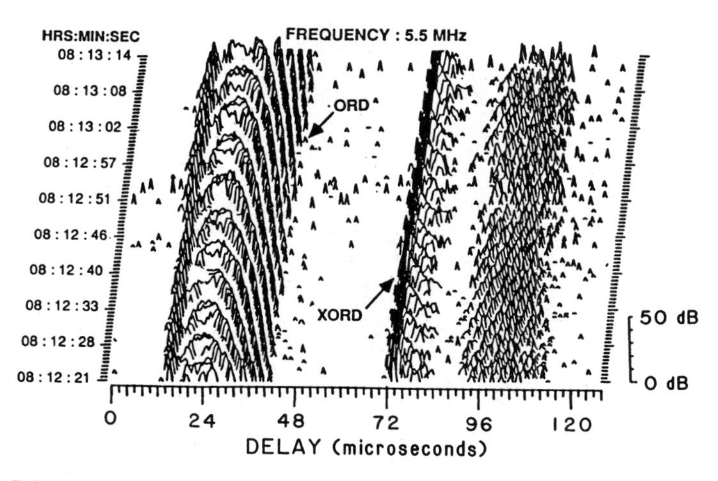

Fig.4-50. Coherent pulse response for a midlatitude channel for a transmitter frequency of 5.5 MHz. The plot is an isometric one showing time history (HRS:MIN:SEC) of the coherent pulse response (amplitude vs delay in μsec). (Courtesy of L. Wagner, NRL)

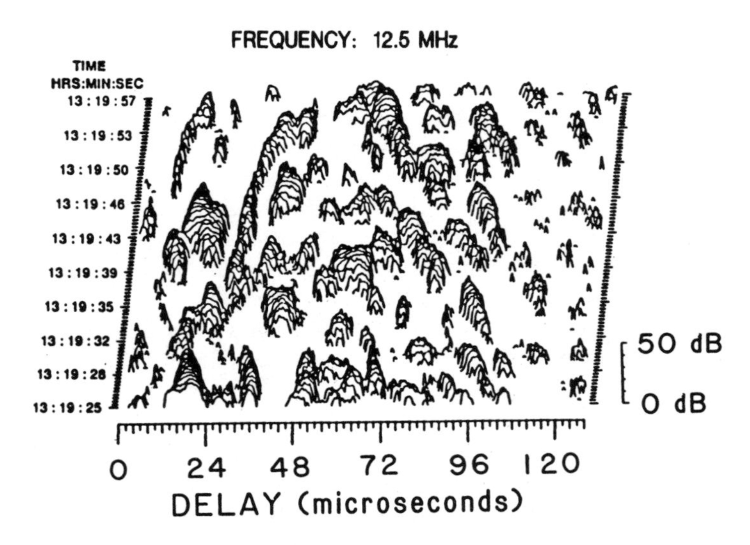

Fig.4-51a. Coherent pulse response for a transauroral channel at a frequency of 12.5 MHz. The in-phase component of the mode pulse response is shown. The response is associated with a one-hop high ray, diffuse return. (Courtesy L. Wagner, NRL)

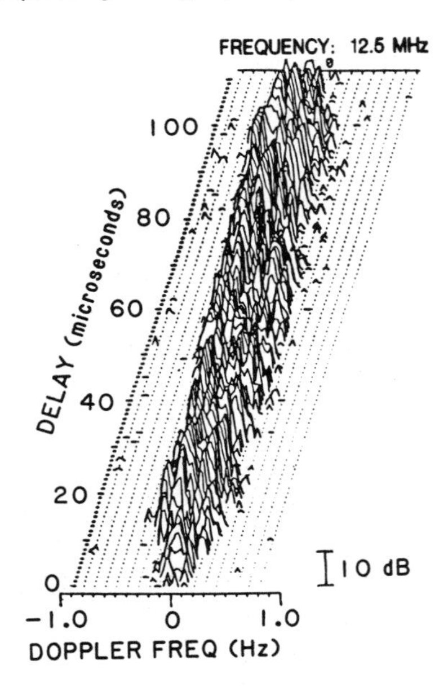

Fig.4-51b. Scattering function for a transauroral channel at a frequency of 12.5 MHz. The corresponding in-phase component of the mode pulse response is given in Figure 4-51a. The response is associated with a one-hop high ray, diffuse return. (Courtesy L. Wagner, NRL)

Nickisch [1990] has examined the propagation on extended random media at HF, and has indicated that the transionospheric propagation theory describing the effects of stochastic media on electromagnetic waves has a degree of validity. This is true despite the fact that practical transionospheric frequencies are substantially greater than 30 MHz. The theory, as outlined by Wittwer [1979, 1982], was developed to explain satellite scintillation phenomena, but it should be valid for highly oblique trajectories at HF for which the ratio of the MOF (as well as the operating frequency) to foF2 is large. Nickisch describes a multiple-screen/diffraction method for examination of extended media, nonuniform electron density, and nonuniform plasma velocity. The Nickisch [1990] approach is an attractive one for study of auroral paths, and for explaining the peculiarities of wideband probe results. The theoretical development builds upon previous work by Knepp [1983] and SRI investigators [Basler et al.,1987, 1988].

Basler et al. [1985, 1987, 1988] have developed a wideband HF channel probe. Figure 4-52 contains two samples of data obtained over a single-hop polar path utilizing the F-layer. The first sample shows both the high and low rays, influenced by a rather significant Doppler spread, and the scattering function exhibits a weak parabolic behavior. The second sample, obtained several hours later and at only a slightly different carrier frequency, exhibits a teardrop pattern for both modes in the scattering function. This is interpreted by Nickisch [1990] as suggestive of a two-stream plasma flow pattern.

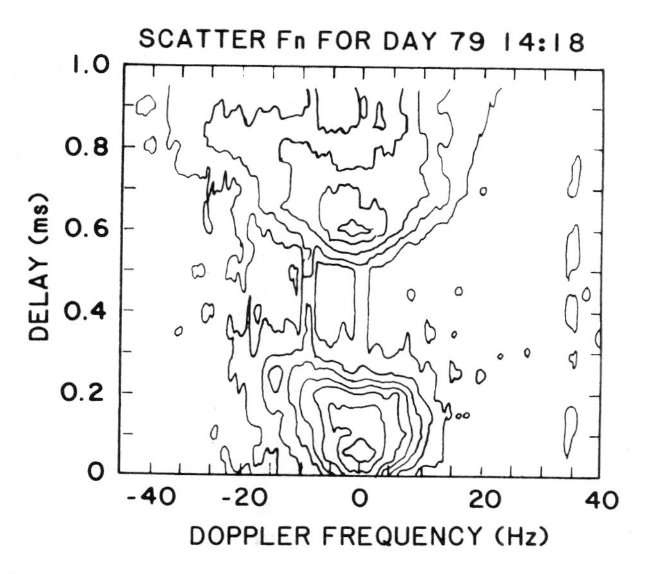

Fig.4-52a. DNA HF Channel Probe data. Both high and low rays are shown. Data obtained on March 20, 1985 at 1418 UT and for f = 10.57 MHz. The probe was developed by Basler et al. [1985]. Curve obtained from Nickisch [1990] appearing in IES'90.

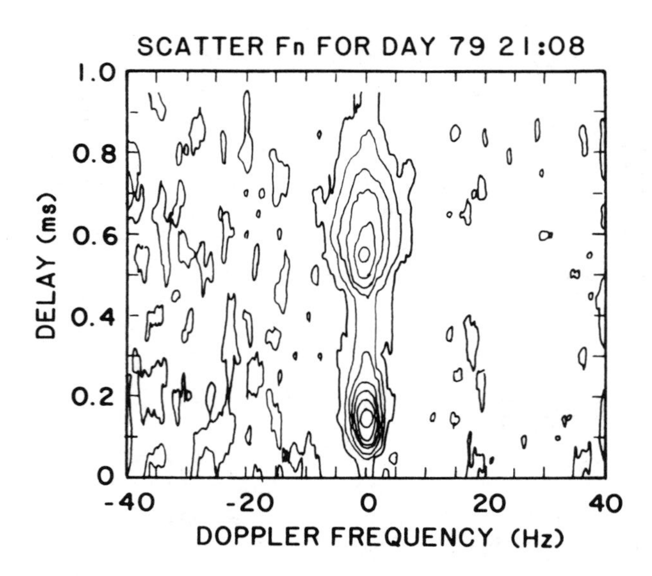

Fig.4-52b. DNA HF Channel Probe data. Both high and low rays are shown. Data obtained on March 20, 1985 at 2108 UT and for f = 10.265 MHz. The probe was developed by Basler et al. [1985]. Curve obtained from Nickisch [1990] appearing in IES'90.

4.13.4 Fading Categories

Discussions of HF fading may be found in CCIR Reports 266-6 [1986m] and 304-2 [1986n]. A number of causes of fading have been identified and they include those listed in Table 4-8.

Fading categories 1, 2, and 4 in Table 4-8 are related to interference between rays, and may be loosely referred to as interference fading. The fastest type of fading (category 1 is sometimes termed flutter fading, and is usually encountered during disturbed periods or for paths transiting the auroral or equatorial (tropical) zones. Flutter fading is associated with spread F conditions.

Fading may further be characterized as selective or nonselective depending upon the bandwidth of the waveform being utilized. It is easy to associate fading categories such as absorption or MUF-failure as essentially nonselective processes. But selectivity is actually a relative term. For example, if we have a multihop situation with a multipath spread of $\delta T_m = 1$ ms, then interference between the two signals would introduce nulls in the signal envelope at a spacing of 1000 Hz (i.e., δT_m^{-1}). Thus in-band diversity requires that the signal bandwidth be in excess of 1 kHz for the stated example. The correlation bandwidth is simply δT_m^{-1} [Maslin, 1987], which increases with decreasing multipath. Selective fading is therefore not always a negative feature. It may be used to achieve some amount of diversity gain. For digital systems, if the symbol duration δT_s is greater than the multipath spread δT_m, any fading which is observed is nonselective. While the presence of multipath may also cause intersymbol interference (ISI), it is only the temporal variation of the multipath which gives rise to selective fading.

TABLE 4-8. CAUSES OF FADING

CATEGORY OF FADING ENCOUNTERED	FADE PERIODS
1. Motion of small-scale inhomogeneities associated with micromultipath	< 1 sec
2. Ionospheric motion in connection with multihop/multilayer multipath	1 - 10 sec
3. Faraday effect, or rotation of the plane of polarization of the radiowave.	0.1 - 2 min
4. Slow fading associated with lens type irregularities in the ionosphere	10 - 60 min
5. Temporal fluctuations in ionospheric absorption (example: SWF)	5 - 30 min
6. Variation in the ionospheric support at a specified frequency. (This corresponds to so-called MUF-failure.)	irregular

In the context of fading, we are principally interested in the first two categories listed in Table 4-8, the categories dealing with ionospheric motion. Fading is observed to be more rapid at the higher range of the HF band since a specified (fixed) amount of ionospheric motion corresponds to a greater phase shift in that region.

If the propagation geometry (and other conditions) is such that only one mode of propagation is available, we find that the impulse response of the channel is somewhat lacking in personality and the reflected signal is actually a composite formed from the superposition of a host of subsignals from a large number of moving reflectors in the ionosphere. This situation leads to diffraction fading and the distribution of amplitudes is Rayleigh in nature. An example of diffraction fading from a single layer is shown in Figure 4-53.

Diffraction (interference) fading is associated with periods which may be as small as a fraction of a second. The form of the amplitude (fading) distribution is of some interest in the design of receivers. The conventional way to describe short-term fading is through the amplitude distribution function $P(V_o)$. It is related to the probability density function $p(V_o)$ by the following equation [CCIR, 1986m]:

$$P(V_o) = \int_{V_o}^{\infty} p(V)\, dV \qquad (4.53)$$

where V is the received signal envelope voltage, and V_o is a specified signal amplitude. In Equation 4.53 the function $P(V_o)$ represents the probability of finding a signal V greater than V_o. Popular probability distribution functions include the Nakagami-Rice function, the Rayleigh function, the normal (Gaussian) function, and the Nakagami-m distribution.

Figure 4-54 shows the distribution function for the Nakagami-Rice case. This is a convenient model since it assumes that the contributory signals to be a steady sinusoidal component and a Rayleigh component possessing a uniform phase probability. We take V_1 to be the rms voltage of the steady component, V_n to be the rms value of the random component, and V to be the received signal envelope voltage divided by $\surd 2$. Two special cases are of interest. The first corresponds to large values of (V_n/V_1) and the second corresponds to small values of this ratio. In the former case the Rice-Nakagami density function reduces to the Rayleigh density function, and in the latter case it reduces to the normal function (provided $V \approx V_1$) which we usually rewrite to conform to a log-normal recipe. This simply means that we express the received signal level with respect to a reference level in decibels. From Figure 4-54, we see that the distribution of received signal is roughly symmetrical (and Gaussian) for $(V_n/V_1) < 0.1$ (or at the -20 dB level or less). On the other hand, we note that if $(V_n/V_1) > 2$ (or about 6 dB), then the distribution is Rayleigh in nature.

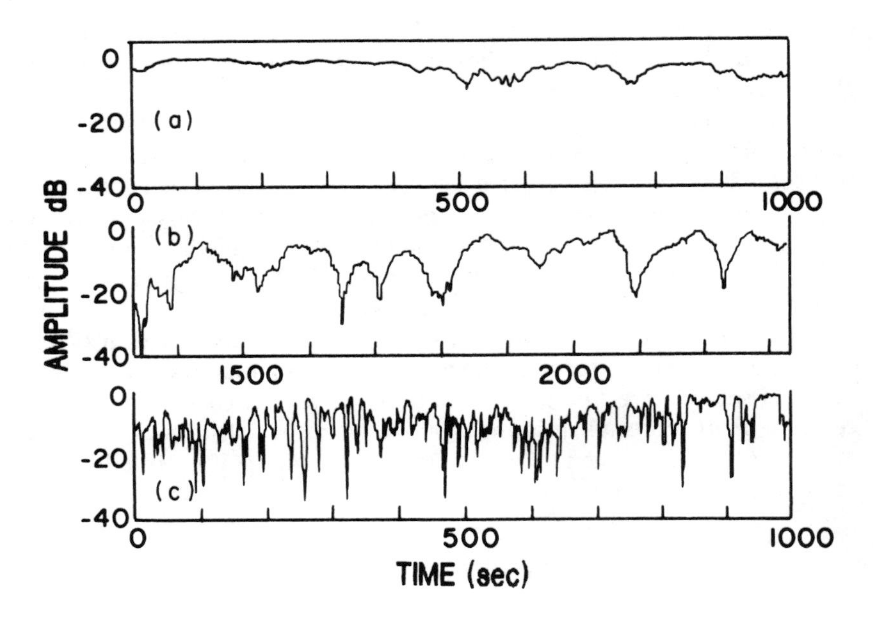

Fig.4-53. Diffraction fading in the ionosphere. (a) Little. (b) Moderate. (c) Severe

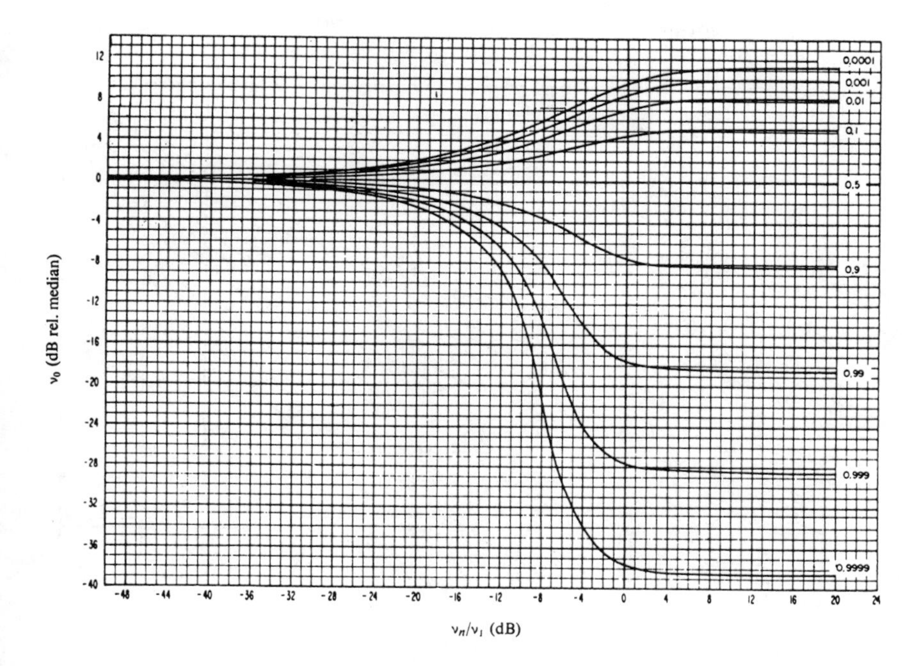

Fig.4-54. Distribution function corresponding to a Nakagami-Rice probability density function [CCIR ,1986m].

Another useful distribution is the so-called *m*-distribution [Nakagami, 1960] which is a one parameter function. For values of *m* which are large, the distribution is Ricean; for values of *m* near unity, the *m*-distribution is Rayleigh in nature. The Nakagami-*m* distribution has also found application in the description of radiowave scintillation statistics in the transionospheric case.

A practical parameter known as the fading range has been defined. It is the difference, (in dB) between the signal levels exceeded 10 % and 90 % of the time. It has been shown that Rayleigh statistics typically apply for short intervals whereas log-normal statistics are more appropriate for analysis intervals of the order of an hour. On the other hand, if the signal levels are quite high, there may be a large specular (single-mode) component involved, and the statistics may be best represented by the log-normal formalism. Often one can represent signal fading in terms of a (known) median component which is log-normal, and an (undetermined) instantaneous component which is thought to be Rayleigh distributed. Figure 4-55 [Picquenard, 1974] is useful in this representation.

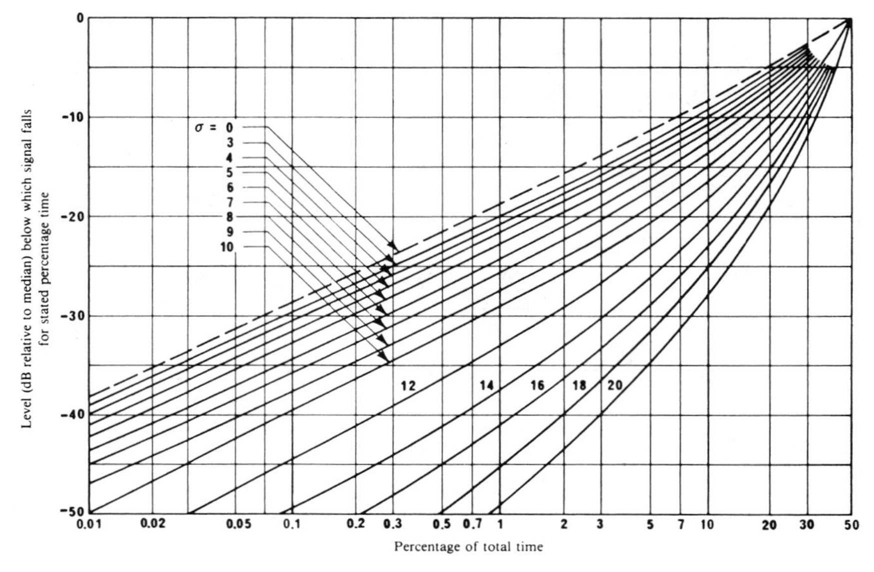

Fig.4-55. Level below which a signal decays when the instantaneous signal strength has a Rayleigh distribution and the daily median values have a log-normal distribution with a standard deviation of σ dB. [CCIR,1986m]

4.13.5 Fading Rates

Fading rates depend upon a number of factors including the degree of ionospheric turbulence and the ratio of the transmission frequency to the MUF. From Report 266-6 [CCIR, 1986m], we may represent the fading rate by the equation:

$$R(\tau) = R_o \exp(-\tau^2/2\tau_o^2) \tag{4.54}$$

where τ_o represents the coherence (time constant) of the HF channel.

The coherence (correlation) bandwidth of the HF channel is given by the inverse of the coherence (correlation) time. Under conditions implicit in Equation 4.54, it is found that the power spectrum of the channel is a Gaussian distribution:

$$P(f) = k \exp(-\tau_o^2 f^2/2) \tag{4.55}$$

where the standard deviation is the correlation bandwidth $1/\tau_o$, and k is a constant. Fading time constants τ_o of from a few seconds to almost a minute have been observed over midlatitude paths; with the smallest values being typically observed when multimode (and especially multihop) propagation is evident. Fading may be far more rapid and intense for paths transiting the auroral zone, but transequatorial fade rates are perhaps the most intense.

Conditions for slow nonselective fading have been discussed in CCIR Report 266-6 and in Stein [1966]. We have also mentioned a few requirements at the beginning of this section. Recall that nonselective fading may be introduced by factors which are independent of multipath including; solar-induced short-wave-fades (SWFs), large-scale focusing irregularities, and possibly MOF-failure[43]. On the other hand, any fading which occurs is nonselective if the information pulse (symbol) length δT_s is *greater* than the multipath spread δT_m. The Doppler condition for slow fading is that the information pulse length be much *less* than the reciprocal of the doppler spread of the scattered signal. For a simultaneous condition of slow *and* nonselective fading, both conditions must be satisfied. In short, we want the pulse to be long enough to overlap all the multipath components, while at the same time, the pulse should be short in comparison with the characteristic times for significant phase fluctuation. The conditions indicated above may be written as follows:

$$BW_d \, \delta T_m = SF \ll 1 \qquad (4.56)$$

where BW_d is the doppler spread and SF is the Spread Factor [Stein, 1966]. In essence it is the media equivalent of the time-bandwidth-product.

It is convenient to associate δT_m^{-1} with a frequency [Stein and Jones [1967]. This frequency function is said to be virtually equivalent to various terms such as selective-fading-bandwidth, correlation bandwidth, and coherence bandwidth. (For a single mode exhibiting a time spread standard deviation σ_T, the coherence bandwidth is $1/2\pi\sigma_T$. See Section 4.14.1.)

Fade rate characterizations are covered in CCIR Report 304-2 [1986n]. At low latitudes, fade rates may be quite variable depending upon the layer which is providing ionospheric support. Flutter fading, which is closely correlated with spread-F-induced scintillation in the equatorial zone, may range between 10/min [Davies and Barghausen, 1967] and 180/min [Carman et al.,1974]. The strongest class of fading is associated with sporadic E, and the equatorial variety may generate rates of as high as 300/min. Normal E and F region fade rates may be as high as 10/min. Midlatitude fade rates are considerably smaller than those generally observed in the equatorial region. The rates become large once more for transauroral paths.

----

43. We use the term MOF-failure instead of the often used term MUF-failure for sake of clarity. They are both the same if the MUF is taken to be an instantaneous value, as obtained from an oblique-incidence ionosonde (OIS). The term MOF is preferred since it is unambiguous.

4.13.6 Intersymbol Interference

As has been indicated above, significant problems may be encountered over HF communication circuits if multipath conditions exist. Major effects include a spread in signal arrival time and frequency, leading to wave interference and signal fading. On a coarse time scale, we are typically concerned with interference between discrete modes of propagation such as 1-hop and 2-hop F modes which may be of the order of several milliseconds. Delays of this amount will limit signaling rates to a few hundred baud. This limit is specified to avoid the multipath-induced overlap of adjacent symbols, and the phenomenon is called intersymbol interference or ISI. The signaling rate limitation is not to be confused with the source rate which may be used.

4.13.7 Mitigation Techniques

A number of schemes have been developed to compensate for the impact of multipath upon HF system performance. Most of the schemes attack the symptom (i.e., fading) rather than the cause. A specific class of countermeasures to mitigate against fading is based upon diversity. Another class of countermeasures avoids multipath by appropriate frequency selection (i.e., propagation near the MOF) or antenna (pattern) selection. Diversity methods will be addressed first.

4.13.7.1 *Types of Diversity*

There are a number of diversity schemes which have been attempted with varying degrees of success. They are listed in Table 4-9.

TABLE 4-9. LIST OF DIVERSITY SCHEMES

1. Space
2. Frequency
3. Angle-of-Arrival
4. Polarization
5. Time
6. Advanced DSP: *Rake* and Equalization

Barratt and Walton [1988] have noted that propagation effects can be grouped into two classes, those which give rise to loss and those which give rise to distortion. The three effects which may cause distortion are ionospheric dispersion, ionospheric parameter nonstationarity, and multipath. The interference fading effects arising as they do from relative motion of multipath structures, will tend to generate system waveform disturbance. This is because the bandwidth of this class of effect is an appreciable portion of the signal bandwidth.

Diversity may be visualized as a procedure whereby information, which has become disguised or scrambled, may be partially or fully retrieved. Full retrieval may necessitate the use of an inverse *mapping* algorithm from which the system has an ability to compensate for the distortion introduced by some physical process. Other measures, not as sophisticated, allow for retrieval based upon data redundancy.

4.13.7.2 *Space Diversity*

Space diversity can be best understood using the notion of a spatial correlation function which describes the relationship between two cw signals which are derived from spaced antennas. The correlation is defined to be unity when the spacing $d = 0$, and common modeling schemes suggest an exponential decay in accordance with the relation:

$$c(d) = \exp[-d^2/2\delta^2] \tag{4.57}$$

where δ is a parameter which depends upon the canonical ionospheric scale of interest. Note that when the separation between the two antennas is equivalent to the canonical scale, then the correlation $c(d = \delta)$ is no more than about 0.61. If $d = \delta\surd 2$, then $c = 1/e \approx 0.37$; and for larger distances the any advantage of diversity is vanishingly small. To get a feel for the relationship between correlation and diversity gain, consider the following. It has been shown that two independently fading signals exhibit a diversity gain of about 15 dB at the 99.9% reliability level; this corresponds to a large separation between the antennas. As the separation between the antennas is reduced to $d = \delta$ for which $c = 0.61$, the diversity gain is still as much as 13 dB [CCIR, 1986m]. The correlation distance is defined to be $\delta\surd 2$.

The next question, of course, is: What is the structure size (or the correlation distance) for representative HF paths? Measurements in the United Kingdom over a large range of HF frequencies and path lengths (excluding medium range and NVIS) have shown that the correlation distance varies between 10 and 25 times the transmitted wavelength. Sub-HF measurements in the United States indicate similar (but somewhat higher) values: 12 to 46 times the wavelength. There is some evidence to suggest that correlation distance is fade rate dependent with smaller distances associated with faster fades. This has implications not only for space diversity but also for time diversity. In general, for situations in which the specular component is dominant, the correlation distance will be distended in comparison with a diffuse scatter from a disturbed ionosphere (i.e., substantial random component). Thus antennas with fixed separations will exhibit better diversity performance in a pathological environment than under benign conditions.

In some instances, several distinct fade rate components may be resolved in the course of observation. These may be related to independent scattering processes; and consequently may possess separate correlation distances.

4.13.7.3 *Angle-of-Arrival Diversity*

This technique may take advantage of the fact that different modes arrive at different elevation angles (for a specified azimuth). Figure 4-56 illustrates the technique. It is worth noting that this diversity is macroscopic in nature allowing (or disallowing) receiver processing of entire propagation modes.

Angle-of-arrival diversity is accomplished through use of highly directional antennas which may point in specified directions, or by automatically adjustable antenna arrays. SNAP-based systems, or steerable-null-array-processors, are used to steer an antenna null in the direction of an unwanted signal such as a jammer. Such systems could also be exploited to avoid a specified multipath component, in addition to satisfying their ECCM function.

Angle-of-arrival diversity schemes may be applied in either or both the elevation and azimuth planes. In the elevation plane, we exploit independent modes; hence we sometimes use the term *mode diversity.* In the azimuth plane, we may exploit any available modes, even ones corresponding to the same reflecting layer. In this situation, the term *mode diversity* is not appropriate; an alternative term which may be used is *path diversity.*

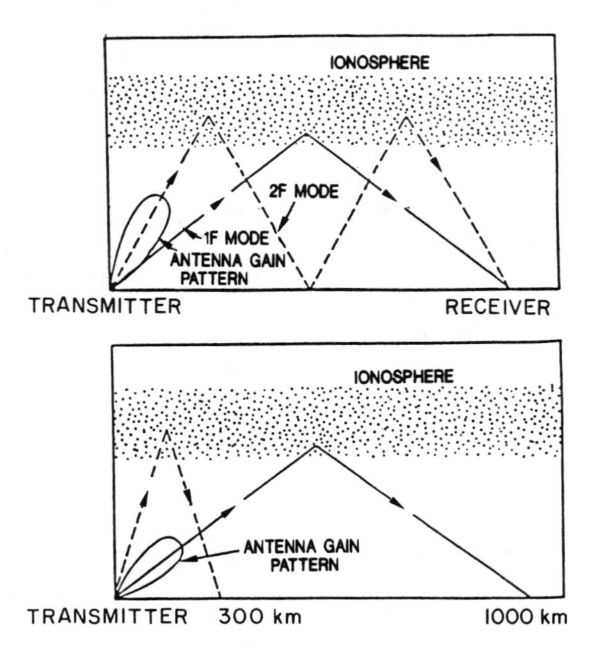

Fig.4-56. Examples by which angle-of-arrival diversity may be accomplished.

4.13.7.4 *Polarization Diversity*

In this case signals are received by differently polarized antennas. The process relies on the notion that orthogonal polarizations behave independently. It is known, for example, that ordinary and extraordinary modes of propagation, having different refractive indices in the magnetoplasma, will travel along different paths in the ionosphere. Thus the so-called O- and X-modes are candidates for diversity combining. Similarly, we may consider other orthogonal pairs such as vertical and horizontal polarization. Measurements made in both the U.K. and the U.S. have confirmed the possibility of diversity gain [Grisdale et al., 1957]. Between the frequencies of 6 and 18 MHz, the polarization diversity gain was equivalent to the space diversity gain achieved for an antenna separation of between 240 and 480 km.

4.13.7.5 *Frequency Diversity*

Frequency diversity is a process in which information is transmitted (and received) on different frequencies nearly simultaneously. This scheme may compensate for selective fading. The frequency correlation distance in Hertz is inversely proportional to the maximum time delay between modes or δT_m (the multipath spread). Since the multipath spread may be as much as 5 ms, the frequency correlation distance may be as low as 200 Hz. Near the MOF, of course, the multipath spread is less and so is the selective fading probability. In the vicinity of the MOF (corresponding to a multipath reduction factor of 0.6 to 0.8) the frequency correlation distance ($1/\delta T_m$) may be of the order of 10 kHz corresponding to a delay spread of about 100 microseconds. At the MOF, ray dispersion (rather than multipath) begins to limit the frequency correlation distance. On the other hand, these high values of frequency correlation distance, being highly dependent upon the coupling of propagation and time-varying ionospheric conditions, may not be practical. In general, information is transmitted in a band of frequencies, and the extent of selective fading may be seen by examination of the correlation between selected pairs of spaced tones in the tone set.

It is noteworthy that frequency diversity gain may only be obtained in a parallel-tone system if the tones are regarded as potential candidates for redundancy, and not exploited solely to achieve higher overall data rates in an ISI environment.

The transmission of multiple tones may be used to compensate for intersymbol interference (ISI), since, to achieve a given high data rate, each of the tones carries digital data at a relatively low baud rate. Medium speed modems such as KINEPLEX, which uses a parallel arrangement of tones, was designed to compensate for fading due to multipath by lengthening the signaling element (or baud duration). The method, although effective in principle, may not achieve the desired result if two nearly equal amplitude multimode components are encountered. In this case, ISI itself is no problem but fading still may still limit system performance.

4.13.7.6 *Time Diversity*

This category includes coding, interleaving, and simple signal repetition schemes. The idea is to add a degree of redundancy. This topic will be covered in Chapter 7.

4.13.7.7 *Advanced DSP Techniques*

The best known schemes involving advanced Digital Signal Processing (DSP) include *Rake* and adaptive equalization. Implicit diversity may be achieved in these countermeasures to multipath and fading. Indeed it is possible to design modems in which the BER is actually reduced in the presence of multipath. This is because the modem resolves the individual (and independently-fading) multi-mode components, and combines them for diversity gain. Equalization and *Rake* methods are sometimes called multipath diversity techniques (see Chapter 7).

4.13.7.8 *Multipath Avoidance Measures*

As we have seen the usual measures to counter the fading problem involve diversity in one form or another. Such measures directly attack the problem. The strategy taken in another class of techniques involves circumvention or avoidance of the problem.

The most obvious strategy for avoidance for a fixed link involves frequency selection near the MUF to reduce the amount of multipath spread. This corresponds to the selection of large values for the multipath reduction factor, MRF. In networked systems, use of connectivity between distant nodes rather than the more obvious short links will increase the MUF values; and this approach will increase the MRF if frequencies close to the MUF are available.

Multipath avoidance is most easily achieved in a real-time environment if a sounder is available since frequencies for which multipath is present may be clearly seen.

4.13.8 Concluding Comments

An excellent source book on fading phenomena and modeling is a special publication *Data Communication by Fading Channels* [Brayer, 1975] which is a collection of reprints, largely from IRE or IEEE journals. It contains material on fading and multipath effects, diversity techniques, the performance of practical systems, channel modeling, error control coding, and related topics of interest. For fading, papers by McNicol [1949], Grisdale et al. [1957], and Koch and Petrie [1962] are noted. Another special collection of papers published by the IEEE Press *Communication Channels: Characterization and Behavior* [Goldberg, 1976] is of special interest in connection with the following topic on channel modeling. Section IV: "HF Ionosphere" is recommended.

4.14 CHANNEL MODELING

The HF channel scattering function, which is a representation of time delay spread against Doppler spread, has been experimentally examined for the wideband case [Wagner et al., 1988; Basler et al., 1988]. (The topic was briefly considered in Section 4.13.3, and Figure 4-52 is a sample scattering function for the wideband channel.) Various models are also available for use in estimating channel properties. An early model [Watterson, 1970] has found restricted use for analysis of narrowband channels. More recently, efforts by Vogler and Hoffmeyer [1989] of ITS, Barratt and Walton [1988] of MITRE , and Malaga [1985] have resulted in a better understanding of the wideband HF channel. Knowledge of the ionospheric multipath structure has been greatly enhanced through investigation by wideband probes and sounders. An FMCW-type sounder has been developed by Salous [1989], and a wideband HF system study has been carried out by MITRE [Perry et al., 1987; Perry and Rifkin, 1985]. In addition, Rome Air Development Center (RADC) has investigated a variety of adaptive probing techniques [Haines and Weijers, 1985], and SRI has developed a set of channel probe parameters [Price, 1983] for disturbed media. The most sophisticated experimental investigation of the HF channel for bandwidths of the order of 1 MHz has been conducted by L.S. Wagner and his colleagues at NRL (See section 4.13.3 and the reference list at the end of the chapter). Here we shall first review channel modeling concepts, and then discuss channel simulators.

4.14.1 Channel Modeling Concepts

Our understanding of HF has been largely limited to narrowband processes. Recently, because of the interest in wideband spread-spectrum signaling, more information about the channel is becoming available for channels which may be characterized by arbitrary bandwidths.

As far as the HF channel is concerned, two forms of propagation effects are noteworthy, losses and distortion. Losses includes those associated with normal r^2 spreading loss, focusing, ducting, and similar effects discussed in Section 4.11, including absorption loss. Polarization mismatch discussed by Davies [1990] is a form of loss which causes flat fading in narrowband systems, but may be highly frequency dependent for wideband signaling. This introduces distortion of the waveform, the other major effect to be considered.

Major sources of distortion include [Barratt and Walton,1987]: time variation of channel parameters, multipath, and signal dispersion. We have already discussed multipath in Section 4.13, and the impact of temporal variability is clear when those variations represent a significant fraction of the signal bandwidth. In general it may be said that the multipath introduces two forms of impairment depending upon whether or not the multipath delay is *intermodal* or *intramodal*. Intermodal delays are due to multimode and multihop propagation, and give rise to intersymbol interference. Intramodal delays are due to geomagnetic field effects, range spread due to ionospheric inhomogeneities, and the dispersive property of the ionospheric medium (independent of any structure present). Thus intramodal delays cause distortion of the pulse and will place limits on channel bandwidth.

Dispersion is the result of the variation of signal velocity with frequency. Thus, individual frequency components transmitted at a time t_0 will arrive at a destination within a time envelope defined by $t_0 + T + \delta t$, where T is the median time delay, and δt is the time delay spread of the individual components. Dispersion has a greater impact on wideband systems than narrowband ones and is generally felt to be responsible for a limitation in the coherence bandwidth. Indeed, a nonlinear phase and group delay properties of the channel transfer function will distort wideband waveforms significantly, and this may be observed both as a broadening of the received pulse *and* a reduction in its amplitude. The magnitude of dispersion is usually defined in units of μsec/MHz. Dispersion is considerably greater for short skywave paths than for long-haul paths, and for narrowband channels it ranges from as much as 500 over short paths to as little as 1-10 over long paths. We observe less dispersion on E region modes than F layer modes; and ground wave propagation exhibits virtually no dispersion. The magnitude of dispersion depends critically on the mode and hop number involved as well as the proximity to the MOF. It also exhibits a frequency dependence, a factor of significance principally for wideband channels. Wagner and Goldstein [1985] have indicated that dispersion can be considerable for signals with bandwidths in excess of 100 KHz.

Another form of signal distortion is that which is introduced by Faraday rotation or polarization interference, a phenomenon discussed earlier in the chapter. If the transmission bandwidth is equal to or larger than the polarization bandwidth (but not large enough to resolve the two magnetoionic modes responsible for the effect) considerable distortion and fading may be introduced. The largest value for the polarization bandwidth occurs in the neighborhood of the crossover of the O- and X-modes, a phenomenon easily detected on conventional ionograms. This would suggest that operation near this point would be prudent in most instances. We shall return to this matter shortly. The smallest polarization bandwidth is observed to occur near the MOF, and is least for the high ray. As indicated in Section 4.5.1 of this chapter, there is a general monotonic decrease with frequency in the polarization bandwidth for the 1F2 low ray. It is interesting to note, however, that a departure from a monotonic behavior arises in the lower part of the transmission band below the nominal crossover frequency, where the polarization bandwidth decreases again. To determine the optimum transmission frequency for establishment of the largest value for BW_c while minimizing polarization interference, we should locate and examine two frequencies: the frequency for least dispersion *and* the crossover frequency. They are not always the same. By definition, we cannot resolve the two modes at the crossover frequency, so any interference distortion still present there would have to be tolerated. Although polarization interference will increase away from this point, pulse distortion may be reduced. We can have the best of both worlds if our transmission bandwidth is sufficiently high to resolve the interfering modes (one of which must have a minimum in pulse dispersion). If the mode separation is 5 μsec or more, then a 200 kHz signal would be necessary to resolve the two waves. Information of the type described here, and derived from wideband probes, allows determination of the maximum (polarization interference-free) communication bandwidths which may be supported without invoking equalization or similar techniques.

Because of the interest in wideband signaling, some of workers have examined the possibility of deriving wideband channel information from a large database of narrowband measurements. Hausman et al. [1988], using data from Wagner's wideband probe, show a good correspondence between inferred wideband properties and direct measurements.

The channel bandwidth BW_c(MHz) has been shown to be related to the group path delay slope by the following expression [Wagner and Goldstein, 1985]:

$$BW_c = 1/(|d\tau/df|)^{\frac{1}{2}} \tag{4.58}$$

where $(|d\tau/df|)$ is the absolute value of the group-path-delay slope (in μsec/MHz). This suggests that the channel bandwidth may be estimated rather conveniently. Salous [1989] using a wideband FMCW probe over a short link found values of $(|d\tau/df|)$ for 1F2 layer O- and X-modes as well as the Es layer. They ranged between 16 and 528 μsec/MHz for the 1F2 O-mode, 16 and 144 μsec/MHz for the 1F2 X-mode, and 16 and 40 μsec/MHz for the Es mode. The implication is that the Es layer, with its smaller values of $(|d\tau/df|)$ and therefore larger channel bandwidth will generally support higher bandwidth communications. Taking $(|d\tau/df|) \approx 16$, we have $BW_c \approx$ 250 kHz; occasionally this condition existed for all three modes analyzed, but it was most consistently realized for the Es mode. Taking the other extreme for which $(|d\tau/df|) \approx 528$ μsec/MHz, we find $BW_c \approx 44$ KHz.

Let $E(t)$ be the response of the HF channel to a signal $E_0(t)$. If $h(\tau,t)$ is the (time-variant) impulse response function, then according to Bello [1963], the (received) response for a linear channel will be:

$$E(t) = \int_{-\infty}^{\infty} h(\tau,t)\, E_o(t-\tau)\, d\tau \tag{4.59}$$

If $E_o(t) = \delta(t)$, a Dirac impulse, then h is actually the received field justifying its designation as the impulse response. The Fourier transform of h is the transfer function H. It is useful to have a statistical description of the channel response. This leads to the notion of a channel scattering function. Under suitable assumptions the channel scattering function is the delay-Doppler power spectrum of the received signal as modified by the channel [Nickisch, 1986, 1990].

Figure 4-57 shows the channel scattering function for an hypothetical situation of 3 ionospheric paths (labeled A,B, and C) resolved in both time delay and Doppler shift. Characterization of the HF channel in this way is central to modeling approaches. In this figure we note that the Doppler shift and spreads are relatively small, and the situation is like that which would be experienced over a benign midlatitude path where E, F1, and F2 modes might be resolved in time delay. Figure 4-58 shows the Doppler spectra for two long paths: one for midlatitude paths and the other for transauroral paths. We note the obvious doppler spread enhancement for propagation over the auroral circuit. This is a result of the rapid motion of ionospheric irregularities within the auroral channel.

The channel scattering function may exhibit a number of modes, and each one may possess considerable time delay and Doppler spreads. (Note that the mean Doppler shift f_d and mean path delay t are not fundamental problems for system design. They are mainly important for signal acquisition and are generally compensated for through use of Doppler tracking and synchronization circuitry [Perry et al., 1987].) The following relations are worth noting:

$$f_c = 1/(2\pi\sigma_\tau) \tag{4.60a}$$

$$\tau_c = 1/(2\pi\sigma_d) \tag{4.60b}$$

In these relations, σ_τ and σ_d are the standard deviations for the time delay and Doppler spreads respectively. Equation 4.60a relates the coherence bandwidth f_c to the inverse of the time delay spread; Equation 4.60b relates τ_c to the inverse of the Doppler spread.

Over a 2000 km path with a wideband (i.e., 1 MHz instantaneous BW) system, Perry et al. [1987] have obtained the *channel* transfer function and the *channel* impulse response. Figure 4-59 shows two interesting distortion features. The first is exhibited in the transfer function, plotted on the left, where a number of nulls are observed over the instantaneous bandwidth. These are related to the interference between the O and X-modes. In the plot on the right, we see the dispersion enhancement from 1 μsec to about 25 μsec.

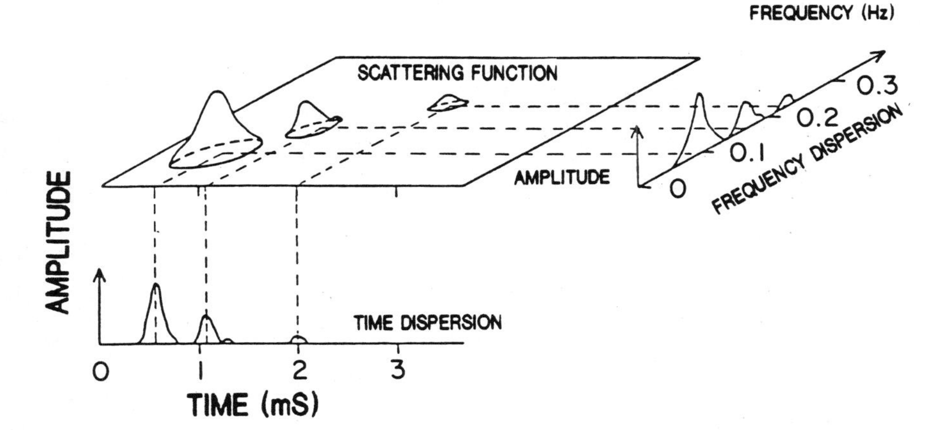

Fig.4-57. Channel scattering function. This illustration suggests that the ionosphere supports three distinct modes of propagation for which both the mean time delays and Doppler shifts are isolated and appear resolved.

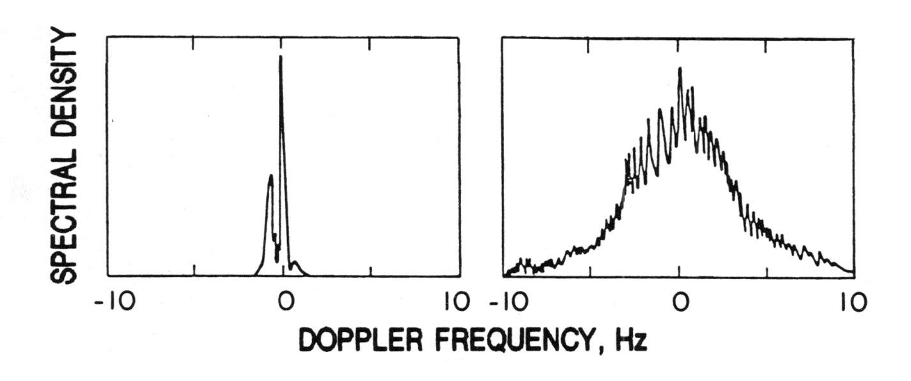

Fig.4-60. Doppler spectra for two long paths, each terminating in Palo Alto California. [Left] Midlatitude path, to Ft.Monmouth, New Jersey. [Right] auroral path, to Thule, Greenland. (Courtesy T.Croft [1982], SRI; after unpublished report by R. Vincent)

4.14.2 HF Channel Models and Simulators

HF channel models are designed to represent the ionospheric channel at the frequency (and waveform characteristics) of interest. These models, in turn, are represented by simulators. The latter is a term usually reserved for hardware devices or software which will generate repeatable and realistic test conditions. In Table 4-10 we list a number of channel models and simulators.

TABLE 4-10. CHANNEL MODELS AND SIMULATORS

Model Reference	Type
Goldberg, Heyd, and Pochmerski [1965]	Stored Ionosphere
Bray, Lillicrap, and Owen [1947]	NB Simulator
Law, Lee, Looser, and Levett [1957]	NB Simulator
Freudberg [1965]	NB Simulator
Di Toro, Hanulec, and Goldberg [1965]	NB Simulator
Walker [1965]	NB Simulator
Clarke [1965]	NB Simulator
Chapin and Roberts [1966]	NB Simulator
Adams and Klein [1967]	NB Simulator
Zimmerman and Horowitz [1967]	NB Simulator
Packer and Fox [1969]	NB Simulator
Watterson, Ax, Demmer, and Johnson [1969b]	NB Simulator
CCIR Report 549-2 (Annex II) [1986o]	Recorder-Reproducer
Perl [1984]	Simulator
Nesenbergs [1987]	WB Model (*)
Barratt and Walton [1988]	WB Simulator (*)
Vogler and Hoffmeyer [1988]	NB & WB Model (*)
Nickisch [1986]	Stochastic Model
Malaga [1981, 1985]	WB Model (*)

* Proposed Model NB = Narrowband WB = Wideband

4.14.2.1 *Watterson Model*

The stationary Gaussian HF channel behavior may be represented schematically as in Figure 4-60. The input signal is fed into a delay line from which the signal may be extracted at certain adjustable taps. For a given tap, the signal is subject to modification in both amplitude and phase by a tap-gain function G_i which is complex and random. In general there are as many taps as resolvable modes in the model. According to CCIR 549-2 [1986o], G_i may be represented as the sum of *two independent complex (bivariate) Gaussian ergodic random processes, each with zero mean values and independent real and imaginary components with equal r.m.s. values that produce Rayleigh fading.*

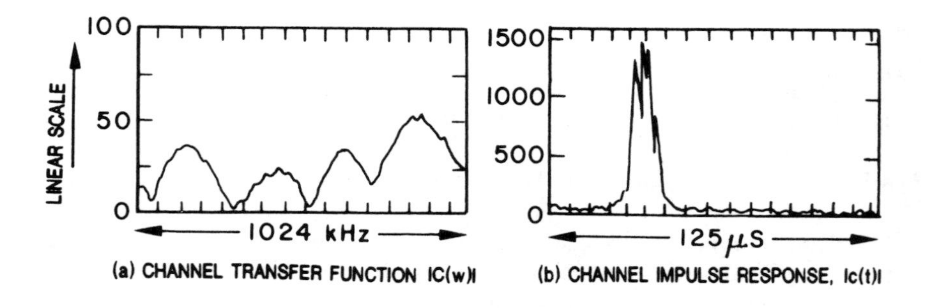

Fig.4-59. Measurements between Eglin AFB and Bedford, MA (2000 km). [Left] Channel transfer function. [Right] Channel impulse response. (From Perry et al. [1987])

$$G_i(t) = G^*_{ia}(t) \exp(j2\pi f_{ia} t) + G^*_{ib}(t) \exp(j2\pi f_{ib} t) \quad (4.61)$$

where a and b represent the two possible magnetoionic components, and the exponentials allow for Doppler shifts for the components. Now, of course, it must be recognized that each $G_i(t)$ is associated with a nominal time delay τ_i. Thus the transfer function for a single mode is given by:

$$H_i(f,t) = G^*_i(t) \exp(j2\pi f_i t) \exp(-j2\pi f\tau_i) \quad (4.62)$$

where we have suppressed the presence of two unresolved magnetoionic modes. The functions $G^*(t)$ are quite important since, if properly normalized, the autocorrelation properties may be used to deduce the coherence time(s) for the individual mode(s). In fact the lag time for which the autocorrelation of $G^*(t)$ drops to $1/e$ is precisely the coherence time [Malaga, 1985].

The tap-gain functions must also be characterized by spectra. For each function $G_i(t)$ there is a spectrum $F_i(f)$ which is actually the sum of the component magnetoionic spectra, with the variable f representing frequency shift. Six parameters may be used to specify the functions $G_i(t)$ and $F_i(f)$: the attenuations for each component (A^*_{ia} and A^*_{ib}), frequency shifts (f_{ia} and f_{ib}), and the frequency spreads ($2\sigma_{ia}$ and $2\sigma_{ib}$). The spectra of the tap-gain functions are written as:

$$F_i(f) = \sum_{j=a}^{b} [A^*_{ij}(2\pi)^{\frac{1}{2}} \sigma_{ij}]^{-1} \exp\{-(f-f_{ij})^2/2\sigma_{ij}^2\} \quad (4.63)$$

which may be represented in Figure 4-61. Under certain conditions the equations 4.61 and 4.63 may be simplified. These conditions arise when we may ignore the distinction between the two magnetoionic modes; i.e., the component spectra are approximately equal and the time delay between the two components is smaller than $\frac{1}{4}$ B_s^{-1}, where B_s is the signal bandwidth. For a 3 kHz signal spectrum, this corresponds to 80 μsec or so. For a 1 MHz spectrum, the maximum allowable delay separation is $\frac{1}{4}$ μsec.

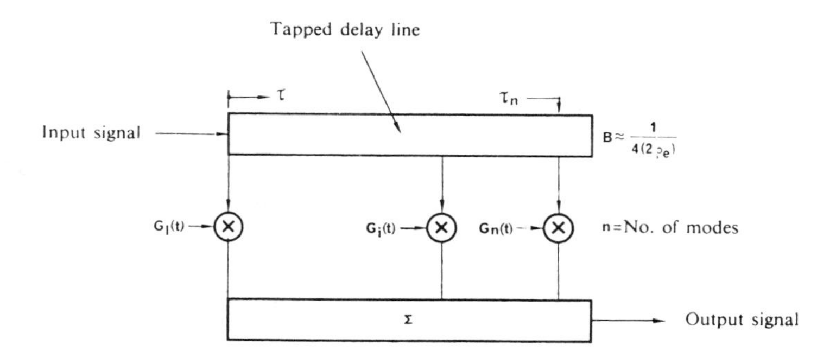

Fig.4-60. Schematic diagram of a Tapped-Delay-Line (TDL) model for the HF channel. [CCIR Report 549-2, 1986o]

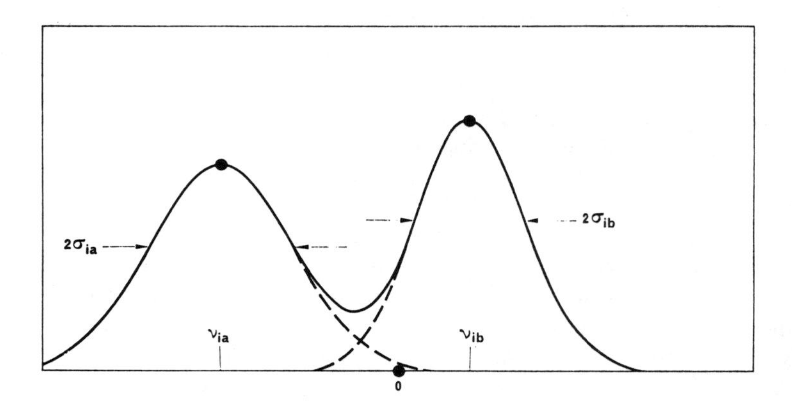

Fig.4-61: Two component Gaussian scatter spectra. (From CCIR Rpt. 549-2 [1986o].)

The validity of the Gaussian-scatter model was confirmed under some specialized conditions by Watterson et al.[1969a,1970]. Furthermore with the inclusion of a specular component, Watterson et al.[1969b] have developed a simulator which could be used in laboratory experiments. It should be noted that the so-called Watterson model is applicable only under conditions of channel stability and stationarity. Unfortunately Watterson and his colleagues did not have a climatological model (nor an analytical model) which could provide estimates of the number of anticipated paths, the corresponding path delays, the mode amplitudes, the Doppler shifts and the Doppler spreads.
Such a model has been developed by Malaga [1985]. In any case, for use in the Watterson model, the following limits are deemed applicable:

- Channel bandwidth ≈ 12 kHz or smaller
- Channel is time stationary
- Channel is frequency stationary
- Channel has low delay dispersion
- Low rays only considered

The transfer function for the narrowband HF channel (NBHF) must account for all of the time-varying frequency-independent tap-gain multipliers, $G^*_i(t)$, from Equation 4-61. Suppressing the explicit representation involving unresolved magnetoionic components, the transfer function for the narrowband channel is given by [Malaga,1985]:

$$H(f,t) = \sum_i A_i\ C_i \exp \{j(2\pi)[f_i t - f\tau_i) \qquad (4.64)$$

where the A_i, f_i, τ_i are constants, the $C_i = C_i(t)$, and all modes are summed.

Equation 4-64 is virtually equivalent to Equation 4.62, and similar versions have been published by others [Vogler and Hoffmeyer, 1988]. The HF channel may be represented by a linear, time-varying transfer function or filter.

4.14.2.2 *Wideband Modeling*

The growth in interest in HF spread spectrum communications, and wideband HF in particular, has occurred as a result of a number of advantages possessed by these methods which would appear to eliminate the negative attributes of the more conventional narrowband HF channel. For this purpose we shall regard 12 kHz as a convenient breakpoint between narrowband and wideband systems. Strictly speaking the breakpoint is blurred; it is certainly system-dependent, environmentally-sensitive, and is the point where ionospheric dispersion and the range spreading of ionospheric inhomogeneities become significant for the operation of the system.

Only a few wideband HF (WBHF) models have been developed, and validation is incomplete as of this writing. Malaga [1985], Nesenbergs [1988], and Barratt and Walton [1987] have proposed models for the wideband channel; and Vogler and Hoffmeyer [1988] have started development of a combined WBHF and narrowband (NBHF) model. These efforts are listed in Table 4-10.

The wideband transfer function may be given by a generalization of the Watterson model for which the narrowband *constants* A_i and τ_i are assumed to be functions of frequency, and C_i is taken to be a function of frequency as well as time. The frequency shift f_i is a constant, just as in the narrowband model. There are a number of other possibilities, but without a significant amount of data upon which to base alternatives, the choices just stated appear to be adequate. A discussion of the significance of the frequency-dependent terms in the WBHF model may be found in Nesenbergs [1988]. The most obvious factors of importance for the WBHF model are the delay dispersion τ_i and the complex amplitudes C_i. The C_i's, being functions of time and frequency, reflects the range dispersion associated with ionospheric inhomogeneities (i.e., scattering).

4.14.3 Multipath Models and Applications

Multipath spread is an important feature in HF signaling since, in the absence of equalization, it limits the maximum rate at which symbols may be transmitted without encountering intersymbol interference. The multipath spread δT_m is inversely proportional to the coherence bandwidth. If the signal bandwidth is greater than the channel coherence bandwidth, then it is possible for the receiver to resolve more than a single multipath component (or mode). This is certainly an advantage since the possibility for diversity gain arises. This gain may only be realized , of course, if the two resolved modes are fading independently. The number of modes Γ which may be resolved is roughly:

$$\Gamma \approx 1 + B_s/B_c \qquad (4.65)$$

where B_s and B_c are the signal and channel bandwidths respectively, and $B_c = 1/\delta T_m$. Malaga [1985] has shown that the BER of a low-data-rate system is reduced as the number of resolved multipath channels is increased. Said another way, it is possible to reduce the power-gains-product $P_t G_t G_r$ and still maintain a fixed bit error rate BER if we can take advantage of the implicit diversity made available through multipath resolution. This suggests that a MOF-seeking architecture for HF signaling is precisely the wrong choice in some instances. We shall return to this matter in Chapter 6. It would appear that a viable multipath model would be quite useful for studying the implicit diversity gain probability for circuits of interest. A sounder system or probe can provide this information in near-real time, and might be of consequence in LPI system operation.

4.15 REFERENCES

Adams, R.T., and M.S. Klein, 1967, "Simulation of Time-Varying Propagation by Computer Control", *IEEE Conf. on Comms. (ICC'67)*, page 74.

Aikens, A.J., R.S. Tucker, and A.G. Chapman, 1944, Appendix A in "HF Sky-Wave Transmission over Short or Moderate Distances Using Half-Wave Horizontal or Sloping Antennas" by A.J. Aikens, Final Report NDRC Project C-70, National Defense Research Council, Washington DC.

Albrecht, H.J., 1957, "Investigations of Great Circle Propagation Between Eastern Australia and Western Europe", *Geof.Pura e Applicata*, 38:169-180.

Albrecht, H.J., 1959, "Further Studies on the Chordal-Hop Theory of Ionospheric Long-Range Propagation", *Arch. Met. Geoph. Bioklim.*, 11:383-391.

Ames, J., T. Croft, T. Adams, and J. Johnston, 1977, "High Frequency, Low-Data-Rate, Low-Probability-of-Intercept Communications Using Skywave", Stanford Research Institute Report 7-4367, Project 5816 (prepared for NADC, Warminster PA), Menlo Park CA.

Appleton, E.V. and W.J.G. Beynon, 1948, "The Application of Ionospheric Data to Radio Communications", Dept.of Scientific and Industrial Research, Special Report No.18, His Majesty's Stationary Office, London.

Appleton, E.V. and W.R. Piggott, 1954, "Ionospheric Absorption Measurements During a Sunspot Cycle", *J.Atmospheric Terrest. Phys.*, No.5, p.141.

Assis, M.S., 1971, "A Simplified Solution to the Problem of Multiple Diffraction over Rounded Obstacles", *IEEE Trans. Ant. & Prop.*, AP-19:292-295.

Barghausen, A.F., J.W.Finnley, L.L.Proctor, and L.D. Schultz, 1969, "Predicting Long-Term Operational Parameters of High Frequency Skywave Telecommunication Systems", ESSA Technical Report. ERL 110-ITS-78, US Dept. Commerce, Boulder CO.

Bailey, D.K., 1959, "The effect of multipath distortion on the choice of operating frequencies for high frequency communication circuits", *IRE Trans Antennas & Propagation*,AP-7:398.

Bain, W.F., 1963, "Abnormal Signal Strengths from Soviet Transmitters", Page Communications Engineers, Inc., presented at *Ninth National Communications Symposium*, PCE-R-1152-0034A, Utica, NY.

Bain, W.F., 1965, "Unusual HF Propagation Modes as Implied by Observations of Transmissions from the USSR and Red China", R-1152-0047A, presented at *URSI Symposium* (October 1964), University of Illinois.

Baker, D.C., and S. Lambert, 1989, "Range Estimation for SSL HFDF by Means of a Multiquasiparabolic Ionospheric Model", *IEE Proc*, 136, Pt.H, (2):120-125.

Barratt, J.J. and T.L. Walton, 1988, "A Real-Time Wideband Propagation Simulator for the High Frequency Band", In *Effect of the Ionosphere on Communication, Navigation, and Surveillance Systems (IES'87)*, ed. J.M. Goodman, USGPO 1988-195-225, available through NTIS,Springfield VA.

Barrick, D.E., 1970, "Theory of Ground-Wave Propagation Across a Rough Sea at Dekameter Wavelengths", Res. Report., Battelle Memorial Institute. (AD 865 840).

Barrick, D.E., 1971a, "Theory of HF and VHF Propagation Across the Rough Sea, 1, the Effective Surface Impedance for a Slightly Rough Highly Conducting Medium at Grazing Incidence", *Radio Science*, 6:517-526.

Barrick, D.E., 1971b, "Theory of HF and VHF Propagation across the Rough Sea, 2, Application to HF and VHF Propagation Above the Sea", *Radio Science*, 6:527-533.

Basler, R.P., G.H. Price, R.T. Tsunoda, and T.L. Wong, 1988, "HF Channel Probe",In *Effect of the Ionosphere on Communication, Navigation, and Surveillance Systems (IES'87)*, ed. J.M. Goodman, USGPO 1988-195-225, available through NTIS, Springfield VA.

Bergman, C.W. and C. Barnes, 1967, "An Improved Three-Frequency Dipole", Research Memorandum, SRI Project 6183, Stanford Research Institute, Menlo Park CA.

Bello, P., 1963, "Characterization of Randomly Time-Variant Linear Channels", *IEEE Trans.Comm.Systems*, reprinted in *Communication Channels: Characterization and Behavior*, edited by B. Goldberg, IEEE Press, New York, NY.

Berry, L.A., 1978, "User's Guide to Low Frequency Radio Coverage programs", Office of Telecommunications Report 78-247, Boulder CO.

Berry, L.A. and M.E. Chrisman, 1966, "A Fortran Program for Calculation of Ground Wave Propagation over Homogeneous Spherical Earth for Dipole Antennas", NBS Report 9178, Boulder CO.

Blomquist, A., 1975, "Seasonal Effects on Ground-Wave Propagation in Cold Regions", *J. Glaciology*, 15:73.

Blomquist, A. and L. Ladell, 1975, "Prediction and Calculation of Transmission Loss in Different Types of Terrain", In *Electromagnetic Wave Propagation Involving Irregular Surfaces and Inhomogeneous Media*, ed. A.N. Ince, NATO-AGARD-CP-144, paper No. 32, Tech. Ed. & Reprod. Ltd., UK.

Bold, G.E.J., 1969, "Power Distribution near the Antipode of a Short-wave Transmitter", *J. Atmospheric Terrest. Phys.*, 31:1391-1411.

Bold, G.E.J., 1972, "The Influence of Chordal Paths on Signals Propagating Near the Antipode of an HF Radio Transmitter", *IEEE Trans. Ant. & Prop.*, AP-20:741-746.

Boithias, L., 1987, *Radio Wave Propagation*, McGraw-Hill Book Company, New York.

Booker, H.G. and R. Lugananni, 1978, "HF Channel Simulator for Wideband Signals", NOSC TR 208, Naval Ocean Systems Center, San Diego CA.

Bradley, P.A., and J.R. Dudeney, 1973, "A Simple Model of the Vertical Distribution of Electron Concentration in the Ionosphere", *J.Atmospheric Terrest. Phys.*, 35:2131-2146.

Bray, W.J., H.G. Lillicrap, and F.C. Owen, 1947, "The Fading Machine and its Use for the Investigation of the Effects of Frequency-Selective Fading", *J.Inst.Elec.Engrs.*, London, 94, Part IIIA: 283-297.

Brayer, K. (Editor), 1975, *Data Communication via Fading Channels*, IEEE Press, New York, NY.

Breakall, J.K., G.J. Burke, and E.K. Miller, 1985, "The Numerical Electromagnetics Code (NEC)", in *Proceedings of the 6th Symposium and Technical Exhibition on Electromagnetic Compatibility*, Zurich, pp.301-308.

Brune, J.F. and J.E. Reilly, 1975, "Compact HF Antenna", R&D Technical Report ECOM 4366, US Army Electronics Command, Ft. Monmouth NJ.

Brune, J.F. and B.V. Ricciardi, 1979, "Modern HF Communications for Low Flying Aircraft" in *Special Topics in HF Propagation*, pp. 4-1 to 4-15, NATO-AGARD-CP-263, Tech. Edit. & Reprod. Ltd., London UK, .

Budden, K.G., 1961, *Radio Waves in the Ionosphere*, The University Press, Cambridge, UK.

Budden, K.G., 1964, *Lectures on Magnetoionic Theory*, Gordon and Breach, New York.

Budden, K.G., 1985, *The Propagation of Radio Waves*, Cambridge University Press, New York NY.

Budden, K.G. and G.W. Jull, 1964, "Reciprocity and Non-Reciprocity with Magnetoionic Rays", *Can. J. Phys.*, 42:113.

Burgess, B., 1979, "The Role of HF in Air-Ground Communications", in *Special Topics in HF Propagation*, NATO-AGARD-CP-263, pp. 2-1 to 2-6, Tech. Edit. & Reprod. Ltd., London UK.

Burke, G.J. and A.J. Poggio, 1977, "Numerical Electromagnetic Code (NEC) - Method of Moments", Part III: User's Guide, Lawrence Livermore Technical Document 116, under contract to NOSC (MIPR-N0095376MP) and the AF Weapons Lab (Project Order 76-090).

Burke, G.J. and A.J. Poggio, 1981, "Numerical Electromagnetics Code (NEC) -- Method of Moments", Volume 1, NOSC Tech. Doc. 116, Vol.1 (developed by Lawrence Livermore Laboratory under contract), (See Burke and Poggio [1977].)

Burke, G.J., 1983, "Numerical Electromagnetics Code - User's Guide Supplement for NEC-3 for Modelling Buried Wires", Lawrence Livermore Laboratory, Livermore CA.

Burrows, C.R., 1936, "Existence of a Surface Wave in Radio propagation", *Nature*, 138:284.

Burrows, C.R., 1937, "Radio Propagation over a Plane Earth, Field Strength Curves", *B.S.T.J.*, 16:45-77 and 16:1203-1236.

Busch, H.F., 1963, "Project Yo-Yo Field Experiments", Technical Report developed for Advanced Research Projects Agency, ACF Industries, Hyattsville MD.

Carman, E.H., M.P. Heran, and J. Röttger, "Simultaneous Observations of Fading Rates on Two Transequatorial HF Radio Paths", *Austral.J.Phys.*, 27:741-744.

Carrara, N., M.T.deGiorgio, and P.F. Pelligrini, 1970, "Guided propagation of HF Radio waves in the Ionosphere", *Space Sci.Rev.*, 11:555-592.

CCIR, 1978, "CCIR Report 373: Definitions of Maximum Transmission Frequencies", in *Recommendations and Reports of the CCIR, 1978; Vol. VI: Propagation in Ionized Media*, ITU, Geneva.

CCIR, 1986a, *Recommendations and Reports of the CCIR, 1986; Vol. V: Propagation in Non-Ionized Media*, ITU, Geneva.

CCIR, 1986b, "CCIR Report 714-1: Ground-Wave Propagation in an Exponential Atmosphere", In *Recommendations and Reports of the CCIR, 1986; Volume V: Propagation in Non-Ionized Media*, ITU, Geneva.

CCIR, 1986c,"CCIR Report 715-2: Propagation by Diffraction", In *Recommendations and Reports of the CCIR, 1986; Volume V: Propagation in Non-Ionized Media*, ITU, Geneva.

CCIR, 1986d, "CCIR Recommendation 368-5: Ground-Wave Propagation Curves for Frequencies Between 10 KHz and 30 MHz", in *Recommendations and Reports of the CCIR, 1986; Volume VI: Propagation in Ionized Media*, ITU, Geneva.

CCIR, 1986e, "CCIR Report 229-5: Electrical Characteristics of the Surface of the Earth", in *Recommendations and Reports of the CCIR, 1986; Volume VI: Propagation in Ionized Media*, ITU, Geneva.

CCIR, 1986f, "CCIR Recommendation 527-1: Electrical Characteristics of the Surface of the Earth", in *Recommendations and Reports of the CCIR, 1986; Volume VI: Propagation in Ionized Media*, ITU, Geneva.

CCIR, 1986g, "CCIR Report 236-6: influence of Terrain Irregularities and Vegetation on Tropospheric Propagation", in *Recommendations and Reports of the CCIR, 1986; Volume VI: Propagation in Ionized Media*, ITU, Geneva.

CCIR, 1986h, "CCIR Report 1008: Reflection from the Surface of the Earth", in *Recommendations and Reports of the CCIR, 1986; Volume V: Propagation in Non-Ionized Media*, ITU, Geneva.

CCIR, 1986i, "CCIR Report 725-2: Ionospheric Properties", in *Recommendations and Reports of the CCIR, 1986; Volume VI: Propagation in Ionized Media*, ITU, Geneva.

CCIR, 1986j, "CCIR Recommendation 373-5: Definition of Maximum Transmission Frequencies", in *Recommendations and Reports of the CCIR, 1986; Volume VI: Propagation in Ionized Media*, ITU, Geneva.

CCIR, 1986k, "CCIR Interim Method for Estimating Skywave Field Strength and Transmission Loss at Frequencies between the Appropriate Limits of 2 and 30 MHz", CCIR 252-2, in *Recommendations and Reports of the CCIR: Vol. VI: Propagation in Ionized Media*, ITU, Geneva.

CCIR, 1986L, "CCIR Report 250-6: Long Distance Ionospheric Propagation Without Intermediate Ground Reflection", in *Recommendations and Reports of the CCIR, 1986; Volume VI: Propagation in Ionized Media*, ITU, Geneva.

CCIR, 1986m, "CCIR Report 266-6: Ionospheric Propagation and Noise Characteristics Pertinent to Terrestrial Radiocommunications Systems Design and Service Planning", in *Recommendations and Reports of the CCIR, 1986; Volume VI: Propagation in Ionized Media*, ITU, Geneva.

CCIR, 1986n, "CCIR Report 304-2: Fading Characteristics for Sound Broadcasting in the Tropical Zone", in *Reports and Recommendations of the CCIR, 1986; Vol. X (part 1): Broadcasting Service (Sound)*, ITU, Geneva.

CCIR, 1986o, "CCIR Report 549-2: HF Ionospheric Model Simulators", in *Reports and Recommendations of the CCIR, 1986; Vol. III: Fixed Service at Frequencies below about 30 MHz*, ITU, Geneva.

Chang, H., 1971a, "Waveguide Mode Theory of Whispering Gallery Propagation in the F-Region of the Ionosphere", *Radio Science*, 6:475-482.

Chang, H., 1971b, "Whispering Gallery Propagation in the E-region of the Ionosphere at HF and VHF", *Radio Science*, 6:465-473.

Chapin, E.W., and W.K. Roberts, 1966, "A Radio Propagation and Fading Simulator Using Radio Frequency Acoustic Waves in a Liquid", *Proc.IEEE*, 54(6):1072.

Christinsin, A., 1986, "Compendium of High Frequency Radio Communications Articles", unpublished technical report, prepared by DCS/Logistics, Air Force Communications Command.

Chvojková, E., 1965, "Analytical Formulae for Radio Path in Spherically Stratified Ionosphere", *NBS J. Res.*, 69D:453-457.

Chvojková E., 1974a, "Radio Path Formula:1- Theory, Circumterrestrial Echo and Other Singularities", *Geomag. & Aeron.*, 14:46-53 (English Edition).

Chvojková E., 1974b, "Radio Path Formula:2- Practical Application"s, *Geomag. & Aeron.*, 14:392-399 (English Edition).

Chvojková E., 1976, "Multiple Propagation Paths Between Satellites Situated in the Ionosphere below the F2-Layer Peak", *J.Atmospheric Terrest. Phys.*, 38:329-331.

Clarke, K.K., 1965, "Random Channel Simulation and Instrumentation", *First Annual IEEE Communications Conf*, Boulder CO, pp. 623-629.

Croft, T., 1982, private communication

Croft, T.A. and L. Gregory, 1963, "A Fast, Versatile, Ray Tracing Program for IBM 7090 Digital Computers" Technical Report No.82, Stanford Electronics Lab, Stanford CA.

Croft, T.A. and H.Hoogasian, 1968, "Exact Ray Calculations in a Quasi-Parabolic Ionosphere", *Radio Science*, 3(1):69-74.

Croft, T.A., 1972, "Sky-Wave Backscatter: A Means for Observing our Environment at Great Distances", *Rev. Geophys.Space Phys.*, 10:73.

Davies, K., 1965, *Ionospheric Radio Propagation*, NBS Monograph 80, USGPO, Washington DC.

Davies, K., 1969, *Ionospheric Radio Waves*, Blaisdell Publishing Company, Waltham MA.

Davies, K., 1990, *Ionospheric Radio*, Peter Peregrinus Ltd, London.

Davies, K. and A.F. Barghausen, 1964, "The Effect of Spread F on the Propagation of Radio Waves Near the Equator", in *Ninth Meeting of Ionospheric Research Committee*, AGARD, Copenhagen, Denmark.

DeMinco, N., 1986, "Automated Performance Analysis Model for Ground-Wave Communication Systems", NTIA Report 86-209, Dept. of Commerce, Boulder CO.

Deygout, J., 1966, "Multiple Knife-Edge Diffraction of Microwaves", *IEEE Trans. Antennas & Prop.*, AP-14:480-489.

Di Toro, M.J., J. Hanulec, and B. Goldberg, 1965, "Design and Performance of a New Adaptive Serial Data Modem on a Simulated Time-Variable Multipath HF Link", *First Annual IEEE Communications Conf.*, Boulder CO, p.770.

Doeppner, T.W., .H. Hagn, and L.G. Sturgill, 1973, "Electromagnetic Propagation in a Tropical Environment", *J. Defense Research, Series B, Tactical Warfare*, 4B(4):353-404.

Dougherty, H.T., 1969, "An Approximate Closed-Form Method of Solution for the Diffraction of Radio-Waves by Irregular Surfaces", *Dissertation Abstracts*, Engineering B. Order No.69-4358.

Dudeney, J.R. 1978, "An Improved Model of the Variation of Electron Concentration in the Ionosphere", *Proc. IEE*, 40:195-203.

ECAC, 1975, "ECAC Calculator Program #11-4: Groundwave Path Loss Model", Electromagnetic Compatibility Analysis Center, Annapolis MD.

ECAC, 1984, "Field Antenna Handbook", prepared for the Joint Chiefs of Staff by J.A. Kuch, IITRI, ECAC -CR-83-200, DoD Electromagnetic Compatibility Analysis Center, Annapolis MD.

ECAC, 1988, "Catalog of Computer Program Abstracts, Vol. 1: Analysis Capabilities", ECAC Handbook 81-047, Electromagnetic Compatibility Analysis Center, Annapolis MD.

Egli, J.J. and R.B. Lacy, 1952, "System Study of Short Range Radio Transmissions Via the Ionosphere", Report 181-25, Project 132A, Coles Signal Laboratory, Fort Monmouth NJ.

Elkins, T., 1979, "Recent Advances in HF Propagation Simulation", paper No.21, in *Special Topics in HF Propagation*, NATO-AGARD-CP-263, V.J. Coyne (Editor), Tech. Edit. & Reprod. Ltd., London UK.

Engel, J.S., 1967, "High Reliability HF Communications", Final Report, Page Communications Engineers Inc., Washington DC.

Epstein, M.R., 1967, "Computer Prediction of the Effects of HF Oblique-Path Polarization Rotation with Frequency", Technical Report 139, Stanford Electronics Laboratories, Stanford University, Stanford CA.

Epstein, M.R. and O.G. Villard Jr., 1968, "Received Polarization of Ionospherically-Propagated Waves as a Function of Time and Frequency", Technical Report #145, Stanford Electronic Laboratories, Stanford University, Stanford CA.

ESEIA, 1984, "Ground-Wave Propagation Charts; Short Title: GW Book", ESEIA Propagation Proj.3-2, Hdqtrs USA Electronic Systems Engineering Installation Agency, Ft. Huachuca AZ 85613-5300.

Fenwick, R.B. and O.G. Villard Jr., 1963, "A Test of the Importance of Ionosphere-Ionosphere Reflections in Long Distance and Around-the-World HF propagation", *J.Geophys.,Res.*, 68:5659-5666.

Fishchuk, D.I. and Ye Ye Tsedilina, 1980, "Propagation of Short Radio waves on a Mid-Latitude Round-the-World Hop Path", *Geomag.& Aeron.*, 20(3):313-317.

Flock, W.L., 1987, "Propagation Effects on Satellite Systems below 10 MHz: A Handbook for Satellite Systems Design", 2nd Edition, Ref. Pub. No. 1108(2), NASA, Washington DC.

Freudberg, R., 1965, "Laboratory Simulator for Frequency Selective Fading", *First Annual IEEE Communications Conf.*, Boulder CO, pp. 609-614.

Fulks, G.J., 1981, "HF Ground Wave Propagation over Smooth and Irregular Terrain", DNA 5796F, Mission Research Corporation, Santa Barbara CA.

Fulton, B., O.Sandoz, and E. Warren, 1960, "The Lower Frequency Limits for F-Layer HF Propagation", *J. Geophys. Res.*, 65:177-183.

Furutsu, K., 1962, "Effect of Ridge, Cliff and Bluff at a Coastline on Groundwaves", *J. Radio Res.Labs*, Vol.9, No.41, Japan.

Furutsu, K. and R.E.Wilkerson, 1971, "Optical Approximation for ther Residue Series of Terminal Gain in Radio-Wave Propagation over Inhomogeneous Earth", *Proc. IEE*, 118(9):1197-1202.

George, P.L., 1971, "The Global Morphology of the Quantity $\int N \nu dh$ in the D- and E-Regions of the Ionosphere", *J. Atmospheric Terrest. Phys.*, No.33, p.1893.

George, P.L., 1972a, "HF Ray Tracing of Gravity Wave Perturbed Ionospheric Profiles", NATO-AGARD-CP-115, paper 32, April.

George, P.L. (1972b) "Dynamic Tilt Correction and its Application to Short Range Single Station Location Systems", Dept. of Defence, Australian Defence Service, Weapons Research Establishment, WRE-Technical Note-1571 (AP), Box 2151, GPO, Adelaide S. Australia, 5001.

Gerson, N.C., 1968, "Ray Tracing Over a Trans-Equatorial Path. Scatter Propagation of Radio Waves", in NATO-AGARD-CP-37.

Gerson, N.C., 1979, comments made at NATO-AGARD conference on *Special Topics on HF Propagation*, AGARD-CP-263, V.J. Coyne (Editor), Tech. Edit. & Reprod. Ltd., London UK.

Gibson, A.J., and P.A. Bradley, "A New Formulation for Above-the-MUF Loss", in *Fifth International Conf. on HF Radio Systems and Techniques*, IEE, Savoy Place, London, UK.

Goldberg, B. (Editor), *Communication Channels: Characterization and Behavior*, *IEEE Press*, New York.

Goldberg, B., R.L. Heyd, and D. Pochmerski, 1965, "Stored Ionosphere", *Fist Annual IEEE Communications Conf.*, Boulder CO., pp. 619-622.

Goodman, J.M., 1979, "Geomorphology of the HF Breakthrough Phenomenon", Paper No. 14, in *Special Topics in HF Propagation*, NATO-AGARD-CP-263, Tech. Edit. & Reprod. Ltd., London UK.

Goodman, J.M. and J. Aarons, 1990, "Ionospheric Effects on Modern Electronic Systems", *Proc. IEEE*, 78(3):512-527.

Goodman, J.M. (ed.), 1983-88, Unpublished notes for AFCEA course 104, *Military Uses of the HF Spectrum*, Armed Forces Communications and Electronics Association, Fair Lakes Court, Fairfax VA

Grisdale, G.L., J.G. Morris, and D.S. Palmer, 1957, "Fading of Long-Distance Radio Signals and a Comparison of Space and Polarization Diversity Reception in the 6-18 Mc/s Range", *Proc. IEE*, 104B(13):39-51.

Grossi, M.D. and B.M. Langworthy, 1966, "Geometric Optics Investigation of HF and VHF Guided propagation in the Ionospheric Whispering Gallery", *Radio Science*, 1:877-886.

Grossi, M.D. and B.M. Langworthy, 1968, "Short-Wave Ionospheric Whispering Gallery", *IEEE International Convention Digest*, No.88, March 18-21, New York NY.

Gurevich, A.G. and E.E. Tsedilina, 1975, "Trapping of Radiation in the Ionospheric Duct during Scattering on Artificial Inhomogeneities", *Geomag. & Aeron.*, 15:713-715.

Gurevich, A.G. and E.E. Tsedilina, 1985, *Long Distance Propagation Of HF Radio Waves*, Springer, Heidelberg, Federal Republic of Germany.

Hagn, G.H.,1964, "Orientation of Linearly Polarized HF Antennas for Short-Path Communication via the Ionosphere near the Geomagnetic Equator", Report on SRI Project 4240, AD 480592, Stanford Research Institute, Menlo Park CA.

Hagn, G.H. and K.A. Posey, 1966, "Survey of Literature Pertaining to the Equatorial Ionosphere and Tropical Communications", Special Report on SRI Proj. 4240, AD 486800, Stanford Research Institute, Menlo Park CA.

Hagn, G.H., J.E. Van der Laan, D.L. Lyons, and E.M. Kreinberg, 1966, "Ionospheric Sounder Measurement of Relative Gains and Bandwidths of Selected Field-Expedient Antennas for Skywave Propagation at Near-Vertical Incidence", Special Technical Report No. 18, SRI Project 4240, AD 489537, Stanford Research Institute, Menlo Park CA.

Hagn, G.H. and G.E. Barker, 1970, "Research-Engineering and Support for Tropical Communications", AD-889-169, Final Report, Contract DA-36--039 AMC-00040(E), SRI Project 4240, Stanford Research Institute, Menlo Park CA.

Hagn, G.H., 1973, "Electrical Properties of Forested Media", in *Workshop on Radio Systems in Forested and/or Vegetated Environments*, conference held in 1973, documentation edited by J. Wait, R.H. Ott, and T. Telfer, Technical Report ACC-ACO-1-74 (AD 780 712)

Hagn, G.H., B.M. Sifford, and R.A. Shepard, 1982, "The SRICOM Probabalistic Model of Communication System Performance", SRI Int. Fin. Rpt. on Project 3603, contract NT-81-RC-1601, SRI International, Menlo Park CA.

Hagn, G.H. and J.C. Gaddie, 1984, "Medium and High Frequency (MF and HF) Ground Electrical Parameters Measured during 1982 at Seven Locations in the United States with the SRI OWL Probe Kit", presented at URSI Commission A, National Radio Science Meeting, Boulder CO., (abstract on p.78)

Hagn, G.H., 1985, (A correction to analytical curves found in Sailors [1984] were made in 1985 and provided to the author of this book by private communication [Hagn,1990]).

Hagn, G. and L. Harnish, 1986, "Measurement Techniques for HF Tactical Antennas", *Proceedings of the DARPA-AFCEA-IEEE Conference on Tactical Communications: the Next Generation*, Fort Wayne, Indiana.

Hagn, G.H., 1988a, "HF Ground and Vegetation Constants", unpublished lecture notes from AFCEA course 104, *Military Uses of the HF Spectrum*, Armed Forces Communications & Electronics Association (Fairfax VA) and SRI International (Arlington VA)

Hagn, G.H., 1988b, "HF Near-Vertical-Incidence-Skywave (NVIS) Communications", unpublished lecture notes from AFCEA course 104, *Military Uses of the HF Spectrum*, Armed Forces Communications & Electronics Association (Fairfax VA) and SRI International (Arlington VA)

Hagn, G.H. 1990, "Estimation of In-Situ Soil and Vegetation Electrical Properties at HF and VHF Using the SRI OWL Kit", (undated technical documentation on the OWL kit), SRI International, provided to the author by private communication.

Haines, D.M., and B. Weijers, 1985, "Imbedded HF Channel Probes/Sounders", *IEEE Military Comm. Conf.*, Boston MA, paper No. 12.1.

Hall, M.P.M., 1979, *Effects of the Troposphere on Radio Communication*, Peter Peregrinus Ltd.,for IEE, London and New York.

Haselgrove, J., Report of Conference on the Physics of the Ionosphere, London Physical Society, 355, 1954

Hausman, C.L., D.R. Uffelman, and T.L. Walton, 1987, "Wideband High Frequency (HF) Skywave Channel Parameters", in *Effect of the Ionosphere on Communication, Navigation, and Surveillance Systems (IES'87)*, J.M. Goodman (Editor-in-Chief), USGPO, order through NTIS, Springfield VA

Hayden, E.C., 1979, "Delineation of Constraints Imposed by Propagation Factors at HF on Jamming of Ships Communications", SouthWest Research Institute Report, Project No. 16-4312, Final Technical Report under contract M00039-75-C-0481, prepared for NAVELEX (now defunct), Navy Dept., Washington DC.

Ho, C.T. and G.M Ranvier, 1963-1964, Two Reports: "A Folded Dipole Antenna for VNN Naval District-Command Junk Communications" and "A 1/4 Wave Vertical Antenna for Command Junks of the RVNAF", Combat Development and Test Center, Saigon , Vietnam.

Holmes, S.K., 1979, "Near Vertical Incidence Skywave for Military Application", USA-CEEIA , Fort Huachuca AZ.

Hortenbach, K.J. and F. Rogler, 1979, "On the Propagation of Short Waves over Long Distances: Predictions and Observations", *Telecomm. J*, 46:320-329.

IEEE, 1991, "IEEE Standard Definitions of Terms for Radio Wave Propagation", IEEE Std 211-1990 (revision of IEEE Std 211-1977), IEEE, New York, NY 10017.

Ippolito, L.J., 1986, *Radiowave Propagation in Satellite Communication*, Van Nostrand Reinhold Publishing Co., New York.

Jones, R.M., 1968, "A Three-Dimensional Ray Tracing Computer Program". *Radio Science*, 3(1):93-94.

Jones, R.M., and J.J. Stephenson, 1975, A Versatile Three Dimensional Ray Tracing Computer Program for Radio Waves in the Ionosphere, Rpt. 75-76 (PB 248856), Office Telecomm., Dept. of Commerce, Boulder CO.

Jones, T.B. (Editor),1969, *Oblique Ionospheric Radiowave Propagation at Frequencies Near the Lowest Usable High Frequency*, AGARD-CP-13, Technivision Services, Slough, UK.

Jursa, A.S. (Editor), 1985, *Handbook of Geophysics and the Space Environment*, AFGL, Air Force, NTIS, Springfield, VA.

Kelso, J.M., 1964, *Radio Ray Propagation in the Ionosphere*, McGraw-Hill Book Company, New York.

Kelso, J.M., 1968, "Ray Tracing in the Ionosphere", *Radio Science*, 3(1):322.

Koch, J.W., and H.E. Petrie, 1962, "Fading Characteristics Observed on a High Frequency Auroral Radio Path", in *NBS J.Research*, 66D (March-April), reprinted in *Data Communications via Fading Channels*, edited by K. Brayer, p.41, IEEE Press, New York, 1975.

Lane, G. and A. Richardson, 1990, "Super Modes and IONCAP", paper contained in *HF-MAP Newsletter*, (Winter Edition), Naval Research Laboratory, Washington DC 20375-5000.

Law, H.B., F.J. Lee, R.C. Looser, and F.A.W. Levett, 1957, "An Improved Fading Machine", *Proc. IEE*, 104B:117-147.

Lawrence, R.S., C.G. Little, and H.J.A. Chivers, 1964, "A Survey of Ionospheric Effects on Earth-Space Propagation", *Proc. IEEE*, 52:4-47.

Levine, P.H., 1978, "Coverage Estimates for Tactical LPI Communications Systems Analysis", Megatek Informal Rpt. (available through Naval Ocean Systems Center, San Diego CA).

Logan, J.C. and J.W. Rockway, 1986, "The New MININEC (Version 3): A Mini-Numerical Electromagnetic Code", Naval Ocean Systems Center, NOSC Tech. Doc. 938, AD-A181 682, available through NTIS, Springfield VA 22161.

Lighthill, M.J., "Group velocity", *J. Inst. of Math. and its Appl.*, 1, 1, 1965

Lloyd, J.L., George Haydon, D.L. Lucas, and Larry Teters, 1978 , "Estimating the Performance of Telecommunication Systems Using the Ionospheric Transmission Channel, Volume I: Techniques for Analyzing Ionospheric Effects Upon HF Systems", (unpublished ITS draft report distributed by Hdqrs. US Army CEEIA for comment).

Lucas, D.L. and G.W. Haydon, 1966, "Predicting Statistical Performance Indices for High Frequency Ionospheric Telecommunication Systems", ESSA Tech.Rpt. EIR 1-ITSA 1, US Government Printing Office, Washington DC 20402.

Lustgarten, M.N. and J.A. Madison, 1977, "An Empirical Propagation Model (EPM-73)", *IEEE Trans. Electromag. Compat.*, Vol. EC-19, No. 3.

Malaga, A., 1985, "A Characterization and Prediction of Wideband HF Skywave Propagation", *IEEE Military Communications Conference*, paper No. 12.5, Boston, MA.

Maslin, N.M., 1979, "HF Communications to Small Low-Flying Aircraft" in *Special Topics in HF Propagation*, NATO-AGARD-CP-263, pp. 3-1 to 3-13.

Maslin, N.M., 1987, *HF Communications: A System Approach*, Plenum Press, New York and London.

McNamara, L. and R. Harrison, 1985-86, "Radio Communicators Guide to the Ionosphere", a series of articles published in *Australian Electronics Monthly*.

McNicol, R.W.E., 1949, "The Fading of Radio Waves of Medium and High Frequencies", *Proc.IEE*, Part III, 96, 517-524.

Millington, G. 1949, "Ground-Wave Propagation over an Inhomogeneous Smooth Earth", *Proc. IEE*, Part III, No. 96, p.53.

Millman, G., 1967, "A Survey of Tropospheric, Ionospheric, and Extraterrestrial Effects on Radio Propagation Between the Earth and Space Vehicles", in *Propagation Factors in Space Communications*, W.T. Blackband (Editor), NATO-AGARD Conference Proceedings, Technivision, Maidenhead, England; also General Electric Report TISR66EMHI, published 1965.

Milsom, J.D., 1985, "Exact Ray Path Calculations in a Modified Bradley-Dudeney Model Ionosphere", *IEE Proc. H. Microwaves, Antenna & Propagation*, 132:33-38.

Mitra, A.P., 1963, "Recombination Processes in the Ionosphere", in *Advances in Upper Atmospheric Physics*, B. Landmark (Editor), Pergamon Press and the MacMillan Co., New York, London.

Mögel, H., 1934, "Kurzwellenerfahrungen im Drahtlosen Überseeverkehr von 1926-1934", *Telefunken-Zeitung*, 15(67):23-39.

Moore, E.J., 1958, "Performance Evaluation of HF Aircraft Antenna Systems", *IRE Trans.on Ant. & Prop.*, AP-6:254-260.

Muldrew, D.B. and R.G. Maliphant, 1962, "Long-Distance One-Hop Ionospheric Radio propagation", *J.Geophys.Res.*, 67:1805-1815.

NBS, 1967, Technical Note No.101, Nos. 1 and 2, (AD687820 and AD687821), available through NTIS, Springfield VA.

Neilson, D.L., 1968, "Ray Path Equations for an Ionized Layer with a Horizontal Gradient", *Radio Science*, 3:101.

Nesenbergs, M., 1988, "Modeling of Wideband HF Channels", in *Scattering and Propagation in Random Media*, NATO-AGARD-CP-419, edited by K.C. Yeh and A.N. Ince, Specialised Printing Services Ltd., Loughton, Essex, UK.

Nickisch, L.J., 1986, "Stochastic Effects on Oblique HF Propagation", Final Report, Contract DNA001-84-C--0253, Mission Research Corporation, Santa Barbara CA.

Nickisch, L.J., 1990, "Propagation Effects in Extended Random Media", in *Effect of the Ionosphere on Radiowave Signals and System Performance, IES'90*, USGPO, available through NTIS, Springfield VA.

Niessen, K.F, 1937, "Zue Entscheidung den Zwischen den Beiden Sommerfeldschen Formeln fur Fortpflanzung von Drahtlosenwellen", *Ann. Phys. Lpz.*, 29:585.

Norton, K.A., 1936, "The Propagation of Radio Waves Over the Surface of the Earth and in the Upper Atmosphere, 1, Ground-Wave Propagation from Short Antennas", *Proc. IRE*, 24:1367-1387.

Norton, K.A., 1937, "The Propagation of Radio Waves over the Surface of the Earth in the Upper Atmosphere, 2, the Propagation from Vertical, Horizontal and Loop Antennas over a Plane Earth of Finite Conductivity", *Proc. IRE*, 25:1203-1236.

Norton, K.A., 1941, "The Calculation of Ground Wave Field Intensity over a Finitely Conducting Spherical Earth", *Proc. IRE*, 29(12):623-639.

Otten, K.W., 1962, "Design of Reliable Long-Distance Air-to-Ground Communication Systems Intended for Operation under Severe Multipath Propagation Conditions", *IRE Trans. on Aerospace and Navigational Electronics*, ANE-9(June):67-68.

Packer, R.J., and J.A.S. Fox, 1969, "A Simulator of Ionospheric Propagation of Amplitude Modulated Signals", Res. Dept. Rpt. No. 1969/24, pp. 1-6. British Broadcasting Corp., Kingswood Warren, Tadworth, Surrey, UK.

Perl, J.M., 1984, "Simulator Simplifies Real Time Testing of HF Channels", *Defense Electronics*, 16(8):103-108.

Perry, B.D., 1988, "Interference and Wideband HF Communications", In *Effect of the Ionosphere on Communication, Navigation, and Surveillance Systems (IES'87)*", ed. J.M. Goodman, U.S. Government Printing Office, 1988-195-225, available from NTIS, Springfield VA.

Perry, B.D., E.A. Palo, R.D. Haggarty, and E.L. Key, 1987, "Tradeoff Considerations in the Use of Wideband Communications", *IEEE Int'l Conf. on Comms.*, Seattle, WA, paper No. 26.2.

Phillips, M.L., and W. Abel, 1958, "F-Layer Transmission on Frequencies Above the Conventionally Calculated MUF", Project Earmuff, Final Rpt. U.S. Army Signal Corps, Contract DA-36-029-SC-72802, USA Signal R&D Labs, Ft. Monmouth NJ.

Picquenard, A., 1974, *Radio Wave Propagation*, Halsted Press, John Wiley and Sons, New York, NY.

Piggott, W.R., 1959, "The Calculation of the Median Sky Wave Field Strength in Tropical Regions", DSIR Radio Research Special Report 23, Her Majesty's Stationary Office, London.

Price, G.H., 1983, "High Frequency Channel Description", DNA-TR-82-196, SRI International, Menlo Park CA.

Pinson, J.M., J. Mellor, W.P. Leach, and R.. Cleland, 1977, "Nap-of-the-Earth Communications (NOE COMMO) System", TCATA Test report FM-320, RCS ATCD-8, Headquarters TRADOC Combined Arms Test Activity, Fort Hood TX.

Radio Science, 1968, special issue on raytracing, Vol.3, No.1.

Rao, N.N., K.C. Yeh, M.Y. Youakim, K.E. Hoover, P. Parhami, and R.E. DuBroff, 1976, "Techniques of Determining Ionospheric Structure from Oblique Radio Propagation Measurements", RADC-TR-76-401, Univ. Of Illinois, Urbana.

Reilly, M., 1990, private communication.

Reilly, M.H. and E.L.Stroble, 1988, "Efficient Ray Tracing Through a Realistic Ionosphere" in *Effect of the Ionosphere on Communication, Navigation, and Surveillance Systems*, J.M. Goodman (editor-in-chief), US Govt.Print.Office, 1988-195-225, available through NTIS, 5245 Port Royal Rd., Springfield VA 22161. (also published in *Radio Science* in May/June 1988 Issue).

Rhoads, F., 1990, private communication, ("IONCAST" program and documentation).

Rice, S.O., 1951, "Reflection of Electromagnetic Waves from Slightly Rough Surfaces", In *Theory of Electromagnetic Surface Waves*, ed. M. Kline, pp.351-378, Interscience & Dover, New York.

Rishbeth, H. and O.K. Garriott, 1969, *Introduction to Ionospheric Physics*, Academic Press, New York.

Rockwell-Collins, 1977, "HF Communications Data Book", Doc. No. 523-0767157-002217, Collins Telecommunications, Rockwell International, Cedar Rapids Iowa 52406.

Roy, T.N., D.B. Sailors, and W.K. Moision, 1987, "Surface Wave Model Uncertainty Assessment", Tech Rpt, NOSC TR 1199, Naval Ocean Systems Center, San Diego CA.

Sailors, D.B., 1984, "Tactical Decision Aids for HF Communication", Technical Document 782, Naval Ocean Systems Center, San Diego CA.

Salaman, R.K., 1962, "A new Ionospheric Multipath Reduction Factor (MRF)", *IRE Trans. Comm. Syst.*, CS-10:221-222.

Sales, G., 1979,"Scatter Injection/Ducted Mode Radar" in *Special Topics in HF Propagation*, NATO-AGARD-CP-263, V.J. Coyne (Ed.), paper No.28.

Salous, S., 1989, "Measurement of Narrow Pulse Distortion over a Short HF Skywave Link: Es and F2 Summer Results", *Radio Science*, 24 (4), 585-597.

Schmidt, O., 1936, "Neue Erklärung des Kurzwellenumlaufes um die Erde, *Z.Tech.Phys*, 17(11):443-446.

Schmidt, A.R., 1960, "A frequency Stepping Scheme for Overcoming the Disastrous Effects of Multipath Distortion on High Frequency FSK Communication Circuits", *IRE Trans. Comm. Syst.*, CS-8(March):44-47.

Schultz, L.D. and R.M. Gallet, 1970, "Normal Ionospheric Absorption Measurements", ESSA Prof. Paper 4, US Dept. Commerce, Washington DC.

Sedgwick, J., 1985, "Managing Valley to Valley Communications", BR Communications Technical Note No. 7, BR Communications, Sunnyvale CA 94088.

Shirley, J.R., 1952, "The Shirley Aerial - A Vertically Beamed Antenna for Improved Short Distance Sky Waves", Report 8/52, Operational Research Section, Far East Land Forces, Great Britain.

Sommerfeld, A., 1909, "The Propagation of Waves in Wireless Telegraphy", *Am. Physik*, 28:665-736.

Southworth, M.P., 1960, "Night-time Equatorial Propagation at 50 MHz: First Results from an IGY Amateur Observing Program", *J.Geophys.Res.*, 65:601-607.

Stein, S., 1958, "The Role of Ionospheric Layer Tilts in Long-Range High Frequency Radio propagation", *J. Geophys. Res.*, 63:217-241.

Stokke, K.N., 1975, "Some Graphical Considerations on Millington's Method for Calculating Field Strength over Inhomogeneous Earth", *ITU Telecomm. J.*, Vol.42, No.III.

Tamir, T., 1977, "Radio Wave Propagation along Mixed Paths in Forest Environments", *IEEE Trans. Ant. & Prop.*, AP-25:471-477.

Taylor, A.H., and L.C. Young, 1928, *IRE*, "Studies of High Frequency Radio Wave Propagation", May issue, pp. 561-578.

Teters, L.R., J.L.Lloyd, G.W. Haydon, and D.L. Lucas, 1983, "Estimating the Performance of Telecommunication Systems Using the Ionospheric Transmission Channel, IONCAP User's Manual", US Dept.of Commerce, NTIA Report 83-127, available through NTIS, Springfield VA (PB84-111210).

Tichovolsky, E.J.,1984, "HF Propagation Using Adiabatic Invariant Theory", RADC-TR-84-59, Rome Air Development Center, Hanscom AFB, MA.

Toman, K., 1979, "High Frequency Ionospheric Ducting - A Review", *Radio.Sci.*, 14:447-453.

Toman, K. and D.C. Miller, 1977, "Computation Study of Long-Range High Frequency Ionospheric Ducting", *Radio Sci.*, 12:467-476.

Tsedilina, E.E., 1975, "Round-the-World Radio Wave Propagation in Ionospheric Wave Ducts", *Geomag.& Aeron.*, 15:371-374.

Tupper, B.C. and G.H. Hagn, 1978, "Nap-of-the Earth Communication Program for U.S. Army Helicopters", Final Report SRI Project 4979, AVRADCOM TR-76-0868-F, SRI International, Menlo Park CA

Utlaut, W., 1962, "Siting Criteria for HF Communication Centers", NBS Technical Note 139, Dept. of Commerce, Office of Technical Services, Washington DC. (govt. report available through NTIS, Springfield VA).

Vogler, L.E., and J.A. Hoffmeyer, 1988, "A New Approach to HF Channel Modeling and Simulation. Part 1: Deterministic Model", NTIA Report 88-240, U.S.Dept.Commerce, Boulder CO.

Wagner, L.S., 1977, "Communications Media Analysis - HF", NRL Memo Rpt. 3428, Naval Research Laboratory, Washington DC 20375-5000.

Wagner, L.S., and J.A. Goldstein, 1985, "High Resolution Probing of the HF Ionospheric Skywave Channel", *Radio Sci.*, 20(3):287-302.

Wagner, L.S., J.A.Goldstein, and W.D.Meyers, 1988, "Wideband Probing of the Trans-Auroral HF Channel", in *Effect of the Ionosphere on Communication, Navigation, and Surveillance Systems (IES'87)*, ed. J.M. Goodman, USGPO 1988-19-225, available from NTIS, Springfield VA.

Wait, J.R., 1964, "Electromagnetic Surface Waves", *Advances in Radio Research*, Vol. 1, pp.157-217.

Wait, J.R. et al. (Editors),1974, *Workshop on Radio Systems in Forested and/or Vegetated Environments*, USACC Technical Report, No.ACC-ACO-1-74 (AD 780712), Springfield VA.

Walker, W.F., 1965, "A Simple Baseband Fading Multipath Channel Simulator", *Radio Sci.*, 1(7):763-767.

Watterson, C.C., G.G. Ax, L.J. Demmer, and C.H. Johnson, 1969a, "An Ionospheric Channel Simulator", ESSA Tech. Memo. ERL-TM-ITS-198, Boulder CO.

Watterson, C.C., J.R. Juroshek, and W.D. Bensema, 1969b, "Experimental Verification of an Ionospheric Channel Model", ESSA Tech. Rpt. ERL-112-ITS-80, Boulder CO.

Watterson, C.C., J.R. Juroshek, and W.D. Bensema, 1970, "Experimental Confirmation of an HF Channel Model", *IEEE Trans. Comm. Tech.*, Vol. COM-18, No. 6.

Weyl, H., 1919, "Ausbreitung Electromagnetischer Wellen ubereinem Ebenen Leiter", *Ann. Phys. Lpz.*, 60:481.

Whale, H.A., 1969, *Effects of Ionospheric Scattering on Very Long Distance Radio Communication*, Plenum Press, New York.

Wheeler, J.L., 1966, "Transmission Loss for Ionospheric Propagation Above the Standard MUF", *Radio Sci.*, 1(11):1303-1308.

White, Ian, 1990, "Antenna Design on a PC", *Electronics & Wireless World*; (also refer to a previous paper by I. White in the same journal: Dec 1989).

Whitten, R.C. and I.G. Poppoff, 1965, *Physics of the Lower Ionosphere*, Prentice-Hall Inc., Englewood Cliffs N.J., pp 131-133.

Woyk, E., 1978, "Ray Tracing Theory and Mirage Occurrence Conditions", *Applied Optics*, 17:2108.

Yefimuk, S.M., B.E. Lyanoy, and V.A. Pakhotin, 1982, "Long Range Propagation of Radio Waves from an Emitter Located in the Ionosphere", *Geomagnetism and Aeronomy*, 22(3):347-349.

Zimmerman, M.S., and J.H. Horowitz, 1967, "A Flexible Transmission Channel Simulator", *International Communications Conf.*, Minneapolis, MN.

Zennick, J., 1907, "Uber die Fortpflanzung Ebener Electromagnetischer Wellen einer Ebener Leiterflache und ihre Beziehung zur Drahtlosen Telegraphie", *Ann. Phys. Lpz.* 23:846.

5

PERFORMANCE PREDICTION METHODOLOGIES

The present interests me more than the past and the future more than the present.

Benjamin Disraeli[44]

5.1 CHAPTER SUMMARY

For those involved in communication technology development and applications related thereto, variabilities associated with the HF channel constitute a significant challenge. To amateurs and other HF devotees, the relatively unpredictable nature of skywave signaling is both compelling and exciting. The exhilaration of communicating "round-the-world" with relatively simple equipment has not been diminished by the advent and growth of satellite technology. Truly the mechanism chiefly responsible for providing such a capability, namely ionospheric channeling, is nature's gift.

In this chapter we shall begin by discussing the need for predictions of HF communication performance. Next we note the relationships between hindcasting, nowcasting, (short-term) forecasting, and (long-term) predictions. We shall review existing HF performance prediction models from the standpoint of the errors which arise in both the ionospheric models which are exploited and the propagation methods which are used. A peripheral theme is the ultimate recognition that HF system performance predictions agree best with reality when prediction models are updated with measured data. This is not surprising, since ionospheric variability is substantial, and typically only median representations of ionospherically-dependent parameters admit to the modeling process. Several approaches by which variability may be accounted for or *tracked* are no doubt possible, but this suggests that the spectrum *planning* process be made more flexible. The process of model updating is examined only briefly in this chapter but is covered fully in Chapter 6. In the present chapter we stress unadulterated prediction methods.

Although we shall stress mainstream ITS prediction methods (such as HFMUFES and IONCAP) and internationally-sanctioned CCIR procedures,

---

44. Benjamin Disraeli (1804-1881), Prime Minister of Great Britain who is his early life was a student of the law and a novelist. This quotation, taken from chapter 24 in *Lothair*, reflects the farsighted views held by the author-statesman. (From *The Home Book of Quotations* by permission of the Vail-Ballou Press, Binghampton NY.)

other models may also provide useful results. Space does not permit a complete discussion of the various models (The IONCAP *User's Manual* alone consists of roughly 200 pages). Nevertheless, we shall strive to provide the reader with a general summary of the attributes of major models, providing a rather complete reference list for further in-depth study. A section on small micro-computer methods is also provided. We conclude with sections dealing with ongoing efforts for improvement in long-term predictions, a commentary on international cooperation, and future developments.

5.2 INTRODUCTION

Geophysical predictions are needed in many sub-fields including: energy, climate, water resources, environment, and the upper atmosphere. Essays of a general nature are contained in a study commissioned by the National Academy of Sciences [NAS, 1978]. The rationale for predictions contained in that collection is still valid today. Above all, predictions are a guide for future planning whether the plans are related to the telecommunications requirements of military/commercial activities or the humane needs associated with natural disasters such as earthquakes or hurricanes. Predictions may rely upon natural laws of physics which lend themselves most readily to theoretical description, or they may be based upon the tendencies exhibited in archival data with the latter leading to the development of quasi-empirical or climatological models. Predictions have improved over the years as a result of two primary factors: the evolution of computers (along with advanced computational methods) and the development of advanced sensors. The accessibility of space with the advent of satellite platforms has led to an enormous advance in our global perspective, especially valuable in the areas of weather forecasting. Satellites have provided a unique collection of scientific data which has augmented our basic understanding of cause-and-effect. This is also true for the complex of geoplasma phenomena which are most easily monitored from the vantage point of space. Radio methods for earth-space and terrestrial-skywave telecommunications are clearly affected by ionospheric phenomena in a manner which is dependent upon the frequency involved. We find that the magnitude of HF propagation effects provides a good index of intrinsic ionospheric variability. This is not only a distraction for users of the HF spectrum but it also introduces an unfortunate vulnerability. Predictions allow one to cope with this problem, and in certain instances, allow one to compensate. It should be noted, since HF is the most vulnerable to the widest range of ionospheric effects, that a major component of ionospheric remote sensing technology has been dominated by HF probes and sounding systems.

Before beginning our general discussion it is of interest to note that the area of HF predictions is no better (but potentially worse) than our knowledge of the underlying ionosphere. Furthermore our capability to predict ionospheric effects is only moderate as noted in the foregoing NAS study.

Interestingly enough, this is to be compared with *high* capability in the area of numerical weather prediction, according to the NAS study. Given the popularly-held negative perception of weather forecasting efficacy, whether justified or not, a moderate ionospheric prediction capability doesn't seem all that satisfactory. It should be remarked, however, that acceptability (unlike accuracy) is a relative term. A method which provides *instantaneous* poor accuracy may, or may not, be acceptable for long-term planning, but will be totally unacceptable for real-time adjustment of optimum working frequencies over an HF link.

There are circumstances which will lead to relatively accurate short-term predictions of HF system performance. Typically these circumstances involve the process of model updating by incorporation of *live* data from sensors which probe the temporal and spatial neighborhoods of the wanted-path control points[45]. At issue is the definition of neighborhoods in the foregoing statement. In the context of HF skywave propagation any sensor, including an oblique-incidence-sounder, which permits ionospheric characterization at the control point may be an effective probe. Conventional wisdom suggests that forecasts lose validity in tens of minutes (the order of magnitude of atmospheric gravity wave periodicities) and if the probe information is either incomplete or if it is displaced from the control point (or points) by a few hundred kilometers or more. Other conditions also place limits upon forecasting performance. For example, the update probe data will be subject to its own scaling errors, and algorithms for converting raw data from the probe into useful update information may be imprecise. Nevertheless, in principle, it is possible to prepare forecasts which will are both accurate and useful. This topic is discussed further in Chapter 6.

Methodologies for preparing long-term predictions of circuit performance, while admittedly inaccurate because of ionospheric variability, do provide useful products for users of the HF spectrum. The presumption made is that (short-term) variability has been properly bounded under the propagation regimes or geophysical conditions for which long-term prediction are desired. It is this set of products which we are addressing in the current chapter.

---

45. *Control point* is a term which flows naturally from the mirror model of HF skywave propagation. In view of the fact that most of the refraction experienced by a *reflected* mode is in the neighborhood of the ray trajectory apogee, exclusive of any high ray modes (see Chapter 4), convenience suggests that the *control point* should refer to the midpoint of the (presumed) great circle trajectory. Accordingly, midpath ionospheric properties which are reckoned at some appropriate height are assumed to control the propagation. Factors which will render the *control point* notion invalid include: strong tilts and gradients, dominance of the *high ray*, above-the-MOF modes, non-great-circle modes, and sundry scatter modes. Another difficulty is the azimuthal insensitivity of the control point approach, a fact which certainly impacts the capability to associate data derived from nonorganic sounders with operational HF paths. This is especially troublesome when the sounder path and the wanted path are virtually orthogonal, even when the control points are common (i.e., paths form a cross in plan view).

5.3 REQUIREMENTS: Predictions and Spectrum Management Guidance

5.3.1 General Broadcast Requirements

It is generally recognized that HF is the most precarious radiofrequency regime in terms of skywave propagation effects, and this personality may result in either positive or negative features in connection with broadcasting and point-to-point operations. As we have seen, considerable flexibility may be required in selecting the optimum set of system parameters to succeed in reaching a particular destination. Of all categories of users, those involved in HF broadcasting may be confronted with the greatest challenge. Aspects contributing to this include: (a) requirement for a distended signal *laydown* pattern to cover widely-separated reception centers, and (b) a means to compensate for skip zone variations for specified receivers even when the diurnal period of transmission is restricted. As we have seen from Figure 4-39, skip distance variability may be considerable. Military broadcast services provide for enhanced performance through incorporation of frequency management techniques; and spectral use efficiency (as a percentage of the MUF-to-LUF envelope) is improved by use of diversity which may partially compensate for fading and intersymbol interference. Given the fact that listeners of civilian broadcasts are disadvantaged (without access to sophisticated radio equipment and with no real-time feedback capability), we recognize that the broadcasting community has an obvious need for a credible long-term prediction capability in order to offer (and advertise) a reasonable set of broadcast channels to potential listeners in designated reception areas. The successful transmission of programs using the shortwave band must account for a number of parameters, including: location of source transmitter, time and duration of the transmission, and the specified reception area for the program. In addition, major attention should be paid to phenomenological parameters of the propagation medium, such as the ionospheric heights and critical frequencies, which ultimately determine the broadcast coverage for a specified frequency. Although predictions are beneficial for other users of the HF spectrum, including tactical- and strategic-military communication services, federal and state emergency communication networks, military affiliated radio systems, and even the radio amateur service, predictions are almost an imperative for the civilian broadcast community.

Coverage prediction will depend upon an ability to predict the ionospheric conditions. These predictions typically involve the exploitation of models of ionospheric structure which are coupled to some appropriate radiowave propagation algorithm. As indicated in Chapter 3, the ionosphere is typically modeled by spatial and temporal functions and some external parameters reflecting solar and magnetic activity control. Usually, the geography for the prediction problem is known, and the ionospheric conditions are prescribed by the model, after one or more input control parameters have been specified.

This places too much burden on a single parameter like the sunspot number, and the result is often unsatisfactory if precision is required. Nevertheless, requirements-driven predictions will rely upon some equivalent solar activity index in most applications for some time to come. Models which have been developed to serve the needs of the point-to-point service are not too satisfactory for analysis of broadcast coverage, and most models fall into this category. This is not simply an I/O issue but involves some rather non-trivial alterations in computer code to make point-to-point models deliver data efficiently for use in development of area coverage maps. Such computer methods are not generally available to the public but certain military activities reportedly use such approaches routinely in the analysis of signal laydown for specified transmitters and in the construction of signal-to-interference contours. The Voice of America has developed a broadcast coverage mapping capability in connection with output from a CCIR computer method termed *HFBC-84* [Lane, 1990].

For guidance in future operations, a measure of ionospheric support variability is also needed. Beyond this, variability in received signal level (or alternatively the basic transmission loss) is required. Models of ionospheric variability expressed in terms of the upper and lower deciles for both the transmission loss and the MUF are available in CCIR publications such as Report 252-2 [CCIR, 1970] and its supplement [CCIR, 1982a] for specified conditions. It is interesting to note that the CCIR MUF variability tables are largely based upon data obtained in the early sixties [Barghausen et al., 1969] [Davis and Groome, 1964]. The CCIR [1986a] has published another field strength variability model specific to the needs of broadcasters. Even so, significant deviations from CCIR *suggestions* have been observed [Gibson and Bradley, 1987] [Fox and Wilkinson, 1986]. The exploitation of existing data banks along with the certification of additional data sets which provide variability information is clearly an important endeavor in the performance prediction process.

5.3.2 Military and Related Requirements

Operational requirements of military users often lead to simplifications of the established main frame procedures in order to provide spectrum guidance in a more accessible manner. This is especially true for tactical commanders who may not have access to real-time sounding information. Tactical frequency management systems, while they may allow for incorporation of real-time data for decision-making in the field, typically *default* to predictions which may be derived from the long-term models similar to IONCAP. One of the primary weaknesses of the larger models is related to Input/Output limitation and a general user *unfriendliness*. The PROPHET system [Rose, 1982] is a good example of a resource management tool which originally exploited simplicity to provide tailored products to the user. Steps have been taken in recent years to improve the models organic to PROPHET and similar systems while retaining user-friendly features for the tactical user. Within the US Department of Defense, the Electromagnetic Compatibility Analysis Center (ECAC) is the catalyst for development of service-specific HF prediction programs. Over the years, a number of developmental systems have emerged with clever names such as FAS-HF, APES, and CHIMPS. Qualified readers may be able to obtain more information about these and other HF performance prediction models by contacting ECAC (North Severn, Annapolis, Maryland 21402-1187).

For many years the U.S. Army has published communication charts and the U.S. Navy has published a document called the NTP 6 Supp-1 [1990] *Recommended Frequency Bands and Frequency Guide.* This guide is currently based upon IONCAP methodology and the actual recommendations are based upon sunspot number ranges specific to a particular year. The range of sunspot numbers for a specified year are based upon long-term running averages reckoned near the publication date and, consequently, may not precisely match currently required conditions. NTP 6 Supp-1 has two methods which are available for users. Both employ look-up tables to retrieve MUF and FOT data. The first method is for users who are communicating over *arbitrary* maritime paths, while the second is tailored for use by communicators terminating at established Communication Stations (COMMSTAs) or Communication Units (COMMUs). The second method also provides an estimate of the antenna radiation angle for the circuit involved. Figure 5-1 is a sample chart taken from NTP 6 Supp-1 for the Naval Communication Area Master Station for the Atlantic region (NAVCAMSLANT) located near Norfolk, Virginia. As may be seen, each region is defined by an alpha-numeric index such as A1 or H4 for example. The radials originate from Norfolk and are drawn every 40 degrees beginning at geographic north, resulting in nine sectors. Range zones extend to 4400 nautical miles (nearly 8000 km), which includes multi-hop paths. The initial range zone A extends from 0 to 400 nmi...; thereafter the zones are progressive in groups. Beyond A and out to 800 nmi. (viz.,B,C,and D), the steps are 200 nmi.; range zone E is from 1000 to 1400 nmi.; and F is from 1400 to 2000 nmi. Finally G, H and I cover the range between 2000 and 4400 nmi. in steps of 800 nmi. An operator locates the region to which he belongs from the chart and finds the appropriate table for the period of operations. The tables are developed for calendar quarters to retain some seasonal information. The sunspot numbers which correspond to the calendar quarter calculations are preset by long-term predictions of same. For those requiring tailored predictions and additional information about NTP 6 Supp-1 contact should be made with the Naval Electromagnetic Spectrum Center (NAVEMSCEN, Washington Navy Yard, Washington DC).

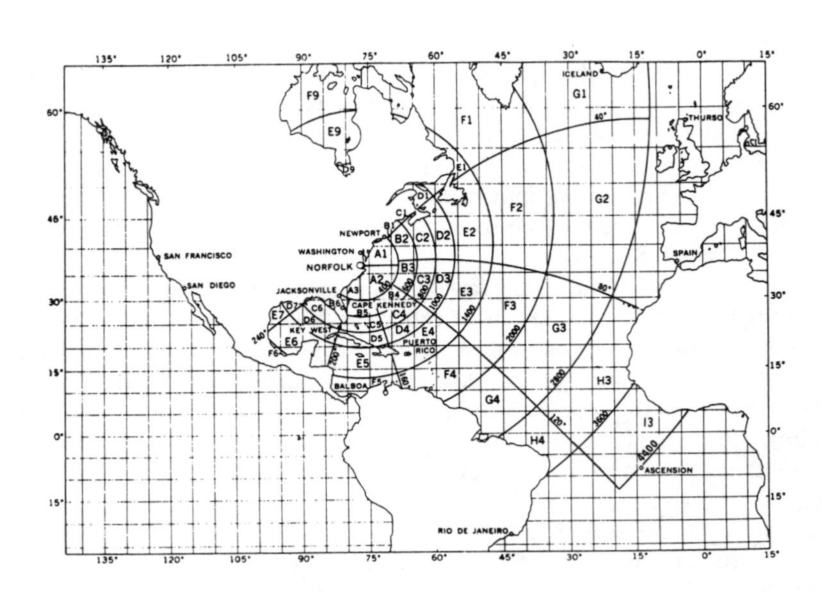

Fig.5-1. Map of the assembly of HF coverage zones for the region under control of the Naval Communication Area Master Station Atlantic (NAVCAMSLANT) which is located in Norfolk, VA. (From NTP 6 Supp-1 [1990], provided courtesy of P. Blais, NAVEMSCEN.)

5.3.3 HF Amateur Radio Needs and Prediction Approaches

Radio amateurs have contributed significantly to our understanding of sky-wave propagation over the years. More importantly, amateurs provide an invaluable service during periods of time when normal telephonic or satellite communication systems are disabled. Such periods may occur during massive earthquakes, floods, hurricanes, and other natural disasters. In some instances, amateur communications provide the only information channel to/from stricken areas until more traditional systems are restored. On occasion, HAMs may serve as relays of information about ships in distress; and may be our window to news concerning political activities or civil unrest in foreign lands.

The frequency prediction *requirements* of radio amateurs are obviously less formalized than those of the civilian broadcast and military sectors. Still, the interest is there. All amateurs must pass an examination before a license will be issued authorizing radio station operation, and many thousands have done so. Such licensing is required to encourage discipline among users in view of the long-distance propagation properties of shortwave signals. Within the USA, the proponent for amateur radio is the American Radio Relay League (ARRL) which also provides assistance to those individuals interested in becoming HAMS. Depending upon the license category, amateurs may make contacts in International Morse Code (or keyed CW), participate in 2-way voice communication with other amateurs, transmit slow-scan television (SSTV) images, transmit facsimile (FAX), transmit radio-teletype (RATT), and exploit the recent packet radio technologies. The amateur sub-bands (within the HF frequency band) are given in Table 5-1. These allocations are decided upon at World Administrative Radio Conferences (WARCs) which are held about every twenty years under the aegis of the International Telecommunications Union (ITU). For the amateur radio service, each member nation represented in the ITU has the responsibility to establish rules for use within its jurisdiction, but consistent with ITU decisions. Within the USA, the Federal Communications Commission (FCC) is the governing body for amateur radio. Within each wavelength band, there is a mode usage plan (i.e., CW, RTTY, SSB, SSTV, FAX and voice) for each specified category of amateur license [ARRL, 1987].

HAMs are avid listeners for signals and broadcasts emanating across the HF band, and there is certainly no lack of sources. Outside of North America alone, there are over 500,000 licensed radio amateurs. Generally speaking, there is no restriction on the process of listening to shortwave signals (which in some instances might be considered private communications) provided the contents of such intercepted signals are not divulged to third party, published, and/or used for personal gain. A *Shortwave Directory* of HF signals is available [Grove, 1989], and up-to-date information is to be found in *The Monitoring Times*, a monthly magazine. Topical articles dealing with clandestine radios, pirate stations, and features dealing with shortwave broadcasting may be found in *Popular Communications*.

TABLE 5-1

AMATEUR BANDS BETWEEN 3 AND 30 MHZ

Wavelength Band Designation (meters)	Frequency Range (MHz)
160	1.80 - 2.00
80	3.50 - 4.00
40	7.00 - 7.30
30	10.10 - 10.15
20	14.00 - 14.35
15	21.00 - 21.45
12	24.89 - 24.99
10	28.00 - 29.70

Predictions have often taken the form of propagation charts to be found in periodicals such as *CQ Amateur Radio* [e.g., Jacobs, 1990] which like *Popular Communications* is another publication of CQ Communications Inc. Master long distance (DX) propagation charts provide information about the "openings" likely to arise for a listener given the season and solar activity level, based upon an assumed fixed value of the effective radiated power (ERP) of 1 kW. The charts indicate predictions of the most likely two-hourly time blocks at which hearability will be possible at up to six frequency bands (in the wavelength notation more conventional with HAMS: 10, 15, 20, 40, 80 and 160 meters). Short skip predictions for the US and Canada have also been made available. The method involves computation of a four-level propagation index based upon a computer analysis of a large number of paths. This index is intended to be indicative of the number of days of the month a particular band and listening area will be accessible at a specified time block. Thus, provided there are transmissions available to be considered, the propagation index is related to a reception probability. Signal quality estimates (related to S-meter reckonings in 3 dB increments) are also predicted using these approaches. For a discussion of these approximate methods, the reader is referred to a book by Jacobs and Cohen [1979] entitled *The Shortwave Propagation Handbook*.

Some computer programs are available to assist Hams in scheduling their activities. Base(2) Systems [Saginaw, MI] has developed propagation aids for the radio amateur using a simplified code for estimating the LUF and the MUF (i.e., a version of MINIMUF). Two products of interest include Band-Aid and MUFMAP, both of which will execute on a PC or compatible computer. Another Ham-oriented product, IONOSOND, is available through W1FM [Lexington, MA]. One of the more popular programs is MINIPROP™ by W6EL Software [Los Angeles, CA]. It provides estimates of signal level and other amateur-specific utilities and runs on a PC. It will be discussed in Section 5.14 where microcomputer prediction programs are examined.

Finally for those interested in amateur radio, many books and documents available either through the ARRL or through local bookstores. A listing of materials, tapes, and educational material may be found in *Uncle Wayne's Bookshelf*, a feature in the magazine *73 Amateur Radio*.

5.3.4 The Spectrum Management Process

Several methods are used for spectrum planning. The ITU has long recognized that the HF skywave channel is a valuable resource, and one of the ITU's technical arms, the CCIR, has developed methods which can be applied by various administrations for optimization of communication and broadcast performance, while limiting the potential for interference with other users. These methods represent the best the community can achieve in the long-term prediction of ionospheric behavior. The various recipes which describe the processes by which radiowaves interact with the ionosphere are not ultimately as critical as is the ionospheric definition in the prediction process.

The ITU, created in 1865 at the Paris International Telegraph Convention, is now composed of 163 nations. Objectives of the ITU are promulgated and maintained through the *International Radio Regulations*. These regulations are updated through agreements reached at the World Administrative Radio Conferences (WARCs). The WARC is one of six major entities comprising the ITU. The period between WARCs is at least 10 years but may be as much as 20 years. The most recent meeting was held in 1979 in Geneva. Another agency within the ITU structure is the International Frequency Registration Board (IFRB) which serves as the official agency for registering the date, purpose, and technical properties of frequency assignments made by member countries. Technical arms of the ITU include the CCIR (International Radio Consultative Committee) and the CCITT (International Telephone and Telegraph Consultative Committee). The CCIR provides much of the guidance to the ITU for outstanding technical issues. Officially this guidance takes the form of published *Recommendations*. The HF prediction methods suggested by the CCIR, therefore, are quite significant for establishing *Recommendations* for spectral planning by the ITU. These documents are taken up at the WARCs and may lead to reallocation of the radio spectrum. This is of considerable importance to all member nations.

In the United States, the Federal Communications Commission (FCC) and the National Telecommunications and Information Administration (NTIA) of the Department of Commerce jointly regulate use of the radiofrequency spectrum. NTIA is responsible for government use and the FCC is responsible for regulation of private use services. Within the government, the Interdepartment Radio Advisory Committee (IRAC) oversees government use of the radio spectrum, and resolves outstanding issues. Each government department, having a member in the IRAC, establishes its own procedures consistent with IRAC decisions. The U.S. Department of Defense, for example, places authority for policy establishment and guidance in the Joint Chiefs of Staff (JCS). The U.S. Military Communications Electronics Board (USMCEB) develops procedures for implementing the JCS guidance. This includes the assignment of frequencies for areas not appropriate for the Commanders-in-Chief (CINCs), who have their own special frequency assignment responsibilities. All DoD components participate in a record system for all frequency resources, and notification is given when a frequency is no longer required. This will make the frequency available for reassignment to other components. Intracommand frequency requirements are passed from the commander to the USMCEB if new assignments are sought. Outside of the United States and if host countries agree, intracommand frequencies may be locally assigned by the commander under certain conditions.

The whole process is rather cumbersome. It is geared to spectral use based upon 1960s technology. Spread spectrum technology and the concepts of frequency pooling, resource sharing, and networking should influence the process in the future. To examine the impact of new spectrum management schemes, it is necessary to request a suite of frequencies on a temporary basis. It has been the experience of the author that such requests are generally approved if it may be shown that little or no interference will be created by the test or experiment.

5.4 RELATIONSHIPS BETWEEN PREDICTION, FORECASTING, NOWCASTING, AND HINDCASTING

The term *prediction* has a rather elusive meaning, depending upon the nature of the requirement for knowledge[46] about the future. In the case of the ionosphere, a distinction is made between long-term predictions and short-term predictions. Long-term predictions of ionospheric behavior may typically be based upon climatological models developed from historical records for specified solar and/or magnetic activity levels, season, time of day, geographical area involved, etc. Very often, the ionospheric prediction is itself based upon a prediction of the solar activity level. In short, the long-term prediction process relies upon the recognition of loosely established tendencies as they relate to relatively simple (and extraterrestrial) driving parameters, and the result is usually an estimate of median behavior. Two sources of error occur in long-term predictions, one arising because of an imprecise estimate of the driving parameter, such as sunspot number, and the second arising from ionospheric variability which is not properly accounted for in the model. Given these difficulties, it may appear surprising that the process can yield useful results, and yet it often does. Long-term predictions are necessary in HF broadcast planning and in other spectrum management activities where significant lead times are involved. Short-term predictions involve time scales from minutes to days. The term *forecast* is sometimes used to describe those prediction schemes which are based upon established cause-and-effect relationships, rather than upon simple tendencies based upon crude indices. In the limit, a short-term *forecast* becomes a real-time ionospheric assessment or a *nowcast*. In the context of HF communications, real-time-channel-evaluation (or RTCE) systems, such as oblique sounders, may be exercised to provide a *nowcast*. Such procedures are useful in adaptive HF communication systems. The term *hindcast* is sometimes used to describe an *after-the-fact*

---

46. *Knowledge of the future* appears to be a contradiction in terms. Given the variability of the ionosphere and the observation of the considerable variability in the MUF and field strength, it is anticipated that future values of HF system parameters cannot predicted with great accuracy. Prediction systems should be evaluated in terms of the success achieved in bounding the the parameter variation over selected epochs. In bounding, we imply the *least-upper-bound*.

analysis of ionospherically-dependent system disturbances. Solar control data are usually available for this purpose, and this may be augmented by ionospheric observation data. Figure 5-2 shows the relationship between the various prediction epochs.

The error associated with any prediction method is critically dependent upon the parameter being assessed, the lead-time for the prediction, and other factors. One of the most important parameters in the prediction of the propagation component of HF communication performance is the maximum electron density of the ionosphere, since this determines the communication coverage at a specified broadcast (or transmission) frequency. The ordinary ray critical frequency, given by the term foF2, may be directly related to maximum F2 layer electron density, and foF2, together with the effective ray launch angle, will determine the so-called Maximum Usable Frequency (or MUF) for a specified transmission distance. Thus, the ability to predict foF2 or the maximum electron density of the ionosphere by a specified method is a necessary step in the prediction of HF system performance if skywave propagation is involved.

The next section discusses the general use of ionospheric models in the present-day prediction process.

5.5 ON THE USE OF IONOSPHERIC MODELS FOR PREDICTION

The nature of ionospheric variability is quite complex, since it arises from temporal and geographical variabilities in upper atmospheric chemistry, ionization production and loss mechanisms, particle diffusion and electrodynamical phenomena. As indicated earlier, general tendencies are fairly well modeled, and much of the variability is understood from a physical point of view. Unfortunately, an understanding of cause-and-effect does not always translate into a prediction capability.

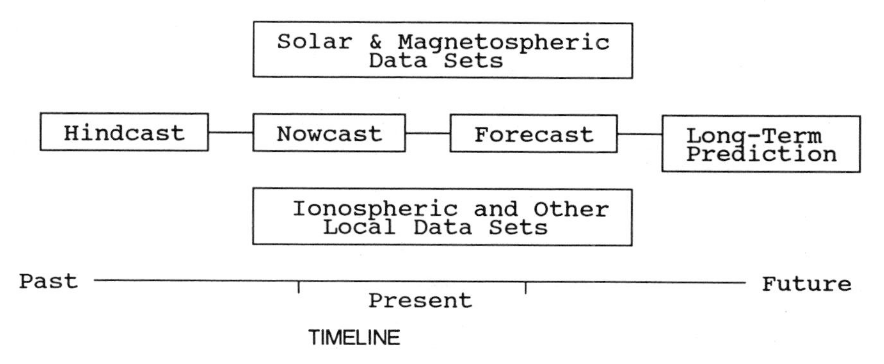

Fig.5-2. Relationships among prediction, forecasting, assessment, nowcasting, and hindcasting.

Because the sources of disturbance cannot be adequately monitored at their points of origin and as they propagate, prediction algorithms are inefficient. An additional complication arises as a result of distortion and attenuation experienced by the propagating disturbance. Moreover, the science which allows us to translate the physical processes in control at the disturbance source to other geographical regimes and times is incomplete. Figure 5-3 depicts the hierarchy of ionospheric disturbances; Table 5-2 provides an estimate of time duration and occurrence frequency for each class of disturbance.

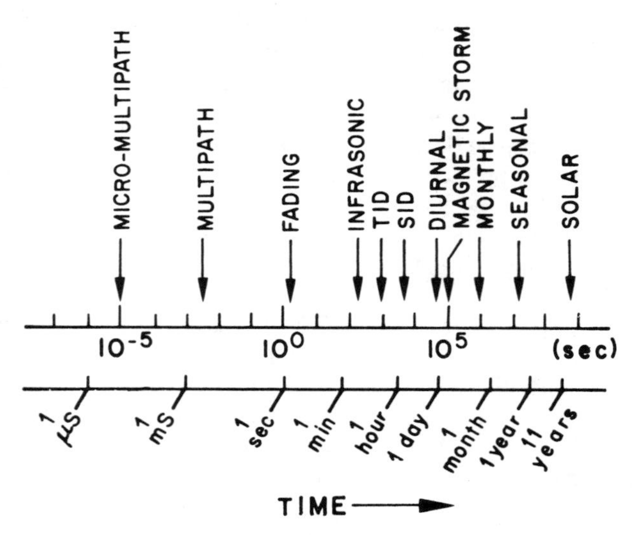

Fig.5-3. Hierarchy of Ionospheric Disturbances

Several models of varying degrees of complexity have been crafted for the purpose of making ionospheric or propagation predictions, or for use in theoretical studies. The historical development of prediction methods until the middle 1950s is given in an account by Rawer [1975], and post World War II activities are summarized by Lucas [1987]. A survey of ionospheric models has been provided by Goodman [1982], following a review by Kohnlein [1978]. Additional information of a general nature may be found in a report by Bilitza [1990] and further insight may be derived from selected technical surveys [Secan, 1989; CSC, 1985]. Unfortunately the survey reports have not been distributed widely. A mini-review of models has been published by Rush [1986]. The paper by Rush includes pure ionospheric models but stress is placed on propagation methods which are in current use and under development.

TABLE 5-2
TEMPORAL VARIATIONS OF HF EFFECTS §

EFFECT	TIME PERIOD {seconds in ( )}	FREQUENCY {Hertz}
Solar Cycle	11 years (3.5×10^8)	2.9×10^{-9}
Seasonal	3 months (7.9×10^6)	1.3×10^{-7}
Diurnal Cycle	24 hours (8.6×10^4)	1.2×10^{-5}
Large-Scale TID	1 hour (3.6×10^3)	2.8×10^{-4}
Short-Wave-Fade	0.5 Hour (1.8×10^3)	5.6×10^{-4}
Small-Scale TID	10 Minutes (6×10^2)	1.7×10^{-3}
Infrasonic Waves	1 Minute (60)	1.7×10^{-2}
Faraday Fading	0.1 - 10 Seconds	10 - 0.1
Interference Fading	0.01 - 1 Second	100 - 1

§ The equivalent frequencies are also provided. A spectral decomposition of the effects will demonstrate a rather featureless continuum for periodicities smaller than a day (or frequencies larger than 10^{-5} Hz). Low frequency terms, being related to well-defined source terms, will cause that part of the spectrum to be discrete.

Some of the models which have been used recently include those of Bent et al.[1975], the International Reference Ionosphere (or IRI) [Rawer et al., 1978, 1981], and the Ching-Chiu model [Ching and Chiu, 1973; Chiu, 1975]. Of more interest to the HF community are models which use the bottomside properties of the ionosphere which influence the skywave propagation most directly. The models which are largely based upon the very substantial data base derived from vertical-incidence sounders are the ones of choice. For several years much effort has been directed toward the analysis of this data base and in the development of suitable mapping techniques and numerical methods for predicting ionospheric properties. Global maps of ionospheric properties have been published, and these data form the basis for many semi-empirical and climatological (statistical) models of the ionosphere. These ionospheric models will play the role of submodels in relative large HF performance prediction codes. We shall return to prediction modeling in Section 5.6.

The U.S. Air Force has developed a class of ionospheric models which are designed to accommodate the insertion of *live* ionospheric data from satellites, terrestrial sensors, and solar observatories. The first model was the so-called Air Force 4-D model [Tascione et al., 1979]. The most recent one is the ICED model [Tascione, 1988], which uses an *effective* sunspot number and a geomagnetic Q-index, the latter being associated with in-situ satellite data describing auroral characteristics. The effective sunspot number used in ICED is based on near-real-time ionospheric measurements derived from a worldwide network of vertical-incidence-sounders; the effective number being that value which, if it were to have occurred, would provide the best match between data and model. The effective sunspot number used in ICED is reminiscent of the *T-index* developed by the Australians [IPSD, 1968] as a replacement for the running 12-month average sunspot number, but the number is more closely related to the real-time pseudoflux concept developed by NRL workers [Goodman et al., 1983, 1984]. Exploitation of this scheme allows for the incorporation of dynamical ionospheric behavior. The model should therefore be applicable to HF broadcasting predictions, and should be particularly appropriate for the modeling of high latitude effects. The topside profile is modeled rather simplistically in ICED, and improvements could include incorporation of multiple scale heights above the F2 peak and a correction for a plasmaspheric contribution to the TEC at great heights. However, these matters are more relevant to considerations of transionospheric propagation. The manipulation of models to derive forecasting information is covered in Chapter 6 which stresses real-time and near-real-time assessment of the propagation path for solution of the *nowcasting* problem.

Recent work by Anderson et al. [1985] has covered the calculation of ionospheric profiles on a global scale in response to physical driving parameters, such as the underlying neutral composition, temperature, and wind; the magnetospheric and equatorial electric field distributions; the auroral precipitation pattern; and the solar EUV spectrum. A subset of these parameters has been used in profile calculations for the development of a semi-empirical, low-latitude ionospheric model (SLIM) [Anderson et al., 1985, 1987] [Sojka and Schunk, 1985]. This kind of approach is computationally very intensive, but the use of coefficient maps from these calculations, which depend on the appropriate parameter values, appears feasible. The Fully Analytical Ionospheric Model (FAIM) [Anderson et al., 1989] uses the structure and formalism of the Chiu model with coefficients fitted to the SLIM model profiles. The development of such programs is required to eliminate the use of oversimplified driving parameters in prediction models and to describe completely the chain of events involved in the solar wind-magnetosphere-ionosphere-atmosphere system. Brief descriptions of SLIM and FAIM are contained in a report by Bilitza [1990].

5.6 THE INGREDIENTS OF SKYWAVE PREDICTION PROGRAMS

The primary purpose of an HF performance prediction model is to provide an estimate of how well a system will work under a given set of circumstances. Typically this translates into some measure of system reliability (see Section 5.11). The components of a complete skywave performance prediction model should include: full documentation (including basis in theory, user's guide, I/O interface data, and machine-specific information), a *user-friendly* preprocessor routine which enables the analyst to set up a computation strategy efficiently, the underlying ionospheric submodel structure, the database or coefficients upon which the ionospheric submodel depends, the noise and interference submodels with associated databases, the antenna and siting factor submodels and their databases, procedures or rules by which propagation is treated, and a set of output products (for each method or option). These major components are shown in Figure 5-4.

Models typically require inputs path geometry (terminal locations in geomagnetic and geographic coordinates), day of year (or month/season), time of day (or some time block), plus an index set to *drive* the ionospheric personality (i.e., solar and possibly magnetic activity). In addition, certain terrain and siting information, antenna configuration/type, and other forms of system data are necessary. Because of the well-established diurnal and seasonal variabilities of the ionosphere, it is not surprising that time-of-day and month (or equivalent) are required as input parameters. Moreover, time block and seasonal data inputs along with receiver location are needed to deduce atmospheric noise, galactic noise, and man-made interference levels. Noise considerations are covered briefly in Section 5.8.

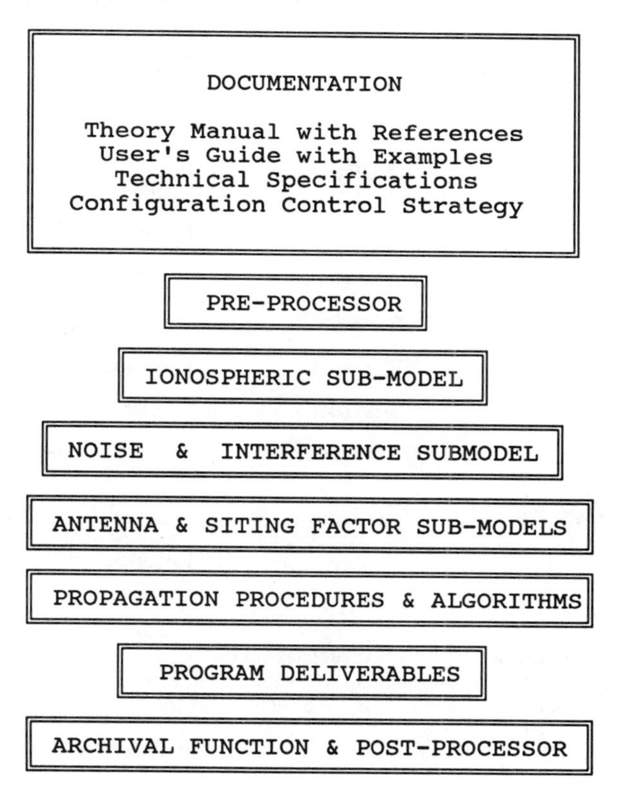

Fig.5-4. Major Components of a Complete Skywave Prediction Program

5.7 BRIEF SYNOPSIS OF PREDICTION MODELS

Propagation prediction models have been developed over the years, and many have incorporated features shown in Figure 5-4. Table 5-3 is a listing of various models and the appropriate references.

TABLE 5-3

SKYWAVE PROPAGATION PREDICTION MODELS

MODEL NAME	ORIGINATOR	REFERENCE
SPIM Method	SPIM:France	Rawer[1952] Halley[1965]
CRPL Method	CRPL:USA	NBS Circular 462[1948]
DSIR Method	Appleton Lab Slough, UK	Piggott[1959]
USSR Method	Soviet Acad. Sci.:USSR	Kasantsev[1947,1956]
FTZ Model	Deutsche Bundespost	Ochs[1970]
§ CCIR-252-2	CCIR/ITU	Rpt.252-2 [CCIR,1970]
§ CCIR-252-2 Supplement	CCIR/ITU	Rpt.252-2 Supplement [CCIR,1982a]
§ CCIR-894-1	CCIR/ITU	Rpt.894 [CCIR,1986a]
HFBC84	WARC/ITU	ITU [1984a]
ITSA-1	ITS-Boulder	Lucas and Haydon [1966]
ITS-78	ITS-Boulder	Barghausen et al. [1969]
HFMUFES4	ITS-Boulder	Haydon et al. [1976]
IONCAP	ITS-Boulder	Teters et al. [1983]
AMBCOM	SRI	Hatfield [1980]
RADARC	ITS-Boulder NRL-Wash.DC	Lucas et al. [1972] Headrick et al. [1971] Headrick and Skolnik [1974]

§ The CCIR Secretariat (ITU, Geneva) retains computer codes for CCIR-252 (HFMLOSS for mainframes); CCIR-252 Supplement (SUP252 for mainframes); and CCIR Report 894 (REP894 for mainframes and micros). See Section 5.14 for a discussion of microcomputer methods.

5.7.1 Historical Development

Current methodology for HF performance prediction evolved gradually, beginning with uncoordinated studies by workers from many countries and organizations. Serious work to establish prediction methods began in earnest during World War II because of the obvious military communication requirements. The earliest methods by the Allies, Germany, and Japan were of the graphical type to speed analysis, because computer methods were not available. The long distance methods used by Germany and those used by the Allies [IRPL, 1943] form an interesting contrast. (The Interagency Radio Propagation Laboratory, IRPL, was a forerunner to the Central Radio Propagation Laboratory, CRPL, at Boulder, Colorado).

In Germany, long distance propagation was analyzed by examination of each mode and path independently. According to an account by Rawer [1975], short paths assumed 1E, 2E, 1F and 2F mode possibilities while for long paths multiple F layer modes alone were considered. At each reflection point (or *control point*, see Footnote No.45) the MUF was deduced by extraction of a value of foF2 for that point (from crude maps) and the appropriate MUF-factor was applied. The overall MUF was logically determined as the lowest of the set of subhop MUFs for each path to be reckoned. Because of nose extension, scatter effects, and the possibility of ducted or chordal mode propagation, this approach, while intuitively pleasing, was pessimistic. The American long path approach, influenced by a more global perspective, used modified control point method which accounted for only two *mirror points* along the great circle path linking communication terminals. These two control points were 2000 km from the communication terminals. This produced a rather optimistic result.

In the period during World War II and after, sounding *networks* were established to provide a basis for the construction of better maps from which foF2 and MUF variation with latitude (and longitude) could be assessed. As previously indicated, significant equatorial anomalies were discovered through examination of this data [Appleton, 1946]. Following World War II, the French organization SPIM was established, while in the United States the agency IRPL became known as CRPL. Both SPIM and CRPL continued the development of more analytical methodologies to replace simpler procedures. Significant improvements in mapping resulted from the incorporation of a modified dip latitude concept to account for geomagnetic control of the ionospheric parameters [Rawer, 1963]. By 1950 Gallet of SPIM developed a mapping technique which soon became part of a computerized method for developing MUF maps. By the early 1960s, Gallet had moved to the United States where he joined with Jones in formulating a basis for the current method for mapping ionospheric parameters [Jones and Gallet, 1962].

5.7.2 Commentary on Selected Models

Models which stem from methods developed by Department of Commerce scientists at Boulder Colorado include ITSA-1, ITS-78, HFMUFES-4, IONCAP , and RADARC. These methods have influenced the design of other prediction models. The CCIR has developed methods for estimating field strength and transmission loss based upon empirical data, and a computer method for propagation prediction was developed for the WARC-HFBC under the aegis of the International Frequency Regulation Board, an organ of the ITU. For more information, the reader is referred to the following: Report 252-2 [CCIR, 1970] and its Supplement [CCIR, 1982a] (both previously cited and published separately), as well as Report 894-1 [CCIR, 1986a] and Recommendation 621 [CCIR, 1986b] which are contained in the 1986 "Green Book" [CCIR, 1986c]. Methods have also been developed in the United Kingdom, Canada, France, the USSR, and India. Many of these have been listed in Table 5-3. Small microcomputer models are discussed in section 5.14. A brief synopsis of selected computer models follows:

ITSA-1 : [Lucas and Haydon, 1966]. This model was developed by the Commerce Department. At the time it was published it represented one of the first computer methods for exploiting augmentations in the underlying ionospheric and geophysical databases. Probably the first computerized method was a program called MUFLUF, which was developed by the Central Radio Propagation Laboratory, a forerunner to the ITSA organization at Boulder. The ITSA-1 model superseded MUFLUF soon after publication. ITSA-1 did not include separate D or F1 layers, and sporadic E was not accounted for. In this program the concepts of circuit reliability and service probability were introduced. MUF variability data were included.

ITS-78 (HFMUFES) : [Barghausen et al., 1969] [Haydon et al., 1976]. ITS-78 actually represents a series of codes developed at ITS-Boulder beginning with ITS-78, and culminating with HFMUFES4. These programs did not include an F1 layer but do include sporadic E. Most of the features of ITSA-1 were included, but with revised F-layer ionospheric data.

IONCAP : [Teters et al., 1983] [Lucas, 1987]. The latest in a string of *main frame* programs developed by ITS and its predecessor organizations. The following improvements over previous ITS models are contained in IONCAP. (1) a more complete ionospheric description; (2) modification in loss equations; (3) empirical adjustment to Martyn's Theorem; (4) revised loss statistics to account for Es and above-the-MUF losses; (5) new methodology for long-distance modeling; and (6) revision to antenna gain models. Unfortunately the documentation to IONCAP is incomplete. A *User's Guide* has been distributed. A draft theory manual has been prepared.

RADARC: [Lucas et al., 1972] [Headrick et al., 1971] [Headrick and Skolnik, 1974]. This program was promoted by the Naval Research Laboratory for use in analyzing the performance of OTH radar facilities. It is a close relative of IONCAP and HFMUFES, however the computational strategy is tailored to provide information along specified radials (and arbitrary distances) from a transmitter rather than for point-to-point communication paths.

FTZ [Ochs, 1970] : This model was developed by the Deutsche Bundespost. It includes an empirical representation of field strength. This method is based upon observations of signal level associated with a large number of circuit-hours and paths, with the majority of the paths terminating in Germany. Since data were obtained without accounting for the individual modes which may have contributed to the result, the model is not fully satisfactory for arbitrary antennas (and patterns). Nevertheless for long distance communication where elevation angles are minimized, the model is quite useful. Furthermore computations require a limited amount of machine time making the FTZ model a valuable method for preliminary screening of a large number of paths.

CCIR 252-2 : [CCIR, 1970]. This model termed *CCIR Interim Method for Estimating Skywave Field Strength and Transmission Loss Between Approximate Limits of 2 and 30 MHz* was initially adopted by CCIR at the 1970 New Delhi plenary. It was the first of three computer methods for field strength prediction which were sanctioned by the CCIR.

CCIR 252-2 Supplement: [CCIR, 1982a]. A field strength prediction method entitled *Second CCIR Computer-based Interim Method for Estimating Skywave Field Strength and Transmission Loss at Frequencies Between 2 and 30 MHz.* The method is more complex than the method of CCIR 252-2 in a number of respects, and the machine time required reflects this additional complexity. A major change is the consideration of longitudinal gradients for the first time. A computer program was completed in 1987.

CCIR 894-1: [CCIR, 1986a]. To assist in the WARC HF Broadcast Conference, a rapid computational method was documented as CCIR Rpt. 894. This document was the result of CCIR Interim Working Party (IWP 6/12) deliberations to produce a prediction program for use in planning by the HF broadcast service. This program is a simplification of CCIR 252-2 (or equivalently IONCAP) but incorporates the FTZ approach for long distance applications. The IONCAP approach is used for paths less than 7000 km, FTZ is used for paths greater than 9000 km, and a linear interpolation scheme is applied for pathlengths in between.

HFBC84 : [ITU, 1984a]. This is a computer code based upon Report 894. An improved estimate of field strength is obtained by a taking the antenna gain (of appropriate broadcast antennas) into account when selecting modes to be included in the calculations. HFBC84 provides the analyst with an practical procedure for mapping the coverage of a specified broadcast antenna. Such a coverage pattern is given in Figure 5-5.

AMBCOM : [Hatfield, 1980]. This program was developed by SRI International in connection with work supported by the Defense Nuclear Agency, and it is a companion program to NUCOM, another propagation program specific to the nuclear environment. One difference between the ITS series of programs and AMBCOM is that the latter employs a 2-D raytrace program while the former programs use virtual methods. In addition, AMBCOM contains within its ionospheric submodel structure a considerable amount of high latitude information including improved auroral absorption models. This should provide for an improved prediction capability for paths through the high latitude region or within its neighborhood. The model allows insertion of up to 41 ionospheric data points along the paths of interest. This capability should make AMBCOM highly suitable for a detailed analysis of links or coverage areas in situations in which the underlying ionosphere is well sampled. The 2-D approach used in AMBCOM is a relaxation of the ionospheric specification requirements implicit in the use of full 3-D methods, but provides a more realistic explanation of coverage than simple (and artificial) virtual methods. A major distinction between AMBCOM and virtual methods used by the CCIR is that the ionosphere defines the path of the ensemble of rays in AMBCOM, whereas a predetermined path is used to define the effective part of the ionosphere (i.e., the *control point*) in the virtual or *mirror* methods. Because of added complexity, the program is generally slower than simpler models. Because AMBCOM uses raytracing and will operate against large electron density gradients, it will predict asymmetric hops and unconventional modes. AMBCOM documentation is not as widely distributed as IONCAP or the CCIR methods. Details on AMBCOM are available by writing SRI International, Attn: Director, Geosciences and Engineering Center, 333 Ravenswood Avenue, Menlo Park CA 94025-3493.

5.7.3 Nuclear Effects Considerations and Models

HF communications are strongly affected by high altitude nuclear detonations and the effects (HANE) have been widely modeled. The effects are examined in terms of phenomenological regimes which depend upon height and yield of burst. Key features of most analyses involve separation of regimes into those which are either prompt or delayed, and the propagation aspects into those which are absorptive or refractive. A summary of weapons effects on communication systems may be found in a book by Glasstone and Dolan [1977].

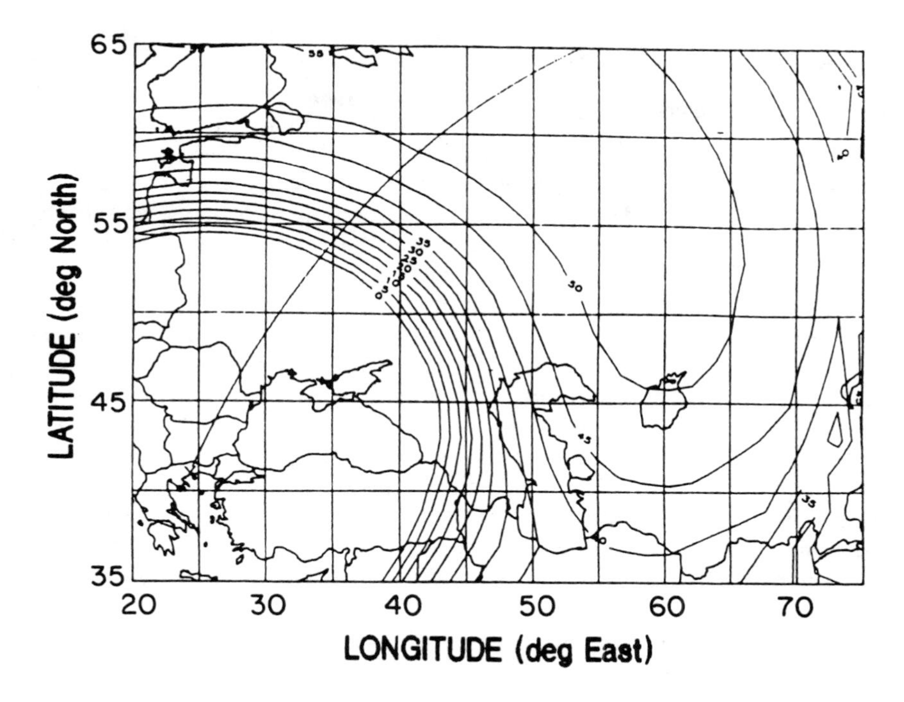

Fig.5-5. Coverage Diagram using HFBC84 for a transmitter located in Kavalla, Greece. The contours constitute lines of constant field strength in dBu. A 250 kW transmitter and a curtain array are assumed. The radio frequency was 11.855 MHz, and the time is 2000 UT during March of 1985. (From Rush [1986])

One of the principal difficulties in prediction of nuclear effects on HF systems is in assessment of late-time effects. These include the rather fascinating influence of disturbance-generated TIDs, as well as spread-F resulting from enhanced striation development. The time history of the radioactive debris is influenced by upper atmospheric winds, leading to a rather poor forecasting capability for debris-related effects beyond a few hours. Absorption effects are the most well understood provide the burst information and debris cloud position and dimensions is characterized. Two phenomena of interest to the analyst include the so-called *beta patch* and the *gamma ray shine*. These descriptive terms describe respectively: (a) an intense and localized absorption zone due to high energy electrons, which map down from the debris to the lower ionosphere along field lines; and (b) the relatively distended absorption blanket arising from x and gamma radiation, which enhances ionization for regions within optical line-of-sight of the debris cloud. Typically, one is little concerned with low-yield low-altitude tactical effects (LANE) as far as the ionosphere is concerned. This is because the prompt effects are largely localized to the region of detonation, and the debris possesses insufficient energy for the cloud to ascend and stabilize above the critical radiation stopping altitude. Thus the D-region is not illuminated and ionospheric absorption will not be encountered. High-altitude detonations (and essentially all yields) will produce significant effects.

The nuclear environment has been observed to include enhanced sporadic E and well-defined *plume-like* features which may introduce yet another mechanism for communication at HF and low-VHF region of spectrum. These enhanced features are sometimes referred to as *bomb-modes*, and (when present) may allow connectivity to be restored more rapidly since absorption is reduced as the radiofrequency is raised. Other possibilities might include the utilization of the enhanced ionospheric scatter modes at VHF and UHF for transmission of low baud rate messages and facsimile. Further strategies, also involving a departure from skywave HF, include: (a) the exploitation of reduced atmospheric noise during nuclear disturbance in connection with the robust features of a ground wave network; and (b) utilization of meteor burst communication (MBC). The combination of HF and MBC has a lot of charm since these systems are essentially complementary. HF skywave is a good long-haul service, while MBC is essentially a short-distance technique. Additionally, MBC possesses low-probability-of-intercept (LPI) features (see Figure 5-6).

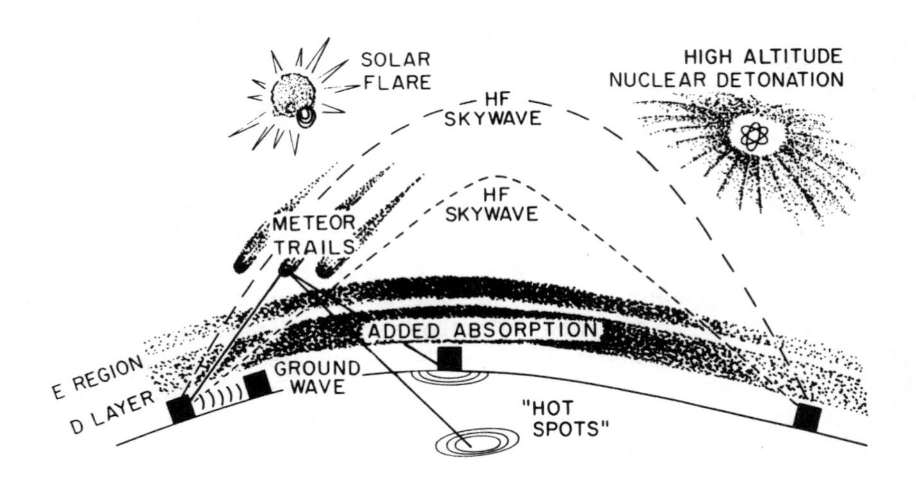

Fig. 5-6. Meteor burst communication geometry

Development of an understanding of the nuclear environment has required a considerable amount of modeling since the advent of the nuclear test ban treaties. Experiments have also been conducted in order to simulate some aspects of low altitude and exoatmospheric detonations. Chemical detonations have been used to simulate shock-induced gravity wave effects, chemical releases such as barium have been examined to assess the motion of ionized plasma, and chemical reagents such as water vapor and hydrogen have been studied for their *quenching* properties. Furthermore, engineering solutions have been sought through development of a variety of diversity measures which rely upon a fairly good picture of the HF channel attributes. Thus studies of the HF channel have been undertaken in the high latitude zone, containing sub-regions which provide the system engineer with no doubt the most favored mimic of nuclear phenomena.

There are a number of computer codes which are specifically dedicated to the analysis of weapons effects on HF communication systems. Two of these are HFNET and WEFCOM. HFNET enables the user to analyze the impact of high- and low-altitude nuclear detonations on specified HF networks, and WEFCOM (Weapons Effects on F-Region Communication Systems) allows the user to evaluate the impact of nuclear disturbances and jamming on the HF channel itself. Another code, SIMBAL, examines weapons effects on VLF, LF, *and* HF links. Details concerning these and related codes may be obtained by corresponding directly with the Defense Nuclear Agency, Washington DC 20305-1000.

5.7.4 Ionospheric Data Used in Prediction Models

The parameters used in major prediction models are the same in many instances and the data sets which represent a given parameter may also be the same. Nevertheless, the manner in which the data are used can lead to extraordinary differences in detail. Fortunately, for purposes of deriving an intuitive *feel* for the various influences on HF propagation/performance, most models are adequate. Indeed, if updating is possible, then many differences may be unimportant, except to the purist.

Table 5-4 is based on a previously unpublished review by Lucas [1987]. It summarizes some of the most important ionospheric parameters, and indicates the models which incorporate the specified data sets.

The parameters listed in Table 5-4 were defined in Chapters 3 and 4; for convenience, selected ionosonde characteristics and their definitions are listed in Table 5-5. The original papers, indicated in the table, review how each parameter is derived and over what period of time the empirical data were assembled. The references in Table 5-3 indicate the specific usage of parameters in a given model.

Statistical distributions are required for certain ionospheric parameters for at least two reasons. First, parameters such as foF2, foEs and hF2 fail to follow Chapmanlike rules, a fact which makes prediction of the average behavior of these parameters less successful than it might otherwise be. Secondly, departures from the mean are perceived to be random variables, and not subject to the prediction process, at least in the deterministic sense. The sporadic E layer and the F2 layer are obvious candidates for statistical treatment. Statistical distributions for foF2 [Lucas and Haydon, 1966] and Es [Leftin et al., 1968] are available.

The only ionospheric height which is explicitly computed in listed computer prediction models is hF2. Still, the variability in hF2 arising from unpredictable sources, such as TIDs, is a significant fraction of the mean diurnal variation. Typically Shimazaki's formula (or some derivative) is employed for estimation of the mean value in hF2, but other approaches may also be used.

TABLE 5-4

IONOSPHERIC PARAMETERS, DATA SOURCES, AND MODELS

Models	Layer Criticals				Layer Heights				Layer Semi-Thickness			
	foEs	foE	foF1	foF2	hEs	hE	hF1	hF2	yEs	yE	yF1	yF2
RADARC	1	2	3	4	k	k	e	5	k	k	k	6
IONCAP	1	2	3	4	k	k	k	5	k	k	k	6
ITS-78	1	2	n	13	n	k	n	5	k	k	n	7
ITSA-1	n	8	n	4,12	n	k	n	5	n	k	n	6
HFBC84	n	8	n	4	k	k	n	9	n	n	n	n
CCIR-252	1	2	n	4	k	k	n	5,11	n	k	n	7
AMBCOM	1	2,10	n	13	k	k	f	5	k	k	f	7

KEY

1. Leftin et al.[1968]
2. Leftin [1976]
3. Rosich and Jones [1973]
4. CCIR [1966a]
5. Shimazaki [1955]
6. Lucas and Haydon [1966]
7. Leftin et al.[1967]
8. Knecht [1962]
9. Lockwood [1984]
10. Hatfield [1980]
11. Leftin [1969]
12. Jones and Gallet [1962]
13. CCIR [1970]

e. determined empirically
f. filled in between E and F2
k. constant
f. not applicable or undefined layer

(Information in this table is based in part upon unpublished material from from Lucas [1987].)

TABLE 5-5.

IONOSONDE PARAMETERS AND DEFINITIONS

Ionosonde parameter	Definition
foE	Critical frequency of the ordinary ray component of the normal E layer. It is the frequency which just penetrates the ionospheric E layer. It is proportional to the square root of Nmax for region E.
h'E	Minimum virtual height of the E layer. This is determined at the point where the ionosonde trace becomes horizontal.
foEs	Critical frequency of the ordinary ray component of the Es (sporadic E) layer.
h'Es	Minimum virtual height of the sporadic E layer, and reckoned at the height where the trace become horizontal.
fbEs	The blanketing frequency for the Es layer. This corresponds to the lowest ordinary wave frequency for which the Es layer allows penetration to a higher layer; i.e., begins to become transparent.
foF2	The critical frequency of the ordinary wave component of the F2 layer. It is proportional to the square root of Nmax for the layer. It is the frequency which just penetrates the F2 layer.
foF	Critical frequency of the ordinary wave component of layer F1. The ionosonde frequency which just penetrates the F1 layer.
h'F2	Minimum virtual height of the F2 layer. It is measured at the point where the trace becomes horizontal.
h'F	Minimum virtual height of the night F layer and the day F1 layer. Again, it is measured at the point where the F trace involved becomes horizontal.
h'F1	Minimum virtual height of the F1 layer, measured at the point where the F1 trace becomes horizontal.
h'FF2	Alternative tabulation of the minimum virtual height of the F layer. It corresponds to the minimum virtual height of night F layer and the day F2 layer. Again it is measured at the point where the appropriate traces become horizontal.
hpF2	Virtual height of the F2 layer corresponding to the frequency f = 0.834 foF2. Based upon a parabolic layer approximation.
M(3000)F2	Ratio of MUF(3000)F2 to the critical frequency foF2.

A new mirror height method having similarities to the Shimazaki approach has been used in HFBC84 [Lockwood, 1984]. The basis for hF2 estimation in the CCIR-252 model is virtual height data (i.e., *h' FF2*) from Leftin et al.[1967] and Leftin [1969]. Recognizing that hF2 is simply the (nonvirtual) height of the F2 maximum, $h_{max}F2$, the Shimazaki relation says:

$$h_{max}F2 = 1490/\mathrm{M(3000)F2} - 176 \tag{5.1}$$

We recognize that M(3000)F2 is MUF(3000)F2 ÷ foF2, and that it is proportional to the secant of the ray zenith angle ϕ. If the layer descends, it is apparent that the secant of ϕ will increase. Consequently M(3000)F2 increases as layer height decreases, and vice versa. This fact is reflected in Equation 5.1. Taking $h_{max}F2$ to be a nominal 300 km, then M(3000)F2 is nearly 3.2. Under this condition, dh/dM(3000)F2 ≈ -150 km. Hence an increase in M(3000)F2 of 0.2 will correspond to a height reduction of 30 km.

Maps of foF2 and M(3000)F2 have been of major importance in HF propagation prediction for years. They are used in various ionospheric models to provide a global distribution of electron density and F2 layer height in other applications. The CCIR [1966a] model, documented as CCIR Report 340-1, consists of an *Atlas of Ionospheric Coefficients* defining foF2 and M(3000)F2, plus actual maps of the parameters EJF(zero)F2 and EJF(4000)F2, when have been defined in previous chapters. The CCIR [1970] model, termed *Supplement No.1*, is an update of the Report 340 which replaces the CCIR [1966a] foF2 *Oslo* coefficients with new ones which better fit the existing data base. Improvements included replacement of the linear dependence of foF2 on sunspot number by a polynomial dependence, and a Fourier representation of the annual variation so that any day could be examined in terms of its surrounding monthly median. The CCIR [1970] coefficients were conceived by Jones and Obitts and are sometimes referred to as the *New Delhi* coefficients. Screen or *phantom* points were required over sparsely sounded oceanic areas for both the *Oslo* and *New Delhi* coefficients. Early versions of ITS-78 (HFMUFES) used *Oslo* coefficients which were reproduced on red computer cards. Later versions used the *New Delhi* coefficients reproduced on blue cards. Thus the terms *red deck* and *blue deck* are sometimes used in references.

Recently, as indicated in Chapter 3, there have been steps to improve the ionospheric coefficients. Within the URSI community (Working Group G.5) Rush et al.[1983, 1984] developed a new coefficient set based upon more fundamental theory. The extensive data base assembled by Rush and his colleagues included *new* data points deduced using a method developed by Anderson [1981]. As pointed out by Rush et al.[1989], in order not to depart too significantly from established CCIR recommendations and long-term prediction methodology, consistency with the structure of the CCIR [1966a] Jones-Gallet coefficient set was required. Fox and McNamara [1986, 1988] have continued the work and have proposed a final set of coefficients. Fox and McNamara organized their data in terms of the T-index rather than in terms of sunspot number, they included more foF2 data in the analysis, and they sought consistency with independent data derived from the Japanese topside sounder ISS-B. They also used methodology in which the coefficients were of higher order at low latitudes than the CCIR/URSI maps. This provides more detail at lower latitudes. The new approach is the basis for a new set of coefficients used by the Australian agency IPS. The improvement over the original set is more than satisfactory. To achieve consistency with the standard format of existing internationally-sanctioned maps, the IPS coefficients were transformed by URSI to coincide with the existing number of coefficients. This process had the effect of degrading the output from the IPS approach somewhat, but consistently smaller residual errors have been noted when compared with the CCIR maps. Ultimately Rush et al.[1989], including the IPS group, have published an update of the foF2 coefficients. Since this revised set, also termed the *1988 URSI coefficient set*, has the same structure as the earlier *1966 CCIR coefficient set* used in IONCAP, an upgrade of IONCAP climatology is straightforward. Figure 3.26 in Chapter 3 shows the difference which arises in the foF2 contours when the two coefficient sets are switched.

Several terms which have been used to describe the various CCIR coefficients. As noted above, the first set to be published as a separate booklet by the CCIR is due to Jones and Gallet [1962], and was approved by the CCIR at its 1966 plenary held in Oslo, Norway. When used in early versions of ITS-78, the coefficient set was reproduced on red cards. The CCIR took note of an alternative coefficient set at its 1970 plenary held in New Delhi, India. This set, developed by Jones and Obitts [1970], and published by the CCIR in 1971, was an improvement in a number of areas over the previous set, was only recommended for use in short-term predictions. Differences in the two sets of coefficients were described in CCIR Report 340 as revised in 1983. A summary of existing coefficient sets is given in Table 5-6. (See Footnote No.50 appearing in the references, Section 5.19.)

TABLE 5-6

VARIOUS SETS OF IONOSPHERIC COEFFICIENTS

Coefficients Epoch	Authors	Plenary Session Location	Computer Deck Designation	Usage
CCIR 1966	Jones-Gallet	Oslo	Red Deck	Long-term
CCIR 1971	Jones-Obitts	New Delhi	Blue Deck	Short-term
URSI 1988	Rush et al.	N/A	N/A	long-term

where N/A means not applicable.

5.8 NOISE AND INTERFERENCE

5.8.1 Relevant Documentation

Noise and interference are major factors in the performance of HF systems. Because of the large number of users, spectral congestion is a major issue at HF; and channel occupancy rates become staggering at night (See Figure 1-3). The established models for describing the effects on radio systems include CCIR 322, CCIR 258-4, CCIR 342-5, and CCIR 670. The current version of CCIR 322 (322-3 approved at the Dubrovnik plenary meeting held in 1986) is based upon work of Spaulding and Stewart [1987]. These investigators have also developed an updated noise model for use in IONCAP. Details of the improvements currently in CCIR 322-3 have been reported by Spaulding and Washburn [1985]; and a discussion of issues relative to Report 258-4 [CCIR, 1982] are found in a report by Spaulding and Disney [1974]. Relevant noise models and CCIR references are given in Table 5-7.

TABLE 5-7

NOISE MODELS AND RELATED CCIR REPORTS [47]

CCIR No.	TITLE OF REPORT	REFERENCE
322	On Atmospheric Noise from Lightning and Galactic Noise	Published as separate booklet. [CCIR, 1964]
322-3	Characteristics and Applications of Atmospheric Radio Noise Data	Replaces CCIR 322 and its supplements. Published as a separate booklet. [CCIR, 1986d]
258-4	On Man-Made Radio Noise	XVIth Plenary Session, Dubrovnik, Vol.VI pp.207-214. [CCIR, 1986e]
342-5	Radio Noise Within and Above the Ionosphere	XVIth Plenary Session, Dubrovnik, Vol.VI pp.214-228. [CCIR, 1986f]
670	Worldwide Minimum External Noise Levels	XVth Plenary Session, Geneva Vol.I, p.224. [CCIR, 1982c]
825	On Channel Occupancy	XVth Plenary Session, Geneva. [CCIR, 1982b]
413	On Receiver Operating Noise Threshold	XIth Plenary Session, Oslo. [CCIR, 1966b]

47. Based upon the deliberations of the XVIIth Plenary Assembly held in Dusseldorf, Germany in 1990, certain revisions may have been made in CCIR documents of record. Particular attention should be given to Volumes I (*Spectrum Management Techniques*) and Volume VI (*Propagation in Ionized Media*). See footnote on page 15.

A NATO-AGARD meeting [AGARD, 1987] devoted to the effects of electromagnetic noise on military systems contains several papers of interest. Special attention is directed to a review article by Spaulding [1987] who examines currently used noise and interference models, and discusses the performance of basic modulation systems.

5.8.2 The System Noise Figure Concept [48]

To estimate the impact of external noise sources on system operation, it is necessary to establish the pre-detection signal-to-noise ratio. Figure 5-7 schematically represents a generic receiver system from input to output, the noise factor and the signal-to-noise ratio associated with the receiver, and the location at which these parameters are reckoned. The system noise factor is given by [Spaulding and Stewart, 1987]:

$$f = f_a + (L_c - 1)(T_c / T_o) + L_c(L_t - 1)(T_t / T_o) + L_c L_t (f_r - 1) \tag{5.2}$$

where f_a is the external noise factor given by $p_n / kT_o b$, F_a is the external noise figure given by $10 \log_{10} f_a$, p_n is the available noise power from a lossless antenna, L_c is the antenna circuit loss (input power/output power), T_c is the temperature (°K) of the antenna and neighboring ground, L_t is the transmission line loss (input power/output power), T_t is the temperature of the transmission line, T_o is the reference temperature (°K), and f_r is the noise factor of the receiver (°K). The noise *figure* in dB is simply $F_r = 10 \log_{10} f_r$. To avoid confusion capital letters are used when discussing the noise figure as well as other terms which may be expressed in decibels, and lower case letters are used when dealing with receiver and antenna noise factors.
The noise power in watts is simply:

$$n = f k T_o b \tag{5.3}$$

where k is Boltzmann's constant $= 1.38 \times 10^{-23}$ J/°K, $T_o = 288$°K, and b is the noise power bandwidth of the receiving system. For an antenna and transmission line which may taken to be lossless, then the overall system noise figure F is approximately the sum of F_a and F_r. Recognizing that $10 \log_{10} kT_o = -204$, we may rewrite Equation 5.3, specifically for the external noise component, in a convenient decibel form:

$$P_n = F_a + B - 204 \tag{5.4}$$

where P_n is in dBW and F_a and B are expressed in dB (where B is in dB-Hz).

---

48. The noise factor is designated by the letter f in this discussion, and should not be confused with the radio frequency. To avoid the possibility of misinterpretation, *f(MHz)* is used to denote the frequency for instances which appear warranted. This comment is specifically relevant to Figure 5-7 and Sections 5.8.2 through 5.8.5.

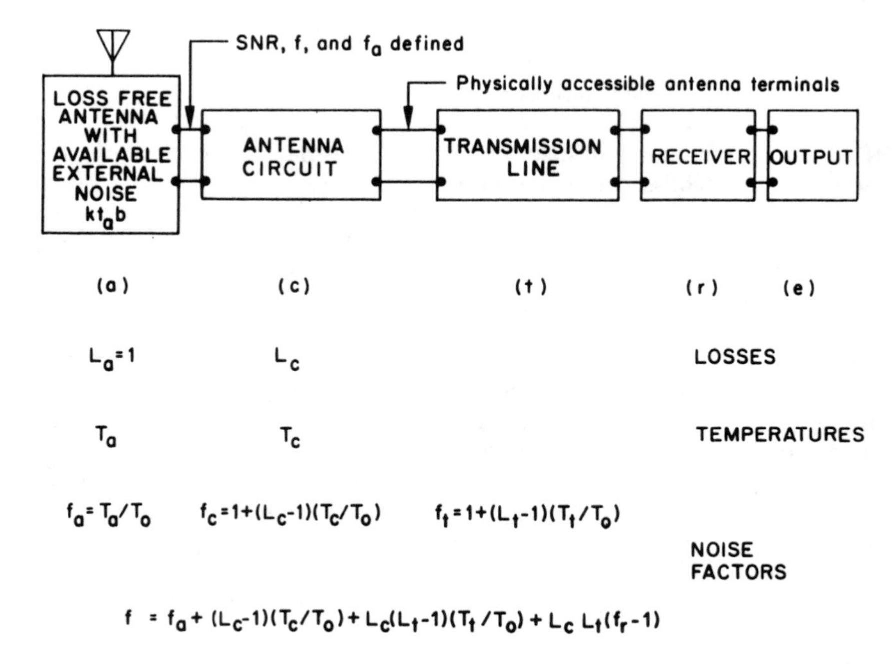

Fig.5-7. Generic receiver system concept, illustrating the locations at which signal and noise parameters may be reckoned. *fa* is the external (antenna) noise factor and *fr* is the receiver noise factor. (From Spaulding and Stewart [1987].)

Another way to represent the external noise factor f_a is as a temperature, where f_a is taken to be the ratio of antenna temperature (resulting from external noise) to T_o.

For specified antennas, it is possible to obtain an expression for the field strength in dB (above 1 μV/m). Such expressions take the form:

$$E_h = F_a + 20 \log_{10} f(\text{MHz}) + B - \Gamma_A \tag{5.5}$$

where Γ_A is a constant dependent upon antenna type and configuration. For a short grounded vertical monopole Γ_A = -95.5 dB. Thus the noise figure (or factor) is a fundamental parameter since it defines for a specified antenna configuration and noise bandwidth the noise level with which the desired signal must compete. We shall now examine the major sources of noise and therefore F_a.

5.8.3 Noise Models and Data

Noise at HF has three major components: atmospheric, galactic, and man-made noise. Another category of noise sources are associated with intentional interferers (jammers). These latter sources will not be discussed here. Figure 5-8 gives the range of expected values for noise. Several features in the figure are of interest. First we see that except for business areas, galactic noise would appear to dominate in the upper half of the HF band. At midband and below, man-made sources become quite important as the galactic component suffers a cutoff because of the high pass filter properties of the ionospheric plasma. Depending upon conditions, atmospheric noise caused by lightning has an enormous range, and may become the dominant noise source, especially in the lower part of the HF band.

5.8.3.1 *Atmospheric*

The major cause of atmospheric noise is lightning strokes which produce broadband noise, and which arise during thunderstorms. Clearly this suggests a preferred source and time distribution for the atmospheric noise contribution. Atmospheric noise, like desirable HF signals, obeys the same physical laws, and may propagate over considerable distances beyond the line of sight. Noise originating in the opposite hemisphere or from sources across the day-night terminator are major contributors to F_a. Even though the events are isolated and of short duration, the composite result, as reckoned from a given receiver may be characterized as quasiconstant for any specified hour. The long distance propagation characteristic of HF has the effect of populating the time domain with signals from the global distribution, but with each individual source being constrained by its own LUF-MUF band-pass filtering operation. Receiver latitude plays an important role at HF. In fact noise is considerably reduced as the latitude increases commensurate with an average increase in distance from the low latitude source regions. Regions where noise is most severe include the African equatorial zone, the Caribbean area, and the East Indies. No account is provided in existing models for effects from a localized source distribution, and azimuthal information is not available because of the manner in which the database (comprising the CCIR 322 model) was generated. Clearly, local noise is important, and its omission will lead to underestimates for anticipated external noise especially during the summertime rainy season. On the other hand, actual antennas may have nonuniform patterns in the bearing (and elevation) plane; and this will modify the noise distributions. Highly directive antennas may yield optimistic or pessimistic values for the observed F_a.

Sailors and Brown [1982] have developed a minicomputer atmospheric noise model using simplified methods. With the advance of computer technology, code simplification is no longer a practical necessity.

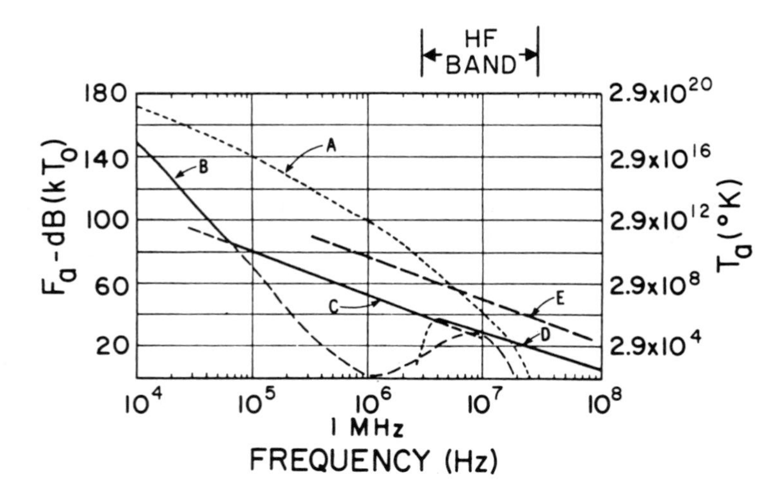

Fig.5-8. Noise figure (dB/kT_0) (LH-scale), and antenna temperature,° K (RH-scale). A: Atmospheric noise from lightning, value exceeded 0.5% of the time; B: Atmospheric noise from lightning exceeded 99.5% of the time; C: Man-made noise at a quiet receiving site; D: Galactic Noise; E: Median business-area, man-made noise. (From Report 670, [CCIR, 1982c].)

5.8.3.2 *Galactic*

Figure 5-9 shows the effective temperature of an antenna which is receiving galactic noise. Galactic (or cosmic) noise originates outside the ionosphere, but for signals to be received at an earth terminal, ionospheric penetration is necessary. Signals in excess of the overhead critical frequency may be received; however if antennas (such as vertical monopoles) have limited gain in the vertical direction, then available lower frequencies will not effectively contribute. Rules for ionospheric penetration imply that the available cosmic noise distribution will always be confined to a small iris near the zenith direction when operating near the critical frequency. As the radiofrequency f exceeds f_c by a large amount, the iris will become distended being defined by a dimension $\phi \approx \sec^{-1}(f/f_c)$.

5.8.3.3 *Man-made*

Man-made noise is not only influenced by the population density, but it also depends upon the technological sophistication of the society. Attempts to relate manmade noise and population density have not been entirely successful, although Lucas and Haydon [1966] have provided an estimate of how population might be used in the prediction of the noise. Propagation may be by either skywave or groundwave methods. Primary sources are local ones, including nearby ignition noise, neon lights, and various electrical equipment.

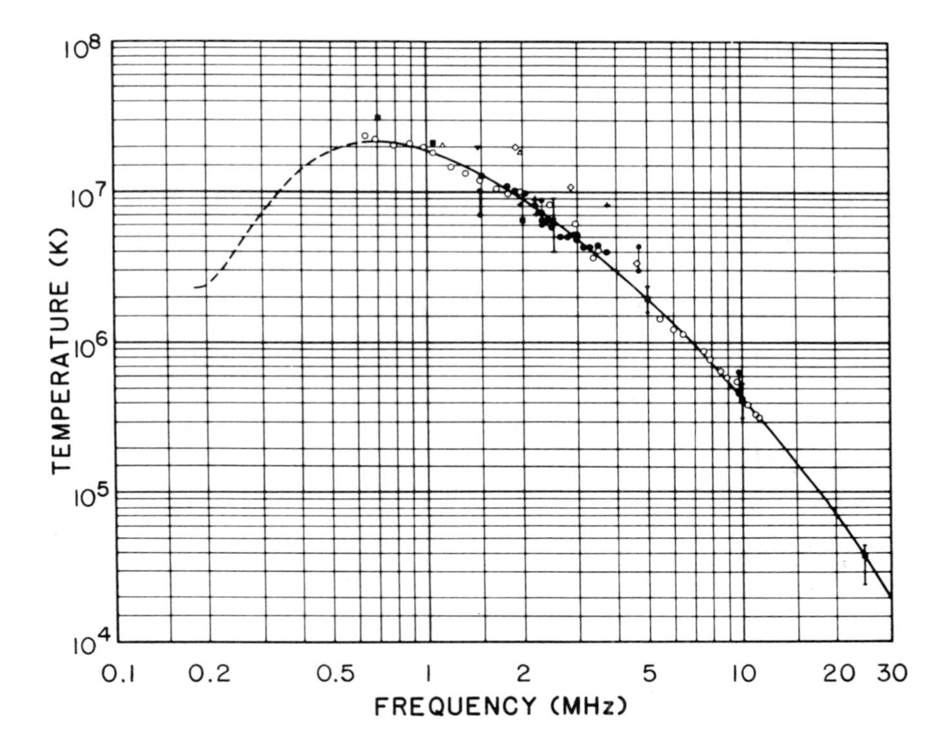

Fig.5-9. Galactic noise as a function of frequency. Data points shown on the plot correspond to experimental results reported by a number of investigators. (Report 342-5, [CCIR, 1986f].)

Figure 5-10 provides a glimpse of residential noise variability across the RF spectrum. We note that the upper and lower deciles differ by approximately 15 to 25 dB throughout, and median values range between roughly 60 dB (at 3 MHz) and 30 dB (at 30 MHz).

A sample man-made noise distribution, expressed in terms of F_a, is given in Figure 5-11 at a frequency of 20 MHz for springtime morning conditions in a residential area. It is seen that the upper-to-lower decile range is about 15 dB. The two log-normal distributions tend to represent the data, one above and one below the median [Spaulding and Disney, 1974]. Galactic and atmospheric noise sources have also been observed to exhibit log-normal distributions.

The man-made noise model described in the earliest versions of CCIR 258 was based upon RF noise measurements originally made by ITS concentrating on sites in the USA. The most recent version of the report, CCIR 258-4 [1982] has been improved by the addition of more modern data, notably data obtained from the USSR. Man-made noise, expressed in terms of F_a, is given in Figure 5-12.

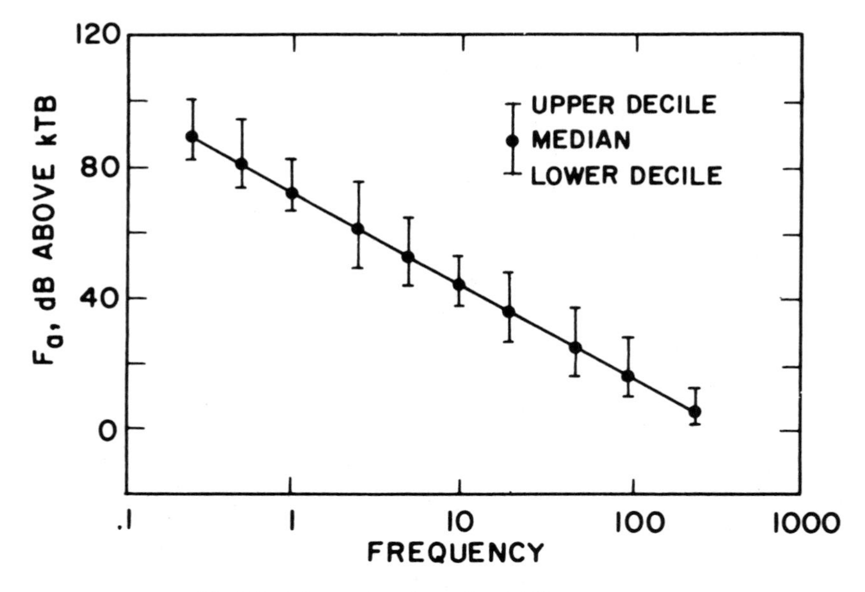

Fig.5-10. Man-made noise and variability.

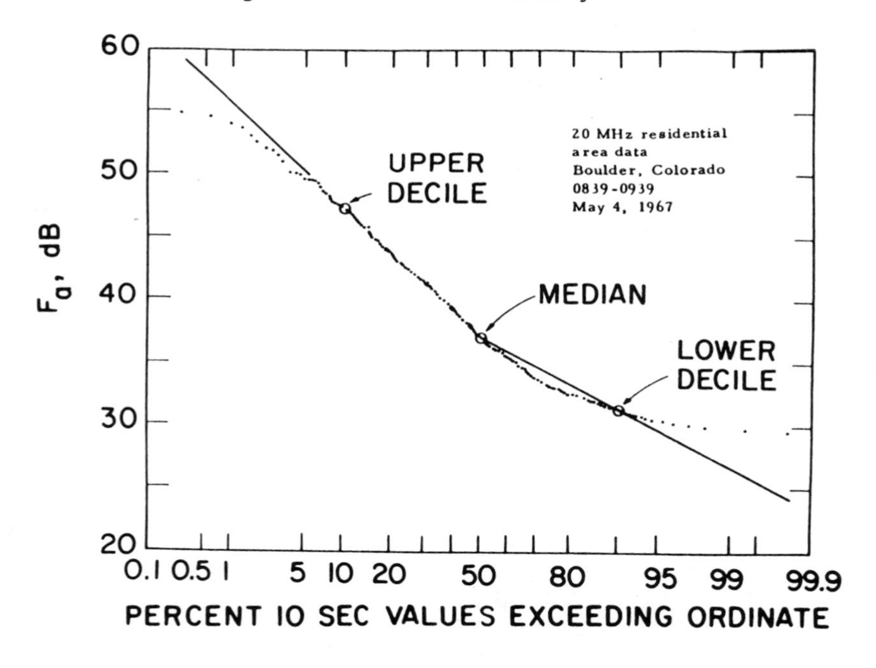

Fig.5-11. Noise distribution at 20 MHz, for spring and morning conditions. Residential area near Boulder, Colorado (From Spaulding and Disney [1974].)

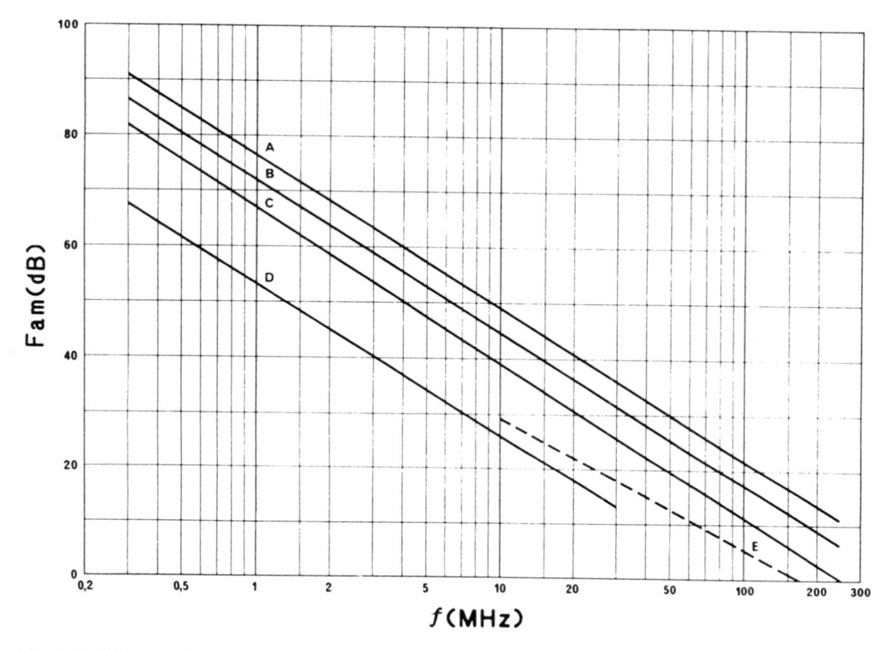

Fig.5-12. Man-made noise levels. A: business, B: residential, C: rural, D: Quiet rural, E: Galactic. (From Report 258-4, [CCIR, 1982e].)

5.8.4 The CCIR 322 Noise Model

Implementation of CCIR 322 may be illustrated through the use of three charts. The first chart (actually one of many) shows contours of F_{am}, the median value of the external noise, for a specified *local time* block at a frequency of 1 MHz. A second chart permits translation of the 1 MHz noise figure medians to the frequency desired. A third chart is used to obtain frequency-dependent statistical information. To obtain an estimate of the atmospheric noise level under specified conditions, we must select an appropriate seasonal map and read the 1 MHz noise estimate for the receiver location dictated.

Figure 5-13 is a CCIR map for the winter season for the 0000-0400 local time block. At Washington DC the value of median 1 MHz noise (i.e., F_{am}) is about 70 dB above $kT_o b$. The next step is to shift this result to the appropriate frequency to be utilized. We do this with companion Figure 5-14. We see that a family of curves is displayed showing the frequency variation of F_{am} but parametric in terms of the 1 MHz value (already obtained). Locating the 70 dB curve, we slide to the frequency of interest. Assuming 10 MHz is that frequency, we see that F_{am} is 35 dB above $kT_o b$. From Figure 5-15 we may deduce noise variability statistics for the frequency of interest.

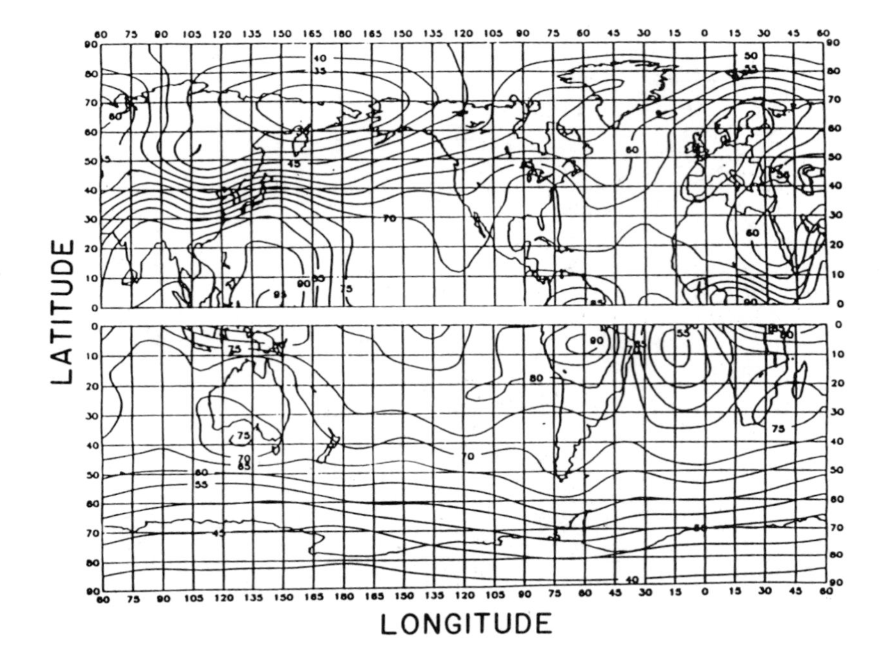

Fig.5-13. World map of atmospheric noise at 1 MHz. Winter. Time block: 0000-0400 LT. Values in dB above kT_0b (From CCIR 322-3 [1986]).

With respect to availability of computer codes from the CCIR Secretariat, the set of new noise coefficients associated with Report 322-3 [1986] is termed NOISEDAT and is applicable for microcomputer application. The program NOISY containing the older coefficients is available for mainframe computer versions of CCIR-322-2 [1964]. Understandably, a number of organizations such as NRL and VOA/USIA have modified the existing mainframe code to accommodate the newer coefficients. It should be noted that they way atmospheric noise, galactic noise and man-made noise are combined has also been modified.

5.8.4.1 *Combination of Noise Sources*

Spaulding and Stewart [1987] have described how each of the noise sources should be combined to estimate system performance effects. At one time propagation prediction methods simply took as the largest of the atmospheric, man-made, and galactic sources as the composite. Over the years this approach has been modified, and a decidedly more attractive method of combining the three sources and obtaining the composite distribution function has been the result.

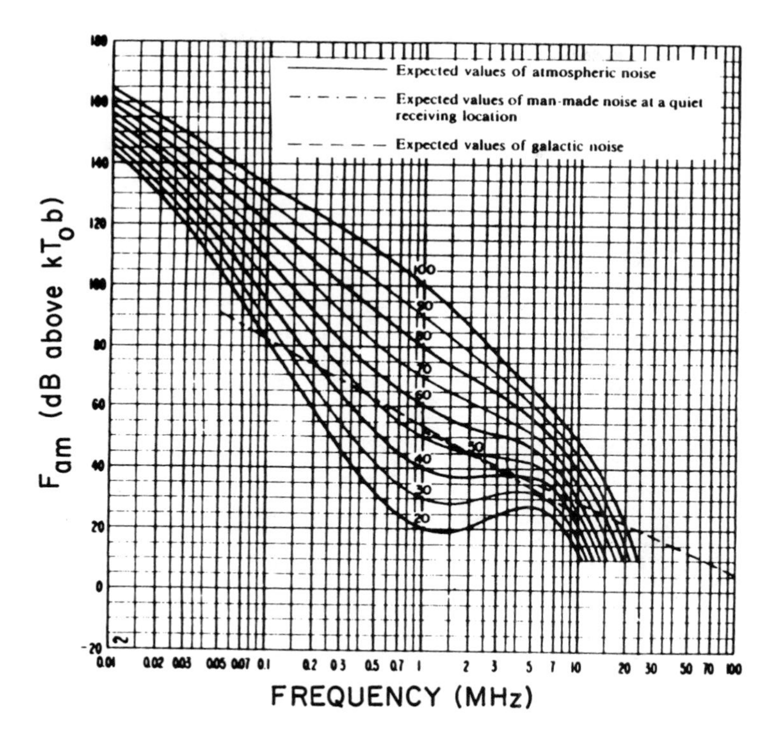

Fig.5-14. Variation of radio noise with frequency. Winter, time block: 0000-0400 LT. Values in dB above kT_0b. (From CCIR 322-3 [1986].)

5.8.4.2 *IONCAP Implementations*

A version of IONCAP containing the newest noise model approach contains the following subroutines: AOIS1, which computes the 1 MHz atmospheric noise levels for 2 adjacent 4-hour time blocks by calling NOISY; NOISY, which uses supplied Fourier coefficients to compute the 1 MHz atmospheric noise value; GENFAM which computes the atmospheric noise at the appropriate frequency as well as variability data; and GENOIS, which combines all of the sources of noise including atmospheric, galactic, and man-made. The subroutine GENOIS has been modified from earlier versions of IONCAP. The new method is also contained within the PC implementation IONCAST developed by NRL [Rhoads, 1990].

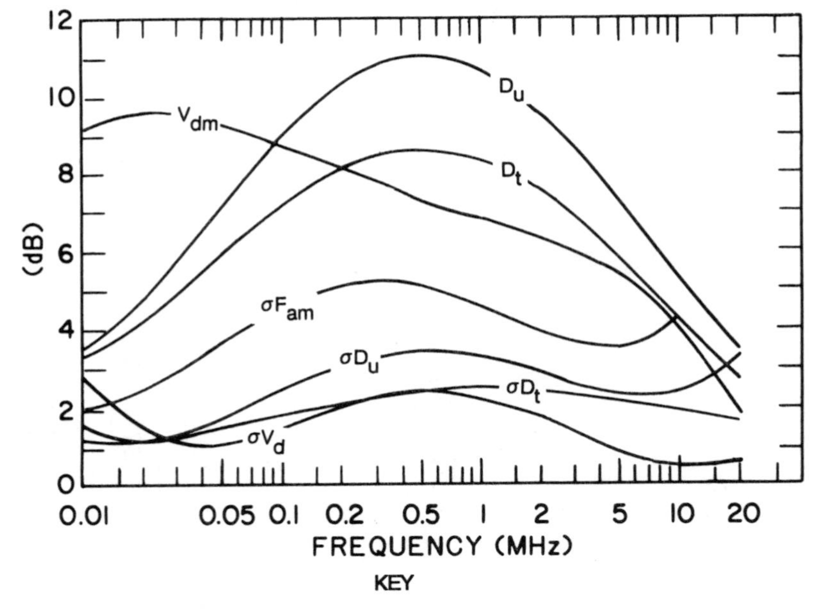

KEY

σF_{am}	:	Standard deviation of values of F_{am}
D_u	:	Ratio of upper decile to median value, F_{am}
σD_u	:	Standard deviation of values of D_u
D_t	:	Ratio of the median value, F_{am}, to lower decile
σD_t	:	Standard deviation of value of D_t
V_{dm}	:	Expected value of the median deviation of average voltage. (Values shown are for a bandwidth of 200 Hz.)
σV_d	:	Standard deviation of V_d

Fig.5-15. Noise variability. Winter, time block: 0000-0400 LT. (From CCIR 322-3 [1986].)

5.8.5 Channel Occupancy and Congestion

As indicated in Chapter 1 (Section 1.3.3), spectral congestion is a major influence in the performance of HF circuits. Figure 1-3 exhibited the nature of congestion for both daytime and nocturnal conditions across the entire HF band. Examination of 50 kHz spectra shows how specified congestion levels appear for daytime and nighttime. Dutta and Gott [1982] suggest the definitions of congestion index Q and the voice channel availability appearing in Table 5.8.

TABLE 5-8. CONGESTION, AVAILABILITY, AND OCCUPANCY

CONGESTION (Q): The probability of randomly finding within each 50 kHz spectrum a 100 Hz frequency interval in which the average interference level exceeds a defined threshold. (original definition modified slightly)

VOICE CHANNEL AVAILABILITY: The probability of finding a 2.5 kHz spectral window where the interference level, measured within contiguous 100 Hz intervals throughout the 2.5 kHz window, is always below a defined threshold level. (This condition is not as highly correlated with congestion as might be supposed)

CHANNEL OCCUPANCY: The fraction of measurement time in which the interference level exceeds a defined threshold.

The definitions in Table 5-8 are tailored to examine the interference statistics of signals which are typically rather narrowband relative to a canonical voice channel.

A definition for spectrum occupancy and its measurement has been discussed by Spaulding and Hagn [1977]. In any case, occupancy is a general term which indicates the level of unavailability of a specified channel (or group of channels within a specified bandwidth), under the condition that exceedance of a given interference threshold implies unavailability. A number of measurement programs have provided information about the distribution of man-made interference and spectral congestion [Wilkinson, 1982] [Dutta and Gott, 1982] [Gibson et al., 1985] [Tesla, 1985] [Moulsley, 1985] [Gibson and Arnett, 1988]. Attempts at modeling spectral occupancy have been made for both narrowband and wideband architectures [Laycock et al.,1988] [Perry and Abraham, 1988]. It would appear that the data on occupancy are highly structured and that separate submodels may be required for each category of user including: fixed , various classes of mobile users, broadcasters, and radio amateur. This would be expected since utilization statistics for the allocated bands depends upon the nature of the population (use).

5.8.5.1 *Wideband Model of Congestion*

In the wideband arena, we take note of work by Perry and his colleagues at MITRE Corporation in which the interference levels have been analyzed in connection with a 1 MHz bandwidth direct-sequence pseudo-noise (DS-PN) spread spectrum waveform. A model has been described by Perry and Rifkin [1987] and Perry and Abraham [1988]. Figure 5-16a shows the logarithm of the occupancy plotted against the interference power. The occupancy here is essentially the probability that a given interference level is exceeded. The model implies a log-log relationship between the minimum noise power level (p_{min}) in the resolution bandwidth, and a certain maximum power (p_{max}) which is simply the power of the largest interferer.

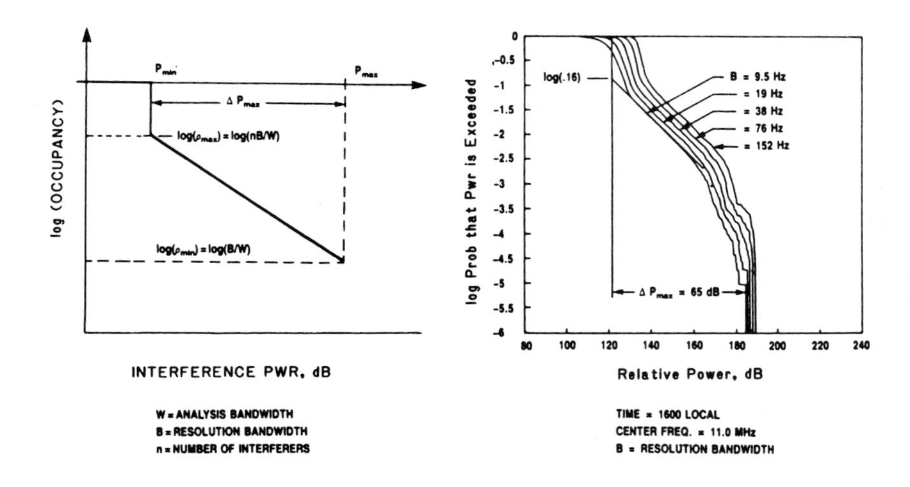

Fig.5-16: [Left] Model of interference distribution in WBHF case. B is the resolution bandwidth, W is the analysis bandwidth, and n = number of interferers. [Right] Spectral distribution of interference during the day at a center frequency of 11.0 MHz. (From Perry and Abraham [1988].)

We have:

$$p_{min} = kTBf_{am} \tag{5.6}$$

where the thermal power spectral density kT is -174 dBm/Hz (or -174 dBm + 34.8 dB = - 139.2 dBm in a 3 kHz channel) and f_{am} is the mean external noise figure which may be deduced from standard CCIR Report 322 analysis or measured. The occupancy under both minimum and maximum conditions are given by:

$$\varsigma_{min} = B/W \tag{5.7a}$$

$$\varsigma_{max} = nB/W \tag{5.7b}$$

where W is the analysis band, B is the resolution bandwidth, and n is the number of resolution cells occupied by interference. Thus the dynamic range of the of the interference distribution is given by:

$$\blacktriangle P_{max} = 10 \log_{10}(p_{max}/p_{min}) \tag{5.8}$$

The value of $\blacktriangle P_{max}$ depends upon the resolution bandwidth which is developed by Discrete Fourier Transform (DFT) processing. Figure 5-16b shows how the occupancy values depend upon B. A departure from the model is observed near P_{min} and P_{max}, and improved versions of the model have been developed. For North America, $\blacktriangle P_{max}$ is between 59 and 78 dB for $B = 9.5$ Hz. The minimum congestion occurs in the vicinity of the largest interferer and the maximum congestion occurs near the noise floor. Perry and Abraham [1988] estimate that the highest signal in a 1 MHz band will be of the order of 10 dB over P_{max}. Recognizing that the noise bandwidth (kTB) at 1 MHz is ≈ 50 dB greater than it is at 9.5 Hz [i.e., 10 $\log_{10}(10^6/9.5)$], they estimate the required dynamic range for the receiver as -50 dB + 10 dB + $\blacktriangle P_{max}$. For $\blacktriangle P_{max}$ = 70 dB, then the required dynamic range is 30 dB. Representative values for ς_{max} and ς_{min} are 0.2 and 10^{-5} respectively.

5.8.5.2 *Narrowband Models of Congestion*

Over a period of five years, scientists at the University of Manchester have made congestion measurements [Gott et al.,1982] [Laycock et al.,1988], and have developed a 1 kHz bandwidth model for HF spectral occupancy. Of some interest is the spatial correlation of congestion measurements. It has been estimated that occupancy models should admit to a significant correlation for distances of 500 km or so [Dutta and Gott, 1981].

In connection with short-range high latitude communication studies in northern Greenland, Ostergaard [1988] has concluded that CCIR 322 [1964] predicted noise levels which were much too high. This is no doubt partially a result of the increased attenuation of atmospheric noise arising from absorption zones within the high latitude region which have not been accounted for properly. Also the data sets from which CCIR 322 was constructed are sparse in the high latitude zone. Ostergaard discloses that the performance of these high latitude HF systems between 2-6 MHz is solely determined by local man-made noise and/or receiver noise. It was found, surprisingly, that the band between 2 and 5 MHz is virtually interference-free in the summertime. As a consequence it would be prudent for system designers to pay more attention to the intrinsic receiver noise figure of operational radios and man-made noise suppression techniques for high latitude communication involving low HF and NVIS approaches.

5.8.5.3 Quasi-Minimum Noise (QMN)

A convenient design criterion for HF receiving systems, especially those deployed in a shipboard environment, is embodied in the Quasi-Minimum Noise or QMN concept. The QMN is that level of shipboard noise which is expected to be exceeded the majority of time and in most geographical areas. Recall that the thermal limit for noise (i.e., kT_0) in a 3 kHz channel is - 139 dBm. For a typical HF receiver, the noise factor may be well in excess of unity. In fact, typical noise figures (in dB) may be of the order of 20 dB, which implies that the limiting noise level is of the order of -119 dBm. This is not the end of the story since external noise components will raise the noise floor.

CCIR Report 322 is the generally accepted international model for external noise contributions, but there are a number of simplified expressions which will suffice for *back-of-the-envelope* calculations. Royce [1981] and others have established the QMN level as a function of frequency *f*, based upon noise measurements performed on U.S. Navy ships:

$$\text{QMN (dB)} = 60.3 - 27.3 \log f(MHz) \tag{5.9}$$

By comparison, Report 670 [CCIR, 1982c] stipulates the worldwide minimum expected noise (figure) to be:

$$\text{WMEN (dB)} = 51.8 - 22.7 \log \text{f(MHz)} \tag{5.10}$$

Engineers from Harris Corporation [1990] have utilized Equations 5.9 and 5.10 and have identified two other expressions using the noise model resident in IONCAP (i.e., CCIR 322), representing rural and urban conditions. The IONCAP rural values are decidedly higher than either QMN and WMEN, not surprising since the CCIR model of F_{am} is representative of median behavior.

Equations 5.9 and 5.10 may be used to establish the limits on receiver sensitivity under selected conditions. Matched antennas tend to deliver much more external (atmospheric) noise power than that which is generated by the receiver subsystem. By deliberately mismatching the antenna so that, as a function of frequency, it corresponds closely to the QMN characteristic, it will be possible to ensure that the noise power delivered by the antenna (i.e., the QMN) is no greater than the self-generated noise, the latter being the ultimate limit. This is an optimum situation, since we are only interested in the SNR for establishing performance criteria. The deliberate mismatching process is carried out in such a way as to produce a flat spectrum of noise output power versus frequency. The use of small (nonresonant) antennas may also be useful in reducing the power (noise + signal) delivered to the receiving subsystem, but the mismatching may still be recommended to optimize the SNR performance and reduce noise pickup for nearby interference sources.

5.8.6 Noise and Interference Mitigation

It should be obvious that noise and interference will have a profound influence on the performance of HF systems. Interference, which may be quite severe in the industrialized sections of the world, can dominate other sources. Such domination is not totally unexpected in the upper part of the HF spectrum, and is even consistent with the CCIR 322 and 258 noise models. We anticipate that atmospheric noise will assume a dominant role in the lowest portion of the HF band, certainly at middle latitudes. Nevertheless, high latitude observations have clearly indicated that this expectation is not observed. Within the auroral zone and possible the cap region other factors act to limit atmospheric sources relative to local man-made signals. A wideband (WBHF) approach allows the receiver to discriminate naturally against narrowband sources while retaining an intrinsic processing gain. Still it has been found that a few narrow-band and relatively high-power signals may even vulgarize WBHF performance. Interference excision techniques are powerful measures which when applied in a wideband environment may enable the maximum advantage of WBHF to be achieved. To determine the advantage which will accrue from such a strategy, it is necessary to model the WBHF channel to determine what might be lost by noise (and elementary frequency band) excision. Remarkably low power level requirements, of the order of several milliwatts, may be enabled for low data rate transmission (say 100 bps) for some skywave paths through use of this technique.

One strategy for coping with interference is simply to avoid it. Indeed, spectrum occupancy meters may be monitored by operators, and along with data derived from sounders, it is possible to avoid occupied channels within the LOF to MOF frequency profile. However, operational occupancy monitors may have insufficient bandwidth resolution relative to the spectral "holes" which are sufficient for some applications. Moreover, manual determination of spectral holes may be inconsistent with the dynamic behavior of channel occupancy. We say *may* because the statement is dependent upon the category of use. Clearly the situation is different in the amateur bands than in the broadcast bands.

Dutta and Gott [1982] have explored the application of congestion information to HF operation and Doany [1981] has examined the impact of congestion on various FSK formats with arbitrary levels of diversity. If M diversity tones are transmitted (with at least a 1 kHz separation between adjacent tones), then the probability that at least two of them will be received free of interference is:

$$P(2, M) = 1 - Q^{M} - M(1-Q)\,Q^{M-1} \tag{5.11}$$

where Q is the congestion index, using the definition given in Table 5-8. The availability of at least two tones will allow a degree of frequency diversity to be accommodated. For $Q = 50\%$ and $M = 6$, then $P(2, M) = 0.9$. Since $Q = 50\%$ implies relatively heavy congestion, it is clear that a sixfold diversity will greatly improve performance under adverse conditions.

Another use which can be made of occupancy measurements is that of passive sounding. Occupancy statistics for skywave signals are strongly dependent upon ionospheric channel behavior. A continuously updated data base of channel occupancy, suitably circulated around an HF network, may provide an alternative to active sounding. No operational system has been deployed using this philosophy.

5.8.7 Effect of Noise on System Performance

Excision and avoidance strategies are not always possible. In most applications for which an HF system must coexist with the noise background, we simply recognize the error rates and attempt to minimize them by selecting appropriate diversity measures. The vulnerability to noise and interference, like multipath fading, is likewise a function of the modulation format selected.

Figure 5-17 shows how the bit error rate (BER) depends on the degree of noise impulsiveness and the presence or absence of diversity as a function of SNR. *waterfall* curves like this have been constructed for various channel conditions and modulation formats. In this case the fading channel was characterized by a Rayleigh distribution and the modulation was non-coherent FSK.

It is noteworthy that noise limits the performance of HF systems for the lower values of SNR, but propagation effects (such as multipath) cause the theoretical improvement in BER for high values of SNR to flatten out. In short, there is a BER floor below which it is not possible to descend since symbol decisions which are corrupted by intersymbol interference and selective fading may be little influenced by increases in the *wanted* signal level. Figure 5-18 bears this out.

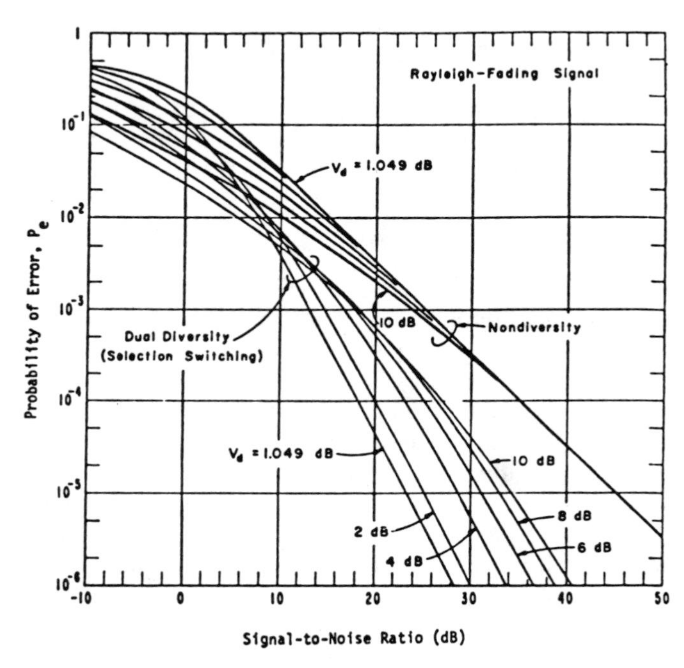

Fig.5-17. Probability of bit error for slow flat Rayleigh fading signal for a NCFSK system, for dual diversity and nondiversity reception (Spaulding [1976]).

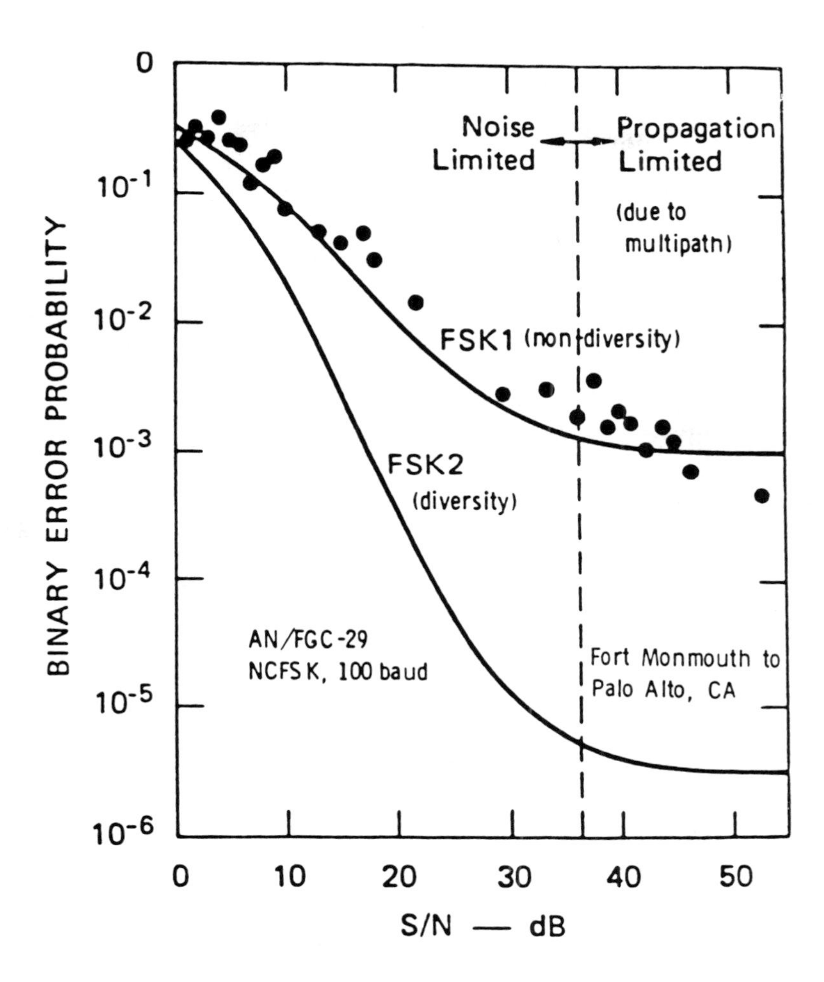

Fig.5-18. System performance as a function of signal-to-noise ratio. (From Hagn [1988].)

5.9 ANTENNA CONSIDERATIONS IN PREDICTION

We have previously discussed the importance of ground constant data in the estimation of antenna performance (viz., Chapter 4: Section 4.4.5). In Section 4.4.6 we briefly examine siting considerations and antenna section. In this section we shall sketch some of the antenna considerations which are important in the prediction process. The reader is referred to CCIR Report 891-1 [1986g] for a rather complete summary.

5.9.1 On Antenna Directivity and Gain

First, it is important to realize that the directivity of an antenna is the same whether it is transmitting or receiving. This does not mean that the same antenna structure will accommodate the transmit and receive processes identically. The ultimate feature of interest for communications is the power gain which is the product of the antenna directivity and the antenna efficiency. Still directivity is a fundamental property of an antenna and is quite important in an analysis of system performance.

The electrical current distribution in an antenna may be used to deduce the antenna directivity (i.e., its pattern). Unfortunately it is not quite that simple. Nearby ground and objects will influence the ultimate pattern. The complete solution requires that the direct radiation from the antenna be combined with the reflected radiation. Thus, some way of accounting for ground reflection coefficient is needed, along with an accounting for the contribution of nearby scatterers (other than the antenna components). Polarization is an important parameter in the analysis. For horizontal polarization, the voltage reflection coefficient for flat perfectly conducting ground is -1 regardless of the angle of incidence. This simplifies the computation of elevation patterns since the horizontal current element has a perfect image corresponding to current flow in the opposite direction. Voltage reinforcements and rarefactions arise as a function of elevation angle with the reinforcements occurring whenever the path difference between the direct and reflected rays are different by ½ wavelength. For an imperfect conductor, the antenna pattern of a horizontal antenna is modified only slightly. For all practical purposes, the use of horizontal antennas allows the analyst to ignore siting factors in the consideration of antenna directivity. This is especially true for situations in which low launch angles are required.

For vertical polarization, associated with a vertical current element, one finds that the complexity of the situation increases. With regards to the reflection coefficient, we note that it is +1 for highly conducting ground such as over sea. Because of this different algebraic sign (as compared to the horizontal case), reinforcement of direct and reflected waves should actually occur at the lowest elevation angles where the path differences are negligible. That is the good news. The bad news is that the phase for the complex voltage reflection coefficient switches to 180 degrees at the so-called Brewster angle at certain elevation angles for poorly (or even finitely-) conducting ground. At the Brewster angle, which occurs at angles of incidence of ≥ 89 degrees over sea and about 75 degrees over land (i.e., Report 891-1, [CCIR, 1986g], the reflection coefficient diminishes significantly. Since the angle of incidence is the complement of the antenna elevation, the use of vertical antennas over land will impact the performance at precisely the launch angles which may be required for long-haul communication. The use of a ground screen (comprised of a set of radials or a mat) may mitigate this negative consequence however. Even so, the length of radials may need to be 50 wavelengths or so to be useful in low elevation applications [Anderson, 1965]. At 20 MHz, the wavelength is 15 meters, so the radial length would need to be 750 meters. This is hardly a practical solution for tactical long-haul use, but permissible for fixed site applications.

5.9.2 Active Antennas

Antenna structures used for reception are typically required to be more flexible in terms of the following: transportability, setup and teardown times, physical size, and frequency agility. For the discipline of communication, the primary role of a transmitting antenna is to deliver adequate power to the input of the propagation medium subchannel. The operational requirements will define the directivity and gain requirements. Naturally, the capabilities (and properties) of target receivers must be accounted for, but the design of transmitting antennas need not consider the problem of noise and interference in the vicinity of the transmitter. The typical transmitter system may includes a sizable antenna structure to optimize gain in specified directions. Receive antennas, on the other hand, must be optimized to receive signals in preference to noise, other user interference, and jamming signals. The requirement for a large aperture may be relaxed, since system performance is determined by the SNR. This results in the possibility that small and relatively inefficient antennas may be used to achieve adequate performance.

Tactical deployments often necessitate the use of small antennas for a variety of reasons: limitations in available real-estate, ease of deployment, and the requirement to minimize physical cross section. Active antennas are well suited for such deployments. By use of a radiofrequency (rf) amplifier coupled to a simple antenna such as a monopole or dipole, it is possible to achieve wideband operation having a fixed output impedance even if the physical dimensions are *short*. The advantages are clear. On the other hand, linearity requirements of the active circuitry are important considerations, since intermodulation products caused by the action of strong interfering signals will degrade the performance of such approaches.

From time to time workers are motivated to embark upon comprehensive measurement campaigns to obtain unique data sets or to correct for inadequacies in existing sets. Data sources maintained by the CCIR have significant shortcomings limiting their usefulness, including: lack of geographical coverage, uncertain system specifications, and target misidentification. An example is CCIR Data Bank D, the international source of all available calibrated HF data. As a result of this, the CCIR has proposed an HF field strength measurement campaign to obtain high quality data. The project, which is still in the planning phase, would require the establishment of a number of computer-controlled transmitters (up to 9) and receivers having known characteristics. The receivers would exploit short vertical active antennas.

5.9.3 Publications and Computer Programs

Publications available through the CCIR include the *Handbook on High Frequency Directional Antennae* [CCIR, 1965], the *CCIR Book of Antenna Diagrams* [1978] and the *CCIR Atlas of Antenna Diagrams* [1984]. Other references are given in Section 4.4.6 of Chapter 4. Method-of-Moments (MOM) techniques are currently in vogue and are providing the analyst with a fuller understanding of the complete antenna problem. Programs such as the Numerical Electromagnetics Code (NEC) are being used for solution of practical problems but are suitable principally for main-frame computers. Nevertheless some limited capability microcomputer versions of NEC are being offered (e.g., MININEC and MN as mentioned in Section 4.4.6). For those interested in additional applications which may be analyzed on a microcomputer, several computer programs are being made available through the CCIR Secretariat. program names include: HFARRAYS, HFRHOMBS, HFMULSLW, HFDUASLW, and HFDUASLW1. See Resolution 63.2 [CCIR, 1986i] and Circulars 22 [ITU, 1984a], 23 [ITU, 1984b], and 95 [ITU,1986a,b]) for more details.

Programs such as HFMUFES and IONCAP contain antenna patterns which are utilized in the HF performance calculations. The antenna subroutines in IONCAP include the ITSA-1 set due to Lucas and Haydon [1966] and an optional ITS-78 set due to Barghausen et al. [1969]. Using information extracted from ITS Report No.74 [Ma and Walters, 1969], the IONCAP "theory manual" describes the evaluation of power gain, radiation resistance, and antenna efficiency for antennas which are included in the program. Methods used are somewhat approximate but are useful in most practical applications.

5.10 PREDICTION PROGRAM DELIVERABLES

The output from the most recent ITS-Boulder programs (viz. HFMUFES and IONCAP) are determined by selection of specific methods. IONCAP methods are listed in Section 5.11.3. Some of the more important outputs associated with any main-frame prediction model include those in Table 5-9. Desirable outputs will include data files which are easily re-transmitted, edited, archived, and disseminated. The dissemination recipients could include "local users" operating in a workstation or PC environment (connected via LANs) and remote users connected by common user media (phone lines/modems). Graphical output is also needed to augment line printer output. Graphical output is not available in officially-distributed versions of the ITS programs. There is no fundamental reason for this; there was simply no overwhelming requirement for graphics at the time the ITS codes were developed. Graphical output displays are available in some codes such as HFBC84 as well as many of the small microcomputer programs (see Section 5.12).

TABLE 5-9. PREDICTION MODEL CATEGORIES

(1) Intermediate Ionospheric Data
- Ionograms and N_e Profiles
- foE, foEs, foF1, foF2
- Ionospheric Layer Heights
- Layer Semithicknesses

(2) Propagation Support Information
- MUF, LUF, FOT, fEs, Multipath,etc.

(3) Independent Mode and Path Analyses

(4) Patterns of Selected Antennas

(5) System Performance Metric
- SNR, Signal Level
- Reliability
- Service Probability
- Coverage Areas

5.11 RELIABILITY: Basic Definitions

The performance of an HF radio system depends upon field strength, noise, competing interference, and other factors which may be functions of the system configuration, mode of operation, and the type of service indicated. Everything else being equal, performance in transmission of facsimile will be far superior to the performance in transmission of high speed data. Given the same category of communication service, performance will be generally degraded by reduction in EIRP, by increased background noise, or by the presence of interference. Reliability is a notion which indicates to the system engineer a probability that the system will perform its function under a given set of circumstances. Some pertinent definitions are provided in Table 5-10.

5.11.1 Basic Mode Reliability

The mode reliability, denoted by the term R_m, is given by:

$$R_m = q\,P_{SNR} \tag{5.12}$$

where q is the mode availability, and P_{SNR} is a conditional probability that the required signal-to-noise ratio (SNR) is exceeded, under the condition that the mode exists. Equation 5.12 presumes that mode presence is independent of signal strength. This approach is useful for computer methods which compute the median field strength for a specified mode of propagation under the condition that the reflection (or refraction) condition actually exists (i.e., CCIR method 252-2). Methods CCIR 252-2(Supplement) and 894 (discussed later on) compute median field strength for all time irrespective of a specified mode availability.

TABLE 5-10
DEFINITIONS FOR RELIABILITY AND RELATED TERMS

1. *Reliability* is the probability that a specified performance is achieved.

2. *Circuit Reliability* is the probability that for a given circuit and single frequency a specified performance is achieved.

3. *Reception Reliability* is the probability that for a given circuit and for all transmitted frequencies a specified performance is achieved.

4. *Service Reliability* is the probability that for a specified percentage of the service area and for all transmitted frequencies a specified performance will be achieved.

5. *Mode Reliability* is the probability that for a single circuit and a single frequency a specified performance will be achieved by a single mode.

6. *Mode Availability* is the probability that for a single circuit and a single frequency a single mode can propagate by ionospheric refraction alone.

7. *Mode Performance Achievement* is the probability that for a single circuit, single frequency, and a single mode (which propagates by ionospheric refraction alone) a given performance is achieved.

Thus, the mode reliability consistent with these latter models is just the fraction of time that the SNR exceeds the required value. An expression for P_{SNR} is given in Report 892-1 [CCIR, 1986h] based upon work by Bradley and Bedford [1976].

5.11.2 Circuit Reliability

As indicated in CCIR Report 892-1 [1986h], Liu and Bradley [1985] have developed a general expression for circuit reliability for the general situation corresponding to an arbitrary number of contributing modes. A practical situation involves only two contributing modes, and the resulting expression for reliability will involve expressions for mode availability and mode performance achievement (definitions 6 and 7 in Table 5-10). We have:

$$R_c = q_1 P_1 + q_2 P_2 + q_{12} P_{12} \tag{5.13}$$

where R_c is the circuit reliability; q_1 and P_1 are the mode availability and mode performance achievement for the case when mode 1 is present; q_2 and P_2 are the mode availability and mode performance achievement for the case when mode 2 is present; q_{12} is the probability that modes 1 and 2 are present simultaneously; and P_{12} is the probability that the combination of modes 1 and 2 will lead to a signal-to-noise in excess of some required level. Even though Equation 5.13 is limited to two contributing modes, its evaluation is not necessarily trivial. Other methods for computing reliability are listed in Table 5-11.

TABLE 5-11

VARIOUS RELIABILITY METHODS IN USE

NAME of METHOD or "SYSTEM" USE	REFERENCE
IONCAP	Teters et al.[1983]
HFMUFES	Barghausen et al.[1969]
Liu-Bradley	Liu and Bradley [1985]
CRC-Canada	Petrie [1981]
CCIR Method	CCIR Rpt.892-1 [1986h]
Maslin Method	Maslin [1978]
Chernov Method	Chernov [1969]
HFBC Method	CCIR Rpt.892-1 [1986h]

It is of interest to look at some special cases of an approximate method developed by Liu and Bradley [1985] for which correlation between two contributing mode MUFs may be taken into account, whereas correlation between mode SNRs is ignored. The method presumes that the basic MUF is normally distributed and the SNR is log-normal (i.e., the SNR in dB is normally distributed). Taking the correlation between two modes as c_{12}, and $Q_1 = q_1 + q_{12}$, $Q_2 = q_2 + q_{12}$ where q_1, q_{12}, and q_{12} were defined previously, we have:

$$R_c = Q_1 P_1 + Q_2 P_2 \{1 - [c_{12}^2 + (1 - c_{12}^2) Q_1]] P_1\} \tag{5.14}$$

where $Q_1 \geq Q_2$. For $c_{12} = 0$ and $c_{12} = 1$, obvious simplifications in Equation 5.14 will result. For purposes of planning, one may take E and F1 modes to be fully correlated (i.e., $c_{12} = 1$), E and F2 modes to be uncorrelated, and F1 and F2 modes to be uncorrelated. Also correlation between dual modes from the same layer are taken to be highly correlated but not necessarily unity. For example experience has shown that two F2 modes have a cross correlation coefficient of 0.8 for purpose of reliability calculations. The effect of vanishing correlation between two contributing modes is to limit the maximum reliability which may be achieved.

IONCAP employs a simplistic scheme for estimating circuit reliability. The method involves combining the signal power from all modes under the presumption that the relative phase relationships between the contributing modes are random. The SNR is taken to be the difference between the means for both signal and noise (in dB), while the variance of SNR is simply the sum of the respective variances. The circuit reliability is taken to be the fraction of days (over a month) that the SNR ≥ the required value. Clearly, if a specified mode does not propagate efficiently, the algorithm automatically disables any significant contribution of that mode to the overall reliability. There is no need to account for mode support explicitly. It is noteworthy that the ionospheric variability and mode support is accounted for implicitly (in terms of SNR variability which is part of the IONCAP model). As expected, comparisons of the various methods for circuit reliability show some differences.

5.12 SAMPLE OUTPUT FROM IONCAP

In this section several output options from IONCAP will be exhibited to illustrate products associated with climatological median models. Table 5-12 gives the number of methods available in IONCAP versions 78.03 and 85.04.

The descriptors of IONCAP methods listed in Table 5-12 are fairly self explanatory. The term REL corresponds to reliability and ANG refers to the elevation angle associated with the specified dominant mode. Methods 1 and 2 allow the user to see the underlying ionospheric data which is used in the other methods. A number of graphical and tabular methods which provide various combinations of propagation data such as HPF, MUF, LUF, FOT, ANG, and foEs are available. However, system performance methods set mainframe models apart from microcomputer models which compute only a limited set of parameters, typically only the propagation parameters and possibly a measure of signal strength. Popular methods in Table 5-12 include numbers 17,20 and 25. A complete system performance is accommodated in method 20 and a condensed version of this is found in method 17. Method 25 allows the analyst to examine system effects mode-by-mode. The reliability vs MUF table found in method 24 is also quite useful, while the antenna methods 13-15 are primarily available for reference purposes.

Figures 5-19 and 5-20 show output from methods 10 (MUF-FOT-ANG) and 24 (Reliability table). A truncated version of method 20 output is given in Figure 5-21. Method 17 output is given in Figure 5-22. These curves were taken directly from the NTIA Report 83-127 describing IONCAP [Teters et al.,1983]. The path length for the Boulder to St. Louis link examined is 1301 km and the Universal Time is 1900 hours. The month selected was January and the sunspot number was set at 100. Other input parameters are printed within the heading lines for each method.

In the computation of reliability, the user must specify a number of system parameters as well as required SNR for a specified modulation format and grade of service. These data are found in various communication handbooks. Tables 4 and 5 in NTIA Report 83-127 give data for an assortment of conditions. The reader should not be surprised to see rather enormous values in the tables just referenced since they reflect the required SNR for a signal in the occupied bandwidth versus noise in a 1 Hz bandwidth. To compare signal and noise in a common bandwidth, one must subtract the system (i.e., noise) bandwidth in dB from the tabulated value.

TABLE 5-12. LISTING OF IONCAP METHODS [49]

METHOD No.	METHOD DESCRIPTION
1	Ionospheric parameters
2	Ionograms
3	MUF-FOT (using nomogram method)
4	MUF-FOT graph
5	HPF-MUF-FOT graph
6	MUF-FOT-Es graph
7	FOT-MUF table (full ionosphere)
8	MUF-FOT graph
9	HPF-MUF-FOT graph
10	MUF-FOT-ANG graph
11	MUF-FOT-Es graph
12	MUF by magnetic indices, K
13	Transmitter antenna pattern
14	Receiver antenna pattern
15	Both transmitter and receiver patterns
16	System performance
17	Condensed system performance, reliability
18	Condensed system performance, service probability
19	Propagation path geometry
20	Complete system performance
21	Forced long path model
22	Forced short path model
23	User selected output lines
24	MUF-REL table
25	All modes table
26	MUF-LUF-FOT table (using a nomogram)
27	FOT-LUF graph
28	MUF-FOT-LUF graph
29	MUF-LUF graph
30	Create binary file of variables

---

49. The IONCAP documentation consists of a user's guide [Teters et al.,1983] which outlines the the various methods which may be selected. A government report describing the underlying methodology is yet to be published officially. Figures 5-19 through 5-22 are taken directly from NTIA Report 83-127 *Estimating the Performance of Telecommunication Systems Using the Ionospheric Transmission Channel: Ionospheric Communications Analysis and Predictions Program (IONCAP) User's Manual.*

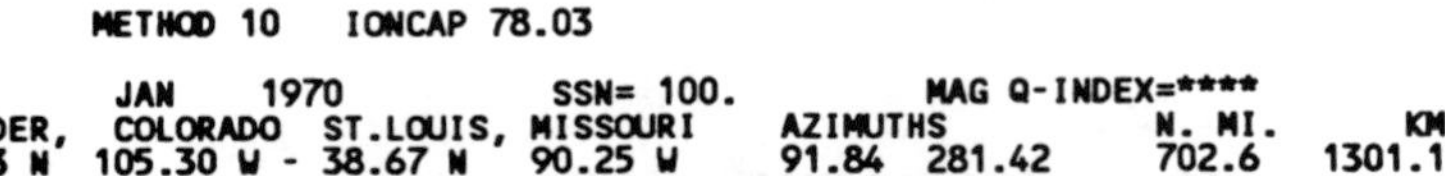

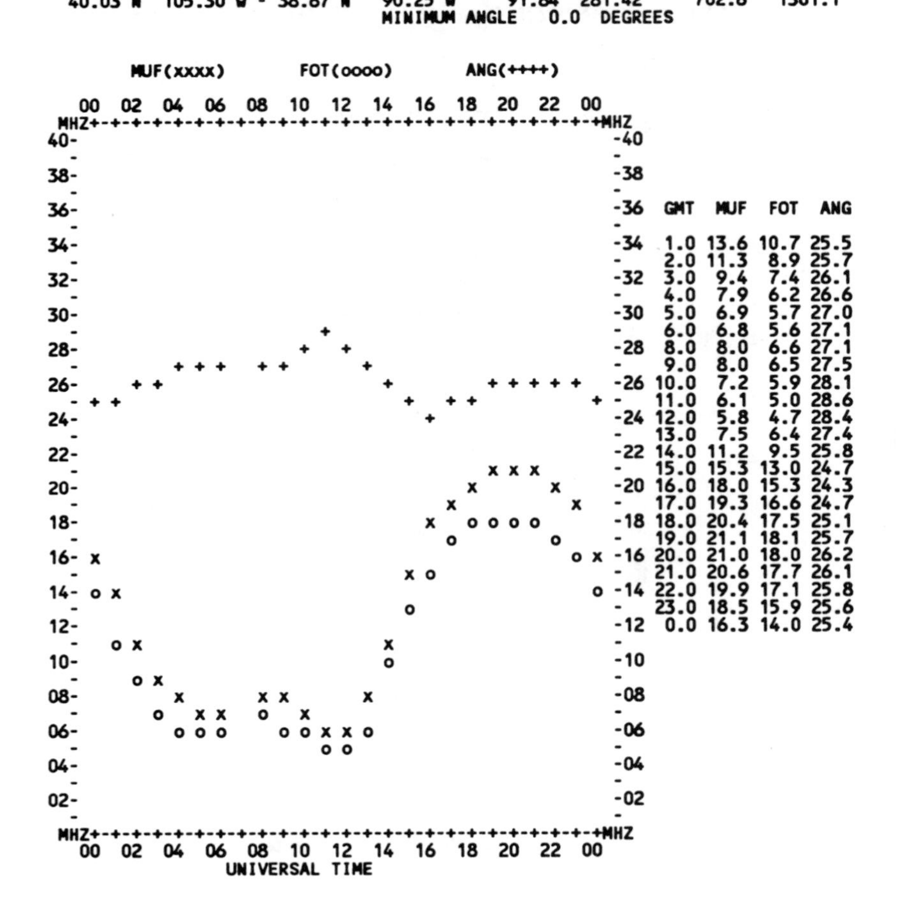

GMT	MUF	FOT	ANG
1.0	13.6	10.7	25.5
2.0	11.3	8.9	25.7
3.0	9.4	7.4	26.1
4.0	7.9	6.2	26.6
5.0	6.9	5.7	27.0
6.0	6.8	5.6	27.1
8.0	8.0	6.6	27.1
9.0	8.0	6.5	27.5
10.0	7.2	5.9	28.1
11.0	6.1	5.0	28.6
12.0	5.8	4.7	28.4
13.0	7.5	6.4	27.4
14.0	11.2	9.5	25.8
15.0	15.3	13.0	24.7
16.0	18.0	15.3	24.3
17.0	19.3	16.6	24.7
18.0	20.4	17.5	25.1
19.0	21.1	18.1	25.7
20.0	21.0	18.0	26.2
21.0	20.6	17.7	26.1
22.0	19.9	17.1	25.8
23.0	18.5	15.9	25.6
0.0	16.3	14.0	25.4

Fig.5-19. Method 10 in IONCAP: (MUF-FOT-ANG). In this method the predicted MUF and FOT (in MHz) are displayed, one value for each hour. The FOT is taken to be 85 % of the MUF. Also plotted is the elevation angle (in degrees) at which the high and low rays converge (i.e., the junction frequency or MUF). It should be noted that the time (i.e., x-axis) is Universal. Thus, for the path selected (i.e., Boulder to St. Louis), sunrise is near the middle of the plot at approximately 1200 hours UT. The model predicts a linear drop in the elevation angle from about 1100 UT until 1600 UT, while the MUF is increasing rapidly. This prediction, based upon empirical data, is consistent with reestablishment of solar control at ionospheric sunrise, with the dominance of photoionization over loss and diffusion terms. This process, because of solar illumination alone, causes the effective F2 layer height to plunge dramatically, after which an equilibrium is reached. (Figure taken from NTIA Report 83-127, [Teters et al.,1983].)

```
                         METHOD 24    IONCAP 78.03

          JAN     1970              SSN = 100.
BOULDER,COLORADO TO ST. LOUIS,MO.               AZIMUTHS              N. MI.      KM
40.03 N  105.30 W - 38.67 N   90.25 W     91.84  281.42         702.6    1301.1
                                        MINIMUM ANGLE     .0  DEGREES
ITS- 1 ANTENNA PACKAGE
XMTR    2.0  TO  30.0  VER MONOPOLE H     0.00 L     -.50 A     0.0  OFF AZ     0.0
RCVR    2.0  TO  30.0  VER MONOPOLE H     0.00 L     -.25 A     0.0  OFF AZ     0.0
POWER =  30.000 KW  3 MHZ NOISE = -150.0 DBW   REQ. REL = .90  REQ. SNR = 55.0

                         FREQUENCY / RELIABILITY
```

GMT	LMT	MUF	2.0	3.0	5.0	7.5	10.0	12.5	15.0	17.5	20.0	25.0	30.0	MUF
8.0	1.0	8.0	.99	1.00	1.00	1.00	.88	.31	.02	.00	.00	.00	.00	1.00
9.0	2.0	8.0	.99	1.00	1.00	1.00	.59	.11	.00	.00	.00	.00	.00	.99
10.0	3.0	7.2	.99	1.00	1.00	.90	.34	.04	.00	.00	.00	.00	.00	.94
11.0	4.0	6.1	1.00	1.00	1.00	.64	.29	.02	.00	.00	.00	.00	.00	.91
12.0	5.0	5.8	.99	1.00	1.00	.52	.31	.03	.00	.00	.00	.00	.00	.91
13.0	6.0	7.5	.79	1.00	1.00	.97	.35	.07	.00	.00	.00	.00	.00	.97
14.0	7.0	11.2	.01	.18	.99	1.00	1.00	.59	.01	.00	.00	.00	.00	.96
15.0	8.0	15.3	.00	.00	.35	.99	1.00	1.00	.99	.50	.03	.00	.00	.96
16.0	9.0	18.0	.00	.00	.07	.99	1.00	1.00	1.00	1.00	.67	.00	.00	[illegible]
17.0	10.0	19.3	.00	.00	.01	.37	.95	1.00	1.00	1.00	.88	.17	.00	.99
18.0	11.0	20.4	.00	.00	.00	.20	.91	1.00	1.00	1.00	1.00	.42	.00	.99
19.0	12.0	21.1	.00	.00	.00	.14	1.00	1.00	1.00	1.00	1.00	.57	.01	.95
20.0	13.0	21.0	.00	.00	.00	.96	1.00	1.00	1.00	1.00	1.00	.55	.01	1.00
21.0	14.0	20.6	.00	.00	.06	.98	1.00	1.00	1.00	1.00	1.00	.45	.00	.99
22.0	15.0	19.9	.00	.00	.38	.97	.99	.99	1.00	1.00	.96	.28	.00	.97
23.0	16.0	18.5	.03	.22	.94	.99	.99	1.00	1.00	.99	.76	.06	.00	.96
.0	17.0	16.3	.35	.80	.98	.99	1.00	1.00	1.00	.75	.35	.00	.00	.96
1.0	18.0	13.6	.63	.94	.99	1.00	1.00	1.00	.90	.65	.34	.01	.00	.99
2.0	19.0	11.3	.77	.98	1.00	1.00	1.00	.89	.57	.21	.02	.00	.00	.99
3.0	20.0	9.4	.93	.99	1.00	1.00	.95	.61	.18	.01	.00	.00	.00	.98
4.0	21.0	7.9	.99	1.00	1.00	1.00	.80	.34	.10	.00	.00	.00	.00	.99
5.0	22.0	6.9	.99	1.00	1.00	.96	.55	.34	.13	.00	.00	.00	.00	1.00
6.0	23.0	6.8	.99	1.00	1.00	.96	.53	.34	.11	.00	.00	.00	.00	1.00
7.0	24.0	7.4	.98	1.00	1.00	.99	.71	.30	.06	.00	.00	.00	.00	1.00

Fig.5-20. Method 24 in IONCAP: Reliability Table. The method by which reliability is estimated is described in an unpublished document *Estimating the Performance of Telecommunication Systems Using the Ionospheric Transmission Channel, Vol.I: Techniques for Analyzing Ionospheric Effects on HF Systems,* by J.L. Lloyd, G.W. Haydon, D.L. Lucas, and L.R. Teters of the Institute for Telecommunication Sciences, Boulder, Colorado (circa, 1978). In essence, the reliability is an estimate of the percentage of days within the month that the available SNR is expected to equal or to exceed a specified requirement. This means that the probability of achieving the required SNR must be computed. A two-sided normal distribution is defined with a mean determined by the predicted mean SNR and with variance deduced from a sum of the variances of the combined noise distribution and the signal distribution. This presumes independence of the noise and signal distributions. Values of parameters just below the heading were default values used by Lloyd and his coworkers in the unpublished *IONCAP theory manual* but are also used in the *IONCAP users manual* [Teters et al.,1983].

```
                              METHOD 20   IONCAP 78.03

         JAN    1970              SSN = 100.
BOULDER,COLORADO TO ST. LOUIS,MO.               AZIMUTHS            N. MI.       KM
40.03 N  105.30 W - 38.67 N   90.25 W        91.84  281.42       702.6   1301.1
                              MINIMUM ANGLE    .0  DEGREES
ITS- 1 ANTENNA PACKAGE
XMTR   2.0  TO  30.0  VER MONOPOLE H    0.00 L    -.50 A    0.0  OFF AZ    0.0
RCVR   2.0  TO  30.0  VER MONOPOLE H    0.00 L    -.25 A    0.0  OFF AZ    0.0
POWER =  30.000 KW  3 MHZ NOISE = -150.0 DBW   REQ. REL = .90  REQ. SNR = 55.0
MULTIPATH POWER TOLERANCE = 10.0 DB   MULTIPATH DELAY TOLERANCE =   .850 MS
```

UT	MUF												
19.0	21.1	2.0	3.0	5.0	7.5	10.0	12.5	15.0	17.5	20.0	25.0	30.0	FREQ
	1F2	1 E	1 E	1ES	2F2	1F2	1F2	1F2	1F2	1F2	1F2	1F2	MODE
	25.7	4.0	4.5	6.6	39.6	22.8	20.0	18.2	18.5	20.1	25.7	25.7	ANGLE
	5.1	4.4	4.4	4.4	5.9	4.9	4.8	4.7	4.8	4.8	5.1	5.1	DELAY
	365.	80.	85.	110.	290.	321.	281.	257.	261.	283.	365.	365.	V HITE
	.50	1.00	1.00	1.00	1.00	1.00	1.00	1.00	.94	.68	.04	.00	F DAYS
	140.	234.	229.	183.	161.	139.	137.	136.	135.	135.	162.	214.	LOSS
	41.	-66.	-57.	-8.	16.	39.	41.	44.	45.	48.	21.	-29.	DBU
	-95	-189	-183	-137	-114	-90	-90	-89	-90	-88	-117	-168	S DBW
	-173	-145	-150	-156	-161	-164	-165	-167	-170	-172	-175	-178	N DBW
	78.	-44.	-34.	19.	47.	73.	74.	77.	80.	84.	59.	9.	SNR
	-6.	106.	96.	44.	16.	-9.	-12.	-14.	-18.	-19.	22.	72.	RPWRG
	.95	.00	.00	.00	.14	1.00	1.00	1.00	1.00	1.00	.57	.01	REL
	.00	.00	.00	.00	.01	.00	.00	.00	.00	.00	.00	.00	MPROB
	.58	.00	.00	.00	.10	.70	.77	.83	.88	.87	.24	.01	S PRB
	16.	3.	3.	3.	3.	5.	3.	3.	3.	7.	25.	25.	SIG LW
	7.	2.	2.	2.	3.	5.	3.	2.	2.	5.	20.	25.	SIG UP
	41.	-66.	-57.	-8.	16.	39.	41.	44.	45.	48.	21.	-24.	VHFDBU
	16.	3.	3.	3.	3.	5.	3.	3.	3.	7.	25.	6.	VHF LW
	7.	2.	2.	2.	3.	5.	3.	2.	2.	5.	20.	8.	VHF UP
	F	E	E	ES	F2	F2	F2	F2	F2	F2	F	F	VHFMOD
	18.	7.	7.	8.	8.	8.	7.	7.	7.	10.	26.	26.	SNR LW
	11.	9.	9.	9.	9.	10.	9.	9.	9.	10.	22.	26.	SNR UP
7.0	7.4	2.0	3.0	5.0	7.5	10.0	12.5	15.0	17.5	20.0	25.0	30.0	FREQ
	1F2	1F2	1F2	1F2	1F2	1F2	1ES	1ES	1F2	1F2	1F2	1F2	MODE
	25.9	21.4	19.8	20.0	27.0	27.0	6.6	6.6	27.0	27.0	27.0	27.0	ANGLE
	5.1	4.9	4.8	4.8	5.2	5.2	4.4	4.4	5.2	5.2	5.2	5.2	DELAY
	368.	301.	278.	282.	387.	387.	110.	110.	387.	387.	387.	387.	V HITE
	.50	1.00	1.00	.99	.47	.05	.14	.07	.00	.00	.00	.00	F DAYS
	123.	120.	119.	118.	125.	146.	168.	192.	224.	225.	227.	229.	LOSS
	53.	46.	47.	52.	47.	31.	15.	-7.	-45.	-45.	-45.	-45.	DBU
	-73	-72	-72	-72	-79	-100	-122	-147	-178	-180	-182	-184	S DBW
	-157	-140	-145	-151	-157	-163	-167	-169	-171	-173	-175	-178	N DBW
	83.	68.	72.	78.	77.	62.	44.	23.	-8.	-8.	-7.	-6.	SNR
	-17.	-5.	-10.	-17.	-10.	9.	25.	57.	70.	70.	69.	68.	RPWRG
	1.00	.98	1.00	1.00	.99	.71	.30	.06	.00	.00	.00	.00	REL
	.00	.78	.94	.00	.00	.00	.00	.00	.00	.00	.00	.00	MPROB
	.77	.55	.69	.85	.69	.30	.12	.04	.00	.00	.00	.00	S PRB
	10.	3.	1.	2.	10.	15.	13.	24.	3.	3.	3.	3.	SIG LW
	5.	3.	3.	1.	7.	17.	25.	25.	8.	1.	1.	1.	SIG UP
	53.	46.	47.	52.	47.	31.	25.	25.	25.	25.	25.	26.	VHFDBU
	10.	3.	1.	2.	10.	15.	6.	6.	6.	6.	6.	6.	VHF LW
	5.	3.	3.	1.	7.	17.	8.	8.	8.	8.	8.	8.	VHF UP
	F	F2	F2	F2	F	F	F	F	F	F	F	F	VHFMOD
	12.	8.	7.	7.	12.	17.	15.	25.	7.	7.	7.	7.	SNR LW
	9.	10.	9.	8.	10.	19.	26.	26.	12.	9.	9.	9.	SNR UP

Fig.5-21. Method 20 in IONCAP: Complete System Performance. The entries are fully described by Teters et al.[1983]. Lines 17-20 were included by Teters et al. as experimental VHF *above-the-MUF* parameters.

```
                              METHOD 17    IONCAP 78.03

         JAN     1970         SSN = 100.
BOULDER,COLORADO TO ST. LOUIS,MO.          AZIMUTHS           N. MI.     KM
40.03 N  105.30 W - 38.67 N   90.25 W      91.84  281.42      702.6    1301.1
                              MINIMUM ANGLE     .0  DEGREES
ITS- 1 ANTENNA PACKAGE
XMTR    2.0  TO  30.0  VER MONOPOLE H    0.00 L    -.50 A    0.0  OFF AZ    0.0
RCVR    2.0  TO  30.0  VER MONOPOLE H    0.00 L    -.25 A    0.0  OFF AZ    0.0
POWER =  30.000 KW  3 MHZ NOISE = -150.0 DBW   REQ. REL = .90  REQ. SNR = 55.0
MULTIPATH POWER TOLERANCE = 10.0 DB   MULTIPATH DELAY TOLERANCE =   .850 MS

 UT  MUF

19.0 21.1   2.0   3.0   5.0   7.5  10.0  12.5  15.0  17.5  20.0  25.0  30.0 FREQ
      1F2   1 E   1 E   1ES   2F2   1F2   1F2   1F2   1F2   1F2   1F2   1F2 MODE
     25.7   4.0   4.5   6.6  39.6  22.8  20.0  18.2  18.5  20.1  25.7  25.7 ANGLE
      .50  1.00  1.00  1.00  1.00  1.00  1.00  1.00   .94   .68   .04   .00 F DAYS
      41.  -66.  -57.   -8.   16.   39.   41.   44.   45.   48.   21.  -29. DBU
      78.  -44.  -34.   19.   47.   73.   74.   77.   80.   84.   59.    9. SNR
      .95   .00   .00   .00   .14  1.00  1.00  1.00  1.00  1.00   .57   .01 REL

 7.0  7.4   2.0   3.0   5.0   7.5  10.0  12.5  15.0  17.5  20.0  25.0  30.0 FREQ
      1F2   1F2   1F2   1F2   1F2   1F2   1ES   1ES   1F2   1F2   1F2   1F2 MODE
     25.9  21.4  19.8  20.0  27.0  27.0   6.6   6.6  27.0  27.0  27.0  27.0 ANGLE
      .50  1.00  1.00   .99   .47   .05   .14   .07   .00   .00   .00   .00 F DAYS
      53.   46.   47.   52.   47.   31.   15.   -7.  -45.  -45.  -45.  -45. DBU
      83.   68.   72.   78.   77.   62.   44.   23.   -8.   -8.   -7.   -6. SNR
     1.00   .98  1.00  1.00   .99   .71   .30   .06   .00   .00   .00   .00 REL
```

Fig.5-22. Method 17 in IONCAP: Condensed System Performance. (From Teters et al.[1983].)

5.13 SIMPLE FIELD STRENGTH AND MUF PREDICTION METHODS

5.13.1 Introduction

There are several programs available for the estimation of HF system performance (including various forms of reliability), and these programs typically take factors such as antenna type, noise and interference, fade allowances, and basic ionospheric properties into account. The ITS developments HFMUFES and IONCAP are well-known *main-frame* programs of a general nature. models of atmospheric noise (Report 322-3 [CCIR, 1986d]), man-made noise (Report 258-4 [CCIR, 1986e]), and ionospheric coefficients (Report 340-5 [CCIR, 1986p] are utilized in these main-frame codes, in one form or another. Other CCIR reports contain information which is important in system performance calculations, and they include Report 266-6 [CCIR, 1986q] on fading and fade allowances, and Report 892 [CCIR, 1986h] on Reliability. Two CCIR methods have been developed for estimating skywave field strength and transmission loss between frequencies of 2 and 30 MHz. These are described in Report 252-2 [CCIR, 1970] and its supplement [CCIR, 1982a]; and computer codes are also available through the CCIR Secretariat. There are a host of computer programs which compute the LUF and the MUF, and a sizable number of codes provide estimates of field strength. Many of these codes are mentioned in the section dealing with microcomputer methods. More than 20 field strength programs have been examined by the CCIR Interim Working Party (IWP) 6/12 in the process of the development of a simple method in preparation for the WARC High Frequency Broadcasting Conference (HFBC) held in 1984.

In the process of foregoing activity by IWP 6/12, a simplified method for estimation of the MUF and field strength was developed and documented as CCIR Report 894-1 [1986a]. It was based upon methods found in CCIR 252-2 and in a second model used by the Federal Republic of Germany (referred to as the FTZ field strength model). The former model (i.e., CCIR 252-2) is used for paths less than 7000 km in length, the FTZ model is used for paths in excess of 9000 km, and an interpolation procedure is employed if the path is between 7000 and 9000 km. The methodology for short- and long-path field strength estimation involved in this simple procedure is similar (but not a replication) of the procedure prescribed for use by IONCAP. CCIR 894-1 describes this simple field strength prediction method, and a companion computer code, REP 894, is available through the CCIR Secretariat. Another model, with certain modifications, HFBC84, was developed and approved at the HF Broadcast Conference. These programs are useful since they are simple and readily admit to the development of area coverage diagrams.

It is instructive to outline the procedures followed in CCIR Rpt.894-1 since many of the previously-outlined phenomena (and propagation *rules*) are involved, and an exposé will serve as a review. The reader is reminded, however that the method is a simplified one, and is but one of a number of approaches in current use. It is not the best model, nor is it the last. Even so, it is sanctioned by the CCIR as a fast method for operational use. Accordingly, given the relatively little space required to discuss 894-1, it is convenient that we give it some attention for purposes of illustration.

5.13.2 Determination of the MUF using CCIR 894-1 Methodology

The ionosphere is generally comprised of regions E, F1, F2, and possibly Es; in CCIR 894 only the E and F2 layers are considered. The basic MUF for a particular path is taken to be the larger of the E and F2 layer basic MUFs. (It will be recalled that the basic MUF is defined on the basis of classical ionospheric considerations to be the high-ray/low-ray junction frequency, whereas the operational MUF is the highest frequency supported by a specified mode. Generally the two should be the same if nose extension arising from side scatter or anomalous propagation is not observed.) For simplicity the ratio of the operational MUF to the basic MUF, called R_{op}, is taken to be unity for E region modes. For F2 modes R_{op} ranges between 1.1 and 1.3 for an EIRP ≤ 30 dBW and between 1.15 and 1.35 for an EIRP > 30 dBW. The higher values are specified for winter-nighttime, and the lowest ratios are specified for summer daytime conditions. A tabulation of the R_{op} values exhibits the symmetry associated with the presumed distribution of above-the-MUF scatter effects.

TABLE 5-13

OPERATIONAL USE OF THE BASIC MUF

Rop	Conditions Specified
1.10	S-D-LP
1.15	E-D-LP, S-D-HP
1.20	S-N-LP, W-D-LP, E-D-HP
1.25	E-N-LP, S-N-HP, W-D-HP
1.30	W-N-LP, E-N-HP
1.35	W-N-HP

KEY

S = Summer
W = Winter
E = Equinox
D = Day
N = Night
LP = Low Power
HP = High Power

R_{op} = Ratio of operational MUF to the Basic MUF

5.13.2.1 *Determination of the E Region MUF*

Recall equation 3.13a for the ordinary ray critical frequency for the E region:

$$\text{foE(MHz)} = 0.9\,[\,(180 + 1.44\,R)\cos\chi]^{0.25} \qquad (3.13\text{a})$$

where R is the 12-month running mean sunspot number, and χ is the solar zenith angle evaluated at the appropriate ionospheric point. In CCIR method 894 we take χ to be an effective solar zenith angle χ_{eff}. Under the vast majority of circumstances $\chi = \chi_{eff}$. When the sun is near the horizon, the situation is more complicated. The following associations are to be made:

$$\chi_{eff} = \chi \qquad \text{for } 0 \le \chi \le 80 \text{ deg}$$

$$\chi_{eff} = 90 - (10.8)^{-1}\exp[0.13(16 - \chi)] \qquad \text{for } 80 < \chi < 116$$

$$\chi_{eff} = 89.907 \qquad \text{for } \chi \ge 116$$

The raw solar zenith angle is deduced at one or more reflection or control points. For paths up to and including 2000 km, the control point (termed CP) is the path midpoint. For paths in excess of 2000 km but less than 4000 km, there are two CPs located 1000 km from the transmitter and receiver terminals. This means that for path lengths only slightly more than 2000 km, the CPs will be quite close together. For paths of almost 4000 km in length, the CPs will be separated by roughly 2000 km, the limiting distance without an intermediate ground reflection. The geographic coordinates of the control points are needed, and these may be deduced using standard spherical trigonometric equations (such as those given in CCIR Report 252, Equations 4-6). With this information, charts similar to Figure 7.21 found in Davies [1965] may be used to deduce χ at the specified CPs; alternatively χ may be calculated. From CCIR 252 [1970], with a change in notation we may write:

$$\cos\chi = \sin \text{Lat}_{CP} \sin \text{Lat}_{sun} + \cos \text{Lat}_{CP} \cos \text{Lat}_{sun} \text{Cos}(15t - 180 - \text{Long}_{CP}) \qquad (5.15)$$

where Lat_{CP} and Long_{CP} the control point latitude and longitude respectively, Lat_{sun} is the subsolar latitude at the middle of the month involved, and the quantity $15t$-180 is the subsolar longitude in degrees. Knowing χ at one or two CPs, it is now possible to deduce the effective value χ_{eff} for use in Equation 3.13a above. Thus one value of foE (or possibly two values in the case of two hops) is (are) obtained. For the case of two E region hops, only the smallest is retained for computation of the basic MUF over the entire path. The E region MUF is found from the following equation:

$$\text{MUF}(D)\text{E} = \text{foE} \sec\phi_{110} \qquad (5.16)$$

where D is the ground range (or great circle distance) associated with the entire path, and ϕ_{110} is the angle of incidence of the ray at a height of 110 km.

5.13.2.2 *Determination of the F Region MUF*

The first step is to deduce the control points involved, not unlike the procedure used in the case of E region modes. However, for the case of the F layer, we take the CP to be the midpoint of the path for $D < 4000$ km. For paths longer than 4000 km, we take CPs to be located 2000 km from each terminal of the path. This holds even for very long paths. It means that for purposes of determining the controlling F region MUF we need only consider the first and last half-hops; the intermediate region is of no consequence in the method. The existence of multiple hops is not necessarily ignored in connection with absorption or other transmission losses encountered in the field strength calculation however. Once the CPs are deduced, we evaluate must determine the overhead MUF and the 4000 km MUF at each CP, and transform this information into the basic MUF for the length of the entire path. The appropriate values of MUF(zero)F2 and MUF(4000)F2 may be extracted from maps in Report 340-5 [CCIR, 1986] for a specified sunspot number (0 and 100) and any of 12 equispaced values of UT. Interpolation is to be used for arbitrary sunspot numbers and times. If foF2 and M(3000) are available, we convert as follows:

$$\text{MUF(zero)F2} = \text{foF2} + \tfrac{1}{2} f_g \quad (5.17a)$$

$$\text{MUF(4000)F2} = \text{foF2}\,(1.1)\,\text{M(3000)} \quad (5.17b)$$

where f_g is the electron gyrofrequency (see Figure 3-9), where convention holds that all frequencies are expressed in MHz.

The required MUF is generally associated with a path length different from 4000 km. As in the case of the E region we have two procedures for computing the F2 layer MUF based upon path length. We shall take up long paths in excess of 4000 km shortly. For paths up to and including 4000 km, the midpath is taken to be the CP, and following the procedure indicated above, we find values for MUF(4000)F2 and MUF(zero)F2 at the CP. Next we convert this information into a MUF(D)F2 value using the following scheme:

$$\text{MUF}(D)\text{F2} = \text{M}(D)\,[\text{MUF(4000)F2} - \text{MUF(zero)F2}] + \text{MUF(zero)F2} \quad (5.18a)$$

where the term M(D) varies with D as follows:

$$\text{M}(D) = 1.64 \times 10^{-7}\,D^2 \quad ,D < 800 \text{ km} \quad (5.18b)$$

and

$$\text{M}(D) = 1.26 \times 10^{-14} D^4 - 1.3 \times 10^{-10} D^3 + 4.1 \times 10^{-7} D^2 - 1.2 \times 10^{-4} D \quad (5.18c)$$

for D between 800 and 4000 km.

Now for paths in excess of 4000 km, there are two CPs, one located 2000 km from the transmitter and the other located 2000 km from the receiver. The MUF associated with the total path is simply the *least* of the two values of

MUF(4000)F2 for the two CPs. The operational MUF for the F2 layer propagation is deduced by forming the product of the basic MUF with the term R_{op}. Finally the overall operational MUF is the *higher* of the MUF(D)E and MUF(D)F2. The 894-1 method suggests that the upper and lower deciles in the operational MUF distribution may be obtained by multiplication of the median operational MUF by factors 1.15 and 0.85 respectively. The OWF thus corresponds to the basic MUF times R_{op} times 0.85.

5.13.3 Determination of Field Strength

The scheme for computing the median field strength is covered fully in Report 894-1 [CCIR, 1986a], and details are found in Report 252-2 [CCIR, 1970] and its supplement [CCIR, 1982a]. We shall provide only the main points in this section.

The first step involves determination of the number of hops along the path in question, and for F layer modes this is obtained by application of the standard formula due to Shimazaki [1955]:

$$h_p(\text{km}) = 1490/\text{M(3000)F2} - 176 \qquad (5.19)$$

Now the next step involves determination of the location of all reflection points (not just the one or two *control points*) used to deduce the MUF). At these points the values of M(3000) and foF2 are found from the CCIR 340 atlas or similar references; foE is deduced from equation 3.13a above. With these data we may now deduce the so-called mirror height h' (as opposed to the Shimazaki virtual height) by the following formula:

$$h' = 358 - (11\text{-}100c_1)\,[18.8 - 320\,c_2^{-5}] + c_1 d[0.03 + 14\,c_2^{-4}] \qquad (5.20)$$

where d is the hop length for an n-hop mode taken to be the total path length D divided by n (i.e. $d = D/n$), c_1 = the larger of 0.04 and M(3000)F2^{-1} - 0.24, and c_2 = the larger of 2.0 and foF2/foE. Now the collection of mirror heights (i.e., the h' values) are averaged to obtain a single representative value.

The radiation (or elevation) angle β is given by the formula:

$$\beta = \tan^{-1}[\cot(d/2R) - (R/R+h')\csc(d/2R)] \qquad (5.21)$$

where h' is taken to be 110 km for the E layer and for F layer modes is found from Equation 5.20 just above. Strictly speaking, a determination of the elevation angle in this manner is appropriate to the path corresponding to the MUF; however in the CCIR 894 method it is used for all frequencies for ease of computation.

Another intermediate parameter required is the so-called virtual (slant) range, p'. This number involves all hops plus the effective radiation angle from Equation 5.21. We have:

$$p' = 2R\sum_{N}[\sin(d/2R)/\cos(\beta + d/2R)] \qquad (5.22)$$

where the summation is over all hops. This is an important number since it will constitute the distance over which spreading loss will assumed to obey the r^{2} rule. In fact the basic free-space transmission loss is taken to be:

$$L_{bf} = 32.45 + 20 \text{Log}_{10} f + 20 \text{Log}_{10} p' \quad (5.23)$$

Losses due to absorption L_i and above-the-MUF propagation L_m need to be considered. The CCIR 252 method for D- and E-layer absorption is applied to each hop and the resultant values are summed to arrive at a composite value. An adjustment must be made for E-layer reflection modes to account for a reduction in absorption owing to reduced absorptive layer penetration. This correction factor L_c is 0 if f exceeds the E-region MUF or for F-region modes. (See Chapter 4 of this book for a general discussion of absorption; additional information may be obtained from the Supplement to CCIR 252-2 [1982a].) Above-the-MUF loss was discussed in Section 4.11.1 of Chapter 4. Repeating Equation 4.52, we have:

$$L_m(\text{dB}) = 130 [f/f_{MUF} - 1]^2 \quad (4.52)$$

where f_{MUF} is the basic MUF for the hop. The value of L_m in the current application is taken to be independent of the number of hops. Nevertheless CCIR 894 limits the loss to a value of 81 dB.

Other losses are necessary to consider. For example, if ground reflection is involved, then losses will be introduced. The ground reflection loss L_g is naturally taken to be zero for a 1-hop path; it is 2 dB for a 2-hop path; and is 4 dB for a 3-hop path.

CCIR Report 252 [1970] contains tables of excess system loss Y_p which is "designed statistically to express the aggregate effects of such phenomena as, the winter anomaly, sporadic-E blanketing, spread-F multipath, off-great-circle propagation, skip distance and horizon focusing, day-to-day variations in layer height, thicknesses, etc." The tables referred to are Tables II and III of CCIR 252; the value of Y_p so found is adjusted by - 9 dB to account for auroral effects [CCIR 894-1, 1986a]. The adjusted excess system loss is termed L_h. Report 894 advises that a another loss, L_z, be included to account for other skywave effects not included. A value of 7.3 dB has been recommended.

Finally we are in a position to estimate the median field strength received by a single specified mode. Before doing so, we must first determine the nature of the path, its total length, and the number of hops and modes to be considered (See Table 5-14).

TABLE 5-14

MODES IN REPORT 894-1 (CONSISTENT WITH CCIR REPORT 252)

Distance Range	Modes
$d < 2000$ km	1E, 1F2, 2F2
$2000 \leq d \leq 4000$ km	2E, 1F2, 2F2
$4000 < d \leq 7000$ km	2F2, 3F2
$7000 < d \leq 9000$ km	3F2 and FTZ [transition region]
$9000 < d$ km	FTZ

The estimation method varies dependent upon the distance of the overall path. For so-called short path considerations ($d < 7000$ km), we have:

$$E_{ts} = 136.6 + P_t + G_t + 20 \operatorname{Log}_{10} f - L_{bf} - L_i - L_c - L_m - L_g - L_h - L_z \quad (5.24)$$

where f is in MHz, G_t is the transmit antenna gain with respect to an isotrope, and the resulting field strength is expressed in dB with respect to 1 μV/m.

For long-path considerations ($d > 9000$ km), the situation becomes more complex and the methodology has an untidy appearance, being based upon a look-up tables and procedures which appear decidedly ad hoc to the superficial reader. Nevertheless the method demands advocacy since it agrees well with the real world. It is based upon a large amount of empirical data and was developed as the so-called FTZ field strength model. We have:

$$E_{tL} = [139.6 - 20 \operatorname{Log}_{10} p'] \, \Gamma(f_M, f_L) - 36.4 + P_t + G_t - L_y \quad (5.25)$$

where p' is computed from Equations 5.21 and 5.22 taking the height = 300 km, and L_y is a long-path excess system loss factor similar to L_z and set equal to -3.9 dB, and $\Gamma(f_M, f_L)$ is given by the expression:

$$\Gamma(f_M, f_L) = 1 - \frac{(f_M + f_H)^2}{(f_M + f_H)^2 + (f_L + f_H)^2} \bullet \left[\frac{(f_L + f_H)^2}{(f + f_H)^2} + \frac{(f + f_H)^2}{(f_M + f_H)^2} \right] \quad (5.26)$$

In this expression f_M and f_L are upper and lower limiting frequencies. The upper limiting frequency f_M is proportional to the basic MUF which is computed for the first and last hops of the path. The least of the basic MUFs is retained as the controlling value but even this is altered to account for diurnal influence and path directionality. The lower limiting frequency f_L is controlled principally by absorption when the path is in daylight. The reader should refer to CCIR Report 894 for details of the long-path model.

For the intermediate distance between 7000 and 9000 km, the field strength is determined by combining the results of Equations 5.24 and 5.25 in the following manner:

$$E_{ti} = E_{ts} + [(D-7000)/2000] \, [E_{tL} - E_{ts}] \quad (5.27)$$

where, as before, the answer is expressed in dB(1 μV/m).

TABLE 5-15. AREAS FOR IMPROVEMENT IN MODELING

1. Maps of ionospheric characteristics, as well as mathematical representations need to be improved. Areas of special need include oceanic zones where vertical incidence sounder data has been sparse.

2. Data from alternative sources needs to be assimilated into the ionospheric data base, in order to assess the ionospheric parameters and their statistical variations over poorly sampled regions. The new data would also provide a consistency check in connection with data derived from routine ionosonde measurements

3. The manner in which the MUF is determined should be examined to see empirical formulas are consistent with observations obtained using oblique-incidence-sounders. Ray tracing methods might also provide the analyst with an improved method if an underlying ionospheric representation is sufficiently precise. Above-the-MUF propagation, nose-extension, and side-scatter effects need to be accounted for more directly. Furthermore, greater care is needed in the treatment of off-great-circle propagation effects.

4. Absorption variability estimates could be improved, especially in the auroral zone

5. Propagation loss predictions for long-distance paths needs to be reconciled with observations; and the transition between so-called long path and short path methods needs to be treated more carefully

6. The treatment of multipath needs to be improved; a general model of multipath needs to be developed, including multi-mode, multi-hop, and scatter modes.

7. Reliability methods should be verified under controlled conditions for which background conditions are well known.

8. Specific features, such as sporadic E and spread F are either imprecisely modeled or ignored in most models. No doubt, basic physical understanding will lead to an identification of observables which will lead to a prediction method. It is anticipated that such methods will be characterized as near-term rather than long-term, however. (Dynamic Model)

9. Tilts and gradients need to be properly modeled so that 3-D ray tracing methods (including magnetic field) may be exploited to their fullest. Macroscopic (large-area) tilts and effects of the termina- should be handled first. Next, steps should be taken to model the gradients/tilts associated with the various circumpolar features and the equatorial anomaly. Finally, an attack must be made in the area of gravity wave-induced TIDs. As in No.8 above, this problem is only meaningful in the context of intermediate and short term modeling. (Dynamic Model)

5.13.4 Comparisons with Data

A number of comparisons have been made between observations and CCIR methods, specifically methods based upon CCIR Report 252 and its supplement. There are also data banks, available through the CCIR Secretariat, which have been used for examination of models (Data Banks C and D; Resolution 63.2 [CCIR, 1986i]). Refer to Report 571-3 [CCIR, 1986j] for a review of these comparisons.

5.13.5 Possible Improvements

There are quite a few improvements which have been, or should be, envisioned to HF communication performance prediction models. A partial list is given in Table 5-15. A number of possible improvements are also presented in CCIR Report 729-2 [1986k].

5.14 SMALL PROGRAMS AND MICROCOMPUTER METHODS

A number of transportable models have been developed to make use of the growing availability of micro- and minicomputers (Refer to Report 1013 [CCIR, 1986L]). The use of electronic mail connection to solar and ionospheric assessment services, such as those provided by NOAA, is placing capability for prediction service in the field where it may be needed, rather than at central establishments with large main frame computers. One of the first micro-computer models to be developed for general use by the public was MINIMUF [Rose,1982a], which is part of the PROPHET family of programs [Rose,1982b]. There have a number of improvements to MINIMUF, the more recent versions being termed MINIMUF 3.5 and MINIMUF85 [Sailors et al., 1986]. Other microprocessor-oriented frequency prediction models have followed: MICROMUF [Bakhuizen, 1984], FTZMUF2 as described by Damboldt and Suessmann [1988a,b], and a series of models based upon algorithms developed by Fricker [SESC, 1988]. Daehler [1990] has developed a MUF-LUF-FOT prediction program having a number of simple models as its basis, but admitting to several update options. Table 5-16 is a compilation of microcomputer methods and corresponding references. A review of various microcomputer methods has been published by Davy et al. [1987]. Field strength models are represented by MINIFTZ4 and the most complete CCIR-sanctioned microprocessor model is REP894. A microcomputer program, developed in accordance with the specifications provided in CCIR Report 1013 and based upon CCIR 894 methodology, is the program MICROP2 [Dick and Miller, 1987].

The Ionospheric Prediction Service of Australia has developed a user friendly microcomputer program called the *Advanced Stand-Alone Prediction System* (ASAPS) [IPS, 1991]. This model exploits the T-index, which was developed by IPS investigators, and draws on a previously-developed *GRAFEX* prediction method [Turner, 1980]. The T-index was discussed previously in Section 2.3.4 of Chapter 2, and earlier in this chapter (viz.,p.330).

Although there is some concern that accuracy may be sacrificed in the development of the microcomputer models, this concern is tempered by the following considerations. First, there have been no in-depth studies as yet which show that the large main frame prediction models significantly outperform their smaller cousins, at least in the prediction of a simple parameter such as the MUF where there is a common basis for comparison. Secondly, in the world of RTCE and ionospheric assessment technology, which may be used for frequent updating of the model input conditions, small microcomputer models may perform quite adequately. This is because temporal updating procedures typically involve the application of scale factors which effectively suppress the physics which may be contained within the more elegant mainframe model. Thus, more rapid temporal updating leads to a convergence in the performance metric of competing models. The same may also be said of spatial extrapolation using models, although in this case an "update" involves the number and location of ionospheric control points used in the extrapolation process. Naturally, one would prefer the flexibility of the larger, more elegant model if the capability to update in either space or time is limited. It should be noted that a number of government agencies and firms are translating the large main frame programs such as IONCAP to run on personal computers, thus making the relative accuracy question mute.

TABLE 5-16

MICROCOMPUTER PREDICTION METHODS AND REFERENCES

MODEL NAME	REFERENCE
FTZMUF2 (foF2 and M3000)	Damboldt and Suessmann [1988a,b]
Fricker (foF2 & hF2)	Fricker [1985]
Compact Ionospheric Model	Clarke [1985]
MINIMUF	Rose [1982a,b]
MICROMUF[1]	SESC [1988] [Bakhuizen ,1984]
MINIPROP[1]	SESC [1988]
MAXIMUF[1]	SESC [1988]
KWIKMUF[1]	SESC [1988]
Gerdes Approach[2]	Gerdes [1984]
EINMUF (MUF-LUF-FOT)[2]	Daehler [1990]
Devereux/Wilkinson Method[2]	Devereux & Wilkinson [1983]
Fricker (Field Strength)	Fricker [1987]
IONOSOND[1]	W1FM [Lexington, MA]
MINIFTZ4 (Field Strength)	Damboldt & Suessmann [1988a,b]
MICROPREDIC	Petrie et al.[1986]
HFBC84 (Micro Version)	Pan and Ji [1985]
HFPC86-CNET Method	Davy et al.[1987]
REP894[3]	CCIR 894 [1986a]
PC-IONCAP (NTIS)	Teters et al.[1983]
IONCAST[4]	Rhoads [1990], NRL-Washington
ICEPAC[4]	Stewart [1990], ITS-Boulder
ASAPS	IPS [1991]

KEY

1 based on Fricker's algorithms
2 approach similar to MINIMUF
3 with CCIR Secretariat: also MICROP2, HFRPC8
4 under development, based upon IONCAP

A microcomputer implementation of IONCAP, termed IONCAST, has been developed at NRL in connection with its high latitude HF propagation program and is fairly widely distributed [Rhoads, 1990]. A separate implementation of IONCAP, called PC-IONCAP, has been developed by ITS and is available through NTIS along with an early version of IONCAP documentation [Teters et al., 1983]. Recently ITS has embarked on the development of an improved PC version, called ICEPAC, which incorporates a better high latitude description [Stewart, 1990].

5.15 COMMENTARY ON SHORT-TERM PREDICTION TECHNIQUES

Short-term prediction methods typically involve the measurement of either an ionospheric or geophysical parameter which is applied to an empirical model or algorithm. We have seen just above that long-term prediction methods provide the system architect and the frequency planner with useful guidance, but that ionospheric variability with time scales of tens of minutes present too much of a challenge. Certainly the ubiquitous median models have no intrinsic short-term forecasting capabilities. One should expect very little correlation between the unfiltered real world and the predictions extracted from a median model. A summary of short-term methods is provided in Report 888-1 [CCIR, 1986m]. Even though long-term models have no capability to assess short-term variability in other than a statistical way, the Achilles heel of short-term forecasting, the reader will discover that long-term models are sometimes used (improperly) by analysts. Milsom [1987] has listed the outstanding problems associated with short term forecasting, and Goodman [1991] has examined ways of coping with short term variability

The derivation of short-term predictions (which we shall hereafter term forecasts) may entail the process of model update with an external geophysical parameter, an ionospheric parameter, or a combination of both. Forecasts which exploit ionospheric measurements for updating purposes are by far more successful. These methods are covered in Chapter 6.

5.16 TOWARD IMPROVEMENT OF LONG-TERM PREDICTIONS

Long-term prediction of ionospheric behavior depends critically upon a reliable representation of past ionospheric data and a known correlation with solar activity, which is the derivative of yet another prediction process. Because of the general lack of a truly accurate representation or model of the ionosphere, which is compounded by the tendency to *drive* these models with a single parameter such as sunspot number, long-term predictions are not dependable. This is because short-term, apparently stochastic disturbance sources or factors, which occur in the actual physical process, are not properly accounted for in the prediction method. Thus, long-term methods for prediction are used to derive coarse guidance. The hope is that they at least reflect the median behavior.

There are long-term tendencies in the solar flux. Recommendation 371-5 [CCIR, 1986n], dealing with the choice of indices for long-term predictions of ionospheric behavior, recommends that predictions which are for dates more than one year ahead of the current period be treated differently than for periods which are less. If predictions are for epochs of more than 12 months in the future, the 12-month running mean sunspot number is to be used for the prediction of all ionospheric parameters, including: foF2, M(3000)F2, foF1, and foE. The 12-month average is employed to average out the shorter period disturbances, which may disguise the long-term tendency of solar flux and its influence on the median ionospheric parameters. For shorter lead times, several indices, including a measure of the 10.7 cm solar flux, as well as the sunspot number, produce equivalent answers in connection with prediction of the parameters foF2 and M(3000)F2. As far as the lower ionospheric parameters foF1 and foE are concerned, it turns out that the 10.7 cm solar flux is the best index for periods up to six months into the future and perhaps even longer. The fact that actual flux (even at 10.7 cm which does not itself interact with the ionosphere) best represents the solar ionization flux, which produces the E and F1 regions of the ionosphere, is well known and is implicit in the CCIR recommendations.

In the design stage, the driving parameters of a prediction model are allowed to take on a range of values, and the system is designed to encompass the results of the calculations. While sunspot number may be an adequate driving parameter for this purpose, it is not optimum for predicting events which will occur in particular days, weeks, or months in the future. Over the past decade, mounting evidence has accumulated [Sheeley et al.,1985] which shows that coronal holes and particular large sunspot groups on the sun are the real sources of high-speed solar wind streams, which feed most immediately into high latitude ionospheric effects and are later felt elsewhere. Observed effects are ionospheric storms, shifted and expanded auroral rings, depressed critical frequencies at mid-latitudes, etc. In other words, a sunspot number which totals all the spots, is too crude a parameter to predict these effects. Instead, the idea would be to view coronal holes and pertinent sunspot regions from the earth, account for the correct number of days for solar rotation to carry these solar features to the central meridian, and then add 2-3 days for the solar wind perturbation to reach the earth. Hence, ionospheric effects could be predicted from solar observations about a week in advance. If one accounts for the fact that several of these solar source features last many solar rotations, then corresponding effects can be confidently predicted to recur every 27-28 days. This is the basis for prediction of effects from solar observations with lead times up to several months. These developments point to the redesign of ionospheric models on the basis of correlating synoptic ionospheric parameter data with a different batch of relevant solar parameters. Shorter-term forecasts (on the scale of hours) may be related to the class of solar flare-related sudden ionospheric disturbances (SIDs), which are associated with bursts of short-wavelength electromagnetic radiation.

Difficulties associated with HF radio circuit performance predictions are outlined in Report 889-1 [CCIR, 1986o]. They are abbreviated in Table 5-17 below, and the list clearly illustrates why HF predictions have mixed reviews.

TABLE 5-17

DIFFICULTIES IN MAKING ACCURATE PREDICTIONS

1. Use of OWF's implies a loss of skywave support 10% of the time
2. Predictions generally ignore storm-time effects.
3. Sporadic E model is not sufficiently accurate.
4. Differences exist between model data bases and observations.
5. The SNR is poorly modeled and an incomplete performance metric.
6. Other user interference is not taken care of properly.
7. Deficiencies listed in Table 5-15 need to be addressed

5.17 A COMMENT ON INTERNATIONAL COOPERATION

It is important for the HF community to recognize the significant number of modern vertical-incidence-sounder (VIS) and oblique-incidence-sounder (OIS) assets which are now becoming available. In many instances the deployment of these assets is coordinated, and compilation of sounder coordinates and other data are available to the potential user. On the international level and under the aegis of Commission G of URSI, an Ionosonde Network Advisory Group (INAG) has been established to coordinate activities, disseminate information, and share new developments in ionospheric characterization by this technique. Paralleling the substantial data base derived from VIS instruments, efforts are now underway to compile similar data from OIS networks. This should be useful in the characterization of HF circuits over ocean areas, where VIS data are sparse, and this should also improve global climatological model predictions in these areas. Two other URSI Commission G working groups deal with ionospheric informatics and mapping/modeling; and these groups participate in the organization of various scientific campaigns in conjunction with other international organizations. In recent years studies involving networks of ionospheric diagnostic devices, including sounders, have been initiated. International cooperation in data collection campaigns and data sharing will no doubt lead to an improvement in our understanding of the ionosphere, as well as in the development of new procedures applicable to a real-time assessment of the global ionosphere. These measures pertain to the automation of data collection, scaling of ionograms, and rapid data dissemination to users. Hopefully these developments may benefit the entire HF community.

5.18 CONCLUSIONS

An overview of prominent ionospheric propagation prediction models was given, and their use for shortwave applications was discussed. The backbone of these programs is a climatological or monthly median ionospheric model, which does not account for short-term variability. Since all such ionospheric models admit to significant errors from this class of disturbance, the choice of prediction model may depend less on the phenomenology embodied in the model and more upon less esoteric matters such as: availability of computer assets, transportability of the model, software maintenance requirements, ease of use, and related issues. This has led to a bifurcation of prediction systems into two classes: one devoted to study of detailed physical processes and long-term planning, and the other driven by short-term tactical requirements. Certain longer period disturbances or features characterized by large geographical scales, may be better described by more detailed models, although a significant empirical component may be involved, and update procedures will be necessitated to improve accuracy significantly. Examples include: day-night transitions, equatorial anomaly regions, high latitude auroral and sub-auroral trough regions, etc. There is also an increasing tendency for large main frame models to be incorporated in microcomputers. Thus, we shall see the most advanced models and methods available to even the most unsophisticated user, and these models will ultimately replace some of the well-known skeletonized models which were developed for microcomputers in the 1970s and early 1980s. In the near term, we anticipate that models such as IONCAP may be incorporated within forecasting systems such as PROPHET as an option. In addition, similar models may be incorporated within advanced modems which employ microprocessors for network management.

Long-term predictions are likely to be required for broadcast planning for some time to come. They are also worthwhile for system studies and planning for military operations. It is unclear to what extent incremental improvements in long-term modeling will provide for anything but small incremental improvements in long-term prediction capability. Computer procedures and display formats may be improved, however, and these cosmetic changes will add value, since they will provide the analyst with a capability to examine the projected data more coherently and in a variety of scenarios. One potential area for long-term performance improvement may arise as the result of a newly-developed scheme for mapping the tendencies of high latitude propagation from 1 week to several months in advance, based upon observation of the evolution of coronal holes and related solar features. There are a number of deficiencies in current modeling approaches and we have identified most of them. Aside from taking more care in representing the ionospheric *personality*, and possible incorporation of 3-D raytracing methods, quantum improvements in prediction capability are not anticipated. The future realm is dynamic modeling.

Long-term modeling approaches may be used to benefit short-term predictions. More dynamic approaches, based on ionospheric soundings, have been discussed. They may be shown to have viability in updating selected prediction models for short-term use. This approach has been found to be particularly useful for local removal of the DC bias errors in ionospheric models, which result from the use of monthly medians and imprecise driving parameters, such as the sunspot number. Updates are particularly relevant for the effective use of adaptive HF schemes. However, a present ionospheric specification decorrelates rapidly when compared with future reality. The update must be performed rapidly. The best application of update for military or civilian broadcast planning may well be in the context of relay station diversity. Thus, the broadcast planner could envision real-time resource management. The resources available in the future may involve backscatter sounder technology, as well as overhead imagery tailored to provide ionospheric *weather* maps. Ionospheric data extracted from the GPS constellation downlink waveforms may be used to provide a more meaningful spatial sampling. These data sources would be coupled to existing assets, such as conventional vertical and oblique sounders and total electron content sensors activated by GPS transmissions. All of this information could be merged with the real-time solar-terrestrial data available through various data services [Joselyn and Carran, 1984]. Relatively high-quality ionospheric information may result from inserting this data into sophisticated ionospheric models which are presently being developed. The possibility exists for the construction of a real-time ionosphere to serve a number of users, not unlike that which has been envisioned by the U.S. Air Force to serve its customers.

Finally, it must be stated that a substantial effort has gone into the general area of ionospheric modeling. From this investment, a considerable amount of insight has been derived, and a number of very interesting methods for performance assessment have evolved. Some of these models include a full range of ionospheric and propagation effects, while others stress simplicity. The modern era will allow selection of the more complex (and complete) models for use in microcomputers as system controllers. Why not? Furthermore these models will have *hooks* allowing real-time update methods to be utilized as the newer sensors become available. In short, prediction methodologies based upon the evolution of long-term median models, and discussed in this chapter, have been an essential catalyst in the development of more dynamic models.

5.19 REFERENCES

AGARD, *Effects of Electromagnetic Noise and Interference on Performance of Military Radio Communication Systems*, NATO-AGARD-CP-420, Proceedings of conference held in Lisbon, Portugal, Specialised Printing Services Ltd., Loughton, Essex, UK.

Anderson, D.N., M. Mendillo, and B. Herniter, 1985, "A Semi-Empirical, Low-latitude Ionospheric Model," AFGL-TR-85-0254, Geophysics Laboratory (now Phillips Laboratory), Hanscom AFB, Massachussets.

Anderson, D.N., M. Mendillo, and B. Herniter, 1987, "A Semi-Empirical, Low-latitude Ionospheric Model", *Radio Science* 22:292.

Anderson, D.N., J.M. Forbes, and M. Codrescu, 1989, "A Fully Analytical, Low- and Middle-Latitude Ionospheric Model", *J. Geophys. Res.* 94:1590.

Anderson, J.B., 1965, "Influence of Sorroundings on Vertically Polarized Log-Periodic Antennas", *Teletenik*, Vol.IX, No.2, pp.33-40.

Appleton, E.V., 1946, "Two Anomalies in the Ionosphere", *Nature* 157:691.

ARRL, 1987, *Tune in the World with HAM Radio*, American Radio Relay League, Newington CT 06111

Bakhuizen, H., 1984, "Program MICROMUF", Radio Nederland Broadcast on 6/17/84.

Barghausen, A.F., J.W. Finney, L.L. Proctor, and L.D. Shultz, 1969, "Predicting Long-Term Operational Parameters of High Frequency Skywave Telecommunication Systems", ESSA Tech. Rpt. ERL-110-ITS-78 (NTIS Doc. No. N70-24144), Boulder, Colorado.

Bent, R.B., S.K. Llewellyn, G. Nesterczuk, and P.E. Schmid, 1975, "The Development of a Highly-Successful Worldwide Empirical Ionospheric Model and its Use in Certain Aspects of Space Communications and in Worldwide Total Electron Content Investigations", in *Effect of the Ionosphere on Space Systems and Communications* (IES'75), edited by J.M. Goodman, U.S. Government Printing Office, Washington DC 20402.

Bilitza, D., 1990, "Solar-Terrestrial Models and Application Software", National Space Science Data Center, World Data Center A for Rockets and Satellites, NSSDC/WDC-A-R&S 90-19, Goddard Space Flight Center, Greenbelt, Maryland.

Bradley, P.A. and M. Lockwood, 1982, "Simplified Method of HF Skywave Signal Mode Reliability", *IEE Second Conference on HF Communication Systems and Techniques*, CP-206, IEE, London.

CCIR, 1964, "World Distribution and Characteristics of Atmospheric Radio Noise", CCIR Report 322, ITU, Geneva.

CCIR, 1965, "Handbook on High Frequency Directional Antennae", ITU, Geneva.

CCIR, 1966a, "Atlas of Ionospheric Characteristics", Rpt. 340-1, in *Recommendations and Reports of the CCIR*, Oslo Plenary, separately published document, ITU, Geneva.[50]

CCIR, 1966b, "Operating Noise Threshold of a Radio Receiving System", Rpt.413, *Recommendations and Reports of the CCIR*, Oslo Plenary, ITU, Geneva.

CCIR, 1970, "CCIR Interim Method for Estimating Skywave Field Strength and Transmission Loss at Frequencies Between the Approximate Limits of 2 and 30 MHz", Report 252-2, in *Recommendations and Reports of the CCIR: Propagation in Ionized Media*, Vol. VI, New Delhi Plenary, ITU, Geneva.

CCIR, 1978, "CCIR Book of Antenna Diagrams", ITU, Geneva.

CCIR, 1982a, "Supplement to Report 252-2: Second CCIR Computer-Based Interim Method for Estimating Skywave Field Strength and Transmission Loss at Frequencies between 2 and 30 MHz", published by ITU, Geneva.

CCIR, 1982b, "On Channel Occupancy", Report 825, ITU, Geneva.

CCIR, 1982c, "Worldwide Minimum External Noise Levels, 0.1 Hz to 100 GHz", Report 670, ITU, Geneva.

CCIR, 1984, "CCIR Atlas of Antenna Diagrams", ITU, Geneva.

CCIR, 1986a, "Propagation Prediction Methods for High Frequency Broadcasting", Report 894-1, *Recommendations and Reports of the CCIR: Propagation in Ionized Media*, Vol. VI, XVIth Plenary held in Dubrovnik, published by ITU, Geneva.

CCIR, 1986b, "Numerical Constants and Interpolation Procedure for the WARC-HFBC Propagation Prediction Method", Recommendation 621, *Recommendations and Reports of the CCIR: Propagation in Ionized Media*, Vol.VI, Dubrovnik Plenary, ITU, Geneva.

CCIR, 1986c, *Recommendations and Reports of the CCIR; Propagation in Ionized Media*, Vol.VI, Dubrovnik Plenary, ITU, Geneva. [This volume is one of 14 "Green Books" published by the ITU.]

CCIR, 1986d, "Characteristics and Applications of Atmospheric Radio Noise Data", Report 322-3, published as a separate booklet (first published in 1963 as CCIR 322-1, but revised as indicated in the 1986 "Green Book": Dubrovnik Plenary), ITU, Geneva.

CCIR, 1986e, "Man-Made Noise", Report 258-4, *Reports and Recommendations of the CCIR: Propagation in Ionized Media*, Vol.VI, Dubrovnik Plenary, ITU, Geneva.

---

50. Supplement No.1 of Report 340-1 was published in 1970 [New Delhi], and a second supplement was published in 1974 [Geneva]. Supplements 1 and 2 were cancelled in 1978 [Kyoto] and replaced by supplement No.3, with the publication appearing in 1980. Report 340-3 was issued in 1980, and it consisted of supplement No.3 and Report 340-1. Report 340 has been reissued at each subsequent plenary meeting: 340-4 in 1982 [Geneva], 340-5 in 1986 [Dubrovnik], and 340-6 in 1990 [Dusseldorf].

CCIR, 1986f, "Radio Noise Within and Above the Ionosphere", in *Reports and Recommendations of the CCIR: Vol.VI, Ionospheric Radio Propagation*, Report 342-5, XVIth Plenary at Dubrovnik, ITU, Geneva.

CCIR, 1986g, "Antenna Characteristics Important for the Analysis and Prediction of Sky-wave Propagation Paths", Report 891-1, in *Reports and Recommendations of the CCIR: Vol.VI, Propagation in Ionized Media*, Dubrovnik Plenary, ITU, Geneva.

CCIR, 1986h, "Computation of Reliability of HF Radio Systems", Report 892-1, in *Reports and Recommendations of the CCIR: Vol VI, Ionospheric Radio Propagation*, Dubrovnik Plenary, ITU, Geneva.

CCIR, 1986i, "Computer Programs for the Prediction of Ionospheric Characteristics, Sky-wave Transmission-loss and Noise", Resolution 63-2, in *Recommendations and Reports of the CCIR: Vol.VI, Ionospheric Radio Propagation*, Dubrovnik Plenary, ITU, Geneva. {See also Mod.1 of Resolution 63-2, dealing with computer programs and their availability), in Conclusions of the Interim Meeting of Study Group 6, Doc. 6/175-E, for period 1986-1990.}

CCIR, 1986j, "Comparisons between Observed and Predicted Skywave Signal Wave Intensities at Frequencies between 2 and 30 MHz", Report 571-3, *Recommendations and Reports of the CCIR: Propagation in Ionized Media*, Vol.VI, Dubrovnik Plenary, ITU, Geneva.

CCIR, 1986k, Developments in the Estimation of Skywave Field Strength and Transmission Loss at Frequencies above 1.5 MHz", Report 729-2,*Recommendations and Reports of the CCIR: Propagation in Ionized Media*, Vol.VI, Dubrovnik Plenary, ITU, Geneva.

CCIR, 1986L, "Microcomputer-Based Methods for the Estimation of HF Radio Propagation and Circuit Performance", Report 1013, in *Recommendations and Reports of the CCIR, 1986: Propagation in Ionized Media*,Vol.VI, XVIth Plenary of the CCIR at Dubrovnik, published by ITU, Geneva.

CCIR, 1986m, "Short-term Forecasting of Critical Frequencies, Operational Maximum Usable Frequencies and Total Electron Content", Report 888-1, *Recommendations and Reports of the CCIR, 1986: Propagation in Ionized Media*, Vol.VI, XVIth Plenary at Dubrovnik, ITU, Geneva.

CCIR, 1986n, "Choice of Indices for Long-Term Ionospheric Predictions", Recommendation 371-5, in *Recommendations and Reports of the CCIR, 1986: Propagation in Ionized Media*, Vol. VI, XVIth Plenary at Dubrovnik, ITU, Geneva.

CCIR, 1986o, "Real Time Channel Evaluation of Ionospheric Radio Circuits", Report 889-1,*Recommendations and Reports of the CCIR, 1986: Propagation in Ionized Media*, Vol. VI, XVIth Plenary at Dubrovnik, ITU, Geneva.

CCIR, 1986p, "CCIR Atlas of Ionospheric Characteristics", Report 340-5, published as a separate booklet, ITU, Geneva.

CCIR, 1986q, "Ionospheric Propagation and Noise Characteristics Pertinent to Terrestrial Radiocommunication Systems Design and Service Planning (Fading)", Report 266, published by ITU, Geneva.

Chernov, Yu A., 1969, "Nadioznost Kanala Radioveshania (Broadcasting Channel Reliability)", *Trudy NIIR*, Vol.1, pp.131-139.

Ching, B.K. and Y.T. Chiu, 1973, "A Phenomenological Model of Global Ionospheric Electron Density in the E-, F1-, and F2- Regions", *J. Atmos. Terrest. Phys.* 35:1615-1630.

Chiu, Y.T., 1975, "An Improved Phenomenological Model of Ionospheric Density", *J.Atmos. Terrest. Phys.* 37:1563-1570.

Clarke, E.T., 1985, "Real-Time Frequency Management in an Imbedded Microcomputer", in IEE Conf.Proc.#245, *Third International Conference on HF Communication Systems and Techniques*, pp.23-25, IEE, London.

CSC, 1985, "Voice of America Propagation Software System Analysis of Available HF Radio Propagation Models", prepared by Computer Sciences Corporation, Falls Church VA, for USIA/VOA under contract No. IA-21633-23 Task 2.

CQ Amateur Radio, monthly periodical, published by CQ Communications Inc., 76 North Broadway, Hicksville NY 11801.

Daehler, M., 1990, "EINMUF: An HF MUF, FOT, LUF Prediction Program", NRL Memorandum Report 6645, (May 18, 1990), Naval Research Laboratory, Washington DC 20375-5000.

Damboldt, T. and P. Suessmann, 1988a, "FTZ High Frequency Sky-wave Field Strength Prediction Method for Use on Home Computers", Forschungsinstitut der DBP beim FTZ, private communication (also referenced as CCIR IWP 6/1 Doc.L14., 1986).

Damboldt, T. and P. Suessmann, 1988b, "A Simple Method of Estimating foF2 and M3000 with the aid of a Home Computer", Forschungsinstitut der Deutschen Bundespost, Darmstadt, FRG., private communication, (also referenced as CCIR IWP 6/1, Doc. 124, 1986)

Davies, K., 1965, *Ionospheric Radio Propagation*, Monograph #80, NBS, Dept.of Commerce.

Davis, R.M. and N.L.Groome, 1964, "Variations of the 3000 km MUF in Space and Time", NBS Rpt. 8498, National Bureau of Standards, Boulder, Colorado.

Davy, P., R. Hanbaba, M. Lissillour, and H. Sizun, 1987, "Microcomputer Based Methods for the Estimation of HF Radio Propagation and Circuit Performance", ICAP-87, *Fifth International Conference on Antennas and Propagation*, Conf.Pub.#274, pp.297-301, IEE, London, UK.

Devereux, E.L. and D. Wilkinson, 1983, "HF Predictions on the Home Computer", *Radio Communication* 59(3):246-248.

Dick, M.I. and B.H. Miller, 1987, "Microcomputer-Based Method for the Estimation of HF Radio Circuit Performance", ICAP '87, *Fifth International Conference on Antennas and Propagation*, Conf.Pub.#274, pp.306-309, IEE, London, UK.

Doany, P., 1981, "A Wideband Frequency Hopping Modem for HF Data Transmission", PhD Thesis, University Of Manchester, UMIST, UK.

Dutta, S. and G. Gott, 1981, "Correlation of HF Interference Spectra with Range", *Proc.IEE*, Part F, Vol.132, No.7.

Dutta, S. and G. Gott, 1982, "HF Spectral Occupancy", in IEE Conference Proc. #206, *Second Conference on HF Communication Systems and Techniques*, IEE, London, UK.

Fox, M.W. and P.J. Wilkinson, 1986, "A Study of the OWF Conversion Factors in the Australian Region", IPS-TR-86, The Ionospheric Prediction Service, Darlinghurst, NSW, Australia.

Fox, M.W. and L.F. McNamara, 1986, "Improved Empirical World Maps of foF2: 1. The Method", Technical Report IPS-TR-86-03, Ionospheric Prediction Service, Sidney Australia.

Fox, M.W. and L.F. McNamara, 1988, "Improved Worldwide Maps of Monthly Median foF2", *J.Atmos,Terrest.Phys.* 50:1077.

Fricker, R., 1981, "Formulae for the Critical Frequency of the F2 Layer", CCIR International Working Party 6/12, Doc. 23, ITU, Geneva.

Fricker, R., 1985, "A Microcomputer Program for the Critical Frequency and Height of the F Layer of the Ionosphere", *Fourth International Conference on Antennas and Propagation*, IEE, pp.546-550.

Fricker, R., 1987, "A Microcomputer Program for HF Field Strength Prediction", IEE Conf.Pub.#274, *Fifth International Conference on Antennas and Propagation*, ICAP'87, pp.293-296, IEE, UK.

Gerdes, N.St.C., 1984, "A Low-Cost Prediction Service for Mobile and Portable Radio Using Ionospheric Propagation", *IEE Conference Publication*, No.238, pp.209-211.

Gibson, A.J. and P.A. Bradley, 1987, "Day-to-Day Variability of HF Field Strengths and Maximum Usable Frequencies", *Fifth International Conference on Antennas and Propagation*, ICAP'87, pp.257-260., IEE, London.

Gibson, A.J., P.A. Bradley, J.C. Schlobohm, 1985, "HF Spectrum Occupancy Measurements in Southern England", IEE Conf. Pub. #245, *Third International Conference on HF Communication Systems and Techniques*, pp.71-75, London.

Gibson, A.J. and L. Arnett, 1988, "New Spectrum Occupancy Measurements in Southern England", IEE Conf.Pub.#284, *Fourth International Conference on HF Radio Systems and Techniques*, pp.159-164, London.

Glasstone, S. and P.J. Dolan, 1977, *The Effects of Nuclear Weapons*, Superintendent of Documents, U.S. Gov. Print. Office, Washington DC 20402.

Goodman, J.M., 1982, "A Survey of Ionospheric Models: A Preliminary Report on the Development of an Ionospheric Model Thesaurus and User's Guide", NRL Report 4830, Naval Research Laboratory, Washington DC 20375-5000.

Goodman, J.M., M.H. Reilly, M. Daehler, and A.J. Martin, 1983,"Global Considerations for Utilization of Real-Time Channel Evaluation Systems in Spectrum Management", *MILCOM '83*, Arlington, Va., Oct 31-Nov 2.

Goodman, J.M., M. Daehler, M.H.Reilly and A.J. Martin", 1984, "A Commentary on the Utilization of Real-Time Channel Evaluation Systems in HF Spectrum Management", NRL Memorandum Report 5454, Naval Research Laboratory, Washington DC 20375-5000.

Goodman, J.M. and M. Daehler, 1988, "Use of Oblique Incidence Sounders in HF Frequency Management," in *Fourth International Conference on HF Radio Systems and Techniques*, Conf. Pub. No. 284, IEE, London, UK.

Goodman, J.M., 1991, "A Review of Methods for Coping with Ionospheric Variability in Connection with HF Systems", *Fifth International Conference on HF Radio Systems and Techniques*, Conf.Pub.No.339, IEE, London, UK.

Gott, G.F., N.F. Wong, and S. Dutta, 1982, "Occupancy Measurements Across the Entire HF Spectrum", NATO-AGARD Conf., *Propagation Aspects of Frequency Sharing, Interference, and System Diversity*, Paris.

Grove, B., 1989, *Shortwave Directory*, Grove Enterprises, Brasstown NC 28902

Hagn, G., 1988, lecture notes on HF noise and interference, AFCEA Course 104, Professional Development Center, AFCEA International Headquarters, Fairfax VA.

Halley, P., 1965, "Methods de Calcul Des Previsions de Point a Point Aux Distances Entre 2500 et 10500 km", Centre National d'Etudes des Telecommunications (CNET), France.

Harris Corporation, 1990, Technical Note 1/90, 221-4520-101, RF Communications Group, Rochester NY 14610.

Hatfield, V.E., 1980, "HF Communications Predictions, 1978 (An Economical Up-to-Date Computer Code, AMBCOM)", *Solar-Terrestrial Predictions Proceedings*, vol.4, pp. D2 1-15, R.F. Donnelly (editor), U.S. Gov. Print. Office, Washington DC.

Haydon, G.W., M. Leftin, and R. Rosich, 1976,"Predicting the Performance of High Frequency Skywave Telecommunication Systems (HFMUFES 4)", OT Report 76-102, Dept. of Commerce, Boulder CO.

Headrick, J.M., J. Thomason, D. Lucas, S. McCammon, R. Hanson, and J. Lloyd, 1971, "Virtual Path Tracing for HF Radar Including an Ionospheric Model", NRL Memo Report 2226, Washington DC, AD 883 463L.

Headrick, J.M. and M. Skolnik, 1974, "Over-the-Horizon Radar in the HF Band", *Proc. IEEE*, 62(6):664-673.

IPS, 1991, "ASAPS: Advanced Stand-Alone Prediction System", Ionospheric Prediction Service, Radio and Space Services, P.O. Box 1548, Chatswood, NSW 2057, Australia. (brochure)

IPSD, 1968, "The Development of the Ionospheric Index T", Report IPS-R11, Ionospheric Prediction Service, Sidney, Australia.

IRPL (Interservice Radio Propagation Laboratory), 1943, "Radio Propagation Handbook", National Bureau of Standards, Washington DC.

ITU, 1984, "World Administrative Radio Conference for the Planning of the HF Bands Allocated to the Broadcasting Service", Report at Second Session of the Conference, General Secretariat , ITU, Geneva.

ITU, 1984a, *Circular 22*, HFARRAYS

ITU, 1984b, *Circular 23*, HFRHOMBS

ITU, 1986a, *Circular 95*, HFMULSLW

ITU, 1986b, *Circular 95*, HFDUASLW

Jacobs, G. and T.J. Cohen, 1979, *The Shortwave Propagation Handbook*, Cowan Publishing Company, 14 Vanderventer Avenue, Port Washington NY 11050

Jacobs, G., 1990, "The Science of Predicting Radio Conditions: Sunspot Cycle Stall Continues", *CQ Amateur Radio*, July, pp.112-114.

Jones, W.B. and R.M. Gallet, 1962, "The Representation of the Diurnal and Geographical Variations of Ionospheric Data by Numerical Methods", *J.Res.Nat.Bur.Standards*, Section D, 66D, pp.419-438.

Jones, R.M. and J.J. Stephenson, 1975, "A Three-dimensional Ray Tracing Computer Program for Radio Waves in the Ionosphere," U.S. Dept. of Commerce, Office of Telecommunications, OT Report 75-76 (PB 248856), Boulder, CO.

Jones, W.B. and D.L. Obitts, 1970, "Global Representation of Annual and Solar Cycle Variation of foF2 Monthly Median, 1954-1958", Telecomm. Res. Rpt. OT/ITSR33, Boulder CO.

Joselyn, J.A. and K.L. Carran, 1985, "The SESC Satellite Broadcast System for Space Environment Services," in *Effect of the Ionosphere on C3I Systems (IES'84)*, pp. 252-254, J.M. Goodman (Editor-in-Chief), U.S.G.P.O., Washington, D.C., 1985-0-480-249:QL-3, available from NTIS, U.S. Dept.of Commerce, 5285 Port Royal Road, Springfield, VA 22161 (AD-163-622).

Kasantsev, A.N., 1947 (translated 1958), "The Absorption of Short Radio Waves in the Ionosphere and Field Strength of the Place of Reception", *Bulletin of Acad.Sci.USSR, Div.Tech.Sci*, No.9, pp1107-1138.

Kasantsev, A.N., 1956, "Developing a Method of Calculating the Electrical Field Strength of Short Radio Waves", *Trudy IRE, Trans.Instit.Rad.Eng.& Elec. of the Acad.Sci.*, USSR, p.2134.

Knecht, R.W., 1962, Lecture No.24, NBS Radio Propagation Course, NBS, Boulder, Colorado.

Kohnlein, W., 1978, "Electron Density Models of the Ionosphere", *Reviews of Geophysics and Space Physics* 16(3):341-354.

Lane, G., 1990, Voice of America, private communication

Laycock, P.J., M. Morrell, G.F. Gott, and A.R. Ray, 1988, "A Model for Spectrum Occupancy", Conf.Pub.#284, *Fourth International Conference on HF Radio Systems and Techniques*, pp.165-171, IEE, London.

Leftin, M., S.M. Ostrow, and C. Preston, 1967, "Numerical Maps of the Monthly Median h'F,F2 for Solar Cycle Minimum and Maximum", ERL Technical Memo #69, Boulder, Colorado.

Leftin, M., S.M. Ostrow, and C. Preston, 1968, "Numerical Maps of foEs for Solar Cycle Minimum and Maximum", U.S.Dept.of Comerce, ERL 73-ITS 63, Boulder, Colorado.

Leftin, M., 1969, "Numerical Maps of the Monthly Median h'F for Solar Cycle Minimum and Maximum", unpublished report, (referenced in ITS-78 documentation [Barghausen et al., 1969])

Leftin, M., 1976, "Numerical Representation of Monthly Median Critical Frequencies of the Regular E Region (foE)", OT Rpt. 76-88, Boulder, Colorado.

Liu, R.Y. and P. A Bradley, 1985, "Estimation of the HF Basic Circuit Relasibility from Modal Parameters", *Proc.IEE*, Vol.132, Part F, 111-118.

Lloyd, J.L., G.W. Haydon, D.L. Lucas, and L.R. Teters, 1978, "Estimating the Performance of Telecommunication Systems Using the Ionospheric Transmission Channel", Vol.1, USACEEIA Technical Report EMEO-PED-79-7, Ft.Huachuca AZ.

Lockwood, M., 1984, "Simplified Estimation of Ray Path Mirroring Height for HF Radiowaves Reflected from the Ionospheric F Region", *Proc.IEEP*, Vol.131, Part F., No.2, pp.117-124.

Lucas, D. and G.W. Haydon, 1966, "Predicting Statistical Performance Indices for High Frequency Telecommunications Systems", Report IER 1-ITSA 1, U.S. Department of Commerce, Boulder, Colorado.

Lucas, D., J. Lloyd, J.M. Headrick, and J. Thomason, 1972, "Computer Techniques for Planning and Management of OTH Radar", NRL Memo Report 2500, Washington DC, AD 748 588.

Lucas, D., 1987, "Ionospheric Parameters Used in Predicting the Performance of High-Frequency Sky-Wave Circuits", Lucas Consulting Inc., Boulder CO 80301, prepared under Navy contract (Attn: Mr. J. Headrick, Code 5320, Naval Research Laboratory, Washington DC.)

Ma, M.T., and L.C. Walters, 1969, "Power Gains for Antennas over Lossy Plane Ground", ESSA Tech. Rpt. ERL 104-ITS 74, U.S. Dept. of Commerce, Boulder CO.

Maslin, N.M., 1978, "The Calculation of Circuit Reliability when a Number of Propagation Modes are Present", CCIR WP 6/1, Doc.79, ITU, Geneva.

McNamara, L.F., 1976, "Short-Term Forecasting of foF2", IPS-TR-33, IPS Radio and Space Services, Dept. Sci., Darlinghurst, N.S.W., Australia.

McNamara, L.F., 1985,"HF Radio Propagation and Communications - A Review", IPS-TR-85-11, IPS Radio and Space Services, Dept. Sci., Darlinghurst, N.S.W., Australia.

Milsom, J.D., 1987, "Outstanding Problems in Short-Term Ionospheric Forecasting", *Fifth International Conference on Antennas and Propagation: A Hundred Years of Antennas and Propagation*, Part 2: Propagation, ICAP 87, IEE Conf. Pub. No. 274, pp. 316-319, IEE, London, UK.

Monitoring Times, periodical, published by Grove Enterprises, Brasstown NC 28902.

Moulsley, T.J., 1985, "HF Data Transmission in the Presence of Interference",IEE Conference Pub.#245, *Third International Conference on HF Communication Systems and Techniques*, pp. 67-70, IEE, London.

NAS, 1978, "Geophysical Predictions", National Academy of Sciences, Washington DC.

NBS, 1948, CRPL Circular 462, Dept of Commerce, Boulder, Colorado.

NTP 6, 1986, "Naval Telecommunication Procedures: Spectrum Management Manual", Commander, Naval Telecommunications Command, 4401 Massachusetts Ave. NW, Washington DC 20390-5290. [Note: Inquiries about this document should be addressed to the attention of Director NAVEMSCEN]

NTP 6 Supp-1, 1990, "Naval Telecommunication Procedures: Recommended Frequency Bands and Frequency Guide", Commander, Naval Telecommunications Command (name defunct in late 1990), 4401 Massachusetts Ave. NW, Washington DC 20394-5000.(stocked at NAVPUBFORMCEN, Philadelphia PA, #0411-LP-179-2350).[Note: this document is published once a year under the control of NAVEMSCEN, a spectrum management center under the aegis of the Naval Computer & Telecommunications Command (NAVCOMPTELCOM)].

Ochs, A., 1970, "The Forecasting System of the Fernmeldetechnischen Zentralamt (FTZ)", in *Ionospheric Forecasting*, NATO-AGARD-CP-49, edited by V. Agy, paper no.43.

Ostergaard, J., 1988, "Short Range Communication Systems, Design and Operation at Very High Latitudes", Conf.Pub.#284, *Fourth International Conference on HF Radio Systems and techniques*, pp.177-181, IEE, London.

Pan, Z. and P. Ji, 1985, "HF Field Strength Measurements in China and their Comparisons with Predicted Values", CCIR International Working Party 6/1, Doc. 253.

Perry, B.D. and R. Rifkin, 1987, "Interference and Wideband HF Communications", in IES'90, *Effect of the Ionosphere on Communication, Navigation, and Surveillance Systems*, J.M.Goodman (Editor-in-Chief), NRL Washington DC (document available through NTIS, Springfield VA).

Perry, B.D. and L.G. Abraham, 1988, "A Wideband HF Interference and Noise Model based on Measured Data", IEE Conf.Pub.#284, *Fourth International Conference on HF Radio Systems & Techniques*, pp.172-175, IEE, London.

Petrie, L.E., 1981, "Selection of a Best Frequency Complement for HF Communications", Communications Research Centre, Dept.Comms.Contract Rpt.#OER80-00339, Ottowa, Ontario, Canada.

Petrie, L.E., G.W. Goudrie, D.B. Ross, P.L. Timleck, and S.M. Chow, 1986, "MICROPREDIC - An HF Prediction Program for 8086/8088-Based Computers," Communications Research Centre Report 1390, Dept. of Communications, Ottawa, Ontario, Canada.

Piggott, W.R., 1959, "The Calculation of Medium Sky-Wave Field Strength in tropical Regions", DSIR Radio Research Special Report No.27, Her Majesty's Stationary Office, London.

Popular Communications, monthly periodical, published by CQ Communications Inc., 76 North Broadway, Hicksville NY 11801-2953.

Rawer, K., 1952, "Calculation of Skywave Field Strength", *Wireless Engineer*, 29:287-301.

Rawer, K., 1963, "Propagation of Decameter Waves (HF-Band)", in *Meteorological and Astronomical Influences on Radio Propagation*, B.Landmark (Editor), Pergamon Press, New York.

Rawer, K., 1975, "The Historical Development of Forecasting Methods for Ionospheric Propagation of HF Waves", *Radio Science*, 10(7):669-679.

Rawer, K., D. Bilitza, and S. Ramakrishnan, 1978, "Goals and Status of the International Reference Ionosphere", *Reviews of Geophysics and Space Physics* 16:178-181.

Rawer, K., J.V. Lincoln, and R.O. Conkright, 1981, "International Reference Ionosphere 79", Report UAG-82, World Data Center A for Solar-Terrestrial Physics, Boulder CO.

Reilly, M.H. and M. Daehler, 1986, "Sounder Updates for Statistical Model Predictions of Maximum Usable Frequencies on HF Sky Wave Paths", *Radio Sci.* 21(6):1001-1008.

Reilly, M.H. and E.L. Strobel, 1988, "Efficient Ray-tracing Through a Realistic Ionosphere," in *Effect of the Ionosphere on Communication, Navigation, and Surveillance Systems (IES'87)*, pp. 407-419, J.M. Goodman (Editor-in-Chief), U.S.G.P.O., Washington, D.C.

Rhoads, F., 1990, documentation on "IONCAST", private communication.

Rose, R.B., 1982a, "MINIMUF: A Simplified MUF Prediction Program for Microcomputers", *QST*, vol. LXVI, 12, 36-38.

Rose, R.B., 1982b, "An Emerging Propagation Prediction Technology", in *Effect of the Ionosphere on Radiowave Systems (IES '81)*, J.M. Goodman (Editor-in-Chief), U.S. Gov. Print. Office, Washington DC 20402.

Rosich, R.K. and W.B. Jones, 1973, "The Numerical Representation of the Critical Frequency of the F1 Region of the Ionosphere", OT Report, Boulder, Colorado.

Rush, C.M., 1976, "An Ionospheric Observation Network for Use in Short-Term Propagation Predictions," *ITU Telecom. J.*, vol. 43, VIII, pp.544-549.

Rush, C.M. and W.R. Edwards, Jr., 1976, "An Automated Technique for Representing the Hourly Behavior of the Ionosphere," *Radio Sci.* 11(11):931-937.

Rush, C.M., M. PoKempner, D.N. Anderson, F.G. Stewart, and J. Perry, 1983, "Improving Ionospheric Maps Using Theoretically Derived Values of foF2", *Radio Science* 18(1):95-107.

Rush, C.M., M. PoKempner, D.N. Anderson, F.G. Stewart, and R. Reasoner, 1984, "Maps of foF2 derived from Observations and Theoretical Data", *Radio Science* 19(4):1083-1097.

Rush, C.M., 1986, "Ionospheric Radio Propagation Models and Predictions - A Mini-Review", *IEEE Trans. on Ant. and Prop.*, vol. AP-34, pp. 1163-1170.

73 Amateur Radio, monthly periodical, published by Wayne Green Enterprises, WGE Center, Forest Road, Hancock NH 03449

Rush, C.M., M. Fox, D. Bilitza, K. Davies, L. McNamara, F. Stewart, and M. PoKempner, 1989, "Ionospheric Mapping: An Update of the foF2 Coefficients", *Telecomm. J.* 56:179-182.

Sailors, D.B., and R.P. Brown, 1982, "Development of a Microcomputer Atmospheric Noise Model", NOSC TR 778, San Diego CA.

Sailors, D., R.A Sprague, and W.H. Rix, 1986, "MINIMUF-85: An Improved HF MUF Prediction Algorithm", NOSC TR 1121, July 1986.

Secan, J., 1989, "A Survey of Computer-Based Empirical Models of Ionospheric Electron Density", NorthWest Research Associates Inc.,Bellevue WA, prepared for Mission Research Corporation, Santa Barbara CA under sub-contract.

SESC, 1988, download from SESC Public Bulletin Board,

Sheeley, N.R.Jr., C.R. DeVore, and J.P. Boris, 1985, "Simulations of the Mean Solar Magnetic Field During Sunspot Cycle 21," *Solar Phys.* 98:219-239.

Shimazaki, T., 1955, "Worldwide Daily Variations in the Height of the F2 Maximum Electron density of the Ionospheric F2 Layer", *J.Radio Res. Labs,* Japan 2(7):86-97.

Sojka, J.J. and R.W. Schunk, 1985, "A Theoretical Study of the Global F Region for June Solstice, Solar Maximum, and Low Magnetic Activity," *J. Geophys. Res.* 90(A6):5285-5298.

Spaulding A.D. and R.T. Disney, 1974, "Man-Made Radio Noise: 1. Estimates for Business, Residential, and Rural Areas", OT Report 74-38, Office of Telecommunications, Dept. Commerce, Boulder CO.

Spaulding, A.D. and G. Hagn, 1977, "On the Definition and Estimation of Spectrum Occupancy", *IEEE Trans. EMC* 3:269-280.

Spaulding, A.D. and J.S. Washburn, 1985, "Atmospheric Radio Noise: Worldwide Levels and Other Characteristics", NTIA Rpt.85-173, ITS, Boulder CO.

Spaulding, A.D. and F. Stewart, 1987, "An Updated Noise Model for Use in IONCAP", ITS, Boulder CO, NTIA Rpt.87-212, PB87-165007-AS, (available through NTIS, Springfield VA).

Stewart, F., 1990, private communication

Szuszczewicz, E.P., E. Roelof, R. Schunk, B. Fejer, R. Wolf, M. Abdu, J.Joselyn, B.M. Reddy, P. Wilkinson, R. Woodman, and R. Leitinger, 1988, "SUNDIAL: The Modeling and Measurement of Global-Scale Ionospheric Responses to Solar, Thermopheric, and Magnetospheric Controls," in *Effect of the Ionosphere on Communication, Navigation, and Surveillance Systems (IES'87)*, pp.321-330, J.M. Goodman (Editor-in-Chief), U.S. Gov.Print. Office, Washington DC, No. 1988-195-225, available through NTIS, Springfield VA.

Tascione, T.F., T.W. Flattery, V.G. Patterson, J.A. Secan, and J.W. Taylor Jr., 1979,"Ionospheric Modelling at Air Force Global Weather Central", *Solar-Terrestrial Predictions Proceedings*, vol. 1, pp. 367-377, ed. by R.F.Donnelly, U.S.Gov.Print.Office, Washington DC 20402.

Tascione, T.F., 1988, "ICED - A New Synoptic Scale Ionospheric Model", in *Effect of the Ionosphere on Communication, Navigation, and Surveillance Systems (IES'87)*, pp. 299-309, J.M. Goodman (Editor-in-Chief), U.S.Gov.Print.Office, Washington, DC. (available through NTIS, Springfield VA).

Tesla, D.D., 1985, "A Method of Channel Occupancy Monitoring in Adaptive HF Systems", Conf.Pub.#245, *Third International on HF Communication Systems and techniques*, pp.76-79, IEE, London, UK.

Teters, L.R., J.L. Lloyd, G.W. Haydon, and D.L. Lucas, 1983, "Estimating the Performance of Telecommunication Systems Using the Ionospheric Transmission Channel: Ionospheric Communications Analysis and Predictions Program (IONCAP) User's Manual", NTIA Report 83-127, NTIS Order No. N70-24144, Springfield VA.

Thomason, J., G. Skaggs, and J. Lloyd, 1979, "A Global Ionospheric Model," NRL Memo Report 8321, Washington DC, AD-000-323.

Turner, J.F., 1980, "GRAFEX Predictions", *Solar-Terrestrial Predictions Proceedings: Volume 4: Prediction of Terrestrial Effects of Solar Activity*, p.D2-85, U.S. Dept. of Commerce, Boulder CO.

Wilkinson, R.G., 1982, "A Statistical Analysis of HF Radio Interference and its Application to Communication Systems", IEE Conference Publication #206, *Second Conference on HF Communication Systems and Techniques*, IEE, London, UK.

Wright, J.W, 1982, "Global Real-Time Ionospheric Monitoring," in *Effect of the Ionosphere on Radiowave Systems, (IES'81)*, J.M. Goodman (Editor-in-Chief), U.S.Gov.Print.Office, Washington DC, available through NTIS, Springfield VA.

6

REAL-TIME-CHANNEL EVALUATION

ITS USE IN SHORT-TERM FORECASTING AND SPATIAL EXTRAPOLATION SCHEMES

What things may be by themselves we know not, nor need we care to know, because after all, a thing can never come before me otherwise ... than as a phenomenon.

Immanuel Kant [51]

6.1 CHAPTER SUMMARY

The concept of Real-Time-Channel Evaluation (RTCE) has been discussed in CCIR Report 889-1 [CCIR, 1986a]. That report suggests that the three stages for implementation of HF frequency management are: long-term forecasting (alternatively, comprised of *long-term prediction* and *long-term forecasting*), short-term forecasting (or simply *forecasting*), and RTCE (*nowcasting*), where the parenthetical terms are other alternatives used in this book. We have examined long term prediction in Chapter 5, and the nature of variability has also been outlined. We also introduced the term *hindcasting*, referring to an assessment of the recent past by using prediction and forecasting methods. In this chapter, however, we shall examine RTCE (or nowcasting) as a means for achieving an improvement in short-term and long term forecasting capabilities.

We begin the chapter by discussing RTCE concepts in general terms. Specific classes of RTCE are identified and examples are provided for each class. They include: oblique-incidence sounding (OIS), channel evaluation and calling (CHEC), vertical-incidence sounding (VIS), backscatter sounding (BSS), frequency monitoring (FMON), pilot-tone sounding (PTS), and error counting system (ECS). The VIS and OIS schemes are covered in dedicated sections of the chapter (viz., 6.3 and 6.4 respectively); whereas CHEC, PTS, and ECS are revisited in chapter 7. We shall spend a little more time on BSS

---

51. Quotation from *Critique of Pure Reason* (1781) by Immanuel Kant. The verse was originally cited by Prof. M. Darnell in his 1975 paper "Channel Estimation Techniques for HF Communications", published by NATO-AGARD [Blackband, 1975]. (Reprinted by Courtesy of Prof. Darnell, by permission Her Majesty's Stationary Office, London, UK.)

(in Section 6.2.2.2) and FMON (in Section 6.2.2.3), with the latter being presented in the dual contexts of communication and direction-finding (DF). The support for DF is given added emphasis (in Sections 6.2.2.3 and 6.3.7). The implications of sounders for HF frequency management is covered in considerable detail. The various model updating schemes are described, with special attention given to exploitation of sounder networks.

6.2 AN INTRODUCTION TO RTCE CONCEPTS

Before beginning, a brief comment on terms is worth reemphasizing. Report 889-1 [CCIR, 1986a] uses the phrase *long-term forecasting* as the first stage of frequency management. In practice, this long-term process involves an estimation of median parameters. It could involve the long-term prediction of solar indices or the incorporation of long-term trends in solar activity using certain classes of current data, such as coronal hole persistence. In this book, we have reserved for the word *forecasting*, whether long or short, that class of techniques which utilize current information (from the solar-terrestrial environment) as a basis. The operable word is the suffix *cast*. If no information is *cast*, either backward or forward in time, we have a pure prediction. As indicated in Section 6.1 above, the CCIR usage of long-term forecasting is actually a consolidation of long-term prediction and long-term forecasting. Being long-term in nature, the word prediction may be used if the context is clear. We propose to reserve the term forecasting to imply any estimate (or prediction) of the future which is based on a current observation. Thus, long- or short-term forecasts are distinguished mainly by the diminution in influence of the observable being applied to derive the forecast. In summary, forecasts extrapolate reality into the future by some appropriate algorithm. Thus forecasts carry local information (including variability) into the future. This has both good and bad features. A long-term prediction, on the other hand, is typically based upon an estimate of the future value of an external index (or set of indices such as K_p or R_I) which is generally used to drive a median representation. The equivalence of Nowcast and RTCE is obvious.

As has been indicated in Chapter 5, short-term forecasting (or simply forecasting) schemes may be successful if the methods allow for prediction of the onset, duration, and magnitude of events such as (i) Short-wave-fades (SWF), (ii) geomagnetic storm-related MOF diminution, and (iii) enhanced high latitude interactions. Other forms of variability such as those which are related to medium- and small-scale TID are not predictable in the traditional manner. Such disturbances present a challenge for practical RTCE systems as well; the more so if the RTCE observable corresponds to a path which is different from the communication path requiring assessment. Although no forecasting method can completely correct for the inadequacy of median representations, real-time measurements from satellites have been shown to reduce HF system outages significantly [Rothmuller, 1978]. This outage reduction was made possible through use of the PROPHET mini-computer terminal which processed the real-time (1-8 A) x-ray data downlinked from SOLRAD HI, and, under appropriate circumstances, provided an alert to the operator and frequency management subsystem. This process allowed partial but timely compensation for the SWF absorption and related effects to be introduced.

The first person to properly define RTCE was Professor Darnell [1978] who has examined various aspects of the subject in recent years [Darnell, 1975a, 1975b, 1979, 1983]. In the first of two 1975 papers, which were given at an AGARD specialists conference in Athens, Greece, Darnell suggests that an appropriate view for HF communicators is found in the philosophy of Immanuel Kant (1781) which simply says that detailed understanding of phenomena is unnecessary as long as results that matter may be characterized properly and adequately. As displeasing as this view may be to the physicist, it is the view which must be taken by the communicator for several reasons. First, the mission of communication must have highest priority without recourse to intermediate (albeit elegant) steps which provide (unnecessary) generality. Second, the process of channel characterization is direct, efficient, and may be more easily engineered into the communication system than may more fundamental approaches. Finally, the basic physics is incompletely understood anyway. In short, a system which is controlled by an appropriate RTCE process favors the design philosophy that the end result does justify the means for achieving it.

Darnell defines RTCE as a stage in the total frequency management system for HF. It appears in CCIR Report 889-1 and we repeat it below:

DEFINITION OF RTCE [52]

> Real Time Channel Evaluation is the term used to describe the processes of measuring appropriate parameters of a set of communication channels in real time and employing the data thus obtained to describe quantitatively the states of those channels and hence the capabilities for passing a given class, or classes, of communication traffic.

Typically RTCE is an ingredient in an adaptive HF communication system (See Chapter 7). Clearly RTCE implies that ultimate parameters of the channel which are important for successful communication are to be monitored. Underlying geophysical or propagation parameters such as solar flux, parameters which *drive* the channel, need not be measured. However, central to the successful exploitation of RTCE methods is the application of observables in a timely fashion. Channel evaluation measurements which are not at

---

52. Definition due to Darnell [1978]. A full discussion of RTCE principles may be found in the CCIR Report 889-1 published by the ITU [CCIR, 1986a].

least near-real-time do not contribute to RTCE, but may be useful as a benchmark for system analysis and the accumulated results may be maintained as a data base for median model development.

We have mentioned the necessity for timeliness in application of RTCE observables without yet indicating the nature of the observables themselves. The essential set of observables must be parameters of the channel which will define the set of waveforms likely to be used for communication. In other words, the observable should provide guidance about data rate or information rate, etc. In order to provide this guidance effectively, the *answers* should be in engineering units which are appropriate for the job at hand. To us the words in CCIR 889-1, the RTCE observable should be error rate for digital data, and intelligibility level for voice.

Figure 4-57 depicts a typical channel transfer function which is in essence a time delay versus doppler display. As indicated in Chapter 4, the time delay results from two factors: the first being related to the distended nature of the medium itself, and the second being caused by the frequency-dependent signal distortion. The frequency dispersion arises from a composite doppler shift of the refracting medium and doppler spread introduced by the *random* motion of scatterers.

Frequency and time dispersion introduce different problems. The former is responsible for time-selective fading, while the latter is responsible for frequency-selective fading. Intersymbol interference (ISI), which was identified in Chapter 4, is caused by multipath and can lead to an irreducible error rate regardless of SNR. A countermeasure to ISI is constructed by transmitting data in parallel channels at low data rates. Thus, in any given channel, the time between adjacent symbols may have a much greater probability of exceeding the multipath spread.

The three primary classes of RTCE are given in Table 6-1 along with a few examples.

TABLE 6-1. THE CCIR CLASSES OF RTCE

Class One: Remote Transmitted Signal Preprocessing
- a. Oblique-incidence sounder (OIS)
  - 1. Pulse Type
  - 2. Chirp Type
- b. Channel Evaluation and Calling (CHEC)

Class Two: Base Transmitter Signal Preprocessing
- a. Vertical-incidence Sounding (VIS)
- b. Backscatter Sounding (BSS)
- c. Frequency Monitoring (FMON)

Class Three: Remote Received Signal Processing
- a. Pilot Tone Sounding (PTS)
- b. Error Counting System (ECS)

6.2.1 Class 1 RTCE:

6.2.1.1 *Oblique-Incidence Sounding (OIS)*

The best known form of this class is the oblique-incidence sounder (OIS) which is discussed fully in Section 6.4. Several network schemes have been examined to provide a basis for a common user approach. Recently the Japanese have fielded such a system [Kuriki and Takeuchi, 1986]. The earliest known system of this class was developed in the United States and deployed in the Pacific basin. CURTS [Probst, 1968], standing for Common User Radio Transmission System, used a network of pulse sounders which were shared by members of the communication network. Communication connectivity was improved by maintenance of shared records of time and frequency dispersion, noise, interference, and measured SNR. The CURTS system was not simply a network of sounders. It was a rather complete communication network which incorporated RTCE (Class 1) to the fullest extent possible, considering the technology available at the time. The reader is referred to additional papers by Ramsay et al.[1967] and Felpenin et al.[1968]. The sounder sub-system within CURTS is sometimes referred to as an *out-of-band* system since the sounders transmit in unassigned channels. This might give rise to interference with other users. The use of short duration pulses limits this problem to one of manageable proportions. A replacement of the pulse sounders within CURTS with those based upon a chirp waveform, while not reducing the use of unassigned channel space, does permit a significant reduction in transmitter power.

6.2.1.2 *Channel Evaluation and Calling (CHEC)*

Another form of Class 1 system developed in the 1960s was CHEC, standing for Channel Evaluation and Calling. This Canadian system [Stevens, 1968] was an in-band sounding system. This means that the sounding *channels* were actually those (and only those) which were assigned for the transmission of communication traffic. Whereas CURTS was designed principally for use by a fixed network of communication nodes, the CHEC system was designed to eliminate difficulties encountered in mobile-to-base station communications. The concept involves base station transmission on selected channels. This transmission is composed of a *cw* portion and a coded portion. The latter consists of a calling code and information on base station interference in the channel being used. Each remote listens for its call and decodes the base station interference level for all channels for which transmissions are provided. In addition, the CW portions of the transmissions are processed for SNR. Propagation is taken to be reciprocal, but account is taken of the different antennas and noise levels for the base station and remote. A special purpose processor is used to compute the best channel for communication back to the base station. A microprocessor implementation of CHEC has also been developed by the Canadians to make the system more automatic. This system is called RACE standing for Radiotelephone with Automatic Channel Evaluation [Chow et al.,1981; McLarnon, 1982].

6.2.2 Class 2 RTCE

The distinctions between Class 1 and Class 2 RTCE systems are rather minor in terms of the waveforms utilized. The principal difference is that in Class 2 RTCE, all processing is done at the base station. The class is characterized by the fact that all activity (i.e., preprocessing, transmission, reception, and final processing) are done at a single station.

The Class 2 RTCE systems include vertical-incidence sounding (VIS), backscatter sounding (BSS) and frequency monitoring (FMON). The first method is the best known, the second is the most complex, and the third class 2 method is the simplest and probably the oldest form.

Even though FMON is basically a simple procedure, it presumes knowledge about the emitter being monitored. Desirable emitter characteristics include: transmitter location, transmit power, antenna gain, antenna pattern, antenna efficiency, polarization, transmitted waveform, local ground constants, and other siting factors. A knowledge of all of these characteristics is not essential for derivation of useful frequency management information. However, transmitter location *is* essential. Accordingly, in our discussion of FMON, we will outline some aspects of direction finding (DF). The discipline of DF is a separate and distinct activity, but may be viewed as a complementary set of data to facilitate the association of the measured properties of unknown signals with specified emitter locations. A passive remote sensing system (such as FMON) must possess information about the transmitter, a matter which is simplified for cooperative links, but it is much more complicated for noncooperative targets. We shall take a slight detour to discuss the DF problem in Section 6.2.2.3.

6.2.2.1 *Vertical-Incidence Sounding (VIS)*

The vertical-incidence sounder (VIS) has been the preferred method for providing information about the overhead ionosphere since the advent of ionospheric investigation. Other methods may provide more detailed information, or may explore regions inaccessible to the ground-based system, but VIS has no peer when cost and simplicity are considered. Virtually all mean models of the lower ionosphere are based upon data derived from VIS systems. In recent years, the oblique-incidence sounder (class 1 RTCE) has replaced the VIS approach for many real-time applications. Still for NVIS geometry and specified tactical HF communication service, VIS is quite important. Discussion is continued in Section 6.3.

6.2.2.2 *Backscatter Sounding (BSS)*

It is well established that neither the ground nor the ionosphere behave as smooth mirror reflectors. For oblique skywave paths, ground backscatter has been used to determine ionospheric properties [Clark and Peterson, 1956; Valverde, 1958]. A survey of the ground backscatter phenomenon is found in a paper by Dieminger [1960], and direct backscatter of HF radar signals from land, sea, and ice surfaces has been reported by Hagn [1962]. Ionospheric motion was monitored by the BSS technique from a sight near Sterling, VA in the late 1950s [Tveten, 1960].

Recently, the Soviet Union has proposed the use of BSS as an economical way to determine sea surface conditions, detect icebergs, tsunami, and ships, and for investigation of the ionosphere [USSR Acad. Sci., 1990]. A prototype Soviet system is termed Bazis-3-32 and is located in Izmiran. The USSR has proposed selling the technology under the name "ICEBERG". The use of BSS for remote sensing of sea surface conditions has also been carried out by agencies in the United States and elsewhere.

Backscatter sounding is also a method which has found use in remote assessment for specific Over-the-Horizon Radar (OTHR) applications [Headrick and Skolnik, 1974]. The French have also examined the concept as an aid in analysis of OTH radar paths [David et al.,1976]. One of the problems in OTHR targeting and tracking is the lack of registration because of the poor ionospheric information between the radar and the target area. This results in a nontrivial transformation of target-echo time-delay to ground range using estimates of the ionospheric height of reflection possibly derived from median models. Fortunately, if engineered properly, it is possible to operate an OTH radar as a BSS system. In this mode, backscatter sounding may provide necessary ionospheric data, although the conversion of a backscatter ionogram into useful information is itself a difficult task.

Typically virtual ray tracing methods are used in OTHR propagation models such as RADARC. Path loss calculations using virtual methods have been shown to compare well with the Jones-Stephenson method [Thomason et al.,1979]; and they perform adequately for OTH radar predictions provided adjustments are made for loss factors such as those introduced by sporadic E.

To be observed at the radar, a target or a segment of ground clutter must scatter energy back toward the radar, so ray re-tracing is implicit in an assessment of detectability. Unfortunately there are a number of competing (surface) scatterers in the practical case. To make matters worse, the ionosphere may not be modeled accurately. In fact, it is well known that ionospheric tilts and gradients along the radar great circle path will introduce virtual (reflection) height variations. Such variations can be significant in HF communication as well as HF radar situations. Superimposed on this systematic large-scale picture are small-scale fluctuations of modest time duration which are not predictable (See Section 3.8 dealing with TID effects). Figure 6-

1 shows a set of nocturnal ionograms for a single hour at four distinctly different latitude zones.

For a specified zone, each trace in Figure 6-1 corresponds to a fixed hour but a different day of the month. Thus we are observing day-to-day variability in virtual height versus transmission frequency at vertical incidence. There are also hour-to-hour variations to contend with. All of this points unmistakably to the requirement for a real-time frequency management system for OTHR systems.

Figure 6-2 is a sample backscatter ionogram for 1-,2-, and 3-hop returns. Figure 6-3 is a stylized copy of the ionogram, suitably annotated to identify important features. We note immediately the long time delays associated with specified ground ranges. This is because we are dealing with a two-way path. The BSS method has been examined by Croft [1967, 1972]. Recently Headrick [1987] has examined BSS in connection with HF broadcasting, a field which many feel is perfectly suited for the use of backscatter sounding as a means to estimate signal strength *laydown* in the targeted (broadcast) areas.

Caratori et al. [1988] have described a set of backscatter ionogram inversion methods. They claim that assessment over a radius of 3000 km is possible, and that the technique is directly applicable to improvement of HF communications over the assessed zone. Using the Australian JINDALEE radar, Lees and Thomas [1988] have examined several ionospheric features including: TIDs, the terminator, normal geographical variations, and even meteor trails.

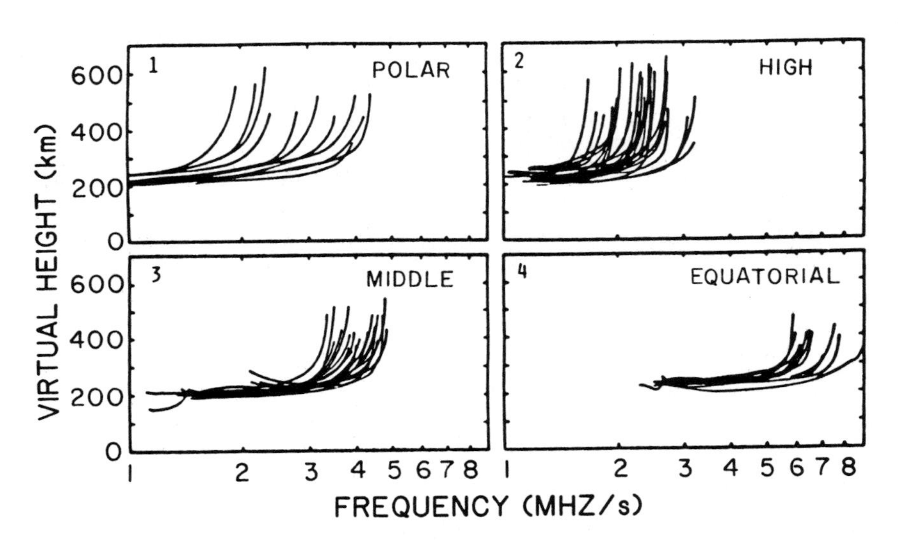

Fig.6-1. Typical nighttime distributions of $h'(f)$ for: (1) polar, (2) high, (3) middle, and (4) equatorial latitude regimes. (Courtesy J.M. Headrick, Naval Research Laboratory.)

They argue strongly for radar backscatter systems as a means to cover large areas with a resolution sufficient to permit successful frequency management of highly oblique paths which are not supported by more conventional sounder systems. Anderson and Lees [1988] have listed among the attributes of JINDALEE the system's doppler resolution of 0.01 - 0.1 Hz as compared to typical ionospheric doppler shifts of 0.1 to 2 Hz. This suggests that the system may provide reasonable estimates of ionospheric layer motions. The spatial resolutions are stated to be of the order of 5- 50 km, and these cells may be assessed rapidly through the parallel processing of multiple beams. Consequently it is possible to obtain coarse maps of *instantaneous* phase path variations over a communication area. Methods for estimating the magnitude of signal *laydown* in the coverage areas have been well established, and may be verified by analysis of backscatter returns. It would appear to the author of this book that a marriage of BSS yielding broad-brush area coverage, with OIS yielding more specific information about given paths, would be useful. It is unfortunate that BSS systems require such a large capital investment given the wealth of information which may be derived. Even so, a number of OTH radar systems have been deployed or are in various stages of development. The primary missions for these system are radar surveillance, but the newer systems have the capability to provide another view of the ionosphere and could be effective RTCE assets. One system has been design specifically for ionospheric studies and to support HF radiowave propagation predictions [Le Saout and Bertel, 1988].

The world's first HF OTH radar, called MUSIC, was developed at NRL in the fifties and was used to detect nuclear explosions at distances of 1700 miles [Gebhard, 1979]. A further development, called MADRE, located at Chesapeake Bay Division of NRL, was the first OTH radar capable of detecting aircraft and missile targets to a distance of over 2000 nautical miles. MADRE has formed the basis for all current OTH radar designs.

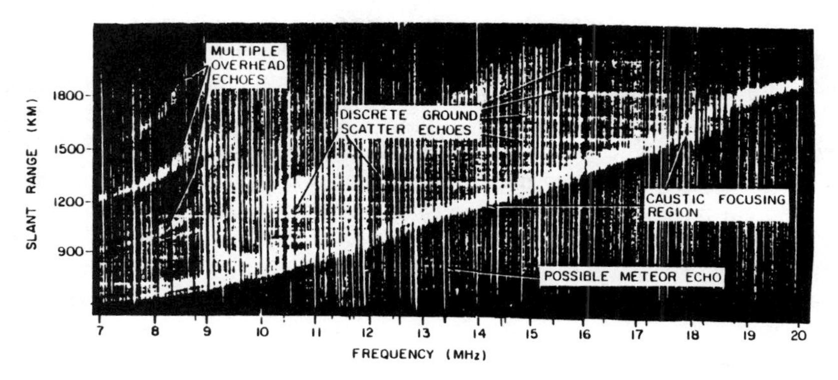

Fig.6-2. Sample backscatter ionogram, showing 1-, 2-, and 3-hop echoes in a time delay (range) versus frequency format. (Courtesy of J.M. Headrick, Naval Research Laboratory.)

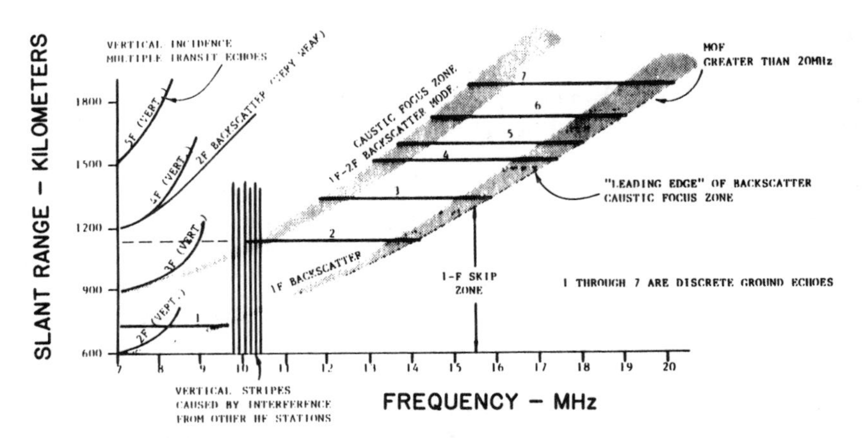

Fig.6-3. Oblique backscatter ionogram. Details corresponding to Figure 6-2. (Courtesy of J.M. Headrick, Naval Research Laboratory.)

Headrick [1990a] examines the current developments in OTH radar, and a large amount of technical information may be found in Chapter 24 [Headrick, 1990b] of the *Radar Handbook* edited by M. Skolnik [1990]. Table 6-2 is a listing of some of the major systems under various stages of development and/or operation. All the listed systems use ionospheric bounce for detection of long range targets. Another HF OTH system, being developed in the United Kingdom, uses surface waves instead of skywaves for detecting low-flying targets and cruise missiles over distances of the order of 200 miles from the radar. This system takes advantage of the reduced propagation losses suffered by ground waves when the ground is (salty and highly-conducting) seawater. See Chapter 4 for more details on ground (or surface) wave propagation.

In conclusion, the contribution of HF radar to the fields of ionospheric specification and modeling has great potential and could be substantial. However, because of the limited number of systems, and the lack of their availability for scientific investigation, contributions to the ionospheric data base have been relatively limited. Special-purpose HF radars have been developed to examine the high latitude ionosphere [Greenwald, 1985], and such systems have provided much information about physics of the various circumpolar features such as the high-latitude trough, the auroral zone, and the polar cap. The paper by Croft [1972] should be read by the serious specialist as a starting point. The various ionospheric propagation effects have been covered by Millman [1978] in considerable detail. Millman suggests that ionospheric errors in HF OTHR may be minimized by using ground-based transponders and swept-frequency sounders.

TABLE 6-2

OVER-THE-HORIZON (OTH) RADAR SYSTEMS

FACILITY	AFFILIATION/LOCATION	REFERENCE
MADRE	NRL/Chesapeake Bay MD	[Headrick and Skolnik, 1974]
JINDALEE	DSTO/Central Australia	[Sinnott, 1986]
JORN	Australia	JINDALEE Operational Radar Net
CNET	CNET/Lannion-France	[Le Saout and Bertel, 1988]
CRIRP	Peoples Republic of China	[Peinan and Du Junho, 1988]
OTH-B	U.S. Air Force	[C^3I Handbook, 1988]
ROTHR	U.S. Navy	[Headrick, 1990a]

As would be expected, the status of many radar and telecommunication systems, which have been developed with military requirements being the principal motivation, is constantly changing. This situation is certainly true in the case of OTH radar. This means that political decisions, rather than scientific ones, may alter any progress which may be made in this area. Possible civilian applications of HF OTHR technology could include the remote sensing of sea state, and the prediction of radio program broadcast quality in specified coverage zones. Aside from the obvious use of OTHR for military target surveillance, it may also be exploited synergistically with other military systems which require ionospheric data from denied areas.

6.2.2.3 *Frequency Monitoring (FMON)*

The frequency monitoring technique is one in which the receiver at location A is scanned in order to detect signals of known origin at location B. Using reciprocity, frequencies which can be received at A are assumed to be capable of receipt at B, if the terminal equipments were reversed. If A and B are nodes in a communication network, and if the test signals from B consist of a range of frequencies, we have the basis for a crude channel probe for frequency management purposes. In the more general situation, the receiver at A is scanning a set of beacons or check targets which are not part of the communication network. Frequently we wish to communicate to a specified node C given FMON information from emitters at B_i, $i = 1,2,...,n$. Provided the target signals can be identified (by call sign or other signature), then it may be possible to estimate the efficiency of propagation between A and C. Precise estimates suggest the requirement for additional information: transmitter power, transmitter antenna gain, and displacement of node C from the emitter population $\{B_i\}$. A succeeding generation system could reduce the requirement for location information if precise direction finding (DF) capability is organic to the FMON system. This implies the existence of antenna structures and algorithms which will permit geolocation.

A text which provides a convenient starting point for those interested in HF direction finding was produced by Gething [1978]. An updated version emphasizing superresolution techniques has been published [Gething, 1991]. Adolf Paul [1985] has examined the concept of passive monitoring with a view toward improvement of HFDF products, and the concept of check target utilization has been examined by Elliott et al.[1990]. A major difficulty in emitter location is not hearability. The problem is mainly associated with ionospheric spatial variability which introduces refractive variability and considerable geolocation uncertainty. Several schemes may be used to locate an HF emitter. In one type of system, geographically dispersed but coordinated sites provide azimuth (or bearing) data to net control in response to a request for the observed bearing of a designated target. These returned lines of bearing are processed at net control by using an appropriate fix algorithm. The confidence of each fix so obtained is dependent upon many factors; some are site-specific and others depend upon ionospheric effects. A classic curve illustrating bearing error variance was developed by Ross [1947] and is reproduced in Figure 6-4. This empirical curve is only indicative of what might be expected for midlatitudes under *nonpathological* conditions.

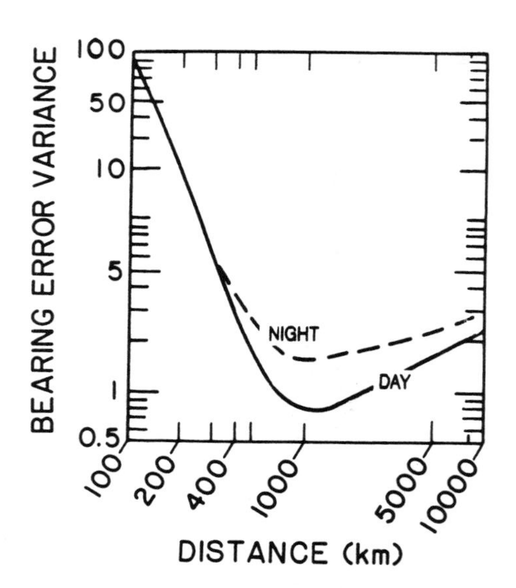

Fig.6-4. The Ross curve exhibiting bearing error variance as a function of distance from the DF site. The minimum variance is obtained at approximately 1000 km, and is greater at night than during the day. (From Ross [1947].)

Figure 6-5a is a sample time history of measured bearing to a fixed transmitter located at middle latitudes. The fluctuations observed are real and represent fluctuations in the ray trajectories which occur during the period of observation. Figure 6-5b shows the nature of azimuthal fluctuations if the emitter is close to the receiver. These rapid and relatively large variations make geolocation of close-in emitters an interesting challenge indeed.

A distinct method for HF emitter geolocation at a single site, called single-site-location (or SSL) involves the measurement of both elevation and azimuth of the incoming signal. Elevation data will enable an estimate of target range to be made assuming the number of ionospheric hops is known.

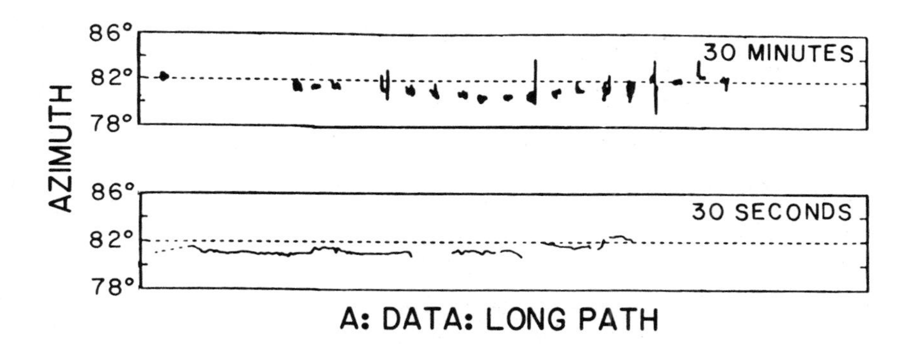

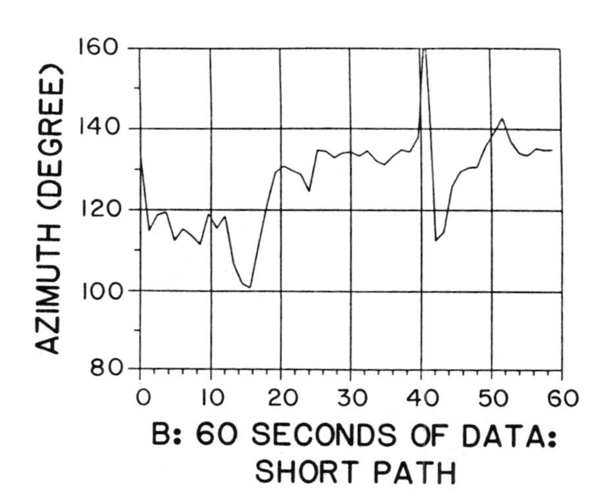

Fig.6-5. (a) Angle-of-Arrival (bearing) fluctuations observed over a 1500 km E-W path in continental United States [Goodman, 1990]. (b) Bearing fluctuations of WWV (5 MHz) at a distance of 86 km from Brighton, Colorado [Paul, 1985].

An estimate of ionospheric height is also necessary for successful application of the SSL method. The SSL method also requires information about the number of hops involved, since an elevation-azimuth pair does not provide a unique answer. Because of the fact that HFSSL systems are required for tactical military operations for which the skywave paths are virtually NVIS in nature, overhead tilts may present a serious problem. Sounders or equivalent systems are essential to the success of SSL. The problem is that they are necessary but are not always sufficient.

From the point of view of extracting RTCE information from signals of interest, it should be noted that a single emitter has the potential to present multiple signals to a receiver. Since these signals represent different paths through the ionosphere, they preserve a *memory* of properties which are likely to be independent of one another. From a communication point of view, if we are able to resolve the paths, we can exploit the path independence properties to achieve a degree of diversity gain. From the point of view of ionospheric assessment, the isolation and analysis of the family of signals from the known emitter will provide an augmented assembly of *control points*. Under certain conditions one may be able to associate specific properties to these *extra* points. We shall return to this later on when we discuss OIS systems and the analysis of oblique ionograms.

Figure 6.6 contains data from a set of oblique ionograms for a path between Skaggs Island, California and Peterson AFB, Colorado. Also given in the figure is a companion display showing the relative signal level for a preselected frequency of 13.45 MHz on the ionogram sweep. The ionograms with start times at 1700 and 1705 GMT both exhibit a strong one-hop low-ray return from the F2 layer at the selected frequency with minor multipath returns above and below. These correspond to a one-hop high-ray signals and E-layer returns. A strong sporadic E component, so evident at 16 MHz and above, is not a factor at 13.45 MHz. A new feature appears on the 1710 and 1715 GMT ionograms; a strong F2 high ray begins to dominate and the O- and X-mode components emerge as resolved components. Although underlying layers do provide some contribution to the multipath, ionograms at 1720 and 1725 GMT exhibit two dominant F2 layer returns, one corresponding to the low ray and the second corresponding to the high ray. The O- and X-mode components are no longer resolved (i.e., separated in time delay) at this time.

Bearing and elevation angle distributions derived from an experimental HFDF array are given in Figures 6-7a and 6-7b respectively. The data was accumulated over roughly one-half hour, whereas each ionogram was a single 2-32 MHz sweep accomplished in less than 5 minutes. Two aspects are worth noting. First, we observe significant dispersion in the elevation plane but with a preponderance of signals being detected within two elevation cells. In actuality four populations are observed: a discrete pair and a dispersed pair. Only one population is evident in the azimuthal plane. This indicates that the signal bifurcation (and spreading) is largely associated with high and low rays.

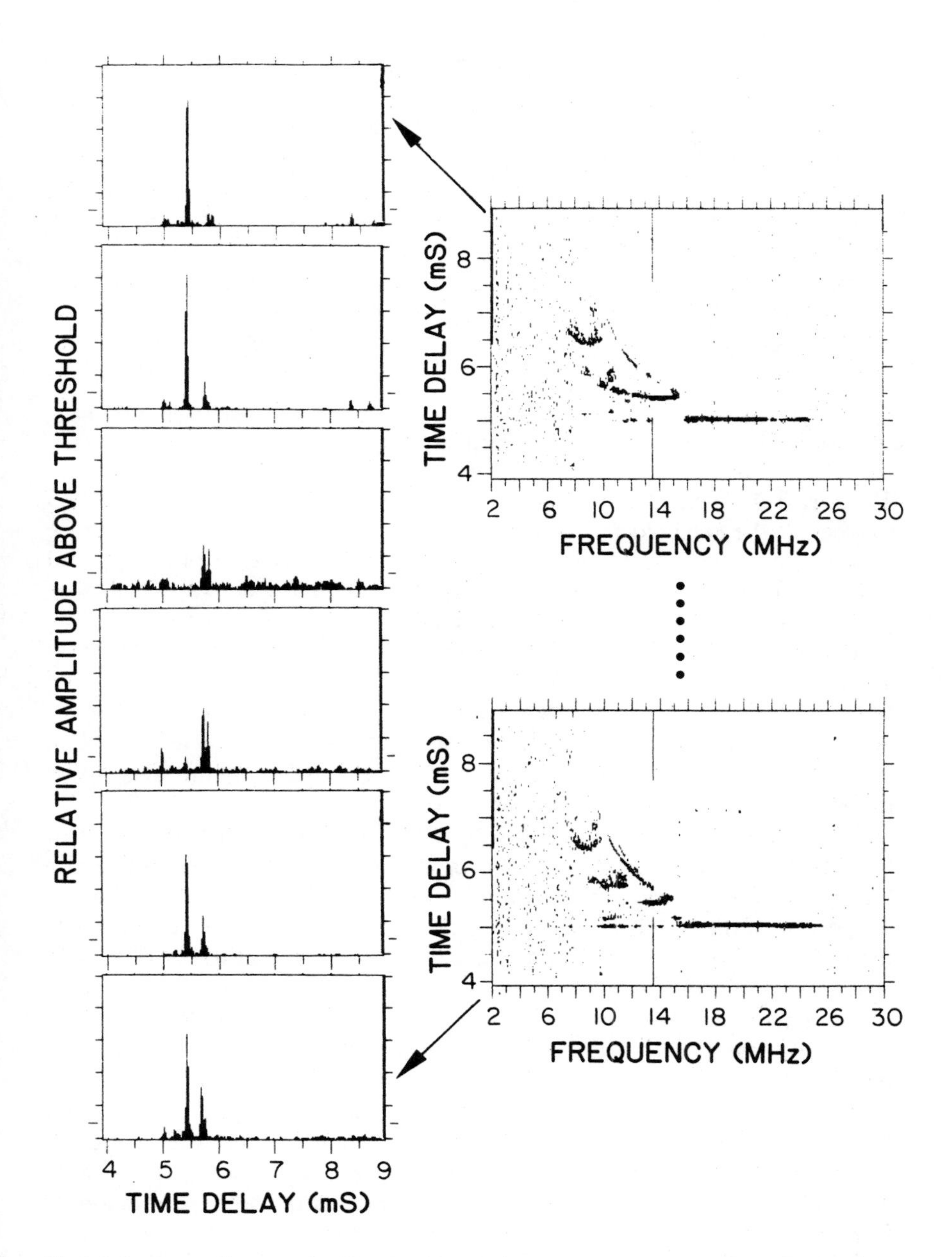

Fig.6-6. Sequence of consecutive ionograms and corresponding multipath spread diagrams with five-minute spacing showing time variability of mode structure at a fixed frequency. Only the starting and ending ionograms are displayed. The multipath spread diagrams are determined for a frequency of about 13.5 MHz. (From Goodman [1990].)

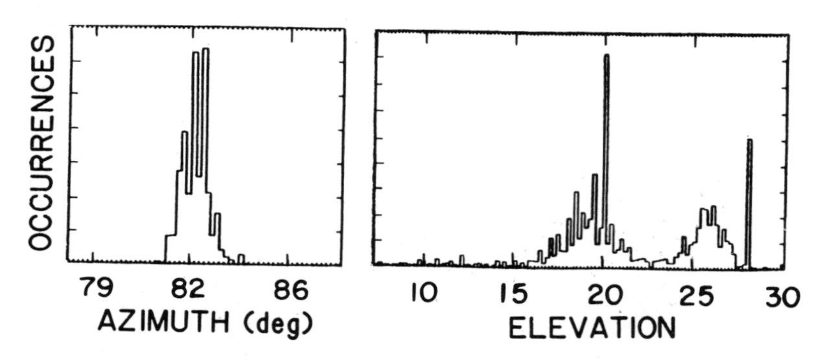

Fig.6-7. (a) [Left] Azimuthal distribution of incoming skywave signals. (b) [Right] Elevation distribution of signals showing multimode propagation. (From Goodman [1990].)

These rays are reflected from the F2 layer, being separated by roughly 5 degrees at the selected frequency. At the MOF of roughly 15 MHz, there would be no such separation in elevation. In fact this is where the high and low rays merge to form the junction frequency or MOF [Goodman, 1990].

It is some interest to note that for an emitter at fixed frequency $f <$ MOF, a reduction in MOF will reduce the time-delay separation between the high and low rays at f. Thus in the late afternoon , as the solar zenith angle is increasing and the MOF is likely to be decreasing, we would expect a gradual convergence of the two signals until such time as $f =$ MOF. Naturally as $f >$ MOF, all support is lost. The reverse situation is true in the early morning hours following sunrise. For more details, refer to Chapter 4.

In conclusion, we find that FMON techniques are the simplest form of RTCE, and with the presumption of propagation reciprocity, may provide the HF operator with a reasonable estimate of signal *laydown* in the region of interest, with the proviso that the distance between the FMON emitter and the communication node of interest is not too great. The utility of FMON in assessment and forecasting of communication performance relies on additional factors, such as knowledge of noise and interference at both ends of the link. This information is not always easy to obtain and disseminate. Consequently, the primary use of FMON as an RTCE scheme is associated with its value in estimating the frequency limits for ionospheric support. MOFs and LOFs cannot be deduced directly unless the ensemble of candidate emitters is finely partitioned over the HF spectrum. Generally speaking, the FMON emitter should be within a few hundred kilometers of the desired node or terminal, although the separation will more or less depend upon circumstances. This maximum separation is related to the correlation distance of ionospheric fluctuations. We shall return to the issue of correlation distance in Section 6.6 which deals with sounder update and extrapolation schemes.

6.2.3 Class 3 RTCE

Pilot-tone sounding and error counting are major forms of Class 3 RTCE. In this class, the channel stimulus and the desired information are multiplexed and transmitted over the HF channel. At the receive end, the signal is demultiplexed, and the channel conditions are derived from an analysis of the recovered stimulus component of the signal. This is used to drive a signal processing method which is appropriate for reconstructing the desired message or data traffic from the corrupted information component of the received signal. The class 3 form of RTCE has been tested on a number of links and it has been demonstrated that significant improvement in signal integrity may be achieved by incorporation of these rather simple methods.

6.2.3.1 *Pilot-Tone Sounding*

Betts and Darnell [1975] have discussed this method, and noteworthy theoretical contributions have also been made by Bello [1965]. The fundamental concept is one in which low-amplitude cw tones are inserted at convenient spectral locations associated with the transmitted waveform. Possibilities include the insertion of tones in (unused) channels contiguous to the communications traffic channels, or insertion of the tones within the spectrum of the data stream itself.

The method is both simple and powerful. Parameters which may be estimated from an analysis of the CW tones include: signal amplitude, SNR, signal phase, Doppler shift, Doppler spread, and time delay spread caused by multipath effects. There are a few concerns about the method, however. One of the pacing issues is the extent to which information derived from pilot tones may be indicative of data errors for specified modulation formats. This concern has been more or less resolved by experiments conducted by Betts and Darnell [1975]. These workers have shown that FSK data errors are well predicted by pilot tone phase errors over single and multiple-hops provided fade rates were slower than 1 Hz and the signal-to-interference ratio is large. The second of these caveats is necessary since pilot tone analysis is only appropriate to the wanted signal; spurious interference is a parameter whose effect must be analyzed or compensated for separately. Another potential difficulty of the pilot tone method is associated with an inability to *predict* MOF failure, since it is, after all, not a sounder. Naturally one could insert cw tones throughout the entire HF spectrum and simulate a sounder system, but this defeats the purpose.

A channel quality sounder, the CQS-100, based upon pilot-tone principles has been developed by Andrew Corporation. Information about this device may be found in a paper by George and Halligan [1985].

6.2.3.2 *Error Counting System (ECS)*

In error counting the channel stimulus is designed to replicate the data traffic so that an analysis of the received stimulus will be equivalent to an analysis of the information bit stream. Alternatively, overhead may be reduced by utilizing the traffic channel directly to derive RTCE information. The data quality or data error rate would be used in a subsystem to determine if frequency changes would be prudent. Darnell [1978] has discussed some possible approaches.

6.3 VERTICAL INCIDENCE SOUNDING (VIS)

The vertical-incidence sounding (VIS) form of RTCE is undoubtedly the most familiar, and has provided most of the information we have today concerning lower ionospheric structure. In this section we shall describe briefly the equipment which is used in this technique, including the modern digital systems. Next we shall discuss some methods of data interpretation, beginning with a comment on the product, i.e., the ionograms themselves. We have previously listed the primary ionospheric parameters which may be extracted (see Chapter 4), and we shall cover these and related ionogram features more completely in this chapter. This topic is followed by a section on applications in the context of RTCE for communication purposes. Discussion of the VIS archival data base is provided in Section 6.3.4, and discussion of the possibilities of linking VIS systems and alternative RTCE instruments in a network configuration is contained in Section 6.7.

6.3.1 Description of the Instruments and Operational Procedures

Given the number of years that vertical-incidence ionospheric sounders (usually called ionosondes but which we abbreviate as VIS) have been used, it is not surprising to find a considerable amount of source material describing the system types and specialized hardware used in specific implementations. Thorough coverage of the full set of system components would require too much space, so we shall cover various evolutionary developments and consider the general concepts. Background may be found in Mitra [1947] and Davies [1965]. Recent information is available in books by Hunsucker [1990] and Davies [1990]. The Air Force Handbook [Jursa, 1985] contains a nice summary of current systems and analysis approaches. The reader is also directed to a group of publications from the World Data Center A in Boulder dealing with various aspects of the subject. References to these publications will be made at appropriate points in the text, but special attention is drawn to the *URSI Handbook of Ionogram Interpretation and Reduction* [Piggott and Rawer, 1972, 1978], and a series of bulletins published by the Ionospheric Network Advisory Group (INAG).

Following the work of Breit and Tuve and others, investigators began to examine the time delay (or virtual height) of ionospheric layer echoes at the transmitted frequency. Since the time delay of these echoes varied with frequency, an examination of the effect over a span of frequencies was carried out. This led to a system in which the transmitter and receiver were both synchronously scanned. Beginning about 1932, the first measurements of the virtual height of reflection, h', versus frequency were made. The $h'(f)$ curves became known as ionograms. One of the first internationally-coordinated scientific campaigns, the Polar Year in 1932, provided the circumstances for proving the merit of this new instrument. In ionospheric investigation the importance of continuous observation cannot be overstated. A station located at Slough in England has provided ionospheric researchers with continuous and regularly-scheduled ionograms since about 1935. This corresponds to five complete solar cycles (six maxima and five minima having been observed). By 1970 more than 40 countries operated sounders, and about 130 of the instruments were in operation around the world.

Analog systems naturally dominated the field until the 1970s at which time hybrid analog/digital systems appeared leading to the emergence of fully digital features in modern instruments of the 1980s. The older systems consisted of several famous versions including the U.S. Department of Commerce C-2 through C-4 sounders, the CSIRO and IPS systems in Australia, the Cossor system in Canada, and Union Radio Systems affiliated with the U.K., to name a few. These systems all swept through the sensible HF band of interest for vertical propagation beginning at about 1 MHz and terminating between 16 and 25 MHz depending on the system. All used high power. The C-4 system, for example, had a transmitter power of 25 kW. The height range was selectable between 500 and 1000 km, and the sweep time was 27 seconds. The ionograms produced by these systems are therefore quite useful in analysis of short-term ionospheric fluctuations with periods in excess of 2 minutes. (Note that the current brand of chirp waveform systems have sweep times of about 5 minutes limiting the resolution to disturbances having periods in excess of 10 minutes. Fortunately most of the ionospheric fluctuations of interest to frequency managers are not of the short-term variety.) Virtually all of the systems display the ionograms on a logarithmic frequency scale. This is an aid to analysis.

Modern digital sounders began to appear in the 1970s and 1980s. Like the situation which led to the improvements in HF communication systems, the digital sounder emerged as a direct result of new device and microprocessor technology. Several systems are listed in Table 6-3. In this table we exclude models of sounder systems which are used for OIS purposes. Oblique-incidence sounders, such as the AN/TRQ-35 (Chirpsounder™), the DGS-256, and a RRL (Japanese) system, will be discussed in Section 6.4. It is worth noting that the Chirpsounder™ technology was first applied to an OIS system, but was modified for VIS operation. The reverse is true for the Digisonde™.

TABLE 6-3

LIST OF RECENTLY-DEVELOPED IONOSPHERIC SOUNDERS

NAME OF SYSTEM[53]	MANUFACTURER OR AGENCY	REFERENCES
Dynasonde	NOAA/SEL Boulder CO-USA	[Wright and Pitteway,1982a,b] [Grubb,1979]
Digisonde™	ULCAR, Lowell MA,USA	[Bibl et al.,1981] [Bibl and Reinisch,1978] [Bulletin 48; INAG,1986]
IPS-5A	IPS Radio and Space Services Sidney,Australia	[Fox and Blundell,1989]
VIS Chirpsounder™	BR Communications Sunnyvale CA,USA	[Barry,1971] [Spec.sheet; BRC,1980]
Skysonde™	Andrew Government Systems Garland TX,USA	[Spec.Paper; AGS,1990a]
Tiltsonde™	Andrew Government Systems Garland TX,USA	[Spec.Paper; AGS,1990b]
610M1	SRI International Menlo Park CA,USA	[Spec.Sheet; SRI,1982] [Bulletin 3; IDIG,1982]
IPS-42	KEL Aerospace Australia	[Spec.Sheet; KEL, 1982] [Bulletin 3; IDIG,1982]
RRL Sounder	Radio Research Lab Tokyo,Japan	[Spec.Data; RRL, 1986] [Bulletin 48; INAG,1986]
Portable Ionosonde	ULCAR[54] Lowell MA,USA	[Haines et al., 1989]

---

53. Digisonde is a registered trademark of the University of Lowell Center for Atmospheric Research (ULCAR), Chirpsounder is a registered trademark of BR Communications, and both Skysonde and Tiltsonde are registered trademarks of Andrew Government Systems.

54. The portable ionosonde has been developed in cooperation with U.S. Army CECOM located at Ft. Monmouth NJ, U.S.A.

Approximately 55 of the C3/C4 type vintage sounders were constructed in the United States, while Union Radio in the United Kingdom built 37 ionosondes. Many of these are no longer used or are in marginally-operational condition. Nevertheless, some of these older systems do continue to operate, faithfully providing data for scientific studies, but the newer systems are beginning to dominate. For example, the C3/C4 sounder located at College, Alaska, and active continuously since 1942, has been replaced by a DGS-256. The Australian KEL systems, specifically the KEL/IPS-42, enjoyed rapid growth in the early 1980s, and the ULCAR DGS-256 (which replaced the DGS-128) exhibited strong sales in the latter part of the 1980s. Also it is noted that Andrew Corporation has developed two vertical incidence sounder systems: one a 3-channel sounder called Tiltsonde™, and a second called Skysonde™ having a military designation AN/TRQ-40.

6.3.1.1 *Digisonde™ (DGS)*

A significant number of digisondes have been deployed as replacements for the older analog systems. Additional deployments include those associated with the Air Weather Service, which plans to establish a network of stations to satisfy its mission objectives. Some research facilities include a DGS-256 installed aboard an AFGL KC-135 airplane for high latitude studies, a DGS-256 installed for the Army at Ft. Monmouth for communication studies, and at least one DGS-256 which is being used by AFGL scientists for ground-based studies. The DGS-256 has the ability to operate in a number of ways: VIS, bistatic OIS (see Section 6.4), BSS [Hunsucker and Delana, 1989], and as an ionospheric drift measurement system. We shall comment principally on the VIS scheme of operation in this section.

The full DGS-256 system is housed in a standard 19" rack, it includes a 7-element antenna receive array, a transmit antenna, a 10 kW pulse transmitter, and appropriate matching transformers, cables, peripherals (including a remote terminal) and software for ionogram scaling (ARTIST). The system employs phase-coherent spectrum integration of the quadrature signal samples. This enables the delivery of relatively high SNR, and Doppler spectra may be recovered and analyzed for estimation of ionospheric motions. Since the DGS-256 includes an array of seven antennas configured to form a narrow beam, there are options available to determine the Angle-of-Arrival of the downcoming ionospheric echoes. This determination is done on a pulse-to-pulse basis.

An important feature of the DGS-256 system is the use of polarized antennas enabling the separation of O and X modes. This is an important factor in the successful operation of the ARTIST autoscaling algorithm which is configured to run on an IBM PC/AT. The sounder operation is design to adapt to local conditions, with the system parameters under software control. For example, it is possible to automatically adjust the transmitter power to the minimum level required to achieve an acceptable SNR. At the receive end, the AGC is digitally controlled for each frequency which is used in the sounding process. Figure 6-8 shows a DGS-256 digital ionogram obtained under quiet benign conditions, and Figure 6-9 corresponds to spread-F conditions.

A rather complete description of the DGS-256 ionospheric sounder system in Volume 2 of the Handbook describing the World Ionospheric/Thermospheric Study (WITS) [Reinisch et al.,1989]. The general DGS-256 deployment plan was described in Bulletin 48 [INAG, 1986]. As of the latter part of 1989, the full *network* of Digisondes consists of a group of about 40 units currently positioned, under construction, or planned. Of these units, nineteen are under the aegis of the U.S. Air Weather Service. We shall return to the network aspects in Section 6.7.

Although the Digisonde can provide useful HF channel information to assist in radio communications, it may also be used as a scientific device. A single DGS-256 system, capable of monitoring the Doppler shifts and angles-of-arrival of downcoming signals, may extract properties of atmospheric gravity waves (such as wave amplitude, phase, wavelength, and velocity). This is a cost effective method compared with an alternative analog scheme whereby a number of stations would be operated simultaneously. The DGS-256 may also operate in an OIS mode. Figure 6-10 contains three successive oblique ionograms for a path from Millstone Hill, MA to Goose Bay, Labrador.

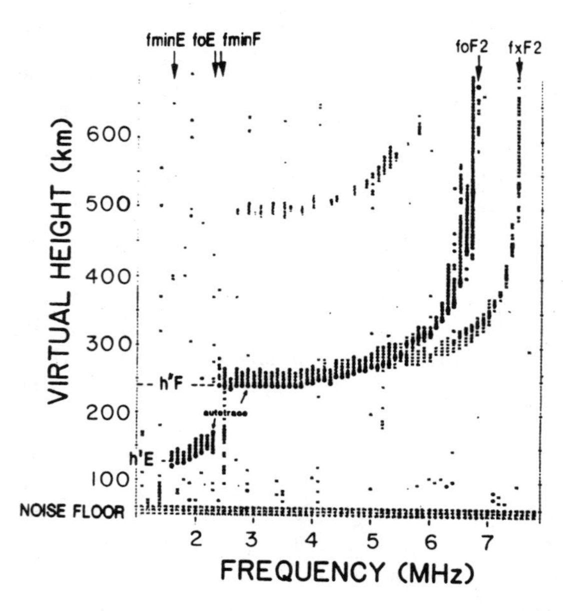

Fig.6-8. DGS-256 ionogram obtained at Goose Bay, quiet conditions, 3-15-83, 1700 AS. (From INAG Bulletins 40/41, Reinisch [1983]).

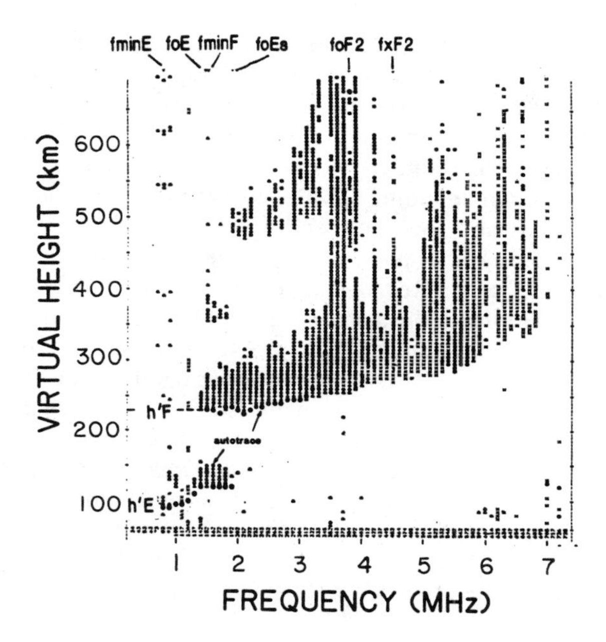

Fig.6-9. DGS-256 ionogram obtained at Goose Bay, disturbed conditions, 3-15-83, 1829 AST. (From INAG Bulletins 40/41, Reinisch [1983].)

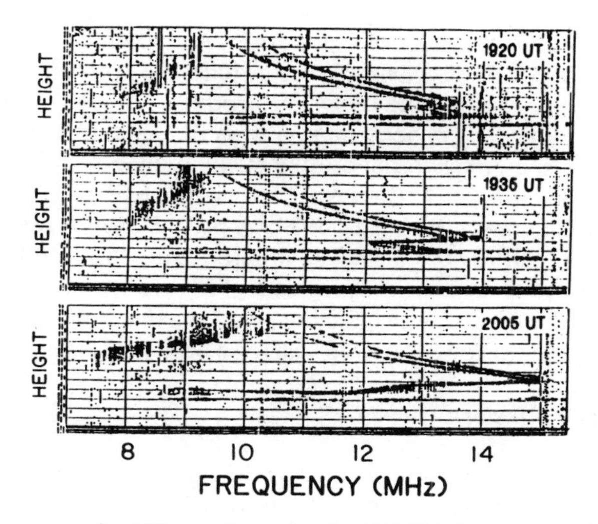

Fig.6-10. Three consecutive OIS recordings using the DGS-256. Data were obtained in the afternoon, on August 24, 1989, over a path between Goose Bay, Labrador and Millstone Hill, MA. (From Reinisch et al. [1989].)

To satisfy U.S. Army tactical communication requirements, the Army Communications-Electronics Command at Ft. Monmouth, NJ, has cooperated with ULCAR in the development of a portable ionospheric sounder [Haines et al., 1989]. This system emulates most of the features of the DGS-256 in a small package and at substantially reduced power. The following observables are available for each sounding frequency: virtual height, amplitude and phase, doppler shift and spectrum, and wave polarization. Standard pulse sounders require 5-50 kW of peak power to overcome the atmospheric noise in the receiver passband. (Recall that receiver noise is not a factor in the lower part of the HF band; the system is external noise limited by possibly as much as 50 dB.) The situation is made even worse by the fact that coherent pulse sounders require a relative large receiver bandwidth to resolve ionospheric heights effectively, but this is inconsistent with reducing the external noise influence. The portable pulse sounder was designed to resolve these problems by employing pulse compression and Doppler integration, thereby achieving about 30 dB of processing gain. Tests were run comparing the DGS-256 (10kW, 50 meter vertical rhombic transmit antenna, high gain receive array) with the portable system (150 watts, 12 meter vertical log-periodic transmit antenna, single crossed dipole active receive antenna with 2 meter dipole length). Although the portable did not perform as well, the resultant ionograms were generally satisfactory.

6.3.1.2 *KEL IPS-42 and Related Equipment*

Approximately 52 ionosondes of the IPS-42, 4A, and 4B class have been installed [Bulletin 37; INAG, 1982]. These systems are not intrinsically digital, but additional units such as the DBD-43 extend their capabilities so that data can be recorded digitally rather than simply on film. Still the IPS-42/DBD-43 complement is sometimes termed a digital ionosonde system. Other KEL products include the KEL-46 Data Analyzer and the KEL-47 Central Processing System [Bulletins 40/41; INAG, 1983]. A program which runs on an IBM PC/XT called TIDPLOT is also available [Bulletin 48; INAG, 1986]. This program operates on a series of ionograms which are stored on IBM formatted disks as ASCII records which are dumped from the DBD-43. An automatic scaling algorithm called SMARTIST has also been developed.

6.3.1.3 *Japanese Sounding System*

A sounding system has been developed by the Japanese which may be used in both VIS and OIS modes [Bulletin 48; INAG, 1986]. We shall return to the Japanese system when we discuss sounder networking issues in Section 6-7.

6.3.1.4 *BRC Vertical-Incidence Chirpsounder™*

BR Communications has developed a Vertical-Incidence Chirpsounder™ System (VIS series) which is similar from their standard line of OIS chirp waveform products. The principles of operation will be discussed when we cover the OIS system developed by BRC. A central feature of the system is its peak output power level, a relatively modest 80 watts.

6.3.1.5 *Skysonde™ and Tiltsonde™*

Andrew Antenna Proprietary Ltd. (AAPL) of Australia, a subsidiary of the Andrew Corporation, a U.S. firm, developed the Skysonde™ vertical-incidence sounder as a component of an army tactical HFDF system called Dragonfix (AN/TSQ-164). The VIS component is nomenclatured as the AN/TRQ-40 by the military. Skysonde™ is basically an FMCW system which operates at 5 to 20 watts; however the frequency scan may be monotonic in frequency or may be operated in a frequency hopping mode for an additional LPI capability. The FMCW function is carried out over a set of spot frequencies. The sweep at each spot frequency is composed of 512 steps of 125 Hz for an effective width of 64 kHz, yielding a height resolution of 2.4 km. A new spot frequency is sampled every 2.56 seconds, and the spot frequency separation is set at 0.1 MHz. A full ionogram is obtained from the exhaustion of a list of spot frequencies, either linear or hopped with no repeat. An ionogram is typically obtained every 15 minutes. Skysonde™ also operates in a fixed frequency beacon mode.

The Tiltsonde™ is similar in many respects to the Skysonde™ system since it generates ionograms. In addition, it measures synoptic and local tilts in the ionosphere which are generally associated with TIDs. The tilt measurement function is accommodated through use of a 3-channel architecture to measure the angle-of-arrival of downcoming signals. This approach is also embodied in Andrew's Skyloc™ HFDF system. Typical daytime and nighttime ionograms from the Andrew sounder system are given in Figure 6-11.

6.3.1.6 *South African Advanced FMCW Ionosonde*

Poole [1985] describes an advanced chirp ionosonde which is a modified version of the BR Communications Chirpsounder™. He makes the distinction between that which is commonly referred to as an *advanced sounder* and a *digital sounder*. Poole suggests that although the output from analog sounders may be digitized (through the process of A-to-D conversion), thus taking on the appearance of a fully digital system, such procedures, though useful, are not truly as advanced as systems which preserve phase information. The truly advanced systems are digital, to be sure, but other features are possible with phase preservation and polarization discrimination.

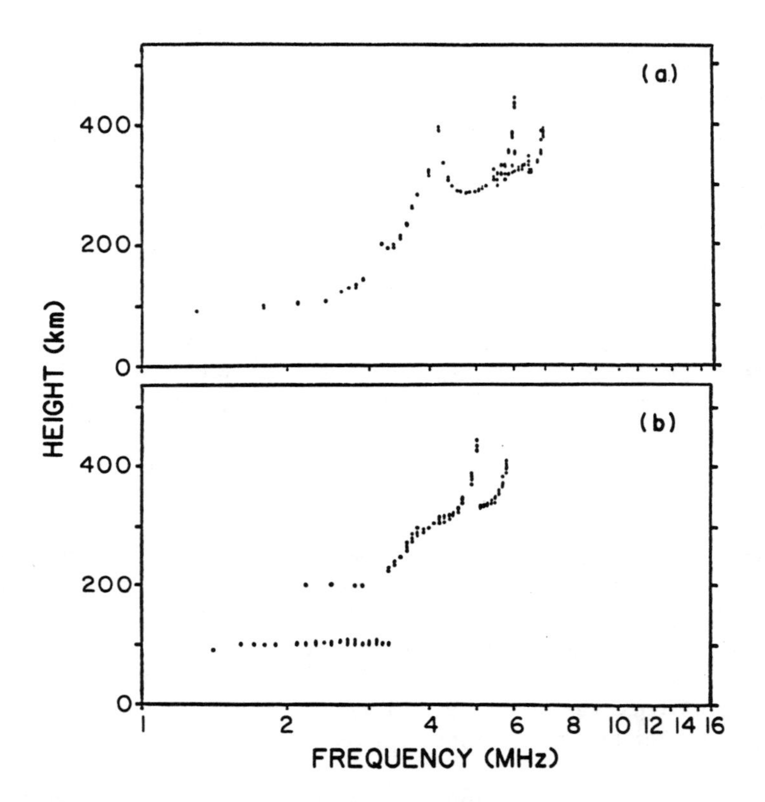

Fig.6-11. a: Daytime ionogram from Skysonde showing cusps at the critical frequencies: foE, foF1, foF2, and fxF2. b: Nighttime ionogram showing 1 and 2-hop Es, and O mode and X mode 1-hop F2 layer propagation. (Skysonde is a registered tradename of Andrew Corporation.)

For example, an advanced system will allow distinctions to be drawn between O- and X-mode signals independent of the time delay patterns involved. Other output information would include the Doppler signature and the angle-of-arrival of the incoming signal. Poole describes a generalized echo from a sounder as a multi-dimensional variable E:

$$E(f, t) \;=\; E(R', V_d, \theta_{NS}, \theta_{EW}, \phi_P, \phi_T, P) \tag{6.1}$$

where R' is the virtual range, V_d is the phase (Doppler) velocity, θ_{NS} is the arrival angle in the N-S direction, θ_{EW} is the arrival angle in the E-W direction, ϕ_P is the polarization angle, ϕ_T is the polarization ellipse tilt angle, P is the echo power, f is the transmitter frequency, and t is the time. Figure 6-12 is a set of parameters obtained from an advanced (digital) chirp ionosonde installed at Grahamstown, South Africa. The virtual range $R' \approx h'$ if we take $\theta_{NS} = \theta_{EW}$ suggesting that the ray trajectories are exactly zenithal. Thus the top curve in Figure 6-12 is just a standard ionogram.

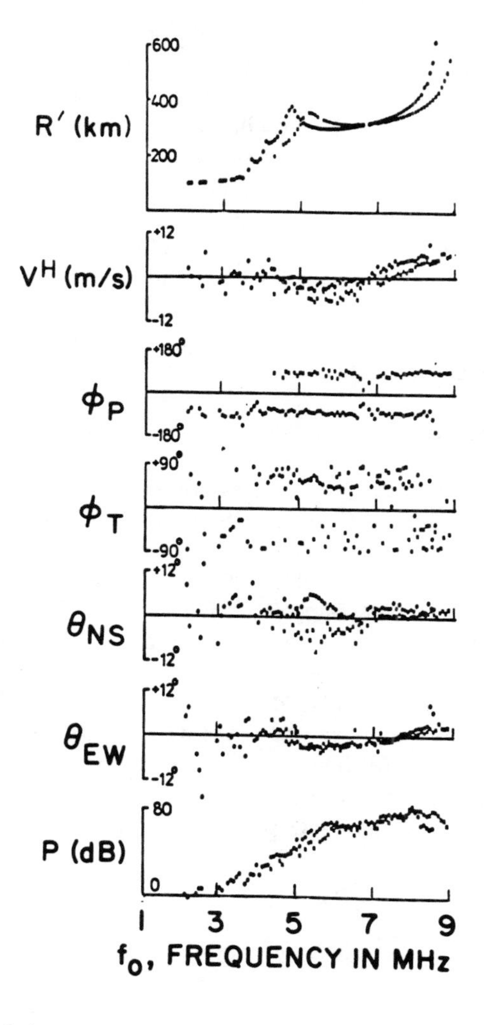

Fig.6-12. A series of plots exhibiting parameters of the echo variable *E* versus *f*, the probing frequency. (From Poole and Evans, 1985].)

One of the negative attributes of chirp methods arises because Doppler frequency shifts cannot be easily measured. The system designed by Poole and his colleagues appears to have ameliorated this problem. Using a result due to Wright and Pitteway [1982a],

$$V_g = V_d + f_0 \; d/df_0 (V_d) \tag{6.2}$$

where V_g is the signal group velocity, we may convert measured phase velocity to group velocity. This is an indirect method which may be validated by comparison with a direct method in which the time rate of change of R' is deduced from a fixed frequency sounding. A fixed frequency sounding approach is provided as an option in the advanced sounder system. Figure 6-13 is a sample plot exhibiting obvious oscillations with a frequency of 3/minute. These are more rapid than free atmospheric internal gravity waves, and fall within the regime of micropulsations and infrasonic waves.

The last plot in Figure 6-12 shows the gradual rise in received power which undoubtedly results from the greater antenna efficiency at the higher frequencies. For a complete discussion of the attributes of the advanced sounder system, refer to Poole and Evans [1985]. They discuss the detailed modifications made to a BRC sounder which in the final implementation yields a number of additional capabilities.

6.3.1.7 *Dynasonde (Advanced Ionospheric Sounder, AIS)*

NOAA/ERL has developed an advanced ionospheric sounder system called the Dynasonde [Wright and Pitteway, 1979a, 1979b] which through suitable programming can be made to reconcile the sampling procedures used in standard ionosondes and the so-called Kinesonde systems also developed at NOAA. The Kinesonde was one of the first truly digital sounding systems which enabled scientists to extract information about ionospheric drift velocities, echo scintillation parameters, and echo location [Wright et al.,1976a, 1976b; Paul et al., 1974; Wright, 1974].

The Dynasonde and some rather novel data processing procedures have been developed by J.W. Wright and his colleagues [Wright and Pitteway, 1982a]. Of interest is a Dopplionogram which has been used to study sporadic E [Wright, 1982b] and gravity waves [Wright and Pitteway, 1982b].

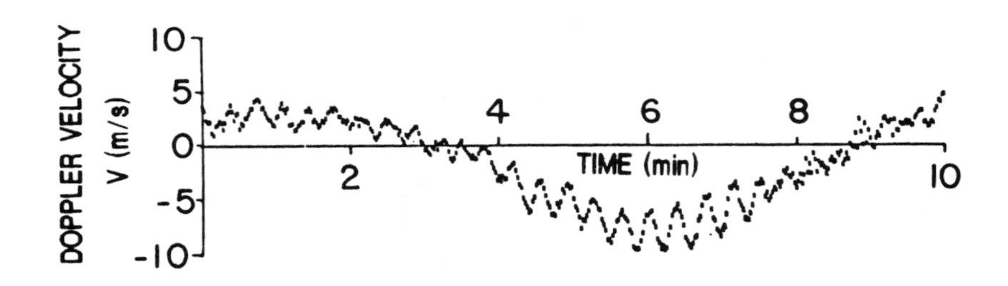

Fig.6-13. Fixed frequency ionogram showing rapid fluctuations (From Poole and Evans [1985].)

The Dynasonde has three operating modes: Ionogram, Kinesonde (K mode), and Basic Measurement Set (B mode). The B mode is a compromise between the Ionogram and K modes which permits satisfactory frequency and time sampling. Besides the Dopplionogram which supplies motion information, other options are also possible in the B mode. One is the Gonionogram which provides information about echo location.

Using the NOAA digital ionosonde Paul [1984,1988,1990] has been able to examine ionospheric variability. He finds omnipresent F-layer fluctuations in a set of data obtained during the solar maximum period 1980-1981 at a site in Colorado. Using MUF(3000)F2 as an effective indicator of variability rather than foF2, because of the ease with which it may be scaled from ionograms, Paul [1989] finds oscillations similar to those exhibited in Figure 6-14 to be quite common. The zenith distance in kilometers is deduced through a measurement of the direction of arrival of the downcoming echoes along with virtual height at the frequency being used. Figure 6-15 shows the spread of apparent echo locations.

Ionospheric F-region tilts are suggested by the data shown in Figure 6-15, but because of refraction effects, measurements of arrival angle cannot be translated directly to reflection points. This problem does not exist in the case of sporadic E reflections since the Es patches are relatively thin and no appreciable refraction arises to distort the computation of tilt angle at the reflection height. Sporadic E variability was a major component of the Paul study. Figures 6-16 and 6-17 shows how the top sporadic E frequency and its apparent position vary. Clearly well-defined and strong gradients are observed. The distinction between Es and F layer fluctuation is most obvious in the time domain. F layer fluctuations tend to be periodic (and suggestive of gravity waves as a source), while Es fluctuations are strongly aperiodic.

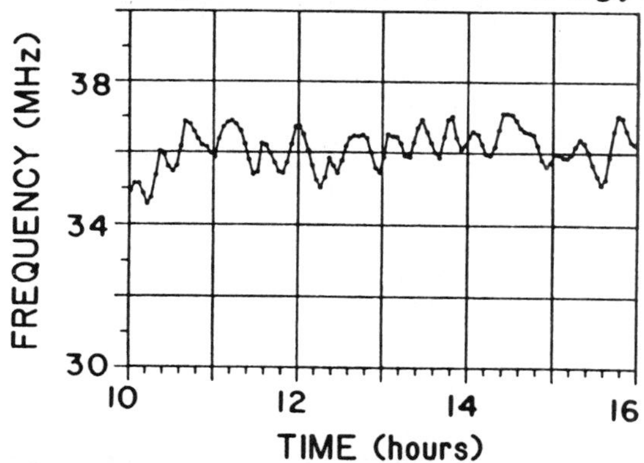

Fig.6-14. Variation in the parameter MUF(3000)F2 . Winter 1981. Solar maximum conditions. The parameter MUF(3000)F2 is scaled from standard VIS ionograms. (From Paul [1989].) This parameter, is actually an estimated value of the *instantaneous MUF* which would be termed an *estimated MOF*, or EMOF, for consistency with definitions used in this book. (See Sections 4.7.2, 6.3.2.1, and 6.3.2.2 for more information.)

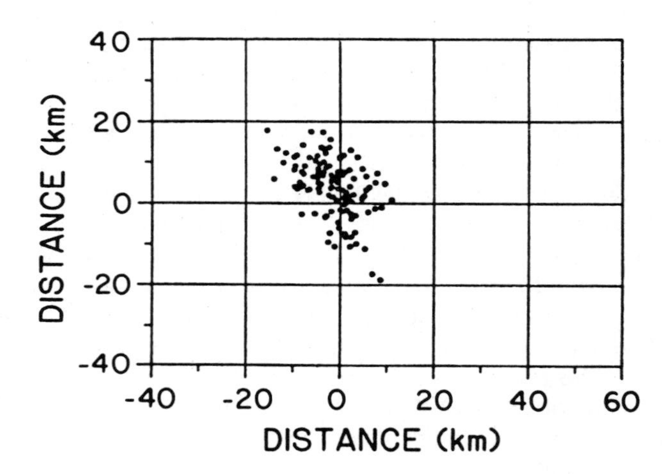

Fig.6-15. Deviations from zenithal propagation for the same data set as Figure 6-14. This corresponds to F-layer echoes (From Paul [1989].)

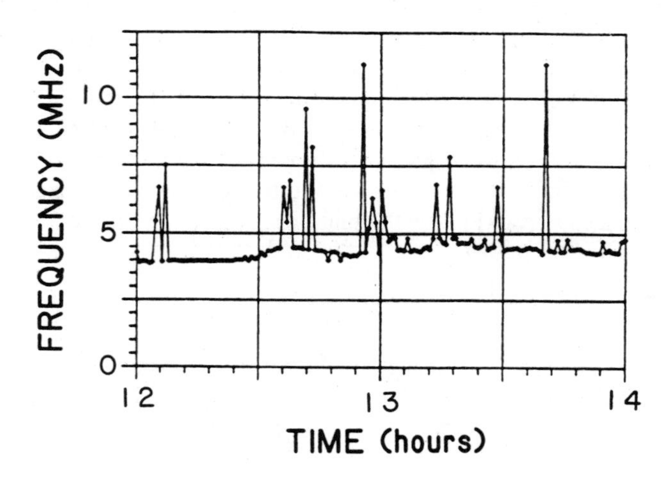

Fig.6-16. Sporadic E top frequency showing intermittent behavior. (From Paul [1989].)

6.3.2 Interpretation of VIS Data

URSI has published UAG-23, a *handbook* dealing with the discipline of vertical-incidence ionogram interpretation and reduction [Piggott and Rawer, 1972]. Subsequently UAG-50, a high latitude supplement, was published [Piggott, 1975]. This was followed by UAG-23A, a revision of the first four chapters in the original document [Piggott and Rawer, 1978].

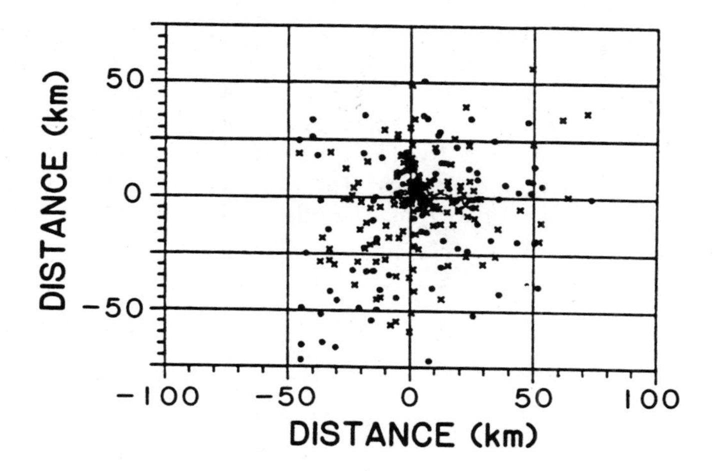

Fig.6-17. Apparent position of the Es Layer. Since Es patches are thin, the AOA is directly indicative of the tilt of the reflecting surface. (From Paul [1989].)

The URSI *handbook* and its revision, published as UAG-23 and UAG-23A by the NGDC/NOAA of the Department of Commerce in behalf of WDC-A for Solar Terrestrial Physics, is the definitive document as far as ionogram interpretation is concerned. Even so, reference must be made to the *Atlas of Ionograms*" [UAG-10; Shapley, 1972] for examples of ionograms, including both benign and pathological cases. These documents are augmented from time to time by articles in the *INAG Bulletins* also available through the WDC-A in Boulder, CO [Conkright, 1990].

6.3.2.1 *Parameter Definitions and Conventions*

Below we provide definitions (and pertinent remarks) either consistent with or directly sanctioned by INAG and contained in UAG-23. For instances in which UAG-13 and relevant CCIR documents differ, such as in the definitions of MUF and related terms, the CCIR versions are preferred (see Section 4.7.2 in Chapter 4.)

a. *Top Frequency of an Ionogram Trace* (*ft*). The highest frequency at which a clear, almost continuous trace is obtained.

b. *Blanketing Frequency* (*fb*). This is the lowest frequency at which a layer begins to become transparent. It depends upon the sensitivity of the ionosonde, but it is usually identified by the appearance of echoes from an overlying layer.

c. *Critical Frequency* (fc). The highest frequency at which the layer reflects and transmits equally; it lies between the top and blanketing frequencies. Use of this definition is generally inconvenient except when it applies to the Es layer. In practice, and especially for F layer traces, it is the frequency on an ionogram at which the slope of the trace becomes vertical, and the virtual height becomes effectively infinite. In fact for a thick layer all three *characteristic* frequencies (ft, fb, and fc) are identical. There are separate critical frequencies fc for each trace on an ionogram. The relationships among fo. fx, and fz follow:

$$fx - fo = fx\,fg\,(fx + fo)^{-1} \approx fg/2 \qquad (6.3a)$$

$$fo - fz = fz\,fg\,(fz + fo)^{-1} \approx fg/2 \qquad (6.3b)$$

$$fx - fz = fg \qquad (6.3c)$$

where fg is the gyrofrequency, and approximations hold for $fo \gg$ fg.

d. *Propagation Modes*. There are two (characteristic) magnetoionic modes of propagation in the ionosphere because of the presence of the uniform geomagnetic field, as discussed in Chapters 2-4. However, as a result of coupling processes three distinct traces may be observed. For ionogram analysis, these three traces are conveniently expressed in terms of their conditions of reflection. We use lower case letters (o, x and z) to symbolize traces and upper case letters (O, X and Z) to symbolize modes of propagation. In addition, we append the suffix "-mode" (e.g., X-mode, if modes are intended) in order to eliminate any possible confusion with the magnetoionic parameters (X, Y and Z). We have:

o trace, O-mode: where the parameter $X = 1$

x trace, X-mode: where $X = 1 - Y$

z trace, Z-mode: where $X = 1 + Y$.

A schematic representation of the three traces for the situation of a single layer is given in Figure 6-18. Clearly, in the no-field case, there is only a single trace corresponding to the O mode since $Y = 0$. Definitions of the parameters X, Y, and Z are provided in Chapter 4. The reader is cautioned that Z (i.e., upper case z) is a term which expresses the ratio of the collision frequency to the radio frequency, and the z trace corresponds to the *third* magnetoionic component which develops because of additional retardation of the X-mode (and trace) near fg. Notice that the impact of closeness to fg is not exhibited on the o-trace, but strong interaction is shown on the x and z traces.

e. *Conventions Used in the Identification of Modes.* The established procedure is to define all layer parameters by the ordinary wave component. Thus the standard ordinary wave critical frequencies foF2, foF1, and foE are used for the F2, F1 and normal E-layers respectively. For sporadic E, the ordinary wave top frequency ft = foEs; and the lowest ordinary wave frequency at which Es becomes transparent is fbEs. The reader should consult UAG-23 for details.

Other frequency specifications are important in the detailed analysis of ionograms. For example, *fmin*, is the lowest observed frequency.

f. *Heights of the Layers.* Two height categories of height are to be considered, virtual and true. Only the virtual heights are directly scaled from a standard ionogram, although the virtual and true heights may be approximately equal for concentrated thin layers (such as sporadic E) which suffer little or no retardation.

The minimum virtual height is of interest since it should represent a frequency at which signals reflected from the designated layer are least retarded. When a trace is horizontal, the condition for scaling a minimum virtual height is rigorously met. Values of minimum virtual height are all denoted by h' followed by the trace in question. For example h' *F2* and h' *Es* are associated with the *o*-traces for the F2 region and the Es layer respectively.

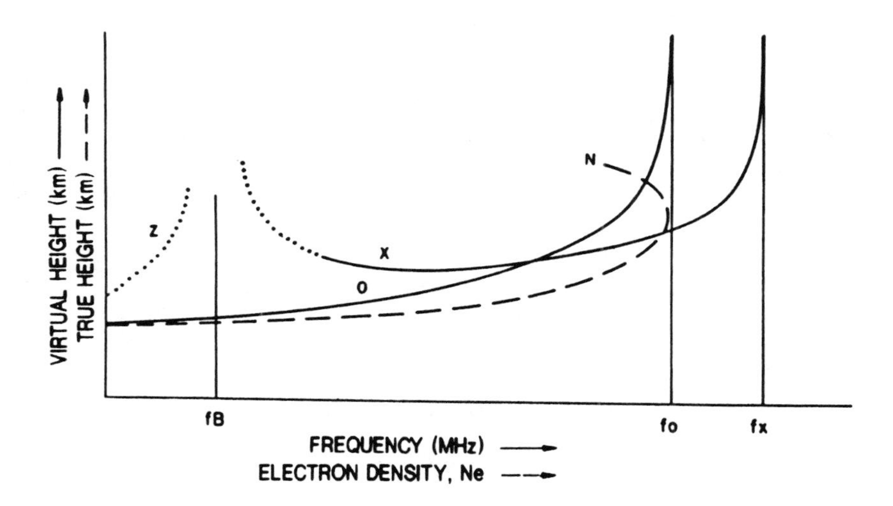

Fig.6-18. The schematic representation of an ionogram for a single layer. The solid lines refer to the radio (sounding) frequency, and the (large) dashed line refers to the electron density. Note that the x and z traces suffer the greatest retardation at the gyrofrequency fg (From UAG-23A [Piggott and Rawer, 1978].)

The terms *hc* and *hmF2* are reserved for methods in which the layer is described by a best-fit parabola, and the *o*-trace near foF2 is fitted to the theoretical *h' f* curve which results. If underlying ionization is disregarded, then the term *hmF2* is used; if considered, then *hc* is used. UAG-23A recommends that the term *hmaxF2* be used to refer to advanced computer methods which exploit both the O-mode and X-mode information.

Figure 6-19 is a schematic representation of a typical daytime ionogram showing the various minimum virtual heights. The schematic in Figure 6-20 shows the standard frequency and height parameters for an ionosphere composed of Es, E, F1, F2 layers and for both one- and two-hop propagation.

g. *MUF Factors.* A convenient distance for use in long-haul HF communication performance analysis (as well as for the analysis of a standardized MUF) is 3000 kilometers. The term MUF(3000) = MUF(3000)F2 when it refers to the F2 layer and can be easily scaled from vertical-incidence ionograms. These terms, upon normalization by *fo* (e.g., foF2) collectively are called *MUF Factors*, or *M-Factors* when normalized, and have some value in converting VIS data into effective maximum transmission frequencies for oblique paths. Additionally, these terms have been shown to be a useful ionospheric index. (See Chapter 4 for definitions of maximum transmission frequencies.)

A convenient height parameter may be deduced if we assume that the layer is parabolic and that there is no underlying ionization.

$$hpF2 = h' \ (f = 0.834 \text{ foF2}) \qquad (6.4)$$

The virtual height *hpF2* corresponds to the frequency $f^* = 0.834$ foF2. There is no physical significance to f^*; it serves only as a way to estimate the height of a equivalent parabolic layer. Serious errors may be introduced if the parabolic layer assumption is invalid. Accordingly, its general use is not recommended. The parabolic layer hypothesis has also been used to estimate the layer height when M(3000)F2 data are available. In fact, a successful empirical relationship has been developed by Shimazaki [1955] which relates *hpF2* with M(3000)F2 where M(3000)F2 = MUF(3000)F2/foF2. Some caution should be observed in using the Shimazaki formula, especially at high latitudes. Moreover, it gives too high an estimate during summer months because of excessive retardation associated with enhanced underlying ionization. We have:

$$hpF2 = -176 + 1490/\text{M(3000)F2} \qquad (6.5)$$

where *hpF2* is in km.

For additional discussion on the MUF-Factor for arbitrary skip distances, see Section 6.3.2.2.

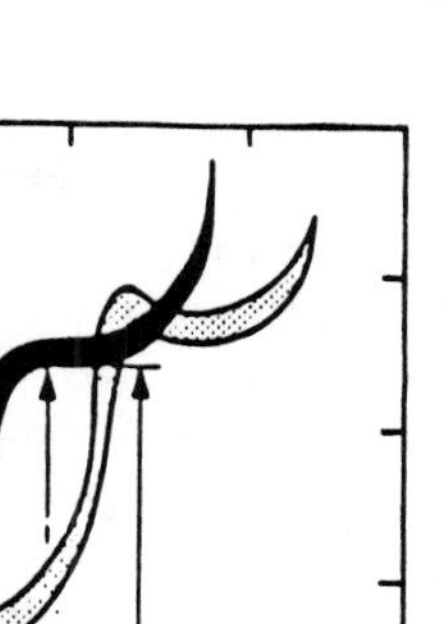

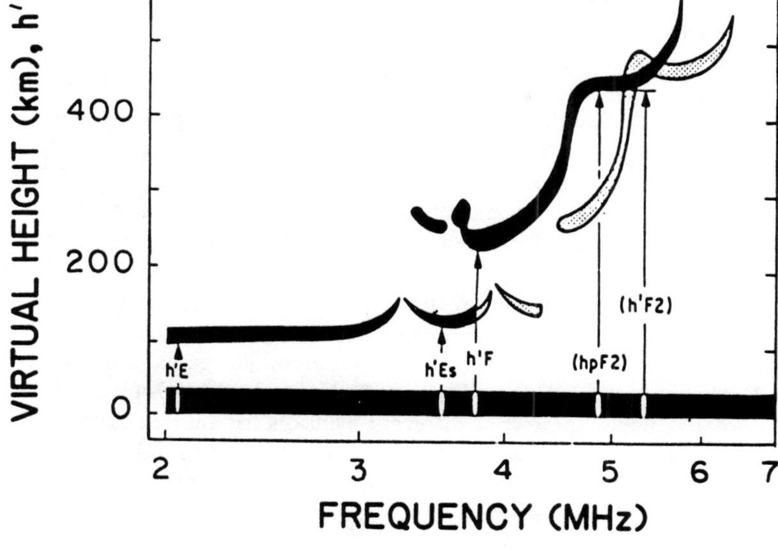

Fig.6-19. Schematic of minimum virtual heights. The dark curves correspond to the O mode while the lighter curves correspond to the X mode. Three layers are exhibited: F2, normal E and sporadic E. (From UAG-23A, a revision of chapters 1-4 of UAG-23 [Piggott and Rawer, 1978].)

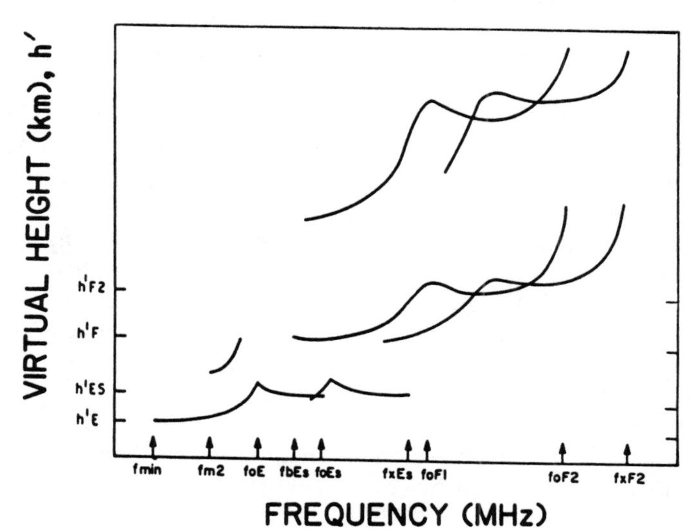

Fig.6-20. Schematic of standard height and frequency parameters. The O and X modes for the Es, F1 and F2 layers are shown. (From UAG-23A, a revision of chapters 1-4 of UAG-23; [Piggott and Rawer, 1978].)

6.3.2.2 *Skip Slider Method and MUF-Factor*

It is observed that the MUF(3000) values are scaled routinely from ionograms using a standard transmission method developed by Smith [1939]. Standard transmission curves are described in Davies [1969]. The concept has led to a simple skip slider method which assumes a simplified propagation model. Taking an ionogram which has a logarithmic frequency scale (from left to right), we construct a plot of M-Factor versus virtual height. (In this instance the M-Factor is plotted in the same logarithmic scale units but runs from right to left.) For additional information on the skip-slider method, see Leid [1967].

The use of the skip-slider method is illustrated in Figure 6-21. The rationale for the method derives from the requirement for a quick (albeit approximate) method for scaling vertical-incidence ionograms to achieve an estimate of maximum usable frequencies for specified oblique paths, not just for 3000 km. Following the general outline provided in the AGARDograph 104 edited by Leid (but using somewhat different symbols for the variables), we may relate the virtual height h' to skip distance d (i.e., ground range) and the angle of elevation β as follows:

$$h' = (d/2) \tan \beta \tag{6.6}$$

The skip slider is obtained by plotting $1/(\sin \beta)$ versus h' in accordance with the right-to-left log-plot procedure mentioned above. In fig 6-21a, two transmission curves are drawn, one for 1000 km and another for 2000 km. The abscissa is, of course, $1/\sin \beta$, which is the MUF-Factor scale. The corresponding values of β are also shown. Figure 6-21b is the schematic of a simple ionogram, which displays h' versus frequency, with the frequency plotted in conventional left-to-right fashion on a log scale.

There are several ways the skip-sliders such as those in Figure 6-21a can be used. If we specify the frequency of operation and have a large number of transmission curves on the slider, we can determine the skip distance d involved. This is done by placing the initial point (i.e., $1/\sin \beta = 1$) of the slider over the specified frequency and locating the curve which comes closest to being tangent to the ionogram trace. On the other hand if we have a fixed frequency of transmission and a fixed skip distance, we can determine the virtual height(s) for the propagation path. In this case, we pick the transmission curve for the specified distance on the slider and then place the initial slider point over the specified frequency on the ionogram trace as in the previous example. This is shown in Figure 6-21c for $d = 1000$ km and $f = 20$ MHz. Here we note that the transmission curve on the slider intersects the ionogram trace in two positions, one corresponding to the low ray and the second corresponding to the high ray. The 1000-km MUF itself is somewhat higher than 20 MHz, and this may be determined graphically by moving the 1000-km slider to the right such that the transmission curve is just tangent to the ionograms trace.

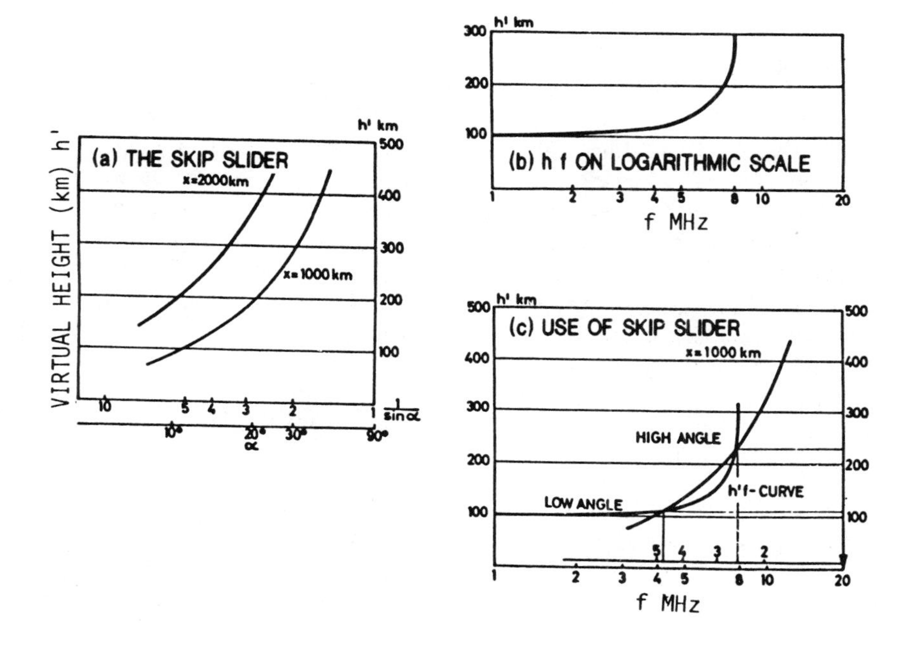

Fig.6-21. The skip-slider method. (a) skip-slider. (b) *h'* (*f*) plotted on a log scale. (c) Use of the skip-slider. The intersection of the ionogram in (b) and the 1000 km slider in (a) yields two points, corresponding to the high (angle) ray and the low (angle) ray. (From Lied [1967].)

At this point the high and low angle rays converge and this condition prevails at the so-called junction frequency. It is remarked that the MUF determined in this way is referred to as the Equivalent Junction Frequency (EJF) in older CCIR documents. In any case, it is still just a rough approximation.

The relationship between the 3000 km M-Factor and virtual height is given in Table 6-4.

TABLE 6-4. M-FACTOR AND THE REFLECTION HEIGHT

M-FACTOR (3000-km)	VIRTUAL HT. (*h'*,km)	M-FACTOR (3000-km)	VIRTUAL HT. (*h'*,km)
4.55	200	2.69	500
4.05	250	2.40	600
3.65	300	2.20	700
3.33	350	2.04	800
3.08	400		

6.3.2.3 *Pathological or Unusual Ionograms*

The most serious difficulty encountered in scaling ionograms occurs when nonstratified conditions are encountered. Tilts, gradients, and traveling disturbances (TIDs) immediately come to mind. Other ionogram observables include spread F and Lacuna phenomena. Spread-F is caused by ionospheric inhomogeneities, which are typically field-aligned, and may be of two types: range spread and frequency spread. The lacuna is a height gap in the ionogram possibly caused by an interaction with plasma instabilities in the region [UAG-23A; p.55]. There are many examples of these phenomena in the references cited, especially UAG-10 and UAG-50. Certain high latitude phenomena which produce interesting ionogram traces. For instance the G condition for which foF1 > foF2 may be observed. Figure 6-22 shows some of the features which are observed at an auroral zone station. FLIZ (for F2 Layer of the Irregular Zone) is a feature of the high latitude trough and ridge which changes in time.

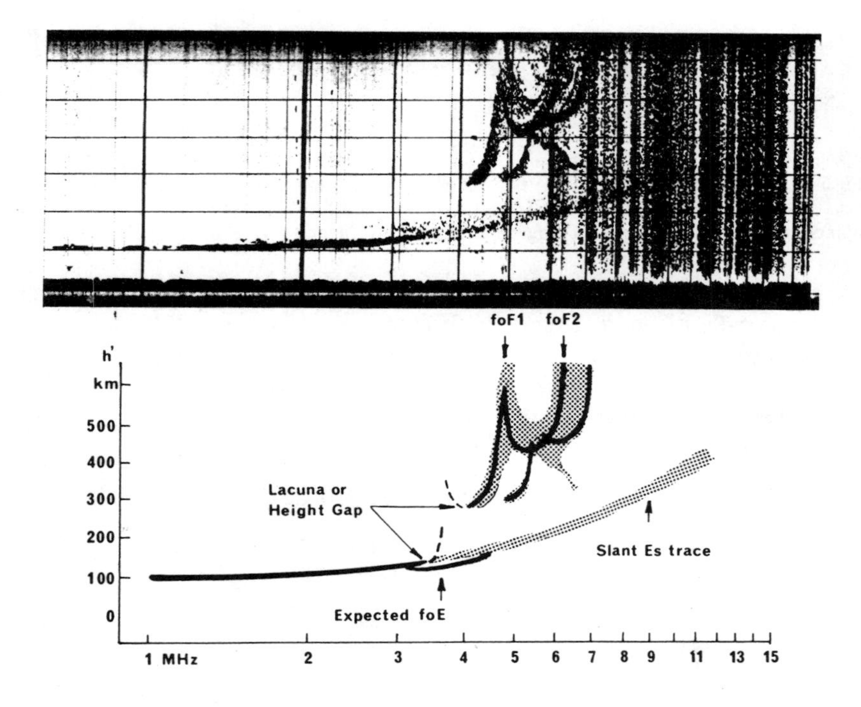

Fig.6-22. Typical daytime ionogram at high latitudes. The station was Narssarssuaq, and the ionogram was recorded on June 15, 1969 at about 4 PM local time. (From UAG-23A, a revision of chapters 1-4 of UAG-23; [Piggott and Rawer, 1978].)

6.3.3 Ionogram Inversion

The transformation of ionograms into actual electron density profiles is essential for many scientific studies, and it is important in many applications which rely on precise layer height data. In this category we include HF coverage predictions, Time-Difference-of-Arrival (TDOA) applications, mode determination for signal sorting in HFDF applications, etc. Kelso [1953] developed one of the earliest methods which was model-independent but it was limited to the O-mode and was computationally slow. Several model-dependent methods allow the determination of certain layer parameters such as peak electron density and layer height [Appleton and Beynon, 1940]. Subsequently the effects of the magnetic field were added and shapes other than simple parabolas were allowed. Reviews of some of the early methods are found in a paper by Thomas [1959] and in the classic book by Kelso [1964]. One important class of inversion methods is characterized by taking the ionosphere to be made up of horizontal laminae. Methods by Budden [1955], Jackson [1956], and Titheridge [1959a,b] comprise this class.

Two fundamental problems associated with ionogram inversion include the so-called valley problem and the uncertainty associated with layer maximum height determination. Both of these difficulties arise from the fact that the E-F valley and the F2 peak (for example) are unobserved by ionosonde, because of the breakdown of monotonicity in the former case and because of an indeterminant value for virtual height in the latter case. The valley problem is not just one which has implications within the valley itself (i.e., distribution of electron density in the valley) but also affects the determination of true heights in overlying layers if valley parameters are estimated or modeled inaccurately. This problem has been addressed by Titheridge [1985].

The POLAN (for POLynomial ANalysis) program is documented in UAG-93 [Titheridge, 1985]. Although such methods formerly required main frame computers, there should be no problem with use of POLAN in a microcomputer environment. UAG-93 indicates that POLAN requires only 40 kbytes of memory on a PDP-11. This is an attractive feature given the current trend in autoscaling methods which exploit microprocessors with capability less than some personal computers. POLAN is quite versatile and may be used for detailed analysis by the serious ionospheric specialist. On the other hand, POLAN may be treated as a black box to use the language of UAG-93. In this black box mode, the only data required is raw virtual height data, magnetic dip angle, and the gyrofrequency. This mode is also user-friendly. Optimized default procedures are supplied in this mode, the program automatically corrects inconsistent input data sets, and the operator is notified of any changes. Finally all results are obtained in a one pass analysis. The program has been compiled under FORTRAN 4 but may be run under FORTRAN 77 with only minor modifications. Copies of all POLAN component programs are available on magnetic media from WDC-A in Boulder CO.

There is a premium on development of compact programs which allow ionogram inversion to be performed on portable machines with acceptable accuracy and speed. An early method for use with microcomputers was reported by Paul [1977]. A simplified version of POLAN, called SPOLAN, has been developed. However no difficulty has been found in the application of POLAN in a minicomputer environment [Titheridge, 1985].

6.3.4 Automatic Ionogram Scaling Systems

As may be inferred from Section 6.3.2, the process of ionogram scaling and interpretation is substantial. Manual methods have evolved over the years for scaling purposes, and the interpretation of ionograms is largely the province of ionospheric specialists. The digital ionogram has led naturally to digital computer methods for scaling and analysis. The laborious manual methods are now being replaced by what is referred to as autoscaling. The advantage of such a technology over previous methods is enormous, since it opens the possibility for real-time data recovery, communication assessment, and short-term forecasting in the field rather than in the laboratory.

There are several autoscaling systems which have been developed and the principal ones are listed in Table 6-5.

TABLE 6-5

IONOGRAM SCALING SYSTEMS

SYSTEM NAME	DEVELOPER	REFERENCE
ARTIST	ULCAR Univ. of Lowell Lowell MA-USA	[Reinisch and Huang,1983] [Reinisch et al., 1983]
CRL SYSTEM	Communication Research Lab Tokyo, Japan	[Suzuki,1988]
SMARTIST	KEL Aerospace Australia	[McCue and Gilbert,1988]
IPS AUTO- SCALING SYSTEM	IPS Radio and Space Services Chatswood, NSW Australia	[Fox and Blundell,1989] [Wilkinson, 1991]

6.3.4.1 *ARTIST.*

The ARTIST system is a well developed software system. It is part of the AN/FMQ12 Digital Ionospheric Sounding System (or DISS) which is basically a Digisonde 256. Strictly speaking, ARTIST (Automatic Real Time Ionogram Scaler with True height) is an automatic scaling algorithm for the DGS-256. It features software for automatically scaling and inverting ionograms leading to the presentation and analysis of true height profiles of electron density. Development includes other advanced features. Work is ongoing and government reports are periodically released describing the progress [Tang et al., 1988a, 1988b]. A sample ARTIST product is provided in Figure 6-23.

6.3.4.2 *IPS Autoscaling System.*

Fox and Blundell [1989] describe the IPS approach which is designed to mimic the activity of an automatic scaler. The software operates only on the ordinary trace, but a properly formatted ionogram derived from any system can be scaled with the IPS system. Figure 6-24 is a midlatitude ionogram for which autoscaling has been applied. The system evolved using the principles embodied in development of expert systems, with the system being trained to scale midlatitude ionograms. The ultimate goal of the IPS system is to provide the Australian region with a real-time model of the ionosphere of major interest to the HF community.

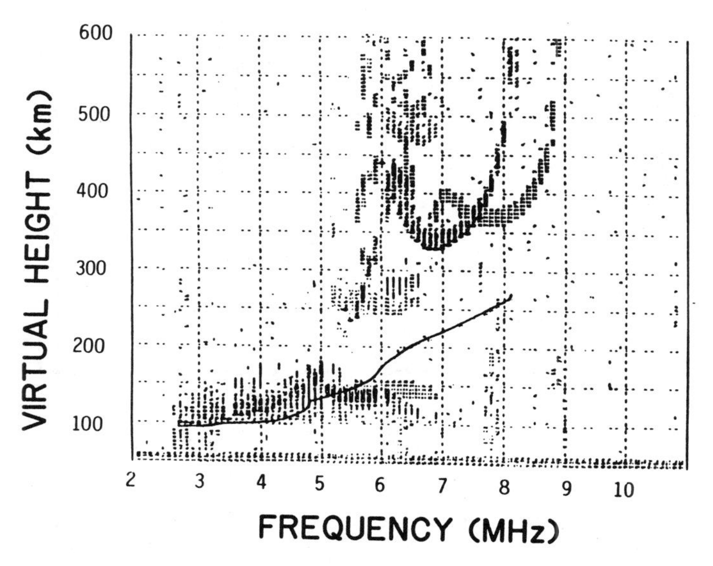

Fig.6-23. Sample ionogram with ARTIST autoscaling. (From Tang et al.,[1988a].)

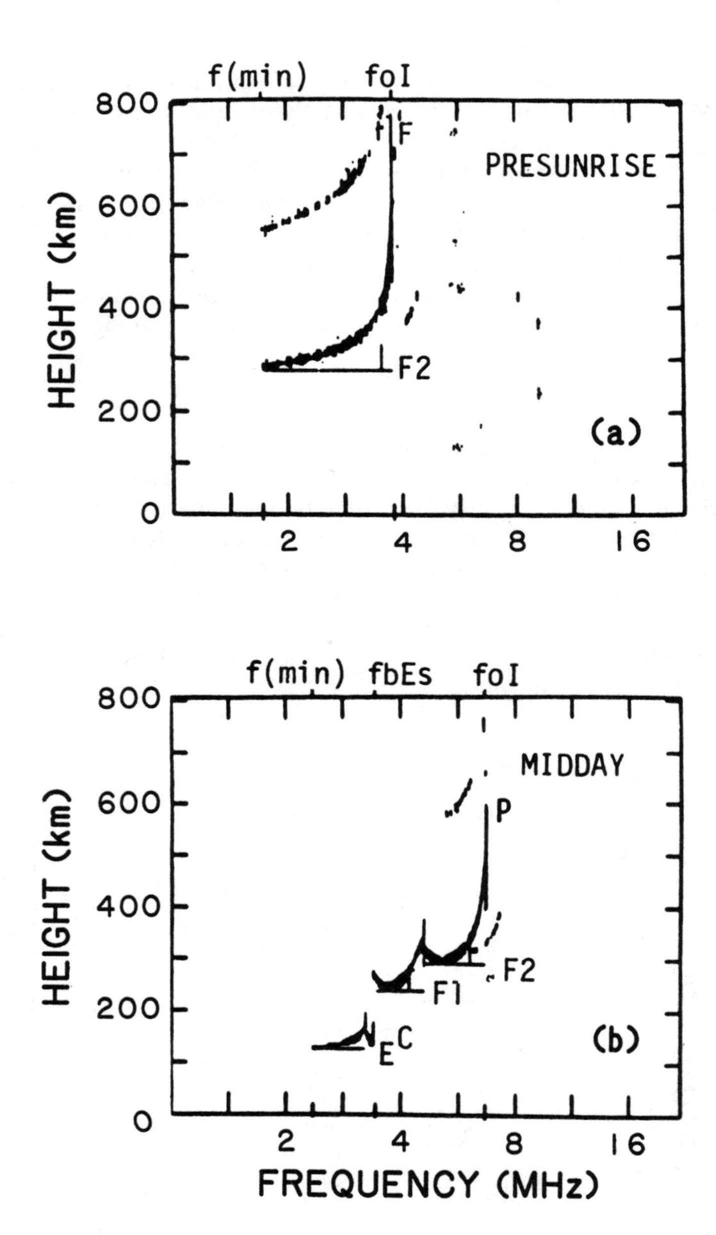

Fig.6-24. Scaled midlatitude ionograms using the IPS system. The station was Camden, Australia. (a) February 8, 1988; 5:30 LT.; foF2 = 3.80 MHz. (b) September 8, 1987; 13:30 LT.; foF2 = 6.71 MHz. (From Fox and Blundell [1989].)

6.3.5 The VIS Data Base and its Use

As has been indicated in Chapters 1 and 2, the National Geophysical Data Center (NGDC) maintains an archive of sounder data (in scaled form), and the technical characteristics of individual sounder stations, and other solar-terrestrial monitoring stations, may be found in a comprehensive directory [Shea et al.,1984]. The raw ionograms are stored at World Data Centers: WDC-A (Boulder), WDC-B (Izmiran, USSR), WDC-C1 (Tokyo), and WDC-C2 (Slough, UK). The Boulder center maintains a record of French and Indian data as well as data from the Western Hemisphere. The index for Western Hemispheric ionograms is found in UAG-85 [Conkright and Brophy, 1982], a document published by the NGDC. A combined international catalog has also been published [UAG-91; Conkright et al.,1984]. See Section 2.5.3 for a discussion of prediction services and products, and Section 2.5.4 for more information on data archives and publications.

Not all stations with ionograms archived in the various World Data Centers are still active. Studies typically take advantage of subsets of the total amount of archived data to satisfy specific needs. Workers at NRL [Martin, 1990] have developed a data base consisting only of those stations which have provided data continuously since April of 1979 covering two solar maxima. The raw data were obtained from two sources: (1) monthly tapes from Air Force Global Weather Central (AFGWC) in Omaha, NB through December 1987, and (2) the USAF Environmental Technical Applications Center (ETAC) in Asheville, NC from January 1988 through December 1990. The NRL data base consists of the following scaled parameters for up to 45 stations in the Northern Hemisphere and up to 7 stations in the Southern Hemisphere: foF2, fmin, foEs, and M(3000). The station list, referred to as the Air Weather Service (AWS) list, is shown in Table 6-6, and the locations are given in Figure 6-25. Figure 6-26 displays a group of sounders organized under the international MONSEE program.

Examining the AWS distribution of sounders, looking down from the North Pole, we obtain a different perspective. To see things more clearly we plot the stations in geomagnetic coordinates, and we overlay a Feldstein auroral oval at UT = 00 hours under the assumption that $Q = 3$ (see Chapter 2 for a discussion of the Feldstein oval). The result, in Figure 6-27, shows that most of the stations are subauroral, although two are within the polar cap and three are within the oval itself. Borderline stations move in and out of the various regions as a function of time because of possible temporal variations in Q as well as earth rotation effects. Figure 2-23 in Chapter 2 showed the diurnal effect, and Figure 2-25 showed the effect of an increased magnetic activity. Since the Kindex and Q are correlated, an increase in Q will expand the Feldstein oval and drive it equatorward. The net effect is to increase the number of auroral stations.

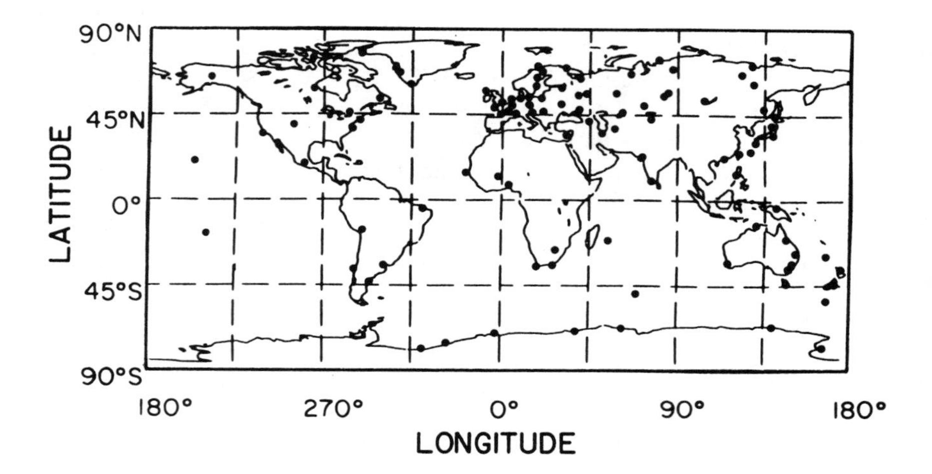

Fig.6-25. Worldwide distribution of VIS sounders in the AWS data base.

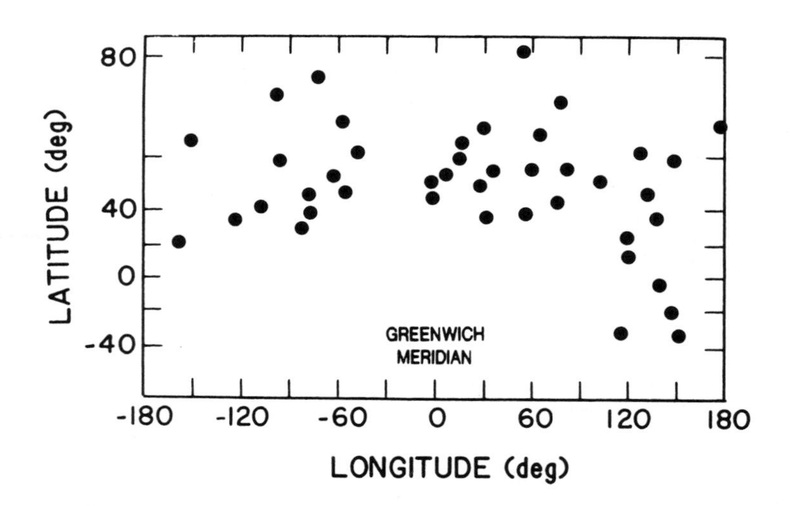

Fig.6-26. Distribution of VIS stations in the MONSEE program. MONSEE, *Monitoring of the Sun-Earth Environment*, operated under the auspices of the Scientific Committee on Solar-Terrestrial Physics (SCOSTEP), which is affiliated with the International Council of Scientific Unions (ICSU).

TABLE 6-6. LIST OF IONOSONDE STATIONS (AWS)

STATION	LATITUDE	LONGITUDE	MLAT	MLONG
Lycksele, Sweden	64.62	18.76	61.24	100.61
Uppsala, Sweden	59.80	17.60	56.26	97.02
Slough, UK	51.50	-0.57	48.30	79.18
Thule, Greenland	77.50	-69.20	86.02	38.82
Narsarssuak, Greenland	61.20	-45.45	67.45	44.45
Poitiers, France	46.57	0.35	42.19	78.35
Kuhlungsborn, Germany	54.12	11.77	50.35	90.00
St. Peter, Germany	8.38	8.38	50.65	87.29
Nicosia, Cyprus	35.17	33.28	28.45	104.98
Krenkel, USSR	80.63	58.05	74.71	144.82
Dickson, USSR	73.50	80.40	67.84	155.68
Murmansk, USSR	68.00	33.00	63.88	113.97
Salekhard, USSR	66.55	66.70	61.59	141.51
Tunguska, USSR	61.58	90.00	56.48	162.46
Yakutsk, USSR	62.00	129.60	55.71	199.74
Cape Schmidt, USSR	68.90	-179.50	64.20	235.02
Magadan, USSR	60.00	151.00	53.21	218.07
Moscow, USSR	55.50	37.30	51.00	112.01
Sverdlovsk, USSR	56.70	61.10	51.99	134.06
Tomsk, USSR	56.50	84.90	51.50	157.09
Irkutsk, USSR	52.50	104.00	47.00	176.00
Khabarovsk, USSR	48.50	135.10	41.56	206.12
Kiev, USSR	50.72	30.30	46.10	104.55
Alma-Ata, USSR	43.25	76.92	37.97	148.72
Ashkhabad, USSR	37.90	58.30	32.66	129.98
Okinawa, Japan	26.16	127.48	18.79	198.24
Wakkanai, Japan	45.23	141.41	37.96	211.87
Akita, Japan	39.43	140.08	32.03	210.74
Taipei, Taiwan	25.00	121.20	17.64	192.03
Tokyo, Japan	35.70	139.50	28.22	210.12
College, Alaska	64.90	-147.80	64.85	261.23
Ottowa, Canada	45.40	-75.90	57.23	358.57
Argentia NAS, Canada	47.30	-54.00	55.52	29.18
Goose Bay, Labrador	53.30	-60.33	62.89	22.81
Kenora, Canada	49.79	-94.49	60.85	330.63
Churchill, Canada	58.80	-94.20	69.70	329.29
Resolute, Canada	74.70	-94.90	83.92	312.49
Vandenberg, California	34.73	-120.57	40.62	302.10
Wallops Island, Virginia	37.90	-75.50	49.83	358.27
Boulder, Colorado	40.00	-105.30	49.14	317.93
Patrick AFB, Florida	28.48	-80.55	40.63	350.33
Bermuda, UK	32.22	-64.41	42.95	12.89
Kahului, Hawaii	20.80	-156.50	21.55	270.27
Manila, Philippines	14.60	121.00	6.74	191.58

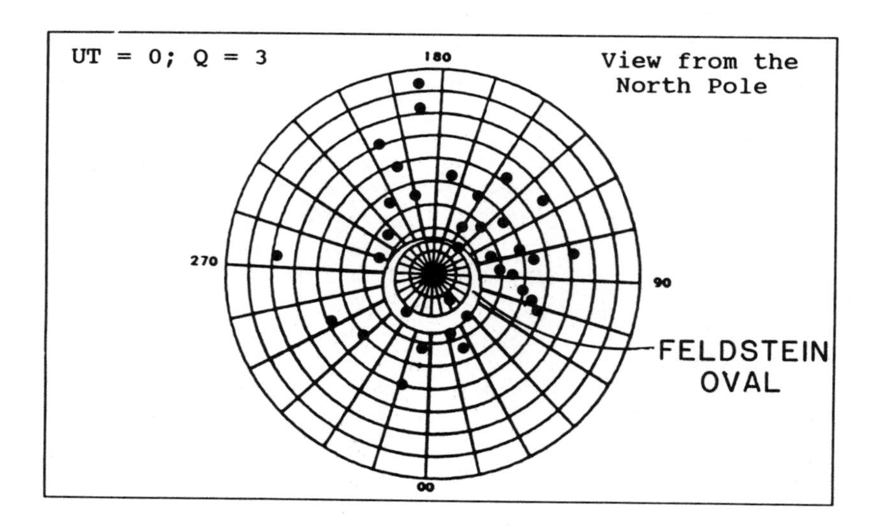

Fig.6-27. Distribution of AWS stations in geomagnetic coordinates. North pole view. The Feldstein oval depicted is for a Q index of 3.

6.3.6 Use of VIS Data in HF Communication

The availability of vertical-incidence sounding data (VIS) has led to the development of long-term prediction models which are used for system design as well as for spectrum planning. *The Global Atlas of Ionospheric Coefficients* contained in CCIR Report 340 [CCIR, 1986b] and a new set of URSI coefficients [Rush et al.,1989] provide a means to summarize knowledge obtained during many years of investigation using VIS instruments. Most models used for ionospheric investigation are largely based upon VIS data sets, including some noteworthy new models such as ICED, which is being developed by the Air Force [Tascione et al.,1988]. Although some algorithms representing transionospheric propagation effects are based upon total electron content data, the small number of polarimetry sites and the diminishing number of VHF satellite beacons suggests that the relatively vast VIS data base will be useful in filling in the gaps. Indeed, the ICED program uses the CCIR data for bottomside modeling and combines this with another topside model to obtain the full ionospheric effect.

In the area of HF communication, the CCIR data is used in virtually all performance prediction models (see Chapter 5). This class of data is also of useful as a real-time decision aid. We have already noted that VIS belongs to

RTCE Class 2 (see Section 6.2.2). The most important use of *live* VIS data is for short-range communications to support Nap-of-the-Earth requirements and other situations which preclude ground wave propagation over short distances. These requirements are usually satisfied by the use of Near-Vertical-Incidence-Skywave (NVIS) systems which were discussed in Section 4.6. The VIS systems provide the NVIS system operator with the existing overhead critical frequencies and the allowable frequency range which will provide the usual *umbrella* coverage without a skip zone. If close-in skywave jamming is to be avoided, and the friendly terminal is at a great distance, then VIS assessment may be used to place the available jamming frequencies within a skip zone by appropriately exploiting a normal skywave method. It turns out that a single VIS instrument may be sufficient for frequency management over a rather large area of up to 300 x 300 km^2. We shall return to this matter when we discuss the extrapolation of VIS and OIS data (and RTCE data in general) in Section 6.6.

6.3.7 Use of VIS Data in SSL and Related Disciplines [55]

One of the systems for which VIS data is virtually an essential ingredient is the HF Single-Site-Location System (HF-SSL). This system must be configured to provide estimates of the position of fixed terrestrial transmitters, the emissions of which are only available over a skywave circuit. Ionospheric information is required to associate the measured arrival angle(s) of the target signal with the proper source (location) of the emission.

In order to design an optimum SSL system it is necessary to take account of several factors. First, we must recognize the practical realities of an operational system itself, and we must describe the system quantitatively in terms of a system metric. (We shall say more about the system metric shortly.) Next, we must recognize limitations imposed by science, technology, cost, and schedule, all of which may be important in the real world. Architectural considerations are generally driven by a requirement or a set of specifications for the equipment. One of these design specifications might well be a requirement for the system to perform demonstrably better than the Ross curve

---

55. In this section we shall depart significantly from the subject of HF communication to concentrate on HF Single-Site-Location (SSL) techniques. The use of real-time remote sensing data to *adjust* or *update* ionospheric mean models is an imperative for certain HFDF applications while for communication applications, such as frequency management, it will provide incremental performance gain. For the HF communications, we may only be concerned about the channel scattering function and signal strength available for demodulation. For HFDF, and HF-SSL, one is certainly concerned about these properties since they characterize the signal hearability, but directionality information is also necessary for emitter geolocation. For readers having limited interest in HF-SSL, Section 6.3.7 may be bypassed without significant loss in continuity.

[Ross, 1947], for example (See Figure 6-4). This may or may not be achievable if factors such as cost and schedule are unrealistic, or if the system is unsophisticated. However, in this section we are more interested in fundamental limitations, such as those imposed by the medium or by algorithmic design. Even current computational speed limitations are not of major concern (within limits). Nevertheless considerations of overall system architecture do bring together a number of established engineering and science disciplines as well as evolving ones such as Artificial Intelligence (AI), computer technology, and remote sensing. The general ideas expressed in this section are based upon unpublished work of the author and Polkinghorn [1990].

The system metric is a quantitative description of the relative merit of achieving a specific state of operational performance within a set of constraints. The system metric has two different forms which are associated with the design and real-time operation of the system. A typical system specification includes a system metric in the form of a list of criteria the system must meet. A value of a system design is then binary. The system is acceptable if it meets all the specification requirements, and unacceptable if it fails to meet one or more of the requirements. An example of such a tradeoff is observation time versus array size versus DF accuracy. After the system is built, one can automatically manage the system using the system metric. There are a number of management strategies (involving system parameter tradeoffs) which could be invoked, but these are beyond the scope this book. The information required to formulate a system metric includes: specification of the signals of interest, their modulation formats, transmission times, number of simultaneous signals, available real estate, set-up times, the number of SSL sites, and their geographical distributions. In evaluating alternative architectures against the system metric, one must include system-wide budgets for accuracy, time delay, dynamic range, location error budgets, and a built-in automatic management system to assure that these parameters stay within bounds. One of the most important results of such an analysis is a list of deficient technologies which should be advanced to achieve the desired level of system performance. The most likely technological problem areas which will influence system performance are: wavefront analysis, DF algorithm design, ionospheric measurement, ionospheric parameter extrapolation, and ray tracing.

In order to determine the location of a terrestrial emitter, several pieces of information are required. The predominant methods involve the measurement of the Direction-of-Arrival (DOA) of the target signal, and the methodology is commonly referred to as Direction Finding (or DF). There are other methods (i.e., geolocation algorithms) which do not require directional information and these include the following : Time-Difference-of-Arrival (TDOA), Frequency-Difference-of-Arrival (FDOA), amplitude DF, and polarization DF. In the *directional* category, one may either measure the total angle (i.e., azimuth and elevation) of the incoming signal from a single receiving station (the so-called SSL method), or one can measure a set of bearings (i.e., azimuths) from several widely-spaced receivers. In both methods, the ionosphere is involved for other than Line-of-Sight (LOS) signals; however in the SSL approach the ionospheric personality may exhibit a more profound influence on the result.

With current SSL technology, the error in location estimation is of the order of 10% of the ground range between (emitter) transmitter and receiver. An accounting of (ionospheric) electron density variability, which ultimately controls skywave refractivity and radio ray direction, is perhaps a more important ingredient in the improvement of SSL performance than is an improvement in instrumental measurement accuracy.

The manner in which most existing SSL systems treat the ionosphere has a substantial bearing on the accuracy which may be achieved. Most systems employ Vertical-Incidence-Sounders (VIS) using a nondirectional antenna in order to determine overhead ionospheric layer heights and densities. In reality, NVIS propagation admits to rather significant departures in actual ray trajectories from the vertical direction. This leads to possible misinterpretations of the ionospheric parameters and of the location to which these parameters should be assigned, unless, of course, proper attention is given to the problem. Time-varying ionospheric tilts which are associated with neutral driving forces in the thermosphere such as acoustic-gravity waves, and macroscopic tilts arising from global influences, day-night terminator effects, and other climatological large-scale features are potential problems for NVIS systems, and especially for SSL systems which operate in an NVIS domain. The most significant problem arises from relatively small scale structure, or small scale Traveling-Ionospheric-Disturbances (TIDs) which are ionospheric tracers to the underlying neutral gas motion. To correct for dynamic layer tilt effects in the NVIS domain, it is necessary to monitor (and understand) the nature of the TID instantaneous structure and its evolution. Thus one must also consider extrapolation rules and the discipline of ionospheric forecasting.

Spatial extrapolation of electron density data is especially important for NVIS (or tactical) SSL systems. The extrapolation distances may be relatively small when compared with long-haul (or strategic) systems, suggesting a reduced ionospheric extrapolation error. Despite this, small-scale features which are partially smoothed-out (or filtered) for the long-haul case dominate the NVIS geometry and possess extremely small correlation distances. Indeed, solution of the overhead tilt problem may present the greatest challenge in the development of the ultimate objective NVIS HF-SSL system architecture.

In the single-site-location (or single-station-location) method, for which independent estimates of emitter location must be ascertained, it is necessary to measure the orientation of the arriving emitter signal and to transform that information into a location estimate. This transformation method may be simple or complex depending upon the degree of accuracy required and the amount of information available. As has been indicated, the difficulty in this process is the requirement for precise and timely ionospheric information. This information is needed for a procedure we call ray retracing. Ray retracing is not conceptually different from ray tracing. For ray tracing, the application might be associated with the general problem of assessing the communication coverage for a broadcast service, or possibly as an aid for determining the type of antenna to use for a particular service area. Ray retracing, on the other hand, emphasizes point-to-point applications; and it is particularly relevant to SSL.

Three pieces of information are involved in the general skywave propagation problem. The first two correspond to the source (transmitter) and sink (receiver), and the third corresponds to the ionospheric subchannel. In principle, information from any two of these three will allow determination of the properties associated with the remaining one. Thus, if the source and sink are known and possibly controlled (as in a sounder), we may derive ionospheric subchannel information. In the SSL problem, we assert that the ionospheric information is available, and we compute a specified emitter property (i.e., location) given an observation of a signal property (i.e., ray orientation) at the receiver. It is the unequivocal requirement for ionospheric structural information which makes SSL so distinctive in comparison with netted HFDF systems for which ionospheric data is regarded by some as only *nice to have*.

Aside from the simplistic median features which are rather well modeled, there are intrinsic variabilities. These are best known as departures from median behavior, an example of which is shown in Figure 6-28.

Another way to examine variability is to examine how a parameter such as foF2 (as observed from a VIS instrument) compares with mean model predictions. In Figure 6-29 we see the variability over a month during which a large magnetic disturbance occurred. The raw data at Boulder was compared with four models: IONCAP, ICED, RADARC, and MINIMUF. Note that a large positive error bias occurs at midmonth. At this time a large F-layer diminution was evident, and none of the models, as configured, could cope with this effect. Even discarding this main phase storm phenomenon, a residual error is present in the data. This residual error is associated with uncompensated diurnal fluctuations, large-scale structures, and TIDs, These features and even smaller variations, not observed as a result of ionosonde undersampling, will map into angle-of-arrival fluctuations. Consequently the primary challenge facing SSL is to detect and correct for these effects, generally in real time.

6.3.7.1 *Ionospheric Limitations of the SSL Method*

One of the more interesting ideas to emerge from studies undertaken by Australian workers to partially compensate for TID and tilt effects is a scheme called *dynamic tilt correction*. George [1972, 1976], while working with the Australian Defence Scientific Service, presented a novel method whereby tilt effects could be reduced for NVIS paths associated with ranges out to roughly 150 km. This method has never been fully implemented in any existing system even though anecdotal information suggests that it is a viable concept.

The more familiar concept of static tilt correction is one in which a local (say overhead) tilt is presumed to apply for all space of interest. It is well known that this method will fail for emitters which are sufficiently displaced from the control point that the ionospheric sensor is assessing. If we use ionospheric data from one location and apply it to another region, we refer to the process as extrapolation. The maximum extrapolation distance depends upon: (1) how rapidly the medium decorrelates, or (2) the scale size of the irregularities present in the zone of extrapolation.

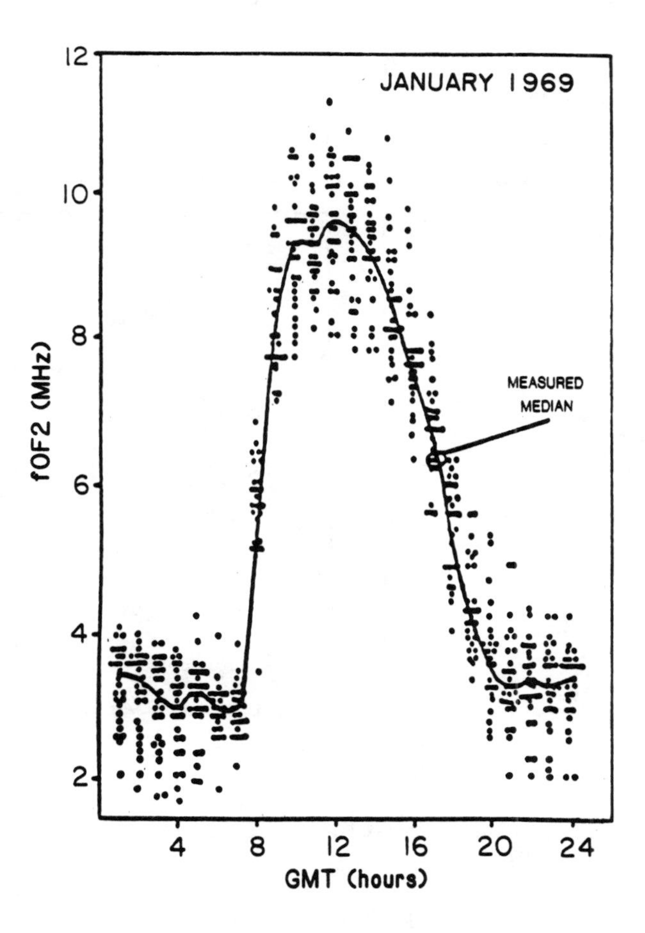

Fig. 6-28. Ionospheric variability at the Slough Observatory.

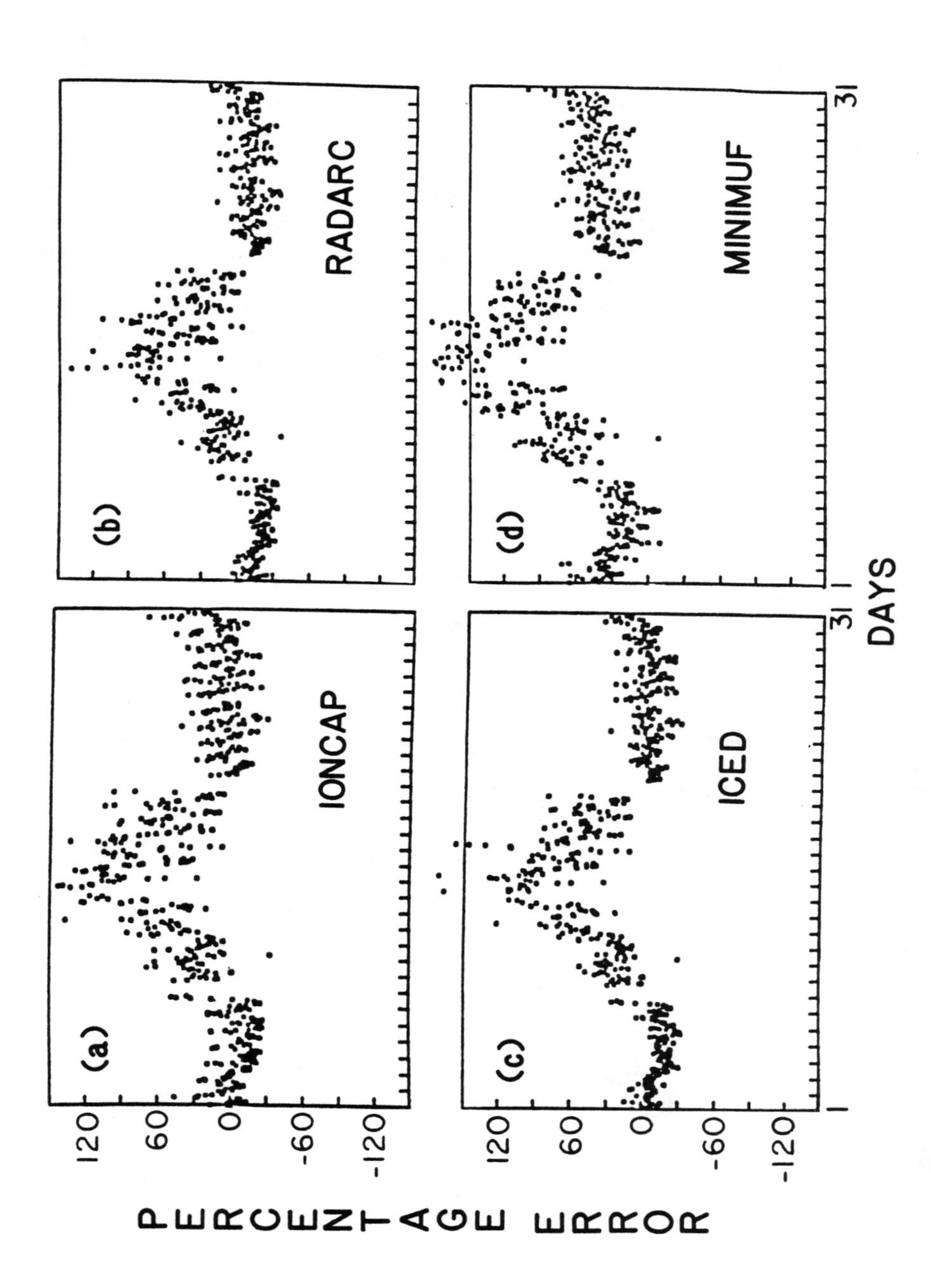

Fig.6-29. Percentage error in model predictions for four ionospheric models in July 1982 at Boulder, Colorado.

A considerable amount of information is available to describe statistically the efficacy of extrapolation. Rush and Gibbs [1973] and Rush [1972, 1976] have examined the spatial and temporal decorrelation of foF2; while Klobuchar and his colleagues at AFGL [Klobuchar and Johanson, 1977] have examined the correlation properties of larger features associated with the total electron content.

Departures from benign (or median) ionospheric behavior are usually viewed as detrimental situations. There is, however, a class of disturbances which lead to improvement in our ability to extrapolate data from place to place. If we were to cross-correlate disturbance amplitudes at two displaced locations, then we would observe higher correlations during severely disturbed days than during quiet days. Put another way, the correlation distance increases during geophysical events which are globally-driven rather than local. Examples include major magnetic storm (foF2 diminution during the main phase, and a short-lived enhancement in foF2 during the initial positive phase). Thus tilt correction schemes would be expected to be sensitive to these effects. In summary, it might be argued that static tilt correction schemes will be less effective during quiet times than during disturbed periods.

McNamara [1988a, 1988b] of Andrew Corporation has described various approaches to ionospheric compensation, and concluded that of the three factors which control SSL performance (viz., angular measurements, ionospheric specification, and ray retracing), ionospheric specification remains the most elusive. A book by McNamara [1991] addresses many of these issues.

Ionospheric tilts introduce bearing errors and elevation errors, with the latter form translating into errors in the estimated range to the emitter. Gething [1978] has treated this subject rather generally. Here, we will simply state the well-established expressions for the effect of static tilts on the accuracy of elevation measurements, emitter range estimation, and bearing measurements. We shall also limit the discussion to thin layers, and we shall ignore the second-order effect of earth curvature. For transverse tilts, the bearing error for N-hops is:

$$\text{B.E.} = (2h/D)\, N_h^2\, \Theta_T \tag{6.7}$$

where B.E. is the bearing error, h is the height of the layer, D is the range to the emitter, N_h is the number of hops, and Θ_T is the transverse tilt angle. Since the error increases with the square of the number of hops, there is a premium on acquisition of bearing data from emitters which are located within 1-hop of the DF site. It is also recognized from Equation 6.7 that the bearing error increases dramatically with decreasing range D for a fixed tilt.

For longitudinal tilts, the effect is to introduce a difference between the transmitted elevation angle β_T and the received elevation angle β_R. This difference is simply twice the longitudinal tilt angle Θ_L, or

$$\beta_R = \beta_T + 2\Theta_L \tag{6.8}$$

If there are no tilts in the longitudinal direction, then the transmitted and received elevation angles are the same, the range error (R.E.) is zero, and the range to an emitter is easily determined from the measurement of the elevation angle. Below we provide equations for the estimated emitter range D (for the no tilt case), and the R.E. for a nonvanishing longitudinal tilt.

$$D = 2\,h \cot \beta_R \tag{6.9}$$

$$\text{R.E.} = 2\,h \Theta_L \csc^2 \beta_R \tag{6.10}$$

But what are the effective ionospheric tilts we might expect? At HF, the impact of tilts is frequency dependent. In other words, for a given ionospheric representation, sufficiently separated probing frequencies will sample different portions of the electron density profile since the depth of ray penetration is an increasing function of frequency. Tilt amplitudes are generally height dependent, and it follows that different probing frequencies will encounter different layer tilts. Thus, in general, a specification of the three-dimensional electron density distribution is not sufficient for estimating ray deviation effects. Ray tracing is required as a general rule, and the probing frequency must be specified.

The depth of ionospheric penetration increases as the radiofrequency is increased. Up to the limit for which the radiowave is no longer reflected from the ionosphere, we find that tilt effects increase with increasing depth of penetration. As we have seen in Chapter 4, polarization rotation also increases as a function of frequency provided the frequency is below the MOF. Once penetration is achieved, such as in earth-space communication, the Faraday effect is proportional to $1/f^2$.

The largest ionospheric tilts will be observed near the MOF, resulting in a maximization in the magnitude of the bearing and elevation angle variances for that frequency regime. Furthermore, a *MUF-seeking* system must also contend with a relatively large amount of polarization fading. On the other hand, signal levels may be enhanced in the neighborhood of the MOF, because of *MOF-focusing* at the junction frequency of the high and low rays, and the probability of multimode interference is reduced.

The bearing deviation depends upon several factors. For a given ionospheric model (i.e., electron density profile, either measured or assumed), a superimposed electron density gradient will generate a bearing error deviation which has marked frequency dependence for a given emitter range. It is found that bearing errors for a target at 200 km may be 10 degrees or more at the lowest propagating frequencies, but may approach 50 degrees for the highest propagating frequencies. Generally speaking, the natural partitioning of measurements into azimuth and elevation components (clearly appropriate for an oblique environment) should likely be replaced by another coordinate system in the NVIS environment, at least in connection with the estimates of system performance. This is because bearing errors become enormous for observation of near-zenithal emitters given the existence of even the smallest of tilt effects, and have little physical significance.

One of the several difficulties associated with geolocation by the SSL method is departure from ionospheric homogeneity. The difficulty is not a fundamental one, but arises out of a lack of information concerning the ionospheric structure. Structural inhomogeneities and layer tilts which cannot be measured or predicted may be treated statistically, but deterministic corrections are clearly impossible. Classes of inhomogeneities may be defined in terms of characteristic time and spatial scales.

Time-varying and uncompensated ionospheric tilts present the ultimate challenge to SSL systems in terms of their ability to approach a geolocation accuracy consistent with the HFDF measurement precision. The sources of these tilts include natural, but dynamic, macroscopic features such as the high latitude trough, the auroral oval and other circumpolar features, the equatorial anomaly, and the day/night terminator. To some extent these features may be modeled, and compensation schemes involving Real-Time-Ionospheric Specification (RTIS) may be successful under specified conditions. A major additional source of tilts is associated with atmospheric gravity waves which have enhanced amplitudes at ionospheric heights. This class of disturbance is often termed an ionospheric wave or a TID, the latter being an acronym for Traveling-Ionospheric-Disturbance. TIDs are actually the ionospheric tracers of the underlying neutral gas motion. The sources of atmospheric waves are typically within the lower atmosphere, and may be the result of meteorological events as well as instabilities originating within the auroral oval.

Two groups of atmospheric waves will propagate and produce ionospheric effects. The first group is a set of modes having temporal scales of roughly 8 minutes or less; the second group consists of all waves having longer periods. The latter group is the more important since it consists of waves which have the largest amplitudes and may travel considerable distances without total extinction. These are called acoustic gravity waves (AGW). The former group consists of waves of smaller amplitude which dissipate quite rapidly with distance; they are termed infrasonic waves.

Generally speaking, MOF variations depend upon both critical frequency (maximum plasma frequency) and ionospheric virtual (mirror) height fluctuations. Fortunately, for tilts associated with acoustic gravity waves (the tracers of which are TIDs), critical frequency and height fluctuations are not fully independent. Generally speaking, height fluctuations are anticorrelated with electron density or critical frequency fluctuations under the presumption that the underlying neutral gas motion is governed by adiabatic compression and expansion. Figure 6-30 shows the general pattern for foF2, h'F2, and M(3000)F2 [Paul, 1988].

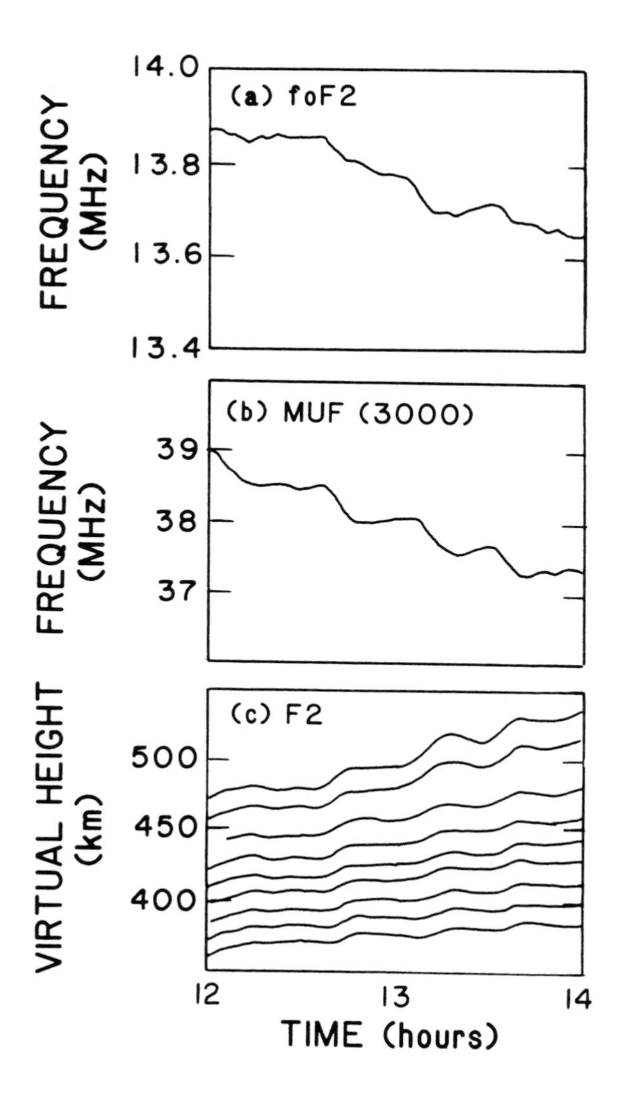

Fig.6-30. Gravity wave-induced fluctuations in foF2, MUF3000, and the virtual height contours. (From Paul [1988].)

Quasiperiodic angle-of-arrival (AOA) fluctuations are often observed, although there are many other instances for which the fluctuations appear to be irregular. Complex patterns may arise for short time scales as a result of multipath fading effects. Some noisy patterns in AOA are thought to be the result of a superposition of TIDs which have unrelated amplitudes and phases as well as different waveforms, since they derive from different sources.

6.3.7.2 *Ionospheric Modeling Support*

A number of SSL system prototypes have used models to provide ionospheric information either in the absence of sounder data or as an adjunct. Some system developers have used IONCAP as a means to develop static tilt information in connection with their analysis of SSL data [Goodman and Uffelman, 1983]. Several climatological models are available for this purpose. It must be recognized that models cannot reflect the dynamics of a given situation and they provide no direct information about AOA variance. A detailed discussion of ionospheric models is found in Chapter 3; propagation factors are described in Chapter 4; and propagation prediction models are discussed in Chapter 5.

6.3.7.3 *Algorithms Used in SSL Determination of Emitter Range*

One may easily estimate the range to a specific emitter given the ray elevation angle. In essence, one exploits the simplicity of well-known ionospheric propagation relations (Martyn's theorem, the Breit-Tuve theorem, and the secant law) to develop an expression linking the measured elevation angle and the ground range. IONCAP or a similar model provides the ionospheric information which is needed in the equation for estimating the ground range d to the emitter:

$$d = 2R\{\cos^{-1}[R\cos\beta \,/\, (R + h')] - \beta\} \tag{6.8}$$

where R is the radius of the earth, h' is the virtual (mirror) height of the ionosphere derived from a computer-generated vertical-incidence ionogram (e.g., from IONCAP), and β is the elevation. This simple result is only an approximation, and admits to significant errors for the longest ranges. However, if we ignore the earth's magnetic field, it provides reasonable estimates of range within the NVIS regime. It has the advantage of being computationally fast. It is called the classical method by some workers.

For quite simple models of the ionosphere, it is possible to determine analytically the range to an HF emitter. This analytic ray-tracing procedure was originally developed by Croft and Hoogasian [1968], but operational improvements have recently been made [Milsom, 1985]. It is termed the Quasi-Parabolic (QP) method and its successful operation requires an admixture of simple ionospheric layer representations (parabolas, linear segments, etc.) which do not necessarily arise in nature. An improved ionospheric model has been developed by Dudeney [1978] but this increased level of sophistication requires numerical ray tracing. To avoid numerical raytracing complexities while retaining some degree of ionospheric realism, Multiple Quasi-Parabolic (MQP) models have also been developed [Baker and Lambert, 1988, 1989]. The success of the general QP approach had been limited to a rather unrealistic ionospheric model [Bradley and Dudeney, 1973] but MQP appears to have a broader application and retains the elegance and speed of an analytic process. Unfortunately it works only if the magnetic field effects can be ignored.

Only a few years ago, conventional wisdom suggested that numerical procedures would be impractical for tactical HF-SSL systems because of the computational load involved, even though such approaches would be more suitable in terms of the magnetoionic complexities which could be addressed. Under these conditions, real-time systems would only permit the use of fast analytic algorithms. Unfortunately, these algorithms are typically based upon unrealistic models of the ionosphere, a fact which places a greater burden on real-time updating procedures to *reacquire* the requisite accuracy. Emerging DSP technology and the enormous increases in processing speed have now made algorithmic simplifications for the sake of speed alone to be unnecessary. While processing speed is very important, even the most elegant computational schemes, even those which involve consideration of the magnetic field, need not be slow.

The check-target procedure is one in which information derived from a known transmitter, of fixed frequency, is utilized to improve the AOA estimates for an unknown emitter. Departures in the position of the known transmitter (or check target) are monitored and these angular departures from the known position are applied to the unknown as a correction. The check target is essentially single-frequency sounder. It has the advantage over a sounder in that the fixed frequency revisit time can be smaller. For effectiveness, the check target (used for AOA calibration) must be closer to the desired target than the sounder (used for channel evaluation) would be from the desired target. Therefore, at least in principle, both temporal and spatial sampling errors (attributed to ionospheric specification by conventional methods) are minimized if check targets are used and positioned properly.

The concept of *closeness* between emitter and check target is an important one. The spatial separation aspect is obvious, but *closeness* also recognizes that ray trajectories are critically dependent upon the frequency of transmission and the electron density function $N(x,y,z;t)$. The check target frequency should not depart too significantly from the emitter frequency, especially in the event of a distended multipath giving rise to selective fading conditions. This is especially important in the neighborhoods of critical frequencies of the MOF where frequency dispersion is pronounced.

Temporal separations should likewise be limited to the differential time-delay difference corresponding to the differential ranges involved. However, given the difficulty of tracking ionospheric microstructure, temporal separations of between milliseconds and seconds will be sufficient for most applications of interest. Even so, there may be a need to sample and average-out rapid AOA fluctuations which would be expected to be at least as fast as the fade rates encountered.

In the NVIS application, an array of check targets, distributed around the SSL system would provide a picture of the local overhead environment corresponding to the probe frequencies involved. A set of swept-frequency sounders could also serve as check targets, and would provide more frequency flexibility.

A major difficulty in the check target method is that the *closeness criterion* cannot generally be satisfied. Given the existence of *smart* and remotely programmable check targets having a communication capability, one could develop a system whereby selected check targets could be directed to transmit at a specified frequency of interest and at a time of interest. In this way two of the closeness criteria could be satisfied. The remaining spatial criterion is dependent upon the location of the unknown emitter and the distribution of check targets. Nevertheless, the check target systems, having transmission and reception capabilities, and being configured with appropriate antennas, should be able receive the unknown emitter by groundwave and skywave, subject to polarization constraints. In some instances, one (or more) of the check targets systems may detect emitter signals propagating by groundwave, a fact which may establish proximity to the nearest check target. This check target would be remotely programmed (from the SSL) to transmit at the emitter frequency. This concept is but one of many which might be considered.

6.3.7.4 *Role of Sounders and Related Equipments*

Vertical-Incidence-Sounders (VIS) have played a strong role in the growth of SSL technology, both as system components and as auxiliary instruments upon which ionospheric characterizations could be made for data sorting purposes. Both applications are important, the first for the operational system, and the second for system design and analysis.

The concept of vertical incidence is a rather misleading when applied at HF. In view of the marked departure from rectilinear propagation for each of the two magneto-ionic modes, especially for rays directed initially toward the zenith, the designation for a *vertical-incidence-sounder* (VIS) should more properly be *vertically-directed-sounder* (VDS) to be more consistent with the physics of the problem. The point is that the VIS (or VDS for the purist) instruments sample a region above the device the location of which is dependent upon the properties of the region being sampled. In short, a first-order VIS analysis tells us what the electron density of the lower ionosphere is, but not precisely where it is. This is a fundamental concern in the application of sounders to the HFSSL problem. The application of antenna arrays in the sounding process could assist in the resolution of this problem.

The signal set available to the SSL system provides a potential resource for use in Real-Time Ionospheric Specification (RTIS). In this connection, the role of passive sounding in future architectures should be considered.

6.3.7.5 *Concluding Remarks on SSL Applications*

It has been indicated that the fundamental weakness in SSL systems is the lack of an adequate ionospheric specification. Another consideration is the availability of an efficient and accurate radiowave propagation algorithm for the geometry involved. For NVIS conditions, ionospheric tilts and gradients provide interesting challenges for SSL design engineers. Several dynamic methods for tilt correction have been suggested to ameliorate this problem.

There are a number of aspects to the geolocation problem. One component focuses on the capability of the system to correctly measure the direction of arrival (DOA) of the incoming signal. This capability naturally depends upon signal parameters such as field strength, duration of signal, and the modulation format, as well as the background interference and noise environment. The efficacy of a measurement of bearing and elevation depends also on the method by which the SSL system determines the most likely result (including the receive array as well as the processing algorithm). A common procedure for arriving at an answer is to analyze the wave surface for planarity. It may be shown that a condition for proper determination of the DOA for a discrete source of RF is that the associated wavefront be linear across the array. By discarding all measurements (i.e., cuts) which do not pass a specified planarity criterion, we will be assured that the consolidated results are associated with discrete modes of propagation. But we are not done. Next we should examine all discrete modes for evidence of source correlation. We should also endeavor to exploit any redundant signals (i.e., discrete multi-mode replicas of the source emitter). Finally, we must have the capability to "ray retrace" through the skywave path and provide a geolocation result. This requires an efficient ray tracing algorithm and a system which faithfully reproduces the ionospheric properties in the regions of interest [Reilly, 1990].

Short of infinite wisdom in the realm of prediction, the successful exploitation of ray retracing to determine source location will depend upon an accurate specification of the refractive properties of the medium. Therefore a sensor for assessing the ionospheric properties is critical for SSL applications, especially under NVIS conditions. The most immediate candidate for the specification scheme of choice is a probe or sounder system which is controlled by the system. For the NVIS geometry, vertical incidence sounders come to mind immediately. It is by no means clear, however, that the conventional swept-frequency sounder is optimum for this purpose. Other approaches might include a channel probe concentrating within the frequency neighborhood(s) of the tasked frequency(ies). The consideration of multiple approaches also has merit, and this approach might involve the exploitation of data from the emitter itself as well as check targets, uncooperative signals of opportunity, data from OTH radar systems, satellite in-situ measurements, topside sounders, or other exotic schemes. In short, we must keep an open mind in our visualization of the ultimate solution to media specification.

6.4 OBLIQUE INCIDENCE SOUNDING

6.4.1 Background

Oblique-incidence-sounding (OIS) has been carried out for a number of years and the work of Hatton [1961] details some of the issues which emerged quite early in the investigation of HF skywave communication circuits. The Canadian Defence Telecommunications Establishment (DTRE) developed an OIS program in 1954 whose primary purposes were to (1) examine the theoretical foundation of oblique propagation at HF, and (2) improve optimum working frequency predictions. DTRE employed a swept-frequency pulse-type sounding system to gather data over several paths during the years from 1954 to 1960. Prior to that time conventional wisdom incorrectly suggested that the following statements were true:

1. [Untrue statement follows] *The F2 layer uniquely determines the MOF for a fixed circuit.* [True statement follows] The F1 layer MOF will occasionally exceed the F2 layer MOF, especially when *hF1* and *hF2* differ significantly, or when substorm MOF diminutions are pronounced. Moreover, Es ionization may control the situation under conditions of blanketing.
2. [Untrue statement follows] *MOF observations agree well with estimated MOFs computed from midpath vertical-incidence soundings.* [True statement follows] Consistently one finds the observed values of MOF to be higher than those which would be estimated from an analysis of midpath data.
3. [Untrue statement follows] *The limit for 1 one-hop F2 path is 4000 km as determined by a static virtual geometry model.* [True statement follows] The canonical 4000 km limit is based upon a nominal F2 layer virtual height of reflection of 300 km. If hF2 exceeds 300 km, then the limit will increase as well. The converse is also true.
4. [Untrue statement follows] *High-angle propagation is not a significant factor in practical HF communication scenarios.* [True statement follows] Although high-angle (Pederson ray) paths are generally ignored with some justification, being subject to significant dispersion, their contribution may to HF connectivity may not always be insignificant. For example, a high-angle ray may dominate a low ray when the latter suffers from obstruction or antenna gain diminution near the horizon.
5. [Untrue statement follows] *Sporadic-E ionization leads to deleterious effects.* [True statement follows] Sporadic E will limit the range of coverage in some instances, and will also introduce scatter and absorption effects. Nevertheless, enhanced Es may lead to improved communication due to elevated MOFs and reduced dispersion of the propagating mode. The primary problem with the exploitation of the Es mode, in the view of frequency managers, is its unpredictability.

6. [Untrue statement follows] *The LUF is primarily a function of absorption factors.* [True statement follows] D-layer absorption and screening effects of lower ionospheric layers are important factors in the determination of the lowest transmission frequency. More precisely, the LUF is determined by a careful analysis of propagation effects and system factors including noise figure in order to ascertain the frequency for which the SNR is adequate to produce a predetermined system reliability.

Studies by Hatton and others have contributed greatly to our understanding of the nature of the oblique incidence sounding and its role in communication assessment and remote sensing of the ionosphere. The OIS method, unlike the VIS scheme, lends itself to examination of more than one ionospheric control point from a single location; and multiple sites will lead to considerable sampling efficiency provided enough receivers are available. It therefore provides the investigator (and communicator) with an assessment of wide-area coverage for an HF system. One of the earliest studies employing multiple sites has been discussed by Lomax [1966] and Nielson [1968]. They describe a 1962 series of VIS and OIS measurements within the Pacific basin. Seventeen OIS paths were monitored and comprehensive observations of spread F patterns were detected. (Spread F is a naturally-occurring phenomenon in the neighborhood of the magnetic equator exhibiting a diurnal maximum in the nocturnal period between local sunset and sunrise the following day.) Lomax observed nose extension effects as well as significant spread echoes which extended the effective MOF to as much as 52 MHz, although normal propagation modes generated much smaller MOFs.

Investigations of oblique-path assessment as an aid to HF frequency management continued with the CURTS experiments [Gould and Vincent, 1962] and other activities [Jull et al.,1965]. The use of OIS for short-term forecasting has been discussed by Petrie and Hunsucker [1967] and Ames and Egan [1967]. We shall return to short-term forecasting methods later.

A synopsis of early studies in which oblique-incidence sounding was employed to characterize radio links may be found in CCIR Report 249-6 [CCIR, 1986c] while operational considerations have been discussed in CCIR Report 357-2 [CCIR, 1986d]. Recent activities within the U.S. DoD and elsewhere are covered in Sections 6.4.3 and 6.6 of this book.

Recently, several organizations have made oblique propagation measurements using OIS instruments such as Digisondes [Reinisch et al., 1989], Chirpsounders™ [Reilly and Daehler, 1986; Daehler et al.,1988], and other types of pulse sounders [Davé, 1988]. Most of these experiments have concentrated on the MOF. The FMCW waveform associated with the experimental OTHB radar in Maine has also led to determinations of the MOF similar to those which could be extracted from a bistatic sounding system (see Section 6.2.2). These data sets have been used to examine the efficacy of various models including: IONCAP, RADARC, AMBCOM, HFMUFES, and MINIMUF (see Chapter 5). The MOF data studied by investigators at the Naval Research Laboratory [Uffelman, 1982; Uffelman et al.,1982] and previously cited work reported by the NRL group has been largely directed toward the development of model update schemes using sounder data as an update parameter.

One of the operational issues often discussed in connection with oblique sounding is the notion of sounding in one direction while applying derived results to an oppositely-directed path. In military practice, mobile units such as ships, aircraft, and jeeps may be required to transmit only as needed in order to limit intercept probability and EW countermeasures. This procedure is sometimes referred to as emission control (EMCON). One strategy might involve the use of short burst-type transmissions in order to reduce *on-air* time. In this instance, the frequency management process, which concerns itself with the selection of optimum transmission frequencies, may be accommodated through use of predictions or through the reception of OIS signals from remote transmitters. Parameters such as the MOF, which summarize tendencies in the gross refractive index over a specified path, would be expected to be reciprocal (see Figure 6-31). This means that ionospheric support will be provided up to MOF (associated with two points) whether signals are transmitted or received. It does not mean that the instantaneous signal strength is necessarily reciprocal, nor does it imply that the SNR (and the system performance) will be reciprocal.

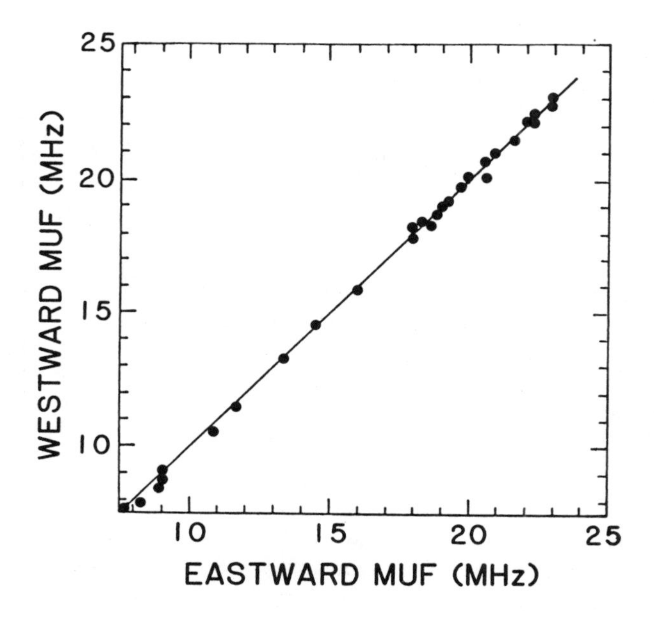

Fig.6-31. Comparison of the MOF for West-to-East and East-to-West paths. (Courtesy BR Communications.)

6.4.2 Description of the Chirpsounder™ Instrument and Method of Operation

The OIS system most favored in connection with military communication requirements is the ubiquitous Chirpsounder™ system, with the earliest versions having the military designation AN/TRQ-35. This system is the defacto standard for Allied sounder operations and official descriptive document is ACP-191 [1987, 1989, 1991]. This document contains the following information: definition of terms, equipment description, operational procedures, station cognizant authorities, and a list of transmit stations. Details about existing stations including station properties, readiness data, and transmitter start times should consult the latest edition of ACP-191. As of this writing more than 80 fixed-site sounder transmitters are reported to be in operation. There are also more than 20 mobile transmitters. Since receivers and spectrum monitors (the other components of the AN/TRQ-35 complement) are passive devices, no attempt is made to control or track specific deployments. Additional (unlisted and unidentified) transmitters are detected from time-to-time, and some transmissions associated with embassy operations, military testing, and short-term R&D activities have been encountered.

BRC has developed a lighter and more flexible version of the Chirpsounder™ system which is backward compatible with the AN/TRQ-35. The following are components of the Chirpsounder™ system along with a brief descriptions.

6.4.2.1 *Chirpsounder™ Transmitter*

All deployed models of the Chirpsounder™ transmitter emit a linear FMCW chirp signal over a selectable range between 2-16 or 2-32 MHz. The sweep period in either case is 4 minutes and 40 seconds. The AN/TRQ-35 operates at 10-100 watts and the newer model TCS-5 is rated at 150W (average power and PEP). The TCS-5 doubles as a communications transmitter (1.6-30 MHz for SSB and CW modes) but the power is limited to 100 Watts in the sounding mode.

A Chirpcomm™ mode may be used to transmit short messages. Chirpcomm™ may be used as an orderwire, but it has been used as a convenient way to identify specified transmitters. It is composed of a 40 character message transmitted redundantly (i.e., 63 times) over the 2-30 MHz sweep interval.

The following definitions are extracted from ACP-191:

Real Time. This term corresponds to UTC (Universal Coordinated Time) as derived from a Universal Time Standard (UTS) such as WWV, WWVH, or BBC. Clocks maintained within the Chirpsounder™ system are synchronized with a UTS.

Sweep Time. Each sounder transmitter is assigned a set of times which define the frequency of occurrence of each sweep and/or the temporal separation between sweeps. These are called sweep times. Examples might include 00, 15, 30, and 45 minutes after the hour, or every five minutes, or any other combination of non-repetitive 5-minute designations but not exceeding 12/hour. However only 4/hour is recommended. These are not actual transmission times. (See Start Time and Transmit Time below).

Start Time. This term defines the temporal lag between transmission of the waveform and the assigned sweep time. To obtain the start time, one must specify the specific minute allotment (start minute) corresponding to the geographical region to which the transmitter belongs as well as a start second. The start seconds are assigned by cognizant authorities within the region involved. The start time is thus the start minute + the start second. ACP-191 recommends that only even start seconds be used to further reduce interference and to enhance the facility for receivers to synchronize to the desired transmitters.

Transmit Time. This term corresponds to the actual time of sounder waveform transmission. The transmit time = sweep time + start minute + start second. The transmit time, so defined, is independent of the hour of the day. The real-time of transmission must account for the hour as well.

Sweep Range. The AN/TRQ-35 system may provide a sweep from either 2-16 MHz or 2-30 MHz.

Scan Rate. The rate at which the sounder scans the HF spectrum depends upon the Sweep Range. For a sweep range of 2-30 MHz, the scan rate is 100 kHz/sec. At the 2-16 MHz, the rate is 50 kHz/sec.

Scan Interval. The time elapsed between successive transmissions from a specified transmitter. This is typically 15 minutes.

6.4.3 Applications of OIS Data: A General Commentary

The Oblique-Incidence-Sounder (OIS) is one of several instruments which provide an indication of the personality of the ionosphere. As a diagnostic tool for specification of HF channel properties, the OIS system has few competitors when relevant factors such as availability of technology, relative simplicity of operation, directness of application to the HF propagation assessment problem, and cost are considered. Specifically OIS technology has provided one method by which HF frequencies may be selected to satisfy certain optimization conditions related to relative signal level and the observed multipath structure. The OIS method belongs to a set of Real-Time-Channel-Evaluation (RTCE) schemes which are principally applicable to fixed HF links characterized by a MUF-Seeking architecture. It is not practical to sound all possible paths in a large communication network, but the benefits from sounding may still be achieved if selected paths are probed and the results are extrapolated to geographically nearby paths. Section 6.6

presents results from several studies which were designed to measure the correlation of ionospheric properties as functions of temporal and spatial variation. In addition, the process of updating well-known models such as IONCAP or MINIMUF is examined in the context of forecasting efficacy.

Information other than the LOF and MOF may be extracted from OIS devices, even though conventional systems do not exploit these possibilities fully. The information may include any or all of the items listed in Table 6-7. Workers at the Naval Research Laboratory (NRL) have examined applications of OIS in support of various U.S. Fleet and U.S. DoD exercises with primary emphasis on extrapolation of data to denied (or unsampled) regions. Early efforts were directed toward the use of OIS data to engage in what may be politely termed propagation tactics [Goodman et al.,1982]. Subsequent work emphasized the communications area more directly [Uffelman and Harnish, 1981, 1982; Uffelman, 1982; Uffelman et al.,1982]. Algorithms and techniques have been emphasized recently, and the issues of temporal and spatial decorrelation of OIS data have been addressed [Reilly and Daehler, 1986; Daehler et al.,1988]. System issues have been discussed by Goodman et al.[1983, 1984] and other system applications of OIS data have been published as well [Goodman and Uffelman, 1983]. NRL retains a substantial data base of OIS records [Goodman, 1984; Goodman and Daehler, 1989]. These data, scaled for the MOF and other parameters, are being deposited in the National Geophysical Data Center in Boulder for access by scientists and telecommunication specialists. Specialists interested in the raw ionograms may secure selected copies by corresponding with NRL directly.

Conventional oblique sounders such as those which employ the chirp waveform indicate the nature of the narrowband HF propagation channel. Noise, interference, and channel occupancy are determined by other devices. In a tactical military environment using an HF link, it is apparent that OIS over that link will indicate the range of frequencies which will propagate. Under the condition that available frequencies are clear of interference and fall within this range, the ingredients for successful communication are achievable. The specification by OIS devices of the range of allowable propagating frequencies over the path being sounded is important because it reduces link establishment time and provides a means to optimize communication performance over time. There is no argument on this point. In fact the U.S. DoD experimented with the idea of sounder networking approaches in the 1960s. Presently OIS has a number of competitors in the context of a cooperative environment for which handshaking is possible. This aspect of the situation was not true in the 1960s. The literature is replete with candidate systems which employ embedded in-band approaches or which involve protocols dependent upon the assessment of other measures of channel efficacy such as bit error rate, signal-to-noise ratio, multipath spread, channel occupancy, etc. Nevertheless, one need only pay casual attention to the field to be impressed with the emphasis placed on earlier OIS approaches by the military services.

TABLE 6-7

INFORMATION FROM OBLIQUE-INCIDENCE-SOUNDERS

1. Frequency band(s) which will support propagation over the path being sounded.
2. Frequency band(s) which exhibit limited multipath conditions.
3. Identification of modes and hop conditions at frequencies of interest.
4. Identification of the LOF and MOF for each propagation mode.
5. Identification of the overall MOF for the complete ionogram.
6. Identification of the overall LOF for the complete ionogram.
7. Estimation of the signal strength for specified modes at specified frequencies.
8. Multipath spread and estimated maximum allowable signaling (baud) rate to avoid ISI.
9. Estimation of the selective-fading or coherence bandwidth as a function of frequency of transmission.
10. Estimation of elevation spread of incoming signals.
11. Estimation of channel bandwidth from group-path-delay slope.
12. Determination of maximum hop rates as a function of frequency from an estimate of pulse stretching (dispersion).
13. Estimates of the magnitude of horizontal gradients or tilts for those situations in which multiple hops may be identified along a great circle path possessing a monotone gradient.
14. Identification of modes which are difficult to predict: (sporadic E, spread F, mixed modes, ducted/chordal modes).
15. Extraction of ionograms suitable for raytracing to support advanced applications such as HFDF, OTH radar, etc.
16. Extraction of ionospheric data for real-time model update.
17. Compilation of ionospheric data for long-term predictions and development of improved models.

6.4.4 On the Interpretation of Oblique-Incidence Ionograms

The detailed interpretation of ionograms has long been the province of a small group of ionospheric specialists. However the needs of communicators are different from those of the physicist, emphasizing channel properties rather than physical properties. For example, the communicator may simply need the MOF without regard to the median time delay associated with the path traversed. The ionosphericist needs the time delay information to invert the ionogram and derive an ionospheric electron density profile. Intermode and intramode delay spreads are important to the communicator as well, but detailed ionospheric understanding is not essential.

A handbook is available to assist in the interpretation of vertical-incidence ionograms (see Section 6.3.2). Unfortunately, no such document is available for OIS data interpretation. Reference material includes: Chapter 4 in the classic NBS monograph 80 [Davies, 1965], Chapter 6 in the newest book by Davies [1990], and selected papers which contain atlases of data for illustration (e.g., Lomax [1966], Davies and Barghausen [1966], and Nielson [1966]). Even so, a definitive document specific to the interpretation of oblique-incidence ionograms is sorely needed. Such a handbook would no doubt include sample OIS output for a variety of geometries, geographies, seasons, solar activities and geomagnetic conditions. Furthermore the document should be augmented with companion synthesized ionograms derived from raytrace calculations to provide insight. Steps should also be taken to select some records for which the control points are close to active vertical-incidence sounder sites. This would allow the rather considerable information available on VIS data interpretation to be exploited. Naturally if the ionospheric profile is known along the ray trajectory from any single method or by any combination of ionospheric diagnostics, then a realistic raytracing analysis can be performed to assist in the interpretation of pathological cases.

Although the scaling of OIS data and its conversion to N(h) profiles are not well established procedures, several methods are under development. Digital methods are most efficient, and some digital OIS data are now available [Wright and Kressman, 1983] for purposes of testing. An indirect reduction method is used by Gething [1969], and direct methods are used by George [1970], Smith [1970], and Rao [1973]. Rash and Gledhill [1984] have compared these methods with actual data from high latitude paths. As anticipated, tilt effects are important.

Automatic scaling methods for OIS such as DORIS (Digital Oblique Remote Ionospheric Scaling) are being developed [Kuklinski et al.,1988]. Such methods would be essential ingredients in a real-time frequency management system based upon oblique sounders.

6.5 OTHER FORMS OF RTCE

A number of alternative forms of RTCE data which differ from those already mentioned may be considered. In some instances the RTCE data is delivered directly (as in a wideband channel probe [Wagner et al., 1988]); in other instances the RTCE data must be developed from purely ionospheric data. A few new examples are given in Table 6-8 along with those previously mentioned. A book by Hunsucker [1990] identifies many of the methods, and Part 2 in a book by Giraud and Petit [1978] provides an interesting sketch of ionospheric measurement techniques. The reader may also refer to selected articles in the WITS Handbook (Volume 2) edited by C.H. Liu [1989].

TABLE 6-8
RTCE METHODS AND IONOSPHERIC DIAGNOSTICS

METHOD	DELIVERABLES AND REFERENCES
OIS[1]	Propagating modes; Table 6-7, Section 6.4
VIS[1]	NVIS propagating modes and $N_e(h)$ profiles; Table 6-3, Section 6.6
WB Channel Probes[1]	Channel properties; Sections 4.13.2 and 4.14
Fixed Frequency Sounding and HF Doppler Technique[1]	TID observations; also Jones [1989]
BSS and Coherent Backscatter Radar[1]	Area coverage patterns, ionospheric structure/dynamics, gravity waves; Section 6.2.2; also Greenwald [1985]
CHEC[1]	Assigned channel properties; Section 6.2.1
FMON and HFDF[1]	Emitter signal strength and location; Sections 6.2.2 and 6.3.7
Noise and Channel Occupancy Measurements[1]	Passive measurement of channel properties, and other-user interference; Section 5.8.6
Pilot-Tone Sounding[1]	In-band sounding of assigned channels; Section 6.2.3
Error Counting[1]	Direct measurement of BER in traffic channel; Section 6.2.3
Topside Sounding[1]	N_e profile above hF2; [Pulinets,1989]
Incoherent Scatter[2]	N_e, T_e, T_i, plasma motion versus height
Trans-Ionospheric Methods[2]	[Klobuchar,1989], [Aarons,1990], [Jursa,1985]
- Faraday polarimeters	TEC
- Group-path-delay/GPS	TEC
- Differential doppler	TEC
- Scintillation	Ionospheric inhomogeneities
Rocket and Satellite Probes for in-situ Measurement	[Szuszczewicz,1989], [Giraud and Petit,1978]
- Plasma probes[2]	Electron density and temperature
- Mass Spectrometry[2]	Ionic constituents
Indirect Methods[1]	[Davies, 1965]
- Riometers	D-layer absorption
- Magnetometers	Ionospheric disturbances
Optical Methods[2]	Images of plasma properties, aurora in the visible band; ionosphere in UV band; McCoy et al. [1987]

KEY
1. Utilizes the HF band.
2. Method belongs to the class of RTIS techniques; channel properties which are equivalent to RTCE products are computed.

RTIS: Real-Time Ionospheric Specification
RTCE: Real-Time Channel Evaluation
HFDF: High Frequency Direction Finding

6.6 SPATIAL AND TEMPORAL EXTRAPOLATION OF RTCE DATA

6.6.1 A Review of Reported Methods

A prediction scheme based on solar observations, once developed, may be the basis for improvements in short-term, as well as long-term forecasts. On the other hand, the ionosphere (and the HF channel) responds to factors other than the sun; and improvement for short-term forecasts has been clearly shown to result from the direct application of ionospheric observations. This should probably be no surprise. The general area of short-term forecasting and spatial projection of ionospheric data is discussed in Report 888-1 [CCIR, 1986e]. This report contains a comprehensive set of references detailing a variety of short-term forecasting approaches. Of interest in the current context is work related to the use of real-time foF2 or MOF measurements to improve near-term predictions. Nevertheless the methodologies are applicable to the general use of other RTCE deliverables with certain adjustments. Considerations for short-term updated forecasts, based on foF2 measurements by vertical-incidence ionosondes, may be found in Rush [1976] and Rush and Edwards [1976]. Also of interest are papers by McNamara [1976, 1985]. It is observed that midlatitude foF2 measurements can be used to improve median model predictions for foF2 at a point removed in space and time out to approximately 500 km in a north-south direction, out to approximately 1000 km in an east-west direction, and ahead by times on the order of one hour. The preceding numbers depend on environmental conditions, and would be influenced by time-of-day, season, solar activity, magnetic activity, geomagnetic latitude, and possibly other factors. Further information is given in the cited references. A severe requirement for an updating system based on ionospheric sounders (or similar devices) is that it be capable of collecting the ionospheric data and distributing it to the prediction modules, where processing and forecasting functions are accomplished, all within the space of an hour or so. This requirement is more likely to be met when predictions and measurements can be confined to smaller geographical areas. The question of data distribution is central in any communication system involving a network of sounders. Various schemes have been suggested including the use of a separate set of robust HF order wires for network management and data dissemination, modulation of the sounder waveforms to provide for data transmission between nodes, the use of satellites for connectivity, and even meteor-burst communication.

As indicated, several methods of forecasting have been attempted over the years. No single method is a panacea. The choice of method should depend upon the application, the forecasting performance required, the ease of implementation, and the overall environment within which the forecasting system must perform. Table 6-9 contains a summary of methods which have been moderately successful.

TABLE 6-9

SUMMARY OF SHORT TERM FORECASTING METHODS

METHOD	REFERENCES
1. Use of a weighted 5-day mean of foF2 to estimate near-term and daily values of foF2. The same strategy can be used in connection with the operational MUF.	Rush and Gibbs [1973]
2. Compute the departure between foF2 and the running 15-day median over a period of time to deduce a trend line. This trend is extrapolated into the future to derive a forecast.	McNamara [1976]
3. Forward projection of the difference between the current value of foF2 and the monthly median prediction.	Wilkinson [1979]
4. The predicted value of the MOF at a future time is computed as the sum of the long-term mean for that time and a term formed from the product of the MOF autocorrelation value between current and future times and the difference between current values of MOF and the long-term mean. The forecast value is: $MOF(t+\tau) = Ave\ MOF(t+\tau) + C(t,\tau)\ [MOF(t) - Ave\ MOF(t)]$ where t is the current time, τ is the lead time for the forecast, and $C(t,\tau)$ is the autocorrelation between t and τ.	Ames and Egan [1965] Ames and Egan [1967]
5. The predicted value of the MOF is deduced in logarithmic form: $\log MOF(t+\tau) = \log \langle MOF(t+\tau)\rangle + k\ [\log MOF(t) - \log \langle MOF(t)\rangle]$ where t is the current time, τ is the forecast lead time, the brackets "<>" suggest that a running 5-day median is taken, and k is essentially the correlation coefficient.	Petrie and Hunsucker [1967]
6. Same method as No. 5 above except that the 5-day running median is replaced with long-term MUF forecasts.	Petrie and Hunsucker [1967]
7. An ionospheric index is deduced from observation by a set of ionosonde stations in the neighborhood of the desired terminal. This index is used in connection with monthly median maps of foF2 to estimate the value of foF2 at the desired terminal. This index is projected forward in time.	McNamara [1979] Wilkinson [1986]
8. A value of MOF is deduced from a mean model of the ionosphere which is adjusted at the current time to be consistent with a current measurement of the MOF. The process typically will involve computation of a pseudoflux (sunspot number) which force model and observations to agree. The model (along with the derived pseudoflux) are used to characterize the future. A PC implementation for MUF,LUF,FOT predictions is available.	Reilly and Daehler [1986] Goodman and Reilly [1988] Goodman & Daehler [1988a]

A rather novel real-time prediction scheme was described by Jones et al.[1978]. It involved the examination of broadcast transmitters located in the general area where forecasting is required. The scheme uses frequency monitoring (FMON) and is a Class 2 RTCE method (see Section 6.2). The underlying principle exploits the generally monotone increasing MOF which exists in the post-sunrise AM period, and the generally monotone decreasing MOF in the PM period (in the neighborhood of sunset). The application is fairly straightforward. For example, in the morning sector, the basic idea is to monitor the time at which a working frequency is first heard and compare this to the predicted time that the MUF is first exceeded. In the evening one observes the time at which the signal is lost. The signal is heard either early or late depending upon whether or not the observed MOF pattern is above or below the predicted MUF curve. At each of these transitions we have a revised estimate of the predicted MUF. In short, we have a twice-daily update scheme using a single frequency. By monitoring multiple frequencies, we may develop additional pairs of update times. As was indicated in Section 6.2, it is essential that we know the location of the signals being used for this method. Otherwise we do not know the region to which the updated model refers.

Beckwith and Rao [1975] were probably the first investigators to predict the MOF for an unsounded path by use of a median model updated with MOF data from another (control) path. The method involved adjustment of the midpath foF2 for the control path so that the observed MOF and predicted MUF agrees. This adjusted value of foF2 is then used for the unsounded path. RMS errors for a full 24 hour diurnal cycle and for each of the 4 six-hourly periods were computed. All-day RMS errors were roughly 3.5-4.0 MHz without update and about 0.8 MHz with update. The smallest RMS errors and the least update improvement were observed in the early morning. The largest errors and greatest improvement were noted near sunset.

The percentage improvement in prediction performance *PI* is given by the following relation:

$$PI = 100\,[S_o - S_p] / S_o \tag{6.9}$$

where S_o represents RMS fluctuations in data, and S_p corresponds to the RMS prediction error. For a linear prediction model [Gautier and Zacharisen, 1965], if we let c = the correlation of data fluctuations, then:

$$PI = 100\,[1-(1-c^2)^{\frac{1}{2}}] \tag{6.10}$$

Note that $c = 1$ means that there is no prediction error, the *PI* is 100%, and all data fluctuations are translated directly from the sampled space to the unsampled space.

Temporal correlation tends to be quite variable and dependent upon the parameter being examined as well as on the environmental conditions. Naturally one would expect the influence of traveling ionospheric disturbances (TIDs) to be quite significant for F region parameters (such as hF2 and foF2) and for periods commensurate with one-half the wave period of the disturbance. TID periods range from minutes (at the infrasonic end of the spectrum) to several hours (for the largest atmospheric gravity waves). The largest amplitude and most coherent disturbances (having scales of 1000 km or so and periods of an hour or more) arise following major substorm events and are relatively rare. The most significant contribution to temporal variability is from a superposition of intermediate- and small-scale waves (having dimensions up to several hundred kilometers and periods of an hour or less) which are uncorrelated and arise from random source locations. This family of waves will influence the maximum ionization densities and the heights at which the maxima occur in a quasirandom fashion. Regardless of the source of variability, the consequences are evident and sometimes striking. Various studies have provided some feel for the average decorrelation time. Generally a value of one hour is taken for daytime conditions although longer durations have been noted. Nighttime and solar minimum values tend to be smaller since the error which must be corrected (i.e., the variance) is less under those conditions. Some of the forecasting error introduced by temporal variability may be compensated for by use of a real-time model. This would eliminate the component of variance associated with observed *DC* bias errors and data trends which may be tracked. After removal of the low frequency terms in this way we are still left with an ostensibly irreducible error from the TIDs. Accordingly, prudence would suggest that correlation should not be expected to exist for longer than an hour. This has implications for the sampling rate required for model updating. The key to the situation is clearly dependent upon the amplitude of TIDs relative to the low frequency terms.

Figure 6-32a is an example of how the ionospheric support varies for three paths terminating at a receiver located at Norfolk, VA. The geometry is given in Figure 6-32b. Two complete days are exhibited for which the MOF, FOT Band, and LOF are displayed. (These parameters, and especially the LOF, are influenced by parameters of the instrument being used, although the MOF is consistent with accepted definitions.) The FOT Band is defined to be the band for which signals are observed to exceed a factory-set threshold and to possess no multipath. Therefore the FOT band should represent a reasonable band for communication.

An interesting test of HF model update has involved the use of the NOSC program MINIMUF, which was discussed in Chapter 5. Aside from the specification of parameters such as terminal locations, day-of-year, and time of day, the MINIMUF program is driven principally by sunspot number or its equivalent. It was thought by workers at NRL that if one could force a MINIMUF estimate of the MUF to agree with the sounder-derived value of MOF, then one could generate an effective sunspot number or pseudoflux number which might be of use for some time in the future. Surely the pseudoflux should be better than a fixed value of predicted solar flux. Tests of this hypothesis were generally positive. Subsequently tests involving IONCAP were conducted with similar results. Figure 6-33 shows how the updated MUF prediction errors vary with time delay between update and prediction. The message is clear. If a model update scheme is to be used, use the result as soon as possible. After a few hours, another update is required to retain some improvement over a simple model prediction.

a.

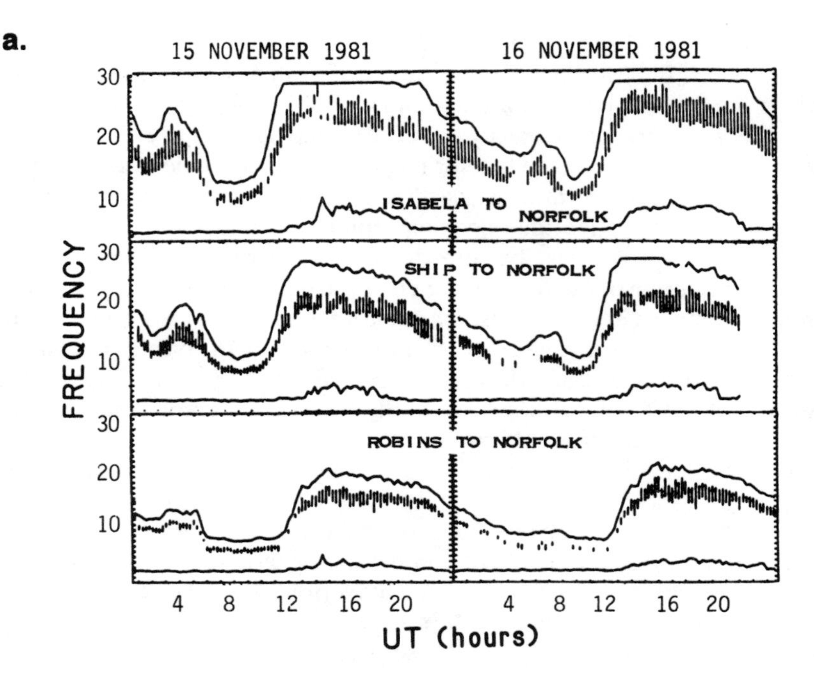

b.

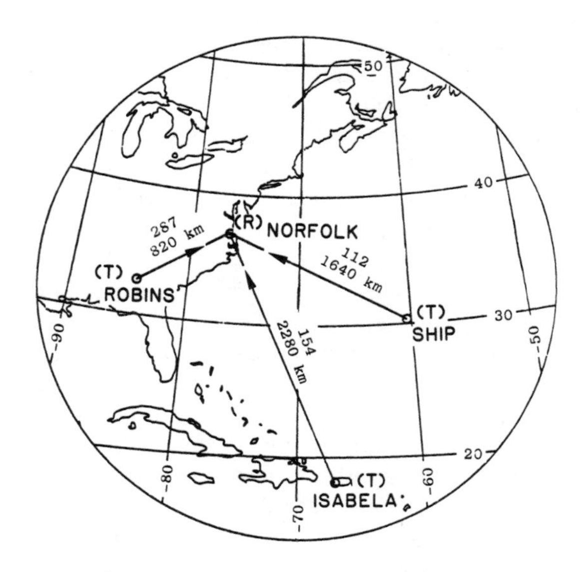

Fig.6-32. (a) Measured MOF, LOF, and FOT band for November 15-16, 1981. The longest path is given on the upper curve, and the shortest is given on the lower curve. (b) Geometry of the situation (From Uffelman and Hoover [1984]; NRL Report 5246)

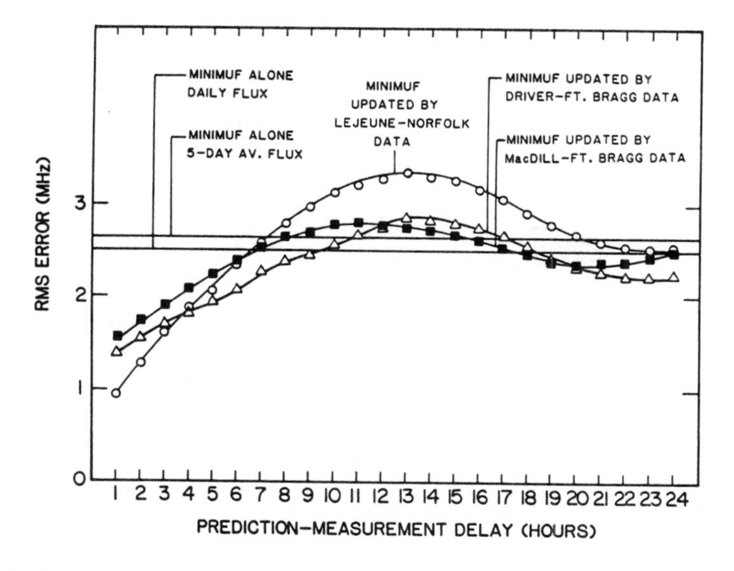

Fig.6-33. RMS error obtained between the observed MOF and that which is determined by a number of techniques involving MINIMUF. Any climatological (median) model will have a similar behavior: rapid increase in error in the first few hours, followed by saturation for a 12 hour lag, and a partial recovery after 24 hours. Data were obtained by Daehler [1984].

Models like MINIMUF or IONCAP provide median values for the MUF so one would not expect a sounder measurement of the MOF to agree with the model estimate of the MUF even if the model performed perfectly. Perfection in this context means that the median of the observations coincides with the model predictions which are, after all, representations of median nature. Examination of foF2 for all hours over a month at any specified site will clearly demonstrate the problem. Not surprisingly MINIMUF documentation frequently specifies a 2-4 MHz RMS error in MUF prediction (compared with actual MOF). This error is reduced rather substantially through use of the sounder update procedure but it is not eliminated entirely. This is because of problems such as the presence of omnipresent traveling ionospheric disturbances and the impact of ionospheric storms. There are situations for which a model driven by a single value of sounder-derived pseudoflux or even sunspot number may be adequate in terms of its ability to provide estimates of MOF for an entire day. Figure 6-34 illustrates this point. Indeed, in this example, the performance of IONCAP (in terms of RMS error) is remarkably insensitive to changes in the assumed value of the sunspot number. However we are not usually this fortunate, and the weight of evidence points to a rather rapid deterioration in prediction performance for those models which are driven by current values. Certainly after a few hours, short-term prediction performance is little better than if long-term methods were used.

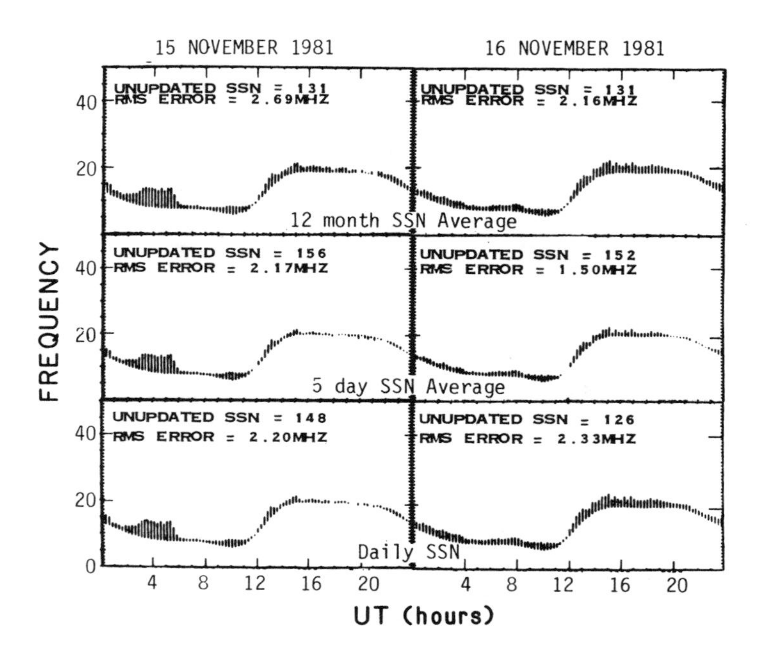

Fig.6-34. A comparison between MOF values and IONCAP predictions for a midlatitude circuit. The vertical bars are the differences between the model prediction and observed value. (See Figure 6-34.)

Nevertheless Uffelman et al.[1984] have shown that the all-day RMS error may occasionally approach the absolute minimum achievable by a single (all encompassing) update of sunspot number or pseudoflux. This minimum RMS error may be visualized as the best fit achievable by sliding the model prediction both horizontally (i.e., time slip compensation) and vertically (i.e.,sunspot number compensation) with respect to a plot of observed MOF versus time. The major factors involved in such occasionally good performance no doubt include: high correlation within a few hours of the update, enhancement in correlation in the neighborhood of 24 hours, and most importantly, non-disturbed conditions. The RMS error for a selected half-day period would be larger than the all-day value.

Magnetic storms provide the most strenuous test of the model update procedure. Figure 6-35 is a comparison of a MINIMUF model prediction based upon an a priori estimate of sunspot number with a set of actual measurements. The agreement is poor, and performance with any static propagation model (including IONCAP or HFMUFES) would be little better. It is of some interest to determine the number of updates that are required by the model input *driver* in order to keep the model prediction within 1 MHz of reality. This is shown in Figure 6-36.

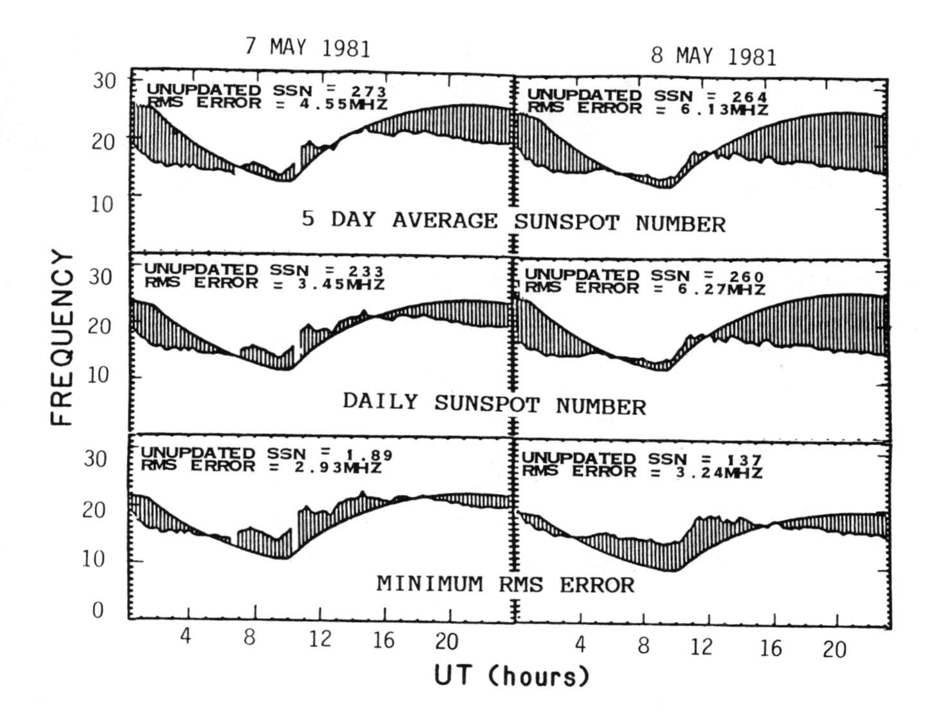

Fig.6-35. MOF observations compared with MINIMUF model predictions for a mid-latitude path between Hurlbert Field FL and Norfolk VA (1176 km) on 7-8 May 1981. The driving parameter for the model was [top] 5-day average sunspot number, [middle] the daily sunspot number. The best the model could do was given in the lowest plot [bottom]. This was a time of very high solar and magnetic activity. (From Uffelman and Harnish [1982]; NRL Report 4849; data were obtained during the SOLID SHIELD exercise.)

Clearly magnetic storms necessitate more rapid updates. The update rate suggested in this instance is much greater than that implied in Figure 6-35 which was obtained under relatively quiet conditions.

Models other than MINIMUF and IONCAP have been used to provide forecasting performance improvement in the update mode. For example, a single effective sunspot number for the ICED model [Tascione, 1988] may be used to fit the model predictions to foF2 data obtained from several vertical-incidence sounders. This same effective sunspot number is used at later times and in other locations. A very similar procedure could be followed using a network of oblique-incidence sounders to provide the consolidated effective sunspot number. In this instance, the properties derived from the OIS instruments would be converted to effective values of foF2 at the control points; thereafter the ICED update procedure would proceed as before. Remark that in the OIS approach, we obtain the observed junction frequency between the high-ray and low-ray modes on the oblique ionogram, which we shall call the maximum observable frequency, MOF. In the cases for which nose extension on the ionogram trace causes a departure between the highest frequency observed and the defined MOF, we simply excise the anomalous extension from consideration. This procedure is justified, since the extension is typically due to side-scatter or related effects, and is not accounted for in available models. The MOF values defined in this way can now be used to determine an effective sunspot number or an effective 10.7 cm. flux as input to the prediction model.

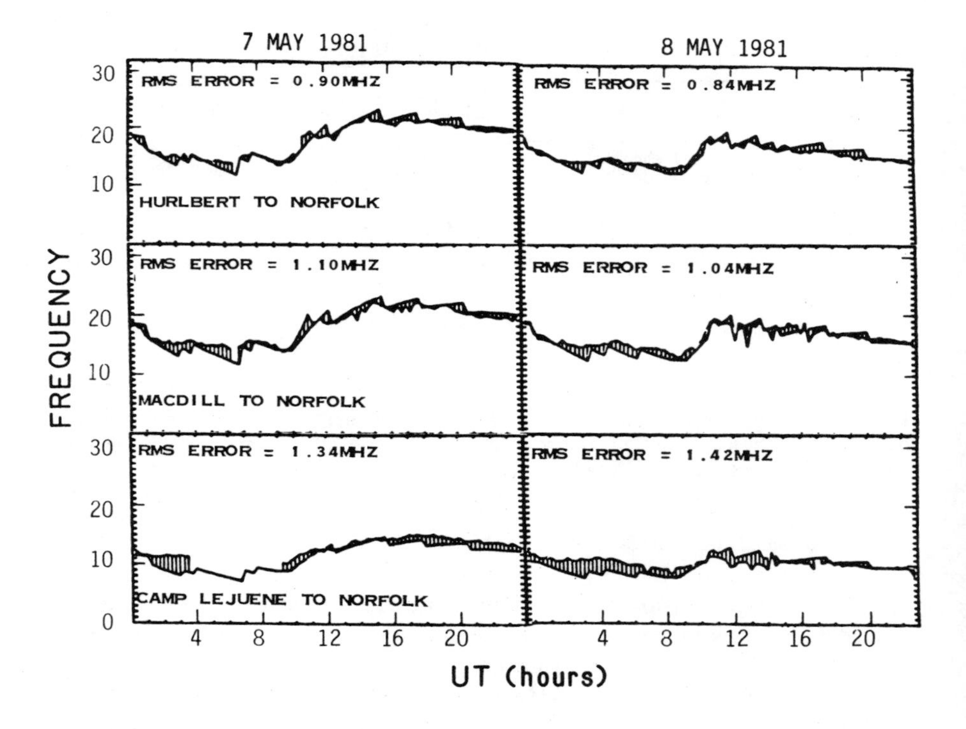

Fig.6-36. Comparison of updated model predictions with MOF observations where deviation is never allowed to exceed 1 MHz. The path is the same as in Fig.6-37. The discontinuities are locations where the deviations reached 1 MHz necessitating an update. The update rate was roughly 1 per hour. The top curve is for 7 May and the bottom curve is for 8 May, 1981. (From Uffelman and Harnish [1982]; NRL Report 4849; data from the SOLID SHIELD exercise.)

6.6.2 Some Current Extrapolation Concepts

Interest in this field has mainly centered on extrapolation of channel properties to denied areas; i.e., the noncooperative environment. For the military, this naturally leads to the concept of propagation tactics for avoidance of threats such as intercept or jamming. It also has implications in HF communications networking since success in extrapolation will reduce the number of node pairs which must participate in sounding.

The basic idea is simple. To estimate an HF propagation parameter for a location (or control point) where data are unavailable, one must assemble data in neighboring regions and extrapolate such data as appropriate to the control point in question. This naturally brings up the issue of spatial persistence or spatial correlation in the process of evaluating the efficacy of a particular scheme of data extrapolation. Some guidance has been obtained from the studies of Rush [1976] and Zacharisen [1965] in connection with foF2, which is helpful, but extrapolation of the MOF has not been studied as extensively.

6.6.2.1 *The Analysis of foF2 Data*

In the analysis of forecasting methods, it is convenient to describe the geographical limits of performance improvement in terms of a confidence ellipse, or more properly, a correlation ellipse. Much effort has been directed toward the definition of such an ellipse in connection with the parameter foF2. Perhaps the most comprehensive, and certainly most cited, studies of foF2 variability were performed by Rush and coworkers [Rush and Gibbs, 1973; Rush et al.,1974; Rush, 1976]. They examined the spatial and temporal correlation of foF2 for the epoch of the International Geophysical Year (IGY) for 32 ionosonde stations located between latitudes of 22 and 71 degrees North. Other workers have examined foF2 variability for other solar epochs, but the results have not been published in the open literature.

An important parameter to consider in correlation studies is the short-term variability which has a relatively low amplitude. Correlation of smoothed data sets, which are only subject to intermediate and long-term variabilities, yields results which are too optimistic. This is because the average diurnal, seasonal, and solar epochal variations are well-behaved, and are highly correlated as well, provided obvious allowances are made for differences in solar zenith angle, local time, etc. Otherwise, successful median models could not be constructed. Since the spectrum of foF2 variability exhibits its highest amplitude components at the lower frequencies (viz., periods of years to days) and since these components are correlated over wide geographical areas, even unfiltered data will exhibit correlation properties which are misleading and not always useful for purposes of short-term forecasting. The object is to examine the correlation of foF2 data which has been passed through a high-pass filter, under the assumption that the low frequency components may be accounted for in any reasonable long-term model.

The general conclusions reached by Rush may be stated as follows: the correlation of foF2 variability decreases with station separation but depends upon geographical latitude, season, and time of day. Correlation tends to be lowest between 2200-0300 LMT especially during winter months. Correlations tend to be greater along the east-west baseline (since local time variations are accounted for in the Rush analysis) and are least for north-south paths. Using a relationship derived by Gautier and Zacharisen [1965], the appropriate spacing for a reasonable percentage improvement to reduce the uncertainty in foF2 variability by application of remote data was obtained. For a 50% improvement, the north-south separations range between 250 and 500 kilometers, and the east-west separations range between 500 and 1000 kilometers. The set of variability data were obtained by forming the difference between the locally observed values of foF2 and the long-term median values. Rush's analysis was based upon the notion of local time compensation. In essence, this amounts to sliding the local time axis at each observatory so that they coincide. This approach, while it does properly characterize the performance of algorithms which correctly model consistent diurnal features, cannot be used to characterize area-wide events which are organized by universal time. Such events might be large-scale TIDs propagating equatorward.

6.6.2.2 *U.S. Navy Extrapolation Experiments*

The unique opportunity to examine extrapolation ideas was afforded by TEAMWORK 80 [Goodman et al.,1982] which involved three subauroral paths. Although the test was of short duration, it was found that a single control path could be used to derive the Maximum Observable Frequency (MOF) on the two other paths with acceptable accuracy. To do so required the elimination of nose extension effects arising from auroral echoes (see Figure 6-37). The process of extrapolation naturally involves the use of some rule or algorithm which expresses the normally anticipated deviation of the MOF on the sounded path from the unsounded path. This algorithm would be expected to depend upon path control point separation. Additional factors such as solar activity, magnetic activity, geographic and geomagnetic coordinates, season, and time of day would also be important considerations in the ultimate deviation. Most model representations include diurnal, seasonal, and solar activity in their recipes. The NRL TEAMWORK 80 study emphasized the use of such representations in either MINIMUF or IONCAP for purposes of performance evaluation. There is, of course, nothing sacrosanct about such selections. Of some interest to those involved with prediction of HF propagation is the duration of acceptability of a channel sounding. It is taken for granted that a sounding will provide a relatively accurate picture of the real-time-channel; and it should therefore provide the frequency manager with better real-time guidance than static models even if the later are updated with global parameters such as solar flux or sunspot number (which have not been reconstructed on the basis of ionospheric data).

Figure 6-38 shows the difference between the modeled MUFs and the observed MOFs for two of the TEAMWORK 80 paths: Kolsaas, Norway and Soc Buchan, Scotland to the USS Mt. Whitney. Clearly a significant (negative) bias error exists as does as an expected random error. An apparent error in the sunrise/sunset terminator was resolved by requiring MINIMUF to accommodate a two-hour time slip.

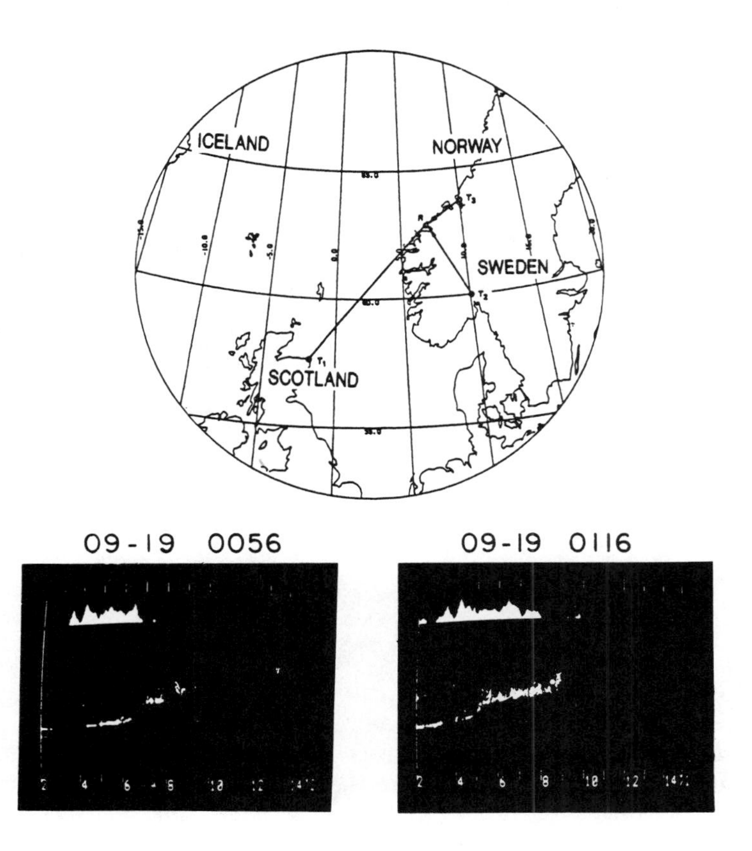

Fig. 6-37. (a) [top] Geometry of the TEAMWORK 80 exercise. (b) [bottom] Nose extension of the ionogram trace for the path between Kolsaas Norway and the USS Mt. Whitney (340 km) (T_2). Notice that the O-X splitting typical of a classical NVIS path is evident. The scatter component appears to extend from the X-mode MOF. The MOF decreases between 0056 GMT and 0116 GMT (also local) while the upper limit of the scatter remains fixed. (From Goodman and Uffelman [1982]; NRL Report 4953)

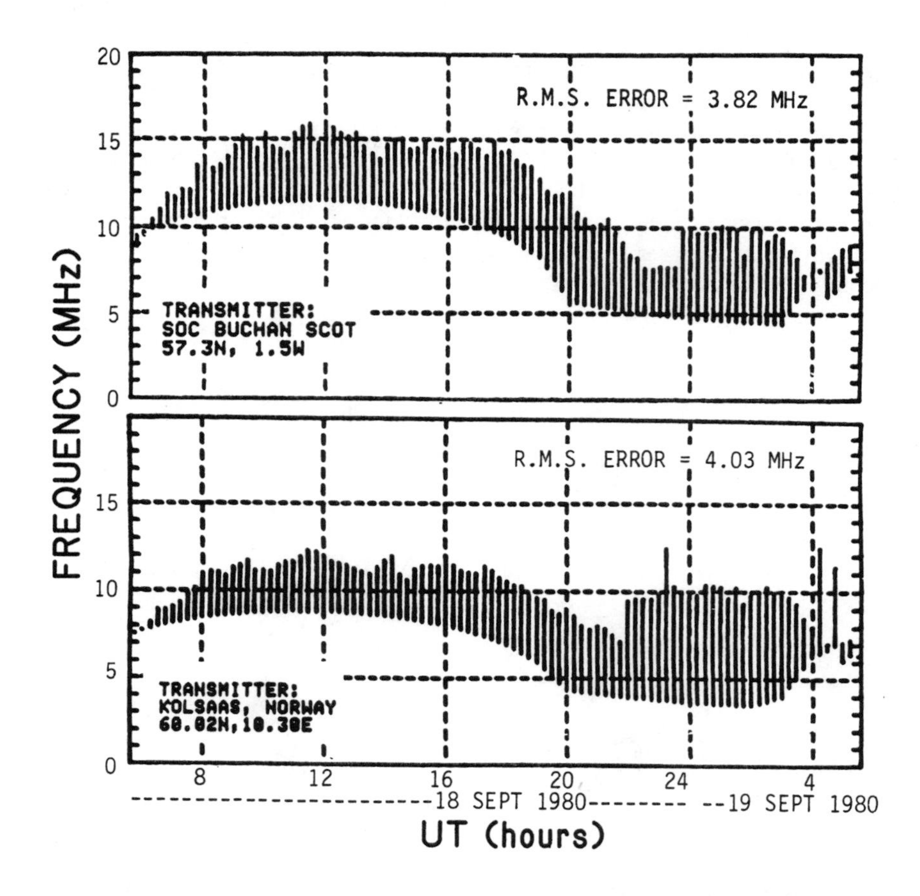

Fig.6-38. Difference between sounder-derived MOF and MINIMUF prediction. A sunspot number of 112 is used. (a) [top] Soc Buchan to Mt. Whitney. (b) [bottom] Kolsaas to Mt.Whitney. The vertical bars represent errors (differences between model and the observation. The smooth lower boundary is the model prediction. The ragged upper curve is the observed data. (from Goodman and Uffelman [1982].)

The model predictions given in Figure 6-38 were obtained by using the sunspot number for the day in question (i.e., $R = 112$). The sunspot number necessary to provide the best fit to the data is significantly higher. Using the Kolsaas Norway to Mt. Whitney path as a control, one obtains a fairly good fit to the data after removing the side scatter effects in the midnight sector. The pseudosunspot number for this control path was determined to be 205. This value was applied to the longer path directly (Soc Buchan to Mt. Whitney), and the results were equally good. Figure 6-39 shows the MOF (and updated-MUF) patterns for both paths.

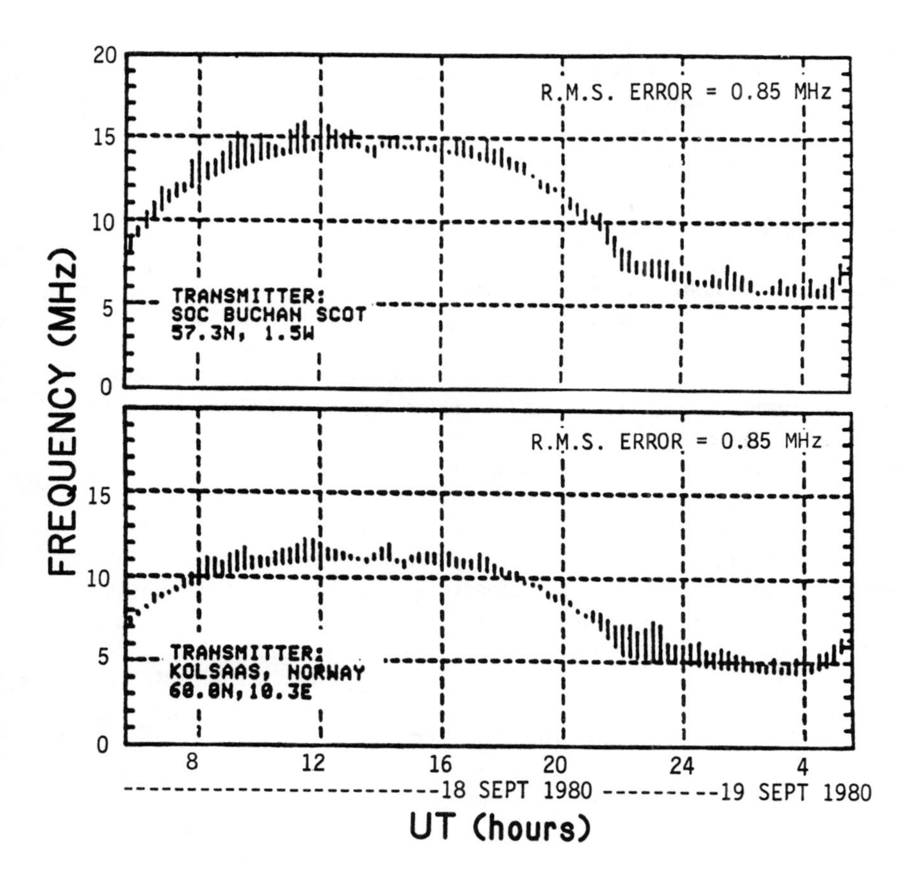

Fig.6-39. Comparison of updated models with actual data. The top curve is the Soc Buchan to Mt. Whitney path. The bottom curve is the control path (Kolsaas, Norway to USS Mt. Whitney). (See Figure 6-38.)

6.6.2.3 *Exploitation of Multiple, Spaced, and Simultaneous OIS Transmissions*

A different use of sounder data for updating a climatological model has been developed recently. It is based on the simultaneous availability of multiple OIS data sets at a single receiver site. For every path which is sampled in this way, we may determined information at one or more control points depending upon the number of hops involved. For the short path case (i.e., length ≤ 4000 km), the association of OIS data with specific control points is rather clear. It is simply the midpoint of the path. For long paths (i.e., length ≥ 4000 km), complications arise. There are two ways this data may be used. The most obvious procedure is to convert the OIS time delay versus frequency data into effective electron density profiles. Procedures for doing this have been developed, but the profiles are somewhat disguised as a result of the path obliquity. In any case one has the possibility of developing a three-dimensional picture of the electron population distribution over the region of interest.

6.6.2.4 *Exploitation of Long-Path Data*

A few comments are in order concerning the utility of long-path data (i.e., lengths ≥ 4000 km), as well as multi-hop data on short-path ionograms (i.e., lengths ≤ 4000 km). If paths have lengths which preclude the appearance of a trace for the first hop (on the ionogram), complications arise. Assuming the path length is between roughly 4000 and 8000 km, the MOF which is observed (corresponding to the lowest order mode on the ionogram) may not be assigned unambiguously; it may be associated with the midpoint of either of two legs. This ambiguity may be resolved by (1) application of climatological models to estimate the subhop which should have the least MOF and therefore possess frequency control, or (2) by the examination of independent data sets which may provide guidance concerning ionospheric horizontal gradients. The same procedure may allow additional data to be extracted from short-path ionograms (0 ≤ length ≤ 4000 km). If the total path length is L, then the first hop MOF will be associated with the midpoint of the path $L/2$; the second hop MOF may be associated with either $L/4$ or $3L/4$; the third hop MOF may be associated with $L/6$, $L/2$, or $5L/6$; etc. Naturally, the MOFs decrease with the increasing number of hops since the path lengths and ray zenith angles are smaller. Since the highest order hop is essentially an NVIS path, the tendency of the MOF family will approach the overhead critical frequency for some control point along the path. But which point? If a vertical incidence sounder is located at the base station, we can compare the high order MOF tendencies with foF2. If they are the same, or nearly so, we may assume that the gradient is monotonic and directed away from the base station.

6.6.2.5 *The Pseudoflux Concept*

Another way to exploit the OIS data sets is to generate a set of pseudofluxes, one for each identified control point, thus forming a map of pseudoflux contours. This approach has been suggested by Goodman et al.[1983, 1984]. Because of the (short-term and small-scale) ionospheric variability which remains following model application, this map of pseudofluxes will also exhibit variability. Naturally, this map of pseudofluxes can no longer be interpreted physically as a solar driver; rather it is an ionospheric index map. Of course, the pseudoflux maps are only defined for a limited region over which real-time OIS data may be extracted. The maps may be used in a manner similar to those which are contained within the *Atlas of Ionospheric Coefficients* [CCIR, 1986b].

An obvious updating option is to determine the effective flux for the sounder control point (SCP) which is the nearest neighbor to the ionospheric control point (ICP) on the unsounded path. This would be used in the prediction model of choice to obtain a MOF prediction on the unsounded path. Without any updating method of this kind, typical MOF prediction errors on unsounded paths are in the range of 2-4 MHz at midlatitudes and for undisturbed conditions. With a fresh update (i.e., within about one hour), the upper bound of MOF prediction error drops to within 1 MHz, if the SCP is within about 300 km of the ICP of the unsounded path. A second option for updating offers further improvement under some circumstances. If a triangle of SCPs can be found which is small enough and surrounds the ICP in question, and the updates at all three SCPs are fresh in the preceding sense, then a pseudoflux for the ICP can be found by linear interpolation of the three surrounding pseudofluxes to the ICP position. The value of this approach has been examined in a study by Reilly and Daehler [1986].

As has been suggested previously, the ionospheric electron density is best organized in terms of geomagnetic coordinates, at least in the F region.
It has also been indicated that east-west correlation distances are larger than north-south correlation distances. Since geographic latitude lines make only small angles with respect to geomagnetic latitude lines, a portion of the enhanced east-west correlation may be associated with a form of geomagnetic field control. Reilly et al.[1990] have found that an improvement in forecasting capability may be developed for large disturbances, under the condition that extrapolation is along lines of geomagnetic latitude. They also suggest that two or more pseudofluxes might be necessary to drive state-of-the-art forecasting models, one for each quasi-independent ionospheric parameter of interest. For example, one might need to apply a separate pseudoflux for determination of foF2 and hF2 (in an ionospheric application or for NVIS propagation), or possibly even three for the E-, F1-, and F2-region MOFs (for oblique communication circuit applications). Such schemes are reminiscent of more complex procedures in which ionospheric layers are computed through consideration of the actual solar flux components. The necessity for use of the pseudoflux concept to deduce E and F1 MOFs over unsounded paths is not entirely clear under benign conditions since these layers are rather Chapman-like in nature and exhibit less drastic TID effects. These facts result in acceptable predictability at least in terms of E and F1 layer critical frequencies. The use of sounders to detect sporadic E over a wide area might lead to a pseudoflux recipe for extrapolation, but this is a matter for additional research.

Daehler [1990] has developed a prediction program called EINMUF which exploits the pseudoflux concept as well as more conventional approaches in conjunction with the MINIMUF-85 MUF prediction method (see Chapter 5).

6.6.2.6 *Correlation of Pseudoflux Indices*

Through analysis of pseudoflux on a global basis, made possible by a multiplicity of OIS data from multiple sites, it is possible to generate maps of pseudoflux index. This map would be similar to the CCIR maps of EJF(zero)F2 (or equivalently foF2). For a specified model which is used to deduce the pseudofluxes, a figure of merit for the model may obtained through an analysis of irregularities appearing on the pseudoflux map. For a perfect model, a single value of pseudoflux will apply to the entire map, and no contours will be observed. Typically, this is not the case. Moreover, the correlation between pseudoflux values which are spatially separated is not unity. This arises because separated points possess ionospheric properties which exhibit a non-vanishing independence. This is caused by TIDs and other time-varying effects which possess their own spatial decorrelation properties.

The correlation of pseudoflux has a direct bearing on the design of a sounder network. Daehler [1984] has shown that the pseudoflux is 0.9 or higher provided the distance between control points is less than 700 km. For control point separations less than 350 km, the correlation averages 0.98 (see Figure 6-40a). It is emphasized that this result corresponds to a midlatitude region and for a relatively restricted period of time. Figure 6-40b gives the geometry of the experiment from which the data were obtained.

6.7 A NETWORK APPROACH TO RTCE (with Emphasis on Sounding)

This section discusses the importance of sounder networks for the improvement of HF communication and related disciplines which depend upon a real-time understanding of ionospheric structure and dynamics. Other than HF communications, the disciplines include HF direction finding, HF Over-the-Horizon (OTH) radar, and a host of earth-space activities in the VHF-SHF domain. In the last case we include UHF satellite communication activities such as the Navy FLTSATCOM which has a geostationary space component, the aging VHF/UHF Transit navigation system which consists of a family of satellites in polar orbit, and the NAVSTAR/GPS Global Positioning System constellation which operates at L-band (1.2 and 1.5 GHz). Indeed any system which depends upon or is constrained by the ionosphere may benefit from information which may be derived from sounder data. The benefits are increased through networking and the near-real-time availability of data sets. Earth-space activities derive useful information from sounder networks because, to first order, the total electron content (TEC) of the ionosphere is largely controlled by foF2, or more properly the electron population within the F2 layer. We shall return to this later.

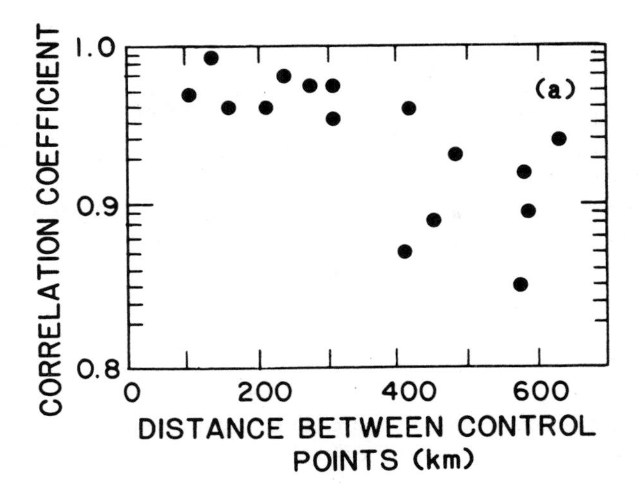

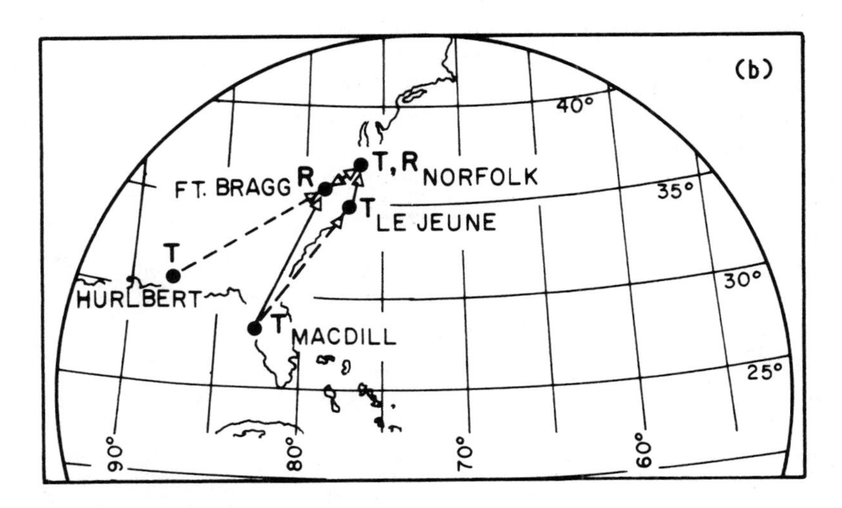

Fig.6-40. (a) [top] Decrease in the correlation coefficient with distance between control points associated with eighteen pairs of paths in the SOLID SHIELD exercise. The control points are path midpoints. (b) [bottom] Geometry of the SOLID SHIELD exercise. Receivers and transmitters are identified by R and T respectively. (From Daehler [1984]; NRL Report 5505.)

6.7.1 Overview of Sounder Networks: Old and New

There have been a number of system tests and experimental campaigns which have involved the assimilation of OIS (and VIS) data sets. Data from many of these campaigns has been organized for estimation of system-wide performance under existing propagation/ionospheric conditions.

6.7.1.1 *U.S. Navy System*

The U.S. Navy procured and deployed a number of pulse sounder systems in the sixties. These systems, comprised of AN/FTP-11 transmitters and an AN/UPR-2 receivers, were deployed on specified ships and Naval Communication stations (COMSTAs). Barker-code pulse compression techniques were used to increase the effective transmitter power (and SNR) while still retaining adequate pulse resolution. The FTP-11 transmitter power was roughly 30 kW. A full scan of 80 discrete frequencies between 2 and 32 MHz (and thus a full ionogram trace) was accomplished in 16 seconds.

6.7.1.2 *CURTS*

The CURTS system (Common User Radio Transmission System) was developed for DCA to make more effective use of existing HF assets. This system was briefly discussed in Sections 6.2.1.1 and 6.4.1. For this system, a pulse sounder network was established in the Pacific basin (see Figure 6-41). The sounder waveform was constructed so that pulse presence could be ascertained, signal strength could be estimated, and time-delay spread could be measured [Daly et al.,1967]. The CURTS sounder swept 120 discrete frequencies between 4 and 32 MHz in 24 seconds, and the nominal peak transmitter power was 15 kW.

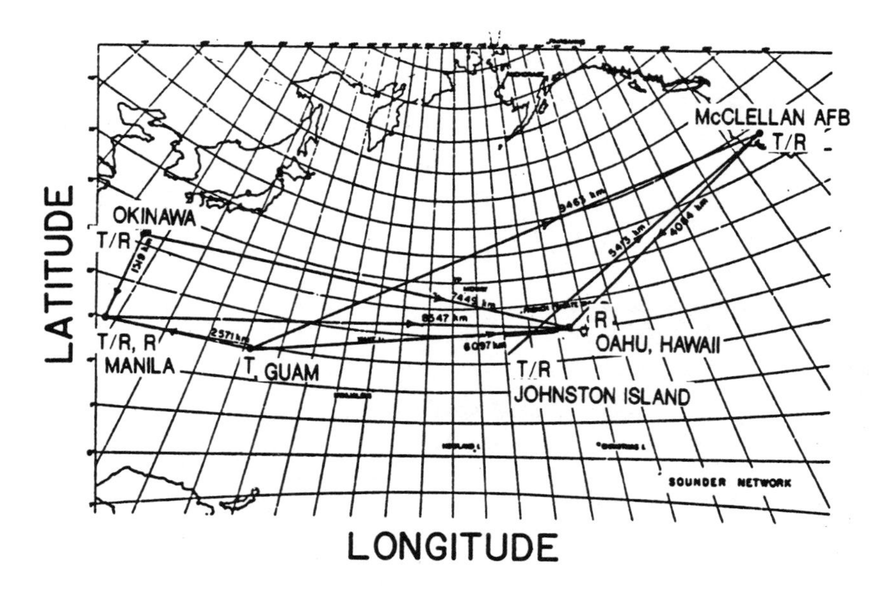

Fig.6-41. CURTS sounder network. (From Daly et al. [1967].)

CURTS was one of the most striking successes attributed to the exploitation of oblique sounders [Jelly, 1971]. The original concept [Probst, 1968; Stevens, 1968] was based upon the real-time assimilation of sounding information along with noise, interference and performance data for all assigned frequencies. A central processor would utilize these data to develop consistent predictions of short-term performance and optimized frequency selections/changes for controlled links. The Pacific test involved a network consisting of six transmitters and two receivers with 120 frequencies being sounded. Between sounding sequences, samples of receiver output were obtained to assess the external noise environment at all assigned frequencies. CURTS employed an out-of-band sounding subsystem. In other words, sounding was not performed at the assigned frequencies for communication. (The assumption is made that this out-of-band sounding will cause negligible interference since the pulses are short and revisits to specified non-assigned channels are infrequent. To estimate the signal strength at the various assigned frequencies, the two nearest neighbors above and below the assigned frequency were processed to obtain a median value. A combination of the measured noise and interference and the median signal strength estimate was used to develop of channel quality index for each assigned frequency based upon a bit error rate analysis. In this manner it was possible for the central processor to rank the assigned channels, with the results being passed to a (human) technical controller for use in frequency change decisions. According to reports, 98% of the frequency changes dictated by the CURTS system gave rise to immediately acceptable traffic as compared to 70% without the aid of CURTS. Despite the advantages which might have accrued from the deployment of CURTS as part of the U.S. Defense Communications System, it was not funded owing to the view that SATCOM was thought to be a better investment at that time.

Even though CURTS was never deployed as an operational system, the lessons learned did find application in a number of systems. The ranking of assigned channels based upon a channel quality index is now central to various in-band systems such as SELSCAN™ developed by Rockwell-Collins. Even the notion of a DoD-wide OIS network to facilitate selection of frequency channels for operational trunks has gained some acceptance.

6.7.1.3 *SRI Sounder Network during FISHBOWL*

For the FISHBOWL nuclear test series in the Pacific in the early 1960s, another complex of pulse sounders was established to examine the impact of High Altitude Nuclear Effects (HANE) on HF trunks in the burst areas. Sounder coverage for the test series is shown in Figure 6-42. The OIS network was established by Stanford Research Institute (SRI). It consisted of four transmitters and eight receivers; 29 propagation paths were monitored including 12 single-hop paths. The equipment consisted of model 902 and 903 sounders developed by Granger Associates. The Granger sounder scanned 160 discrete frequencies between 4 and 64 MHz at two vastly different pulse lengths every 20 seconds (100 μsec and 1.6 msec). This allowed the delivery of both coarse and relatively high resolution ionograms for every scan interval. The very extensive OIS coverage was augmented by a group of vertical-incidence sounders, and the combined data base facilitated the study of HANE-generated TIDs [Lomax and Nielson, 1968].

6.7.1.4 *NOSC Data Base Used in Development and Testing of MINIMUF*

Engineers at the Naval Ocean Systems Center (NOSC) have assembled a large amount of pulse sounder data which has been in the development and verification of their MUF prediction program MINIMUF. This database consists of OIS ionograms derived from the Navy Tactical Sounder System (or NTSS, consisting of the FTP-11 and UPR-2), the Granger sounders, and some others. Figure 6-43 shows the family of oblique paths from that complex of sounders which were studied by engineers at the Naval Ocean Systems Center (NOSC). The analysis of the resulting data ultimately formed the basis for MINIMUF, a Navy MUF prediction algorithm [Rose et al.,1978]. One of the earliest versions of the program was verified for 23 different transmission paths which had lengths ranging from 800 to almost 8000 km. Over the years, more data sets have been included in upgrades of MINIMUF.

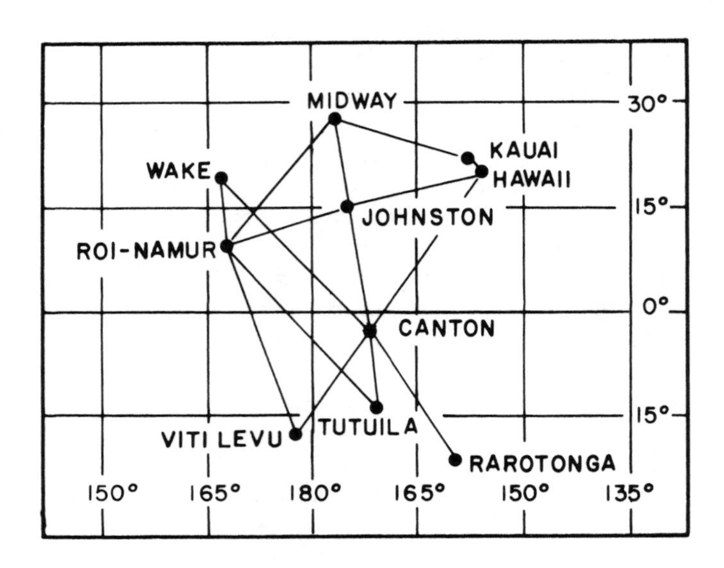

Fig.6-44. SRI sounder network employed during high altitude nuclear tests in the Pacific. (From Lomax and Nielson [1968].)

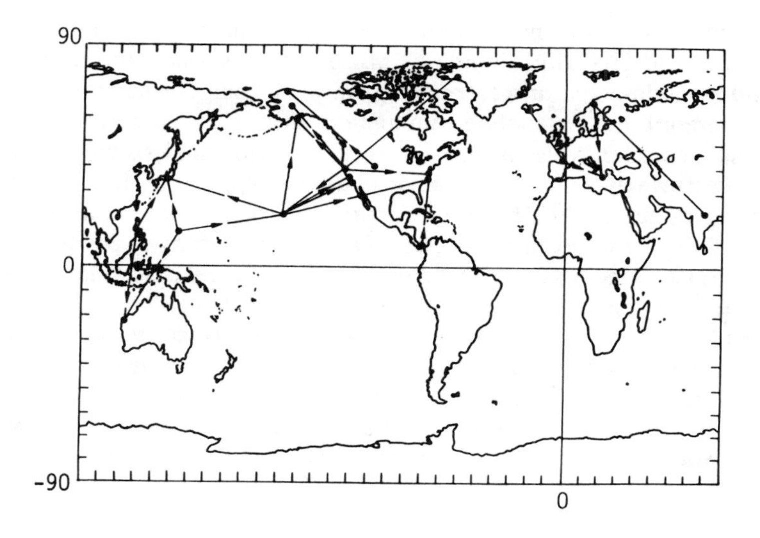

Fig.6-43. Path segments used in the NOSC data base of OIS data. (From Rose [1982]; NOSC unpublished report: *196 Path Months of HF Oblique Incidence Sounder Data.*)

6.7.1.5 *Other OIS Data Bases*

Other special purpose tests have been conducted by NRL, and a sizable data base of MOFs has been formed as a result [Goodman and Daehler, 1988].

Although campaigns of the foregoing type have provided a wealth of information about HF propagation effects and have led to some interesting modeling concepts, the sounder networks (or paths) which provided the data were not organized as part of an operational system. Only CURTS was structured so that operational considerations could be assessed.

6.7.2 Proposed Sounder Networks

6.7.2.1 *European VIS Network*

Milsom [1983] has reviewed HF skywave frequency management for the European environment. He concludes that the existing European VIS *system* (whose data is archived at the various World Data Centers) is not adequate to effectively evaluate HF propagation conditions effectively in western Europe. Figure 6-44 illustrates the projected forecasting coverage for the network in 1983 and a modified coverage based upon a slight reorganization of the sounders and the deployment of several new ones. Such forecasting coverage is fairly good over western Europe. This is independent of the known and more recent deployments of other sounding systems such as the Digisonde™ DG-256) and OIS Chirpsounder™. Milsom discusses how the sounder data could be assimilated at a forecasting/network control center, and suggests that meteor burst or landline communication might be used for net control and data transfer. Milsom's network configuration analysis was based upon the same principles of VIS data correlation as applied by Rush [1976] who first suggested the development of an ionospheric observation network as an aid to short-term forecasting. (The correlation issue was discussed briefly in Sections 6.6.1 and 6.6.2). The dimensions of the correlation ellipses shown in Figure 6-44 are only intended to be representative. They may shrink or expand depending upon time-of-day and solar-terrestrial conditions. The existence of structure will reduce the dimensions while large disturbances will expand the size of the ellipses. Simple modeling procedures may serve to make the ellipses larger, thereby suggesting a reduced number of sounders which would be required to assess the propagation environment over a specified area.

6.7.2.2 *Global Network of Dynasondes*

In connection with work on the Dynasonde sounder system, NOAA scientists [Wright, 1975, 1977; Wright and Paul, 1981] proposed a global digital ionosonde network. Although the primary sensor complement in the system would consist of VIS devices (presumably of the Dynasonde type), the total system requirement for media sampling necessitates the addition of an OIS complement.

Fig.6-44. Proposed European VIS network. (From Milsom [1983].) On the right is the existing coverage (circa 1982); On the left is the revised coverage obtained by modest additions and the movement of several sounders.

The deployment suggested by Wright and Paul [1981] was partially based upon the existing VIS network, but modifications and additions were constrained by practical as well as geophysical considerations. The network topology was not significantly influenced by geopolitical factors however. About 90 VIS instruments are specified for the total system with unsampled areas being addressed by other sensors such as OIS systems.

An open ocean region cannot be sampled using the VIS method alone unless a sensor is in the required neighborhood. On the other hand, OIS, comprised of noncolocated transmitter and receiver terminals, senses a limited region of the ionosphere which is roughly midway between the two terminals for the case of a one-hop path. The Wright-Paul concept calls for about six VIS instruments to cover the European region and supplemented by OIS for intermediate distances. This composite provides a spatial sampling resolution roughly commensurate with the setup suggested by Milsom.

Using the DGS-256 as the basis, scientists at ULCAR have proposed a *GLOBAL HF NET* [INAG, 1986; Bulletin No.48]. This network would be comprised of 20 nodes, all within a 1-hop propagation distance from each other. Both vertical and oblique soundings would be made. Soundings would be scaled automatically and measurements of noise would be made in order to . Each node in the system concept would have the capability to communicate with up to 4 other nodes. Figure 6-45 shows the network topology.

Neither the Milsom nor the Wright-Paul concepts explicitly address how the VIS and/or OIS sensors are to be integrated into a real-time system. The development of a centralized global data (fusion) center would certainly be a formidable undertaking, if it were to be operated as a real-time network. Although there is scientific interest in making global data sets more readily available, it is not clear that there is an overwhelming need for users to access global data on a real-time basis. However, national and regional military interests may require real-time access of OIS and VIS data sets within a particular theater of operations. Moreover, it is possible to envision the transfer of skeletonized data sets between various regions. Nevertheless non-real-time access of global data is clearly valuable for scientific and military purposes. Applications include improved modeling and hindcasting to explain system malfunctions or propagation anomalies. Convenient methods for accessing archived data sets are being developed by the National Geophysical Data Center (NGDC) , NOAA, and the Department of Commerce in Boulder, CO.

6.7.3 Existing Systems and Resources

Several systems have been introduced which may provide HF users with information about channel behavior over oblique paths without the necessity for procurement of additional transmitters. In each of these instances, a network of specially-designed transmitters has been established and receivers are positioned to provide for assessment of designated paths.

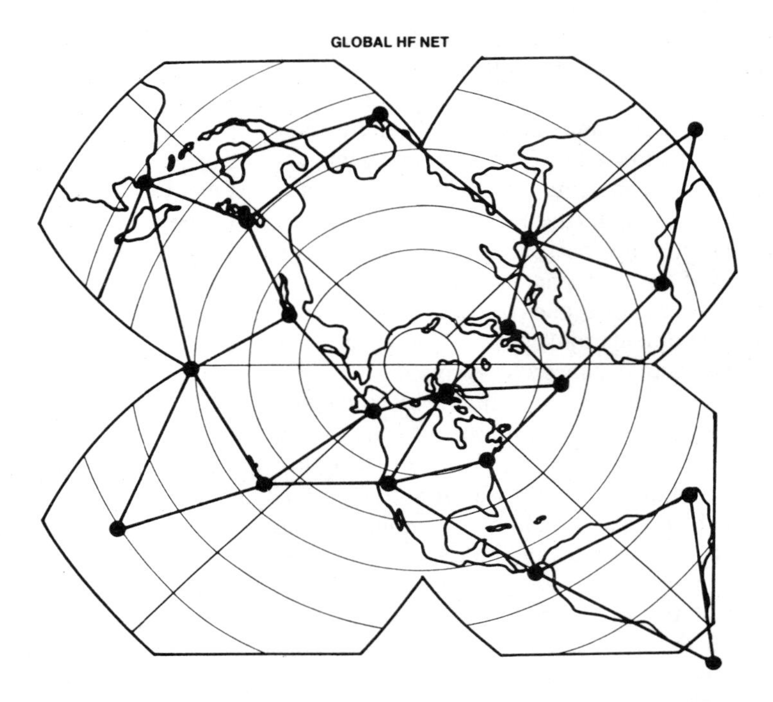

Fig.6-45. Proposed worldwide sounder network based upon DGS-256 sounding systems. This is referred to as *Global HF Net.* (From Bulletin 48, INAG [1986].)

In the networks which have been established, the transmitted waveforms are processed by compatible receivers to obtain the ionospheric channel information. Since many HF skywave applications depend principally upon ionospheric structure near the control point of the path (independent of path orientation to first order), one may simply move receiver systems to appropriate locations to obtain the control point distribution that is desired. We shall briefly describe three such networks: the AN/TRQ-35, (or Chirpsounder™) the DGS-256 (or Digisonde), and the Japanese sounder networks. The DGS-256 operates in both the OIS and VIS modes. An alternative procedure for retrieval of summary data is available from designated DGS-256 sites. Scaled data sets for selected sites are to be assimilated within the SESC real-time data base as well as within the SELDADS system. The real-time data is available to qualified users through the satellite broadcast service and it may be extracted through a computer-to-computer file transfer service.

A discussion of operational sounding systems of the OIS type is contained within Report 357-2 [CCIR,1986d] while the ionospheric propagation issues are more fully covered in Report 249-6 [CCIR, 1986c].

6.7.3.1 *The Chirpsounder™ Network*

The currently existing network of Chirpsounder transmitters is listed in *Ionospheric Sounder Operations* which is published by the U.S. DoD and referred to as ACP-191 [JCS, 1987, 1989, 1991]. The principles of operation were discussed in Section 6.4.2. To avoid mutual interference, the world is divided into four regions each having a central authority for the assignment of transmission start times. Each area is allotted a specific start-time minute. Within this minute allotment, specified stations within an authority are assigned start-time seconds, usually an even second specification with odd seconds being used for guard time. Hourly sweep time sets are also assigned in increments no smaller than 5 minutes. A sample sweep time set might be 00, 15, 30, and 45 minutes. If the start time minute/second were 3 minutes and 16 seconds, then the actual transmit time in this instance would be 03:16, 18:16, 33:16, and 48:16 for each hour of operation. Given that four areas may accommodate 30 transmitters, no more than 120 transmitters are permitted according to this allotment/assignment strategy. Figure 6-46 gives the global distribution of Chirpsounder™ transmitters. Figure 6-47 and 6-48 show the distributions for North America and Europe respectively.

The necessity for use of sounder networking as a means for reducing the number of sounder transmitters has been examined in several papers, and concrete suggestions have been made to bring this about [ARFA, 1990; Josephson, 1990]. The U.S. DoD policy for *Ionospheric Sounder Operation in the US&P* is consistent with ACP-191 ground rules, but lays the foundation for two distinct subnets of Chirpsounder transmitters. The two subnets would be organized on the basis of two distinct classes of sounder operations: common user and special operations. Spadework for the DoD policy was performed jointly by designated government representatives serving on the *United States Military Communications Electronics Board (USMCEB) Working Group on Oblique Incidence Sounders*, chaired by representatives of the Naval Electromagnetic Spectrum Center (NAVEMSCEN). The U.S. Air Force retains responsibility for maintenance of ACP-191. As it stands, both the DoD policy on sounders and ACP-191 refer only to placement and operation of the Chirpsounder™ systems. This does not preclude the possibility of incorporation of other sounder systems in the future.

The special operations subnet, although constrained to follow the requisite frequency management rules stipulated by applicable frequency regulations, would be free to operate with directional antennas using the minimum transmission cycle time of 5 minutes. The common user subnet, on the other hand, would normally operate with a transmission cycle time of 15 minutes and, to provide the greatest amount of coverage, would make use of omnidirectional antennas. A Data Fusion Center (DFC) would be the focus for the assimilation, analysis and temporary storage of OIS data, and dissemination of tailored products to designated subscribers.

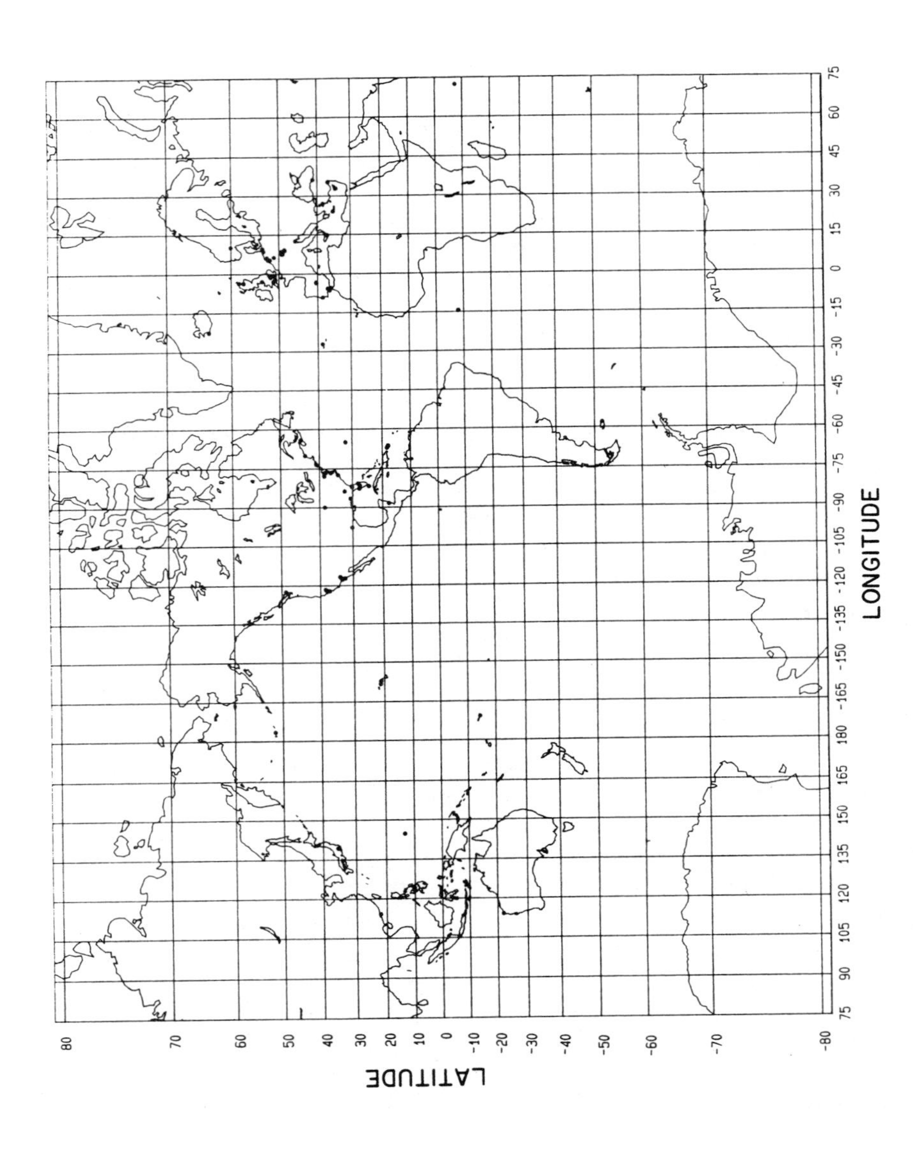

Fig.6-46. Global Distribution of OIS Chirpsounder transmitters (excluding mobiles). The distribution shown is current as of 1991. The numbers associated with the small circles correspond to the station number designations given in ACP-191. (From ACP-191 [1991]. Chirpsounder is a registered tradename of BR Communications.)

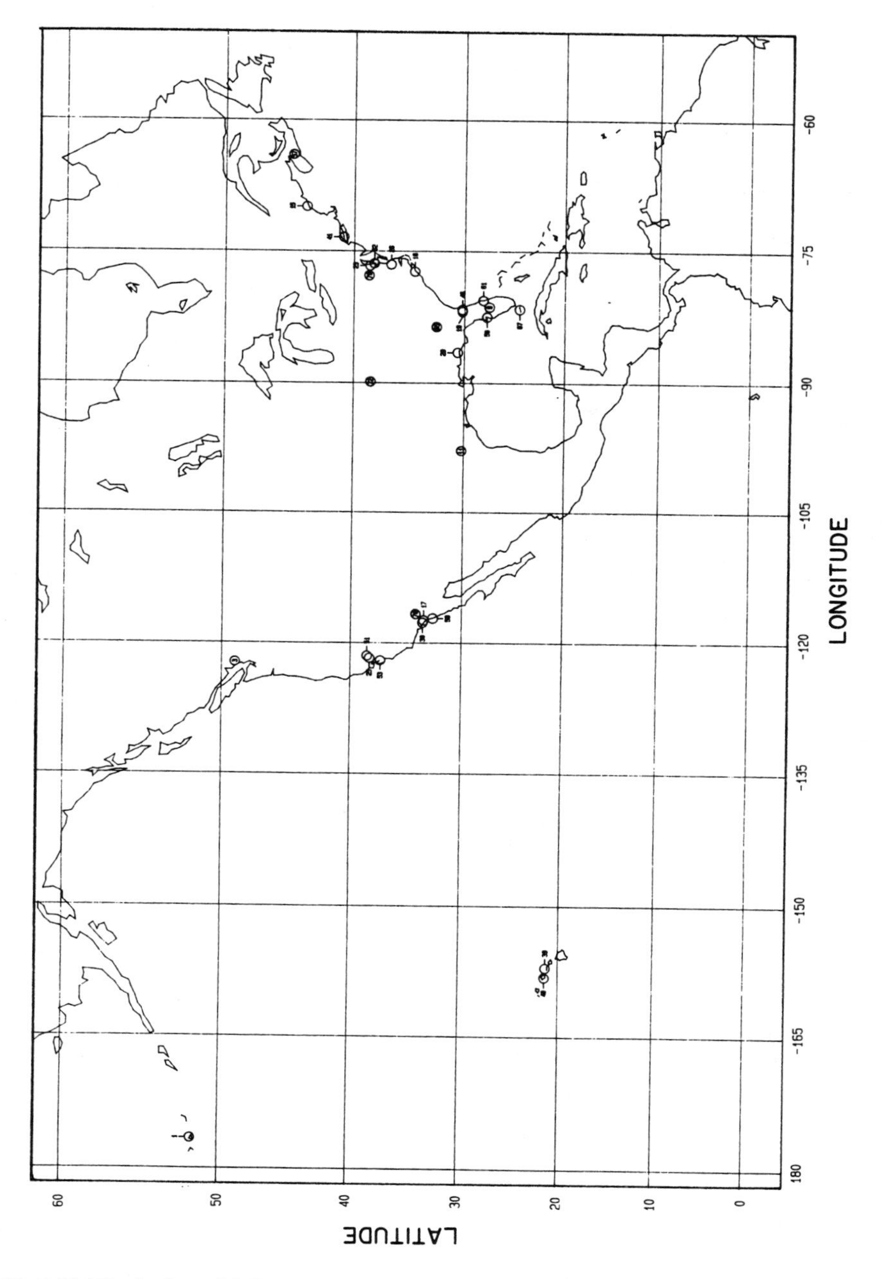

Fig.6-47. Distribution of OIS Chirpsounder transmitters in North America (excluding mobiles). The distribution shown is current as of 1991. The numbers associated with the small circles correspond to the station number designations given in ACP-191. (From ACP-191 [1991]. Chirpsounder is a registered tradename of BR Communications.)

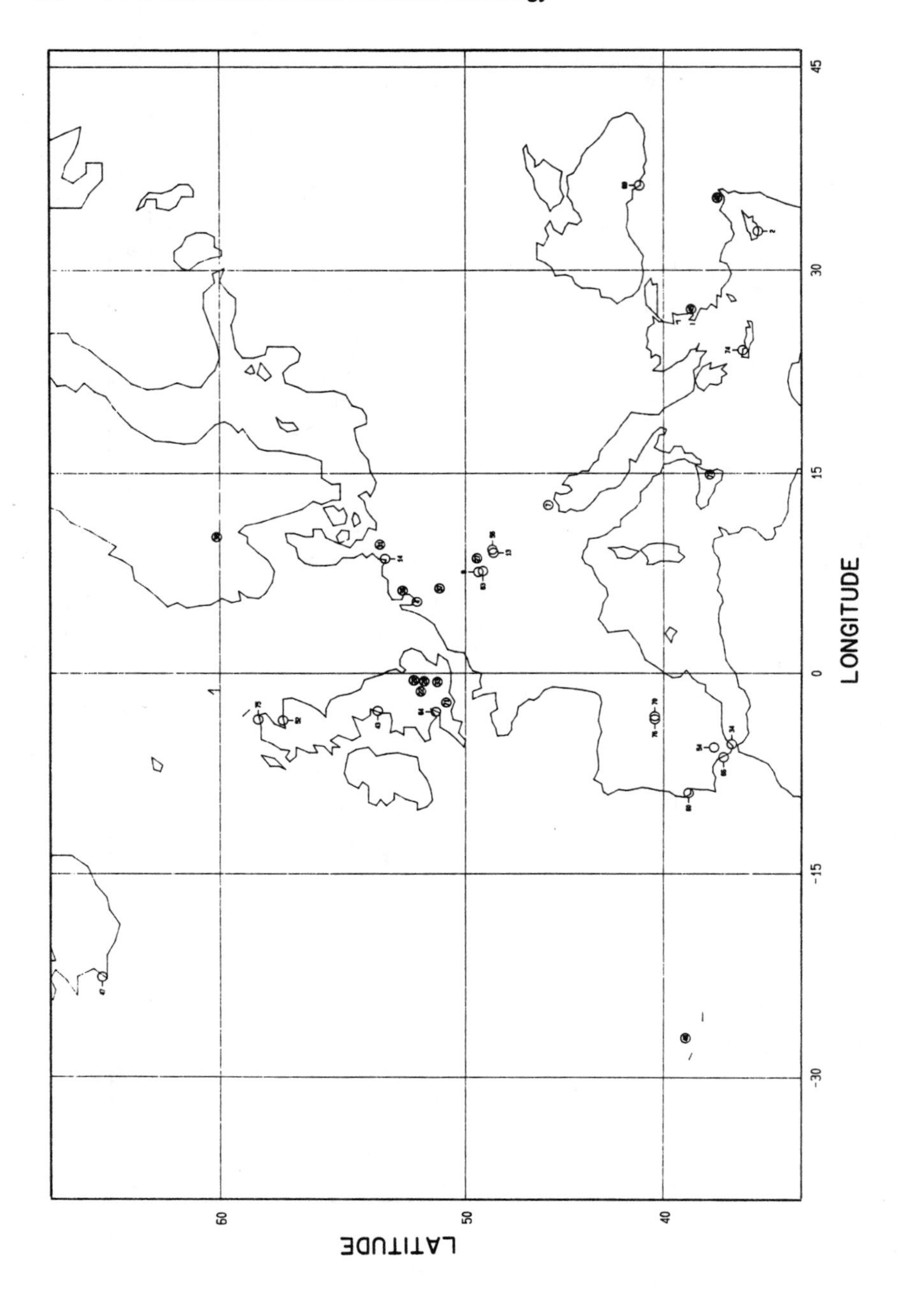

Fig.6-48. Distribution of OIS Chirpsounder transmitters in Europe (excluding mobiles). The distribution shown is current as of 1991. The numbers associated with the small circles correspond to the station number designations given in ACP-191. (From ACP-191 [1991]. Chirpsounder is a registered tradename of BR Communications.)

Van Troyen and Van de Capelle [1988] have used Chirpsounders™ for a study of a potential diplomatic network at HF; for example, connectivity between Bangkok and ten Far East embassies. The proposed star network would normally require one transmitter at each of the ten outstations and a receiver at the base station. Based upon sounder studies between Brussels and Athens, the authors suggest the use of only a single transmitter which would be shared among all of the outstations, housed for one month at each location to obtain monthly summaries. The single transmitter should be revisited every 10 months. Apart from a cost savings which is obvious, there are some disadvantages. Real time information is only available at one site at any time. Although some forecasting improvement does arise from spatial extrapolation of current data to the other (unsampled) sites, it is clear that frequency predictions at the nine (unsampled) outstations will not be much better than climatological estimates obtained from standard long-term programs such as IONCAP. Granted, over time, summaries of the actual paths involved should produce a median regional representation which will be superior to a global model. Indeed the proposed scenario will lead to a regional model in the neighborhood of Bangkok. It is fairly well established that special-purpose regional representations perform better than global versions in the business of developing accurate predictions. Even so, there an inadequacy in the suggested approach unless all seasons are uniformly sampled for each path under consideration. Since the revisit time is almost a year, there might be a tendency for specified paths to be sounded during a single season. This will limit the value of the regional model being developed. The use of a default model such as IONCAP, updated with a sounder-derived pseudoflux index, could be exploited for predictions until such time as the regional model based upon Chirpsounder™ data is mature and tested for reliability.

6.7.3.2 *The Digisonde™ Network*

The Digisonde™ was discussed in Section 6.3.1. We will limit the remarks here to networking issues. A proposed network concept to provide global frequency management was shown in Figure 6-45 [INAG Bulletin 48, 1986]. Figure 6-49 illustrates the current and planned network of Digisondes as of September 1989. The system will eventually have a dial-up modem capability to retrieve data from individual ionosonde stations. Scaled data sets are available through Air Force channels and/or SESC in Boulder, CO. As indicated in Chapters 2 and 3, solar-terrestrial and ionospheric data may be retrieved in a timely fashion from an SESC service called SELDADS. This service (now called SELVAX) has been upgraded to allow computer-to-computer transfer of data sets. As the VIS ionosonde network ages, individual units are being replaced by modern digital sounders such as the DGS 256.

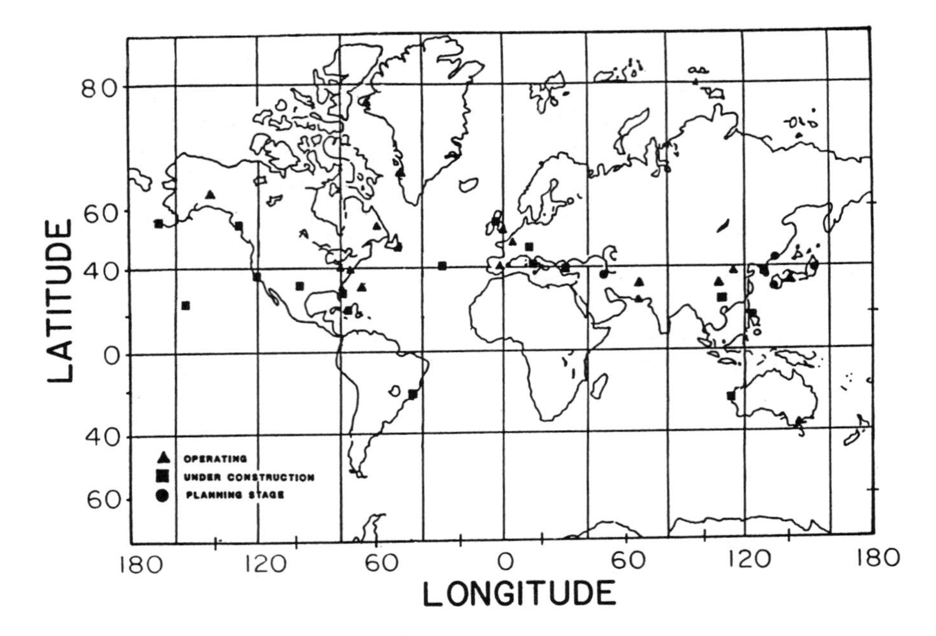

Fig.6-49. Global network of DGS-256 (Digisonde) ionospheric sounders. (From Reinisch et al.[1989]; in the *WITS Handbook*, Vol.2 [Liu, 1989]. Digisonde is a registered tradename of ULCAR, Lowell University, MA.)

6.7.3.3 The Japanese Network

The Japanese have deployed a network of sounders within Japan [Wakai, 1981]. Little information is available about this network, but it reportedly is composed of five transmitters/receivers. Figure 6-50 shows the network topology [INAG Bulletin 33, 1981]. The system can operate in the VIS and OIS modes. Any node (i.e., receiver) can produce five ionograms without waveform contention. According to available documentation [CCIR, 1986d], a small portable receiver unit is available for use by authorities who do not belong to the network.

6.7.4 Leverage Afforded by OIS Deployments

A distinct advantage provided by OIS strategies for environmental assessment is that each base station may sample more than one independent control point. In fact, assuming sufficient numbers of receivers are available at a fixed station, one may sample at least as many control points as there are sounder transmitters. In truth, one may obtain information at many more control points than this since multiple hop paths are associated with multiple control points. But, the control point information for multiple hop paths is generally ambiguous. We may observe the value of the composite MOF for a multiple hop path but, without some analysis, we do not know the subhop to which it should be assigned. This is important if channel data is to be extrapolated in space. In fact, spatial extrapolation of channel information extracted from multihop paths may be impossible. Figure 6-51 shows the paths (with lengths less than three hops) linking a Puerto Rican site with existing transmitters. Figures 6-52 and 6-53 are similar plots for receivers located at Washington, DC and Keflavik, Iceland respectively.

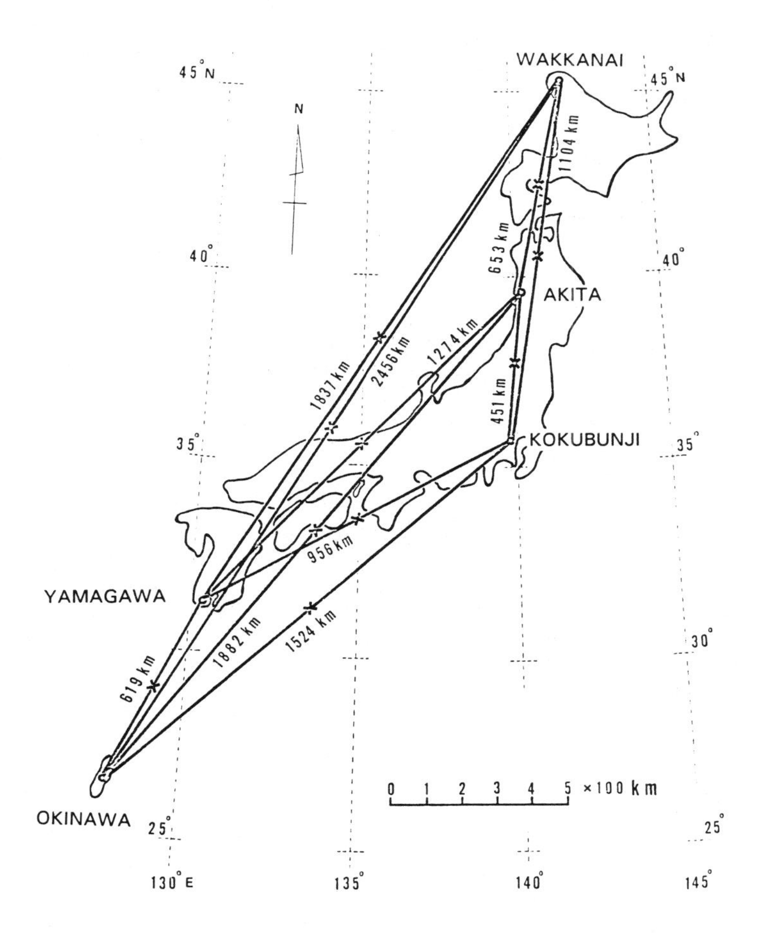

Fig. 6-50. Ionospheric sounding network in Japan. Both vertical and oblique sounders are involved. The cross marks on the paths specify path midpoints, or F2 layer control points. (From Wakai [1981]; INAG Bulletin 33.)

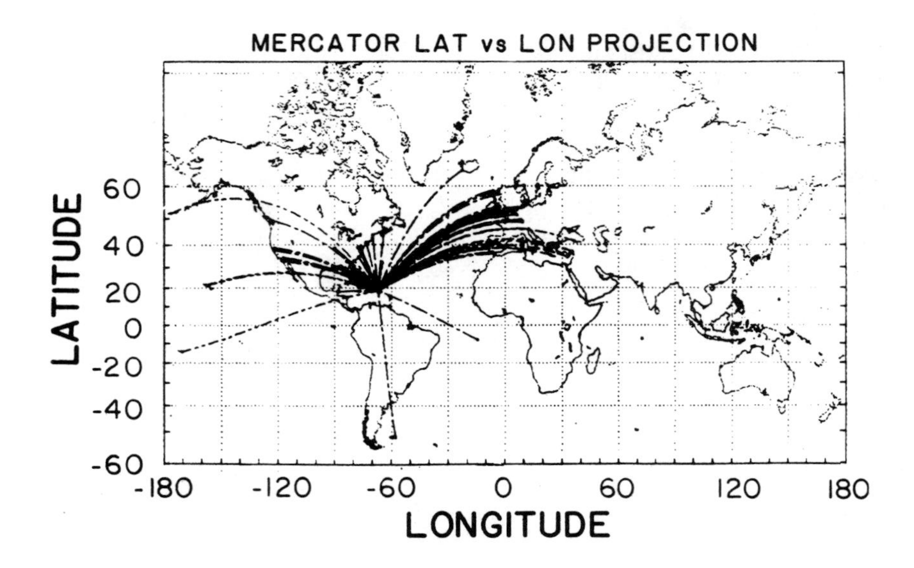

Fig.6-51. Paths linking a receiver located in Puerto Rico with Chirpsounder transmitters which are listed in ACP-191 [1991]. (Chirpsounder is a registered tradename of BR Communications.)

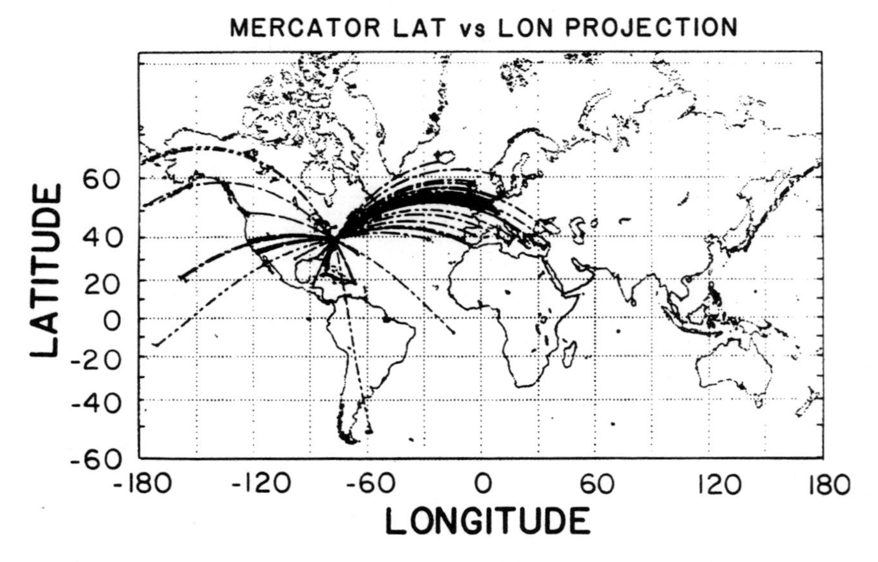

Fig.6-52. Paths linking a receiver located in Washington, DC to Chirpsounder transmitters listed in ACP-191 [1991]. (Chirpsounder is a registered tradename of BR Communications.)

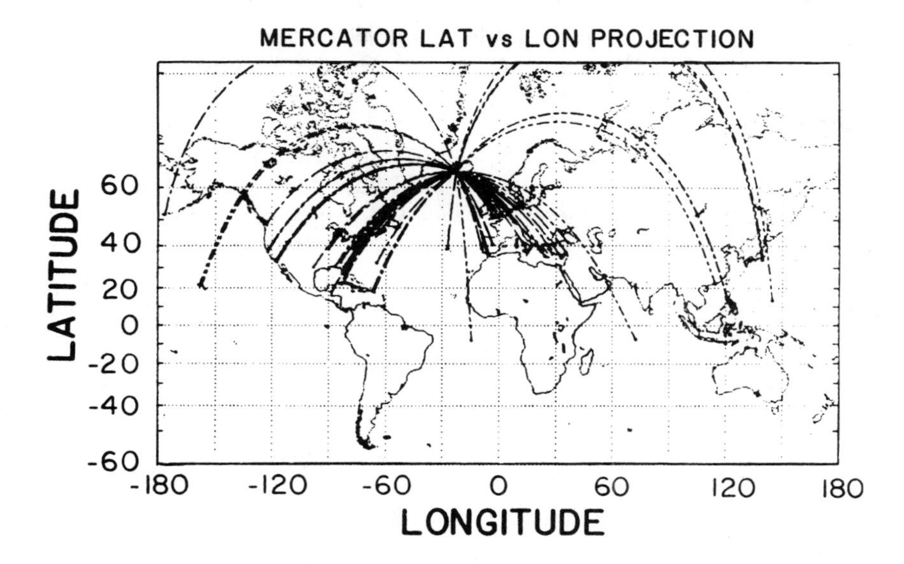

Fig.6-53. Paths linking a receiver located in Keflavik, Iceland to Chirpsounder transmitters listed in ACP-191 [1991]. (Chirpsounder is a registered tradename of BR Communications.)

A path is nominally a single-hop path if it allows presentation of at least the first hop on an oblique-incidence ionogram. We have seen that higher order hops may be observed, and may provide additional information. For an ionogram which displays four distinct hops (including the first), we will have unambiguous information at the midpath control point, with the possibility that redundant information may extracted from the middle leg of the three-hop path. If the path length is L, then the second hop will deliver a MOF which is assigned to either $L/4$ or $3L/4$. The third hop will deliver a MOF which is assigned to $L/6$, $L/2$, or $5L/6$. For the fourth hop, the assignment is $L/8$, $3L/8$, $5L/8$, or $7L/8$. By careful analysis, one may exploit the fact that the controlling MOF on a multihop path is always the smallest one. For an arbitrary long path, a single hop may not be possible. This occurs when the path length is much greater than about 4000 km. Ionospheric data derived from such paths cannot be associated with a single control point unambiguously (see Section 6.6.2.4).

If we can associate the observed MOF (or other ionospheric information) with a particular subhop along a path, and if receiver and transmitter systems are colocated, it is clear that the number of control points sampled will be $\frac{1}{2} N(N\text{-}1)$ where N represents the number of common receiver/transmitter sites. According to the rules specified in ACP-191, N is limited to 120, worldwide.

Nevertheless this suggests that 7140 paths may be evaluated. That's the end of the good news. The bad news is multifold: far too many receivers would be required to analyze all of the paths, many of the paths are multihop with limited hearability, the multi-hop paths introduce potentially ambiguous ionospheric data (as indicated above), and the existing distribution of transmitters is nonuniform. Given political reality and geographical constraints, there is little hope for making the overall distribution a uniform one. Moreover, we generally restrict consideration to those paths which allow propagation by a single-hop; and there are typically only a handful of sounder transmitters located within 4000 km of existing receivers (and vice versa).

It should be emphasized that the negative consequences associated with analysis of multihop oblique ionograms centers on the difficulty of an unequivocal assignment of ionospheric properties to a specific control point or set of control points. If it is required to assess the communication channel over a multihop operational path which is similar to a sounded path, there are definite advantages to be obtained by embarking on a thorough analysis of the sounder path. In this case, any ionospheric data might be useless for extrapolation purposes, but the path-specific channel information could have considerable value. In practice, however, it is not always possible to have coincidence of radio paths and sounder paths.

Figure 6-54 shows the one-hop path segments which could be sounded if receivers were to be located at each existing transmitter site. The control points for these paths are provided in Figure 6-55. The two major regions are near the eastern continental United States (CONUS) and Europe. An expanded version of the path segments for the United States is found in Figure 6-56, and the control points are shown in Figure 6-57.

It is sometimes important to know the azimuthal and range distribution of sounder systems for purposes of updating. There was never any intention by administrators of the sounder systems to do other than provide point-to-point sounding information between specified terminals. Naturally, this strategy is not consistent with the desire for uniform sounder coverage, as stipulated in the present context. Unplanned quasiuniform coverage may occur in certain instances; e.g., at sites located in southeastern CONUS, or in central Europe. In most instances, however, the coverage is quite nonuniform. Figure 6-58 shows the distribution of all Chirpsounders™ reckoned from Washington DC. We note that the distribution of paths with lengths less than 4000 km are more evenly distributed in bearing than are the paths having lengths greater than 4000 km.

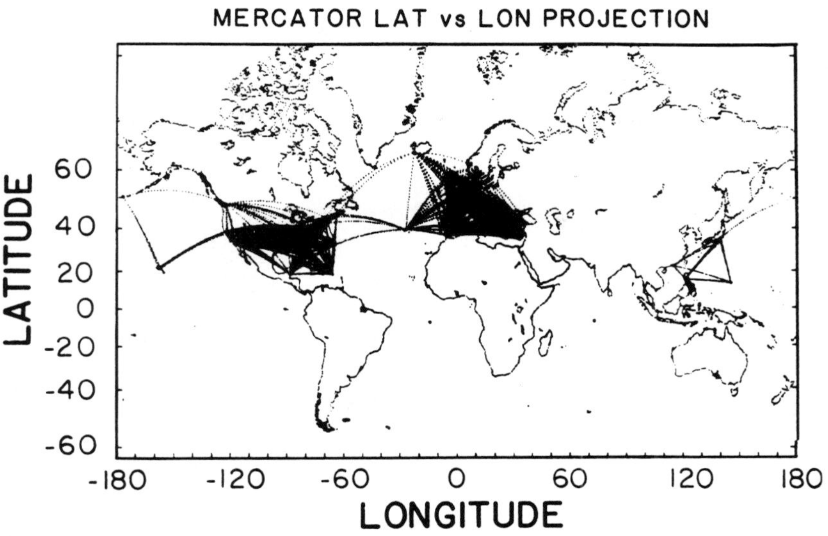

MERCATOR LAT vs LON PROJECTION
LATITUDE
60
40
20
0
-20
-40
-60
-180
-120
-60
0
60
120
180
LONGITUDE

Fig.6-54. Single-hop paths linking Chirpsounder transmitters and receivers. It is assumed that compatible receivers are deployed at all stations which house the transmitters. The control points for these paths are given in Figure 6-55. (Source for coordinate data: ACP-191 [1991]. Chirpsounder is a registered tradename of BR Communications.)

MERCATOR LAT vs LON PROJECTION
LATITUDE
60
40
20
0
-20
-40
-60
-180
-120
-60
0
60
120
180
LONGITUDE

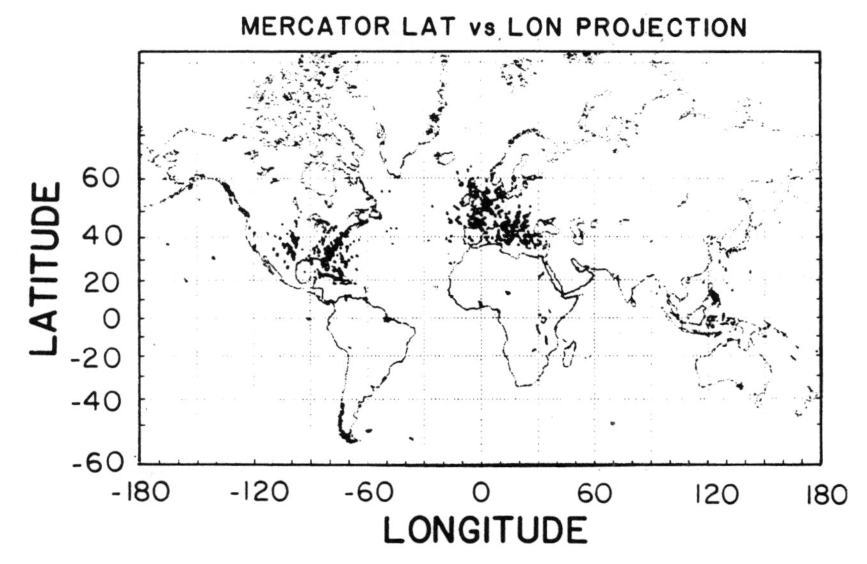

Fig.6-55. Display of the control points associated with the set of one-hop sounder paths shown in Figure 6-54. (Source for coordinate data: ACP-191 [1991].)

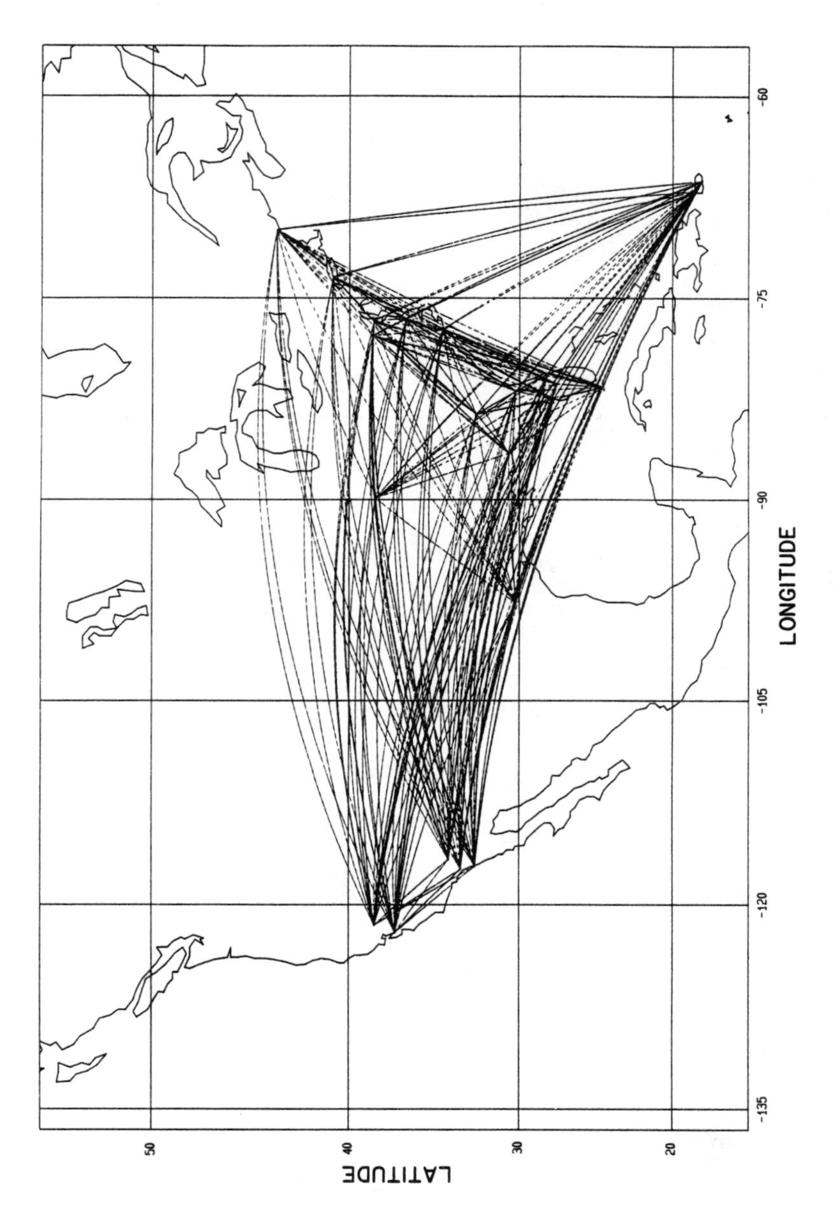

Fig.6-56. One-hop paths linking Chirpsounders in the forty eight contiguous states. Control points are shown in Figure 6-57. (Source for coordinate data: ACP-191 [1991]. Chirpsounder is a registered tradename of BR Communications.)

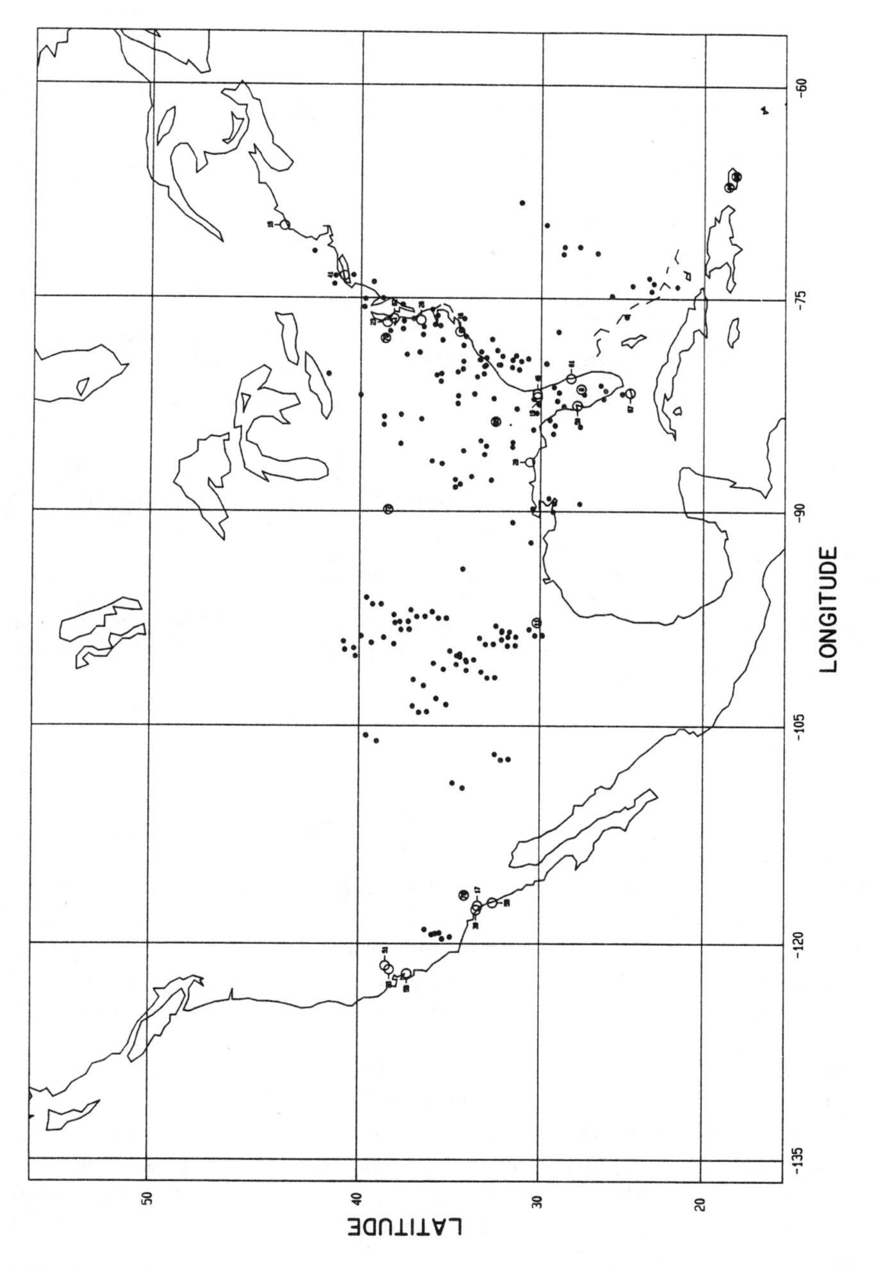

Fig.6-57. Display of the control points associated with the single-hop paths in Figure 6-56. (Source for coordinate data: ACP-191 [1991].)

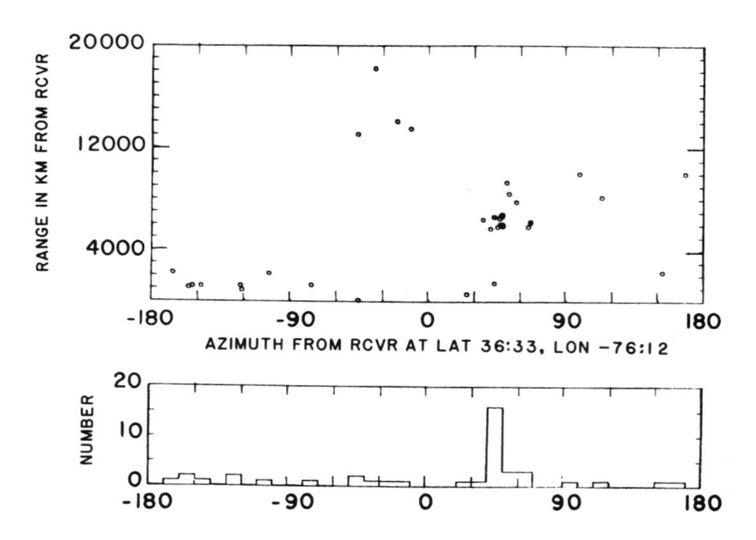

Fig.6-58. Single-hop paths terminating in the Washington, DC area. The upper plot shows the sounder range (km) from the receiver, while the lower plot gives the number of sounders in ten degree azimuth bins. There is clearly a clustering of sounders in the 30 - 60 degree sector. These constitute Chirpsounders deployed in the European region. (Chirpsounder is a registered trade-name of BR Communications.)

6.8 CONCLUDING REMARKS

In this chapter we have examined RTCE schemes emphasizing sounding techniques. Other forms of RTCE which are components of adaptive HF systems will be considered in the next chapter. It is hoped that the reader will now have a better appreciation for the range of RTCE schemes which are available and the types of channel variability which these schemes will expose.

With the emergence of new technologies and the increased availability of microprocessors with advanced DSP capabilities, real-time assessment and short-term forecasting schemes appear feasible for operational users. Predictions will always be important for system design and planning at HF, and a biproduct of RTCE technology growth will be an expanded data base for the development of more realistic models. Still the future of HF lies in the development and proper application of RTCE in modern HF radios and other HF systems. We have addressed this promise in the current chapter. In the following chapter we will look at certain implementations of RTCE and advanced features.

6.9 REFERENCES

Aarons, J., 1990, "Forecasting Morphology and Dynamics of F-Layer Irregularities", in *Effect of the Ionosphere on Radiowave Signals and System Performance*, IES'90, J.M. Goodman (editor-in-chief), USGPO, available through NTIS, Springfield, VA.

ACP-191, 1987, "Ionospheric Sounder Operations", published for the Chairman of the USMCEB under direction of the Joint Chiefs of Staff for use by Armed Forces of the United States and other users of U.S. communications facilities, Pentagon, Washington, DC 20301-5000. (Revisions have been published in 1989 and 1991.)

AGS, 1990a, Andrew Government Systems specification sheet on the Skysonde™ Vertical Incidence Ionosonde, Andrew Corporation, Garland, TX.

AGS, 1990b, Andrew Government Systems specification sheet on the Tiltsonde™ Three Channel Vertical Incidence Ionosonde, Andrew Corporation, Garland, TX.

Ames, J. W., and R.D. Egan, 1965, "Short Term Prediction of the Optimum HF Communication Frequency", *1st Ann. IEEE Commun. Conf.*, conference record, pp. 657-659.

Ames, J. W., and R.D. Egan, 1967, "Digital Recording and Short-term Prediction of Oblique Ionospheric Propagation", *IEEE Trans. Ant. and Prop.*, AP-15(3):382-389.

Anderson, S.J. and M.L. Lees, 1988, "High Resolution Synoptic Scale Measurements of Ionospheric Motions with the Jindalee Radar", in *The Effect of the Ionosphere on Communication, Navigation, and Surveillance Systems*, J.M. Goodman (editor-in-chief), USGPO, available through NTIS, Springfield VA.

Appleton, E.V., and W.J.G. Beynon, 1940, "Application of Ionospheric Data to Radio Communication Problems", *Proc. Phys. Soc.*, 52(1):518.

Baker, D.N., and S. Lambert, "Multiparabolic Ionospheric Model for SSL Application", *Electronics Letters*, 24:425-426.

Baker, D.N., and S. Lambert, 1989, "Range Estimation for SSL HFDF Systems by Means of a Multi-Quasiparabolic Ionosphere Model", *Proc. IEE*, 136(H):120-125.

Barry, G.H., 1971, "A Low Power Vertical Incidence Ionosonde", *IEEE Trans.Geosci.Electron.*, GE-9(2):86-89.

Basler, R.P., and T.D. Scott, 1973, "Ionospheric Structure from Oblique-Backscatter Soundings", *Radio Sci.*, 8(5):425-429.

Beckwith, R.I., and N.N. Rao, 1975, "Real-Time Updating of Maximum Usable Frequency Predictions for HF Radio Communication", *IEEE Trans. on Comms.*, Feb. issue, pp. 286-288.

Bello, P.A., 1965, "Some Techniques for the Instantaneous Real-time Measurement of Multipath and Doppler Spread", *IEEE Trans. Comm. Tech.*, 13(3):285-292.

Betts, J.A., and M. Darnell, 1975, "Real-time HF Channel Estimation by Phase Measurements on Low-level Pilot Tones", in *Radio Systems and the Ionosphere*, NATO-AGARD-CP-173, W.T. Blackband (editor), Tech. Edit. and Reprod. Ltd., London, UK.

Bibl, K., and B.W. Reinisch, 1978, "The Universal Digital Ionosonde", *Radio Sci.*, 13(3):519-530.

Bibl, K., B.W. Reinisch, and D.F. Kitrosser, 1981, "Digisonde 256, General Description of the Compact Digital Ionospheric Sounder", University of Lowell, Center for Atmospheric Research, Lowell, MA.

Blackband, W.T. (editor), 1975, *Radio Systems and the Ionosphere*, NATO-AGARD-CP-173, Tech. Edit. and Reprod. Ltd., London, UK. (Also available through NTIS, Springfield, VA.)

Bradley, P.A., and J.R. Dudeney, 1973, "A Simple Model of the Vertical Distribution of Electron Concentration in the Ionosphere", *J. Atmos. Terrest. Phys.*, 35:2131.

BRC, 1980, BR Communications Specification sheet on the VIS series Vertical Incidence Ionospheric Chirpsounder™ Systems, Barry Research Corporation, Sunnyvale, CA.

Budden, K.G., 1955, "A Method for Determining the Variations of Electron Density with Height from Curves of Equivalent Height against Frequency", in *The Physics of the Ionosphere*, The Physical Society, London, UK.

Caratori, J., R. Schwab, and C. Goutelard, 1988, "Real Time Large Scale Modelesation of the Ionosphere with Backscatter Sounding - New Results", in *Fourth International Conference on HF Radio Systems and Techniques*, IEE, Savoy Place, London, UK.

CCIR, 1986a, "Real-Time Channel Evaluation of Ionospheric Radio Circuits", Report 889-1, in *Recommendations and Reports of the CCIR, 1986: Propagation in Ionized Media*, Vol.VI, (XVIth Plenary Assembly in Dubrovnik), ITU, Geneva.

CCIR, 1986b, "Atlas of Ionospheric Coefficients", Report 340-6, in *Recommendations and Reports of the CCIR, 1986: Propagation in Ionized Media*, Vol.VI, (XVIth Plenary Assembly in Dubrovnik), ITU, Geneva.

CCIR, 1986c, "Ionospheric Sounding at Oblique Incidence", Report 249-6, in *Recommendations and Reports of the CCIR, 1986: Propagation in Ionized Media*, Vol.VI, (XVIth Plenary Assembly in Dubrovnik), ITU, Geneva.

CCIR, 1986d, "Operational Sounding at Oblique Incidence", Report 357-2, in *Recommendations and Reports of the CCIR, 1986: Fixed Service at Frequencies Below About 30 MHz*, Vol.III, (XVIth Plenary Assembly in Dubrovnik), ITU, Geneva.

CCIR, 1986e, "Short-term Forecasting of Critical Frequencies, Operational Maximum Usable Frequencies and Total Electron Content", Report 888-1, *Recommendations and Reports of the CCIR, 1986: Propagation in Ionized Media*, Vol.VI, (XVIth Plenary Assembly in Dubrovnik), ITU, Geneva.

Chow, S.M., G.W. Irvine, B.D. McLarnon, and A.R. Kaye, 1981, "Communications for Small Communities in Developing Countries", *Proc. Pacific Telecommunications Conference*, Pacific Telecommunications Council, Honolulu, HI.

Clarke, C., and A. M. Peterson, 1956, "Motion of Sporadic E Patches Determined from High-Frequency Backscatter Patches", *Nature*, 178:486-487.

Conkright, R.O., and H.I. Brophy, 1982, "Catalog of Ionosphere Vertical Soundings Data", UAG-85, National Geophysical Data Center, NOAA, Boulder, CO.

Conkright, R., and M. O. Ertle, 1984, "Combined Catalog of Ionosphere Vertical Soundings Data", UAG-91, National Geophysical Data Center, NOAA, Boulder, CO.

Conkright, R., 1990, (private communication) INAG bulletins are distributed to a mailing list organized by the INAG Chairman through the WDC-A for Solar-Terrestrial Physics, Boulder, CO. The Ionosonde Network Advisory Group (INAG) is under the auspices of URSI Working Group G1 of Commission G.

Croft, T., 1967, "The Interpretation of HF Sweep-frequency Backscatter Soundings to Deduce the Structure of Localized Ionospheric Anomalies", TR 116, SU-SEL-08-029, Stanford Electronic Lab, Stanford University, Stanford, CA.

Croft, T., 1972, "Skywave Backscatter, A Means for Observing our Environment at Great Distances", *Rev. Geophys. Space Phys.*, 10(1):73-155.

C^3 I Handbook, 1988, prepared by editors of *Defense Electronics*, EW Communications, Inc., Palo Alto, CA.

Croft, T., and H. Hoogasian, 1968, "Exact Ray Calculations in a Quasi-Parabolic Ionosphere with No Magnetic Field", *Radio Sci.*, 3(1):69-74.

Daehler, M., 1984, "An HF Communications Frequency Management Procedure for Forecasting the Frequency of Optimum Transmission", Memorandum Report 5505, Naval Research Laboratory, Washington, DC.

Daehler, M., 1990, "EINMUF: An HF MUF, FOT, LUF Prediction Program", NRL Memorandum Report 6645, (May 18, 1990), Naval Research Laboratory, Washington, DC 20375-5000.

Daehler, M., M.H. Reilly, F.J. Rhoads, and J.M. Goodman, 1988, "Comparison of Measured MOFs with Propagation Model Forecasts", in *Effect of the Ionosphere on Communication, Navigation, and Surveillance Systems*, IES'87, J. M. Goodman (editor-in-chief), available through NTIS, Springfield, VA.

Daly, R.F., K.D. Felperin, T.I.Dayharsh, and B.C. Tupper, 1967, "CURTS Phase 2. Automatic Frequency Selection System - Signal Processing, Application, and Evaluation", AD815884, Final Report on Project 5757, DCA contract 100-66-C-0075, SRI, Menlo Park, CA.

Darnell, M, 1975a, "Channel Estimation Techniques for HF Communications", in *Radio Systems and the Ionosphere*, NATO-AGARD-CP-173, Tech. Edit. and Reprod. Ltd., London, UK.

Darnell, M., 1975b, "Adaptive Signal Selection for Dispersive Channels and its Practical Implications in Communications System Design", in *Radio Systems and the Ionosphere*, NATO-AGARD-CP-173, W.T. Blackband (editor), Tech.Edit.and Reprod.Ltd., London, UK.

Darnell, M., 1978, "Channel Evaluation Techniques for Dispersive Communications Paths", in *Communications Systems and Random Process Theory*, J.K. Skwirzzynski (editor), Sijthoff and Noordhoff, The Netherlands, pp.425-460.

Darnell, M., 1979, "An HF Data Modem with In-Band Frequency Agility", in *Recent Advances in HF Communication Systems and Techniques*, IEE Conference Proceedings, Savoy Place, London.

Darnell, M., 1983, "Real-Time Channel Evaluation", in *Modern HF Communications*, NATO-AGARD-LS-127, available through NTIS, Springfield VA 22161.

Davé, N., 1988, "Comparison with Oblique Sounder Data of High Latitude HF Propagation Predictions from RADARC and AMBCOM Computer Programs", in *Effect of the Ionosphere on Communication, Navigation, and Surveillance Systems*, IES'87, John M. Goodman (editor-in-chief), available through NTIS, Springfield, VA.

David, P., C. Goutelard, and J.P. Van Uffelen, 1976, "Procédé de Sélection de la Fréquence Optimale pour une Transmission de données sur canal Ionosphérique", in *Radio Systems and the Ionosphere*, NATO-AGARD-CP-173, W.T. Blackband (Ed.), paper No. 17, NASA Accession No. N77-20303, NTIS, Springfield VA 22161.

Davies, K., 1965, *Ionospheric Radio Propagation*, NBS Monograph 80, USGPO, Washington, DC.

Davies, K. 1990, *Ionospheric Radio*, IEE Electromagnetic Wave Series 31, Peter Peregrinus Ltd., in behalf of the IEE, London, UK. (Also available through IEEE, Piscataway, NJ.)

Davies, K. and A.F. Barghausen, 1966, "The Effect of Spread F on the Propagation of Radiowaves Near the Magnetic Equator", in *Spread F and its Effects upon Radiowave Propagation and Communication*, P. Newman (editor), NATO-AGARDograph 95, Technivision, Maidenhead, England.

Dieminger, W., 1960, "Ground Scatter by Ionospheric Radar", in *Proc. Avionics Research: Satellites and Problems of Long Range Detection and Tracking*, edited by E. Glazier, E. Rechtin, and J. Voge, AGARD Avionics Panel, Copenhagen (1958), Pergamon Press, New York.

Dudeney, J.R., 1978, "An improved Model of the Electron Concentration with Height in the Ionosphere", *J. Atmos. Terrest. Phys.*, 40:195-203.

Elliott, W., Q. R. Black, and Wm. M. Sherrill, 1990, "Parish Task 1: Check Bearing Detection of DF System Faults" (draft), Southwest Research Institute, San Antonio, TX.

Felpenin, K.D., T.I. Dayharsh, J.L. Ramsay, and B.C. Tupper, 1968, "The Operational Validation of the CURTS using 2400 Baud Autodin", Final Report, Part 2, SRI SWM 16353U, Stanford CA.

Fox, M.W., and C. Blundell, 1989, "Automatic Scaling of Digital Ionograms", *Radio Sci.*, 24(6):747-761.

Gautier, T.N., and D.H. Zacharisen, 1965, "Use of Space and Time Correlation in Short Term Ionospheric Predictions", *1st IEEE Ann. Commun. Conv.*, conference record, pp.671-676.

Gebhard, L.A., 1979, *Evolution of Naval Radio-Electronics and Contributions of the Naval Research Laboratory*, Naval Research Laboratory Report 8300, USGPO, Washington DC.

George, P.L., 1970, *J. Atmos. Terrest. Phys.*, 32:905.

George, P.L., 1972, "HF Ray Tracing of Gravity Wave Perturbed Ionospheric Profiles", NATO-AGARD conference proceedings, Wiesbaden, Germany, AGARD-CP-115, paper No. 26.

George, P.L., 1976, "Dynamic Ionospheric Tilt Correction and its Application to Short Range Single Station Location Systems, Weapons Research Establishment Technical Note, WRE-TN-1571(AP), Australia.

George, P.L., and R.J. Halligan, 1985, "Techniques for Real-time HF Channel Evaluation in Aid of Optimized Data Transmission", *Proc. URSI/IPS Conf. on the Ionosphere and Radio Wave Propagation*, D.G. Cole and L.F. McNamara (editors), TR-85-04, Ionospheric Prediction Service, Sidney, Australia.

Gething. P.J.D., 1969, *J. Atmos. Terrest. Phys.*, 31:347.

Gething, P.J.D., 1978, *Radio Direction Finding*, published in behalf of the IEE by Peter Peregrinus Ltd., Southgate House, Stevenage, Herts, England.

Gething, P.J.D., 1991, *Radio Direction Finding and Superresolution*, published in behalf of the IEE by Peter Peregrinus Ltd., London, UK. (Also available through the IEEE, Piscataway, NJ.)

Giraud, A., and M. Petit, 1978, *Ionospheric Techniques and Phenomena*, D. Reidel Publishing Co., Boston.

Goodman, J.M., 1984, "A Description of the NRL Oblique-Incidence-Ionogram Data Base", Memorandum Report 5452, Naval Research Laboratory, Washington, DC.

Goodman, J.M., 1990, unpublished data from an HFDF Experiment conducted by NRL in 1989.

Goodman, J.M., and D.R. Uffelman, 1982, "The Role of the Propagation Environment in HF Electronic Warfare", Report 4953, Naval Research Laboratory, Washington, DC.

Goodman, J. M., D. R. Uffelman, and F. O. Fahlsing, 1982, "The Role of the Propagation Environment in HF Electronic Warfare", in *Propagation Effects on ECM-Resistant Systems in Communication and Navigation*, H. Albrecht (editor), NATO-AGARD conference proceedings, Tech. Edit. and Reprod. Ltd., London.

Goodman, J.M., and D.R. Uffelman, 1983, "On the Utilization of Ionospheric Diagnostics in the Single-Site-Location of HF Emitters", in *Propagation Factors Affecting Remote Sensing by Radio Waves*, NATO-AGARD-CP-345, Tech. Edit. and Reprod. Ltd., UK.

Goodman, J.M., M.H. Reilly, M. Daehler, and A.J. Martin, 1983, "Global Considerations for Utilization of Real-Time Channel Evaluation Systems in Spectrum Management", *Proc. MILCOM '83*, Arlington, VA.

Goodman, J.M., M. Daehler, M.H.Reilly and A.J. Martin, 1984, "A Commentary on the Utilization of Real-Time Channel Evaluation Systems in HF Spectrum Management", Memorandum Report 5454, Naval Research Laboratory, Washington, DC.

Goodman, J.M. and M. Daehler, 1988a, "The NRL Data Base of Oblique-Incidence Soundings of the Ionosphere", Memorandum Report 6337, Naval Research Laboratory, Washington, DC.

Goodman, J.M., and M. Daehler, 1988b, "Use of Oblique Incidence Sounders in HF Frequency Management", in *Fourth International Conference on HF Radio Systems and Techniques*, Conference Publication No. 284, IEE, Savoy Place, London UK.

Goodman, J.M., and M.H. Reilly, 1988, "Shortwave Propagation Prediction Methodologies", *IEEE Transactions on Broadcasting*, June issue.

Gould, R.G., and W.R. Vincent, 1962, "System Concepts for a Common-user Radio Transmission Sounding System (CURTS)", Stanford Research Institute Report, Menlo Park, CA.

Greenwald, R.A., 1985, "High Frequency Radio Probing of the High Latitude Ionosphere", Johns Hopkins, *APL Technical Digest*, 6(1):38-50, available through DTIC, AD-A187 055.

Grubb, R.N., 1979, The NOAA/SEL HF Radar System (Ionospheric Sounder), NOAA Tech. Memo. ERL-SEL-55, NOAA/SEL, Boulder, CO.

Hagn, G., 1962, "An Investigation of Direct Backscatter of High Frequency Radio Waves from Land, Sea Water, and Ice Surfaces", SRI Project 2909, Final Report, Vol.2, Stanford Research Institute, Stanford University, Stanford, CA.

Haines, M., D.F. Kitrosser, B.W. Reinisch, and F.J. Gorman, 1989, "A Portable Ionosonde in Support of Reliable Communications", in *Operational Decision Aids for Exploiting or Mitigating Electromagnetic Propagation Effects*, NATO-AGARD-CP-453, Spec. Print. Serv. Ltd., Loughton, Essex, UK.

Hatton, W.L., 1961, "Oblique-Sounding and HF Radio Communication", *IRE Trans. on Commun. Syst.*, September issue, pp.275-279.

Headrick, J.M. and M.I. Skolnik, 1974, "Over the Horizon Radar in the HF Band", *Proc.IEEE*, Vol. 62, No.6, pp.664-673.

Headrick, 1987, unpublished report on the use of backscatter sounding as an aid to assessing the performance of HF broadcasts, private communication.

Headrick, J.M., 1990a, "Looking Over the Horizon", *IEEE Spectrum*, July issue, pp. 36-39.

Headrick, J.M., 1990b, "HF Over-the-Horizon Radar", in *Radar Handbook*, M. Skolnik (editor-in-chief), McGraw-Hill Publishing Company, New York.

Hunsucker, R.D., 1991, *Radio Techniques for Probing the Terrestrial Ionosphere*, Springer-Verlag, New York.

Hunsucker, R.D., and B.S. Delana, 1989, "First Results from HF Oblique Backscatter Soundings to the Northwest of College, Alaska Using a Modified ULCAR Digisonde D-256", Geophysics Laboratory, GL-TR-89-0135, Hanscom AFB, MA 01731-5000.

IDIG, 1982, Bulletin No. 3, International Digital Ionosonde Group, URSI Working Group G10.

INAG, 1981, Ionospheric Station Bulletin No. 33, Ionosonde Network Advisory Group, Commission G, URSI, distributed by WDC-A for Solar-Terrestrial Physics, Boulder, CO.

INAG, 1982, Ionospheric Station Bulletin No.37, Ionosonde Network Advisory Group, Commission G, URSI, distributed by WDC-A for Solar-Terrestrial Physics, Boulder, CO.

INAG, 1983, Ionospheric Station Bulletins No. 40 & 41 (combined), Ionosonde Network Advisory Group, Commission G, URSI, distributed by WDC-A for Solar-Terrestrial Physics, Boulder, CO.

INAG, 1986, Ionospheric Station Bulletin No.48, Ionosonde Network Advisory Group, Commission G, URSI, distributed by WDC-A for Solar-Terrestrial Physics, Boulder, CO.

INAG, 1991, Ionospheric Station Bulletin No. 56, Ionosonde Network Advisory Group, Commission G, URSI, distributed by WDC-A for Solar-Terrestrial Physics

Jackson, J.E., 1956, "A New Method for Obtaining Electron Density Profiles from *P'-f* Records", *J. Geophys. Res.*, 61(1):107-127.

Jelly, D.H., 1971, "Ionospheric Sounding as an Aid to HF Communications", AD902437, CRC Report 11225, Communications Research Centre, Ottowa, Ontario, Canada.

Jones, T.B., 1989, "The HF Doppler Technique for Monitoring Transient Ionospheric Disturbances", in *World Ionosphere/Thermosphere Study: WITS Handbook*, Vol. 2, C.H. Liu (editor), SCOSTEP, University of Illinois, Urbana. IL.

Jones, T.B., C.T. Spracklen, and C.P. Stewar, 1978, "Real Time Updating of MUF Predictions", in Conference Proceedings, NATO-AGARD-CP-238, Vol. 1, Tech. Edit and Reprod. Ltd., London, UK.

Josephson, J., 1990, private communication of: (a) draft US-DoD Sounder Policy, and (b) draft planning document for sounder utilization in the NATO theater (ARFA).

Jull, G. W., G.E. Poaps, and J. P. Murray, 1965, "Prediction of HF Channel Characteristics using Frequency Sounding Equipment", *1st IEEE Ann. Commun. Conference*, proceedings, pp. 653-656.

Jursa, A.S. (scientific editor), 1985, *Handbook of Geophysics and the Space Environment*, Air Force Geophysics Laboratory, Air Force Systems Command, U.S. Air Force, available through NTIS, Springfield, VA.

KEL, 1982, data on the DBD-43 (digital add-on), DBD-43 (digital remote terminal), the KEL-46 (data analyzer), and the KEL-47 (central processor), KEL Aerospace, Australia.

Kelso, J.M., 1953, "The Determination of the Electron Density Distribution of an Ionosphere Layer in the Presence of an External Magnetic Field", Scientific Report 55, Ionosphere Research Laboratory, Penn State University, PA.

Kelso, J.M., 1964, *Radio Ray Propagation in the Ionosphere*, McGraw-Hill Book Company, New York.

Klobuchar, J.A., 1989, "Modern Total Electron Content Measurement Techniques", in *World Ionosphere/Thermosphere Study: WITS Handbook*, Vol. 2, C.H. Liu (editor), SCOSTEP, University of Illinois, Urbana. IL.

Klobuchar, J.A., and J. M. Johanson, 1977, "Correlation Distance for Mean Daytime Electron Content", AFGL-TR-77-0185, Air Force Geophysics Lab, Hanscom AFB, MA, (ADA048117).

Kuklinski, W.S., K. Chandra, and B.W. Reinische, 1988, "Preliminary Development of Oblique Ionogram Automatic Scaling Algorithm", Univ. of Lowell, ULCAR Scientific Report No. 13, GL-TR-89-0184, Geophysics Laboratory, Hanscom AFB, MA.

Kuriki, I., and T. Takeuchi, 1986, "Vertical/Oblique Incidence Ionospheric Sounding", *J. Radio Res. Labs. (Japan)*, Vol.33.

Lees, M.L., and R. M. Thomas, 1988, "Ionospheric Probing with an HF Radar", in *Fourth International Conference on HF Radio Systems and Techniques*, IEE, Savoy Place, London, UK.

Leid, F. (editor), 1967, *H.F. Radio Communications with Emphasis on Polar Problems*, NATO-AGARD, AGARDograph 104, Technivision, Maidenhead, England.

Le Saout, J.Y. and L. Bertel, 1988, "Antenna Arrays of the CNET Backscatter HF Radar", in *Fourth International Conference on HF Radio Systems and Techniques*, IEE, Savoy Place, London, UK.

Liu, C.H. (editor), 1989, *World Ionosphere/Thermosphere Study: WITS Handbook*, Volume 2, published in behalf of SCOSTEP, International Council Scientific Unions, University of Illinois, Urbana, IL.

Lomax, J.B., 1966, "Spread-F in the Pacific", in *Spread F and its Effects upon Radiowave Propagation and Communication*, P. Newman (editor), NATO-AGARDograph 95, Technivision, Maidenhead, England.

Lomax, J.B., and D.L. Nielson, 1968, "Observation of Acoustic-gravity Wave Effects Showing Geomagnetic Field Dependence", *J. Atmos. Terrest. Phys.*, 330:1033-1050.

Martin, A.J., 1990, Naval Research Laboratory, private communication.

McCoy, R.P., L.J. Paxton, R.R. Meier, D.D. Cleary, D.K. Prinz, K. D. Wolfram, A.B. Christensen, J.B. Pranke, and D.C. Kayser, 1988, "RAIDS: An Orbiting Observatory for Ionospheric Remote Sensing From Space", in *Effect of the Ionosphere on Communication, Navigation, and Surveillance Systems*, IES'87, J.M. Goodman (editor-in-chief), USGPO, available through NTIS, Springfield, VA.

McCue, C.G., and J. D. Gilbert, 1988, "Automatic Ionogram Scaling", *INAG Bulletin 52*, distributed by WDC-A, Boulder CO.

McLarnon, B.D., 1982, "Real Time Channel Evaluation in an Automatic HF Radio Telephone System", *Proc. 2nd IEE Conference on HF Communications Systems and Techniques*, IEE, Savoy Place, London, UK.

McNamara, L.F., 1976, "Short-Term Forecasting of foF2," Tech. Rep. IPS-TR-33, IPS Radio and Space Services, Dept. Sci., Darlinghurst, N.S.W., Australia.

McNamara, L.F., 1979, "The Use of Ionospheric Indices to Make Real- and Near-Real-Time Predictions of foF2 Around Australia", *Solar-Terrestrial Proceedings: Symposium in Meudon, France*, Edited by P.A. Simon, G. Heckman, and M.A. Shea, published jointly by NOAA, Boulder, CO. and AFGL, Hanscom AFB, MA.

McNamara, L.F., 1985, "HF Radio Propagation and Communications - A Review," Tech. Rep. IPS-TR-85-11, IPS Radio and Space Services, Dept. Sci., Darlinghurst, N.S.W., Australia.

McNamara, L.F., 1988a, "Ionospheric Limitations to the Accuracy of SSL Estimates of HF Transmitter Locations", in *Ionospheric Structure and Variability on a Global Scale and Interaction with Atmosphere, Magnetosphere*, NATO-AGARD-CP-441, Munich, Germany.

McNamara, L.F., 1988b, "Ionospheric Modelling in Support of Single Station Location of Long Range Transmitters", *J. Atmos. Terrest. Phys.*, 50(9):781-795.

McNamara, L.F., 1991, *The Ionosphere: Communications, Surveillance, and Direction Finding*, Orbit Foundation Series, Krieger Publishing Company, Malibar, FL.

Millman, G., 1978, "Ionospheric Propagation Effects on HF Backscatter Radar Measurements", in *Effect of the Ionosphere on Space and Terrestrial Systems*, IES'78, J.M.Goodman (editor), USGPO, available through NTIS, Springfield, VA.

Milsom, J.D., 1983, "HF Skywave Links and Frequency Management in Europe", in *Towards Improved HF Communications in the European Envionment*, SHAPE Technical Centre, The Hague, NL (Conference in 1982).

Milsom, J.D., 1985, "Exact Ray Path Calculations in a Modified Bradley/ Dudeney Model Ionosphere", *IEE Proc. Microwaves, Antenna and Prop.*, 132:33-38.

Mitra, S.K., 1947, *The Upper Atmosphere*, The Royal Asiatic Society of Bengal, Calcutta, India.

Neilson, D., 1966, "Oblique Sounding of a Trans-equatorial Path", in *Spread F and its Effects upon Radiowave Propagation and Communication*, P. Newman (editor), NATO-AGARDograph 95, Technivision, Maidenhead, England.

Paul, A.K., 1977, Simplified Procedure for Calculating Electron Density Profiles from Ionograms for Use with Minicomputers", *Radio Science*, 12:119-122.

Paul, A., 1985, "Passive Monitoring of the Ionosphere", in *Effect of the Ionosphere on C^3I Systems*, J.M. Goodman (editor-in-chief), pp.84-91, USGPO, available through NTIS, Springfield, VA.

Paul, A., 1988, "Monostatic Observations of F-Region Structure", in *Effect of the Ionosphere on Communication, Navigation, and Surveillance Systems*, IES'87, J.M. Goodman (Editor-in Chief), USGPO, available through NTIS, Springfield, VA.

Paul, A., 1989, "Ionospheric Variability", TR 1277, Naval Ocean Systems Center, San Diego, CA.

Paul, A., 1990, "Applications of High Accuracy Digital Ionosonde Data", in *Effect of the Ionosphere on Radio Signals and System Performance*, IES'90, J.M. Goodman (Editor-in-Chief), USGPO, available through NTIS, Springfield, VA.

Paul, A.K. , J.W. Wright, and L.S. Fedor, 1974, "The Interpretation of Ionospheric Drift Measurements, VI. Angle of Arrival and Group Path (echolocation) Measurements from Digitized Ionospheric Soundings: The Group Path Vector", *J. Atmos. Terrest. Phys.*, 36:193-214.

Peinan, Jiao, and Du Junhu, 1988, "The Localization technique for the Experimental HF Skywave Radar", in *Fourth International Conference on HF Radio Systems and Techniques*, IEE, Savoy Place, London, UK.

Petrie, H.E., and R. D. Hunsucker, 1967, "Short-term Forecasting for a High-frequency Long-distance Point-to-Point Communication Path", *IEEE Trans. Ant. and Prop.*, AP-15(3):390-394.

Piggott, W.R., 1975,"High Altitude Supplement to the URSI Handbook on Ionogram Interpretation and Reduction", Report UAG-50, World Data Center A for Solar-Terrestrial Physics, Boulder, CO.

Piggott, W.R., and K. Rawer, 1972, "URSI Handbook of Ionogram Interpretation and Reduction", Second Edition, UAG-23, World Data Center A for Solar-Terrestrial Physics, Boulder, CO.

Piggott, W.R., and K. Rawer, 1978, "URSI Handbook of Ionogram Interpretation and Reduction", Second Edition (revision of Chapters 1-4), UAG-23A, World Data Center A for Solar-Terrestrial Physics, Boulder, CO.

Polkinghorn, F., 1990, Naval Research Laboratory, private communication.

Poole, A.W.V., 1985, "Advanced Sounding: 1. The FMCW Alternative", *Radio Sci.*, 20(6):1609-1616.

Poole, A.W.V., and G.P. Evans, 1985, "Advanced Sounding: 2. First Results from an Advanced Chirp Ionosonde", *Radio Sci.*, 20(6):1617.

Probst, S.E., 1968, "The CURTS Concept and Current State of Development", in *Ionospheric Radio Communications*, K. Folkestad (editor), Plenum Press, New York.

Pulinets, S. A., 1989, "Prospects of Topside Sounding", in *World Ionosphere/Thermosphere Study: WITS Handbook*, Vol. 2, C.H. Liu (editor), SCOSTEP, University of Illinois, Urbana. IL.

Ramsay, J.L., K.D. Felpenin, and T.I. Dayharsh, 1967, "A Demonstration of the Consistent Operational Success of the 2400 Baud Autodin Data Transmission Using CURTS Phase II System", Final Rpt., Part 1 ,SRI SWM 16352U, Stanford CA.

Rao, N.N., 1973, *J. Atmos. Terrest. Phys.*, 35:1561.

Rash, J.P.S., and J.A. Gledhill, 1984, "Electron Density Profiles over the Southern Ocean from Oblique Incidence Ionograms", *J. Atmos. Terrest. Phys.*, 46(10):945-951.

Reilly, M.H. and M. Daehler, 1986,"Sounder Updates for Statistical Model Predictions of Maximum Usable Frequencies on HF Sky Wave Paths", *Radio Science*, 21(6):1001-1008.

Reilly, M.H., 1990, "Upgrades for Efficient 3D Ionospheric Ray Tracing - Investigation of NVIS Effects", in *Effect of the Ionosphere on Radiowave Signals and System Performance*, IES'90, J.M. Goodman (editor-in-chief), USGPO, available through NTIS, Springfield, VA.

Reilly, M.H., F.J. Rhoads, J.M. Goodman, and M. Singh, 1990, "Updated Climatological Predictions of Ionospheric and HF Propagation Parameters", in *Effect of the Ionosphere on Radiowave Signals and System Performance*, IES'90, J.M. Goodman (editor-in-chief), USGPO, available through NTIS, Springfield, VA.

Reinisch, B.W., 1983, "The Digisonde 256 - Additional Features", *INAG Bulletin 40/41*, distributed for URSI by WDC-A for Solar-Terrestrial Physics Boulder, CO.

Reinisch, B.W., and H. Huang, 1983, "Automatic Calculation of Electron Density Profiles from Digital Ionograms; 3. Processing of Bottomside Ionograms", *Radio Sci.*, 18:477.

Reinisch, B.W., R.R. Gamache, J.S. Tang, and D.F. Kitrosser, 1983, "Automatic Real Time Ionograms Scaler with True height Analysis - ARTIST", Scientific Report No. 7, AFGL-TR-83, Air Force Geophysics Laboratory (now Phillips Lab), Hanscom AFB, MA. (ADA135174).

Reinisch, B.W., K. Bibl, D. Kitrosser, G.S. Sales, J.S. Tang, Zhao-Ming Zang, T.W. Bullett, and J.A. Ralls, 1989, "The Digisonde 256 Ionospheric Sounder", Chapter 13, in *World Ionosphere/Thermosphere Study, WITS Handbook*, Vol. 2, C.H. Liu (editor), SCOSTEP Secretariat, Univ. of Illinois, Urbana.

Rose, R.B., 1982, "196 Path Months of HF Oblique Incidence Sounder Data", NOSC code 5326, Naval Ocean System Center, San Diego, CA. (Unpublished report courtesy of Bob Rose.)

Rose, R.B., J.N. Martin, and P.H. Levine, 1978, "MINIMUF 3: A Simplified HF MUF Prediction Algorithm", TR-186, NOSC, San Diego, CA.

Ross, W., 1947, "The Estimation of Probable Accuracy of HF Direction Finding Bearings", *J. IEE*, 94, Part IIIA: 722-726.

Rothmuller, I.J., 1978, "Real-Time Propagation Assessment: Initial Test Results", in *Effect of the Ionosphere on Space Systems and Communications*, IES'78, J.M. Goodman (editor), USGPO, available through NTIS, Springfield VA.

RRL, 1982, data sheet on the Radio Research Laboratory (Japan) vertical and oblique incidence sounder network, in *INAG Bulletin No. 3*, page 12.

Rush, C.M., 1972, "Improvements in Ionospheric Forecasting Capability", AFCRL-72-0138, Air Force Cambridge Research Laboratory, Hanscom AFB, MA, (AD742258).

Rush, C.M., 1976, "An Ionospheric Observation Network for Use in Short-Term Propagation Predictions," *ITU Telecom. J.*, 43(VIII):544-549.

Rush, C.M., and J. Gibbs, 1973, "Predicting the Day-to-Day Variability of the Mid-latitude Ionosphere for Application to HF Propagation Predictions", AFCRL-TR-73-0335, Air Force Cambridge Research Laboratory, Hanscom AFB, MA, (AD764711).

Rush, C.M., D. Miller, and J. Gibbs, 1974, "The Relative Daily Variability of foF2 and hmF2 and their Implications for HF Radio Propagation", *Radio. Sci.*, 9:749-756.

Rush, C.M., and W.R. Edwards, Jr., 1976, "An Automated Technique for Representing the Hourly Behavior of the Ionosphere", *Radio Sci.*, 11(11):931-937.

Rush, C.M., M. Fox, D. Bilitza, K. Davies, L. McNamara, F. Stewart, and M. PoKempner, 1989, "Ionospheric Mapping: An Update of the foF2 Coefficients", *Telecomm. J.* 56:179-182.

Shapley, A.H., 1970, "Atlas of Ionograms", UAG-10, National Geophysical Data Center, NOAA, Boulder, CO.

Shea, M.A., S.A. Militello, and H.E. Coffey, 1984, "Directory of Solar-Terrestrial Physics Monitoring Stations", MONSEE Special Publications No. 2, TR 84-0237, AFGL, Hanscom AFB, MA.

Shimazaki, T., 1955, "Worldwide Variations in the Height of the Maximum Electron Density of the Ionospheric F2 Layer", *J. Radio. Res. Labs.* (Japan), 2:86.

Sinnott, D.H., 1986, "The JINDALEE Over-the-Horizon-Radar System", *Proc. Conf. Air Power in Defense of Australia*, Canberra, Australia.

Skolnik, M. (editor-in-chief), 1990, *Radar Handbook*, Second Edition, McGraw-Hill Publishing Company, New York.

Smith, N., 1939, "The Relation of Radio Skywave Transmission to Ionosphere Measurements", *Proc. IRE*, 27:323.

Smith, M.S., 1970, *J. Atmos. Terrest. Phys.*, 32:1047.

SRI, 1982, data sheet on the Model 610M1 Digital Ionosonde, SRI International, Menlo Park, CA.

Stevens, E.E., 1968, "The CHEC Sounding System" in *Ionospheric Radio Communications*, K. Folkestad (editor), Plenum Press, New York.

Suzuki, J., 1988, "Ionospheric Data in Japan for July 1988", Vol. 40, No. 7, Communication research Lab., Min. of Posts and Telecomm., Tokyo.

Szuszczewicz, E., 1989, "In-Situ Measurement Techniques for Ionospheric/ Thermospheric Investigations", in *World Ionosphere/ Thermosphere Study: WITS Handbook*, Vol. 2, C.H. Liu (editor), SCOSTEP, University of Illinois, Urbana. IL.

Tascione, T.F., H.W. Kroehl, and B.A. Hausman, 1988, "ICED - A New Synoptic Scale Ionospheric Model", in *Effect of the Ionosphere on Communication, Navigation, and Surveillance Systems*, IES'87, J.M. Goodman (editor-in-chief), USGPO, available through NTIS, Springfield, VA.

Tang, J., R.R. Gamache, E. Lee, D.F. Kitrosser, and B.W. Reinisch, 1988a, "Status of ARTIST Upgrade", Scientific report No. 12, ULCAR, University of Lowell, GL-TR-89-0183, Geophysics Laboratory, Hanscom AFB, MA.

Tang, J., R.R. Gamache, and B.W. Reinisch, 1988b, "Progress in ARTIST Improvements", Scientific Report No. 14, ULCAR, Lowell University, GL-TR-89-0185, Geophysics Laboratory, Hanscom AFB, MA.

Thomas, J.O., 1959, "The Distribution of Electrons in the Ionosphere", *Proc. IRE*, 47(2):162-175.

Thomason, J., G. Skaggs, and J. Lloyd, 1979, "A Global Ionospheric Model", NRL Memo Report 8321, Washington DC, AD-000-323.

Titheridge, J.E., 1959a, "The Calculation of Real and Virtual Heights of Reflection in the Ionosphere", *J. Atmosp. Terrest. Phys.*, 17(1,2):96-109.

Titheridge, J.E., 1959b, "Ray Paths in the Ionosphere; Approximate Calculations in the Presence of the Earth's Magnetic Field", *J. Atmos. Terrest. Phys.*, 14:50-62.

Titheridge, J.E., 1985, "Ionogram Analysis with the Generalized Program POLAN", Report UAG-93, World Data Center for Solar-Terrestrial Physics, NGDC, NOAA, Boulder, CO.

Tveten, L.H., 1960, "Ionospheric Motions Observed with High Frequency Backscatter Sounders", *J.Res. Nat. Bur. Standards*, D: Radio propagation, 65D(2):115-127.

Uffelman, D.R., L. O. Harnish, and J. M. Goodman, 1982, "The Application of Real-Time Model Update by Oblique Ionospheric Sounders to Frequency Sharing", in *Propagation Aspects of Frequency Sharing, Interference, and System Diversity*, H. Soicher (editor), NATO/AGARD-CP-332, Technical Edit and Reproduction Ltd., London, UK.

Uffelman, D.R., 1982, "HF Propagation Assessment Studies over Paths in the Atlantic", in *Effect of the Ionosphere on Radiowave Systems* (IES'81), J.M. Goodman (editor-in-chief), USGPO, avail. through NTIS, Springfield, VA.

Uffelman, D.R. and L.O. Harnish, 1981, "HF Systems Test for the SURTASS Operation of February 1981", Memo Report 4600, Naval Research Laboratory, Washington, DC. (ADA106747).

Uffelman, D.R. and L.O. Harnish, 1982, "Initial Results from HF Propagation Studies during SOLID SHIELD", Memo Report 4849, Naval Resaearch Laboratory, Washington, DC. (ADA118492).

Uffelman, D.R., L.O. Harnish, and J.M. Goodman, 1982, "The Application of Real-time Model Update by Oblique Ionospheric Sounders to Frequency Sharing", in *Propagation Aspects of Frequency Sharing, Interference, and System Diversity*, H. Soicher (editor), NATO-AGARD-CP-332, Tech. Edit. and Reprod. Ltd., UK. (Also available through NTIS, Springfield, VA.)

Uffelman, D.R., and P. Hoover, 1984, Report 5246, Naval Research Laboratory, Washington, DC.

Uffelman, D.R., L.O. Harnish and J. M. Goodman, 1984, "H.F. Frequency Management by Frequency Sharing Assisted by Models Updated in Real-Time",Memorandum Report 5284, Naval Research Laboratory, Washington, DC.

USSR Acad. Sci., 1990, "Meteorological Ionospheric Station for Backscatter Sounding "ICEBERG", Academy of Sciences, USSR.

Van Troyen, D.L.R. and A.R. Van de Capelle, 1988, "Ionospheric Modelling Based on Oblique Chirp-Sounders", in *Fourth International Conference on HF Radio Systems and Techniques*, Conference Pub. No. 284, IEE, London.

Wagner, L.S., J.A. Goldstein, and W.D. Meyers, 1988, "Wideband Probing of the Trans-Auroral HF Channel", in *Effect of the Ionosphere on Communication, Navigation, and Surveillance Systems*, IES'87, J.M. Goodman (editor-in-chief), USGPO, available through NTIS, Springfield, VA.

Wakai, N, 1981, "Ionospheric Vertical/Oblique Sounding Network in Japan", in *INAG Bulletin No. 33*, URSI, WDC-A, Boulder, CO.

Wilkinson, P.J., 1979, "Prediction Limits for foF2", in *Solar-Terrestrial Predictions Workshop*, Vol. 1, R. F. Donnelly (editor), USGPO, Dept. Commerce, available through NTIS, Springfield, VA.

Wilkinson, P.J., 1986, "Short-term Forecasting Using Daily Ionospheric Indices", *Solar-Terrestrial Proceedings: Symposium in Meudon, France*, Edited by P.A. Simon, G. Heckman, and M.A. Shea, published jointly by NOAA, Boulder, CO. and AFGL, Hanscom AFB, MA.

Wilkinson, P.L., 1991, "The IPS 5A Ionosonde Project", INAG Bulletin No.56, INAG, distributed by WDC-A for Solar-Terrestrial Physics, Boulder CO.

Wright, J.W., 1974, "The Interpretation of Ionospheric Radio Drift Measurements, VII. Diffraction Methods Applied to E-Region Echo Fading: Evidence of a Focusing Model", *J. Atmos. Terrest. Phys*, 36:721-740.

Wright, J.W., 1982, "Application of Dopplionograms to an Understanding of Sporadic E", *J. Geophys. Res*, Volume 87, No. 3 (March issue).

Wright, J.W., 1975, "Development of Systems for Remote Sensing of Ionospheric Structure and Dynamics: Functional Characteristics and Applications of the Dynasonde", NOAA ERL SEL 206, U.S. Dept. of Commerce, Boulder, CO.

Wright, J.W., 1977, "Development of Systems for Remote Sensing of Ionospheric Structure and Dynamics: The Dynasonde Data Acquisition and Dynamic Display System", NOAA ERL SEL 206, U.S. Dept. of Commerce, Boulder, CO.

Wright, J.W., M. Glass, and A. Spizzichino, 1976a, "Interpretation of Ionospheric Radio Drift Measurements, VIII. Direct Comparisons of Meteor Radar Winds and Kinesonde Measurements: Mean and Random Motions", *J. Atmos. Terrest. Phys*, 38:719-729.

Wright, J.W., G. Vasseur, and P. Amayenc, 1976b, "Interpretation of Ionospheric Radio Drift Measurements, VIII. Direct Comparisons Between Field Aligned Ion Drifts by Incoherent Scater and Kinesonde Measurements", *J. Atmos. Terrest. Phys*, 38:731-738.

Wright, J.W., and M.L.V. Pitteway, 1979a, "Real Time Data Acquisition and Interpretation Capabilities of the Dynasonde, 1. Data Acquisition and Real-time Display", *Radio Sci*, 14:815-825.

Wright, J.W., and M.L.V. Pitteway, 1979b, "Real Time Data Acquisition and Interpretation Capabilities of the Dynasonde, 2. Determination of Magnetoionic Mode and Echolocation using a Small Spaced Receiving Array", *Radio Sci*, 14:827-835.

Wright, J.W., and A.K. Paul, 1981, "Toward Global Monitoring of the Ionosphere in Real Time by a Modern Ionosonde Network: The Geophysical Requirements and Technological Opportunity", NOAA ERL, Boulder, CO.

Wright, J.W., and M.L.V. Pitteway, 1982a, "Data Processing for the Dynasonde: The Dopplionogram", *J. Geophys. Res*, 87(A3):1589-1598.

Wright, J.W., and M.L.V. Pitteway, 1982b, "Application of Dopplionograms and Gonionograms to Atmospheric Gravity Wave Disturbances in the Ionosphere", *J.Geophys. Res*, 87(A3):1719-1721.

Valverde, J.F., 1958, "Motions of Large Scale Traveling Disturbances Determined from High Frequency Backscatter and Vertical Incidence Records", Stanford Propagation Laboratory, Stanford University Science Rpt. No.1.

Zacharisen, D.H., 1965, "Space-Time Correlation Coefficients for Use in Short-Term Ionospheric Predictions", NBS Report 8801, available through NTIS, Springfield, VA.

7

ADAPTIVE HF & THE EMERGING TECHNOLOGIES

"We must cut our coat according to our cloth, and adapt ourselves to changing circumstances."

W.R. Inge[56]

7.1 CHAPTER SUMMARY

Adaptive system concepts have existed for many years but specific realizations have only resulted from the development of technological advances. It has been long recognized that adaptivity will permit system performance to be optimized in a temporally and spatially-varying environment. This chapter describes growth in adaptive HF as stimulated by availability of advanced technologies. So-called *robust* system approaches have also been implemented as a consequence of technological improvement although these schemes, which cope with environmental effects by provision of increased margins, are philosophically quite different.

We begin the Chapter by reviewing some definitions of *adaptive HF*[57]. We take note of the fact that a clear definition is rather elusive being relative to the vantage points (or "sandboxes") of the respective authors. However it is clear that *adaptive HF*, when viewed as a process, is distinct from some systems which are intrinsically robust. The ultimate definition advanced herein is compared with others put forward by government agencies and supporting contractors. We note that the term *adaptive HF* is used in so many ways, and that computer searches of keywords such as *modern HF* or *advanced HF* will yield as many truly adaptive concepts as the term *adaptive HF* will yield inappropriate ones. Moreover, a term like *advanced HF* is descriptive of a

---

56. Dean of St. Paul's Cathedral, London (1860-1954). Quote from "Lay Thoughts", in *Home Book of Quotations*, 9th edition, 1964, Vail-Ballou Press, Binghampton, N.Y.

57. At least four terms are used in connection with HF system adaptivity: *adaptive HF*, *robust HF*, *modern HF*, and *advanced HF*. The latter two terms refer to state-of-the-art HF technologies, while *adaptive* and *robust* have more precise meanings. *Adaptive HF* requires the capability of channel trait-tracking coupled with near-real-time system parameter adjustment to optimize system performance. *Robust HF* is designed to overcome disturbances without the need for system parameter adjustment. (See Sections 7.5 and 7.8.6.)

superset, an umbrella term under which *adaptive HF* may be contained. Typically *advanced HF* includes adaptive and robust system concepts without significant distinction. Within this chapter we will examine new technologies which might serve as ingredients to advanced (and perhaps adaptive) HF. Many of these technologies involve component development but others involve propagation disturbance mitigation schemes using signal processing or involve the implementation of new procedures for resource management.

7.2 INTRODUCTION

In the late seventies, HF communication systems experienced a re-examination based upon several factors including the military requirements for a balanced mix of assets to affect reliable, survivable, enduring, and affordable command and control of forces. The *engine* which drove this renewal of interest was the evolution of new equipment and technology. This chapter addresses the latter factor in very general terms, and emphasizes adaptive HF as an architectural option which has been stimulated by the availability of new technology. The reader will note that many of the technological advances are not unique to HF but may apply equally well to other frequency regimes as well as other system types.

A discussion of the rationale for adaptive HF is provided emphasizing some of the identified requirements and current attitudes of the military user. This rationale recognizes the necessity for redundancy in communications and sees HF as a valuable commodity in the event of satellite system failure. Clearly a broad set of independent communications media is an advantage since an adversary would be forced to invest in a wide range of countermeasures.

7.3 BACKGROUND

7.3.1 The History of Adaptive HF

The trials and tribulations associated with HF radio system development are well-known, at least as far as so-called skywave propagation is concerned. In this connection, HF communication is presently perceived to be decidedly influenced, largely in a negative manner, by the ionosphere. In the early years of HF radio development the profound ionospheric influences on the band were recognized, but negative comments were virtually nonexistent since there were few alternatives upon which to base comparisons. Longwave systems were too large and expensive and were somewhat limited in data rate. Direct cable systems were expensive to deploy and maintain, and did not offer the flexibility required by the military or by the mobile user. Finally, satellite systems did not yet exist.

The possibility of high frequency propagation modes enabled the early communication pioneers to develop long-haul BLOS[58] capabilities at power levels which were relatively modest using manageable antenna structures. During this early period two distinct constituencies emerged: one was concerned with the development of HF radio and related technologies, and the other was concerned with upper atmospheric phenomena. Nevertheless, the most important features of the medium were identified before this division became relatively permanent following World War II.

Over the years HF technologists and system operators (including amateurs) have uncovered many of the more salient features of the HF channel, and a majority of these features are directly traceable to a specific ionospheric property such as ionospheric height, critical frequency, etc. At the same time ionosphericists have established the cause-and-effect relationships with respect to the sun-earth system and have developed a myriad of models to describe ionospheric phenomenology. Unfortunately, these early models described median behavior and large variabilities about median behavior have been found to exist. Moreover, such variability was difficult to predict. In short, although the ionospheric personality was fairly well understood in terms of its median climatology, there remained an apparently unknown (and largely deleterious) component.

Nevertheless, with the availability of these median models of channel behavior, it was found that HF communication reliabilities of the order of between 60 and 90% could be achieved, provided a sufficient range of frequencies were available for selection. Actually, it is striking that the reported reliabilities were so high given the fact that the median model estimates accounted for diurnal and solar-induced variabilities in only the crudest fashion. Unfortunately reliabilities less than 50% were often apparent as a result of limited frequency assets and training inadequacies. These low levels of reliability allowed a somewhat negative image of HF to emerge. Clearly measures were required to achieve greater reliability. Certainly it would be necessary to have an adequate number of frequencies to taken advantage of the crude median models (and thereby approach 90% reliability), but it would also be necessary to obtain a measure of the real-time channel in order to

---

58. Beyond-Line-Of-Sight. A number of terms should be noted in connection with the range of communication coverage: LOS, ELOS, BLOS, and NVIS. By convention, principally within the military C^3 community, Line-Of-Sight (LOS) coverage implies distances as great as 30 nautical miles. A more precise (physical) definition for LOS is given in Chapter 4. Extended-Line-Of-Sight (ELOS), ranges between 30 and 300 nautical miles, while Beyond-Line-Of-Sight (BLOS) refers to distances in excess of 300 nautical miles (480 km). *Umbrella* skywave coverage devoid of a *skip zone* is termed Near-Vertical-Incidence-Skywave (NVIS). Therefore skywave coverage may separated into NVIS and BLOS components. Alternatives to NVIS are LOS and ELOS.

account for channel variability about the median. If this could be done successfully, then reliabilities approaching 100% could be realized, at least in principle.

Faced with the challenge of obtaining a real-time measure of ionospheric and HF channel behavior, sounding methods developed by ionosphericists were modified by radio engineers for application to communication problems. In the 1950s and 1960s, special tests were conducted employing these new sounding methods with highly successful results. Optimum transmission frequencies were determined for individual links by sounding the paths and this resulted in an improved network connectivity since the available frequencies could be managed more effectively. Adaptive HF was born. Even so, with the advent of artificial earth satellites and the promise of communication performance far exceeding that which could be achieved by terrestrially-based systems, HF was forced to accept the dubious distinction of *backup* to satellite communication systems.

The recent history of HF (roughly 1960 to 1980), has been dominated by policy decisions based upon negative aspects (i.e., the trials) whereas the positive aspects (i.e., the tribulations) were largely ignored. As a result of this benign neglect, HF assets, including man and machine, were neglected, leading to a significant HF infrastructure deficit. By 1980, a number of factors has led to a modest *rebirth* of HF. In the 1980s, a number of HF modernization efforts were initiated by the military services, and various adaptive HF concepts were implemented in hardware.

7.3.2 Elements of the Adaptive Process

As has been indicated, the ionosphere exhibits considerable variability and this changing behavior is directly related to attributes of the temporally- and spatially-varying HF channel. Some ionospheric states provide better channel conditions than others. The designer is challenged to match system parameters to the existing channel conditions in a manner which is both efficient and optimal. Since the ionospheric state varies temporally, the matching process must be updated periodically or even continuously. The update rate (or duty cycle) is dependent upon the sensitivity of performance to parameter change. For example, if the performance is judged solely on the basis of negligible intersymbol interference at a fixed frequency, we would like to pack as many uncorrupted symbols as possible into the channel. This suggests that an adjustment in the transmission data rate is to be made to account for the deleterious effects of multipath spread. For the fixed frequency case, we find that the update rate would be dependent on the slow diurnal changes of the ionosphere. Unfortunately this approach, while limiting catastrophic failure arising from intersymbol interference, is guaranteed to yield zero throughput during long periods occasioned by lack of ionospheric support, a matter which is ultimately independent of data rate.

The oldest form of adaptation to the HF channel involves frequency selection. HF signals travel decidedly different trajectories at different frequencies. As a corollary to this, a wideband signal will *see* a different channel than will a narrowband signal. By the process of frequency management, it is possible to find and exploit the most appropriate channel conditions. Naturally, one is also able to avoid those frequency domains above-the-MOF and below-the-LOF where support is lacking altogether.

It is clear that the ability to adapt to the HF channel is directly related to the temporally- and spatially-varying nature of the channel itself. Moreover, we must be able to *learn* the channel by some process and adapt at a speed commensurate with the underlying ionospheric state changes. Thus the cornerstone of adaptive HF is a capability to define the current HF channel properties of interest. This may involve in-band sounding, out-of-band sounding, or a host of other schemes for estimating the channel parameters (see Figure 7-1). An equally important component of adaptive HF systems is a capability to respond to acquired knowledge about the channel. This implies system parameter flexibility. It is also important to recognize that the performance is dependent upon the full environment including additional factors such as system noise, channel occupancy, and the electronic countermeasure conditions[59]. Finally we will deliberately exclude from the category of adaptive HF attributes those which correspond to a class of *robust* design attributes. These are typically associated with some low-data-rate military systems which may not have the luxury of an adaptive countermeasure to the HF channel condition.

7.4 COMMUNICATION REQUIREMENTS FOR ADAPTIVE HF

The reader should now be well-versed in the trials and tribulations of HF communications, especially the propagation aspects. The HF skywave mode is utilized to achieve BLOS connectivity, and this mode is hampered by propagation-related disturbances which are often unpredictable. The situation is complicated even more by frequency congestion, archaic frequency assignment and management procedures, and even more serious problems such as deliberate interference (i.e., jamming) and the very special circumstances encountered in a nuclear environment. How does one cope with these problems? HF is often regarded as yesterday's technology. Is it a forgotten resource? What are the civil and military requirements which should make HF, and adaptive HF in particular, an important ingredient in future architectures for telecommunication systems?

----

59. Electronic countermeasures, or ECM, generally implies jamming in the military context.

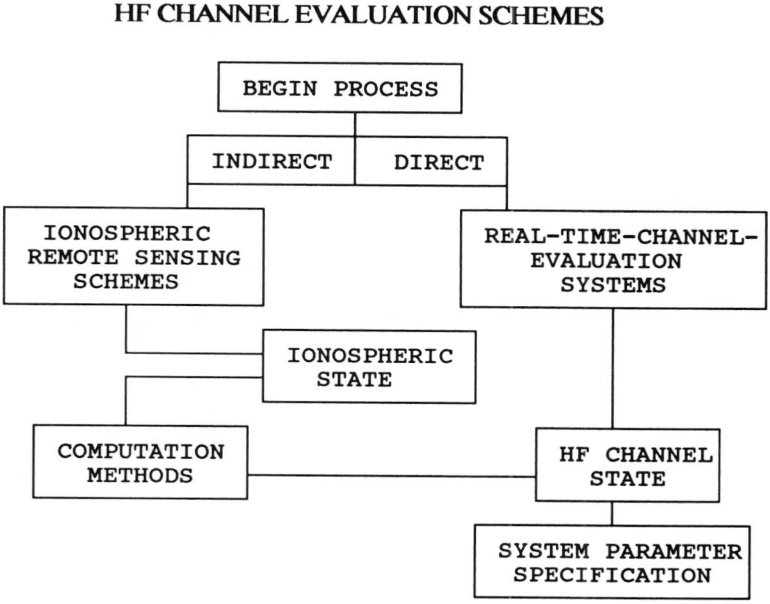

HF CHANNEL EVALUATION SCHEMES

Fig.7-1. Generic procedures for assessing the HF channel. The process may be direct, in which case systems parameters are monitored; or it may be indirect, in which case the ionospheric parameters are measured, with the channel parameters being computed.

Modern military communication requirements have encouraged the development of certain new techniques (some not specific to HF) and stimulated reinvestment in older HF techniques or systems. Validated requirements for improvement in HF communications include [O'Mahony, 1991]: reduction of logistical/operational support, improvement in reliability and capacity, reduction in vulnerability to jamming and intercept, achievement of interoperability, and integration into multimedia networks. Broadly interpreted, this list may be organized into three basic requirements: system reliability, responsiveness, and survivability, as shown in Table 7-1. To an extent, these attributes are virtually consistent with those incorporated in the design of civilian systems.

TABLE 7-1. MILITARY COMMUNICATION REQUIREMENTS

RELIABILITY
RESPONSIVENESS
SURVIVABILITY

7.4.1 Reliability

Reliability implies an ability to provide continuous high quality service to users of the system regardless of the circumstances posed by the total environment. This total environment includes propagation condition, natural interference and noise, ECM, and nuclear effects.

7.4.2 Responsiveness

Responsiveness implies the ability to perform specified time-critical functions consistently without any appreciable reduction in the communication reliability. Typically communication responsiveness requires automatic system capability for timely dissemination of command and control information.

7.4.3 Survivability

Survivability implies the ability to maintain service even in the event of highly disruptive natural or nuclear disturbances as well as an ECM attack. Survivability is the single most important feature which distinguishes a civilian system from a military one.

7.4.4 On Satisfying the Requirements

To satisfy the requirements mentioned above, it is clear that several issues have to be addressed. Initially we need specific information about the environment and also about the requirements themselves. How disturbed is the environment? What is the nature of the threat? How reliable is the system to be, and at what cost? How responsive? How survivable? Is the system to be used in the trans-attack period, is it to be used in the post-attack period, is it to be used for reconstitution of government, or is it to be used simply for peacetime conditions? A block diagram of the salient criteria is given in Figure 7-2.

It is presumed that the system designer has some knowledge of the ionospheric disturbances to be expected as well as an estimate of the jamming threat. This is essential to determine the margins of protection which should be incorporated into the system. For instance, if 10 dB of excess absorption is anticipated, then one may compensate for this eventuality through augmentation of the power gains product ($P_t G_t G_r$) by the same amount. Typically disturbances may be minimized through application of appropriate countermeasures, but usually with some cost. In the example above, we may increase the antenna gains or increase the transmitter power. The negative features of these options are obvious. If we increase the power, we increase the congestion level for other users, and military users will have their likelihood of detection increased. Furthermore, antenna gain enhancement is inconsistent with an omnidirectional requirement specified for some broadcast systems. Another ionospheric disturbance requiring system compensation is multipath, which may introduce fading or may give rise to ISI[60] in digital signaling. One countermeasure for this effect is to reduce the signaling rate so that adjacent symbols do not overlap. There are naturally more elegant methods and these will be discussed later. No matter what the ionospheric disturbance, an a priori estimate of its uncompensated impact on the operating system will enable the system architect to include appropriate counters in the system design. It must always be recognized that the incorporation of system margin as a static design parameter typically fixes the ultimate upper limit of system performance as a price for controlling the lower limit of performance. Adaptive procedures provide for a flexibility in system parameters enabling the optimum performance to be achieved, at least in principle. Of course, some form of channel evaluation and assessment subsystem is required if adaptive HF is to work. If this subsystem fails, then the possibility exists that the adaptive HF system will not perform even as well as a *plain vanilla*[61] radio. It certainly will not be competitive with a robust system approach which is characterized by a static margin for all propagation-sensitive system parameters.

First we must define the total environment. Looking at the set of requirements, we must then play our technical options against the environment to determine the set of options which best satisfy the requirements. Budgetary constraints may limit the options but prudence dictates a set of technical solutions which will somewhat overcompensate for the expected conditions. It is generally easier to delete unnecessary options in the final stage of design than to modify the completed design with add on features.

In the pursuit of improved reliability, responsiveness, and survivability, there are two basic responses. The first concentrates on the methodology of operation, while the second stresses basic system design as the focal point. In the former response, we take a more remedial approach in which training (or retraining) is dictated. Clearly, the successful operation of ordinary (non-adaptive) radios puts a considerable stress on the operator. Improvements in this area are to be made first. Too often, inappropriate frequency selection procedures have led to significant performance degradation in military operations. This is neither the fault of the basic radio nor the operator. Rather it is the fault of management for not providing the level of training required to maintain an adequate degree of military readiness. Clearly this is a zeroth

---

60. Intersymbol Interference. A condition in which adjacent symbols overlap, leading to ambiguities in decisions relative to their states or values. This condition may be the result of multipath propagation and/or signal dispersion.

61. The descriptive term *plain vanilla* HF radio usually refers to a non-adaptive single sideband (SSB) HF radio system. This class of radio appears to be the defacto standard providing minimum capability based upon an examination of specification sheets obtained from vendors.

order measure. A first order enhancement involves automation of operator-intensive functions. At this level, we still have not altered the fundamental properties of the radio.

The second response for improved reliability, responsiveness, and survivability is to redesign the radio. Considering the technical solutions, hardware implementations result from two different approaches. The first involves real-time adaptivity to circumvent disturbances, while the other involves the incorporation of robust waveforms or other strategies which will provide for quasi-continuous immunity from disturbances. In some instances, the two approaches may be combined. The full ensemble of these new system designs should be referred to as *advanced-HF*, although regrettably the term *adaptive HF* is used by many to describe systems composed of both the robust (i.e., nonadaptive) and the truly-adaptive approaches. The most accurate (and descriptive) terms by which all *advanced HF* systems may be organized are *adaptive* and *robust*. We shall be more precise about this shortly.

STEPS IN SYSTEM DEFINITION

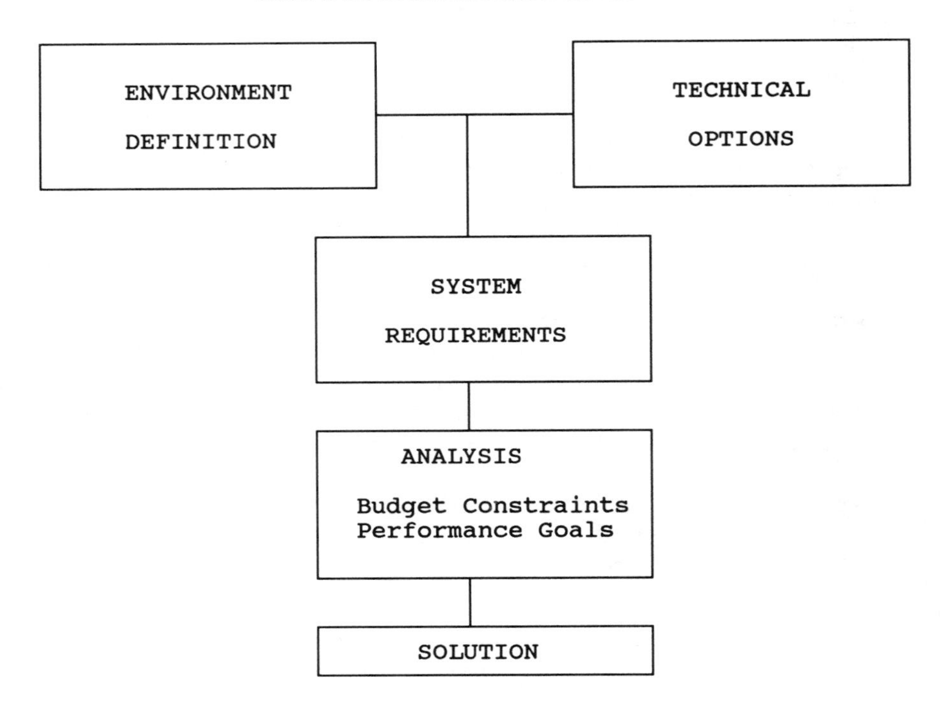

Fig.7-2. Steps in system definition. Recognizing the system requirement, we play the technical options against the environmental conditions. Ultimately, life-cycle cost analysis may dictate the solution.

7.5 A MATTER OF DEFINITION

Adaptive HF has an elusive definition. There have been more than a handful of definitions advanced by government organizations, the contractor community, and radio system specialists associated with academic institutions. There are several common features in these definitions, but many advance corporate interests (and product lines) and this tendency often leads to excess baggage. In arriving at an appropriate definition for *adaptive HF* we shall first look at the various forms of adaptivity. Then we examine RTCE[62] as a cornerstone of adaptive decision making. Finally, we discuss system diversity as a necessary ingredient in the avoidance or mitigation of channel impairment. Without some form of system diversity, there would be no need to invest in RTCE appliques.

An important goal of military communication managers is to achieve interoperability between the armed services for theater and strategic operations. This requires that certain common features must be possessed by all radio equipment. Interoperability could be stipulated at the service level, the national level, or the international level. Ultimately this implies commonality of specified system functional capabilities, and agreements on waveforms, protocols for connectivity, and other factors. Such issues provide the stimulus for standardization agreements between the services (i.e., Military Standards), federal agencies (i.e., Federal Standards), and allied countries (viz.,Standardization Agreements or *STANAG* s between NATO members). Naturally not all radios need possess the same functional capabilities. Still, in order to permit integration of communication assets, care must be taken to insure that necessary adaptive features are accounted for in a rational way. Such a rational approach is contained in FED-STD-1045 and MIL-STD-188-141A which are discussed later on in the Chapter. It must be remembered that *adaptive HF* is realized by the availability of new device and component technologies. These technologies enable use to be made of digital signal processing schemes in advanced receivers which exploit diversity to compensate for channel impairments which are independently characterized by RTCE capabilities. Automation is the key. Harrison [1987] has described 10 levels at which ALE[63] may be achieved and maintained in HF systems. Interestingly enough, the term ALE appears to convey a more general meaning within the HF community, incorporating many *adaptive HF* features to achieve link establishment. Before examining the Harrison concept, we shall

---

62. Real-Time-Channel-Evaluation. RTCE is covered in Chapter 6. A related function for sky-wave channels is Real-Time-Ionospheric-Specification (RTIS).

63. Automatic Link Establishment (see Section 7.6.5). A related term, ALM, stands for Automatic Link Maintenance (ALM).

outline a four tier framework for *adaptive HF.* Later on we shall dissect these four tiers to accommodate the 10 step *staircase* model (see Section 7.6.5). This issue will be revisited at the end of the Chapter (i.e., Section 7.9) in connection with development of Federal and Military standards.

7.5.1 Types of Adaptivity Based Upon Communication Level

There is an hierarchy of adaptive HF characteristics depending upon the level of communication involved. Figure 7-3 indicates the four basic types of adaptivity. Let us examine each one briefly. The first two types are relevant for a single transmission path at one or more frequencies. The final two types are associated with multiple paths. A breakdown is given in Figure 7-4.

7.5.1.1 *First Type : Transmission Adaptivity*

This is the lowest level. Here one is concerned with adaptivity characteristics which are available over a fixed link and at a fixed transmission frequency. Such characteristics include data rate, transmission waveform, coding scheme, transmitter power, antenna pattern, and performance assessment. At this level, it may be possible to match the data rate to be proportional to the maximum rate which the ionospheric channel will support. LPI[64] may be accommodated by adaptive power control. Adaptive null-steering may allow for placement of unwanted signals at nulls within the antenna pattern. However, the concept operates best against groundwave signals or skywave modes which are isolated. Thus we would not expect null-steering concepts to operate properly in a pathological environment such as that which might arise in the vicinity of the auroral zone, or during nuclear disturbances.

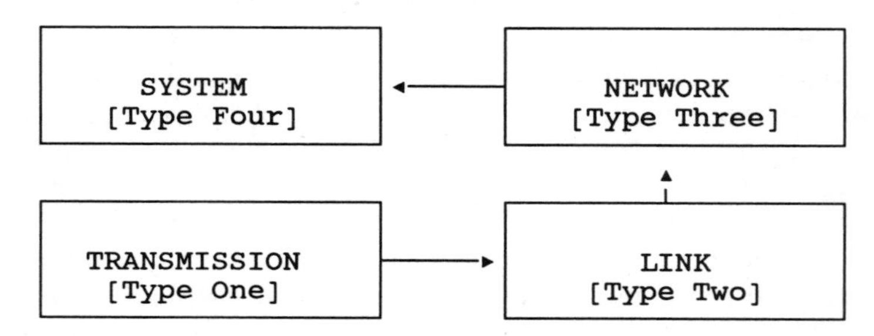

Fig.7-3. Basic levels at which adaptivity may be achieved

---

64. Low Probability of Intercept. LPI is a requirement for a number of critical military communication systems. It may be achieved by power reduction, spread spectrum techniques, and by adaptive beam steering.

7.5.1.2 *Second Type: Link Adaptivity*

In this instance, we are concerned with point-to-point communications. The characteristics at this level include various frequency management and control functions, oblique sounding, channel probing, and spectrum monitoring. RTCE schemes are included in the link adaptivity category. Although general information concerning a coordinated ECM attack may derive from a higher level system component, countermeasures to jamming may be applied at the link level. The logical ECCM[65] schemes involve either spread spectrum or frequency hopping or both. These active countermeasures may be invoked adaptively (i.e., only used when needed) in order to optimize performance. In a sense these schemes allow for the coexistence of wanted and unwanted signals, and should be placed in the *robust* system parameter category. However, in the present context, in which we allow for invocation of the techniques as a conditional response, active ECCM is an attribute within the link adaptivity family. Passive ECCM techniques involve either an assessment or prediction of relative channel availability for friendly links and for the threat. Thus it may be possible to circumvent (rather than coexist with) enemy jamming or interception by appropriate narrowband frequency selection.

7.5.1.3 *Type Three: Network Adaptivity*

At the network level of adaptivity, we are dealing with multinode networks. The characteristics to be included are the following: adaptive routing, flow control, protocol handling, and data exchange between nodes. This type of adaptability provides for the accessibility of alternate routing options in the event of ionospheric disturbances, or if specified nodes become inoperable. Network management activities include those which assess the network status and provide for adaptive reorganization of the network topology.

It is clear that overall HF system connectivity may be improved if the network is comprised of a sufficient number of independent links. We thus think of successful communication in terms of an *end-to-end* process rather than one which is *point-to-point*. A combination of multiple frequencies and paths leads to an even greater gain in connectivity [Rogers and Turner,1985]. This philosophy is implicit in several strategic HF radio system designs (see Section 7.8.6). Adaptive networking concepts have been developed in the United States, the United Kingdom [Goodwin and Reed, 1987] [Goodwin and Harris, 1988], and Canada [Nourry and Mackie, 1987].

---

65. Electronic Counter-Countermeasures. These are techniques which are used to thwart ECM, and principally the jamming threat. Active ECCM may involve frequency hopping, direct-sequence spread spectrum, or both. Passive ECCM may involve propagation tactics.

ADAPTIVITY ATTRIBUTES

TYPE ONE: TRANSMISSION	TYPE TWO: LINK
DATA RATE CODING SCHEME TRANSMISSION WAVEFORM TRANSMITTER POWER ANTENNA PATTERN - Null Steering PERFORMANCE ASSESSMENT - Limited RTCE	FREQUENCY MANAGEMENT CHANNEL SOUNDING FREQUENCY AGILITY - Passive ECCM - Active ECCM SPECTRUM MONITORING PERFORMANCE ASSESSMENT - Full RTCE Complement
TYPE THREE: NETWORK	**TYPE FOUR: SYSTEM**
ROUTING FLOW CONTROL PROTOCOL HANDLING DATA EXCHANGE NETWORK REORGANIZATION PERFORMANCE ASSESSMENT - Network-wide RTCE	MULTI-MEDIA COMMS - Meteor Burst - Bomb Modes - Ground Wave BRIDGE TO ISOLATED NODES RESOURCE MANAGEMENT PERFORMANCE ASSESSMENT - System Assessment

Fig.7-4. Adaptivity attributes organized into the four primary types. These correspond to basic levels of communication: transmission, link, network, and system. The common thread in these adaptivity types is performance assessment.

Shipboard Architectures

McGregor et al.[1985] of NRL have designed two special purpose networking architectures: one for HF Intratask Force (HF-ITF) communications, and the second for HF Ship-to-Shore (HF-SS) communications. The designs are quite different. The HF-ITF exploits groundwave, and the predominant mode of propagation of the HF-SS is skywave. At any one time, the HF-ITF structure is established through execution of a link-cluster-algorithm (LCA) [Wieselthier et al.,1983] which decides the status of network nodes based upon data exchanged between neighboring platforms. Certain nodes may be characterized as either clusterheads, which service a number of ordinary nodes, or gateways, which connect the clusterheads together. The backbone of the HF-ITF is comprised of clusterhead and gateway nodes. The network is designed to reorganize itself continuously and automatically through use of the LCA algorithm, which operates on status messages organized under a Time-Division-Multiple-Access (TDMA) protocol. A frequency hopping[66] code-division-multiple-access (FH-CDMA) is proposed as one scheme for reducing the vulnerability to jamming. An additional ship-to-shore network, the HF-SS [Hauser et al., 1984] is dependent upon skywave connectivity, and is more subject to propagation disturbances. Frequency hopping waveforms are not used in the HF-SS, however the system is made more robust through use of a one-half rate Golay code[67], diversity combining, and ARQ[68]. It features adaptive link control and procedures for handling multi-precedence traffic and the system supports the ANDVT[69] terminal [Garner et al.,1982].

Adaptive Network Synthesis

Concepts of adaptive networking are not always easy to test in the field. This is especially true for system topologies which must be nuclear-capable. As a result, network synthesis studies have been used to examine the efficacy of candidate methods. DePedro [1985] has used two simulation programs , *HFCON* and *HFDELAY*, in conjunction with *IONCAP* to test various hypothetical configurations under benign and stressed conditions. He proposes a backbone configuration with ordinary nodes connected by means of SEL-SCAN™[70], and with backbone nodes linked by sounders for more precise frequency management. Harris Corporation has described a serial modem networking model [McRae, 1985] and have synthesized the proposed network [Boyd and McRae, 1987]. (See Section 7.7.)

7.5.1.4 *Type Four: System Adaptivity*

At the overall system level, the adaptivity includes those functional capabilities which allow multimedia communication. Characteristics include: system level frequency management and control, coordination of resources in a stressed environment, and the determination of bridges to isolated nodes. The incorporation of frequency extension (to low VHF) will improve access to

---

66. Frequency Hopping is one method for achieving bandwidth expansion. The frequencies which are activated in a hopping strategy are generally determined on the basis of a pseudo-random (PN) sequence. (See Section 7.6.6).

67. Coding methods are covered in Section 7.6.7.

68. ARQ, Automatic Repeat Request, is a method for enabling retransmission of messages, data blocks, or packets on the basis of error detection by the receiver (see Section 7.6.7).

69. The ANDVT system is discussed in Section 7.6.8.4. The ANDVT supports data transmission of 2400 bps (uncoded), 1200 bps (coded), and 300 bps (coded with quadruple diversity).

70. SELSCAN is a radio controller developed by Rockwell-Collins.

nonclassical modes of propagation such as meteor burst and nuclear-induced *bomb modes*, and frequency reduction (to MF) will allow access to lower loss groundwave operation. Adaptive frequency extension and reduction as well as other system adaptivity attributes require a network-wide RTCE capability.

7.5.2 The Cornerstones of Adaptivity

It is seen from Figure 7.4 that performance assessment is an ingredient in all four levels of adaptivity. Within every level, it is an essential function, typically accomplished through the analysis and application of some form of RTCE information. The types of RTCE have been discussed in Chapter 6. However, in order to achieve adaptivity there are additional capabilities which must exist, including an ability to be responsive to the RTCE information. This implies automation and microprocessor control. Furthermore it is essential that there exist within the family of system attributes a countermeasure to combat the deleterious channel behavior derived from the RTCE information. Even so, without performance assessment (and RTCE) the concept of adaptive HF would be an academic one. Lacking the suitable prerequisites, the closest approach to adaptive HF is approached through use of forecasting methods (i.e., predictions).

The need for RTCE arises because of the variability in the *total* environment. Table 7-2 lists the environmental parameters which must be monitored so that system parameters may be adapted.

7.5.3 Diversity as a Mitigation Scheme

Radiowave disturbances are typically associated with a specified parameter such as signal strength, polarization, phase, or frequency. If a transmitted waveform is monitored over two circuits which are independently disturbed (with respect to a specified parameter) then a combination of the two received (and distorted) waveforms will lead to a conditional recovery of the original waveform. Diversity combination and kindred procedures comprise the most common methods for mitigation of HF radiowave disturbances. For skywave conditions of signal fading, diversity methods which may be exploited include : time, space, frequency, polarization, and angle (see Figure 7-5).

Most other complex schemes are versions of the five basic types mentioned above. Time diversity is the method most commonly used. *Rake*, for example is an implicit (time) diversity concept which compensates for the fading which arises from multipath spread. Channel equalization, a matched filter method which compensates for ISI, is made adaptive by an algorithm which adjusts the filter characteristics in response to time-varying ionospheric changes. The Automatic Repeat Request (ARQ) method and its variants take advantage of the fact that channel quality is not stationary. Even coding and interleaving schemes exploit the existence of channel diversity in the time domain.

TABLE 7-2. ENVIRONMENTAL VARIABILITIES

PROPAGATION

- VARIABILITIES OF THE SKYWAVE PATH
  - Maximum Observable Frequency (MOF)
  - Lowest Observable Frequency (LOF)
  - Optimum Working Frequency (FOT)
  - Angle-of-Arrival (AOA)
  - Multipath Spread
  - Doppler Shift and Spread
  - Polarization Bandwidth
  - Fade Rate & Amplitude
  - Absorption Loss
  - Skip Distance
  - Instantaneous Channel Bandwidth
- VARIABILITIES OF THE GROUNDWAVE PATH
  - Ground Constants
  - Tropospheric Refractivity
  - Terrain and Obscuration Effects
  - Sea State and Salinity

NOISE ENVIRONMENT

- VARIATIONS IN AMBIENT NOISE
  - Galactic
  - Atmospheric
  - Man-Made
- VARIABILITY IN SPECTRUM OCCUPANCY

THE ECM ENVIRONMENT (Threat)

- JAMMING
- SIGNAL INTERCEPT

The PATHOLOGICAL ENVIRONMENT [Relates to Propagation]

- NUCLEAR EFFECTS VARIABILITIES
  - Absorption Blanket
    - Beta Patch
    - Gamma Ray Shine
  - MOF Diminution
  - Bomb Modes
  - Gravity Wave Effects (Large Scale TID)
- NATURALLY DISTURBED MEDIA
  - Region-Specific (Auroral, Equatorial)
  - Sudden Ionospheric Disturbances (SWF)
- GEOMAGNETIC STORM-RELATED VARIABILITIES

THE BENIGN ENVIRONMENT [Relates to Propagation]

- NORMALLY-OCCURRING IONOSPHERIC VARIATIONS
  - Small and Medium scale TID
- DEPARTURES FROM MEDIAN (Model) BEHAVIOR

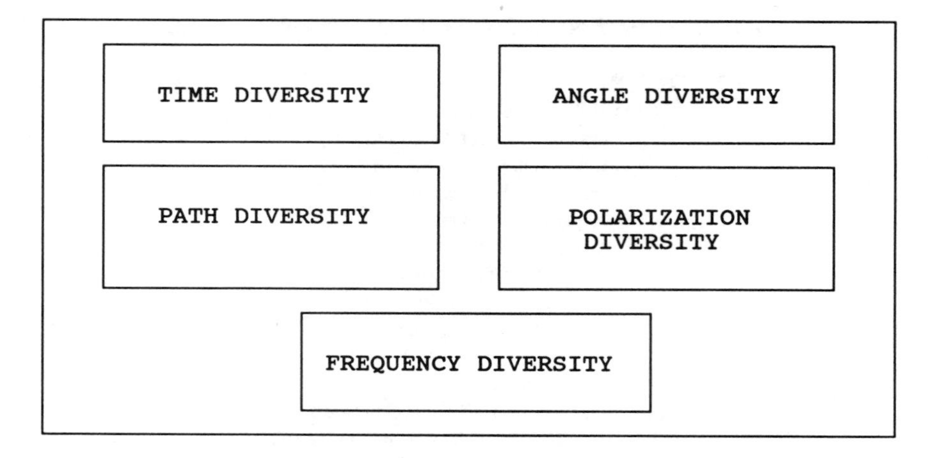

Fig.7-5. Types of diversity employed to mitigate HF channel disturbances.

Angle diversity is exploited through use of a directive antenna, whereas space diversity is exploited by use of spaced antennas. In both instances the generic designation might be simply *path diversity.* Under the same umbrella is *station diversity,* a form of path diversity which is exploited in relay-capable systems.

7.5.4 An Encompassing Definition

If we analyze the common features of Adaptive HF communication systems we find that they resemble those found in a military C⁴I system. The reader is reminded that C⁴I is an acronym standing for: Command, Control, Communication, Intelligence, and Computers. Since communication is one of the *C*s in C⁴I, it may seem odd that Adaptive HF itself tends to mimic the road map for C⁴I. Consider the following.

A commander exercises *command and control* of his forces on the basis of military *intelligence* information (about friend and foe alike), and it is said that *communication* is the glue which holds it all together. The fourth *C* in C⁴I stands for *computers* which recognizes the importance of modern computational decision aids in the entire process. How is this analogous to *adaptive HF* methodologies?

An *adaptive HF* system provides *communication* capability based upon system *intelligence* (of the total communication environment), and *command and control* may be regarded as process by which adaptation is achieved. *Computers* and similar devices in the communication designer's toolkit bind the components of an adaptive system together.

The definition of adaptive HF given in Figure 7-6 is all-encompassing, albeit unwieldy.

```
A fully adaptive HF system typically op-
erates under microprocessor control in-
corporating automation for most operator
-intensive functions, and is capable,
through marriage of a variety of diver-
sity schemes with a set of system status
intelligence functions, of automatically
establishing and maintaining systemwide
operation in an adroit manner, immediate-
ly responsive to time-varying changes in
the total environment including:
system propagation factors, external
noise, channel occupancy, and the ECM
conditions.
```

Fig.7-6. A definition of adaptive high frequency communications. This was obtained from unpublished lecture notes for the George Washington University course GWU-1159, *New HF Communication Systems and Technology* conducted by the author between 1984 and 1991.

Adaptive HF systems generally do not possess (or require) a full complement of diversity schemes. The communication requirements and the projected operational environments will define the variabilities of interest and their relative magnitudes. Cost may ultimately limit the flexibility which is designed into a system. Furthermore the adaptivity attributes built into the system are more relaxed for a benign midlatitude environment than for a pathological transauroral situation or a nuclear-disturbed environment.

Robust HF system attributes reduce or even eliminate the necessity for a system to adapt to the changing environment. More will be said about this later.

Military standards have been promulgated to enable HF communication systems to operate together efficiently. The trend toward standards development, has provided vendors and procurement officials with a road map for achieving a common level of interoperability. This "stairway" approach developed by Harrison [1987] is discussed in both MIL-STD-188-141A and FED-STD-1045. Within 188-141A, the features of *adaptive HF* are described and adaptive communications is defined (see Figure 7-7 and Table 7-3).

```
Adaptive HF describes any HF communica-
tions system that has the ability to
sense its communications environment,
and, if required, to automatically ad-
just operations to improve communica-
tions performance.
```

Fig.7-7. Definition of *adaptive HF* communications in the U.S. MIL-STD-188-141A, *Interoperability and Performance Standards for Medium and High Frequency Radio Equipment.*

TABLE 7-3. ESSENTIAL ADAPTIVE FEATURES[71]

1. Channel (frequency) scanning capability.
2. Automatic Link Establishment (ALE) using an imbedded selective calling and an option to inhibit responses shall be included.
3. Automatic sounding (station-identifiable transmissions). A capability to disable sounding and an option to inhibit responses shall be included.
4. Limited Link Quality Analysis (LQA) for assisting the ALE function:
   a. Relative data error assessment
   b. Relative SINAD[72] (optional)
   c. Multipath/distortion assessment

Recently Smith and Goodwin [1989] defined adaptive HF as "... one which automatically carries out the functions of establishing the communication link(s) and exchanging the message(s) in an optimum manner despite the variations in propagation conditions and the high probability of interference inherent in the HF band."

7.6 TECHNOLOGY: FOUNDATION OF ADVANCED/ADAPTIVE HF

7.6.1 Documentation

During the decade of the Eighties and somewhat before, there has been ample documentation of selected aspects of emerging and state-of-the-art technologies in connection with communication systems. With respect to HF communication systems, the reader is referred to proceedings from a series of topical symposia sponsored by the Institution of Electrical Engineers [IEE-UK, 1979, 1982, 1985, 1988, 1991]. Papers of interest are also found in published records of conferences dedicated to a broader range of communication systems and propagation/media effects on these systems [IEE, 1987]. It should also be noted that the Advisory Group for Aerospace Research and Development (AGARD) of NATO is an excellent source of new data on HF systems and how they are being utilized in the framework of the military environment. Of interest also are certain AGARD Lecture Series publications, and AGARD-LS-145 [AGARD, 1986] gives the reader new information on the propagation impact on modern HF communication systems design. Numerous government reports address state-of-the-art technologies having application at HF as well as research and development activities which

---

71. From military standard MIL-STD-188-141A, 15 September 1988.

72. SINAD corresponds to the signal plus noise plus distortion to noise plus distortion ratio.

promote and exploit emerging technologies. A book by Maslin [1987] provides an overview of the impact of modern technology on HF systems. Information may also be gleaned from various topical conference proceedings and workshops. General developments in HF equipment and systems have been addressed by Wilson [1986], and technological trends in receiver design have been discussed by Rhode [1985]. Under contract to the U.S. Defense Nuclear Agency (DNA), workers from SRI International have documented an adaptive HF radio design [Chang et al.,1980; Ames et al.,1984]. The work, although strongly tied to a specific hardware choice, contains general information for those involved in early stages of adaptive system design.

7.6.2 Technology Areas

A major ingredient in technological enhancement is the infusion of advanced microprocessing equipment, allowing systems to be controlled more effectively and in some cases without significant operator intervention. Major system components such as transmitters, receivers antennas, couplers, processors, and clocks have also been improved over the years; and device technology has led the way. New device and component technologies have enabled system engineers to implement successful robust and/or adaptive strategies for link, system-wide and network applications. New system concepts for ALE, ALM, LPI, and mixed media applications have been promoted within the military. Wideband signaling and spread spectrum approaches are also being examined (see Sections 7.6.6 and 7.6.8).

The major technological areas are indicated in Table 7-4. As may be seen, a number of the technological improvements which relate to function are associated with the *Adaptive HF* attributes discussed earlier in the chapter.

Clearly, all three areas of technology are inextricably related. Device technology is used in the design of system components, and the manner in which components are configured in a system will define the functional capabilities of the system. The growth in device technology has clearly been the *engine* which has driven improvements in HF system components as well as functional capabilities.

Microprocessors are becoming integral components of radios completely modifying the architectures of radio subsystems responsible for the basic processes of transmission and reception. We shall not discuss new devices directly, leaving that task to the various trade journals which are readily available. Instead, we will restrict our attention to system component technologies and functional capabilities with emphasis on the latter. In keeping with the tone of this book, it is proper that we emphasize the functional aspects, which are requirements-driven, over the system components and devices, which are basic building blocks. There exists a considerable body of engineering data dedicated to this topic.

TABLE 7-4. MAJOR ELEMENTS OF TECHNOLOGICAL ADVANCE

NEW DEVICES

- Very-Large-Scale-Integration (VLSI)
- Charge-Coupled-Devices (CCD)
- Very-High-Speed-Integrated-Circuits (VHSIC)
- Surface-Acoustic-Wave (SAW)
- Bubble Memory

COMPONENTS AFFECTED

- Receivers
- Transceivers
- Antennas
- Antenna Couplers
- Power Amplifiers
- Synthesizers

FUNCTIONAL CAPABILITIES
[System Technologies]

- Digital Signal Processing (DSP)
- Speech Synthesis
- Diversity Reception
- Multipath Compensation
- Frequency Agility
- Spread Spectrum (Chirp, DS, and FH)
- Wideband Signalling
- Interference Excision & Reduction
- Steerable Null Array Processors (SNAP)
- Power Control
- Low-Probability-of-Intercept (LPI)
- Imbedded RTCE
- Automatic-Link-Establishment (ALE)
- Automatic-Link-Maintenance (ALM)
- Network Management & Control
- Real-Time Frequency Management
- Expert System and AI applications

7.6.3 System Components

The modern HF radio has achieved a degree of reliability once thought impossible. This improved level of performance has generally been attributed to the influence of digital instruments and techniques; viz., the availability of smaller and more efficient components, the incorporation of *smart* transceivers with imbedded RTCE to adapt to channel behavior, and the utilization of robust waveforms and signal processing to counteract residual impairments.

7.6.3.1 *Power Amplifiers*

The inventory of military HF radios has been dominated by vacuum tube power amplifiers (PAs). These systems are tuned to suppress nonlinear effects and undesirable intermodulation products (IMP), but the process is electromechanically intensive and leads to rather modest retuning speeds. This technology is unreliable and clearly unsuitable for current or future FH-SS applications.

The architectural trends in power amplification have been greatly influenced by the development of RF power transistors. These new systems may be designed to be broadband throughout all stages, making them highly suitable for functions requiring frequency agility.

7.6.3.2 *Receivers and Transceivers*

Until recently, receivers have been characterized by analog devices such as crystal filters, narrowband RF amplifiers and local oscillators, and the operat ion has been manpower-intensive. Operators have been required to monitor preselected channels for noise and traffic, select appropriate frequencies for communication, tune the receiver, check on system performance, and participate in other *overhead* activities to enable communication to be established and maintained over a link or set of links. This reliance on the human operator has often led to poorer communication performance than would be suggested by the channel capacity under specified conditions if system parameters were to be optimized. Digital receiver designs provide a capability to reduce, and in some instances eliminate, human operators from the loop. In short, human operators, previously serving as inefficient proxies for legitimate receiver functions, are left free to make higher level decisions.

Newer digital receivers operate under microprocessor control and exhibit frequency agility. This property enables the radio receiver to shift frequency rapidly and precisely in accordance with an established or computed frequency-time plan. There are some practical difficulties which arise when reception and transmission are simultaneous and the receiver and transmitter units are colocated [Maslin, 1987]. The problem may be ameliorated by supplying sufficient front-end protection for the receiver.

Digital synthesizers are needed to achieve the necessary switching speeds required by modern HF radios. These components are programmable and can be used to mimic the local oscillator (LO) functions of receivers and exciters integral to FH-SS systems. The performance of digital synthesizers may be characterized by the spectral properties (i.e.,purity of synthesized signals) and by the speed with which the unit may be tuned. The tuning speed is dictated by the *settling time* of the synthesizer, and this property clearly determines the maximum hop rate for a FH-SS application or for scanning over preselected channels in the HF band. The spectral *purity* is a factor in receiver *selectivity,* since reciprocal mixing of LO noise sidebands with a large interferer may introduce a competitive *spike* at the receiver intermediate frequency (IF).

The forms of synthesizers produced may be organized into two fundamental categories: indirect and direct. Indirect synthesizers utilize phase-lock loops while the direct synthesizers make use of mixing and filtering schemes. As a consequence the indirect scheme is relatively sluggish, being characterized by a *settling time* of the order of a millisecond. The direct scheme is fast, having a *settling time* of the order of microseconds. Early on, the direct method led to systems which exhibited poor noise floor performance but this has been partially ameliorated in current units. Direct synthesizers are more costly than their counterparts. Research and development is leading to improvement in synthesizer design, but more work still remains to be done. Improved processing is being achieved through incorporation of faster integrated circuits (ICs), and size is being reduced by the use of surface acoustic wave (SAW) technology.

Filters are important components in HF receivers, and the digital variety is generally preferred to the analog type on the basis of both performance and flexibility. This is especially true in the intermediate frequency (IF) and audio frequency sections [Rhode, 1985]. Chebyshev and Bessel filter representations may be implemented, but since the process is under software control, any number of filter types and performance characteristics may be selected.

Aside from agility, the modern receiver must accommodate advanced DSP[73] capability. This DSP capability may allow for processing gain through exploitation of deliberate redundancy in the waveform structure (see Section 7.6.8).

The design of transceivers[74], like receivers, have been strongly dictated by agility requirements, automatic operation, adaptivity and a desire to reduce operator-intensive operations. Many of the previous comments in this section also apply to transceivers. They are typically used in mobile applications, in which compactness would be desirable.

Modern HF transceivers must have the ability to change frequency rapidly in order to compensate for channel variability, at least that part which is directly related to the frequency setting. Thus frequency agility is a necessary attribute, especially in a nuclear environment but it may not be sufficient. The ECM environment may dictate that additional measures be invoked. For example, mitigation of jamming may require that frequency be changed more

---

73. Digital Signal Processing. The term DSP is used to describe a large class of advanced digital methods.

74. A transceiver is a transmitter and receiver combined in a common housing. A transceiver is comprised of common circuit components for both radio transmission and reception. Simplex operation is employed.

rapidly than the variability in the propagation medium would suggest. A technique generally referred to as frequency hopping may be invoked. Transceivers must be able to hop at rates of more than a few hundred hops/second in most scenarios.

7.6.3.3 *Antennas and Couplers*

One of the most important considerations in the setup of a communication link is the specification of the type of antenna to be used. Treatment of antenna theory is beyond the scope of this book, but practical information may be found in various handbooks. Sections 4.4.5 and 4.4.6 suggest additional references. As we have seen, it is difficult to always separate antennas from propagation. Modes which might normally propagate with great efficiency between two terminals could be inhibited if low antenna gain (or a null) is associated with the associated ray launch angle. For the quasi-static environments associated with ground segments employed in ground-to-ground, ground-to-air, and shore-to-ship communications, antenna selection should not represent a formidable problem. However, this does not suggest that the job is always trivial.

Most antennas deployed prior to the 1980's were narrowband in nature, restricting the flexibility in operational scenarios, although log-periodic antennas were available at a number of installations. Multifrequency requirements have led to the introduction of broadband solutions, although the procurement tendencies have included acquisition of multiple narrowband antennas, a factor leading to growth in the size of *antenna farms*. Broadband replacements should assist in elimination of the real estate requirements. There is a clear requirement for relatively small multifrequency antennas for tactical field use as well as for installation of such antennas on rooftops (i.e.,diplomatic communications from embassies) and other small platforms. One solution has been suggested whereby multiple frequencies could be employed with the 4-element conical logarithmic spiral antenna structure (Spira-cone™) without tuning or frequency restriction [Smith and Werner, 1988]. The Spira-cone antenna may be used for transmission and reception, and offers Faraday fade protection since it is a mixed polarization device.

Some care must be taken in the selection and installation of antennas used in tactical situations, and the specification of antennas to be installed on small mobile platforms, helicopters, and fixed wing aircraft present special challenges. Maslin [1987] discusses HF antennas with special emphasis on airborne platforms, and Christinsin [1990, 1991] has considered the type of antenna which should be used under NVIS propagation conditions. Recently, experimental measurements of a helicopter-borne NVIS antenna have been made by Barr et al.[1991] in support of a USMC requirement. Using a number of approaches, including Numerical Electromagnetic Code (NEC) modeling, scale model tests, as well as real airframe measurements, Cox and

Vongas [1991] have examined the consequences of electrically short antennas mounted on helicopters. The issue of electrically small antennas is but one application of the use of modeling prior to *cutting metal*. It is anticipated that computer modeling will play a major role in the design of antennas for use in novel platform environments. See Section 4.6 for more information on NVIS propagation.

Generally one must consider transmitter and receiver antennas separately since the processes suggest different configuration requirements. A transmit antenna is typically larger than a receiver antenna. However, it is possible to use relatively small antennas possessing a modest power-gain product rather effectively, although not efficiently, with the proviso that the overall HF system can compensate for the residual inefficiency. This is accomplished by adaptivity and various DSP measures to achieve processing gain over background noise and interference.

Broadband antennas have a large number of applications, including use in frequency-hopping spread-spectrum (FH-SS) applications. Receive antennas can be made broadband but transmit antennas are typically narrowband. As a consequence, antenna couplers are used to tune transmitter antenna elements. However, tuning speeds are limited if mechanical switches (i.e., relays) are used. This implies the exploitation of alternative solid state devices to increase the switching speeds to accommodate fast frequency hopping applications. Although solid state devices (i.e., PIN diodes[75]) are limited in the level of transmitter power which may be accommodated, being typically far less than 1 kilowatt, modern HF systems do not require enormous levels. Even so, an automatic high power antenna coupler, rated at up to 1 kw and having a range of tuning speeds between roughly 10 and 400 milliseconds, has been developed [Conticello, 1984].

Cross section reduction and mission disguise are needs which have led to the development of conformal antenna concepts. Concealment and hardening requirements for some military applications has led to the development and testing of buried and ground contact antennas. The effectiveness of such antennas is dependent upon the electrical properties of the ground within which the antenna elements are buried or upon which they are placed. Some packages may be rolled out quickly and are referred to as *rapid deployment antennas* (RDAs). RDA systems may be provided as packages which are lightweight, broadband, and omnidirectional in the NVIS mode. These factors make them ideal for a number of tactical HF requirements [Whigham, 1989].

---

75. A PIN (Positive-Intrinsic-Negative) diode converts light energy to current. It is a depletion layer photodiode which operates in essentially the reverse fashion as an LED (Light Emitting Diode). Light entering the device is absorbed by electrons in the valence band of the semiconductor material generating electron-hole pairs. The resulting electric field generates an electric field in a load circuit.

Small antennas will always be needed in tactical military operations. To achieve compactness as well as broadband capability, it is necessary to incorporate active elements in the antenna configuration. Active antennas have been shown to be effective if designed properly to reduce the impact of large interfering signals. This implies the incorporation of active components which are quite linear thereby reducing the amplitude of intermodulation distortion products.

After frequency selection, the most common difficulty encountered in HF communication is antenna selection and deployment. In operations, such as Desert Shield in 1990-91 and Desert Storm in 1991, the inadequacy of training often led to the selection of frequencies which had no possibility of supplying skywave support and the use of field expedient antennas, such as whips, which were clearly inappropriate for the job at hand (i.e., NVIS connectivity). Training of field communicators is clearly necessary to optimize HF system operation. This requirement will be relaxed as ALE radios become part of the operational inventory.

7.6.4 System Technology: Modulation Techniques

The choice of modulation format is critical at HF because of the precarious nature of the propagation channel. For a specified modulation, the performance of the system increases monotonically with signal-to-noise ratio (SNR), but not without limit. Considering digital modulation, one finds that the bit error rate (BER) initially decreases rather sharply with SNR, but a BER floor is ultimately reached irrespective of further increases in transmitter power (and SNR). The first regime is dictated by noise limitation and the latter regime is dictated by propagation (i.e., multipath) limitations. Both forms of limitation may be ameliorated somewhat by the institution of diversity measures. However the degree of performance improvement is dependent upon the robustness of the modulation format employed.

The subject of modulation is an extensive one. Readers may find discussion of the various signaling formats in a book by Tomasi [1987] as well as articles which appear in topical conferences and trade journals. There are two basic signal format options available: analog and digital. Analog signaling is a natural format for voice, it is rather simple to produce, and waveform processing is not difficult. On the other hand, security is not easy to implement, and propagation disturbances may be insurmountable. Examples of analog formats include: single sideband (SSB) and double sideband amplitude modulation (AM). A defacto baseline standard for HF radios compatible with FED-STD-1045 is SSB capability (see Section 7.6.5).

Vintage HF radios, lacking technological improvements developed in the 1980s and early 1990s, are still in use, especially in the military. The modulations employed by these systems is likely to be one of the following: low rate FSK, SSB voice, amplitude modulation (AM), and continuous wave (CW).

At HF, the primary digital schemes are FSK, MFSK, and DPSK [Maslin, 1987], although other formats are used in special cases. Digital methods are preferred since they are amenable to security protection measures, and they perform with adequacy in multipath environments. Digital formatting is not as simple as analog and the processing is not always trivial. Still the advantages outweigh the disadvantages.

FSK and its variants are very popular modulation techniques since they are robust in the face of propagation disturbances. Binary FSK employs two frequencies, or tones, to convey a *mark* (a "one") or a *space* (a "zero). Conventionally the detection process is noncoherent. The minimum frequency separation of binary FSK *mark* and *space* tones is roughly twice the input bit rate. It is interesting to note that some diversity protection against multipath fading is achieved through use of binary FSK signaling. For a multipath delay δt, a tone frequency separation $\delta f = 0.5\ 1/(\delta t)$ will cause a selective fade appearing on one tone of the tone-pair to produce an enhancement on the other. Thus a multipath separation of 4 ms may be accommodated in this way if a tone spacing of 125 Hz is used. Naturally, an immunity to fading caused by significant microscale multipath components would require extremely large tone spacings to insure that at least one of the tones is available for detection. Operation near the MOF should increase the probability that multipath spread will be limited. In this region, where δt may be less than 150 μS, fading will be correlated over an entire 3 kHz channel. Irrespective of the achievement of multipath-induced fade protection in the present context, or the lack of it, the engineer is still faced with a multipath-induced symbol rate limitation. An FSK format for transmission of data at 75 bps will normally employ a rather narrow shift of 85 Hz. For aircraft applications this is too small; a shift of 850 Hz may be used in this instance.

The most serious propagation-related challenge to digital signaling is multipath which may introduce ISI if the baud rate is sufficiently high. Avoidance of ISI can be achieved in a number of ways. One method is to provide a guard time between symbols equal to the largest multipath spread. This places a limit on the maximum symbol transmission rate of the order of 200 symbols/sec under the presumption of a spread of 4 ms. However, the input data (bit) rate may be higher than this. In this connection, M-ary signaling, a typical multistate system which accommodates higher input rates, may be utilized quite effectively. If 8-ary FSK, defined by 8 tones or 3 bits/symbol is employed, then the data rate would be 600 bps. Higher data rates may require the use of multiple carriers. This suggests the development of digital modems with a parallel signaling capability (see Section 7.6.8). It is noteworthy that M-ary signaling involves the transmission of only one tone at a time; parallel tone signaling involves simultaneous transmission. The latter technique is used for transmission of high data rates, whereas the former is used for low to modest data rate transmissions. A good example of a multiple tone system is the U.S. Navy ANDVT [JTCO, 1981] which uses a 16 tone format at 2400 bits/sec for non-secure purposes. However the ANDVT uses a QPSK modulation scheme.

PSK systems require phase tracking and thus may be more vulnerable to pathological ionospheric conditions. DPSK has an intrinsic ability to adjust to changes since it uses the received phase of the previous element as a phase reference. The performance of PSK systems is generally better than FSK, although conditions of noise, interference, and signal fading properties must be considered. The Kineplex system, a 4-phase 20-tone DPSK system, will support the transmission of data at 2400 bps but does not perform well in a frequency hopping mode (see Section 7.6.8.8 and Table 7-11). This is principally because of phase noise introduced at each hop interval.

Some other modulation formats of interest include binary FM, otherwise known as continuous phase frequency shift keying (CPFSK), and minimum shift keying (MSK). A brief discussion of these methods has been given by Maslin [1987]. Other methods include frequency exchange keying (FEK), and coherent frequency exchange keying (CFEK) . A CFEK low data rate modem ,integrated into a 3 watt manpack transceiver, has been successfully tested at high latitudes. Multipath conditions appeared to provide a degree of diversity gain [Atkinson, 1990].

In most modulation formats, detection may be either coherent or noncoherent. Coherent systems generally provide better performance but are more difficult to implement. Differential PSK (or DPSK), for example, has a 3 dB poorer performance than coherently-detected binary PSK (CBPSK) in a fading environment because errors in the former come in pairs. Still, because of the self-*adaptive* tracking capability intrinsic to differential methods, DPSK has an advantage if the propagation channel exhibits significant channel variabilities[76].

There have been numerous systems developed to combat HF channel disturbances with varying degrees of success. Refer to Section 7.6.8 for an identification of specific system implementations of modulation formats. The U.S. Navy has developed a digital modem for those engineers that don't like to make decisions prior to testing, one which permits the implementation of a variety of modulation schemes under software control [NAVELEX, 1979].

---

76. Since DPSK demodulation does not involve the extraction of a carrier phase reference in order to make a decision, it is relatively simple to implement and it is not subject errors associated with acquisition and maintenance of absolute phase reference information. Time variations in received phase which are slower that the twice the symbol period are automatically countered since decisions are based upon a phase comparison between adjacent symbols. If phase jitter is pronounced, a more robust modulation scheme such as NCFSK is required.

7.6.5 System Technology: Functional Capabilities and ALE

In this section we initiate our discussion of functional capabilities which help HF systems operate in pathological environments and under high levels of interference. We stress ALE, which incorporates many of the other capabilities such as imbedded RTCE, polling, selective scanning, automatic handshaking, networking, store and forward, and network management.

It is well known that point-to-point HF radio links are sometimes difficult to establish. A major factor is the selection of an optimum frequency for communication, a daunting problem which may exacerbated by the lack of channel information, frequency flexibility, and related factors. Because of this, operators of conventional systems have a tendency to remain on inefficient channels far too long, possibly taking other unsophisticated measures to compensate for poor system performance (e.g., increasing transmitter power). The link establishment problem is challenging one, and its solution should lead to improvement in spectrum utilization efficiency and a reduction in false starts (i.e., unsuccessful handshakes) which clutter the spectrum with unproductive signals. For many years it has been felt that automation of the link establishment function would be the only way to solve this operational problem.

Table 7-5 provides the standard levels of automated HF systems as defined by Harrison [1985, 1987]. Figure 7-8 depicts the *stairway approach*. As previously mentioned, these HF-ALE concepts have been largely assimilated within various government standards[77,78]. It has been said that the military standard (i.e., MIL-STD-188-141A) is the *engine* which drove the development of the federal standard (i.e., FED-STD-1045). ALE radios have been developed by several vendors and tests have been conducted for compliance with FED-STD-1045. Additional information about current and evolving standards is found in Section 7.9.

Table 7-6 lists the attributes at each level of automation for ALE radios. Harrison [1987] has specified a number of operational rules which are consistent with level attributes. The operational rules appear in Table 7-7, and are cross-referenced with level attributes given in Table 7-6.

----

77. MIL-STD-188-141A: Military Standard, *Interoperability and Performance Standards for Medium and High Frequency Radio Equipment* dtd 15 September 1988, available from Naval Publications and Forms Center, NPODS, 5801 Tabor Avenue, Philadelphia, PA 19120-5099.

78. FED-STD 1045: proposed Federal Standard, *Automatic Link Establishment*, dated 24 January 1990., prepared by ITS, NTIA, Dept. of Commerce for the Office of Technology and Standards, National Communications System.

TABLE 7-5. LEVELS OF HF SYSTEM AUTOMATION

LEVEL 0: Noninteroperable Radio
LEVEL 1: Baseline System
LEVEL 2: Selective Calling and Handshake
LEVEL 3: Scanning
LEVEL 4: Sounding
LEVEL 5: Polling
LEVEL 6: Connectivity Exchange
LEVEL 7: Link Quality Analysis and Channel Selection
LEVEL 8: Automatic Message Exchange
LEVEL 9: Message Store and Forward
LEVEL 10: Network Coordination and Management

7.6.6 *System Technology: Spread Spectrum*

It may seem surprising that spread spectrum technology may be applied within the HF band given the bandwidth limitations which obtain as a result of frequency-dependent distortion. However the basic virtues of spread spectrum still apply. Pickholtz et al.[1982] have examined the origins of the subject, and a tutorial on spread spectrum theory has been published [Scholtz, 1982]. General insight is provided by Dixon [1976] who indicates that three types of modulations are employed in spread spectrum applications: *direct sequence* (DS), *frequency hopping* (FH), and *chirp* or *pulsed-FM*.

Applications of spread spectrum technology at HF have included schemes for avoidance of impairments introduced by multipath such as selective fading and intersymbol interference. A more obvious application is derived from the property of spread spectrum systems to deliver *processing gain* against unwanted background noise and deliberate interference (i.e., jamming). Various CDMA[79] schemes have also been developed. Other applications of spread spectrum include: message screening, signal hiding, and ranging. To understand the manner in which these applications are attacked requires an appreciation for the type of system involved: DS-SS, FH-SS, or chirp.

Even though spread spectrum would appear to offer many benefits at HF, there are negative aspects as well. In some instances they may appear to be bureaucratic in nature, but there are some legitimate technical matters to consider. Spread spectrum signaling bandwidths may need to be limited as the result of special frequency allocation constraints. More fundamentally, the ionosphere may not support large instantaneous bandwidths characteristic of direct sequence systems (see Section 7.6.6.1).

---

79. Code Division Multiple Access. Powerful correlation and code recognition schemes permit the coexistence of multiple codes at the same time. Alternative access schemes include Time Division Multiple Access, TDMA, and Frequency Division Multiple Access, FDMA.

TABLE 7-6. ATTRIBUTES & COMPONENTS OF AUTOMATIC RADIOS

LEVEL 1: Baseline System

1. Experienced Operator
2. Directory
3. Standard Procedures
4. *HF Single Sideband Radio*

LEVEL 2: Selective Call & Handshake
[Operational Rules: 1-6]
1. Baseline System Attributes
2. Digital Signaling
3. Digital Addressing
4. Calling and Handshaking
5. *Digital Modem*
6. *Digital Controller*

LEVEL 3: Scanning
[Operational Rules: 1-7]
1. Level 2 System Attributes
2. Continuously Scanning RCVR
3. Scan-capable XMTR

LEVEL 4: Sounding
[Operational Rules: 1-8]
1. Level 3 System Attributes
2. Pulse Sounding
3. Passive Sound Reception (EMCON)

LEVEL 5: Polling
[Operational Rules: 1-9]
1. Level 4 Attributes
2. Connectivity Data Acquisition
3. Connectivity Data Storage
4. Direct Connectivity Data

LEVEL 6: Connectivity Exchange
[Operational Rules: 1-10]
1. Level 5 Attributes
2. Expanded Memory Matrix
3. Uni/bilateral Connectivity Exchange
4. Group Call Connectivity Exchange
5. Connectivity Transfer by Sounding
6. Indirect Connectivity Data

LEVEL 7: LQA/Channel Selection
[Operational Rules: 1-12]
1. Level 6 Attributes
2. Link Quality Determination
3. Exchange of Link Quality Data
4. Automatic PWR Reduction

LEVEL 8: Automatic Msg. Exchange
[Operational Rules: 1-12]
1. Level 7 Attributes
2. Point-to-Point Message Exchange
3. Orderwire and/or Operator Messages
4. Message Store & Delayed Exchange
5. Channel Cond. Message Exchange

LEVEL 9: Msg. Store & Forward
[Operational Rules: 1-13]
1. Level 8 Attributes
2. End-to-End Msg.Exchange

LEVEL 10: Network Cood.& Mgmt
[Operational Rules: 1-13]
1. Level 9 Attributes
2. Mgmt. of Station Message Loading
3. Net level Propagation Compensation
4. Optimization of Net Reliability

TABLE 7-7. OPERATIONAL RULES FOR HF-ALE RADIOS

1. HF-ALE receive capability is independent and parallel.
2. HF-ALE radios always listen for ALE calls.
3. HF-ALE radios always automatically respond to calls (unless thy are inhibited).
4. HF-ALE radios will not interfere with active channels (exclusive of priority interrupts)
5. HF-ALE radios spend minimum time on channels.
6. HF-ALE radios employ highest levels of mutual capability.
7. HF-ALE radios always scan (unless otherwise occupied for other purposes).
8. HF-ALE stations always sound periodically (unless inhibited).
9. HF-ALE stations always seek (unless inhibited) and maintain track of connectivities.
10. HF-ALE stations always exchange direct connectivities (unless inhibited), and also maintain track of indirect connectivities with other stations.
11. HF-ALE stations always exchange (unless inhibited) and measure the received signal quality of other stations.
12. HF-ALE stations automatically adjust transmitter power levels to the minimum essential level (under presumption that such capability exists).
13. HF-ALE stations always automatically relay messages to other stations (unless inhibited).

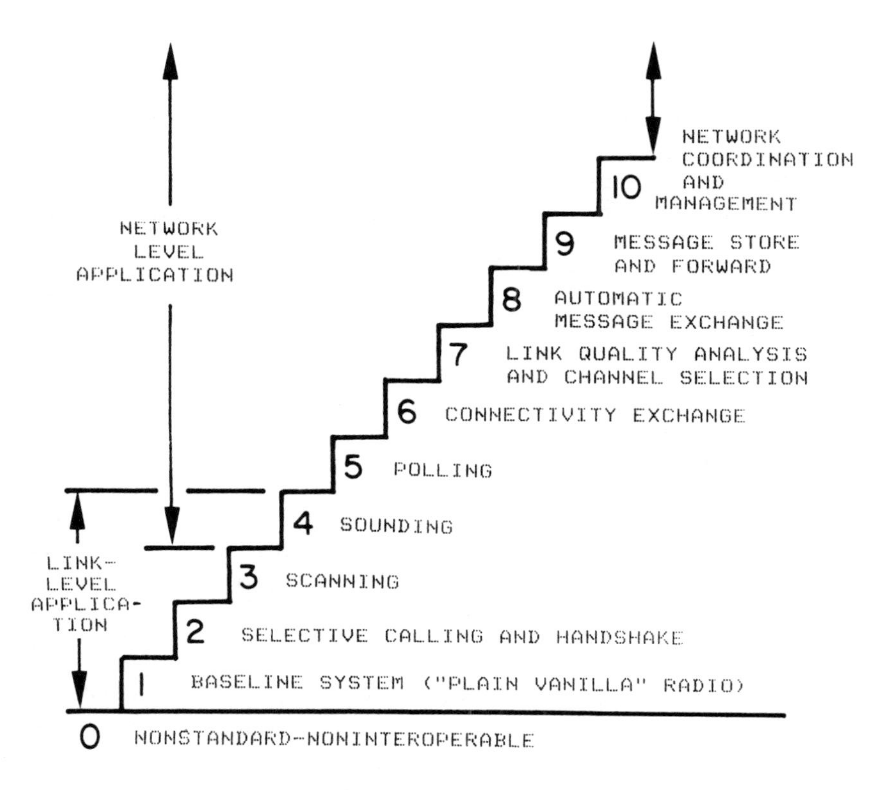

Fig.7-8. Depiction of the levels of HF system adaptivity. (From Harrison [1987].)

Recently Yeh and Liu [1988] have studied ionospheric effects upon HF spread spectrum systems with emphasis upon systems possessing an expanded bandwidth which is *instantaneous*. A comprehensive report by these authors, as appended with a PhD Thesis by Wagen [1988] and a paper on ionospheric dispersion [Yeh and Dong, 1988], provides considerable insight and is highly recommended for those interested in the more detailed propagation aspects. An outline of various spread spectrum modulation methods and the effect of the ionosphere has published by Milsom and Slator [1982].

Spread spectrum signaling has led to more interest in wideband model development (see Section 4.14). More careful examination of the distortion effects introduced by dispersion in the background ionosphere and scattering from ionospheric irregularities has become necessary. The effects are characterized by the dispersion bandwidth f_d and coherence bandwidth f_c respectively (see Section 7.6.8).

7.6.6.1 *Direct Sequence Spread Spectrum: DS-SS*

This process is one in which the bandwidth of the information baseband signal (i.e., *information rate R*) is deliberately expanded (i.e., *spread*) by the application of a digital code sequence whose bit rate (i.e., *RF bandwidth* B_{RF}) is larger than the information rate. The term *Direct Sequence Spread Spectrum* (DS-SS) is used for spread spectrum systems which employ this method of broadbanding. The usual method for implementing this method is through application of a pseudonoise (PN) sequence with a high chip rate. Occasionally DS-SS systems are referred to as *pseudonoise* systems. Also, to distinguish DS-SS from other means for bandspreading, the term *direct spread* is sometimes used. Generally speaking, the DS-SS bandwidth B_{RF} (or B_{SS}) is taken to be twice the code clock rate [Dixon, 1976].

The processing gain for a DS-SS system is simply the ratio B_{RF}/R where R is the information rate. For an information rate R of 1000 bps and a clock rate of 500 kHz (implying $B_{RF} = 1$ MHz), we have a processing gain of 1000 or 30 dB. The central question is whether or not this processing gain can be achieved over ionospheric skywave paths in other than exceptional conditions.

Milsom and Slator [1982] have derived a useful formula for the maximum chip rate R_c (which they refer to as C_r), the square of which is inversely proportional to the group time delay slope, $d\tau/df$.

$$R_c = [\, a / (d\tau / df) \,]^{\frac{1}{2}} \qquad (7.1)$$

where a is a constant which for a 3 dB loss is equal to 1.5 [Sunde, 1961].
It is noted that $(d\tau/df)^{-\frac{1}{2}} = B_c$, the channel bandwidth (see Section 4.14.1 and Equation 4.58). Taking $d\tau/df = 100$ microseconds/MHz, we may compute a maximum chip rate ,under the aforementioned 3-dB loss condition, of roughly 120 kilochips/sec. Clearly higher chips rates are advantageous in order to develop the greatest capability to resolve individual multipath

components and to achieve implicit diversity. At the same time Equation 7.1 suggests that there is a limit to the chip rate owing to dispersion from the background ionosphere. Ironically, this limitation is most severe near the MOF, the frequency region near which conventional narrowband systems are most efficient since multipath is minimized and background noise is reduced.

An increase of the chip rate is also an advantage in amelioration of the impact of polarization effects. Recall from Section 4.5.1, we may define a polarization bandwidth B_F arising from Faraday rotation. Increasing the chip rate will enable the O and X modes to be more easily resolved. A smaller value of polarization bandwidth is associated with more Faraday fading of unresolved modes but a greater time delay separation at a specified carrier frequency. Bandwidth expansion may thus compensate for adverse effects of periodic Faraday fading with sufficiently high chip rates.

7.6.6.2 *Frequency Hopping Spread Spectrum: FH-SS*

This process is one in which the RF carrier frequency is programmed to jump from place to place in discrete steps (i.e., a specified pattern). This pattern of jumps is defined by a code sequence, typically a PN code, which enables transmission to be performed in accordance with a specified frequency-time plan. Frequency hopping achieves the bandspreading advantage against ECM threats but without the dispersion problems which might obtain if an instantaneous bandwidth were to be developed and transmitted through the channel. The frequency hopping process is a generalization of *plain vanilla* FSK signaling. However, the frequency set used in frequency hopping is quite large rather than only two as in binary FSK. Frequency hopping spread spectrum systems (FH-SS), or *hoppers* are generally the preferred method for bandspreading in military systems.

The bandwidth of a FH-SS system is $B_{RF} = m\ B_{channel}$ where m is the number of unique frequency channels which are in the hop set. Clearly the processing gain for a FH-SS system is m, the number of frequencies. In order to achieve a processing gain of 30 dB we must hop over 1000 frequencies. If the information rate is 1 kilobits/sec (as in Section 7.6.6.1), this implies 1 hop per information bit. By hopping faster than the information rate it is possible to build redundancy into the system, allowing more than one opportunity for each symbol to be interpreted in the presence of selective fading or interference. Slow frequency hopping may be vulnerable to follower jammers. Fast hopping may counter this threat by *outrunning* the jammer. Fast frequency hopping does present some design challenges, but will provide predictable protection provided the geometrical situation is defined (i.e., jammer XMTR, friendly XMTR, and victim RCVR separations) [Darnell, 1986].

Because of group path delay variation with carrier frequency, random frequency hopping, driven by a PN sequence, will introduce random time delay fluctuations or time jitter. The range of this jitter is $B_{RF}\ (d\tau\ /df)$ where

$d\tau/df$ is the group path delay slope. Time jitter will introduce hop period variations following transmission through the channel. If μ is the maximum fractional hop period allowed, then the time jitter range should not exceed μ/R_{hop}, where R_{hop} is the hop rate. If the total occupied bandwidth B_{RF} (or $B_{SS} = m B_C$) can be established then we can determine the number of independent channels m assuming a nominal value of 3 kHz for B_C. Taking $\mu = 0.1$. $R_{hop} = 1000$ hops/sec, and $d\tau/df = 100$ microseconds/MHz, we find $B_{SS} = 1$ MHz. This implies that m, the number of frequencies in the hop set, may be as high as about 333 without overlap. More restrictive values for μ or faster hop rates will reduce m.

7.6.6.3 *Chirp Modulation*

Chirp modulatiion is a process in which the RF carrier is swept over a wide band during a specified period of time or pulse interval. The typical system involves the development of a linear FM chirp. The processing gain for a chirp waveform is called the compression ratio (the sweep time × frequency sweep).

A class of oblique-incidence sounders exploit the chirp waveform in order to suppress interference and provide cleaner ionograms for greater ease in interpretation[80]. A rather robust HF communication system has been developed by BR Communications through modulation of a linear FM chirp waveform and the application of redundancy[81]. Another modem employing the chirp technique has been implemented by University of Manchester scientists supported by the MoD in the United Kingdom [Atkinson, 1990].

7.6.7 System Technology: Dealing with Data Errors

The ionospheric channel is not ideal and one should anticipate that data transmission errors will arise. Errors may be introduced in digital systems as a result of noise conditions and signal fading conditions. Burst errors are introduced during impulsive events (i.e., noise spikes) and at the nulls of a signal fading pattern. Gaussian noise tends to introduce random errors. Also, multipath time dispersion will introduce symbol interpretation errors as a result of ISI. Error patterns may also be related to the modulation format used. Specif-

---

80. Chirpsounder is an ionospheric sounding system developed by BR Communications. The military system is the AN/TRQ-35 (see Chapter 6). In recent years a lighter version of the AN/TRQ-35 system has been nomenclatured by the U.S. military services.

81. Chirpcomm is a communication adapter exploiting a chirp waveform modem. It was developed by BR Communications. It has been used as a component of an HF frequency management terminal comprised of a chirpsounder sub-system. The Chirpcomm system may be used for transmitter identification.

ically, if some form of differential modulation is employed, we would expect errors to occur in pairs since adjacent symbols are related. The distribution of errors is important since it provides information vital in the specification of an error control strategy.

To deal with the data error problem, we must develop a strategy for detecting the errors and if necessary, correcting for them. Error detection involves data monitoring and provision for indicating *when* data errors arise. Methods for error detection include: parity checking, exact-count encoding, vertical and longitudinal redundancy checking, and cyclic redundancy checking (CRC). It is important to recognize that the primary goal of error detection is to prevent the occurrence of undetected errors, not correct them. It remains for the specific system to account for detected errors in some way. This might involve retransmission of the suspicious symbols or FEC[82]. Systems which combine the detection and correction functions are called EDAC[83] systems. Tomasi [1987] provides a description of various error detection and error correction approaches. Books dedicated to coding methods for error correction are available [Lin, 1970] [Clark and Cain, 1981], and Grossi [1986] has prepared a tutorial on the subject as part of a NATO-AGARD lecture series publication entitled *Propagation Impact on Modern HF Communication System Design.*

7.6.7.1 *The HF Channel and the Shannon Limit*

Shannon [1948, 1949] has suggested that information may be transmitted over a communication channel at a rate which is proportional to the product of the channel bandwidth and the base 2 logarithm of the signal-plus-noise to noise ratio. The capacity is given by:

$$C = B\log_2(1 + S/N) \tag{7.2}$$

where B is the bandwidth (Hz) and S/N is the signal-to-noise ratio. The Shannon result refers to *symbol rate* and not *transmission rate*, even though there is usually a straightforward relationship between the two. The implications of the Shannon formula has been examined by Bell [1986] and a series of articles by Walters [1989]. Communication will fail if data rates in excess of C are attempted, and the delay time (to implement coding/decoding) may be excessive for systems which approach C. One important class of codes is a block code for which we may define a code rate R_{code} as the ratio of information bits (k) to transmission bits (n). Each message block contains n-k redundant bits and 2^k possible code words. The Shannon result indicates that there exists a code rate R_{code} possessing an arbitrarily high probability of error-free

---

82. Forward Error Correction.

83. Error Detection and Correction.

decoding. Simply put, the probability of decoding error can be reduced to an arbitrarily small level by an increase in the code length n while keeping R_{code} less than C. The Shannon theorem indicates the ultimate limit on baud (or information) rate but does not provide any guidance as to the nature of the code which must be constructed to achieve the result. It does, however, provide code designers with a basis for faith in the pursuit of more efficient codes. In effect. Shannon's work asserts that signal power, signal noise, and bandwidth will set a limit on communication (symbol) rate but not communication accuracy.

Figure 7-9 depicts an HF communication channel in logical form. The source of data may be a computer terminal or even a human being. From this source of information, there exits a sequence of discrete symbols or a continuous waveform. The source encoder transforms the source output into a sequence of binary symbols (i.e., 0's and 1's) for insertion into the channel encoder. The channel encoder develops another sequence, typically longer for redundancy and error control, which is a codeword representation of the information sequence. In some instances the source encoder incorporates bandwidth compression (as in voice), while the channel encoder incorporates bandwidth expansion. Following development of the codeword representation, also a series of 0's and 1's, it must be tailored prior to transmission through the HF channel. This is accomplished by the modulator function, which generates unique waveforms which will represent the states which exit the channel encoder. The channel distorts the transmitted signal which is also vulgarized by noise and interference. At the receiver the entire process is repeated but in reverse direction. Demodulation recovers the transmitted codewords, and the channel decoder recovers the original information sequence. Finally, the application the source decoder provides meaningful data or physical waveforms back to the end user.

Darnell [1986] points out the advantages which accrue from digital encoding, and outlines the schemes which appear to offer the most promise. They include: adaptive filtering, quenched filtering, soft-decision decoding, correlation reception, variable rate decoding, and post-reception processing. Of these, adaptive filtering, quenched filtering, and variable rate decoding are most readily perceived as features to be incorporated in an adaptive HF radio. The adaptive nature of the other schemes is not as obvious, but they are important countermeasures to channel disturbances nonetheless.

Soft decision decoding, which associates *hard decisions* with additional information to be used as weights or confidence levels, can also be viewed as a type of adaptive process in view of the time varying nature of the confidence level information which is typically related to received signal properties.

Adaptive filtering is a process whereby the receiver senses channel interference and adjusts the digital filter parameters to optimize the signal-to-noise plus interference ratio (SNIR). Quenched filtering is a process in which the digital filter is activated (or initialized) only when instructed, and at a time when the greatest probability of optimum SNIR would be expected. Clearly quenched filtering requires time synchronous system operation. Variable rate decoding procedures could be applied for signaling schemes in which the source data is encoded at multiple rates and transmitted as an integrated bit sequence over the channel. Darnell [1983] has described the system control aspects of this approach. Variable rate decoding is related to the code combining strategy which has been suggested by Chase [1973]. The nature of code combining will be covered later on.

7.6.7.2 *Error Detection, Correction, and Control*

Error Detection

The most prevalent methods for error detection include the application of redundancy, exact count encoding schemes, parity checking in various forms, and cyclic redundancy checking.

Character redundancy implies the repeated transmission of data characters. If the strategy is to transmit characters twice, then an error will be detected if received characters are not in pairs. *Message redundancy* implies the repeated transmission of a set of characters. If the set is not received twice in succession, then an error in the message has occurred. Redundancy in these forms could be used to trigger the initiation of an ARQ[84].

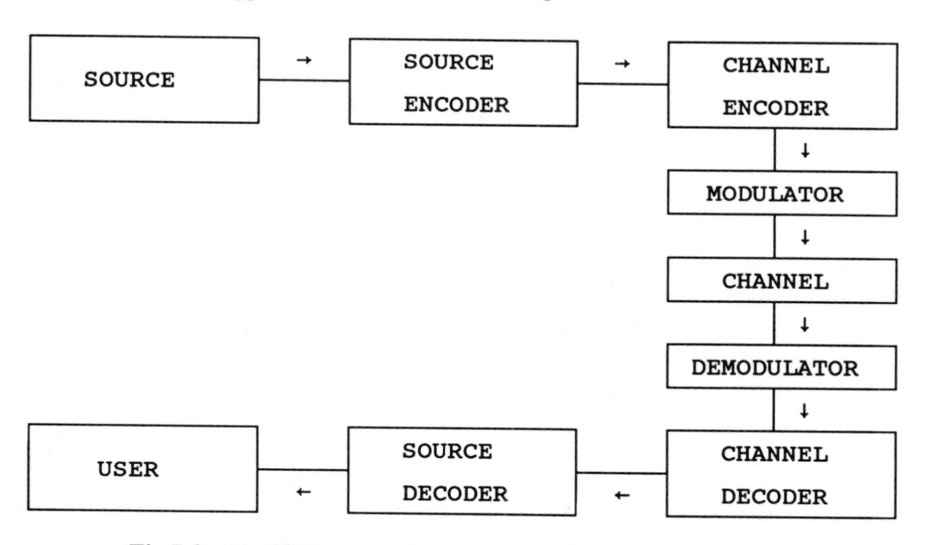

Fig.7-9. Block Diagram of a Generic Data Communication Channel

---

84. Automatic Request for Repeat (of transmission).

Exact counting is a scheme for correcting characters in which a fixed count of 1's representing a character is dictated. If the count is required to be 3, then other counts imply that an error has occurred. Like the redundancy method, exact counting is a natural scheme for initiation of an ARQ.

Parity checking is a rather simple procedure for detecting errors and is used in a number of data transmission systems. Character parity checking involves adding an extra bit (called a parity bit) to each character to insure either an even or an odd number of 1's in the binary code representing the character. This is sometimes called vertical redundancy checking (VRC). A similar procedure can be used in the context of a message block in addition to (or instead of) parity checking on a character-by-character basis. Longitudinal redundancy checking (LRC) requires the maintenance of a specified parity (usually even) for the sequence of information bits organized by bit position. The sequence of bits making up the LRC character which results from this process is called a block check sequence (BCS) and it conveys no information whatsoever. It is only used for error detection. VRC and LRC can be employed either independently or in a coordinated way. Used together almost all errors may be detected.

Cyclic redundancy checking is the most reliable method of error detection. This method, although not simple to implement, will detect about 99.95% of all transmission errors. The most generally accepted code of this type is the CRC-16, which employs 16 bits in the BCS [Tomasi, 1987].

Error Correction

Error correction methods include ARQ and Forward Error Correction (FEC) schemes. Both methods, either individually or in combination as in Hybrid Error Correction (HEC), have been used in system applications.

ARQ is generally viewed to be the most reliable method for insuring the integrity of transmitted messages under most circumstances. Conditions of path non-reciprocity may alter this view. Also , if the environment is changing more rapidly than the ARQ duty cycle, a type of nonreciprocity generated by temporal changes in the channel may be introduced. This suggests that short messages or pieces of messages should be transmitted, especially during disturbed periods. This would reduce the probability that channel impairments will influence the transmission, return, and retransmission paths over the time period required for the handshake. Each ARQ sequence requires a fixed amount of overhead or time delay. Therefore, the use of increasingly smaller message block lengths will eventually become counterproductive. The whole process could be made adaptive by optimizing the ratio of message length to delay time in accordance with path conditions.

It should be noted that ARQ requires that every block of data which is transmitted be confirmed as received with errors or without errors. A positive acknowledgement is termed *ACK* and a negative acknowledgement is termed a *NAK*. Redundancy which must accompany the data stream in an ARQ system is relatively small as compared with systems which correct errors without the necessity for feedback.

FEC detects and corrects errors at the receiver without the necessity to call for retransmission. This implies that sufficient redundancy has been built into the transmitted data to accomplish this. Such methods are not always as reliable as ARQ. On the other hand, FEC may be more advisable in a stressed environment which is changing rapidly since it does not require a message return path (i.e., a handshake) which may prove to be vulnerable or even undesirable. A number of codes are possible in the FEC process. One of the more common methods is based on the Hamming code, which is a linear block code. Still FEC is not perfect, even if a substantial amount of redundancy is engineered into the data stream.

HEC is a hybrid error correction scheme which is a composite of FEC and ARQ. Small error rates tend to favor the use of fixed overhead FEC systems while larger error rates favor ARQ systems. Consequently an HEC system is an optimum approach for transmission over a real channel since errors generally appear in bursts over the short time frame and clusters of bursts over a longer time frame. Random errors are easily corrected by the FEC portion. Interleaving can be used to randomize burst error patterns while clusters of errors, associated with long deep fades, may require message retransmission.

7.6.7.3 *Coding Structures*

No attempt will be made to cover such a rich topic in all of its generality. Error-correcting coding for digital communications has been discussed by Clark and Cain [1981] and an earlier book by Lin [1970] is a useful reference on the subject. A collection of key papers has been compiled [Jayant, 1976] for those interested in acquiring additional background.

General Features

It is found that he most common coding structures are block codes and tree codes. Block codes are "memoryless" since the process involves the mapping of a block of k input symbols into n output symbols without consideration of any previous symbols. Tree codes, on the other hand, are generated by a device having memory. In this method a group of m input symbols are mapped into n output symbols as before, but the n output symbols are determined by an augmented symbol set of dimension $m + v$, where v represents a set of previous symbols. The number $m + v$ is called the constraint length of the tree code.

Codes may also be classified as either linear or nonlinear in nature with linear codes being most common. Linear tree codes are sometimes referred to as convolutional codes, while linear block codes may be called group codes.

Code rates R_{code} are defined as the ratio of the number of input symbols to output symbols. For both block and tree codes, typical values for R_{code} range between 1/4 and 7/8. Low rate codes have the greatest redundancy and are used in some strategic systems requiring the greatest error correction capability with acceptable levels of throughput efficiency. Both types of coding structure allow one to define a minimum distance[85] of the respective codes, d_{min}, a parameter which allows one to quantify error detection and correction capabilities for a specified code. A large value of d_{min} implies an improved error detection and correction capability.

An example of the class of linear block codes is the Hamming code which is prevalent since it permits single errors in a codeword to be corrected with the least number of redundant bits. Multiple errors may not be corrected, however up to two errors/codeword may be detected.

Another class of linear block codes are so-called cyclic codes which are relatively easy to implement with feedback shift registers. This is because every cyclic shift in a codeword is another codeword in the allowable set. Examples include Golay and Bose-Chaudhuri-Hocquenghen (BCH) codes. Codes such as BCH also belong to a general class of polynomial generated codes along with Reed-Solomon codes (RS).

Decoding

The decoding process for linear block codes is optimized in the form of maximum likelihood schemes. Such schemes involve the logical grouping of allowed codewords with disallowed sequences on the basis of the number of bit position differences. This process may be visualized by forming a decoding lookup table. Comparison of input data with the table allows one to associate an arbitrary input sequence with one out of several allowable sequences. In short, the decoder selects as the proper codeword the one which is closest in so-called Hamming distance. The process yields the minimum probability of error in decoding an observed sequence as an allowed codeword out of a number of possibilities. Implementations using lookup tables are not realistic for long block codes. Other block decoding algorithms include the error trapping decoder and the Meggitt decoder [Grossi, 1986].

Convolutional codes may be decoded by exhaustive tree search or sequential decoding methods [Shanmugan, 1979]. In general, the major categories of tree decoding algorithms include: the sequential methods, threshold decoding, table lookup algorithms, and the Viterbi algorithm. The Viterbi algorithm is

---

85. The distance d between any two codewords a and b is the weight of the combination (i.e., the sum) of a and b using modulo-2 arithmetic. The weight w of a codeword sequence is defined to be the number of nonzero elements. Thus $d(a,b) = w(a+b)$. The minimum distance of a code is found to be the minimum weight of any nonzero codeword in the entire code.

actually a maximum likelihood method.

Hard decision decoding or soft decision decoding may be used. In the former method, the decision concerning a codeword is dependent upon a comparison of the bit positions of a received symbol with bit positions for all allowable codeword sequences. No auxiliary information is used in processing the decision. A maximum likelihood decoder is one example of hard decision scheme since the decision is based only upon determination of the codeword candidate which differs from incoming sequence in the smallest number of bit positions. In the soft decision process, other information enters into the algorithm, and the maximum likelihood method is conceptually easy to generalize to accommodate this type of decision process.

Concatenated Codes

Multiple levels of encoding have been used to simplify the encoding/decoding system. Typically two levels of encoding are utilized, an outer code and an inner code, which are linked together in a chain (i.e., concatenated). Outer codes are generally block codes of the Reed-Solomon variety which make decisions at the symbol rate rather than the bit rate. Inner codes include the following as possibilities: orthogonal codes, short block codes, and convolutional codes.

The overall length of a concatenated code is nN bits, where n is the symbol length of the outer block code and N is the inner code block length (i.e., bits/symbol). The outer block code is nonbinary and consists of k K-bit information symbols, $n - k$ check symbols yielding a length of n bits. The inner code consists of symbols of length K bits, and $N - K$ parity bits to yield the N bit block length. The overall code rate is simply kK/nN where the composite code rates are defined in the usual way.

Burst Error Correction Methods

Decoders are generally optimized to compensate for random errors. Randomization of errors may be accomplished through use of interleavers. These devices randomize or scramble the transmitted data stream. At the receive end the data stream is deinterleaved, a process which reorganizes the transmitted data stream but disorganizes the burst errors which may have occurred. The overall error rate is the same. The complete procedure enables more efficient and predictable error correction to be performed. Interleaver structures may be internal or external, depending upon the configuration of the hardware. If the interleaver is imbedded in the coder/decoder hardware, it is termed internal; otherwise the interleaver is termed an external device. External interleaver/deinterleaver setups are used in the implementation of Viterbi decoding, sequential decoding, and BCH decoding. Meggitt decoding structures may use internal interleaver/deinterleaver configurations.

Coding Gain

To establish the merit of a coding system, it is necessary to determine performance against a standard system or coding approach. For a communication system, the ratio of the information bit energy (E_b) to the noise spectral density N_0 which is required to enable a specified level of performance (i.e., bit-error-rate, BER) is an appropriate index to use in such comparisons. The standard method for ascertaining coding gain is to prepare curves of bit error probability for both coded and uncoded situations versus E_b/N_0, and to deduce the difference between the two (along a horizontal line) for a specified error probability at issue. The results are typically given in dB.

Code Combining

A process by which low rate convolutional codes are processed to achieve a composite redundancy sufficient to remove significant numbers of errors. With a feedback circuit this procedure may be made adaptive. In this way a minimum code rate R_{code} can be established which is sufficient to achieve a specified error rate.

7.6.7.4 *Adaptive Coding Processes*

Coding methods by their nature provide varying degrees of protection. Some schemes are more robust in the face of channel disturbance but carry with them a significant overhead burden. Other methods are relatively fast provided the channel is only moderately disturbed, but may fail under virulent conditions. It seems natural therefore to visualize the development of coding schemes which are adaptive in nature. For example, code combining strategies may be made adaptive and such measures are useful in strategic HF systems. This method exploits the same basic code structure while altering the effective code rate. In general one could imagine code structures themselves being selected adaptively depending on the state of the channel [Chase, 1984]. Code combining features a soft decision strategy and a maximum likelihood decoding procedure instead of minimum distance decoding. Maximum likelihood sequence estimation (MLSE) is discussed in Section 7.6.8.7.

7.6.8 System Technology: DSP and Advanced Modems

The actual *brains* of an advanced transceiver is the digital modem. The word *modem* is an acronym standing for modulator-demodulator. Early versions of these devices were used simply for converting digital signals into quasi-analog signals for transmission and reconverting the quasi-analog signals into digital form. Now a number of additional functions have been added as component technology has improved to match the system requirements. Some of the requisite properties of digital modems are provided in Table 7-8.

7.6.8.1 *General Principles*

Monsen [1986] has discussed the implications of the propagation channel upon decisions related to modulation format and coding scheme. Central to any discussion of waveform performance in a channel is a knowledge of the channel properties. Bello [1963] has characterized random time-variant linear channels. This classic paper may be found in a volume entitled *Communication Channels: Characterization and Behavior* [Goldberg, 1976]. The reader is also referred to books by Stein and Jones [1967] and Kennedy [1969] for a general review of communication channel properties and methods of characterization.

Assuming that the ionospheric channel is characterized by a transfer function $H(f,t)$, with t_c and f_c being the coherence time and coherence bandwidth respectively, the following relationships are found to hold:

$$\text{Doppler spread:} \quad \sigma_d\,(\text{Hz}) = (1/2\pi)t_c^{-1} \tag{7.3}$$

$$\text{multipath spread:} \quad \sigma_t\,(\text{sec}) = (1/2\pi)f_c^{-1} \tag{7.4}$$

where σ_d is related to the channel bandwidth (and is a measure of the fading rate), and σ_t approximates the multipath delay spread. Table 7-9 contains some bandwidth conventions frequently used in this book. Also included are definitions of various channel properties and terms of general interest.

Characterizations of the two spread functions (i.e., σ_d and σ_t) may be developed rigorously in terms of the spectral moments in a statistical channel model for which $H(f,t)$ is characterized by a two dimensional correlation function [Bello, 1963]. Design principles for signal processing systems may be based upon less sophisticated notions.

TABLE 7-8. DESIRABLE PROPERTIES OF DIGITAL MODEMS

A. REDUNDANCY; in connection with the process of data stream synchronization.
B. WAVEFORM GENERATION; from original voice signals and/or data sources.
C. WAVEFORM CONVEYANCE; to the transmitter exciter.
C. WAVEFORM ACCEPTANCE; from the receiver.
D. BUILT-IN ERROR DETECTION AND CORRECTION (EDAC).
E. BUILT-IN CHANNEL EVALUATION; possibly achieved through exploitation of well-designed preambles.
F. DATA RATE SUPPORT.

Monsen [1986] defines two constraints which provide guidance: the *learning constraint* and the *diversity constraint.*

With a change in notation for consistency, we have:

$$\text{Learning Constraint:} \quad R >> \sigma_d \tag{7.5}$$

$$\text{Diversity Constraint:} \quad B_s \geq f_c \tag{7.6}$$

Learning Constraint

The learning constraint carries with it the implication that channel properties may be derived or *learned* by some process. Such a process might be in-band sounding, pilot-tone sounding, wideband probing, or some other form of RTCE system. The process must operate with a duty cycle sufficient to allow channel properties to be established and exploited by the system. In short, the channel must be stationary for a suitable reckoning interval. Equation 7.5 defines that interval in terms the data rate R. Coherent systems have an advantage over noncoherent systems if the learning constraint is satisfied. Noncoherent detection must be used to avoid the problem associated noisy data decisions, if the learning constraint is not satisfied.

Diversity Constraint

The diversity constraint deals with the necessity to stimulate segments of the spectrum which are uncorrelated. This approach is the enabling rule for a process termed *implicit diversity.* This is distinct from the concept of *explicit diversity* which includes some of the more obvious types based upon: spaced reception, message repetition, polarization flexibility, and selectivity in frequency, mode, path, and launch angle, etc. For *implicit diversity* to exist, the bandwidth B_s of the signal must be larger than the coherence bandwidth f_c. If the reverse were true, then the entire band would exhibit a simultaneous fading property. As a consequence, no *implicit diversity* would be possible. Although *implicit diversity* finds its application in the context of decorrelation over the frequency domain (i.e., Equation 7.6), there is also a possibility for another form of *implicit diversity* whereby decorrelation in the time domain is exploited. These schemes are possible in a fast fading environment wherein adherence to the learning constraint might appear marginal (i.e., $R \approx \sigma_d$). Clearly this suggests that the individual codewords would require adequate redundancy (i.e., additional check bits) to achieve time diversity gain.

TABLE 7-9

FREQUENTLY USED CHANNEL PROPERTY TERMS

SYMBOL	DEFINITION
$H(f,t)$	Channel Transfer Function
f	Carrier Frequency (Hz) Radio Frequency (Hz)
t	Time (sec)
t_c	Decorrelation Time (sec) Coherence Time (sec)
f_c	Decorrelation Frequency (Hz) Correlation Bandwidth (Hz) Coherence bandwidth (Hz) Selective-fading Bandwidth (Hz)
B_F	Epstein Polarization Bandwidth (Hz)
f_d	Dispersion Bandwidth (Hz)
σ_d	Fading Rate (Hz) Doppler Spread (Hz) [or B_d for cw input]
σ_t	Multipath Spread (sec) [or T_m for impulse]
L	Channel Spread Factor (unitless)
T	Symbol Period (sec)
T_c	Chip period (sec)
$1/T_c$	Chip rate R_c (chips/sec) Code clock rate
B_c	Channel bandwidth (Hz)
B_s	Signal Bandwidth (Hz)
B_{RF}	RF Bandwidth (Hz) Transmission Bandwidth B (Hz)
B_{ss}	Spread Spectrum Bandwidth (Hz) [$\approx 2 \times$ chip rate in DS-SS system] [m $\times$ B_c, m = channels in FH-SS system]
R_{hop}	Hop Rate (hops/sec) {FH-SS system}
R	Transmitted Data Rate (bps) [for a binary input]
R_{code}	Code Rate (unitless)
$1/T$	Transmitted Symbol Rate (symbols/sec)
C	Shannon's Channel Capacity (symbols/sec)

Spread Factor

The product of the doppler spread and the multipath spread is defined as the spread factor[86].

$$\sigma_d \sigma_t = L \qquad (7.7)$$

The significance of the spread factor is that it provides a design criterion for avoidance of ISI as well as pulse coherence. Avoidance of ISI suggests that $T >> \sigma_t$, where σ_t is the multipath spread. Avoidance of pulse distortion suggests that $T << 1/\sigma_d$, where σ_d is the doppler spread.
Thus we have:

$$\sigma_t << T << 1/\sigma_d \qquad (7.8)$$

This implies that the spread factor L must be significantly less than 1 for adaptive systems to function properly. In practice L ranges between 1/200 and 1/2000 [Maslin, 1987].

Low Data Rates and ISI

Since $1/T$ is the transmitted symbol rate, to eliminate ISI it is clear that T should be well in excess of the multipath spread σ_t. At the same time the diversity constraint must be satisfied if any implicit diversity is to be achieved. For an M-ary modulation scheme, groups of $\log_2 M$ bits are transmitted as a symbol which may assume 1 of M possible values. The raw data rate is $R = (1/T)\log_2 M$, which for situations such that M is small implies that R is the order of $1/T$. Since $1/T << f_c$ to avoid ISI, then $R << f_c$. Then surely $R << B$. This means that the bit rate is very much smaller than the available bandwidth. The fact that R must be smaller than the available bandwidth means that a bandwidth expansion technique is required to "excite" the diversity components which are present over the available bandwidth B.

Excitation of diversity components may involve invocation of spread spectrum waveforms. This matter was discussed in Chapter 4. Equation 4.65, for example, indicates that the number of modes Γ which may be resolved is roughly the quotient of signal and channel bandwidths (i.e., $B_s/B_c + 1 = \Gamma$). It is important to recognize that the signal bandwidth is defined by the waveform, whereas the channel bandwidth is defined by the medium and its characteristics.

----

86. A different terminology is adopted by some authors to define doppler and multipath spreads. For example, Stein and Jones [1967] call B_d the doppler spread and T_m the multipath spread. Thus $L = T_m B_d$.

7.6.8.2 *Signal Processing Schemes for HF Communications*

We have indicated that data errors may be minimized by the introduction of coding, while burst errors may be randomized by interleaving approaches so that coding schemes may be more effective. We have also noted that the choice of modulation format makes a difference in the performance of systems in the presence of multipath and noise. Techniques which have been used to reduce the BER and improve HF system reliability have included M-ary transmission (Piccolo system), parallel channel PSK transmission (Kineplex system), and a variety of explicit diversity schemes. Still, in order to increase the data rates in the presence of channel distortion and ISI introduced by multipath conditions, more advanced digital signal processing techniques must be invoked. Serial data transmission has received attention because of a need to retain more efficient bandwidth utilization, but such a transmission scheme generally requires equalization of the channel. In recent years equalization schemes, used successfully for compensation of telephone line distortion, has been a growth industry within the community of HF modem designers.

Wideband signaling schemes realized in FH-SS, DS-SS, or chirp formats furnish another set of challenges to the system designer. A spread spectrum approach, before all else, allows the capability to resolve multipath. A PN sequence (i.e., DS-SS) approach has the potential to resolve multipath components provided they are separated by more than the chip duration. In general, bandwidth expansion without equalization leads to the demodulation of only one multipath component while the other components contribute to the overall noise [Simon et al.,1985; Wagen, 1988].

It is beyond the scope of this book to detail the wide variety of DSP techniques which have been suggested or used in HF communication systems. Two volumes consisting of collections of papers on digital signal processing [Rabiner and Rader, 1972; Oppenheim, 1976] outline many of the features being implemented in current DSP concepts and digital filter design. Generic descriptions of the majority of methods are provided within Sections 7.6.8.3 through 7.6.8.8. Table 7-10 contains a list of the primary DSP schemes in use today. In preparation for a summary of DSP methods in current use, we shall shortly review the implications of parallel and serial data transmission (Section 7.6.8.3). In the remainder of Section 7.6.8 dealing with advanced technology, we shall cover the following topics: equalization structures, the *Rake* equalizer, maximum likelihood detectors, and a summary of selected HF modems.

7.6.8.3 *Serial and Parallel Tone Signaling*

The choice of parallel or serial tone signaling depends largely upon data rate requirements, anticipated multipath and doppler spreads , and the availability of suitable equalization equipment. Since multipath places a limit on signaling rates to the order of 100-200 baud, high bit rates must be enabled by M-ary signaling and/or use of parallel tones.

One of the stated problems in parallel signaling is associated with a *crest factor* limitation. Performance of systems is based upon average power whereas modulators are peak power limited. Parallel transmission systems exhibit large peak-to-mean power ratios in the event tones add in phase. Clipping may ameliorate this problem somewhat, but the *crest factor* loss may be 3 dB worse than it is for a serial transmission format. Nevertheless, in practice other issues are far more compelling in the decision of which way to go.

Disadvantages associated with parallel signaling include its sensitivity to selective fading and its inefficiency in use of bandwidth. On the other hand, serial schemes are generally complex, requiring sophisticated DSP techniques.

7.6.8.4 *Digital Voice Transmission*

Earlier Analog Approaches

The compression of the radiotelephone signal spectrum has been taken up by the CCIR [1986a]. Of particular interest is a system called Linked-Compressor-Expander (or Lincompex) [CCIR, 1986b] which uses an auxiliary narrowband FM channel for system control. The information carried over this auxiliary channel is the degree to which the signal is compressed prior to transmission over the 3 kHz channel. At the receiver the auxiliary data is used to restore the spectral shape of the signal. Analog Lincompex has had good success over noisy and fading channels, and spectrum utilization efficiencies of around 50% have been achieved in some instances.

TABLE 7-10. HF SIGNAL PROCESSING SCHEMES

TECHNIQUE NAME	PROCESS
Parallel-Tone Signaling	Long symbols (bauds) on adjacent frequencies to increase overall data rate while eliminating ISI
Rake Processing	Parallel Matched Filters
Linear Equalization	Distortion reduced by filtering the received signal.
Decision Feedback Equalization	Distortion reduced by filtering the received and past decisions.
Maximum Likelihood Estimation	Message matched filter, Viterbi decoding.

Digital Lincompex

A separate version of Lincompex, called Synchronized-Compressor-Expander (or Syncompex) [CCIR, 1986b], replaces the control channel FM modulation format with a digital control channel. Many versions of Lincompex and Syncompex have been developed. The value of Lincompex systems for HF communications is now well-established [Weller,1982]. Lincompex has been recommended by the CCIR[87], and it is incorporated into a current military standard [88] MIL-STD-188-141A. A recent version of Lincompex, similar in some respects to Syncompex, has been developed by AMAF Industries [AMAF, 1990a]. The AMAF system, called LINK-PLUS™[89], connects in series with the voice path in the transmitter, providing voice compression and a control modulation which is added to the compressed voice signal. The Lincompex demodulator connects in series to the voice path at the receiver, it demodulates the control tone, and regenerates the voice envelope expanding the voice waveform to its original dynamic range. Before delivering the signal to the audio output of the receiver, the control tone is filtered out.

Approach to Digital Voice

Although jamming presents a challenge for the designer of military HF radios, there is also the issue of security. Requirements for transmission of secure data and voice places additional demands on ultimate design of the transceiver. Voice is, of course, an analog process and it must be digitized to introduce privacy through the application of encryption schemes. Another feature associated with such a transceiver is bandwidth compression to reduce the distortion of the transmitted signal. On the surface this would appear to be inconsistent with the requirement to represent voice as a digital data stream.

----

87. CCIR Recommendation 455-1: "Improved Transmission System for HF Radiotelephone Systems", in *Reports and Recommendations of the CCIR*, Vol.III, [CCIR,1986a], unmodified since first publication in 1974.

88. MIL-STD-188-141A: *Interoperability and Performance Standards for Medium and High Frequency Radio Equipment* published in 1988 (see Section 7.8).

89. The LINK-PLUS Model 2200 system, developed by AMAF Industries, uses the companding scheme recommended by CCIR 455-1. Details have been published by Jensen [1982]. The package is actually configured as a communication enhancement module. Recent tests in connection with frequency hopping radios [AMAF, 1990b]. The Joint Interoperability Test Center (JITC) tested several HF radio systems [DISA, 1991]to support requests of the United States Special Operations Command (USSOCOM) and the U.S. Marine Corps. Lincompex, used to improve Autovon connectivity on one phase of the testing, proved to be a valuable technique even under unsettled magnetic activity conditions.

Vocoders

Modulation schemes such as pulse code modulation (PCM) encode voice directly but are inefficient in bandwidth utilization. A vocoder (or Voice-Operated Coder) does not directly encode the human voice. Instead it measures voice parameters and transmits these as a slow digital bit stream. At the receiver the encoded voice parameters are used to synthesize key features of the original voice by using an electric circuit to imitate operation of the vocal cord, throat, and mouth. The vocal cord and the complex vocal cavity are associated with specific resonance patterns in the spectral domain. The resonances in human voice are called *formants*, which are characterized by center frequency, bandwidth, and amplitude. Given these properties, a vocoder system operation is based upon the transmission of digital codes which correspond to designated features of speech. Knowledge of the formant structure allows for reconstruction of speech at the receiver.

Linear Predictive Coding (LPC)

One military approach for transmission of digital voice calls for use of a technique called Linear Predictive Coding (LPC)[90] which examines the instantaneous spectral properties of the voice and encodes the analog signal into data (i.e., numbers) which are used to represent the voice. These numbers are then transmitted as a bit stream. The US Navy uses the LPC algorithm in its ANDVT (Advanced Narrowband Digital Voice Terminal) which employs a 39-tone parallel signaling scheme. Other parallel schemes also exist. To send data at say 2.4 kilobits/sec (required for digital voice) is not possible. To achieve the requisite packing density (about 1 bit per Hz), the data stream is split into a number of substreams which are used to modulate one (or more) of the parallel tones. Encryption is then applied. Synchronous reception is required to successfully demodulated this waveform. De-encryption may then be applied, followed by transformation of the *numbers* to analog speech by inversion of the LPC algorithm. The receiver must be designed to operate such that symbol decisions are made at the requisite time (i.e., at the transition times). Otherwise errors result. Signal distortion, which is the primary cause of timing distortions, is greatest for the wider bandwidths. This is why narrowband processes are desired. The downside of parallel processing is that the transmitter amplifier must be equipped to handle a situation where the parallel tones *add-up* and are transmitted in phase. Consequently there is a *crest factor* penalty suffered for parallel signaling.

----

90. Linear Predictive Coding (LPC) is based upon the assumption that the resonances of the vocal tract may be represented by a linear all-pole filter [Maslin, 1987]. LPC-10 is a standard system which uses 10 poles corresponding to characterization of up to 5 resonances (formants).

Speech Recognition and Processing

There are obvious advantages to speech processing. Such processing schemes include the exploitation of vocoders for bandwidth reduction and allowance for institution of communication security. Limited vocabulary speech recognition schemes have the potential for reducing the bandwidth even more, thereby leading to a significant diminution in competing noise. Words contained within a limited vocabulary may be encoded by a relatively small number of bits. Vocabularies of 128, 256, 512, and 1024 words would require 7, 8, 9, and 10-bit addresses respectively. Assuming a 1024 word vocabulary and a speech rate of 1 word/sec, a bit rate of 20/sec would be required excluding the incorporation of additional check bits for EDAC purposes and engineering bits for synchronization, etc. A composite bit rate of 50-100 bits/sec would be expected to be adequate. When compared with conventional vocoder approaches this data rate represents a significant bandwidth reduction, but it is enormous when compared with analog approaches.

It is possible to visualize the development of a speech recognition system which allows for interpretation into more than one language. In other words one could map one language into another thereby allowing limited vocabulary speech in one language to be *heard* as another language. For example, a military commander could broadcast in English but be heard in French, German, Italian, etc.

Most available vocoders operate with bit rates of 1200-2400 bits/s, and since this rate is marginal under the best of circumstances, one could not expect digital voice modems based upon conventional vocoder principles to operate efficiently under adverse conditions. Darnell [1986] suggests that it may be impossible to derive a satisfactory grade of service in congested environments such as Europe especially during nocturnal hours. A hybrid approach involving speech recognition and synthesis has been proposed by Chesmore and Darnell [1985]. This system has an information rate of 25 bits/s.

According to Maslin [1987] the principal advantages of speech processing include: bandwidth reduction, message compression, audio noise reduction, security and privacy, and radio control.

7.6.8.5 *Equalization Methods*

Distortion associated with data transmission through the ionospheric channel may be considerable. One fundamental difficulty with such distortion coupled with multipath is that it increases the channel *reverberation time*. As a consequence, ISI difficulties may certainly arise if the baud duration is less than the sensible multipath spread. This effect may either be tolerated by using a lower baud rate, or it may be ameliorated through use of equalization. An equalizer is a device, or an equivalent circuit, which compensates for amplitude and time delay variations. The ultimate objective of equalization is the elimination of ISI. Equalization[91] is defined in the second edition of *Telephony's Dictionary* [Langley, 1986] as "the reduction of frequency distortion and/or phase distortion of a circuit by the introduction of networks to compensate for the difference in attenuation, time delay, or both, at the various frequencies of the transmission band." Adaptive equalization is the process of automatic adjustment of an equalizer based upon the *learned* properties of a distorted input in order to produce an undistorted output. How the channel is learned and how the channel information is applied to achieve equalization is not simply a matter of taste. The realization of specific equalizer structures depends critically upon the requirements involved.

As indicated, equalization is one of a number of methods which may be used to compensate for multipath. The basic idea is to cancel the effect of the channel transfer function by producing a distortion-free replica of the transmitted signal. It is crucial that these methods account for the channel variability to be useful at HF. Stemming from approaches applied to non-HF channels, Morris [1983] breaks automatic adjustable equalization technology into two major types: Preset equalization and adaptive equalization. In preset equalization, the tap weights are adjusted by a *training sequence* which simulates the line characteristics. The preset equalizer must therefore *learn* the channel. Adaptive equalizers are based upon a decision-directed concept in which iterative corrections to the tap gains are derived from the output of the equalizer and applied continuously.

There exists a class of equalization processes which successfully remove ISI, an attribute which distinguishes them from the generic *Rake* process which does not. Such equalizers must work considerably faster than a *Rake* equalizer, since they sum the multipath components before pulse compression. In short the processing requirements are defined by the chip rate, the speed at which weighted summation is driven.

Equalization techniques belong to two major groups: linear and decision-feedback. The linear equalizer is comprised of a linear filter structure which operates on the received signal with ISI removal being the principal goal. This operation is followed by data detection. A decision-feedback approach exploits a linear filter, however the detected data sequence is filtered and reinserted into the signal path. The process of feedback accomplishes the task of canceling ISI due to past digits for which decisions have already been made.

---

91. The term equalization is a general one, but some authors choose to differentiate between the various types. This is especially true for *Rake* equalization structures. *Rake* is not specifically designed for removal of ISI, but it is a very efficient form of equalizer which operates optimally if baud rates are not excessive. *Rake* is designed to perform a weighted summation of the multipath components *following* pulse compression. Non-*Rake* equalizers are designed to sum the multipath components *before* pulse compression.

There are a number of equalizer structures. Fixed structures employ matching networks, or filter configurations, based upon known properties of the channel. This scheme is clearly inadequate for the ionosphere, long known for its time-varying dispersive nature and multipath properties. Manually adjustable equalizers might work under very special circumstances, but would not represent a practical solution for HF channel equalization in operational systems. Various equalizer structures have been outlined by Morris [1983].

The tapped delay line transversal filter is a structure which is well suited for exploitation in the process of adaptive equalization. The decision-feedback equalizer structure may be made adaptive through the incorporation of an algorithm which adjusts the filter parameters in a continuous fashion.

Dhar and Perry [1982] and Perry [1983] of MITRE Corporation have examined links for which only a single mode is known to exist, and have successfully applied wideband (1 MHz) adaptive equalization techniques. The goal was the development of an equalization strategy to compensate for intramodal (small-scale) multipath and propagation dispersion. In the late 1970's MITRE started a program to develop a wideband equalizer based upon an R&D program of channel measurements. The methods developed by MITRE suggest that F layer modes, the primary ones used for most HF communication, can be effectively equalized. The obvious applications include protection from jamming and covert communications. Also, fading resulting from intramodal interference is eliminated. These workers use an equalizer which a programmable transversal filter having a bandwidth of 1.024 MHz and an impulse response duration of 125 microseconds, corresponding to 128 taps. Perry and his group use three different algorithms to compute the equalizer complex coefficients (tap weights): inverse filter, matched filter (*Rake*), and Weiner filter. We will comment on the *Rake* filter more fully in Section 7.6.8.6.

MITRE Measurements have suggested that the channel is stable for up to 10 seconds. As a consequence, adaptivity overhead requirements will not place a tremendous stress on the system. A decision feedback approach has been suggested to satisfy the need to equalize the time-varying channel .

7.6.8.6 *Rake Equalizer*

The *Rake* system, as described by Price and Green [1958], was probably the first comprehensive attempt to attack the problem of multipath distortion actively through application of knowledge of the channel. According to Price and Green, "passive" approaches would involve various *explicit diversity* schemes, use of SSB transmission, and coding. Kathryn, nomenclatured AN/GSC-10, was a digital data modem developed by the U.S. Army [Kirsch et al.,1969; Bello, 1965] and used the *Rake* principle. An early type of ionospheric channel simulator based upon an inverse *Rake* concept has been described in a paper by Goldberg et al.[1965].

A *Rake* processor is a matched filter which performs a weighted summation of multipath components of the received signal following pulse compression. The process is more efficient than adaptive equalization. Since it is a spread spectrum scheme, it provides a degree of protection against certain classes of jamming. Also it enables resolved time dispersion to be used as an effective diversity measure against multipath-induced fading. The term *implicit diversity* derives from this property. Stein and Jones [1967] interpret the process as one in which multipath is resolved and time-tagged through use of wideband PN waveform signaling. In short the multipath may be viewed as a sequence of definable echoes in the time domain. These echoes are "raked" together and combined in diversity. *Rake* is not explicitly designed to compensate for ISI. Thus there is a symbol rate restriction of typically 100-200 symbols per second. Parallel *Rake* filters can be employed to handle this problem. Code-division-multiple-access techniques could also be applied provided spread spectrum waveforms are used.

Bello [1988] has examined the performance of *Rake* modems over non-disturbed and wideband HF channels. In Bello's view, the *Rake* technique involves the construction of a filter matched to each received symbol. At the same time this filter may be visualized as the cascade of two filters, one matched to the transmitted symbol and one matched to the channel. The *Rake* operation can also be conveniently reinterpreted as a correlation operation. Many versions of the *Rake* technique exist, and these derive principally from the architectural differences associated with various DSP configurations. An examination of four *Rake* modems has been conducted by Bello [1989], and Wagner et al. [1989] have determined wideband HF channel properties which may be used to estimate the performance of *Rake* designs.

7.6.8.7 *Maximum Likelihood Sequence Estimation*

As indicated in Section 7.6.7.3, MLSE presents an optimum way to decode linear block codes. The process, as exemplified in the Viterbi algorithm, yields the minimum probability of error in decoding an observed sequence as an allowed codeword out of a number of possibilities. A number of workers have examined Maximum Likelihood Sequence Estimation (MLSE) structures in connection with adaptive HF receiver designs [Clarke, 1981] [Magee et al.,1973] [Viterbi, 1967] [Forney, 1972, 1973] [Ungerboeck, 1974] [Hayes, 1975]. MLSE, like equalization, may take advantage of *implicit diversity* gain.

Crozier et al.[1982] describe a 2400 bps modem suitable for HF communications based upon MLSE principles. The adaptive receiver, termed an Ungerboeck receiver, enables the following process to achieve MLSE: (modified) Viterbi decoding in conjunction with adaptive matched filtering. If the channel state is known exactly, the Ungerboeck MLSE receiver is optimal in the limit of white Gaussian noise. Decision-feedback equalizers may also be used for sequence estimation but they are not optimum methods.

7.6.8.8 *Sampling of HF Digital Modulation Systems*

The array of digital radio components including modems, transceivers, system controllers and associated digital devices has grown dramatically during the 1980s and early 1990s. Of particular interest are digital modems and radio systems which exploit the more advanced digital concepts for ionospheric disturbance compensation. Table 7-11 is a sampling of systems which have been developed for use at HF. Not all have been successfully fielded, being conceptual developments only. Some data for construction of Table 7-11 has been extracted from company brochures, but the primary sources include government reports, and papers published in refereed journals or presented at topical conferences. The list is not exhaustive, only indicative.

7.6.9 System Technology: Expert Systems and AI Applications

7.6.9.1 *Introduction*

This topic has been examined by the author [Goodman, 1989] with special emphasis on real-time-decision-aids (RTDAs) such as RTCE devices including sounders. The need to extract current information for immediate application has placed a premium on speed and accuracy, factors which are not totally consistent with human operators or analysts. Improvements are being made in at least two areas: off-line RTCE data interpretation and the application of RTCE data in adaptive systems. Expert or knowledge-based systems may be exploited for such purposes.

On the Development of Decision Aids

The advance of micro-computer technology, the growing sophistication of specified propagation models, and the expanding ability to sense the medium and apply that knowledge in real time is leading to an improvement in system performance predictions making such products more meaningful to the tactical military. At the same time, this real-time capability may be exploited more directly through use of sophisticated systems to adapt the modulation or coding process to an optimum structure. Advanced system architectures and DSP is leading to a fundamental change in the way HF is viewed. There are a number of tools available which can make HF more viable, and the selection rules will depend upon the communication requirements and the nature of the time-varying HF channel.

The maturation of artificial intelligence disciplines should provide the user of advanced C^3I decision aids an ability to manage the plethora of information more effectively. A major factor in this field of activity is the evolution of *self-adaptive* system architectures. In this context, the decision aid may be exploited by other machines or processes with little or no human intervention.

TABLE 7-11. DIGITAL RADIOS AND MODULATION SYSTEMS[a]

SYSTEM	RESPONSIBLE AGENCY [or Developer]	REFERENCES
Adapticom	U.S.Army Electronics Cmd.	[Goldberg, 1966]
ALIS Processor[b]	Rhode and Schwarz	brochure [Rohde and Schwarz, 1991]
Andeft	General Dynamics Corporation	[Porter, 1968]
Artrac	ITT Standard Radio, Sweden	[SjÖbergh, 1984]
Autolink[c]	Harris Corporation	brochures [Harris Corp., 1991a,b]
Chirpcomm[d]	BR Communications	brochure [BRC, 1990]
Codex	Codex Corporation	
Echotel	AEG Research Center, FRG	[Brakemeier and Lindner, 1988]
FRM		[Zimmerman and Harper, 1967]
Harpcall	Zellwegen Telecommunications	[BÜrgisser and Norton, 1991]
-------	Hughes Aircraft Co.[e]	brochure [Hughes, 1989]
Kathryn AN/GSC-10	U.S.Army Electronics Cmd.	[Goldberg, 1966] [Zimmerman and Kirsch, 1967] [Bayley and Ralphs, 1972]
Kineplex	Collins Radio	[Doelz et al.,1957]
Link-Plus	AMAF Industries	brochure [AMAF, 1990a]
--------	Loral Terracom[f]	brochure [Loral, 1990]

----

a. This listing is intended to be representative of many of the products available to assist in the transmission of digital data over HF links. The equipments may also have use for ALE, ALM, and control functions. The list is certainly not exhaustive. Any inclusion (or exclusion) of HF products should not be interpreted as approval (or disapproval) by the author. A complete list is available from a variety of sources, including JMG Associates, 8310 Lilac Lane, Alexandria, Virginia 22308-1943.

b. Rohde and Schwarz also make the HF 850 radio equipment equipped with ALIS, other options.

c. Harris Corporation has developed the RF 3200, RF 5000, RF 7100, and RF 7200 series, the RF 5254B serial tone and the RF 3466T parallel tone modems, and the AN/URC-119(v) and the AN/URC-121(v).

d. BR Communications markets a variety of HF equipments which are compatible with its niche in the HF market, chirp waveform sounders. A current version of the AN/TRQ-35, the Model RCS-6 Chirpsounder receiver (nomenclatured AN/URQ-39) includes Chirpcomm, an other features. Other equipments include the Model 4180 transceiver, the Model FMT-5A adaptive frequency management terminal and the Model 6078 time diversity modem.

e. Hughes Aircraft, Ground Systems Group, has developed the AN/PRC-104B which is light-weight manpack radio. This is part of the U.S. Army IHFR family.

f. Loral Terracom developed the AN/PRC-132 radio used by the Special Operations Forces.

TABLE 7-11. DIGITAL RADIOS AND MODULATION SYSTEMS [Continued from previous page]

SYSTEM	RESPONSIBLE AGENCY [or Developer]	REFERENCES
---------	Mackay Communications[g]	brocures [Mackay, 1991]
Marconi Serial Modem	GEC/Marconi	[Whittemore et al.,1988]
Model 1045[h] Data Controller	Frederick Electronics	[Conti, 1991]
Transcall[i]	Transworld Communications	brochure [Transworld, 1991]
Pathfinder 9000 Series	Sunair Electronics	brochure [Sunair, 1991]
Piccolo	United Kingdom	[Bayley and Ralphs, 1969]
Robust	Magnavox Corporation	[Baker, 1988]
---------	Signatron[j]	brochure [Signatron, 1987]
Stresscom[k]	ITT	[Wetmore, 1988]
VSM[l]	Naval Research Laboratory[m]	[Tate, 1989]
---------	Tadiran[n], Israel	[Shpigel and Perl, 1988]

g. Mackay Communications markets a variety of HF systems including the AN/GRC-223 tactical HF radio set, the MSR 8000 transceiver, the MSR 5050 receiver, and various modems, control units, and associated equipment.

h. Frederick Electronics markets other HF equipment including a data modem, the Model 1102.

i. Transworld makes the Model RT-1616 transceiver, DATACOM terminal, and a variety of HF SSB/FSK products. The "Transworld 100 System" includes the TW100 transceiver which is advertised as ALE-compatible with FED-STD-1045 and MIL-STD-188-141A.)

j. Signatron has developed the S-821 300 bps modem designed for LPI operation, burst transmission, and covert communications.

k. ITT has previously developed the Newlook system, a system designed for U.S. DoD strategic applications. The system is not adaptive, but designed to operate optimally in an environment for which RTCE (and sounding) is either inappropriate or impractical. Newlook/Stresscom are in Section 7.8.6.1.

l. The Variable Speed HF Modem (VSM) is a light, compact HF modem for transmitting information at 2400 bps. The modem is compatible with the AN/USQ-83 and ANDVT TACTERM.

m. The Naval Research Laboratory (NRL) has been involved in HF technology base R&D, along with other government laboratories in the U.S. and other countries. In the United States, noteworthy developments have been made by the Army CECOM, the Naval Ocean Systems Center, and the Air Force Rome Laboratories.

n. Tadaran markets the HF-2000 series of HF equipments. They also have developed a frequency management system, MESA.

In some instances, the application may involve human-like interpretations of RTCE data sets, especially if ionospheric properties are to be ascertained. Knowledge-based processes mimic the human operator in terms of cognitive skill while departing from negative human behavior traits such as fatigue, boredom, and distractions. These factors may lead to errors or logical inconsistencies.

Media Specification and RTCE

It is well recognized that the High Frequency (HF) radio regime (3-30 MHz) is the most precarious band in terms of its interaction with the ionosphere. Ionospheric refractivity exhibits considerable variability in both space and time by virtue of its strong dependence upon the ionosphere. Nevertheless it provides for Beyond-Line-Of-Sight (BLOS) communication connectivity and long-range surveillance potential. The nature of HF skywave propagation is known to be strongly influenced by ionospheric personality, and difficulty in the assessment of these traits (or physical properties) may be encountered in support of quasi-adaptive systems which are designed to allow for near-real-time adjustment of (system) parameters, based upon *trait tracking* and feedback, to provide for an optimization of overall system performance. The difficulties experienced in this assessment stem from several factors including: the nature of ionospheric variability itself, the mission of the system, and the definition of the system performance metric (i.e., measure of success), which defines the parameters of the assessment process. These parameters involve specification of the most appropriate Real-Time-Channel-Evaluation (RTCE) scheme to be employed, the duty cycle associated with RTCE application, and other notions such as model update, and so on.

The role which RTCE plays in the optimization of HF performance has been dealt with extensively in the literature (viz., see Darnell [1986]). It is obvious that RTCE schemes should be imbedded within the HF system itself for maximum effectiveness; nevertheless relatively successful exploitation of non-organic and *out-of-band* ionospheric sounding systems has been well documented. These systems, such as the well-known AN/TRQ-35, provide the system manager with the information which may be used to select the best available frequency for transmission.

With the growth in development of automatic HF radio communication concepts, advanced digital modems, and microprocessor technology, the intrinsic capability to rapidly adjust system parameters to an optimal set has evolved to the point that the ultimate limit in HF system performance may be associated with media specification. Media specification is thus the fundamental limit on the capability to tailor performance to the optimal value.

Trends Toward Automation and Expert Systems

Human operators are the controllers in most operational HF radio systems, but it is evident that modern technological advances are changing the architectures of the next generation systems. Adaptive HF schemes are leading to fundamental improvements in the potential for an improvement in HF communication performance. Automation is the key to effective system management and control, and the incorporation of RTCE is an essential ingredient along with the availability of fully-adaptive system components. These components include those which permit changes to be made in selected parameters at various levels: i.e., the transmission level, the link level, the network level, and the overall system level. Figure 7-10 gives a picture of the decisions which may be involved in the operation of an adaptive HF system.

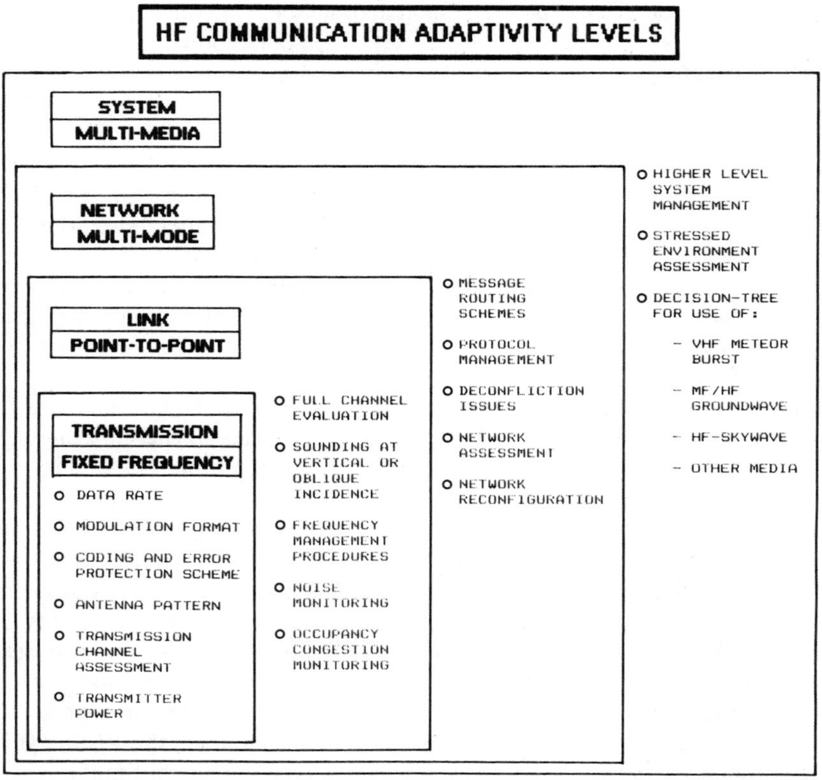

HF COMMUNICATION ADAPTIVITY LEVELS

SYSTEM
MULTI-MEDIA

NETWORK
MULTI-MODE

LINK
POINT-TO-POINT

TRANSMISSION
FIXED FREQUENCY

O DATA RATE
O MODULATION FORMAT
O CODING AND ERROR PROTECTION SCHEME
O ANTENNA PATTERN
O TRANSMISSION CHANNEL ASSESSMENT
O TRANSMITTER POWER

O FULL CHANNEL EVALUATION
O SOUNDING AT VERTICAL OR OBLIQUE INCIDENCE
O FREQUENCY MANAGEMENT PROCEDURES
O NOISE MONITORING
O OCCUPANCY CONGESTION MONITORING

O MESSAGE ROUTING SCHEMES
O PROTOCOL MANAGEMENT
O DECONFLICTION ISSUES
O NETWORK ASSESSMENT
O NETWORK RECONFIGURATION

O HIGHER LEVEL SYSTEM MANAGEMENT
O STRESSED ENVIRONMENT ASSESSMENT
O DECISION-TREE FOR USE OF:
- VHF METEOR BURST
- MF/HF GROUNDWAVE
- HF-SKYWAVE
- OTHER MEDIA

Fig. 7-10. Decision Levels Associated with Adaptive HF Systems. Four basic levels of adaptivity are given in boldface capital letters. Also given is a one-word descriptor (in caps) and a set of attributes for each level.

There are a number of other systems which are available for use as sources of RTCE information. Systems which incorporate Automatic-Link-Establishment (ALE) and Automatic-Link-Maintenance (ALM) require that RTCE data be assessed either implicitly or explicitly. The hierarchy of HF ALE systems has been described by Harrison [1987]. He identifies ten "logical and unique functions which are critical to automatic linking in the HF environment". The so-called standard levels of automated HF systems described by Harrison were depicted in Figure 7-8. Each level of automation proceeds logically and stepwise from the lowest adaptability quotient to the highest. Level one corresponds to a *plain vanilla* (non-adaptive) radio and the ultimate system at level ten corresponds to a full complement of adaptive features. It is noteworthy that level 10 functions would be based upon HF-ALE operational rules designed to duplicate the skill levels of experienced HF operators and system managers. Clearly expert system methodologies apply in the realization of Harrison's model of HF system automation.

7.6.9.2 *Examples of Decision Aids and AI Applications*

Top Level Decision Making

There are a number of initiatives in the services to address the problems of C^2 decision-making. Dacunto [1988] describes current Army C^2initiatives in terms of the Army Command and Control Master Plan (AC2MP), and indicates that the army decision-makers, in order to exercise power effectively, must use information to develop the products of C^2. These products are decisions and directives. Information is thus the key. Command and control processing begins with information acquisition and ends with execution of a directive or decision. It is clear that in systems of this type, decision aids will prove to be essential avenues for facilitating the process, and timeliness is normally a desirable attribute.

Battlefield Spectrum Management

The Army has developed battlefield spectrum management schemes to facilitate the function of communication which is termed the *glue* which holds C^2 together. Within the HF domain, the Army PROPHET Evaluation System (APES) employs a family of frequency and resource management tools based upon propagation (and possibly threat) models which are resident on a PC. This army development, and nearly-equivalent versions of the PROPHET methodology addressing needs of the sister services, has brought many of the tools of decision-making to the ultimate user rather than simply the C^2 manager. This decentralization may introduce risk if procedures are not properly organized in advance but time from decision to execution is reduced by this form of vertical divestiture.

Artificial Intelligence (AI) schemes have recently been described in connection with the battlefield spectrum management (BSM) problem [Morcerf et al.,1987]. In the BSM arena, automation concepts for system control and operation are driven by increased complexity of the newer and more versatile radios, the need to coexist with a proliferated equipment environment, and a requirement to improve reaction time.

Propagation Assessment

The PROPHET system (and its variants) developed by NOSC is one of the first successful examples of a full class of computer-based decision-aids at HF [Rose, 1981]. Richter [1989] has reviewed decision aids in the context of propagation assessment, emphasizing products developed at NOSC. Rose [1989, 1991] has recently explored expert systems as an aid in propagation disturbance assessment, and a new program called the Disturbance Impact Assessment System, DIAS, emphasizes high latitude effects. Rulesets in the DIAS program allow assessment of SID, PCA, ionospheric storms, auroral absorption, auroral Es, and auroral E effects.

Roesler [1990] has developed an HF/VHF resource management expert system called PROPMAN. Propagation rules are established by means of an ongoing data collection effort. The concept includes an accumulation of other solar-terrestrial data derived from the Space Environment Services Center, SESC, in Boulder by means of satellite link or landline. This facet is similar to PROPHET, but the methodology for HF prediction is different.

Sounder Scaling and Propagation Analysis

The problems associated with sounder-aided propagation predictions can be substantial. Sanders [1986] has described a general system dedicated to HF battlefield spectrum management which uses sounders as a basis for prediction improvement. A basic difficulty arises in connection with interpretation of the ionograms, a matter discussed in Chapter 6. Dickson et al.[1991] have developed a DSP enhancement to the Chirpsounder™ receiver called the Radio Oblique Sounding Equipment, ROSE. This system is an add on to the Chirpsounder™ and provides for a color display as well as a significant improvement in resolution.

Ionogram scaling is also a fundamental weakness in the sounder approach. Vertical incidence sounder scaling approaches have been discussed in Chapters 4 and 6. Progress in the area of oblique incidence sounder scaling has been more sluggish, although the Digital Oblique Remote Ionospheric Sensing Program, DORIS, developed at the University of Lowell for the U.S. Air Force [Kuklinski et al.,1989] has shown promise. One area where progress may be achieved is the maintenance of accurate time at the transmitter and receiver. The absence of this capability has lead to an uncertainty in the establishment of absolute values for the ionospheric heights of reflection, which is an impediment to certain model update schemes. Daehler [1990] has proposed the use of GPS clocks at all sounder terminals.

Many ionograms are not clean and must be enhanced in order to develop a picture of the mode structures involved. Following this process, mode identification must be achieved if any ionospheric analysis is stipulated. This is usually the case in HFDF and HFSSL operations, but may be less critical in HF frequency management scenarios. A standard method for mode identification involves a comparison of the observed ionogram with a synthesized ionogram.

HF System Design and Planning

Approaches to HF system design based upon AI disciplines have been discussed by Jowett [1987], and Chesmore [1991] has described aspects of knowledge-based systems as applied to HF system design and control. For the purpose of evaluating electromagnetic compatibility (EMC) of colocated HF systems, a computer-aided design tool called the Communication Engineering Design System, COEDS, has been developed [Alexander et al., 1991]. COEDS may also be used to automate the development of frequency management plans.

7.7 A SAMPLING OF HF SYSTEMS AND NETWORKS

In this section we shall identify a few HF communication systems as examples of current capability. In most instances, the sampling pertains to systems consisting of radios having common design which are organized in a network configuration. These systems are generally deployed to satisfy a specific mission. In one case (i.e., SHARES), the system is composed of an aggregate of dissimilar systems which are organized to satisfy a Federal government requirement in time of emergency. Those included are not a complete set, and there is no implication concerning system efficacy relative to other systems which may not be included in the discussion. Naturally, no sensitive military networks are mentioned. Even so, there are a host of other systems which could be identified, but space will not permit a discussion.

Scope Signal is a U.S. Air Force program to upgrade HF ground station facilities. It constitutes a network and replaces the Giant Talk HF system.

7.7.1 General Commentary on HF Networks

Networking is one of the most powerful means of thwarting spatially-dependent impairments, provided the impairment source exhibits low instantaneous correlation over a subset of links which comprise the network. Networking has value regardless of the switching system used, but a packet switching architecture would generally be preferred over message switching. At HF, the concept is used in strategic systems such as *Newlook* (see Section 7.8.6.1). The goal of networking is to achieve reliable *end-to-end* connectivity, while individual *point-to-point* connectivities may be quite low. The *Telecommunications Dictionary* offers one definition of network as "an organization of stations capable of intercommunication...", but there is an implication that such intercommunication should be accomplished automatically. The incorporation of solid state equipment, robust/adaptive modems, and ALE capability have made automatic relaying functions perform more efficiently. O'Mahony [1991] has examined HF networking and lists the following essential control functions for an ensemble of nodes: management of message queues and priorities, automatic route selection, store-and-forward relaying, control and monitoring of radio equipment, adaptive selection of frequency and data rate, error control, exchange of connectivity and status information, frequency management at all levels, and message I/O. The control functions are similar to those listed in Table 7-6. Existing and proposed standards FED-STD-1045 through FED-STD-1049 deal with networking issues at various levels (see Table 7-17).

The Open System Interconnection (OSI) model for data systems is an internationally accepted standard which allows systems to interconnect and interoperate. The OSI model[92] is composed of three general groups of levels, being ultimately partitioned into seven layers or functional tiers. The three fundamental groups are: a network-dependent group of three layers, a single transport layer, and a group of three application oriented layers.

Several network studies/projects are ongoing. One is the Improved HF Data Network (IHFDN) being developed by Harris Corporation under contract to the USAF Rome Labs. This network [Baker and Robinson, 1991] will handle digital voice and data at rates between 75 and 2400 bps in a 3 kHz bandwidth, will allow Doppler shifts up to 50 Hz, will handle multipath of 5 milliseconds, and will operate in a fading environment. A network simulator, the INS, has been developed to test the system. A layered approach was used in the network design (i.e., the OSI model), and attributes of the MIL-STD-188-110A modem were emphasized.

---

92. OSI is an internationally agreed model for data systems which interconnect. The term "Open" implies that the data systems utilize a set of protocols sanctioned by the International Standards Organization (ISO) and the CCITT. The seven layers are:
    1. Physical Layer: the electrical and mechanical interface.
    2. Link: moves data between connected nodes.
    3. Network: interfaces to packet network, packetizes, reassembles message, flow control.
    4. Transport: moves message from source node to destination node.
    5. Session: establishes, maintains, and terminates the logical link.
    6. Presentation: editing, mapping, translation.
    7. Application: application programs and dialogue.

    (Source: Telephony's *Telecommunication Dictionary*, by G. Langley, 2nd edition, Telephony Publishing Company, Chicago, IL., 1986.)

7.7.2 Agency Coordination: SHARES

Based upon a National Security Council (NSC) recommendation to President Kennedy in 1963, the National Communications System (NCS) was established in order to establish a nationwide communication capability for use by the Federal government in times of crisis. In 1984, by Executive Order 12472, President Reagan established an administrative structure for the NCS whereby the Defense Department (DoD) was the Executive Agent for the system, and the Director of the Defense Communications Agency (DCA) would be the system manager. Twenty three agencies and departments of the U.S. government are presented in the NCS administrative structure.

The NCS does not rely on any specific communication medium exclusively. However, HF radio is a major contributor to the integrated capability. One of the success stories of the NCS is a program called SHARES which stands for *SHAred RESources*. SHARES[93] is an HF radio program which provides a backup capability to the Federal government to pass/exchange emergency information using existing HF assets including radio amateurs. Purposes of SHARES include: enduring backup to vulnerable leased systems, provision for extended HF coverage for use by all agencies, provision of the *flagword* SHARES to identify critical message traffic, standardization framework for operational procedures and message formatting, and provision for a "work-around" for jamming by providing more frequency availability to circumvent the threat. Participation in SHARES is open to all Federal agencies, and the responsibilities include the maintenance of a SHARES Directory, and the participation in readiness training and exercises. The Directory consists of information about member station capability (i.e., operational modes and schedule). Sample modes of operation include those given in Table 7-12.

7.7.3 Cross Fox

The Cross Fox communication system is an LF/HF maritime communication system in support of NATO requirements in the northern and central regions of Allied Command Europe (ACE). The system is designed to operate on the common user principle and is centrally controlled. The system is composed of 28 fixed sites under the control of either of two designated management centers (MCs). Capabilities include broadcast, ship-to-shore, and point-to-point communications.

---

93. SHARES documentation includes: a description of the program (NCS Directive 3-3), a user's manual (NCS Manual 3-3-1) which gives operating procedures and message formats, and the SHARES Directory (NCS Handbook 3-3-1) which provides station ID by state, agency, and call sign.

TABLE 7-12. MODES OF OPERATION FOR SHARES STATIONS

A. Single Sideband (SSB) Voice
B. Radio Teletype (RTTY)
C. Continuous Wave (CW)
D. SSB Voice + RTTY
E. SSB Voice + CW
F. SSB Voice + RTTY + CW
G. RTTY + CW
H. Packet Radio
I. FED-STD-1045 Adaptive
J. SELSCAN[94] Adaptive
K. AUTOLINK[95] Adaptive
L. SELCAL[96] Adaptive

Connectivity will be achieved as a result of a frequency management subsystem currently being developed and tested[97]. The frequency management subsystem (FMS) has been designed for the HF component of Cross Fox, and incorporates data obtained from chirp sounding and channel occupancy receivers located at several designated receiver sites. The FMS also includes computers, and related equipment which are to be located at the designated MC, and the hardware associated with the integrated capability to provide frequency management is referred to as the frequency management facility (FMF). Sounder transmitters are situated at locations of interest to insure coverage in the theater of operations. The FMF, and specifically its component chirp sounder receiver sites, will also take advantage of sounder transmission of opportunity.

The appropriate FMF provides data such as ranked choices of operating frequencies, sites, and antennas to the designated MC as requested. The determination of the optimum frequencies and equipments is based upon a number of procedures: ionospheric propagation prediction, ground wave propagation prediction, real-time ionospheric sounding, and real-time measurements of channel occupancy. The ionospheric model to be employed is IONCAP. The exact procedures to be followed in the final FMS process are being developed.

---

94. SELSCAN is a trademark of Rockwell-Collins Corporation.

95. AUTOLINK is a trademark of Harris Corporation.

96. SELCAL is a feature central to an adaptive HF radio developed by Hughes Corporation.

97. The Cross Fox frequency management subsystem is being developed by Harris Corporation under contract to the Norwegian Defence Communications and Data Services Administration.

7.7.4 Customs Over-the-Horizon Network (COTHEN)

The U.S. Customs Service, together with military and civilian law enforcement agencies, and in concert with a number of federal governments in the American zone, have developed a communication system which has proved to be useful in the *war on drugs.* The system includes the Customs Over-The-Horizon Network or COTHEN[98]. This communications capability is comprised of a network of HF SSB radios possessing ALE capability. As configured, the system is interoperable with other ALE systems which are consistent with federal standards, and is computer-driven [Williams, 1990].

7.7.5 FEMA National Radio System (FNARS)[99]

The U.S. Federal Emergency Management Agency (FEMA) has developed the FEMA National Radio System (FNARS) under a mandate from the President (viz., Executive Order 12472). FNARS is an HF SSB radio system which can transmit both voice and data, and has the capability to operate in both secure and nonsecure modes. The system must be interoperable with HF radio systems utilized by state and local public safety/health networks as well as federal agencies. It is also designed to support networks such the FEMA Switched Network (FSN), and the Mobile Air Transportable Telecommunication System (MATTS). In order to exploit all possible residual communication capacity which may be available during emergencies, it is planned to accommodate the vast resources associated with radio amateurs, including RACES and the Military Affiliated Radio System (MARS)[100].

FEMA will manage, operate and maintain FNARS which will consist of systems located Washington DC (Headquarters), Sperryville VA, 10 Federal/Regional Centers, and 59 State/Territory Emergency Operation Centers. During emergency and crisis situations, all of these specified sites will be interconnected. Additional connectivity will be achieved with the Armed Services, the Coast Guard, and local communication systems (viz., police, fire departments, hospitals, etc.). FNARS Transmitter power ratings range from 1-10 kwatts.

----

98. COTHEN was developed by Rockwell-Collins Corporation, and possesses features contained in SELSCAN, a product of Rockwell-Collins.

99. The FINARS, is based upon technology developed by Harris Corporation.

100. The primary purpose of MARS is to serve as an alternative means of communication during during emergencies. The MARS network consists of about 233 military and 3800 amateur radio stations. MARS is used to send *MARS-grams* and has been used for phone patches.

FNARS radios will include both fixed site and mobile versions, and will have ALE capabilities incorporating LQA and preset channel scanning. Users of the system will be able to make selective calls (i.e., individual or group) and broadcasts.

7.7.6 National Radio Communication System (NARACS)

The National Radio Communications System (NARACS) is a voice and data communications system between Federal Aviations Administration (FAA) Headquarters office in Washington D.C. and regional offices, support aircraft, and specified facilities [DoT, 1989]. During emergency situations the NARACS will provide the minimum essential communications for management, operation and/or reconstitution of the National Airspace System (NAS) in support of FAA/DoT and U.S.DoD missions. During situations which are not classified as national emergencies, NARACS supports crash site investigations, aviation security, and routine day-to-day operations.

NARACS provides an HF SSB capability, and automatic system operation capabilities are patterned on the basis of a SELSCAN™ architecture.

7.8 STRATEGIC HF SYSTEMS AND THE NUCLEAR ENVIRONMENT

The term *strategic HF* commonly refers to HF systems which have been designed to operate in a nuclear-stressed environment.

7.8.1 Introduction and Reference Material

One of the principle reasons HF communication systems have enjoyed a modest capital and R&D investment has been the perception that HF should be included in a balanced mix of C^3 assets which together would have a high probability for survivability in the face of nuclear engagement. There is a school of thought that HF is more vulnerable to nuclear effects than any other component of the suite of communication systems. Although this may be so, depending upon the nature of the communication service required, it is also true that the HF channel tends to repair itself over time. So there is an ultimate survivability, if not continuity, associated with the use of HF in a nuclear-disturbed environment. In a sense, the pathological behavior we attribute to nuclear disturbances are only an exaggerated version of the normal environment. Nevertheless the rapid changes which may be introduced in the trans- and post-nuclear environments will place the greatest stresses on HF systems whether they were designed to be fully adaptive or not. This section we shall outline some aspects of the nuclear environment as they pertain to HF systems.

A survey of nuclear phenomenology and first-order effects on radiowave propagation systems may be found in *The Effects of Nuclear Weapons* by

Glasstone and Dolan [1977]. Radar and radio propagation effects are examined in Chapter 10 of their book. Although necessarily limited in detail, it remains a valuable resource since much of the documentation which pertains to this subject is not available to the public.

Various nuclear effects on the ionosphere and radio propagation have been reported. One of the first available treatments is due to Utlaut [1958]. Subsequently Knapp et al.[1967] reviewed propagation effects, and Crain and Booker [1963] have examined nuclear effects on an HF radiowave circuit. HF communications effects have also been studied by Dayharsh [1965] and Nielson et al.[1967].

Hoerlin [1976] published a summary of United States nuclear test experiences. Local and large area effects were both observed during the active nuclear test period and results have been published, principally in technical reports. The *Argus* experiment [Christofilos, 1959], consisting of three A-bomb detonations in the exoatmosphere, created an artificial electron belt lasting for more than a decade. Although of academic interest, the implications of extremely high altitude shots for HF system operation is not overly critical. Low-altitude and ground-level detonations of modest yield have also been reported, but the primary propagation consequence (i.e., absorption) is largely restricted to line-of-sight links which intersect the fireball. Multimegaton bursts at tropospheric heights or even at ground level may cause ionospheric (D-layer) absorption at late times as the radioactive debris ascends buoyantly beyond the so-called *stopping altitude* of ionizing radiation. Such bursts may also introduce ionization redistribution due to interaction of the ionospheric plasma with the detonation shock wave. The *prompt* effects become important when the detonation occurs above the stopping altitude. Some of the more interesting phenomena were observed during the *Hardtack* series (*Orange* and *Teak* shots) in the late 1950s and the *Fishbowl* series of high altitude nuclear tests in the early 1960s (*Starfish, Checkmate, Bluegill,* and *Kingfish*). These tests, involving detonations over Johnston Island in the Pacific Ocean, were coordinated with ionosonde measurements enabling maps of ionospheric disturbances to be constructed.

Acoustic gravity waves generated by high altitude nuclear detonations were observed by a number of investigators [Kohl, 1964] [Lichtman and Andersen, 1962] [Newman et al.,1966]. These phenomena, which have a unique signature, are associated with large scale traveling ionospheric disturbances (TIDs) which are actually tracers to the underlying neutral gas motion. Nuclear-generated TIDs may play a key role in HF connectivity, not only in the hemisphere of the burst but in the conjugate region as well. The ionospheric response to nuclear-generated TIDs has been documented in papers by Nelson [1968] and Lomax and Nielson [1968].

7.8.2 Coping with the Nuclear Environment.

The nuclear environment is thought to place a greater stress on HF systems than on systems operating at other frequency regimes. Implicit in this view is the assumption that nuclear detonations have a substantial effect on the ionosphere and its structure. The extent of HF system disruption depends crucially upon the details of the ionospheric impact and its time history. This, in turn, is a function of the height and yield of individual bursts, and finally upon the sequencing of multiple bursts. We will not cover actual scenarios in this section for obvious reasons. We shall only cover the salient features which are available in the public domain.

The first step in the coping process involves examination of available test data to determine the range of effects. A second step would involve empirical modeling of effects as extrapolated from available test data. Due to the paucity of test data, this is not always possible. This suggests two additional and parallel steps: (a) the search for natural events which might serve as mimics of nuclear-disturbed conditions, and (b) theoretical analysis and synthesis of effects using computational physics approaches. As in most situations for limited data is available to describe phenomena, hybrid semi-empirical models are the result. The complete picture is quite complex involving as it does a wide range of disciplines including: fluid mechanics, electrodynamics, magnetohydrodynamics, upper atmospheric chemistry, and details of nuclear detonation biproducts as a function of the height and yield of burst. Once an appropriate phenomenological model is specified, a model describing the radio system is a logical next step. Table 7-13 outlines the major input parameters and products associated with nuclear effects models.

As has been suggested, modeling is the principal method of coping with nuclear effects. A number of models[101] such as *HFNET*[Frolli,1979] and *NUCOM*[Nelson, 1976] have been developed and used in a variety of nuclear *laydown* scenarios.

---

101. Major nuclear effects codes are available through the auspices of the U.S. Defense Nuclear Agency, Washington DC. *HFNET* was developed by Mission Research Corporation and was designed to compute nuclear effects for large networks, possibly involving multiple bursts. *NUCOM* was developed by SRI International and provides more detailed system performance information for single links with a variety of receive antenna selections. For *HFNET* the detailed antenna analysis is performed off-line. Other disturbed media codes are generally needed as inputs to the propagation codes (viz., *SCENARIO*). For information about these and similar nuclear effects codes, the reader should contact the Defense Nuclear Agency.

TABLE 7-13
PARAMETERS ASSOCIATED WITH NUCLEAR EFFECTS MODELS

INPUT INFORMATION
- Time of Day
- Season (or Day of Year)
- Solar Epoch (or Solar Activity)[101]
- Geographic Coordinates of Burst(s)
- Geomagnetic Coordinates of Burst(s)[102]
- Altitude and Yield of burst(s)
- Burst Scenario (or Sequencing)
- Noise Conditions Specified
- Interference Conditions Specified
- Propagation Path(s) (or Network Topology)
- Selected Media (HF Skywave, HF Groundwave, HF-LOS, Meteor Burst, Mixed)
- Selected Propagation Modes (All,Mode-specific,Hop-specific)
- System Performance Figure of Merit (or Requirement for Success)
- System(s) Parameters[103]
  - Antenna Types
  - Modulation and Coding Formats
  - Power Gains Product
  - Diversity (Space,Time,Polarization,frequency)
  - Receiver Processing

MODEL TYPES
- Atmospheric
- Ionospheric
- Geomagnetic Field
- Propagation (Stand-alone, Combined with Other Submodels)
- Noise & Interference (CCIR, Other)
- Nuclear Phenomenology Model
- System and Channel Model
  - Voice, Data, TTY, Other
  - Switching Architecture (Message,Packet)
  - Waveform (Wideband,Narrowband,Spread Spectrum,etc.)

OUTPUTS
- Transmission Losses (Fireball, Debris, *Beta Patch, Gamma Ray Shine*)
- Refraction Effects (Layer Penetration, Tilt, Gradient, and Lens Effects)
- Scatter Effects (Enhanced Spread-F, Auroral Clutter, *Bomb Modes*)
- Signal Effects
  - Fading & Scintillation
  - Doppler Spread and Shift
  - Multipath Spread
  - System Performance (BER, Reliability, Throughput)

----

102. The sunspot number (or the 10.7 cm solar flux index) may be entered directly, or it may be derived based upon a projection and the specification of the future date. Climatological models may also be driven by other solar and magnetic activity indices. (see Chapter 2 for details.)

103. Magnetic coordinate specification may be an intermediate calculation derived from the magnetic field model rather than an input parameter.

104. Input parameters may be prespecified by selection of the System Model.

7.8.3 Height-Dependent Effects: Prompt and Delayed

The height and yield of nuclear detonations has a profound influence on the amplitude, geographical extent, and duration of disturbance effects such as absorption-related blackout. If we fix the yield of the burst we may estimate the spatial extent and duration of outages as a function of detonation altitude. The most significant ionospheric impact of nuclear detonation occurs for 1 Megaton shots in the height range 30 - 300 km. For smaller yields the lower limit may approach 60 km, while for larger yields the lower limit may descend to tropospheric levels. The major factor is the energy contained in the buoyantly-rising radioactive debris cloud. In this connection it important to recognize that a *stopping* or *containment* altitude exists for major forms of radiation such as x-rays, gamma rays, beta particles, alpha particles, and other debris ions. X-rays, gammas, neutrons, and debris particles are generated at detonation time and are available to produce *prompt* ionization in the D-layer of the ionosphere. It is required, however, that the radiation originate above the *stopping* altitude for excess D-layer ionization to be produced immediately. *Delayed* D-layer ionization will be produced at the moment the rising radioactive debris transcends the containment altitude. In this case the primary ionization sources include betas and gammas. *Stopping* altitudes for the various species are given in Table 7-14. Ionizing radiation, be it *prompt* or *delayed*, will produce D-region absorption and may be sufficient to cause a blackout of HF circuits which penetrate the enhanced D-layer. The prompt ionization enhancement is relatively short-lived, being roughly an hour in duration. The delayed effects may last for many hours or days, depending upon the intensity, height, and dimension of the debris cloud.

7.8.3.1 *Beta Patch and Gamma Ray Shine*

Ionization enhancement in the D-layer has two forms; one associated with gammas (and x-rays) and called *Gamma Ray Shine*, and the second associated with betas and termed *Beta Patch*. The former produces excess ionization in every part of the D-layer which is within view of the radioactive debris. Thus *Gamma Ray Shine* is not unlike the D-layer absorption introduced by x-ray flares, but with a different vantage point and source phenomenology. *Beta Patches*, on the other hand, are associated with high energy extranuclear electrons and are constrained to travel down (and up) geomagnetic field lines. Since betas do not suffer R^2 spreading, being effectively mapped from the debris cloud to the D-layer interaction point, *Beta Patches* are far more intense layers of excess ionization than those which are associated with *Gamma Ray Shine*.

7.8.3.2 *Acoustic Gravity Waves (AGWs)*

It is well known that AGWs may be generated by a variety of sources and that Traveling Ionospheric Disturbances (TIDs) are the ionospheric tracer to the underlying neutral gas motion. The response of the ionosphere to AGWs has been established by Hines [1960], Nelson [1968] and others. The nature of the AGW signature for a specified nuclear detonation depends, as in the prompt ionization case, on the height and yield of the burst. The nuclear detonation produces a blast wave which progresses in all directions. The component of the shock which propagates into the tenuous upper atmosphere is transformed into a traveling disturbance of the AGW type so familiar to ionospheric workers. These disturbances, observed indirectly by means of a spaced ionosonde network have been found to travel at speeds of the order of 630 meters/sec [Kohl, 1964]. The oscillation period of the disturbance increases slowly with distance from the burst, ranging between roughly 30 and 90 minutes over a propagation distance of 4000 km. The ensemble of speeds, wavelengths, and amplitudes vary as a function of ambient atmospheric conditions as well as nuclear burst properties. Some of more interesting properties of AGWs include: (a) increasing disturbance amplitude with increasing altitude as a consequence of the exponential decrease in gas density with height, (b) gradual dissipation of AGW amplitude in the upper F-layer as a result of an increase in kinematic viscosity, (c) surfaces of constant wave phase which tilt in the direction of motion of the wave, and (d) energy flow upward and parallel to the surfaces of constant wave phase. These AGW properties and their consequences are summarized in Table 7-15.

TABLE 7-14. *STOPPING* OR *CONTAINMENT* ALTITUDES

WEAPON OUTPUT		STOPPING ALTITUDE (km)
A.	*Prompt* Radiation	
	X-Rays	48 - 88
	Gamma Rays	24
	Neutrons	24
	Debris Ions	113
B.	*Delayed* Radiation	
	Beta Particles	56
	Gamma Rays	24

The magnetic field has a profound influence on the electron density distribution in the ionosphere. Poleward motion of an AGW will, through ion-neutral drag forces, drive ionization downward compressing the electron-ion gas thereby increasing the electron density and the local critical frequencies. Equatorward motion will produce the opposite effect, since ions will be driven upward leading to an expansion of the electron-ion gas. For the *Fishbowl* series of tests over Johnston Island, a northern hemispheric site, AGWs influenced a general depletion of F-layer critical frequency in the neighborhood of the burst and in the southern region north of the magnetic equator. As the wave progressed over the equator where the filed lines are horizontal, no rarefaction or compression of the electron-ion gas was observed. However, southward of the magnetic equator, the field lines tilt downward again and significant increases in the ambient electron concentration and the parameter foF2 would be expected and were in fact observed. Due to the large diminution of electron concentration in the neighborhood of the burst, a phenomenon not fully understood, the anticipated increase in electron concentration poleward of the burst is disguised somewhat. The large diminution of foF2 in the burst area, while observed in many high altitude nuclear tests, is currently thought to be a nocturnal phenomenon. Conjugate region enhancements would be expected to occur both day and night.

7.8.3.3 *Large-Scale Diminutions and MOF-Failure*

The implication of large-scale diminutions in foF2 in the burst area is clear. Such events will lead to *MOF failure*, a type of blackout resulting from a lack of ionospheric support unless dynamic frequency management and robust relaying strategies are imposed. This situation may also be countered by use of skywave media alternatives such as VHF meteor burst or netted groundwave when possible.

TABLE 7-15. AGW Properties and Consequences

PROPERTY	CONSEQUENCE
A. AGW amplitude increases exponentially with altitude for energy conservation.	TID amplitudes are greater at F-layer altitudes than at the D and E-layer heights. This is also consistent with an overall increase in ionospheric variability with altitude.
B. Viscous damping of AGW arises as the kinematic viscosity increases with altitude. This form of viscous damping becomes important as the collisional mean free path becomes a significant fraction of the AGW wavelength.	Tendency for a monotonic increase in AGW amplitude with increasing altitude (by property A) is thwarted by loss of AGW energy. This loss is in the form of heat (i.e., disordered motion) which causes the temperature of the neutral gas and the electron-ion plasma to increase throughout the thermosphere. This process is a major factor in the upper atmospheric energy balance.
C. Surfaces of AGW phase tilt forward in the direction of horizontal motion. These surfaces gradually become vertical as viscous damping becomes dominant. (See B.)	Disturbances associated with AGWs (i.e.TIDs) are first observed at F-layer heights. Lower ionospheric effects are seen after some lag time and with reduced amplitude (property A).
D. Energy flow is along the surfaces of constant wave phase.	The *direct* energy sources for AGW are in the lower atmosphere rather than in the magnetosphere or elsewhere. Nevertheless the AGW energy source (i.e., seat of the disturbance) may have developed from magnetospheric particle precipitation or solar cosmic rays, as well as tropospheric events.

7.8.3.4 *Bomb Modes*

These peculiar modes of propagation are associated with reflection from enhanced regions of ionization in the E and F layers. These zones will presumably support ionospheric reflection in the VHF domain thereby mitigating D-layer absorption, a phenomenon which might render an HF system unusable.

7.8.3.5 *Spread F*

Spread-F, a naturally occurring phenomenon, has been observed in a number of nuclear tests. Spread F presents interesting signal processing challenges.

7.8.3.6 *Sporadic E*

Sporadic E ionization has been observed to be enhanced in a number of nuclear tests. These features may be exploited in a manner similar to *bomb mode* ionization if the system architecture admits to adaptivity.

7.8.3.7 *Auroral Arcs*

Striking auroral arcs and auroral displays were observed during the *Starfish* nuclear test. These visual displays gave way to less dramatic but quite substantial radio propagation effects.

7.8.4 Absorption and Refraction Regimes

Most of the pronounced absorption effects associated with HF skywave communication may be identified with the D-region enhancements in electron concentration, and refraction effects are largely associated with variations in E and (especially) F-region ionization. This allows nuclear effects analysis to be conveniently separated into absorption and refraction regimes. In most practical studies, absorption and refraction may be considered independently, although the former may strongly limit the importance of the latter.

Absorption associated with *prompt* ionization sources will cause disturbances in HF circuits for roughly an hour in the neighborhood of the burst, under the presumption, of course, that the appropriate *stopping* altitudes are exceeded and the weapon yield is the order of 1 Megaton or so [Glasstone and Dolan, 1977]. Outages would be most pronounced for NVIS and short-haul links for which lower HF frequencies would be used. The precise definition of burst *neighborhood* depends upon height and yield of the burst, but height is an ultimate constraint on prompt x-ray and gamma ray-induced D-layer ionization.

Delayed absorption effects depend critically on the evolution of the radioactive debris cloud. Since this is difficult to predict with accuracy, late-time effects on HF systems remains an uncertainty in most nuclear effects codes. The *beta patch*, being a rather concentrated zone of enhanced absorption, constitutes the major threat to connectivity for a fixed link.

Refraction effects are quite significant, especially at later times, as AGW wave effects introduce large TID effects including layer tilts, horizontal gradients, and critical frequency variations. Large redistributions of electrons, plume formations, and auroral arcs may introduce *bomb modes* thereby enhancing the capability to communicate at frequencies far in excess of normally propagating HF frequencies. This feature would allow for a reduction in absorption ($1/f^2$) will maintaining connectivity. There are some potential problems associated with large scale critical frequency diminutions which may have the opposite effect, a problem thought to most pronounced at night.

7.8.5 Scaling Rules

The frequency scaling rule for absorption in a nuclear environment is the same as it is for the benign case. We assume that the $1/f^2$ rule applies. The temporal scaling for absorption is difficult since it depends to such a strong degree on specified phenomenological regimes. Fireball attenuation, although intense, obeys a $1/t^2$ rule, while absorption within the D-layer is much slower. Nocturnal absorption follows a $t^{-1.2}$ dependence and daytime absorption has a range between t^{-1} and $t^{-0.6}$.

Neither the frequency nor the temporal scaling rules for refraction are well defined. Detailed raytracing methods are required.

7.8.6 Countermeasures to Nuclear Effects

7.8.6.1 *High Latitude Studies and the Search for Analogies*

As has been indicated previously, *HANE*[104] have analogies in the natural environment, especially under certain disturbed conditions. Auroral disturbances and other manifestations of geomagnetic storm phenomena seem to mimic nuclear effects which have been actually observed. The temptation then is to examine these natural events more carefully to see what lessons can be learned. For this reason, among others, studies of high latitude propagation effects have a special significance since magnetic storm phenomena are most pronounced in the polar, auroral, and high midlatitude regions. We have discussed many of these phenomena in Chapter 2 (Section 2.4.3) and Chapter 3 (Section 3.4). Means to overcome or partially compensate for high latitude effects can usually be applied to the nuclear case. Still, there are some striking differences which remain to be modeled since not all nuclear features can be characterized conveniently with natural analogies.

7.8.6.2 *Menu of Mitigation Schemes*

Table 7-16 contains a list of some generic mitigation schemes which may be used to avoid or ameliorate nuclear-induced disturbances. There are others which could be mentioned. All of the schemes involve diversity in one way or another.

Path diversity, frequency diversity, and relay-capable networking imply the capability to *go around* the nuclear disturbance in some manner. A diversity of nodes (i.e., terminal locations), paths, or propagation modes is necessary to insure this.

----

104. *HANE* stands for High Altitude Nuclear Effects. It is used to distinguish between effects arising from ground-level and tropospheric bursts, termed Low Altitude Nuclear Effects (*LANE*) and those above the troposphere. The most pronounced *HANE* corresponds to a detonation altitude above 30 km but below 300 km.

TABLE 7-16. NUCLEAR EFFECTS MITIGATION SCHEMES

Path Diversity
Frequency Diversity & Agility
Relay-Capable Networking
Increased Transmitter Power
Frequency Extension (f_{max} increase to LVHF)
Time Diversity
Spread Spectrum
Adaptive Antenna
Non-coherent Detection
Ground Wave Propagation

Frequency extension and groundwave propagation correspond to media diversity. Frequency extension typically implies the use of frequencies well into the low VHF (or LVHF) band. This reduces the f^{-2} absorption loss provided the selected LVHF frequency is supported by the medium. Candidate non-standard media include meteor trails and nuclear-induced *bomb modes*. Naturally the use of ground wave modes, especially at low HF and MF, will eliminate ionospheric propagation problems. Since skywave noise may be reduced in a nuclear environment, due to increased absorption, and since groundwave signals are not so reduced, the S/N for groundwave modes should be enhanced. Thus an increased reliability can be achieved for a fixed range. Alternatively, a fixed message reliability can be maintained over a larger coverage area.

Time diversity schemes involve effective data rate reduction and error-correction measures such as powerful low-rate codes, interleaving structures, and other techniques (ARQ, FEC, EDAC) possibly applied in combination. Coding, a time redundancy technique, is especially effective in a slow fading environment such as that which would be associated with AGWs. Mitigation of relatively high fade rates which are encountered in multipath or in a striated plume environment generally requires a reduction in data rate.

Non-coherent detection is more robust than a coherent scheme in a nuclear-stressed environment characterized by a channel having a large doppler spread component. This preference for non-coherent processing is not universal, but it is one way problems introduced by dynamic changes in the medium may be confronted more successfully.

One normally considers an adaptive array as a means for elimination of interfering signals by a null-steering process. The same procedure may also be applied in the context of multipath isolation problems. Hansen and Loughlin [1981] have developed a 4-element adaptive array for the elimination of multipath interference. They used an LMS algorithm which finds the least mean-square difference between the array output and a modulated pilot signal. Used against 1E, 1F1, and 1F2 ionospheric modes, the system could pre-select any given mode and discriminate against the other two with an isolation of the order of 15 dB. Their approach also has the potential for discriminate against interferers by time nulling. In a nuclear environment, adaptive antenna arrays may have a number of problems when trying to isolate wanted modes from unwanted ones, or when trying to null-out an interfering signal or a jammer. Generally speaking the capability of an array to isolate targets depends upon the number of elements in the array. Angular resolution will be degraded by non-linearities in the incoming wave surface, and such non-linearities will arise if the medium defocuses the signal path, or if a large number of modes are simultaneously detected or spread conditions exist. In short, adaptive antenna arrays may fail in their primary function in a multipath environment for precisely the same reason that diversity works.

Although increasing the transmitter power will not generally hurt as far as enhancing the system reliability, it is a poor strategy given the enormous absorption losses to be contended with. A hundred dB or more of attenuation may be encountered if the ray trajectory is at the wrong place at the wrong time. It would be far better to select another path, or another time. The excess transmitter power is not needed for benign conditions, and for disturbed conditions it doesn't provide benefits commensurate with the increase in system cost.

Some advantages of spread spectrum do not automatically obtain in a nuclear-disturbed environment. Frequency hopping waveforms may encounter enhanced time jitter, and direct sequence waveforms may suffer from increased multipath and ISI. The usual applications of spread spectrum include LPI, covert communication, and reduction of vulnerability to geolocation. The nuclear environment presents challenges for spread spectrum techniques as applied to both ECM and ECCM.

7.8.7 Strategic HF System Types

Two so-called *strategic-HF* systems are worth a comment: *Newlook/Stresscom*[105] and *Robust*™[106]. Both systems are designed to provide connectivity in pathological environments such as that which would be expected as the result of nuclear detonations. They are both designed to operate within a network. *Newlook/Stresscom* uses a shotgun approach in frequency management, whereas *Robust*™ uses a sounding approach. As a result, *Newlook/Stresscom* spreads-out the risk in data transmission over the HF band, possibly

----

105. *Newlook* was developed by ITT. The current implementation is called *Stresscom.*

106. Magnavox developed the *Robust HF* system. The reader should not confuse this tradename with the the adjective *robust* (without the capital R).

extended into the LVHF region. *Robust*™, by using a sounding approach, involves more risk since it is possible for a poor channel to have been selected. On the other hand, a good channel selection should enable *Robust*™ to have superior performance over shotgun approaches.

7.8.6.1 *Newlook/Stresscom*

Newlook was designed to permit robust end-to-end communications of a diverse network of nodes, and its architecture made it especially suitable for strategic aircraft applications. It has evolved over the years, becoming more compact with tranceivers replacing separate transmitters and receivers. The current version is dubbed *Stresscom*, but the architecture is not substantially different than the original concept. The system is characterized by the following properties: 4-ary FSK signaling to transmit dibits of information, DS-SS to assist in detection of energy presence, uses a block code structure error correction, frequency hopping for jam protection and processing gain, multiple redundancy through simultaneous transmission over specified frequency blocks, tiered relaying structure based upon unique frequency-time plans for each level, and VHF frequency extension to accommodate bomb modes and/or to mitigate against enhanced D-layer absorption. *Newlook/Stresscom* makes no valiant effort to seek the MOF nor to avoid frequency bands where interference may be bothersome. It is not an adaptive radio system, and is designed to coexist with interference and operate despite propagation disturbances. Tests have been carried out at high latitudes during magnetic storm conditions which mimic some aspects of the nuclear environment.

7.8.6.2 *Robust*™

Robust™ has some of the same features as the *Newlook/Stresscom* system, but architectural differences are required since the requirements are not identical. *Robust*™, for example, exploits convolutional codes and code combining processes to maintain reliability at a high level of data integrity. As such, it possesses an adaptive property. The code combining strategy adds sufficient redundancy to maintain information accuracy, but not necessarily the data throughput rate, at a specified level. This procedure is an attempt to match the data rate to the channel capacity. Thus the *Robust*™ system has both truly robust and adaptive features. Another adaptive feature of the *Robust*™ system is the use of sounding procedures for channel selection rather than shotgun procedures. Perhaps the Magnavox *Robust*™ system should have been trademarked *Adaptive* instead. Indeed, using the definitions favored in this book, *Newlook/Stresscom* possesses more truly robust features.

7.8.6.3 *Chirpcomm*™

Chirpcomm™, a communications adapter to the BR Communications Chirpsounder™, may also be viewed as a candidate for use in a nuclear environment. Despite the fact that it exploits a sounding waveform, *Chirpcomm*™ uses the shotgun method for message transmission. Messages are repeated 63 times as the chirp waveform is swept over the HF band. Message combining increases the probability of correct message receipt.

7.9 TRENDS TOWARD STANDARDIZATION

7.9.1 The Need for Standards

Since the various military and civilian organizations have markedly dissimilar requirements for HF radio, it would be unexpected to find strong architectural similarities between the various systems. Furthermore, with the proliferation of approaches for communication performance improvement, it is not surprising to find that the assortment of system components which comprise a basic HF radio will differ from one vendor to another. Many large organizations such as the military services may procure separate radios to satisfy the local tactical requirements of subordinate activities as well as the organization-wide strategic requirements. Unless the organization is comprised of totally independent activities without mutual requirements, it would be desirable for this process to result in a standardized set of radio attributes which would permit all radios to be interoperable (in a specified set of operational modes). Unfortunately this result will not generally occur without imposition of a set of standards by the controlling organization. This circumstance is, of course, not unique to HF radio systems; it is the consequence of a free market system. Nevertheless, if systems are to be interoperable, it will be necessary to ensure a degree of functional commonality through the promotion of performance and design standards. This must be done without the relaxation of mission requirements resulting from vulgarization of the system specifications. A balance must be drawn between a proposed standard and the requirements-driven specifications.

It is almost axiomatic that system performance enhancements will involve equipment specifications, waveform designs and signal processing upgrades which are more complex than their predecessors. This increased complexity (and diversity of approach) is not without some risks, including the possible introduction of additional failure modes. More distressing is possibility that different manufacturers will develop and market systems which, if procured, will be unable to interoperate. Because of the role which HF may play in the event that satellite, landline, and other terrestrial communication systems are incapacitated , the interoperability of existing and future-planned HF radios becomes an imperative. This requirement extends to international networks and to other situations involving allied nations for which some integration of HF assets would be desirable. In the United Kingdom, Smith [1991] has proposed the development of a set of standards which account for all important characteristics of *Automatically Controlled Radio Systems* (ACRS).

7.9.2 Recent Chronology of Standards Development in the USA

In 1984, the Plans and Programs component of the National Communications System (NCS) tasked the Mitre Corporation to investigate the degree of interoperability among various agencies. Mitre subsequently developed a draft standard waveform for automatic link establishment which would serve as the basis for a Military Standard on *Interoperability and Performance Standards for Medium and High Frequency Radio Equipment* (MIL-STD-188-141A) as well as a Federal Standard on *Automatic Link Establishment* (FED-STD-1045). Additional Federal Standards include: *HF Radio Automatic Networking* (FED-STD-1046), *HF Radio Automatic Store-and-Forward* (FED-STD-1047), *HF Radio Automatic Networking to Multi-Media* (FED-STD-1048), and *HF Radio Automatic Operation in Stressed Environments* (FED-STD-1049).

Within the US, the General Services Administration (GSA), pursuant to the Federal Property and Administrative Services Act of 1949, issues various standards. The relevant HF standards are developed to provide a basis for design and procurement of HF systems. Some of the documents of interest to the reader are included in Tables 7-17 and 7-18.

7.9.3 ALE Radios: Tests for Compliance with the Standards

Several ALE radios have been developed. They include: the Model 1045 by Frederick Electronics, the MD-9189 ALE Processor by Sunair Communications, the RS-5000 Falcon™, by Harris Corporation, and other systems manufactured by Rockwell-Collins, Motorola, MacKay, and Transworld Electronics.

The U.S. Department of Commerce (ITS-Boulder) directed an interoperability test of several FED-STD-1045 HF radios [Smith et al.,1991] employing channel simulators as well as on-the-air tests in the real environment. On-the-air exercises demonstrated that test control difficulties may be encountered as a result of unpredictable propagation disturbances. Tests using a hardware version of the Watterson model channel simulator (i.e., a Harris Model RF-3460) against the ALE radios were far less taxing. On the other hand, they may also have been less realistic.

TABLE 7-17

US FEDERAL AND MILITARY STANDARDS

{FEDERAL}

FED-STD-1003: Telecommunications, Synchronous Bit Oriented Data Link Control Procedures (Advanced Data Communication Control Procedures)

FED-STD-1037: Glossary of Telecommunication Terms

FED-STD-1045: HF Radio Automatic Link Establishment

pFED-STD-1046: HF Radio Automatic Networking[107]

pFED-STD-1047: HF Radio Automatic Store-and-Forward

pFED-STD-1048: HF Radio Automatic Networking to Multi-Media

pFED-STD-1049: HF Radio Automatic Operation in Stressed Environments.

1. link protection
2. anti-interference
3. encryption
4. reserved

{MILITARY}

MIL-STD-188-100: Common Long-Haul and Tactical Communication

MIL-STD-188-110: Equipment Technical Design Standards for Common Long-Haul/Tactical Data Modem

MIL-STD-188-114: Electrical Characteristics of Digital Interface Circuits

MIL-STD-188-141A: Automatic Link Establishment Systems

MIL-STD-188-148: Interoperability Standard for Anti-Jam Communications in the High Frequency Band (2-30 MHz); Classified, not generally available.

---

107. The "p" prefix used for FED-STD-1046 through 1049 indicates that these standards are proposed and not yet fully sanctioned.

TABLE 7-18

INTERNATIONAL STANDARDIZATION DOCUMENTS

{NATO}

STANAG 4203:	Technical Standards for Single Channel HF Radio Equipment
STANAG 5035:	Introduction of an Improved System for Maritime Air Communications on HF, LF, and UHF
STANAG 4285:	Characteristics of 1200/2400/3600 Bits per Second Single Tone Modulators/ Demodulators for HF Radio Links (DRAFT); Classified, not for general distribution

{QUADRIPARTITE}

QSTAG 733:	Technical Standards for Single Channel High Frequency Radio Equipment

{ITU}

Radio Regulations

{CCIR}

CCIR Rec. 455-1:	Improved Transmission System for HF Radiotelephone Circuits
CCIR Rec. 520:	Use of High Frequency Ionospheric Channel Simulators

7.10 REFERENCES

Aarons, J. (editor), 1986, *Propagation Impact on Modern HF Communications System Design*, NATO-AGARD-LS-145, Specialised Printing Services Ltd., Essex, England.

AGARD, 1986, *Propagation Impact on Modern HF Communications System Design*, NATO-AGARD-LS-145, Specialised Printing Services, Ltd., Loughton, Essex, UK.

Alexander, P., J. Holtzman, P. Magis, G. Prescott, and C. Tsatsoulis, 1991, "A Computer-Aided Design System for Frequency Management of HF Communication Systems", in *Fifth International Conference on HF Radio Systems and Techniques*, IEE, Savoy Place, London.

AMAF, 1990a, specification sheet for Model 2200 LINK-PLUS™, AMAF Industries, Columbia, MD.

AMAF, 1990b, press release, AMAF Industries, Columbia, MD.

Ames, J.W., N.J.F. Chang, and T.D. Magill, 1984, "Development and Testing of Adaptive HF Radio Techniques", Report No. DNA-TR-84-379, Defense Nuclear Agency, Contract No.DNA-001-80-C-0253, SRI International, Menlo Park, CA.

Atkinson, S., 1990, "Mil-Tech Datacomms", *Electronics World and Wireless World*, June, pp.489-494.

Baker, M., and N.P. Robinson, 1991, "Advanced HF Digital Networks", *Modern HF and Emerging Technologies*, AFCEA Symposium, Annapolis Chapter, 18-20 November, 1991, Annapolis, MD.

Baker, S., 1988, private communication, lecture notes on Robust HF from AFCEA Course No. 104 (on HF communication systems and technology) organized by J.M. Goodman.

Barden, W., 1989, "Neural Networking", *PCM Magazine*, pp. 69-74.

Barr, E., A.S. Eley and J.H. Choi, "Near Vertical Incidence Skywave Antenna Examination", 1991, Proc. 5th Int.Conf. on *HF Radio Systems and Techniques*, held in Edinborough, Scotland in July 1991; IEE, Savoy Place, London WC2.

Bayley, D. and J.D. Ralphs, 1969, "The Piccolo 32-tone Telegraph System", *Point-to-Point Telecomm.*, 14(2):78-90.

Bayley, D. and J.D. Ralphs, 1972, "The Piccolo 32-tone Telegraph System in Diplomatic Communications", *Proc.IEE*, Paper 6750, 119(9):1229-1236.

Black, W., undated, "HarpCall Test Results and Possible Applications", Technical Note 364, SHAPE Technical Center, The Hague, NL.

Brakemeier, A., and J. Lindner, 1988, "Single-tone HF Data Transmission with 2.4 kbps and 4.8 kbps", in *Fourth International Conference on HF Radio Systems and Techniques*, IEE, Savoy Place, London, UK, pp.339-343.

Bell D.A, 1986, "Channel Capacity: the Meaning of Shannon's Formula", *Electronics and Wireless World*, April, pp.66-67.

Bello, P.A., 1965, "Selective Fading Limitations of the Kathryn Modem and Some System Design Considerations", *IEEE Trans.Comm.Tech.*, COM-13:320-333.

Bello, P.A., 1988, "Performance of Some RAKE Modems over the Non-Disturbed Wideband HF Channel", *MILCOM'88 Conference Record*, IEEE/DoD/AFCEA, IEEE Headquarters, New York, N.Y.

Bello, P.A., 1989, "Performance of Four Rake Modems Over the Non-Disturbed Wideband HF Channel", RADC-TR-89-91 (F19628-86-C-0001), Bedford, MA.

Boyd, R.W. and D.D McRae, 1987, "Simulation of an Automated Robust HF Packet Network", in *MILCOM '87*, IEEE Conference, (14.3.1-14.3.5), pp. 340-344.

Brayer, K. (editor), 1975, *Data Communications via Fading Channels*, IEEE Press, New York, N.Y.

BR Communications, 1990, company brochures describing the Model 4180 system, the XCS-6 Chirpsounder Transceiver, Chirpcomm™, and other products.

Bürgisser, E., and M.H. Norton, 1991, "Robust Low Data Rate Transmission over HF and its Performance on Real Radio Channels", *Modern HF and Emerging Technologies*, AFCEA Chesapeake Chapter Symposium, 18-20 November, 1991, Annapolis, MD.

CCIR, 1986a, "Improved Transmission System for HF Radiotelephone Circuits", Recommendation 455-1, in *Recommendations and Reports of the CCIR*, Vol.III (Fixed Service at Frequencies below about 30 MHz), Geneva.

CCIR, 1986b, "Improved Transmission Systems for Use over HF Radiotelephone Circuits", Report 354-5, in *Recommendations and Reports of the CCIR*, Vol.III (Fixed Service at Frequencies below about 30 MHz), Geneva.

Chang, N.F., J.W. Ames, and G. Smith, 1980, "An Adaptive Automatic HF Radio", Final Report, DNA 5507F, Contract DNA001-79-C-0364, SRI Project No.8710, SRI International, Menlo Park CA.

Chase, D., 1973, "A Combined Coding and Modulation Approach for Communication over Dispersive Channels", *IEEE Trans.Comms.*, Com-21:159-174.

Chesmore, E.D. and M. Darnell, 1985, "A Low-Rate Digital Speech Message Transmission System for HF Engineering Orderwire (EOW) Applications", *International Conference on HF Communication Systems and Techniques*, IEE Conference Pub.No.245, IEE, Savoy Place, London, UK.

Chesmore, E.D., 1991, "Artificial Intelligence Techniques in HF System Design", in *Fifth International Conference on HF Radio Systems and Techniques*, IEE, Savoy Place, London.

Christinsin, A., 1986, *Compendium of High Frequency Radio Communications Articles*, USAF, Department of Defense, USGPO: 1986-652-123/40137.

Christinsin, A., 1990, "Near Vertical Incidence Skywave Yields Reliable HF Communications", *Signal*, November, p.39.

Christinsin, A., 1991, "High Frequency NVIS Vehicular Communications for Command and Control", paper presented at AFCEA Symposium on *Modern HF and Emerging Technologies*, 18-20 November, Annapolis, MD.

Christofilos, N.C., 1959, "The Argus Experiment", *J.Geophys.Res.*, 64:869.

Clark, G.C., and J.B. Cain, 1981, *Error-Correction Coding for Digital Communications*, Plenum Press, New York.

Clark, A.P., 1981, "Detection of Digital Signals Transmitted over a Known Time-Varying Channel", *Proc. IEE*, 128(Part F, No.3):167-174.

Conti, P.S., 1991, Frederick Electronics Corporation, private communication, company brochure.

Conticello, C., 1984, "A Breakthrough in Mobile HF Communication: A New Generation of High Power Antenna Couplers", *Signal*, March, pp.71-76.

Cox, J.W.R. and G. Vongas, 1991, "Comparison of Some Predicted and Measured Characteristics of an RF Loop Antenna Mounted Upon a Helocopter", Proc. 5th Int. Conf. on *HF Radio Systems and Techniques*, venue: Edinborough, Scotland, July 1991; IEE, Savoy Place, London WC2.

Crain, C.M. and H. G. Booker, 1963, "The Effects of Nuclear Bursts in Space on the Propagation of High Frequency Radio Waves Between Separated Earth Terminals", *J.Geophys.Res.*, 68(8):2159-2166.

Crozier, S., K. Tiedemann, R. Lyons, and J. Lodge, 1982, "An Adaptive Maximum Likelihood Sequence Estimation Technique for Wideband HF Communications", *Progress in Spread Spectrum Communications*, Vol.2,, MILCOM'82, IEEE, DoD, AFCEA, pages 29.3-1 through 29.3-9.

Dacunto, L.J., 1988, "Army Command and Control Initiatives", *Signal*, November, pp.63-69.

Daehler, M., 1990, "Oblique-Incidence Sounder Measurements with Absolute Propagation Delay Timing", in *The Effect of the Ionosphere on Radiowave Signals and System Performance, IES'90*, pp.418-421, USGPO, Available through NTIS, Springfield VA.

Darnell, M., 1983, "HF System Design Principles", in *Modern HF Communications*, NATO-AGARD-LS-127, Specialised Printing Services Ltd., Essex, England.

Darnell, M., 1986, "HF System Design", in *Special Course on Interaction of Propagation and Digital Transmission Techniques*, NATO-AGARD-R-744, Specialised Printing Services Ltd., Essex, England.

Dayharsh, T.I., 1965, "HF Communication Effects: Ionospheric and Mode of Propagation Measurements", DASA 1701, Final Report, Contract DA 36-039 SC-87197, Stanford Research Institute, Menlo Park, CA.

DePedro, 1985, "On Adaptive HF Networks in Benign and Stressed Environments", *MILCOM 85*, IEEE.

Dhar, S. and B.D. Perry, 1982, "Equalized Megahertz Bandwidth HF Channels for Spread Spectrum Communications", *Progress in Spread Spectrum Communications*, Vol.2, MILCOM'82, IEEE, New York, NY.

Dickson, A.H., P.C. Arthur, and P.S. Cannon, 1991, "ROSE - A DSP Enhancement to the Barry Ionospheric Sounder", in *Fifth International Conference on HF Radio Systems and Techniques*, IEE, Savoy Place, London.

DISA, 1991, "Test Report AN/TSC-122 Type Communications Central Interoperability"; performed by: Joint Interoperability Test Center (JITC), Fort Huachuca, AZ.; under aegis of: Joint Tactical Command, Control, and Communications Agency of the Defense Informations Systems Agency (DISA); Rpt. 90-SOC-T002/87-MC-T005 dtd. 31 October 1991,

Dixon, R., 1976, *Spread Spectrum Systems*, Wiley-Interscience, New York.

Doelz, M.L. et al., 1957, "Binary Data Transmission Techniques for Linear Systems", *Proc.IRE*, 45:656-661.

DoT, 1989, "Federal Aviation Administration National Radio Communications System High Frequency/Single Sideband", System User's Guide, DoT document No.6610.14, FAA, Washington D.C., July.

Drazovitch, R.J., B.P. McCune, and J. Roland Payne, 1982, "Artificial Intelligence: An Emerging Military Technology", *IEEE Conference*, pp 341-348.

Forgie, J.W., A.J. McLaughlin, and C.J. Weinstein, 1987, "An Intelligent C3 Terminal Architecture", *IEEE Conference*, (5.4.1-5.4.7), pp. 181-187.

Forney, G.D.Jr., 1972, "Maximum Likelihood Sequence Estimation of Digital Sequences in the Presence of Intersymbol Interference", *IEEE Trans. Information Theory*, IT-18(3):363-378.

Forney, G.D. Jr., 1973, "The Viterbi Algorithm", *Proc.IEEE*, 61(3):268-278.

Frolli, M.R., 1979, "HFNET: A Computer Program to Calculate Nuclear Effects on HF/VHF Communications Systems", Volumes I and II, MRC-R-515, Mission Research Corporation, California.

Frost, V., R. Moats, R. Spohn, S. Fechtel, and R. Balasubramaniam, 1987, "The Evaluation of HF Communication Systems Using a Simulation Workstation", *IEEE Conference*, (14.1.1-14.1.7), pp. 327-333.

Garner, J., T. McChesney, and M. Glancy, 1982, "Advanced Narrowband Digital Voice Terminal", *Signal*, November, pp.7-16.

Glasstone, S. and P.J. Dolan, 1977, *The Effects of Nuclear Weapons*, 3rd Edition, Dept.of Defense, Superintendent of Documents, USGPO, Washington DC 20402.

Goldberg B.(editor), 1976, *Communications Channels: Characterization and Behavior*, IEEE Press, New York, N.Y.

Goldberg B., 1966, "300 kHz-30 MHz MF/HF", *IEEE Trans. Comm. Tech.*, Com-14:767-784.

Goldberg, B., R.L. Heyd, and D. Pochmerski, 1965, "Stored Ionosphere", *IEEE Int. Conf. Comm.*, pp.619-622.

Goodman, J.M., 1989, "Decision-Aid Design Factors in Connection with HF Communication and Emitter Location Disciplines", *Operational Decision Aids for Exploiting or Mitigating Electromagnetic Propagation Effects*, NATO-AGARD Conference Proceedings (San Diego), AGARD-CP-453, Specialised Printing Services Ltd, Loughton, Essex, UK.

Goodwin, R.J. and A.P.C. Reed, 1987, "Design Considerations for an Automated HF Data Network with Adaptive Channel Selection", NATO-AGARD-CP-420, AGARD Conference Proceedings (Lisbon), page 12-1 to page 12-11.

Goodwin, R.J., and S.P. Harris, 1988, "The Design and Simulation of an Adaptive HF Data Network", in *Fourth International Conference on HF Radio Systems and Techniques*, IEE, Savoy Place, London, UK, pp. 1-5.

Grossi, M., 1986, "An Introduction to Error Control Coding with Application to HF Communications", *Propagation Impact on Modern HF Communications System Design*, NATO-AGARD-LS-145, Specialised Printing Services Ltd., Essex, England.

Hague, J., 1987, "Improved Coding and Control of HF Systems in a Non-Gaussian Noise Environment", NATO-AGARD Conference, Lisbon, AGARD-CPP-420, pp.16-1 to 16-17.

Hansen, P.M. and J.P. Loughlin, 1981, "Adaptive Array for Elimination of Multipath Interference at HF", *IEEE Trans. Antennas and Propagation*, AP-29(6):836-841.

Harris Corp., 1991a, "Tactical HF Communications: A Systems Approach", *Modern HF and Emerging Technologies*, AFCEA Chesapeake Chapter Symposium, 18-20 November, 1991, Annapolis, MD.

Harris Corp., 1991b, company brochure describing the RF-5000 series of HF radio equipment.

Harrison, G., 1985, "Functional Analysis of Link Establishment in Automated HF Systems", WP86W00015, Mitre Corporation, McLean VA.

Harrison, G., 1987, "HF Radio Automatic Link Establishment Systems", *MILCOM'87*, Conference Record, 1:43-47.

Hauser, J.P., D.N. McGregor, and D.J. Baker, 1984, "Design and Simulation of an HF Ship-Shore Communication Network", NRL Report 8805, Naval Research Laboratory, Washington DC 20375-5000.

Hayes, J.F., 1975, "The Viterbi Algorithm Applied to Digital Data Transmission", *Communications Society*, 13(2):15-20.

Hayes, D.F., 1987, "An Integrated Voice/Data Architecture for HF Tactical Communications", in *MILCOM 87*, IEEE, (2.6.1-2.6.4), pp. 68-71.

Hines, C.O., 1960, *Canadian J. Phys.*, 38:1441.

Hoerlin, H., 1976, "United States High-Altitude Test Experiences", University of California, Los Alamos Scientific Laboratory (LASL), LA-6405, Los Alamos NM.

Hughes, 1989, brochure from Hughes Aircraft Corporation describing the AN/PRC-104B.

IEE-UK, 1979, *Recent Advances in HF Communication Systems and Techniques*, Institution of Electrical Engineers, Conference Digest No.1979/48, London, UK.

IEE-UK, 1982, *HF Communication Systems and Techniques*, 2nd International Conference, Institution of Electrical Engineers, London, UK.

IEE-UK, 1985, *HF Communication Systems and Techniques*, 3rd International Conference, Institution of Electrical Engineers, Rpt.No.245, London, UK.

IEE-UK, 1987, *Fifth International Conference on Antennas and Propagation: A Hundred Years of Antennas and Propagation*, Part 2: Propagation, Conference Publication 274, IEE, UK.

IEE-UK, 1988, *HF Radio Systems and Techniques*, 4th International Conference, Institution of Electrical Engineers, Report No.284, London, UK.

IEE-UK, 1991, *HF Radio Systems and Techniques*, 5th International Conference, Institution of Electrical Engineers, Report No. 339, London, UK.

Jayant, N.S. (Editor), 1976, *Waveform Quantization and Coding*, IEEE Press, New York, NY.

Jensen, R.G., 1982, "Digital Lincompex", *RTCM Assembly Meeting*, Seattle, Washington.

Jerome, A.C., S.P. Baraniuk, and M.A. Cohen, 1987, "Communication Network and Protocol Design Automation", *IEEE Conference*, (14.2.1-14.2.6), pp. 334-339.

JTC3A, 1989, HF Information Exchange Meeting, held at BDM, 20-23 March

JTCO, 1981, "Performance Specifications of the ANDVT Tactical Terminal", Specification No.TT-B1-4210-0087B, Joint Tactical Communication Office, Ft. Monmouth NJ.

Jakubowicz, O.G., 1989, "Neural Networks for Intelligence Analysis", *Signal*, April, pp. 47-54.

Johnson E.E., 1991, "Linking Protection for HF Radio Networks", in *Modern HF and Emerging Technologies*, AFCEA Symposium, Annapolis Chapter, 18-20 November, 1991, Annapolis, MD.

Jowett, A.P., 1987, "Artificial Intelligence in HF Communication Systems", in NATO-AGARD-CPP-420, Lisbon Conference, Paper No. 14., pp. 14-1 to 14-25.

Kennedy, R.S., 1969, *Fading Dispersive Communication Channels*, Wiley-Interscience, a Division of John Wiley and Sons, New York.

Kirsch, A.L., P.R. Gray, and D.W. Hanna Jr., 1969, "Field Test Results of the AN/GSC-10 (Kathryn) Digital Data Terminal", *IEEE Trans.Comm.Tech.*, Vol.COM-17, pp.118-128.

Knapp, W.S., C.F. Meyer, and P.G. Fisher, 1967, "Introduction to the Effects of Nuclear Explosions on Radio and Radar Propagation", General Electric Co., TEMPO, DASA Report 1940.

Knoefel J.O., 1988, "An Expert System for Single Channel Radio Link Planning", *Signal*, November, pp.83-86.

Kohl, H., 1964, "Acoustic Gravity Waves Caused by the Nuclear Explosion on October 30th, 1961", *Electron Density Distribution in the Ionosphere and Exosphere*, NATO and NDRE, John Wiley and Sons, New York (also North-Holland, Amsterdam).

Kuklinski, W.S., K. Chandra, and B.W. Reinisch, 1989, "Progress of DORIS Automatic Scaling", Scientific Report No.15, GL-TR-89-0186, Geophysics Laboratory, Hanscom AFB, Massachusetts 01731-5000.

Langley, G., 1986, *Telephony's Dictionary*, Second Edition, Telephony Publishing Corp., Chicago, IL.

Lichtman, S.W. and E.J. Andersen, 1962, "Ionospheric Effects of Nuclear Detonations in the Atmosphere", *Proceedings of the International Conference on the Ionosphere*, published by the Inst. Physics and the Physical Soc., London, UK.

Lin, S., 1970, *An Introduction to Error-Correcting Codes*, Prentice-Hall Inc., Englewood Cliffs, New Jersey.

Lomax, J.B. and D.L. Nielson, 1968, "Observation of Acoustic-Gravity Wave Effects Showing Geomagnetic Field Dependence", *J. Atmos. Terr. Phys.*, 30:1033-1050.

Loral-Terracom, 1990, company brochure describing the AN/PRC-132 radio.

Mackay, 1991, company brochure describing the AN/GRC-223 radio set and other radio equipments.

Maslin, N., 1987, *HF Communications: A Systems Approach*, Plenum Press, New York, N.Y.

Magee, F. R. Jr, and J. K. Proakis, 1973, "Adaptive Maximum-Likelihood Sequence Estimation for Digital Signaling in the Presence of Intersymbol Interference", *IEEE Trans.Information Theory*, IT-19:120-124.

McGregor, D.N., J.E. Wieselthier, D. J. Baker, A. Ephremides, and J. P. Hauser, 1985, "Networking Concepts for Naval HF Communications", *HF Communication Systems and Techniques*, Conference Proceedings No. 245, IEE, Savoy Place, London, UK.

McRae, D.D., 1985, "Reliable HF and Survivable Communications", *Signal*, August, pp. 75-79.

Milsom, J.D. and T. Slator, 1982, "Consideration of Factors Influencing the Use of Spread Spectrum on HF Skywave Paths", *Second Conference on HF Communication Systems and Techniques*, IEE, Savoy Place, London, UK.

Monsen, P., 1986, "Modern HF Communications, Modulation and Coding", *Propagation Impact on Modern HF Communicatiions System Design*, pp.8-1 to 8-10, NATO-AGARD-LS-145, Specialised Printing Services Ltd., Essex, England.

Morcerf, L.A., K.R. Kontson, and J.E. Clema, 1987, "A Concept for the Application of Artificial Intelligence Technology to Battlefield Spectrum Management", *IEEE Conference*, 5.1.1-5.1.6

Morris, D. J., 1983, *Communication for Command and Control Systems*, Pergamon Press Inc., New York, NY.

Mosier, R.R. and R.G. Clabaugh, 1958, "Kineplex, A Bandwidth-Efficient Binary Transmission System", *AIAA Trans.(Part 1: Communications and Electronics)*, 76:723-728.

NAVELEX, 1979, "HFDM: AN/USQ-83(XH-1)V, The High Frequency Digital Modem, Operation and Maintenance Manual", Sylvania Systems Group, for NAVELEX, under the title "High Frequency Digital Modem, Technical Objectives and Goals", Navy Depart., Washington DC.

Nelson, R.A., 1968, "Response of the Ionosphere to the Passage of Neutral Atmospheric Waves", *J. Atmos. Terrest. Phys.*, 30(5):825-835.

Nelson, G.P., 1976, "NUCOM/BREM: An Improved HF Propagation Code for Ambient and Nuclear-Stressed Environments", DNA 4248T, Final Rpt., Contract DNA 001-76-C-0261, GTE Sylvania, Needham Heights, MA.

Newman, P., B. Jones, and L. McCabe, 1966, "F-Region Irregularities After High Altitude Detonation of July 9, 1962", *Spread-F and its Effects upon Radiowave Propagation and Communication*, P. Newman (editor), NATO-AGARD, Published by Technivision, Maidenhead, England.

Nielson, D.L., J.B. Lomax, and H.A. Turner, 1967, "Prediction of Nuclear Effects on HF Communications", DASA 2035, Final Report, Contract DA-49-XZ-436, Stanford Research Institute, Menlo Park CA.

Nourry, G.R., and A.J. Mackie, 1987, "The Design and Performance of an Adaptive Packet-Switched HF Data Terminal", in *Effects of Electromagnetic Noise and Interference on Performance of Military Radio Communication Systems*, NATO-AGARD-CP-420, Specialised Printing Services Ltd., Loughton, Essex, UK.

O'Mahony, T.P., 1991, "High Frequency Networks: Technology for the 1990s", in *Modern HF and Emerging Technologies*, AFCEA Symposium, Annapolis Chapter, 18-20 November, 1991, Annapolis, MD.

Oppenheim, A.V.(Chairman, Editorial Committee), 1976, *Digital Signal Processing, II*, IEEE Press, New York, N.Y.

Perry, B.D., 1983, "Preliminary Measurement of Reciprocity Effects on Megahertz Bandwidth Equalized HF Skywave Paths", Report M83-7, MITRE Corporation, Bedford MA.

Pickholtz, R.L., D.L. Schilling, and L.B. Milstein, 1982, "Theory of Spread-Spectrum Communications-A Tutorial", *IEEE Trans. on Comms.*, COM-30(5):855-884.

Porter, G.C., 1968, "Error Distribution and Diversity Performance of a Frequency-Differential PSK HF Modem", *IEEE Trans. Comm. Tech.*, Com-16:567-575.

Price, R., and P.E. Green, Jr., 1958, "A Communication Technique for Multipath Channels", *Proc. IRE*, 46:555-579; also in *Communication Channels: Characterization and Behavior*, B. Goldberg (editor), IEEE Press, New York, N.Y.

Rabiner, L.R., and C.M. Rader (editors), 1972, *Digital Signal Processing*, IEEE Press, New York, N.Y.

Rhode U., 1985, "Technological Solutions for HF Communications", in *Third International Conference on HF Communication System and Techniques*, Conference Publication 245, IEE, London, UK.

Richter, J., 1989, "Propagation Assessment and Tactical Decision Aids", in *Operational Decision Aids for Exploiting or Mitigating Electromagnetic Propagation Effects*, NATO-AGARD Conference Proceedings (San Diego), AGARD-CP-453, Specialised Printing Services Ltd, Loughton, Essex, UK.

Rohde and Schwarz, 1991, company brochure describing the HF 850 radio, the ALIS processor, and other equipments.

Roesler, D.P., 1990, "HF/VHF Propagation Resource Management Using Expert Systems", in *The Effect of the Ionosphere on Radiowave Signals and Systems Performance, IES'90*, J.M. Goodman (Editor-in-Chief), pp.313-321, USGPO, Available through NTIS, Springfield VA.

Rogers, D.C. and B.J. Turner, 1985, "Connectivity Improvement Through Path and Frequency Diversity", *HF Communication Systems and Techniques*, Conference Proceedings No.245, IEE, Savoy Place, London UK.

Rose, R., 1981, "PROPHET - An Emerging HF Prediction Technology", in *Effect of the Ionosphere on Radiowave Systems, IES'81*, J.M. Goodman (Editor-in-Chief), pp.534-542, USGPO, 1982:0-367-001:QL3, Washington DC, Available from NTIS, Springfield VA.

Rose, R., 1989, "PROPHET and Future Warfare Decision Aids", *Operational Decision Aids for Exploiting or Mitigating Electromagnetic Propagation Effects*, NATO-AGARD Conference Proceedings (San Diego), AGARD-CP-453, Specialised Printing Services Ltd, Loughton, Essex, UK.

Rose, R., 1991, "DIAS: A High Latitude Ionospheric Disturbance Expect System", in *The First Annual Conference on Prediction and Forecasting of Radio Propagation at High Latitudes for* C^3, Monterey California.

Sanders, C., 1986, "Battlefield Spectrum Management for High Frequency Communications Systems: Sounder-Aided Propagation Predictions", ECAC-CR-86-056, prepared for Communications/Automated Data Processing Center, Ft. Monmouth NJ 07703-5203.

Scholtz, R.A., 1982, "The Origins of Spread-Spectrum Communications", *IEEE Trans. on Comms.*, COM-30(5):822-854.

Schutzer, D., 1987, *Artificial Intelligence*, Van Nostrand Reinhold Company, New York.

Shanmugan, K.S., 1979, *Digital and Analog Communication Systems*, John Wiley and Sons, New York.

Shannon, C.E., 1948, "A Mathematical Theory of Communication", *Bell System Tech. J.*, 27:379-423 and 27:623-656.

Shannon, C.E., 1949, "Communication in the Presence of Noise", *Proc. IRE*, 37:10-21.

Shipley, C., 1989, "What Ever Happened to AI?", in *PC/Computing*, March, pp. 64-74.

Shpigel, A., and J.M. Perl, 1988, "An Adaptive Digital Receiver for HF Communication", in *Fourth International Conference on HF Radio Systems and Techniques*, IEE, Savoy Place, London, U.K., pp. 331-334.

Signatron, 1987, company brochure from Signatron Corporation describing the S-821 modem.

Simon, M.K., J.K. Omura, R.A. Scholtz, and B.K. Levitt, 1985, *Spread Spectrum Communications*, Volume II, Computer Science Press.

Sjöbergh, B., 1984, "Modern High Frequency Practices and Systems", *Communications International*, May Issue.

Smith, G. K. L., 1991, "Automatically Controlled HF Radio Systems - Toward Standardization", Proc. 5th Int. Conf. on *HF Radio Systems and Techniques*, held in Edinborough, Scotland, July 1991; IEE, Savoy Place, London, UK.

Smith, G.K.L., and Goodwin, 1989, "Adaptive HF Management", IEE Conference Digest No. 1989/33, IEE, London, UK.

Smith, G.F., and W.L. Werner, 1988, "The Renaissance of HF Communications", *Signal*, January, pp.55-62.

Smith, P.C., D.R. Wortendyke, C. Redding, and W.J. Ingram, 1991, "Interoperability Testing of FED-STD-1045 HF Radios", NTIA Rpt., Boulder, CO.

Stein, S. and J. J. Jones, 1967, *Modern Communication Principles*, McGraw-Hill Book Company, New York.

Sunair Electronics, 1991, company brochure describing 9000 series of HF radio equipment, including Pathfinder™ ALE system.

Sunde, E.D., 1961, *BSTJ*, XL:353-422.

Tate, D.L., 1989, "The Variable Speed HF Modem Modulator Design", NRL Report 9169, Naval Research Laboratory, Washington DC.

Tomasi, Wayne, 1987, *Advanced Electronic Communications Systems*, Prentice-Hall, Inc., Englewood Cliffs, New Jersey.

Transworld Communications, 1991, company brochures describing the TW100F SSB transceiver and the DATACOM TW9000 Data Communications Terminal.

Ungerboeck, G., 1974, "Adaptive Maximum Likelihood Receiver for Carrier Modulated Data Transmission Systems", *IEEE Trans. on Comms.*, COM-22(5):624-636.

Utlaut, W.F., 1959, "Ionospheric Effects Due to Nuclear Explosions", NBS Report 6050, Project 8520-12-8510, NBS Laboratories, Boulder CO.

Viterbi, A.J., 1967, "Error Bounds for Convolutional Codes and an Asymptotically Optimum Decoding Algorithm", *IEEE Trans. Information Theory*, IT-13(2):260-269.

Wagen, J.F. 1988, "Simulation of HF Waves Reflected from a Turbulent Ionosphere", in *Ionospheric Effects of HF Spread Spectrum Systems* by K.C. Yeh and C.H. Liu, Technical Report UILU-ENG-88-2559, Univ.of Illinois, Urbana-Champaign, IL.

Wagner, L.S., J. A. Goldstein, W. D. Myers, and P. A. Bello, 1989, "The HF Skywave Channel: Measured Scattering Functions for Midlatitude and Auroral Channels and Estimates for Short-Term Wideband HF RAKE Modem Performance", *MILCOM'89 Conference Record*, IEEE Headquarters, New York, N.Y.

Walters, L.V., 1989, "Shannon, Coding and Spread Spectrum", series of four articles in *Electronics & Wireless World*, January, March, April, May issues.

Weller, W.J., 1982, "Lincompex - A Shot in the Arm for HF Radio", *Signal*, January, pp.25-28.

Wetmore, G. 1988, private communication, lecture notes on NEWLOOK and STRESSCOM from AFCEA Course 104 (on HF communication systems and technology) organized by J. M. Goodman.

Whigham, J.A., 1989, "Tactical Low Profile High Frequency Antenna", *Signal*, June, pp.16-17.

Whittemore, M.J., J.C. Currie, L. Mays, and J. P. Gilliver, 1988, "2.4 kbit/s Adaptive Serial Modems for HF Skywave Paths", in *Fourth International Conference on HF Radio Systems and Techniques*, IEE, Savoy Place, London, UK., pp. 326-330.

Wieselthier, J.E., D.J. Baker, A. Ephremides, and D.N. McGregor, 1983, "Preliminary System Concept for an HF Intra-Task Force Communication Network", NRL Report 8637, Naval Research Laboratory, Washington DC 20375-5000.

Williams, R.H., 1990, "The U.S. Customs Service Spearheads the Drug War", *Signal*, December, pp.52-54.

Wilson, Q.C., 1986, "Developments in HF Equipment and Systems: Mobile and Portable Terminals", in *Propagation Impact on Modern HF Communications System Design*, NATO-AGARD-LS-145, Specialised Printing Services Ltd., Loughton, Essex, UK., pp.10-1 to 10-19.

Yeh, K.C. and C.H. Liu, 1988, "Ionospheric Effects on HF Spread Spectrum Systems", Final Technical Report, No.72, UILU-ENG-88-2559, U.S.Army Contract No.DAAB 07-84K-K531, University of Illinois, Urbana, IL.

Zimmerman, M.S. and R.C.Harper, 1967, "FRM - A New Modem Technique for Rapidly Fading Channels", *1967 IEEE International Conference on Communications Digest*, p.69.

Zimmerman, M.S. and A.L. Kirsch, 1967, "The AN/GSC-10 (KATHRYN) Variable Rate Data Modem for HF Radio", *IEEE Trans. Comm. Tech.*, Com-15:197-204.

EPILOGUE

Nature, to be commanded, must be obeyed
Francis Bacon
Novum Organum

Modern telecommunication systems are designed to have channel capacities with the potential to approach the so-called Shannon limit of data rate, and to provide this capability with high reliability over a range of link or network dimensions. This is a lofty objective for the HF band , long known for its intrinsic bandwidth limitations, spectrum crowding, and propagation effects. When all is said and done, the secret of utilizing the HF band effectively first involves the observation or recognition of that which nature provides, followed by full exploitation. In certain sports activities this procedure is sound strategy. In American football, for example, a defensive team may choose to sacrifice short yardage in what is termed a *prevent defense*, a series of tactical retreats, thereby avoiding a catastrophic strategic failure. The offensive team, recognizing this defensive maneuver, may choose to take what is offered gladly, rather than wasting *downs* on near certain failure. Ultimately the small gains may *integrate* to a successful conclusion. So such is the situation in the world of HF, and this mind-set is succinctly expressed in the epigraph by Francis Bacon printed above.

We have provided the reader with considerable information detailing the precariousness of HF vis-a-vis the ionospheric personality. Awareness of this personality has led to the natural development of systems which utilize spectrum above the so-called shortwave band in both the military and civilian sectors since, broadly speaking, plasma refractivity effects decrease with increasing frequency. Intrinsic bandwidths available for communication are also larger at the higher frequencies. Below HF, ionospheric influence is less pronounced because of a reduction in effective penetration of the ionic medium, although not lacking altogether, and low data-rate military strategic systems may exploit this relative insensitivity.

Against this background, one wonders about the future of HF for use in communication and related systems. Why bother? Is there any advantage to the continued use of this part of the spectrum given the apparent problems? The answer is yes, and there are a number of reasons for this. From a military requirements perspective, typical rationales include the following interrelated encapsulations: balanced mix of assets, communications flexibility, and overall system diversity. In the other-government and civilian sectors, exploitation of

HF is dominated by amateur interests and the broadcast community. In these latter applications, HF and its neighboring bands are used for reasons of economy and practicality. Nearly everybody has a standard radio, and many come equipped with shortwave capability.

Any system which requires the transmission of an electromagnetic signal through the ionosphere or uses it as a *mirror* may be affected by refractive properties in the ionosphere which is spatially- and time-varying. Long-range communications, over-the-horizon (OTH) surveillance radars, and related systems which operate in the HF band and its near neighborhood are clearly the most sensitive to changing ionospheric conditions. As indicated just above, these factors account for the tendency for military systems to gravitate away from HF to both the high and lower ends of the radiowave spectrum. The reasons are compelling. At the low end, it is hoped to minimize ionospheric interaction through reduced penetration of the most ionized portion of the medium, while at the high end of the spectrum the ionospheric interaction is minimized as a result of an approach to unity in refractive index as the radio frequency becomes large. It is seen that this strategy, while sound, is not always infallible. Indeed, satellite-borne systems in the UHF/SHF frequency range are by no means safe from degradation or disruption as a result of scintillation problems. Even during apparently benign solar and magnetic activity conditions, ionospheric scintillation may be an important factor.

The fundamental limits of the ionosphere are imposed by two factors which are sometimes difficult to separate: propagation (magnetoionic properties, anisotropy, dispersion) and media effects (including ionospheric structure and dynamics). For HF communication and surveillance systems, detailed structure defines the impulse response, and information about the dynamical behavior is central in development of the scattering function. These data may be extracted from ionospheric probes. Models of the HF channel have been obtained for narrowband and wideband cases, but such representations, while useful for system design, are not generally capable of providing the specific information required for ionospheric compensation in real time. Real-time ionospheric specification is essential to optimize performance. Otherwise system performance goals must be relaxed to accommodate a set of uncompensated media variabilities. In this case models are quite useful in specification of the margins over which an (adaptive) HF radio system must be prepared to adapt. Non-adaptive radios must be sufficiently *robust* to operate in spite of the predicted effects. Both approaches, *adaptive* and *robust*, are being used in state-of-the-art HF radio systems.

Communications systems are vital to the success of various DoD missions. They should be reliable and contain a minimum of errors. For military purposes there are additional requirements of security, and measures may be necessary to avoid jamming as well. The earliest form of long-haul terrestrial communication involved either longwave or HF bands, and to this day civil and military use is considerable. But, for the most part, satellite communication systems provide the bulk of existing and planned-future capability. Even so, HF will continue to have a special role to play, even in peacetime. This is quite evident when one considers the relative advantages of HF during the stressed conditions associated with national disasters, international emergencies, and search-and-rescue operations. There is a vast reservoir of HF amateur radio operators (HAMS) which may serve as a resource under such circumstances. This resource should be nurtured.

An apparently excessive use of the HF spectrum (3-30 MHz) has led to some interesting frequency management challenges. HF is required for tactical and strategic military purposes, while other federal agencies are using the HF band in contingency networks in the event of national disasters. Non-military use abounds. and international broadcasts by organizations such as BBC, VOA, Radio Liberty, Deutsche Welle, Radio Nederland, Radio Moscow, and others are quite well known. HF is also used in connection with Over-the-Horizon radars, many of which are now reaching operational status. Furthermore, HF usage is relatively common in remote or inclimate areas where satellite coverage is inadequate, and it is an important method for communication in developing countries. Ionospheric variability only further exacerbates the situation of spectral congestion against a background of fixed frequency assignments. The growing trend toward use of spread spectrum methods is not without controversy in this regard. Real-time spectral monitors may assist in locating unoccupied potions of spectral space. At the same time, parallel signaling formats which incorporate redundancy may permit collisions to occur on isolated tones while still retaining sufficient diversity to counter fading or increase the effective symbol rate.

HF communication systems differ in many respects, so it is not surprising that the media constraints are system-dependent. For example, analog systems may suffer from fading problems, while digital systems may successfully counter fading but still be confronted with the baud rate limitations imposed by intersymbol interference. Wideband systems have well-known intrinsic advantages over standard narrowband signaling schemes. Disadvantages are associated with the dispersive nature of the ionosphere, a property which distorts short pulse waveforms and places a limit upon available bandwidth for transfer of information over the channel. Modern matched filter techniques such as *Rake*, with or without parallel processing, may be used to achieve implicit diversity thereby compensating for fading effects introduced by multipath propagation. A step forward in digital signal processing (DSP) speed and complexity is embodied in a class of adaptive equalization schemes which enable a resolution of the intersymbol interference (ISI) problem. In these systems, the unsatisfactory limit on baud rate which is imposed by distended multipath echoes is largely removed. This gain is not, of course, without cost. For example there are still diversity and learning constraints which must be satisfied.

Ray tracing methods are finding application, especially in the context of HFDF and OTH radar. *Virtual raytracing* methods are still used in mainframe communication models such as IONCAP but 3-D raytracing is re-emerging in special applications. This is significant, because with the possibility of improved ionospheric specification, a far greater burden will be placed upon the raytracing component in a prediction/assessment algorithm. 3-D raytracing methods are now being applied in instances which require an accurate relation between known launch angles at one end of the ray path and geolocation at the other end.

Future SATCOM systems operating at EHF may partially resolve the data rate, reliability, security, and related limitations imposed by operation at lower frequencies. Even with this projected capability, however, a backup is still required, and HF will continue to operate in this capacity for long-haul communications. In addition, HF will be employed as a method for communication using the near-vertical-incidence-skywave (NVIS) method in regions where line-of-sight (LOS) propagation is not possible or under circumstances in which satellite-based solutions are not practical.

The futility of developing fully self-adaptive systems which must operate over uncooperative links dictates that ionospheric effects will continue to be a major contributor to communication ineffectiveness at HF. An ultimate solution to this problem may involve the deployment of a real-time global ionospheric assessment system.

Another method of communication which is receiving renewed interest is meteor burst communications (MBC), operating in the low VHF band. Civilian interest in this intermittent method of communication has predated military interest principally as a result of the stress upon satellites. One such early system, called SNOTEL, was developed by the Dept. of Commerce for snowpack telemetry of environmental data. Early experimental efforts in Canada, at the Shape Technical Centre in the Netherlands, and recent studies under the aegis of the U.S. Department of Defense have led to an increased interest in the value of MBC in combating ionospheric perturbations arising from solar flares and nuclear disturbances. The footprint associated with MBC also provides a degree of communication privacy and protection from BLOS jamming. Sporadic E scattering may serve to reduce this latter advantage.

In summary, with respect to telecommunications, we note that system architects are designing systems which are not strongly limited by ionospheric interaction. Satellite systems, for example, are gravitating to the SHF and EHF areas, although military users may require some UHF capability for some time. Commercial systems have long ago abandoned VHF and UHF technology in favor of the SHF band in providing point-to-point and broadcast services. Non-satellite strategic connectivity will be provided by exploitation of the ELF and VLF bands which have a limited ionospheric interaction, while emergency groundwave systems at LF are intrinsically impervious to the ionosphere. Nevertheless, we find that HF systems have a considerable appeal in many instances, especially in the case of certain military scenarios. In the face of nuclear effects, or national emergencies, it is possible that HF will be the communication system which can restore communication most rapidly. It is strongly affected by ionospheric effects, but owing to the fact that the ionosphere repairs itself (and is a satellite which doesn't fall down), HF will definitely have a role to play in the future. We find that modern HF system concepts are being formulated and advanced techniques are being used to eliminate many of the negative effects of the ionosphere.

Global assessment of the ionosphere from satellite sensors would have a major impact on HF-OTH radar as well as other systems. Candidates might include the following: the Air Force Trans-Ionospheric-Sensing-System (TISS), DMSP-borne earth-observation UV limb scanners, and possibly data obtained from GPS and Transit waveforms. Other less expensive, and therefore attractive, options include networking of selected oblique-incidence and vertical-incidence sounders. Suggestions have been made to monitor HF fixed-frequency beacons and timing signals to extract ionospheric information over available paths. A combination of all these methods suitably linked by a versatile empirical model should lead to the most objective ionospheric specification for a variety of purposes. But what about forecasting? This is still a challenge. Solar input would appear to have a place here. Otherwise highly adaptive systems would be essential for optimal performance.

There is clearly a need to study the ionosphere and its coupling to the magnetosphere above and the troposphere below in order to more fully develop the insight required to specify the radiowave propagation effects introduced by these various media. This venture must be closely coupled to studies of properties associated with both the sun and the interplanetary medium in order that the potential for an adequate predictive and forecasting technology may be realized in full. Current empirical models of the ionosphere are inadequate except perhaps for system design guidance, and more effort could fruitfully be spent in this area. They are especially poor over oceanic areas and in zones where experimental observations are sparse. Physical models serve to fill this void in some instances but they also provide predictions which are not useful in most operational scenarios. In addition, the more sophisticated scientific models, whether they be empirical or physical in nature, may overburden the computational capacity of an operational system. As a result, the future trend may be directed toward the development of more simplified operational models which may be updated with real-time observables. It is emphasized, however, that agencies which provide funds for such studies should recognize that the scientific models must precede the operational models to fully identify the relationships involved.

We have indicated in the text that decision aids come in many forms and address a variety of problem areas. They may involve open or closed-loop schemes. A canonical open-loop scheme is a sunspot number-driven prediction model such as IONCAP. Closed-loop schemes may involve certain organic RTCE methodologies which are specific to a given system. This class of schemes may obviate human intervention and favor automatic Real-Time-Decision-Aids (RTDA) as necessary drivers for system parameter adjustment in an adaptive system or network. Hybrid approaches involving models which may be updated through application of RTCE observables may provide a viable alternative to the system designer. In any case, an algorithmic form of ionospheric (or HF channel) specification is the central ingredient which is required in modern era decision aids. Future aids will incorporate RTCE to a greater extent than found in current systems. Improvement in ionospheric specification brought about by RTCE incorporation, and the transformation of derived information into parameters which will allow for ionospheric compensation is required for any significant improvement in HF system performance to be realized. In addition to an improved specification of the ionosphere, it may be advisable for state-of-the-art 3D raytracing techniques (including magnetic field effects) to be incorporated into the codes. This statement is most appropriate for HFDF applications. Future decision aids must provide timely assessment of local and remote tilts as well as their predicted behavior. They must provide for a timely and accurate specification of the position of circumpolar features such as auroral arcs, the midlatitude trough, and the polar cap. Also, for operation in the transequatorial environment, a precise assessment of the Appleton anomaly is needed.

Currently there are a number of initiatives directed toward the improvement of global ionospheric models, and work is also being undertaken in the area of HF propagation modeling. Unfortunately, this activity is only loosely coupled with the development (and deployment) of a variety of RTCE devices. Coordination of model studies and RTCE experimentation would be prudent in order to allow for the development of more efficient RTDAs and decision aids.

Most of the regular macroscopic features of the ionosphere are reasonably well understood although certain details remain as perplexing problems to the user community. There have been noteworthy successes in this area of macroscopic modeling especially in the spatial domain, even though noteworthy problems are still encountered in the extrapolation of relatively small-scale features to unsampled areas. Moreover, with respect to short-term prediction or forecasting capability, we encounter far more serious deficiencies. This is because there is absolutely no possibility to interpolate between known observational values; namely, to bound the time for which a forecast is required with known values from the past and the future. This *bounding procedure* is always possible in connection with *spatial forecasting*, or more properly, extrapolation. The only issue is degree of correlation between observables and the precise form which the extrapolation algorithm will take. We must reluctantly conclude that requisite knowledge of the irregular ionosphere is inadequate. This relates to both the basic phenomenology and triggering functions.

As a consequence we typically treat variability as a statistic. This deficiency affects both transionospheric and ionospheric-reflected propagation assessment in profound ways. There are a number of observational approaches which may assist in the amelioration of this problem. A combination of real-time observation (to provide a global *nowcast* or *snapshot*) and a basic understanding of the underlying physical processes (to improve forecasting capabilities) is required.

The problem of ionospheric behavioral prediction has permeated the psyche of HF specialists for decades. To some communications practitioners, HF systems and the accompanying ionospheric *baggage* are to be avoided at all costs. To others, the topic is rich with variety and represents a continuous challenge. Although advances are typically incremental as we learn more about the HF channel personality, the knowledge gained is leading to the solution of more general problem areas, ones which may also require a fundamental understanding of ionospheric processes including a prediction capability. The development of novel adaptive approaches may have application in a variety of fields. Beneficiaries of these advances include not only other HF systems such as OTH radar, but also UHF satcom, satellite radar, and GPS navigation systems as well. Aside from these decidedly practical issues, there is virtue in examination of the ionospheric medium for its own sake. Solar-terrestrial interactions define parameters essential for a complete assessment of the global energy budget. For example, atmospheric acoustic-gravity waves (AGW), which are closely associated with traveling ionospheric disturbances (TID), provide a major if not dominant heat source for the thermosphere. Thus. a definition of the irregular ionosphere and an ability to predict its impact upon generalized radiowave systems in near-real time, facilitated by ongoing HF studies and developments, could be the single most important contribution in the decade of the 90's.

It is hoped that this book has stimulated the reader not only to appreciate the science and technology unique to HF systems, but to appreciate the adventure associated with ionospheric study in general.

INDEX